GEOLOGY OF CONTINENTAL SLOPES

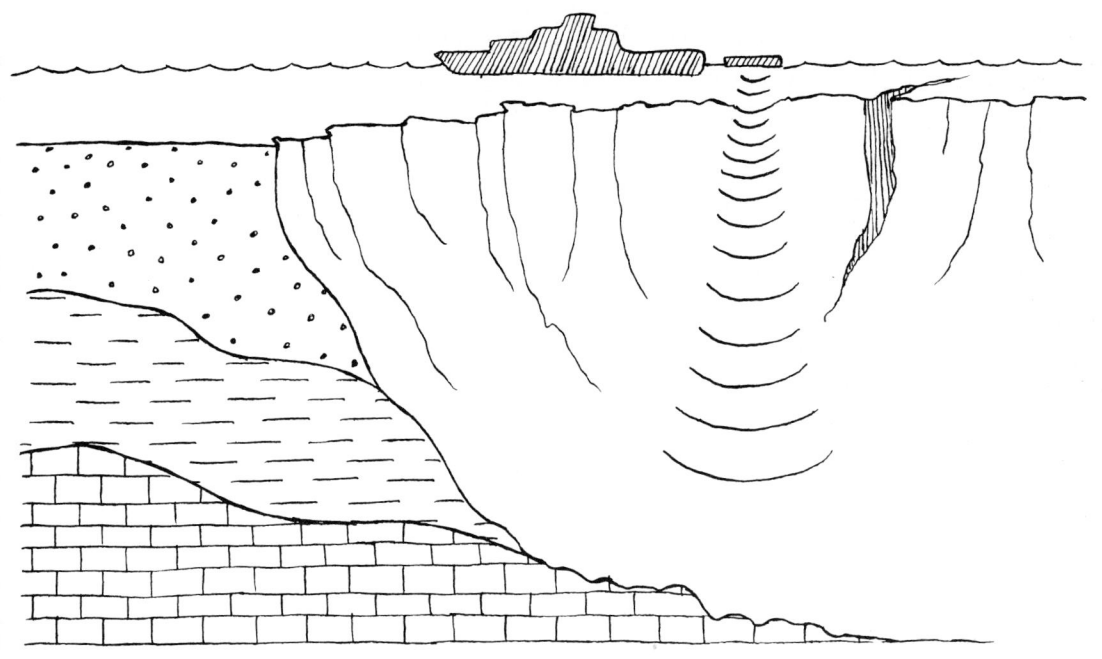

Edited by

Larry J. Doyle
Department of Marine Science
University of South Florida

and

Orrin H. Pilkey
Department of Geology
Duke University

Copyright 1979 *by*

SOCIETY OF ECONOMIC PALEONTOLOGISTS AND MINERALOGISTS
Special Publication No. 27

Tulsa, Oklahoma, U. S. A. August, 1979

PREFACE

Continental slopes are the edges of continental blocks, the zones of change from continental crust to oceanic crust. Here, prograded shelf sedimentary prisms end. Slopes and their associated water columns are also zones of flux through which pass all sediments destined to reach the continental rise and deep ocean basins. They are critical links in the chain of sedimentary processes that eventually carry sediment to the true ocean basin floor. In spite of their importance, until recently continental slopes have been largely ignored when compared with research focused on other provinces of the continental margins and deep sea.

Now, spurred by the recognition that a key portion of the margin has been overlooked and by the extension of hydrocarbon exploration into ever deeper waters, interest in continental slopes has burgeoned. In response to this, we convened a special symposium sponsored jointly by the Society of Economic Paleontologists and Mineralogists and the American Association of Petroleum Geologists at the 1978 meeting in Oklahoma City. This volume, a result of that meeting, is comprised of papers presented at that time with the addition of several papers in order to round out certain subjects.

The book is organized into three major parts. In the first, CONTINENTAL SLOPES: OVERVIEWS, general concepts of slope geology, resource potential, and related engineering problems are dealt with. Since environmental impacts on organisms are so important, a chapter on slope biology is included in this section. Two chapters treat the important subject of down-slope bottom transport mechanisms and their classification, a subject which has caused much confusion in the past. These two chapters cover some common ground but differ significantly in that Lowe presents a fresh quantitative approach to the theory of rheologic behavior regimes while Nardin et al focus on mechanisms of slides, rockfalls, and sedimentologic and stratigraphic character of resultant sediment bodies.

Part 2, MODERN CONTINENTAL SLOPES, is composed of a series of chapters presenting the results of individual current geologic investigations of Atlantic, Gulf, Pacific, and Antarctic continental slopes. Here new insights are provided into slope structure, sediments and their engineering properties, and sedimentary processes which affect interpretation of whole margin systems.

Part 3, ANCIENT CONTINENTAL SLOPES, presents in a series of chapters, current investigations of ancient slope deposits and the interpretation of the processes which were their antecedents. Cook's chapter bridges the gap between recent and ancient, pointing out that study of the past in many cases may be the key to understanding the present.

Finally, we would like to acknowledge Marina Doyle, whose artwork appears on the first cover and cover page, and Phyllis Frothingham and Sharlene Pilkey, who provided major contributions in technical editing.

LARRY J. DOYLE
AND
ORRIN H. PILKEY
Co-Editors

GEOLOGY OF CONTINENTAL SLOPES

Edited by Larry J. Doyle and Orrin H. Pilkey, Jr.

CONTENTS

PART 1. THE CONTINENTAL SLOPE: OVERVIEWS

CONTINENTAL SLOPES . . . . *Arnold H. Bouma*	1
MAN'S ACTIVITIES ON THE CONTINENTAL SLOPE . . . . *Harold D. Palmer*	17
SOME GEOTECHNICAL ENGINEERING PROBLEMS OF UPPER SLOPE SITES IN THE NORTHERN GULF OF MEXICO . . . . *Leland M. Kraft, Jr., Kerry J. Campbell, and M. R. Ploessel*	25
PETROLEUM POTENTIAL OF PASSIVE MARGIN SLOPES . . . . *Mahlon M. Ball*	43
THE BIOTA AND BIOLOGICAL PROCESSES OF THE CONTINENTAL SLOPE . . . . *Gilbert T. Rowe and Richard L. Haedrich*	49
A REVIEW OF MASS MOVEMENT PROCESSES, SEDIMENT AND ACOUSTIC CHARACTERISTICS, AND CONTRASTS IN SLOPE AND BASE-OF-SLOPE SYSTEMS VERSUS CANYON-FAN-BASIN FLOOR SYSTEMS . . . . *T. R. Nardin, F. J. Hein, D. S. Gorsline, and B. D. Edwards*	61
SEDIMENT GRAVITY FLOWS: THEIR CLASSIFICATION AND SOME PROBLEMS OF APPLICATION TO NATURAL FLOWS AND DEPOSITS . . . . *Donald R. Lowe*	75

PART 2. MODERN CONTINENTAL SLOPES

I. ATLANTIC AND GULF OF MEXICO SLOPES

CURRENTS IN SUBMARINE CANYONS AND OTHER TYPES OF SEA VALLEYS . . . . *Francis P. Shepard*	85
STRUCTURE OF THE CONTINENTAL SLOPE OFF THE EASTERN UNITED STATES . . . . *John S. Schlee, William P. Dillon, and John A. Grow*	95
SEDIMENTATION ON THE EASTERN UNITED STATES CONTINENTAL SLOPE . . . . *Larry J. Doyle, Orrin H. Pilkey, and C. C. Woo*	119
GEOTECHNICAL PROPERTIES OF CONTINENTAL SLOPE DEPOSITS—CAPE HATTERAS TO HYDROGRAPHER CANYON . . . . *George H. Keller, Douglas W. Lambert, and Richard H. Bennett*	131
RECENT HISTORY OF MASS-WASTING ON THE UPPER CONTINENTAL SLOPES, NORTHERN GULF OF MEXICO, AS INTERPRETED FROM THE CONSOLIDATION STATES OF THE SEDIMENT . . . . *James S. Booth*	153
DEEP CARBONATE BANK MARGIN STRUCTURE AND SEDIMENTATION IN THE NORTHERN BAHAMAS . . . . *Henry T. Mullins and A. Conrad Neumann*	165
SEDIMENT FACIES OF PLATFORM-BASIN TRANSITION, TONGUE OF THE OCEAN, BAHAMAS . . . . *Wolfgang Schlager and Andrew Chermak*	193

II. PACIFIC AND ANTARCTIC SLOPES

SANTA CRUZ BASIN, CALIFORNIA BORDERLAND: DOMINANCE OF SLOPE PROCESSES IN BASIN SEDIMENTATION . . . . *T. R. Nardin, B. D. Edwards, and D. S. Gorsline*	209
SMALL-SCALE SLUMPS AND SLIDES AND THEIR SIGNIFICANCE FOR BASIN SLOPE PROCESSES, SOUTHERN CALIFORNIA BORDERLAND . . . . *Michael E. Field and Samuel H. Clarke, Jr.*	223
SLOPE AND BASIN BENTHIC FORAMINIFERA OF THE CALIFORNIA BORDERLAND . . . . *Robert G. Douglas and Hal L. Heitman*	231
QUATERNARY SEDIMENTATION ON THE TECTONICALLY ACTIVE OREGON CONTINENTAL SLOPE . . . . *L. D. Kulm and K. F. Scheidegger*	247

SEDIMENTATION ON THE ANTARCTIC CONTINENTAL SLOPE
.... *John B. Anderson, Dennis D. Kurtz, and Fred M. Weaver* 265

PART 3. ANCIENT CONTINENTAL SLOPES

ANCIENT CONTINENTAL SLOPE SEQUENCES AND THEIR VALUE IN UNDERSTANDING MODERN SLOPE DEVELOPMENT
.... *Harry E. Cook* 287

DEPOSITIONAL MECHANICS OF THICK-BEDDED SANDSTONES AT THE BASE OF A SUBMARINE SLOPE, TOURELLE FORMATION (LOWER ORDOVICIAN) QUEBEC, CANADA ... *Richard N. Hiscott and Gerard V. Middleton* 307

SAND TRANSPORT THROUGH CHANNELS ACROSS AND EOCENE SHELF AND SLOPE IN SOUTHWESTERN OREGON
.... *R. H. Dott, Jr. and Kenneth J. Bird* 327

CATSKILL DELTA SLOPE SEDIMENTS IN CENTRAL APPALACHIAN BASINS; SOURCE AND RESERVOIR DEPOSITS
.... *J. Douglas Glaeser* 343

SLOPE SEDIMENTS IN SMALL BASINS ASSOCIATED WITH A NEOCENE ACTIVE MARGIN, WESTERN HOKKAIDO ISLAND, JAPAN .... *George deVries Klein, Hakuyu Okada, and Kiyohiro Mitsui* 359

PART I
THE CONTINENTAL SLOPE: OVERVIEWS

CONTINENTAL SLOPES

ARNOLD H. BOUMA
U.S. Geological Survey
Menlo Park, California 94025

ABSTRACT

The continental slope is defined as the zone extending from the shelf break and terminating at the continental rise where the gradient becomes less than 1:40 or where the slope is bounded by a deep-sea trench or a marginal plateau. Although the slope is commonly the steepest physiographic province of the continental margin, a single simple definition cannot be given. Its width ranges from 20 to 100 kilometers; its upper limit at the shelf break normally starts in water depths between 100 and 200 meters; its lower boundary falls in the depth range 1400 to more than 3200 meters. The slope province can be smooth or terraced, may contain steep escarpments or intraslope basins, or can be very irregular as a result of slumping, fault scarps, or diapirs. Most slopes are cut transversely at many places in their upper part by submarine canyons that connect to deep-sea fans which form constructive sediment bodies on lower slopes, rise, trenches, and parts of the abyssal plain.

Because slopes are located in both active and inactive tectonic settings, the concepts of sea-floor spreading and plate tectonics are critical for interpreting the evolution of the slope. Construction by sediment up- and out-building, slumping and sliding, the influence of sea level fluctuations, reef formation or salt tectonics result in different slopes than are formed by the action of subduction and plate accretion alone. However, the accretionary basins in some active margin settings may have substantial sediment buildup as in passive margin settings.

As the transition zone between the continental shelf and the deep sea, the continental slopes form a major segment of sea-floor studies for scientific, economic and political reasons. Differing tectonic and depositional histories result in a variety of sediment suites and morphology, and in some areas these factors combine to produce conditions favorable for accumulation of economic quantities of oil and gas as well as selected hard minerals.

INTRODUCTION

The continental slope is the central part of the continental margin and is generally located above the transition between continental and oceanic crust. Morphologically, the slope starts at the shelf break. Its lower boundary, however, is less well defined, as it can lie in different physiographic provinces. Half of the world's continental slopes terminate in deep-sea trenches or shallower bordering depressions, the rest in continental rises or deep-sea fans (Curray, 1966). The continental slopes cover an area of approximately 28.7 million km^2, which is 5.6% of the earth's surface (Drake and Burk, 1974). Their lengths total more than 110,000 km (Stanley, 1969 b).

The term "continental slope" was proposed by Wagner (1900) for the entire area between the shelf break and the abyssal floor. Heezen, Tharp, and Ewing (1959, p. 18–19) restricted the term to "that relatively steep (3–6°) portion of the sea floor which lies at the seaward border of the continental shelf." Their definition is still in use and is sufficiently general to define the continental slopes despite their nonuniformity.

The physiographic components of continental margins have characteristics that depend on their tectonic and sedimentary histories. The sedimentary history is closely governed by sediment input and structural deformation. Yarborough (1977) distinguishes three fundamental styles of structural deformation related to evolution of margins and plate tectonics. The first is based on extensional tectonics caused by the break-up of the super continents. The second refers to the deformation of plate boundaries by shear-zone tectonics, while the third relates to subduction that occurs at the leading edge of moving plates.

An understanding of the structural style, geologic history, and types of sediment accumulation on slopes not only suits a scientific curiosity but is critical to the petroleum geologist who has to evaluate source rock and reservoir characteristics, influence of evaporites and related tectonism, maturation levels, effectiveness of seals, and other parameters.

Research by the academic and government oceanographic communities, as well as the petroleum industry, has provided a better, though still incomplete, understanding of the geologic processes of this physiographic province. More effort, especially from the industrial side, has been concentrated on the continental slope of the northern Gulf of Mexico than on any other slope in the world, and even knowledge gained by that effort is far from complete (see Bouma, Moore and Coleman, 1976, 1978).

MORPHOLOGIC CHARACTERISTICS

Nonuniformity of continental slopes makes it impossible to give a sharp definition of this physiographic province. The variation in shape, gradient, and location of the lower boundary makes it necessary to provide a lengthy description if all slopes are to be discussed. Shepard, in his "Submarine Geology" (1948, 1963, 1973), presents a tour du monde which is still the best available. In his third edition (1973), he combines the continental shelf and slope under the term continental terrace because of their close interrelations.

The upper boundary of the continental slope is at the shelf break, a physiographic feature usually easy to define. The break is found in water depths ranging from about 100 to 200 m and may be sharp or gradual, but the increase in slope can generally be found.

The lower boundary typically is gradual and located at depths ranging from 1400 to 3200 m of water, and locally at much greater depth. The most classical boundary is at the change into the continental rise where the gradient decreases to 1:40. This boundary is found in less than half of all continental slopes, however. Locally, the lower slope and the rise may be covered by a deep-sea fan, and where it is, no real lower boundary exists. In places the lower boundary is found in a depression, such as a deep-sea trench, where the continental slope may terminate at a depth of 10,000 m, or a shallower group of bordering depressions such as the southern California Borderland. Where such a borderland or a marginal plateau is present, the slope may be divided into two parts, one on the inner edge and one on the outer edge. The outer-edge slope is referred to as a marginal escarpment (Heezen, Tharp and Ewing, 1959).

The continental slope is a very narrow province, its width ranging from 20 to 100 km. As a consequence, it displays a relatively steep gradient, averaging 4.17° for the upper 1800 m (Shepard, 1973). Off major deltas, this angle may be 1.3°, off fault coasts with negligible shelves about 5.6°, off young mountain coasts 4.6°; the average slope over the upper 2000 m off stable coasts that have no major rivers averages about 3° (Shepard, 1963, p. 298). Within the realm of generalization, it can be stated that continental slopes in the Pacific are steeper than those in the Atlantic, and those in the Atlantic are steeper than those in the Indian Ocean. Steepness greater than 10° may occur in walls of submarine canyons and at places on margins of small basins such as those in the southern California Borderland (Emery, 1960). The steepest slopes observed are off coral islands, where the angle can be 45° down to considerable depth (Fairbridge, 1966).

The slope may be smooth, slightly convex, or irregular on a large or small scale. It may be terraced on the middle part as in the western Gulf of Alaska above the Aleutian Trench (von Huene and Shor, 1969; von Huene, 1972), or consist of an inner and an outer slope as in the southern California Borderland (Emery, 1960; Moore, 1969) or the Blake Plateau (Curray, 1966). The borderland consists of banks and islands with separating deep basins. In borderland regions, the continental slope is off the outer bank where the drop to the deep sea is located. The Blake Plateau is a marginal plateau of intermediate depth with a relatively smooth surface that separates the continental slope into an upper and a lower section. Smaller scale irregularities are produced by mass movement, folding and faulting, diapirism causing intraslope basins and highs, and slope erosion.

Slope trends are remarkably straight or gently curving and do not reflect the irregular nature of the shoreline. Such trends strongly favor structural control and/or smoothing by prograding sediments. Major deltas are an exception. Audley-Charles, Curray, and Evans (1977) recognize four principal types of modern deltas in terms of drainage patterns and tectonic setting: deltas on continental crust, at continental margins, at rifted continental margins, and at marginal ocean basins. Deltas that develop entirely on continental crust generally do not build over extensive marine evaporites that might underly a deltaic pile, nor are they likely to have a well-developed deep-water fan facies. In contrast, most major deltas developed at continental margins have a deep-water fan, commonly larger in volume than that of the shallow-water deltaic prism.

A dominant characteristic of the morphology of continental slopes is the many incisions cut more or less transversely. These submarine canyons may start near the top of the slope or even on the continental shelf. Where they head depends primarily on the width of the shelf and secondarily on the availability of sand and silt-sized material that can be transported by bottom flow from shallow to deep water, thereby causing erosion. These remarkable features are the feeders for sands and silts to deep-sea fans, continental rises and abyssal plains. Submarine canyons are influenced by different dynamic processes and contain in their axes sediments that are different from those normally found on the adjacent uncut parts of the continental slopes (Shepard and Dill, 1966; Whitaker, 1976; Shepard et al, 1977).

Origin and Internal Structure

Modern geophysical techniques and the Deep Sea Drilling Project have provided significant data needed to understand the internal structure and the origin of the continental slope and a base for replacing older theories by new ones. An attempt to review the great body of modern geologic literature falls beyond the scope of this paper. Most of those works can be found in Shepard (1973) and in the many bibliographies in edited publications such as those of Stanley (1969a), Bird and Isacks (1972), Kahle (1974), Burk and Drake (1974a), Bouma, Moore and Coleman (1976, 1978), and McFarlan, Drake and Pittman (1977).

One of the first contributions dealing specifically with the origin of the continental slope is a paper by Dietz (1964), who discussed a number of earlier views, and then proposed a new classification (Table I) (see also Emery, 1965; Fairbridge, 1966; Curray, 1969). Since the slope is part of the continental margin, it can belong to either an active or an inactive (passive) margin, depending on whether or not the margin is associated with young deformation, volcanism, or active seismicity. It should be kept in mind, however, that even passive margins are not undeformed (Burk and Drake, 1974b). Inactive margins are also known as Atlantic-type continental margins, though not entirely restricted to the Atlantic. They border the Arctic and Norwegian Seas, the north and south Atlantic except for some small areas (see below), the Indian Ocean except for the Sunda Sea, the Antarctic except for the Scotia Arc, the Bering Sea, the Sea of Okhotsk, the Sea of Japan, the South China Sea, and parts of the Mediterranean (Heezen, 1974). Tectonically active margins or Pacific-type continental margins (Fisher, 1974) are not all confined to the Pacific Ocean; they occur off the Antilles and South Antilles in the Atlantic and off the Indonesia-Sunda Trench in the Indian Ocean.

Thinning of the continental crust and transition to oceanic crust generally takes place underneath the continental slope. Specifically in active margins, however, this transition can become complex. On passive margins, the transition may lie underneath the shelf if progradation of the slope has taken place.

Emery (1977) distinguishes six distinct types of continental slope (Fig. 1) and mentions that some hard-to-define transitions exist. His first main type (Type A) consists of folded or faulted steps or ridges of rock more or less mantled by sediments. Steps may result from stretching, thinning, and isostatic sinking of continental crust during rifting or from deformation of post-rift sediments by rearrangement of crustal plates after rifting. Emery gives examples of type A slopes in the Atlantic (Fig. 1) in the following places: off southern Greenland, off northern Puerto Rico, in the Caribbean off Venezuela, Colombia and Panama, north slope off the Falkland Plateau, and off the northern and southern sides of the Bay of Biscay.

Slope type B is a progradational type, either as delta growth over the shelf edge or along continental margins with a large supply of detrital material from land. Layers of slope sediments very commonly continue seaward within the continental rise with no or few discontinuities. Good examples of Type B slopes, such as off Nova Scotia, inshore of the Blake Plateau, off Rio de Janeiro, off southwestern Africa, and off much of Europe (Fig. 1), are numerous.

Slope type C represents a continuation of shelves that are built upon pre-rift and initial rift rocks with surface irregularities that are small relative to the thickness of the covering sediments. The margin has been down-warped toward the ocean during the entire period of shelf sedimentation, causing shallow marine sediments to form a wedge with greatest thickness of from several to more than 10 km near the shelf break. The face of the continental slope is more or less continuously eroded by mass movement and bottom-density currents, causing a steepening, possi-

TABLE I—CLASSIFICATION OF CONTINENTAL SLOPES

		Example	Declivity
I.	*Primary* (Structural) Flank of accretionary orogen (usually collapsed continental rise prism)	California (Mesozoic); West coast of South America	Steep Steep
II.	*Secondary* (Structural) Continental Rift Scar	Atlantic coast of Africa (?); Gulf of California	
III. Type	*Secondary* (Modified by Sedimentation) 1. Face of prograded paralic beds 2. Up-lapped Continental Rise 3. Continental Embankment (uplapped continental rise in turn converted by delta foreset beds)	Atlantic coast of USA Parts of Antarctica Gulf Coast of USA, Texas and Louisiana	Moderately steep Gentle Very gentle
	4. Carbonate up-building (atoll-like)	Florida (west coast) Yucatan	Very steep

(After Dietz, 1964)

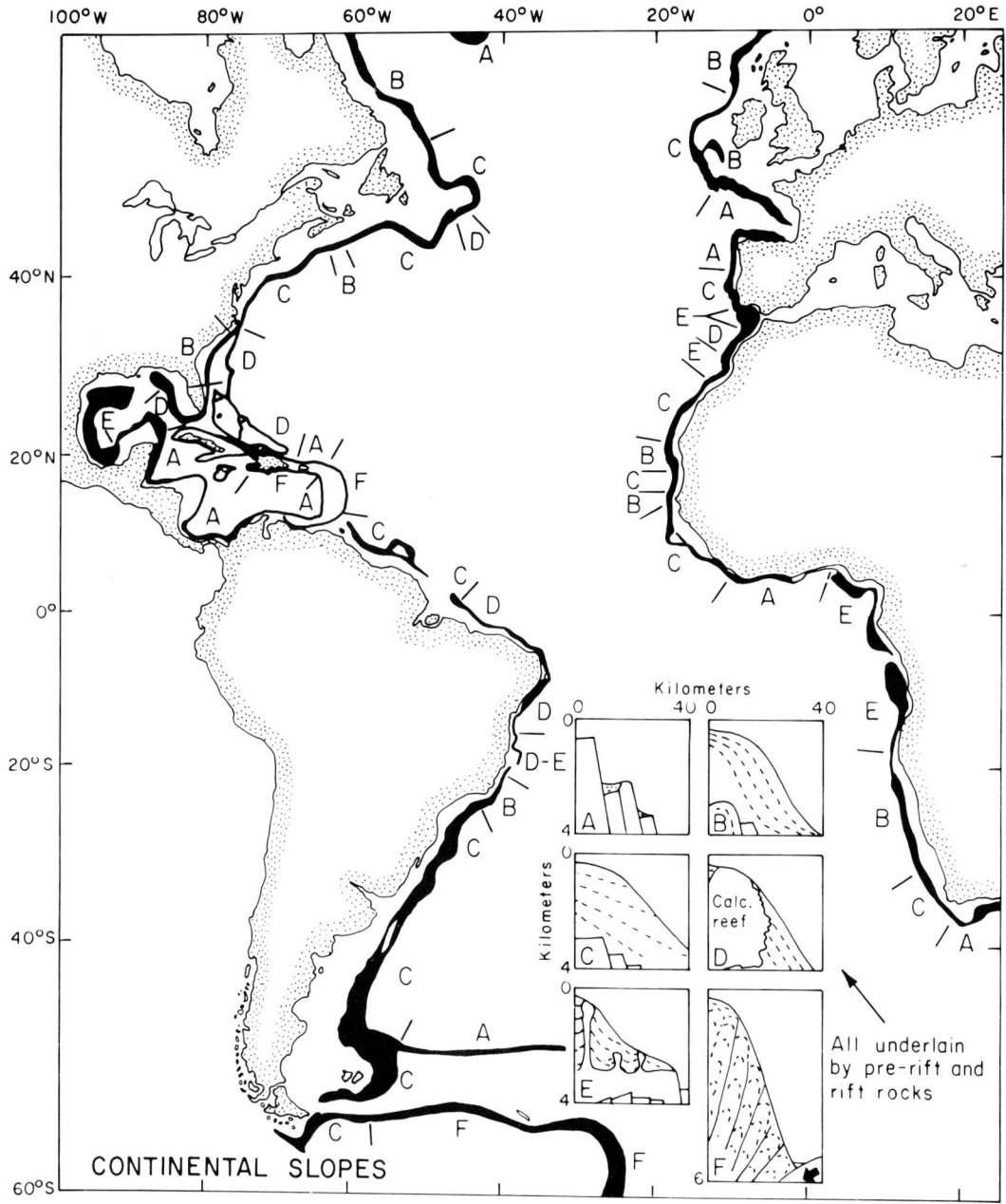

Fig. 1.—Types and distribution of continental slopes in the Atlantic Ocean. Redrawn with permission from K. O. Emery (1977, Fig. 5).

bly a response to a slight rotation during the down-warping. This type of continental slope is the most common in the Atlantic Ocean (Fig. 1).

Slope type D is related to type C in history but differs in that the continental slope is supported by the growth of massive calcareous reefs. The thickness of these reefs requires large subsidence of the underlying crust. The reef may be exposed as a steep outer escarpment or its flanks may be covered by predominantly calcareous sediments. This type slope is restricted to near-tropical waters of the present time, where the input of terrigenous material is low.

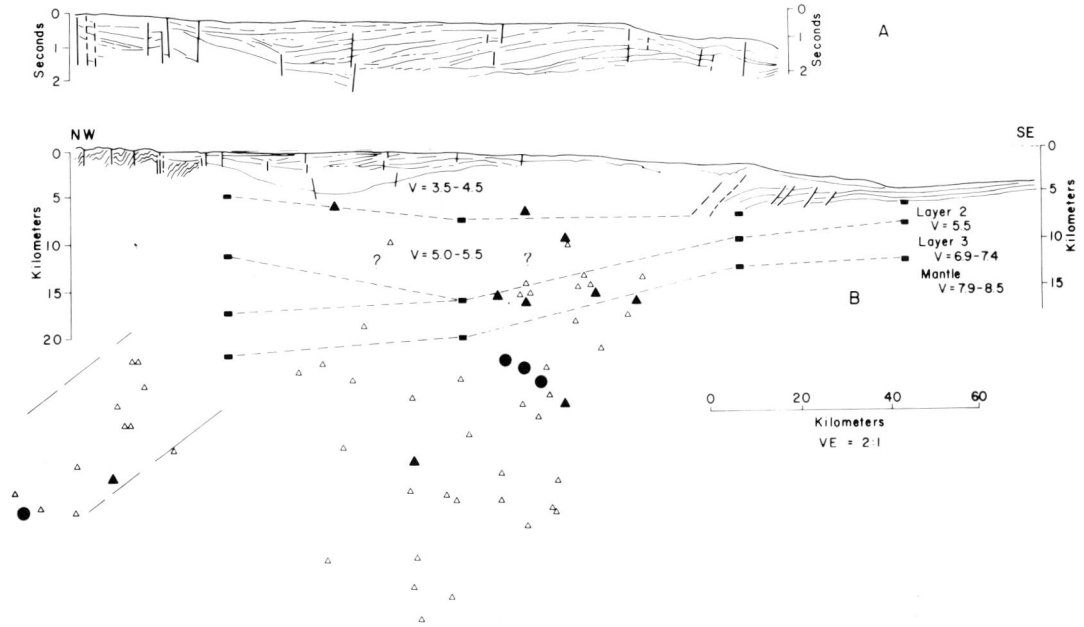

Fig. 2.—Cross section from Kodiak Island to Aleutian Trench, based on 24-fold multichannel record. A) a reflection section. B) depth section. Hypocenters: filled triangle located with more than 50 stations; open triangle located with 10-50 stations; filled circle located from more than 50 stations with depth from $_pp$ phase. Velocity V in km/s. Dashed lines from Shor and von Huene (1972), geology of Kodiak Island from Moore (1969). Reproduced with permission from R. von Huene.

Slope type E is controlled by evaporite layers and their resulting diapirs, commonly an extension of a similar-type shelf. This group includes the mud and shale diapirs, chiefly a product of pro-delta clays. The best known examples are in the Gulf of Mexico, in particular near the Sigsbee Escarpment (see contributions in Bouma, Moore, and Coleman, 1976, 1978), off Angola-Gabon, and off Nigeria (Fig. 1). Sediments become strongly deformed under diapiric action and very irregular bottoms with highs and lows result. Salt diapirism in northwestern Gulf of Mexico is known to prograde seaward, thereby bringing rise sediments to slope depths and slope deposits to shallower locations (Watkins, Worzel, and Ladd, 1976; Watkins et al, 1978).

Emery's last slope type (F) incorporates those structures that are related to convergence. The underthrusting process scrapes off sediments riding on top of the oceanic crust, mixes them with lower continental slope and trench sediments, and deforms the mix into stacked wedges with shearing in between or in any of a number of other arrangements (Fig. 2) (see also Seely and Dickinson, 1977). Although this type can be found in the Atlantic Ocean (south of Puerto Rico and Hispaniola; outer Antilles arc; south of Falkland Plateau), it is far more common in the Pacific.

Both active continental margins and flanks of active island arcs are associated with plate consumption that generates arc-trench systems. Subduction in trenches is reflected by deformation of sediments lying on oceanic crust and can involve part of the upper basement as well. As subduction continues, the deformed rocks form a mass of growing bulk known as the subduction complex (Seely and Dickinson, 1977). Apparently, this complex grows in width as successive increments are added by underthrusting at the trench, and at the same time it tends to grow upward and becomes a positive structural feature. Part of the continental slope in deep-sea trench areas is called the trench inner slope (i.e., from the trench-slope intersection to the first major change in inclination above this intersection) to differentiate it from the trench outer slope, the area between the trench and the seamount chain bordering the adjacent abyssal plain (Fig. 3) as can be observed in the Aleutian setting. This example is actually somewhat unique when considering the rest of the Pacific basin because the trench outer slope may not have a seamount chain and no abyssal plain. At least the lower part of the trench inner slope is generally underlain by the subduction complex, a thick section of deformed abyssal plain and trench deposits containing vari-

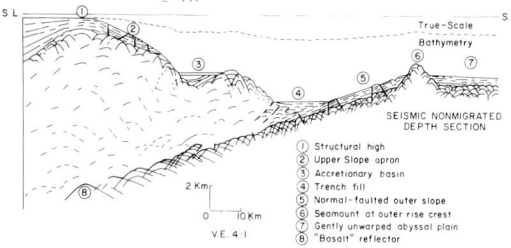

FIG. 3.—Composite diagram of common active margin characteristics. Redrawn with permission from D. R. Seely and W. R. Dickinson (1977, Fig. 6).

able proportions of oceanic crustal pieces (Seely, Vail, and Walton, 1974). This section of lower slope is covered in places by a rather thin veneer of lower slope sediments. It is characterized by a highly irregular surface with accretionary basins and structural bathymetric highs (Fig. 4). The upper part of the inner slope of many trenches is covered by a rather thick sediment apron that has prograded across the edge of the shelf or has been transported by contour currents. Such slopes are found in the eastern Aleutians, Mid-America, Peru-Chile, Tonga-Kermadec, and Japan.

Plate-surface deformation may decrease from the toe of the slope in an upslope direction and faults steepen landward because of the active belt of thrust and folds that influences the abyssal plain and trench sediments scraped off the underthrusting plate. Normally the thrusts dip landward in areas where inner-slope structure can be discerned, and associated folds have seaward vergence. This gives a prevailing landward dip at lower structural levels beneath the slope, causing progressive landward tilting of accretionary basins seen in the steeper landward dips of former slope-basin sedimentary units. Landward vergence is also observed in some cases (Seely, 1977).

The characteristics of the sediments overlying any part of the continental slope are thus constrained by their tectonic and sedimentary histories. Geotechnical properties of the slope's surfi-

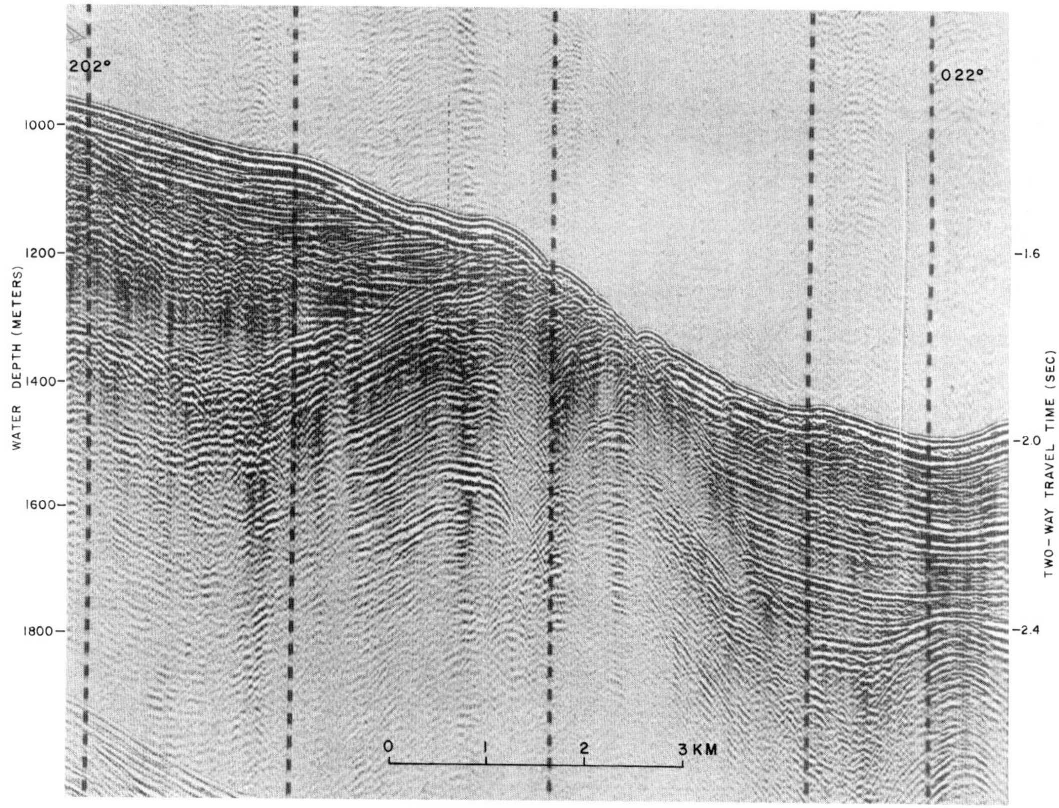

FIG. 4.—Inner trench slope section off Kodiak Island, Alaska, showing spillover from a filled depression into a lower depression separated by a structural high. R/V Sea Sounder, line 292, 1977, 160 KJ Sparker profile.

cial sediments are critical to an understanding of the history of a slope segment.

Sediments and Processes

Sediments.—Mud is the predominant sediment on the continental slope with lesser amounts of sandy and gravelly material. Rock outcrops are common on steep slopes off narrow shelves adjacent to mountainous coasts (Cotton, 1918; Shepard, 1948, 1963 1973; Stanley, 1969a, b; McCave, 1972; Stanley and Swift, 1974). Carbonate sediments are less represented but can be a major constituent in certain regions (Milliman, 1974). In general, the slope sediments are finer than those on the adjacent shelf and continental rise.

Utilization of high-resolution seismic techniques can provide insight into the structural and depositional characteristics of the sediments. Several workers are using the seismic facies models from Sangree and others (1976, 1978) or follow similar approaches. Additional coring and shallow drilling can provide the ground truth and stratigraphy needed to get full value from the seismic results (e.g., Amery, 1976, 1978; Mitchum, 1976, 1978; Sidner, Gartner, and Bryant, 1977, 1978). Damuth and Hayes (1977) demonstrated the value of 3.5 kHz high-frequency precision depth recordings for studying the depositional/erosional processes on the sea floor on the basis of the character of the echo. The echo also reflects the amount of coarse material present. A combination of seismic and coring/drilling is to be expected to provide data needed to understand the sedimentary processes responsible for the materials encountered at any part of sea bottom.

Sea-level Changes.—Several workers stress the significance of sea-level changes as well as tectonic processes as a control of amount and coarseness of sediment supply (Emery, 1968a; Curray, 1977; Vail, 1977; Sidner, Gartner, and Bryant, 1977, 1978). During sea-level lowering, an increase of river runoff takes place and coarse sediments are transported to an advancing coastline that may come close to the shelf break. Stanley, Fenner, and Kelling (1972) stated that transport across the shelf break during low stands of sea level certainly takes place where the shelf is wide. Consequently, an influx of granular material can be moved onto the continental slope under the influence of currents and waves, causing sheet flow or slumping.

During a rising sea level, less material arrives at the shoreline owing to a decrease in gradient of the lower reaches of a river system that results in deposition and reestablishment of a gradient of those lower reaches before sediment can arrive at the shore. Further, the transgressive shoreline will prevent much of the "grainy" material from being transported across the submerged shelf to the shelf break unless a vast growing delta progrades faster than the retreat of the land/sea boundary.

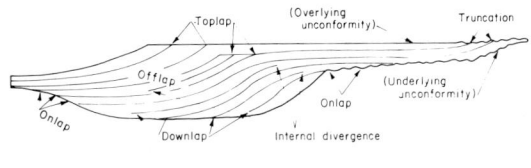
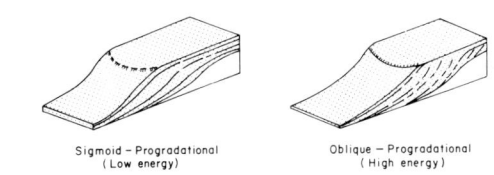

Fig. 5.—Seismic stratigraphic reflection terminations and shelf-margin seismic facies types. Redrawn with permission from P. R. Vail (1977, Figs. 1, 6).

Vail (1977) and Vail et al (1977) recognize depositional sequence boundaries on seismic records by identifying reflections caused by lateral termination of strata, termed onlap, downlap, offlap, and truncation (Fig. 5). Age and areal distribution can be used to determine relative changes in sea level (Fig. 6). Sidner, Gartner, and Bryant (1977, 1978) combined seismic facies with textural and paleontological data and interrelated climatic zones with eustatic sea-level changes and with depositional characteristics for the northwest Gulf of Mexico outer shelf and upper slope (Fig. 7).

Input of Fine-grained Material.—About 8×10^9 tons per year of fine-grained suspended sediment will be deposited on the world's slope and the rise, the greater part going to the large deep-sea fans in front of major supply points of fine-grained materials such as the Ganges, Indus, and Mississippi Rivers (McCave, 1972). Although the transgressive period is lacking activity in a depositional sense, recent investigations clearly reveal that the area slope is far from quiescent.

Several processes take place under the influence of general meteorological and climatological conditions. Pierce, Nelson, and Colquhoun (1972) observed intrusion of colder slope water onto the shelf during winter and spring months, primarily a result of greater density of coastal waters at lower temperatures. Cascading of cold shelf water over the Carolinian Slope surface resulted from this temperature drop. During late summer and early fall, these workers observed intrusions of slope water onto the shelf caused by upwelling

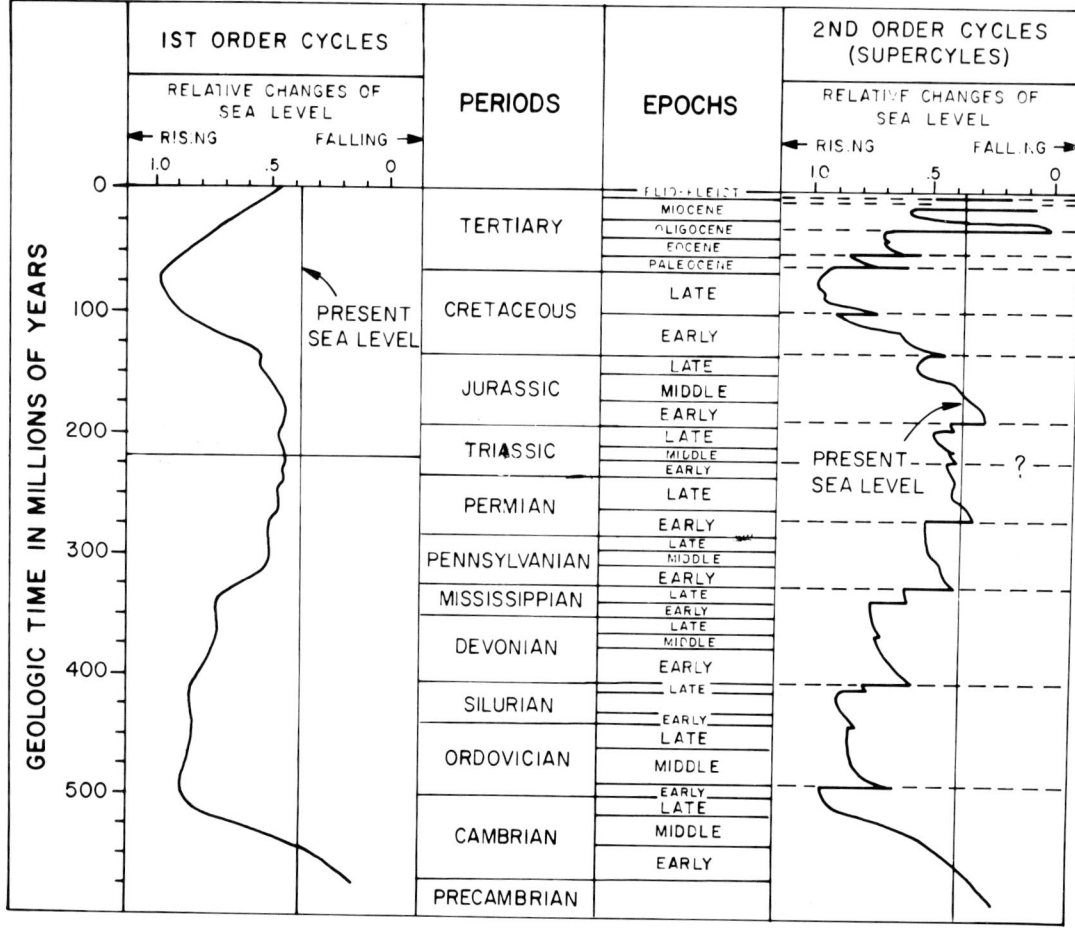

Fig. 6.—Relative changes of sea level based on seismic reflection stratigraphy. Redrawn with permission from P. R. Vail (Vail et al, 1977, Fig. 1).

and thermal stratification that prevented mixing. Such flows are strong enough to move fine bottom sediment or prevent deposition of fine-grained matter.

Bruce (1977) demonstrated the seasonal character of the Somali Current, which develops between May and September during the southwest monsoon in the Indian Ocean. The highest velocities, 350 cm/sec at the surface, were measured during July and August. Such a current is comparable to the Gulf Stream and the Kuroshio, but differs by being shallower (200–400 m) and basically wind driven. The current turns sharply in an offshore direction, and cold upwelling water is found in the turning area. Again, bottom effects were not investigated, but the impact may be considerable.

Internal Waves.—Cacchione and Wunsch (1974), in dealing with realistic periods and amplitudes, showed upslope amplification of near-bottom velocities beneath internal waves onto the continental margin. This will cause incipient movement of sediments of a wide range of sizes found on the outer shelf and upper slope, even prior to breaking of the waves (Cacchione and Southard, 1974). The experiments of Southard and Cacchione (1972) showed that near-bottom velocities in the zone of internal-wave breaking are much greater than the amplified velocities just downslope. It seems likely that breaking internal waves are capable of moving sediments, although direct observations are lacking. Southard and Cacchione (1972, p. 96) suggested that several dominant effects in their experiments might be observed when internal waves impinge onto the continental margin:

"1. Breaking of the waves, with advance of large decaying vortices upslope and return flow of mixed fluid downslope, creating a band of

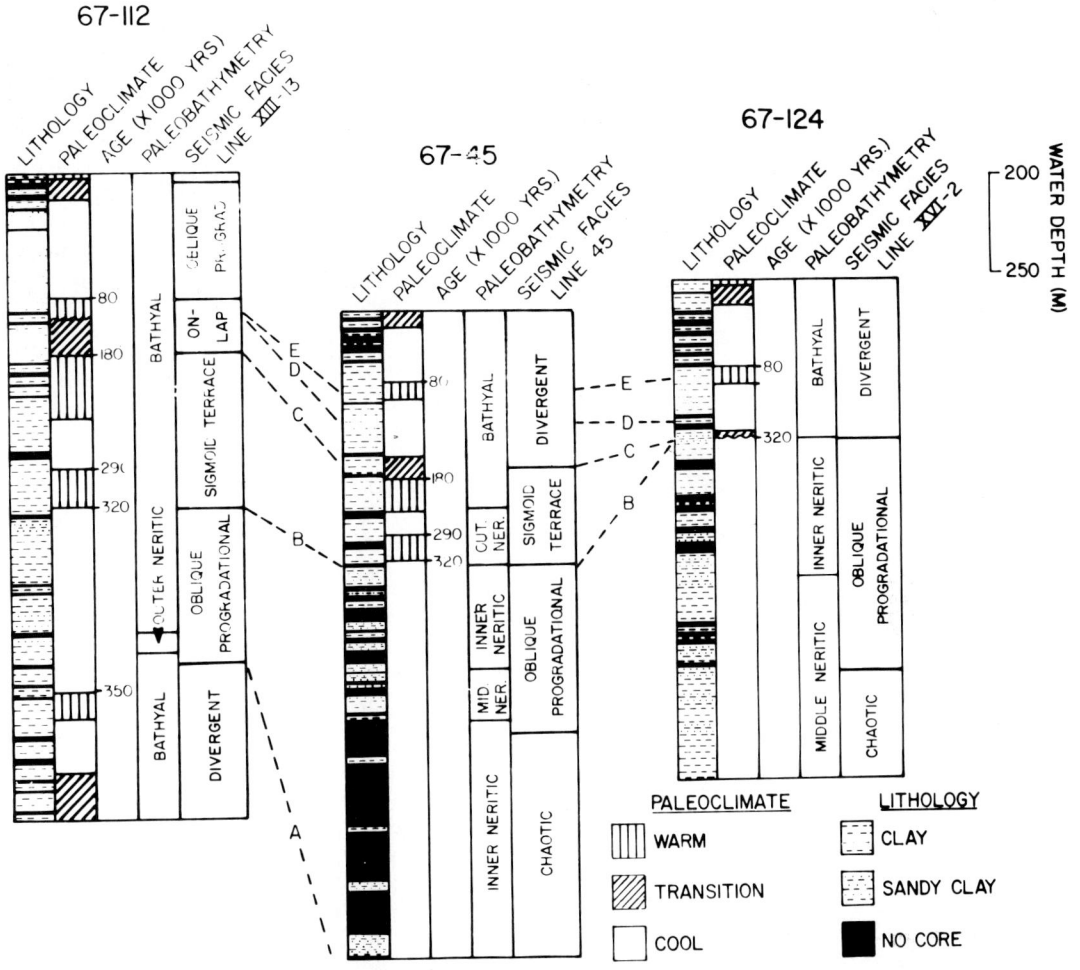

Fig. 7.—Lithologic, chronologic, paleoclimatic, paleobathymetric, and seismic facies summary from three drill sites in northwest Gulf of Mexico. Reproduced with permission from B. R. Sidner (Sidner, Gartner, and Bryant, 1978, Fig. 6).

sediment movement with width at least an order of magnitude greater than the wave amplitude;

2. Abrupt onset of sediment movement at the point of breaking and gradual decrease in intensity downslope;

3. Predominant downslope movement of sediment except near the breaking point, causing deposition in a narrow band just upslope of the point of breaking and erosion in a wide band farther upslope;

4. Generation of downslope-migrating bed forms in the wide band of net downslope sediment transport; and

5. Suspension of fine sediment by the advancing breakers and horizontal transport away from the slope in the mixed layer.

In regard to generation of bed forms, the downslope flow would produce trains of asymmetrical sediment ripples migrating downslope provided only that the current is strong enough to transport at least some sediment as bedload (and that sediment is not coarser than about one-half millimeter). If the maximum downslope velocity, attained not far upslope of the point of breaking, is great enough, dunes or sand waves migrating downslope might also be produced (Simons and Richardson, 1963; Southard, 1971). In regard to suspension of fine sediment, horizontal spreading of mixed fluid away from the slope at the appropriate density level might contribute to the development of turbidity maxima observed in oceanic thermoclines (Costin, 1970)."

Tidal Effects and Submarine Canyons.—Stanley, Fenner, and Kelling (1972) discussed the tidal influence of water masses in transporting coarse sediment from the outer shelf onto the upper slope. Their observations in submarine canyons and the many contributions from Shepard and his co-workers demonstrate that slow progradation of the outer continental shelf is continuing as modern spillover. These modern depositional processes continue to modify the shelf-break environment.

Dill and other investigators demonstrated in the late fifties and early sixties that inner shelf sediments transported along the coast by longshore currents can end up in the heads of submarine canyons. This trapping is particularly critical on narrow shelves when such canyons start close to shore (see Shepard and Dill, 1966).

Although it is generally accepted that turbidity currents originate from shallow-water sediment trapped in the upper reaches of submarine canyons, the less spectacular daily processes in submarine canyons and fan valleys commonly result in a net upward motion under the action of tidal forces that have a concentrated effect in these continental slope incisions. Shepard et al (1977) recently demonstrated the existence of a cycle consisting of a relatively strong upcanyon flow followed by a rapid build-up of a much stronger downcanyon current too weak to register. By their observations, these flows are slow turbidity currents activated by river, floods and favorable storm conditions. They distinguish two types of canyons. One type is present off large rivers with a clear relation to river mouths. The processes here are primarily marine; turbidity currents will be initiated at frequent intervals off such river mouths when large masses of relatively coarse material are introduced onto the sea floor which slopes steeply into deep water. The other type of canyon lies off large deltas that have built across the shelf; they are characterized by a large number of valleys cutting the advancing slopes.

Carbonates.—Although carbonate-containing continental slopes are less common than noncarbonate slopes, their presence warrants more than just their mention. Milliman (1974) has summarized much of the literature about carbonates; detailed discussions and references can be found in his book.

Marine carbonates are divided into shallow water (reef and upper shelf), sublittoral (shelf and upper slope: 20–300m), and deep sea (>200 m). Milliman (1974) states that the percentage of total carbonate in slope sediments averages 5.8% (see also Sverdrup, Johnson, and Fleming, 1942). The average carbonate deposition rate on the slope is 1.5 g/cm^2/1000 yrs. This value may be subject to revision when further data become available.

The sublittoral zone is the transition from shallow to deep sea. Decrease in temperature and amount of light penetration, accompanied by an increase in hydrostatic pressure characterizes this zone. Many present-day transitional carbonates consist of material that was deposited during the last rise in sea level (Milliman and Emery, 1968). Calcium carbonate solubility depends on the temperature of the water. Emery (1968b) suggested that biogenic carbonate deposition increases in lower latitude shelves and correspondingly is more important on the warmer western sides of the ocean than on the eastern sides. Complications arise where there is an influx of terrigenous sediment causing dilution and burial of many carbonate producing organisms. The carbonate content of the upper continental margin is, however, a first-order function of sediment supply and a second-order function of water temperature and corresponding carbonate productivity. One observes a decrease of benthonic foraminifera and a relative increase in planktonics from the outer continental shelf onto the upper slope. In addition one commonly finds echinoid spines in upper slope sediments.

Owing to a greater current activity and larger distances from rivers, many outer continental shelves and uppermost slopes contain deposits that are richer in carbonates than those on the adjacent inner shelf. Encrusting coralline algae, together with corals, bryozoans and barnacles form a prominent facies along the shelf edge. Barnacles tend to decrease toward tropical latitudes, a trend that is likely due to certain fish that graze seemingly continuously on the substrate, preventing settlement of larvae.

Many outer shelves and upper slopes in water depths of 40–100 m are characterized by small topographic highs rising normally not more than 1–5 m above the surrounding bottom allowing them to be covered with various carbonate producing organisms and sediments. Such algal ridge system or reef is discontinuous and tends to parallel the shelf break. The substrate normally is algal that started during low stands of sea level. It is suggested that increased biological productivity at the shelf break due to upwelling is the main reason for their location. Their inception during early-transgression is partly supported by the presence of oolites on many shelves and upper slopes. Modern oolites tend to form in warm, agitated water, generally in depths less than 2 m. The reddish-brown color and the 10–15 thousand-year age based on carbon-14 dating make those brown oolites relict. Their position outside the algal ridge system in depths greater than 163 m, however, is still hard to explain.

Instability.—Instability phenomena, predominantly slumps, are common features on the slope proper. Unstable conditions can result from "rapid" accumulation due to the nearness of a sediment source, sudden impinging forces such as large storm waves or earthquakes, tectonic upwarping resulting from diapiric activity or upheaval of the upper plate in a subduction zone as demonstrated by Hampton and Bouma (1977), and Hampton et al (1978) (Fig. 1). Quantitative information is still far from abundant although an increase in marine geotechnical studies can be noted. Such investigations, together with seismic surveying, side-scan sonar, direct bottom observations, coring, drilling and current and wave measurements, become a high priority task for understanding natural hazards as part of scientific and applied research (Hampton et al, 1978).

ANCIENT CONTINENTAL SLOPES

A study of the present continental margins commonly requires evaluation of their past history as well as an understanding of old margins now completely encased in continental crust (Burk and Drake, 1974a, b). Ancient continental margins in a general sense have been studied since the early days of geology, and during the past few decades the concepts of sea-floor spreading and plate tectonics have found their way into these studies. The body of available literature is overwhelming; some of the more recently edited works provide the more informative papers and bibliographies (see Burk, 1965; Kay, 1969; Bird and Isacks, 1972; Burk and Drake, 1974a; Dickinson, 1974; Dott and Shaver, 1974; Kahle, 1974; McFarlan, Drake, and Pittman, 1977).

Since the continental slope is part of the continental margin and tectonically not a unit by itself, no author has singled out this physiographic province for an in depth publication. The ancient slope is mentioned and sometimes described in sedimentological contributions as part of submarine canyon-deep sea fan studies where channelized facies cut into shaly slope deposits. Specifically worth mentioning are the Italian contributions (Mutti and Ricci Lucchi, 1972, 1975; Mutti, 1977) and subsequent investigations based on their work (e.g., Walker and Mutti, 1973; in Dott and Shaver, 1974; in Whitaker, 1976; Nilsen, 1977; Bouma and Nilsen, 1978).

ECONOMIC ASPECTS AND CONSIDERATIONS

Hedberg and Moody (1977, p. 62) claim that a review of the "probable geology of the continental slopes and rises over the world conclusively shows that the petroleum prospects of these oceanic realms are at least good enough to justify extensive exploration drilling, if, as, and when, there is demand for petroleum at prices adequate to cover the costs of production in these environments" (see also Hedberg, 1970). Weeks (1974) indicated that about 4.6×10^6 km^2 of continental slope may be prospective for hydrocarbons.

Although organic-rich sediments are commonly found in the oxygen minimum zone, which ranges from 200 m to 400 m off Peru (Hafferty et al, 1978), organic-rich deposits do occur at much greater depths on the continental slope (Suess, 1976). We know little in detail about the present slope as the recent studies of the northern Gulf of Mexico intraslope basins have shown (Bouma et al, 1975, 1976, 1978; Trabant and Presley, 1978; McKee et al, 1978) (Fig. 8). Moreover, Dow (1977) notes that continental slopes and rises are commonly the sites for high organic productivity because of nutrients supplied by upwelling and river runoff. Slope sediments average 0.6-1.0% by weight organic carbon, which makes them the most organic-rich continental margin deposits. Appropriate time and temperature conditions are required to convert organic matter to petroleum, which means 2 to 4 km of burial is required for oil generation and 3 to 7 km for gas, depending on the geothermal gradient, rate of accumulation, and age of the source section. One of the major problems is to find reservoir rocks with adequate porosity and favorable structure. Except for carbonate areas we have to search for the submarine canyon-deep sea fan systems, as the slope sediments normally are too clayey to provide sufficient porosity and permeability (see also Yarborough, 1977).

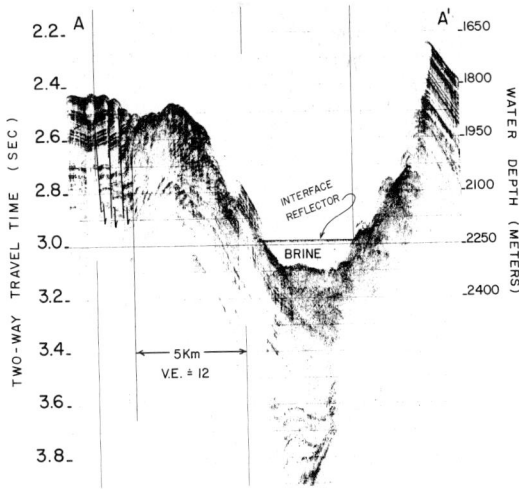

FIG. 8.—High-resolution, 3.5 kHz, seismic profile over the Orca Basin, north central Gulf of Mexico, showing a strong reflector between normal sea water and the brine. Reproduced with permission from P. K. Trabant (Trabant and Presley, 1978, Fig. 3).

In addition to hydrocarbons, a few other minerals of economic interest may be present in certain slope sediments (Cruickshank, 1974). Mineralization associated with subduction can provide metallogenic belts with good core bodies in island arc areas but little research on this topic has been done so thus far. Certain types of nodules and concretions, metalliferous muds, phosphorite, barite, and glauconite may be some of the more economic products that can be found in surficial and subsurface recent and ancient slope areas once the market shows depletion of the more obvious locations.

ACKNOWLEDGMENTS

Larry J. Doyle, Dave A. Cacchione, Robert H. Dott, Jr., Laverne D. Kulm and Kenneth O. Emery offered valuable suggestions on the manuscript. Appreciation also expressed to K. O. Emery, R. von Huene, P. R. Vail, B. R. Sidner, and P. K. Trabant for their permission to use their illustrations.

REFERENCES

AMERY, G. B., 1976, Structure of continental slope, northern Gulf of Mexico, in Bouma, A. H., Moore, G. T., and Coleman, J. M., eds., Beyond the Shelf Break: Am. Assoc. Petroleum Geologists, Marine Geol. Comm. Short Course, v. 2, p. H1–H16.

———, 1978, Structure of continental slope, northern Gulf of Mexico, in Bouma, A. H., Moore, G. T., and Coleman, J. M., eds., Framework, Facies, and Oil Trapping Characteristics of Upper Continental Margin: Am. Assoc. Petroleum Geologists, Studies in Geology, No. 7, p. 141–154.

AUDLEY-CHARLES, M. G., CURRAY, J. R., AND EVANS, G., 1977, Location of major deltas: Geology, v. 5, p. 341–344.

BIRD, J. M., AND ISACKS, B., eds., 1972, Plate Tectonics, selected papers from the Journal of Geophysical Research: Washington, D.C., Amer. Geoph. Union, 951 p.

BOUMA, A. H., MOORE, G. T., AND COLEMAN, J. M., eds., 1976, Beyond the Shelf Break: Am. Assoc. Petroleum Geologists, Marine Geol. Comm. Short Course, v. 2, New Orleans, La., 163 p.

———, ———, AND ———, eds., 1978, Framework, Facies, and Oil Trapping Characteristics of Upper Continental Margin: Am. Assoc. Petroleum Geologists, Studies in Geology, No. 7, introduction.

———, AND NILSEN, T. H., 1978, Turbidite facies and deep-sea fans—with examples from Kodiak Island, Alaska: Offshore Technology Conf., Preprints OTC 3116, p. 559–570.

———, SMITH, L. B., SIDNER, B. R., AND MCKEE, T. R., 1975, Submarine geomorphology and sedimentation patterns of the Gyre Intraslope Basin, northwest Gulf of Mexico: Texas A & M Univ., Dept. of Ocean., Tech. Rept. No. 75-9-T, 163 p.

———, ———, ———, AND ———, 1976, Gyre Basin, an intraslope basin in northwest Gulf of Mexico, in Bouma, A. H., Moore, G. T., and Coleman, J. M., eds., Beyond the Shelf Break: Am. Assoc. Petroleum Geologists, Marine Geol. Comm. Short Course, v. 2, p. E1–E28.

———, ———, ———, AND ———, 1978, Intraslope basin in northwest Gulf of Mexico, in Bouma, A. H., Moore, G. T., and Coleman, J. M., eds., Framework, Facies and Oil Trapping Characteristics of Upper Continental Margin: Am. Assoc. Petroleum Geologists, Studies in Geol., No. 7, p. 298–302.

BRUCE, J. G., 1977, Somali Current: recent measurements during the southwest monsoon: Science, v. 197, p. 51–53.

BURK, C. A., 1965, Geology of the Alaska Peninsula Island Arc and Continental Margin, part 1: Geol. Soc. America Memoir 99, 250 p.

———, AND DRAKE, C. L., eds., 1974a, The Geology of Continental Margins: New York, Springer-Verlag, 1009 p.

———, AND ———, 1974b, Continental Margins in perspective, in Burk, C. A., and Drake, C. L., eds., the Geology of Continental Margins: New York, Springer-Verlag, p. 1003–1009.

CACCHIONE, D. A., 1970, Experimental study of internal gravity waves over a slope: Mass. Institute Technology, Dept. Earth and Planetary Sciences., Rept. 70-6, 226 p.

———, AND SOUTHARD, J. B., 1974, Incipient sediment movement by shoaling internal gravity waves: Jour. Geophys. Res., v. 79, p. 2237–2242.

———, AND WUNSCH, C., 1974, Experimental study of internal waves over a slope: Jour. Fluid. Mech., v. 66, p. 223–239.

COSTIN, J. M., 1970, Visual observation of suspended-particle distribution at three sites in the Caribbean Sea: Jour. Geophys. Res., v. 75, p. 4144–4150.

COTTON, C. A., 1918, Conditions of deposition on the continental shelf and slope: Jour. Geol., v. 26, p. 135–160.

CRUICKSHANK, M. J., 1974, Mineral resources potential of continental margins, in Burk, C. A., and Drake, C. L., eds., The Geology of Continental Margins: New York, Springer-Verlag, p. 965–1000.

CURRAY, J. R., 1966, Continental terrace, in Fairbridge, R. W., ed., Encyclopedia of Oceanography: New York, Reinhold Publ. Corp., p. 207–214.

———, 1969, Shallow structure of the continental margin, in Stanley, D. J., (convener), The new Concepts of Continental Margin Sedimentation: Amer. Geol. Inst. Short course lecture notes,: Washington, D. C., p. JC-XI1-1-22.

———, 1977, Modes of emplacement of prospective hydrocarbon reservoir rocks of outer continental marine environments, in McFarlan, E., Drake, C. L., and Pittman, L. S. eds., Geology of Continental Margins: Am. Assoc. Petroleum Geologists Continuing Education Course Note Series No. 5, p. E1–E14.

Damuth, J. E., and Hayes, D. E., 1977, Echo character of the East Brazilian continental margin and its relationship to sedimentary processes: Mar. Geol., v. 24, p. 73–95.

Dickinson, W. R., ed., 1974, Tectonics and Sedimentation: Soc. Econ. Paleontologists Mineralogists Spec. Pub. No. 22, 240 p.

Dietz, R. S., 1964, Origin of continental slopes: Amer. Scientist, v. 52, p. 50–69.

Dott, R. H., Jr., and Shaver, R. H., eds., 1974, Modern and Ancient Geosynclinal Sedimentation: Soc. Econ. Paleontologists Mineralogists Spec. Publ. No. 19, 380 p.

Dow, W. G., 1977, Petroleum source beds on continental slopes and rises, in McFarlan, E., Drake, C. L., and Pittman, L. S., eds., Geology of Continental Margins: Am. Assoc. Petroleum Geologists, Continuing Education Course Note Series No. 5, p. D1–D37.

Drake, C. L., and Burk, C. A., 1974, Geological significance of continental margins, in Burk, C. A., and Drake, C. L., eds., The Geology of Continental Margins: New York, Springer-Verlag, p. 3–10.

Emery, K. O., 1960, The Sea off Southern California, a modern habitat of petroleum: New York, Wiley 366 p.

———, 1965, Characteristics of continental shelves and slopes: Am. Assoc. Petroleum Geologists Bull., v. 49, p. 1379–1384.

———, 1968a, Shallow structure of continental shelves and slopes: Southeastern Geol., Duke Univ., v. 9, p. 173–194.

———, 1968b, Relict sediments on continental shelves of the world: Am. Assoc. Petroleum Geologists, v. 52, p. 445–464.

———, 1977, Stratigraphy and structure of pull-apart margins, in McFarlan, E., Drake, C. L., and Pittman, L. S., eds., Geology of Continental Margins: Am. Assoc. Petroleum Geologists, Continuing Education Course Note Series No. 5, p. B1–B20.

Fairbride, R. W., 1966, Continental slope, in Fairbride, R. W., ed., Encyclopedia of Oceanography: New York, Reinhold Publ. Corp., p. 205–207.

Fisher, R. L., 1974, Pacific-type continental margins, in Burk, C. A., and Drake, C. L., eds., The Geology of Continental Margins: New York, Springer-Verlag, p. 25–41.

Hafferty, A. J., Codispoti, L. A., and Huyer, A., 1978, JOINT II, R/V Melville Legs I, II and IV, R/V Iselin Leg II: Bottle data March 1977–May 1977, International Decade of Ocean Exploration, Coastal Upwelling Ecosystems Analysis, Data Report 45 (in press).

Hampton, M. A., and Bouma, A. H., 1977, Slope instability near the shelf break, western Gulf of Alaska: Marine Geotechnology, v. 2, p. 309–331.

———, Bouma, A. H., Carlson, P. R., Molnia, B. M., Clukey, E. C., and Sangrey, D. A., 1978, Quantitative study of slope instability in the Gulf of Alaska: Offshore Technology Conf., Preprints OTC 3314, p. 2307–2318.

Hedberg, H. D., 1970, Continental margins from viewpoint of the petroleum geologist: Am. Assoc. Petroleum Geologists Bull., v. 54, p. 3–43.

———, and Moody, J. D., 1977, Petroleum prospects of continental slopes and rises: Am. Assoc. Petroleum Geologists and Soc. Econ. Paleontologists Mineralogists Program and Abs., Washington, D. C., p. 62–63.

Heezen, B. C., 1974, Atlantic-type continental margins, in Burk, C. A., and Drake, C. L., eds., The Geology of Continental Margins: New York, Springer-Verlag, p. 13–24.

———, Tharp, M., and Ewing, M., 1959, The Floors of the Oceans: I. The North Atlantic. Geol. Soc. America Spec. Paper No. 65, 122 p.

Kahle, C. F., ed., 1974, Plate Tectonics—Assessments and Reassessments: Am. Assoc. Petroleum Geologists, Memoir 23, 514 p.

Kay, M., ed., 1969, North Atlantic—Geology and Continental Drift: Am. Assoc. Petroleum Geologists, Memoir 12, 1082 p.

McCave, I. N., 1972, Transport and escape of fine-grained sediment from shelf areas, in Swift, D. J. P., Duane, D. B., and Pilkey, O. H., eds., Shelf Sediment Transport: Process and Pattern: Stroudsburg, Dowden, Hutchinson & Ross, Inc., p. 225–248.

McFarlan, E., Drake, C. L., and Pittman, L. S., eds., 1977, Geology of Continental Margins: Am. Assoc. Petroleum Geologists, Continuing Education Course Note Series No. 5, 128 p.

McKee, T. R., Jeffrey, L. M., Presley, B. J., and Whitehouse, U. G., 1978, Holocene sediment geochemistry of continental slope and intraslope basin areas, northwest Gulf of Mexico, in Bouma, A. H., Moore, G. T., and Coleman, J. M., eds., Framework, Facies, and Oil Trapping Characteristics of Upper Continental Margin: Am. Assoc. Petroleum Geologists, Studies in Geology No. 7, p. 313–326.

Milliman, J. D., 1974, Marine Carbonates: New York, Springer-Verlag, 375 p.

———, and Emery, K. O., 1968, Sea levels during the past 35,000 years: Science, v. 162, p. 1121–1123.

Mitchum, R. M., 1976, Seismic stratigraphic investigation of west Florida slope, Gulf of Mexico, in Bouma, A. H., Moore, G. T., and Coleman, J. M., eds., Beyond the Shelf Break: Am. Assoc. Petroleum Geologists, Marine Geol. Comm. Short Course, v. 2, p. I1–I35.

———, 1978, Seismic stratigraphic investigation of west Florida Slope, Gulf of Mexico, in Bouma, A. H., Moore, G. T., and Coleman, J. M., eds., Framework, Facies, and Oil Trapping Characteristics of Upper Continental Margin: Am. Assoc. Petroleum Geologists, Studies in Geology No. 7, p. 193–224.

Moore, D. G., 1969, Reflection Profiling Studies of the California Continental Borderland: Structure and Quaternary Turbidite Basins. Geol. Soc. America Spec. Paper 107, 142 p.

Moore, G. W., 1969, New formations of Kodiak and adjacent islands, Alaska: U.S. Geol. Survey Bull. 1274-A, p. A27–A35.

MUTTI, E., 1977, Distinctive thin-bedded turbidite facies and related depositional environments in the Eocene Hecho Group (South-central Pyrenees, Spain): Sedimentology, v. 24, p. 107–131.

———, AND RICCI LUCCHI, F., 1972, Le torbiditi dell' Appennino setentrionale—introduzione all' analisi di facies: Soc. Geol. Ital. Mem., v. 11, p. 161–199.

———, AND ———, 1975, Turbidite facies and facies associations, in Examples of turbidite facies and facies associations from selected formations of the Northern Appennines: Ninth Intern. Cong. Sedimentology, Nice, France, Fieldtrip Guidebook A 11, p. 21–36.

NILSEN, T. H., 1977, Turbidite facies and sedimentation patterns, in Nilsen, T. H., Late Mesozoic and Cenozoic Sedimentation and Tectonics in California: San Joaquin Geol. Soc. Short Course, Bakersfield, p. 39–52.

PIERCE, J. W., NELSON, D. D., AND COLQUHOUN, D. J., 1972, Mineralogy of suspended sediment off the southeastern United States, in Swift, D. J. P., Duane, D. B., and Pilkey, O. H., eds., Shelf Sediment Transport: Process and Pattern: Stroudsburg, Dowden, Hutchinson & Ross, Inc., p. 281–306.

SANGREE, J. B., WAYLETT, D. C., FRAZIER, D. E., AMERY, G. B., AND FENNESSY, W. J., 1976, Recognition of continental-slope seismic facies offshore Texas-Louisiana, in Bouma, A. H., Moore, G. T., and Coleman, J. M., eds., Beyond the Shelf Break: Am. Assoc. Petroleum Geologists, Marine Geol. Comm. Short Course, v. 2, p. F1–F54.

———, ———, ———, ———, AND ———, 1978, Recognition of continental-slope seismic facies, offshore Texas-Louisiana, in Bouma, A. H., Moore, G. T., and Coleman, J. M., eds., Framework, Facies, and Oil Trapping Characteristics of Upper Continental Margin: Am. Assoc. Petroleum Geologists Studies in Geology No. 7, p. 87–116.

SEELY, D. R., 1977, The significance of landward vergence and oblique structural trends on trench inner slopes, in Talwani, M., and Pitman, W. C., III, eds., Island Arcs, Deep Sea Trenches and Back Arc Basins: Maurice Ewing Series 1, Amer. Geophysical Union, Washington D.C., p. 187–198.

———, AND DICKINSON, W. R., 1977, Stratigraphy and structure of compressional margins, in McFarlan, E., Drake, C. L., and Pittman, L. S., eds., Geology of Continental Margins: Am. Assoc. Petroleum Geologists, Continuing Education Course Note Series No. 5, p. C1–C23.

———, VAIL, P. R., AND WALTON, G. G., 1974. Trench slope model, in Burk, C. A., and Drake, C. L., eds, The Geology of Continental Margins: New York, Springer—Verlag, p. 249–260.

SHEPARD, F. P., 1948, Submarine Geology: New York, Harper & Brothers, 348 p.

———, 1963, Submarine Geology, (2nd ed.): New York, Harper & Row, 557 p.

———, 1973, Submarine Geology, 3rd ed: New York, Harper & Row, 517 p.

———, AND DILL, R. F., 1966, Submarine Canyons and other Sea Valleys. Chicago, Rand McNally & Co., 381 p.

———, MCLOUGHLIN, P. A., MARSHALL, N. F., AND SULLIVAN, G. G., 1977, Current-meter recordings of low-speed turbidity currents: Geology, v. 5, p. 297–301.

SHOR, G. G., JR., AND VON HUENE, R., 1972, Marine seismic refraction studies near Kodiak, Alaska: Geophysics, v. 37, p. 697–700.

SIDNER, B. R., GARTNER, S., AND BRYANT, W. R., 1977, Late Pleistocene Geologic History of the outer Continental Shelf and upper Continental Slope, northwest Gulf of Mexico: Texas A & M Univ., Dept. of Ocean., Tech. Rept. 77-5-T, 131 p.

———, ———, AND ———, 1978, Late Pleistocene geologic history of Texas continental shelf and upper continental slope, in Bouma, A. H., Moore, G. T., and Coleman, J. M., eds., Framework, Facies, and Oil Trapping Characteristics of Upper Continental Margin: Am. Assoc. Petroleum Geologists, Studies in Geology No. 7, p. 243–266.

SIMON, D. B., AND RICHARDSON, E. V., 1963, Forms of bed roughness in alluvial channels: Trans. Amer. Soc. Civil Eng., v. 128, p. 284–302.

SOUTHARD, J. B., 1971, Representation of bed configurations in depth-velocity diagrams: Jour. Sed. Petrology, v. 41, p. 903–915.

———, AND CACCHIONE, D. A., 1972, Experiments on bottom sediment movement by breaking internal waves, in Swift, D. J. P., Duane, D. B., and Pilkey, O. H. eds., Shelf Sediment Transport: Process and Pattern: Stroudsburg, Dowden, Hutchinson & Ross, Inc., p. 83–97.

STANLEY, D. J., ed., 1969a, The new Concepts of Continental Margin Sedimentation: Short course lecture notes, Amer. Geol. Inst., Washington, D.C., 280 p.

———, 1969b, Sedimentation in slope and base-of-slope environments, in Stanley, D. J., (convener), The new Concepts of Continental Margin Sedimentation: Amer. Geol. Inst., Washington, D.C., Short course lecture notes, p. DJS-8-1-25.

———, FENNER, P., AND KELLING, G., 1972, Currents and sediment transport at the Wilmington Canyon shelfbreak, as observed by underwater television, in Swift, D. J. P., Duane, D. B., and Pilkey, O. H., eds., Shelf Sediment Transport: Process and Pattern: Stroudsburg, Dowden, Hutchinson & Ross, Inc. p. 621–644.

———, AND SWIFT, D. J. P., eds., 1974, The new Concepts of Continental Margin Sedimentation, II: Sediment Transport and its Application to Environmental Management: Amer. Geol. Inst., Short Course Lecture Notes, Falls Church, Virginia, 1155 p.

SUESS, E., 1976, Nutrients near the depositional interface, in McCave, I. N., ed., the Benthic Boundary Layer: New York, Plenum Publ. Co., p. 57–79.

SVERDRUP, H. U., JOHNSON, M. W., AND FLEMING, R. H., 1942, The Oceans, their Physics, Chemistry and General Biology: Englewood Cliffs, Prentice-Hall, Inc., 1087 p.

TRABANT, P. K., AND PRESLEY, B. J., 1978, Orca Basin, anoxic depression on continental slope, northwest Gulf of Mexico, *in* Bouma, A. H., Moore, G. T., and Coleman, J. M., eds., Framework, Facies, and Oil Trapping Characteristics of Upper Continental Margin: Am. Assoc. Petroleum Geologists, Studies in Geology No. 7, p. 303–312.

VAIL, P. R., 1977, Seismic recognition of depositional facies on slopes and rises, *in* McFarlan, E., Drake, C. L., and Pittman, L. S., eds., Geology of Continental Margins: Am. Assoc. Petroleum Geologists, Continuing Education Course Note Series No. 5, p. F1–F9.

———, MITCHUM, R. M., AND THOMPSON, S., III 1977, Global cycles of relative changes of sea level, *in* Payton, C. E., ed., Seismic Stratigraphy—Application to Hydrocarbon Exploration: Am. Assoc. Petroleum Geologists, Memoir 26, p. 83–97.

VON HUENE, R., 1972, Structure of the continental margin and tectonism of the eastern Aleutian Trench: Geol. Soc. America Bull., v. 83, p. 3613–3626.

———, AND SHOR, G. G., JR., 1969, The structure and tectonic history of the eastern Aleutian Trench: Geol. Soc. America Bull., v. 80, p. 1889–1902.

WAGNER, H., 1900, Lehrbuch der Geographie: Hanover, Hahn.

WALKER, R. G., AND MUTTI, E., 1973, Turbidite facies and facies associations, *in* Middleton, G. V., and Bouma, A. H. eds., Turbidites and Deep-Water Sedimentation: Soc. Econ. Paleontologists Mineralogists, Pacific Section, Short Course Lecture Notes, Los Angeles, p. 119–157.

WATKINS, J. S., WORZEL, J. L., AND LADD, J. W., 1975, Deep seismic reflection investigation of occurence of salt in Gulf of Mexico, *in* Bouma, A. H., Moore, G. T., and Coleman, J. M., eds., Beyond the Shelf Break: Am. Assoc. Petroleum Geologists, Marine Geol. Comm. Short Course, v. 2, p. G1–G34.

———, LADD, J. W., BUFFLER, R. T., SHAUB, F. Y., HOUSTON, M. H., WORZEL, J. L., 1978, Occurrence and evolution of salt in deep Gulf of Mexico, *in* Bouma, A. H., Moore, G. T., and Coleman, J. M., eds., Framework, Facies, and Oil Trapping Characteristics of Upper Continental Margin: Am. Assoc. Petroleum Geologists, Studies in Geology No. 7, p. 43–66.

WEEKS, L. G., 1974, Petroleum resources potential of continental margins, *in* Burk, C. A., and Drake, C. L., eds., The Geology of Continental Margins: New York, Springer-Verlag, p. 953–964.

WHITAKER, J. H. McD., ed., 1976, Submarine Canyons and Deep-Sea Fans: Benchmark Papers in Geology Series: Stroudsburg, Pa., Dowden, Hutchinson and Ross, 461 p.

YARBOROUGH, H., 1977, Continental margin types related to plate tectonics and evolution of margins, *in* McFarlan, E., Drake, C. L., and Pittman, L. S., eds., Geology of Continental Margins: Am. Assoc. Petroleum Geologists, Continuing Education Course Notes Series No. 5, p. A1–A8.

MAN'S ACTIVITIES ON THE CONTINENTAL SLOPE

HAROLD D. PALMER,
Associate
Dames & Moore, Washington, D.C.

ABSTRACT:

Man's activities on the continental slopes of the world do not begin to approach the intensity and magnitude of his efforts on continental shelves, yet increasingly we find exploration and development strategies focused upon areas beyond the shelf-break. The primary activity at present involves the search for hydrocarbons, and plans for drilling slope provinces now include sites in water depths in excess of 1500 m. Non-petroleum resources may include old placers, but a greater potential lies in the authigenic minerals which may contain useful elements. Deep-sea benthic fisheries are now under consideration, and advances in trawling and trapping devices may promote a new direction in commercial activity off the shelf. Our exploitation of shelf resources must take into account the impacts and hazards attendant to operating within a dynamic environment. Gravity transport of large volumes of surficial slope sediments is now considered commmonplace, and for activities which rely upon the sea floor for support of structures, pipelines, cables, and other objects the rates and scales of local to regional movements must be understood.

INTRODUCTION

The theme of this volume addresses physical processes acting upon one of the more remote and poorly understood physiographic provinces of the earth, the continental slopes. As Emery and Uchupi (1972) point out,

"the continental slope is the most significant topographic discontinuity of the earth's crust because it marks the general position of the contact between low density rocks of the continents and the high-density rocks of the ocean floor."

They continue to point out that much of this topographic discontinuity has been obscured by extensive progradation of the basal slope and rise deposits, and that there is a blurred transition from the true continental terrain to the oceanic domain. The interest in this brief chapter is not in defining upper and lower bounds to a physiographic province, but rather in forthcoming activities within this region with an eye toward the exploitation of slope resources, or at least the uses to which the slope may be subjected.

THE SLOPE ENVIRONMENT

Previous chapters have provided a sound descriptive basis for our understanding of agents and processes active on the slope, and we must note that with a global perspective we are considering a region with declivities averaging 4° (Fairbridge, 1966), but whose gradients range from vertical in certain coralline areas to slopes where, by definition, the gradient flattens beyond 1:40 at which one passes into the province of the rise. Here, we will consider the slope to occupy that region beyond the shelf-break where slope declivities increase perceptibly, and include both the shelf-break region itself and the lower portion of the slope merging with the continental rise.

As we have seen in preceeding chapters, it is a diverse environment, in places monotonously smooth, in others a rugged and unstable area prone to slumps and other displacements where large-scale movements are not uncommon. Across and within this poorly understood region, the increasing tempo of man's activities demands that we develop a greater understanding of slope characteristics such that allowances may be made in structures and practices which involve the region. The remainder of this chapter will examine activities which affect, or are affected by, the continental slope.

SUBMARINE CABLES

Transoceanic communication is now taken for granted, as satellite links and the numerous submarine cables carry thousands of conversations and teletype messages hourly. But for decades the only means of rapid communication was by cable. The first attempt to lay a transatlantic cable was made in 1857–1858, but the cable failed after a few weeks of operation.

Nearly a decade later, the British vessel "Great Eastern" established the first successful link between Newfoundland and England but the capacity of this early cable was quite limited, permitting a transmission rate of only 3 words per minute! (Sprankle, written comm., 1978). A Pacific cable between San Francisco and Honolulu was completed in 1902, and by 1904 the link had been extended to Manila. Today numerous communications cables pass between major continental and insular centers.

Most submarine cables are of small size, on the order of 5-6 cm outside diameter and although they are armored and include repeaters at intervals along their length, they are still tenuous links when one considers the forces acting in a turbidity current, slump or other mass movement on the slope. Walker (1978) summarized the hazards to cables as being both tensional stresses due to mass motion of the fluid and sediment as it passes the cable, scouring, and sandblasting caused by the abrasive impact of entrained particles. Flow velocities of up to 10 m/sec have been calculated for several turbidity currents which have broken cables, and although this process has eluded direct quantitative measurement, we must assume that such forces are not uncommon on continental slopes. More specific examples may be found in Heezen and Ewing (1955), Heezen (1956) et al (1969) and Krause et al (1970). The renowned and sequential cable breaks associated with the Grand Banks earthquake of 1929 are forever a point of reference in any discussion of turbidity currents, and their counterparts, breaks in African and South Pacific cables, substantiate the power of such phenomena. Although the speed of advance of the Grand Banks current has been revised downward from earlier overestimates to a figure of about 30 km/hr. or 8.3 m/sec., (see discussion by Emery and Uchupi, 1972, p. 296; Shepard and Dill, 1966), the strain induced by displacement of the cables was sufficient to part them.

There is no doubt but that communications cables are subject to breakage due to mass movements of slope materials and this hazard remains, along with snaring and breakage by fishing vessels, as one prime concern of communications agencies. As satellite links become the standard means of intercontinental communications, submarine cables will prove too costly and unreliable to maintain, and they will become a secondary means of passing information between distant points.

WASTE DISPOSAL

If one consults a dictionary, the first definition of *disposal* is generally given as . . . "an orderly or systematic placement, distribution or arrangement" while the fourth meaning is . . . "a discarding or throwing away." We cannot consider ocean disposal as "throwing away" since the waste materials remain, in some form, within the sea, it's inhabitants or its sediments. Throwing away implies no return, or a disappearance, and this is impossible in the physical sense of particulate matter and/or solutions being released in the sea. Such materials are merely recycled, and given the state of our knowledge of marine processes we cannot be certain of the rates or scales of these cycles.

Returning to the first definition of disposal implying orderly or systematic placement ocean dumping may be politically systematic, since a designated time and place may be specified, but we cannot say that such practices are orderly, for the reasons just given. Therefore, since much of the material disposed of is potentially hazardous to marine life and water quality, it is crucial that the site of disposal be carefully selected and that subsequent dumping be carried out within that region set aside for disposal.

Most of the waste materials disposed of at sea consist of dredge spoil, industrial by-products and sewage sludge. Where shelves are broad, such as mid-Atlantic Bight area, deep water disposal beyond the shelf-break is too costly, and dump sites have been designated at a variety of sites on the inner and middle shelf. Only the most toxic and noxious materials (radioactive wastes, pathogenic materials, etc.) are released in these deepwater sites.

Typical of these sites is deepwater dumpsite 106 (DWD 106) on the continental slope 106 nautical miles southeast of Ambrose Light and 90 nautical miles east of Cape Henlopen, Delaware. As pointed out be NOAA (1976), this site is used by more than 25 different dumpers in the New York-New Jersey area to dispose of various industrial chemicals. Wastes are dumped just beneath the sea surface by a barge moving at a speed of 3 to 5 knots. Typical waste materials include: hydrochloric acid by-products, inert ore slurry from production of titanium dioxide pigments, residual sludge from galvanizing and plating operations, liquid wastes from production of textile manufacturing and from etching and photographic processes, water solutions of inorganic salts, and similar materials derived from diverse manufacturing processes.

The major findings from the first NOAA seasonal baseline study in May 1974 were published in 1975. These findings indicate that little if any of the dumped materials reach the bottom in the vicinity of the site. A second seasonal baseline study in July 1975 used a research submersible and a surface vessel. In that study, *Alvin* made 10 dives to depths of from 1,086 to 2,507 m, and box cores, epifauna and water samples were taken, and in 3 dives in the abandoned radioactive dumping area both sealed and leaking drums were encountered (NOAA, 1977). Traces of plutonium and cesium were found in sediments at this site. On the west coast of the United States, a similar radioactive waste disposal site was established in an area 35 miles west of San Francisco near the Farallon Islands. At that site, 47,500 drums of low level radioactive wastes were dumped at depths ranging from 1,000 to 2,000 meters.

Deepwater dumpsites in European waters are used in accordance with the criteria set forth in the Oslo Convention of 1972 and the various London Conventions of the early 1970's. These requirements include stipulations stating that wastes must be dumped at least 190 nautical miles (342 km) from the nearest land and in water depths of at least 12,000 ft (or nearly 3,000 meters). In the United Kingdom, the Conventions are embodied in the Dumping at Sea Act of 1974 and four sites have been approved—two in western approaches to the islands and two off the Iberian Peninsula.

In the United States, the practice of ocean dumping of toxic substances and noxious industrial wastes will be eliminated by law on 31 December 1981. Dumping of dredge spoil will be permitted, but the use of the continental slope as waste repositories will cease by that date. This action, in accordance with Public Law PL 95-153, is a timely and necessary requirement in light of our meager knowledge of slope circulation patterns, life and dynamic processes. With respect to living resources, dumping activities may be most harmful since numerous studies have shown that most organisms assimilate and concentrate trace metals and other toxic substances through "bio-accumulation" in benthic and pelagic food chains. However, in an exhaustive report addressing the question of dredged material disposal beyond the shelf-break, Pequegnat et al. (1978) have concluded that there are several advantages to disposal in the deep sea. Pelogic life, including those species of commercial importance, is greatly reduced in oceanic versus shelf waters. In addition, there is a similar reduction in the numbers of benthic organisms, and broad dispersal of sediment discharged by dumping through a water column hundreds of meters deep implies little direct impact at a local site. Although the economics of long-distance transport (a round trip of nearly 400 km at some eastern U.S. sites) now makes deep ocean disposal unattractive, new techniques and cost-saving improvements in dredging-disposal procedures may change with time as other sites become unavailable.

LIVING RESOURCES

For centuries, commercial fisheries have found the continental shelves and banks to be rich areas for both finfish and shellfish resources. Yet the greatest fishing areas of the world lie above continental slopes and shelf-breaks where upwelling processes bring nutrients into the euphotic zone. Much of this activity has focused upon the pelagic fisheries, and only recently has the aspect of a deep benthic fishery been given serious consideration. In upwelling areas where high productivity prevails in the euphotic zone (0–80 m water depth), it seems at first paradoxical that there may not be an equally valuable benthic fishery beneath these enriched waters. But as Didyk et al. (1978) point out, in upwelling areas the precipitation of abundant organic detritus through the water column promotes rapid oxygen consumption and depletion in the "disphotic" zone (80–200 m+) where dim light and little or no effective plant production prevail. Here, where the resulting anoxic waters intersect the upper continental slope, bottom sediments also become anoxic and benthic populations of commercial importance find little food to sustain commercially abundant populations.

Studies off the Atlantic margin suggest that there may be a significant crab fishery in the upper slope regions of the mid-Atlantic region. Haefner and Musick (1974) and subsequent investigations by Wigley et al (1975) support the contention that both red crabs (*Geryon quinquedens*, Figure 1a) and lobster (*Homarus americanus*) inhabit the upper slope in numbers which represent commercially economic abundance. The

FIG. 1.—A) The deep sea red crab (*Geryon quinquedens*) may become a commercially important species on the upper continental slope. Catches recovered with otter trawls have yielded up to 1588 kg/hour off southern New England (see text). B) Red crab fisheries may be exploited by deployment of crab pots, such as this example. The long dimension measures just over 1 meter (photos courtesy of Virginia Institute of Marine Sciences).

habitat of crab species ranges from 40 to 2155 m, but is principally found in the bathyal zone. Catches shoaler than about 215 m are from the colder water regions in the Gulf of Maine. Catch records during exploration to assess concentrations are often impressive. Meade and Grey (1973) note one catch of 1587.6 kg (3,500 lbs) per hour was made with a standard otter trawl off southern New England. Deployment of crab pots, such as the example shown in Figure 1b, is an alternative to deep water trawling practice. The most comprehensive discussion of red crabs will be found in Wigley et al (1975) who include discussions of survey technique and bottom photographs from a towed benthic sled.

The American lobster has been fished in canyons and on the upper slope for many years. Recent data by Wigley et al (1975) indicate a general reduction in biomass, with the most common occurrence for lobster lying between 412 and 503 m. Off New England, Uzman, et al. (1977), report squid present to at least 300 m depths in the slopes surrounding Veatch submarine canyon. In 1973, approximately 32 million kg (70 million lbs) of the longfinned squid *Loligo pealei* were caught on the shelves and slopes off the northeastern U.S., of which only about 1 million kg was caught by U.S. vessels. Squid are presently under consideration as another commercial fishery for the U.S. fleet. Although this brief review has focused on the eastern coast of the U.S., extensive deep bottom fisheries exist through the world (see Russell-Hunter, 1970; Gulland, 1976).

OIL AND GAS DEPOSITS

From the perspective of both the cost of endeavor and the economic return, the recovery of hydrocarbon resources is the major activity now in progress on the worlds' continental slopes. Estimates of the location of offshore hydrocarbon reserves has been made by Kruger (1978) who polled 6 major oil companies to determine their outlook on the percentage of reserve versus water depth. The consensus was that 65% of all reserves were in depths less than 200 m, the isobath generally accepted as the shelf-break. Thirty percent of global reserves lay between 200 m and 2,500 m, while only 4% were assigned to the continental rise and the remaining 1% of the abyssal plains. The basis for such predictions is provided by reconnaissance geophysics, geology and the structural traps seen, sediment thickness, and other parameters. The most extensive and thickest sections of deepwater sediments flank the coasts of southern Alaska, east coast of the U.S. and Canada, southwestern and eastern coasts of South America, much of the African coast, western India and east of Viet Nam. Other isolated basins west of the United Kingdom and Norway round out the list. With the adoption of a 370 km (200 nautical mile) sovereign state jurisdiction, most of these areas will be in national waters.

Off the eastern U.S., the upper continental slope is believed to contain the highest average percentage of preserved organic matter in surface sediments of the Atlantic margin. Regarding deeper sediments of the margin, a 305 m core drilled into the upper slope of New Jersey contained significant amounts of light hydrocarbons (methane, ethane, propane) which may, in part, have leaked from deeper accumulations (Mattick et al, 1977). Even deeper, beneath the continental rise and adjacent abyssal plain, drilling of the DSDP has confirmed an extensive deposit of Cretaceous black, hemipelagic shale up to 300 m thick. If these deposits exist in the slope, sufficient temperature gradients could have been established to generate significant quantities of petroleum. In a discussion of geothermal gradients in continental rises, Emery and Skinner (1977) note that thermal maturation of organic matter into petroleum occurs between 60° and 150° C for oil and between 100° and 200° C for gas. They conclude that continental rises may contain sig-

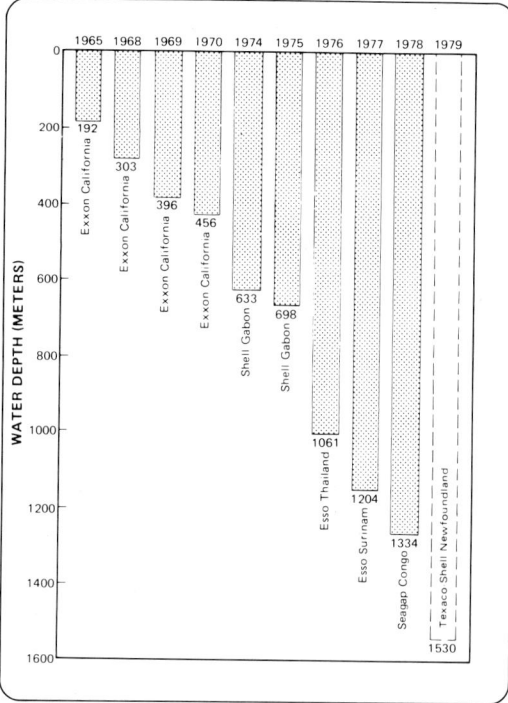

FIG. 2.—The progress of deep water drilling is displayed in this diagram which highlights the increasing capability to explore the continental margin (see text; source, Oil and Gas Journal, February, 1978).

nificant quantities of hydrocarbons, a position, we have noted, not held by several major oil companies. At least for the Atlantic margin of the central U.S. Mattick et al (1977), find favorable stratigraphic and structural evidence to support optimism for finding hydrocarbon deposit beyond the present shelf-break.

There is no doubt but that exploratory drilling will continue into even deeper waters of the continental slope. Lease blocks in many U.S. Sale areas lie at depths over 400–500 m, and the current stratigraphic test well in the Baltimore canyon area is drilling in 805 m of water. Off Eire, the deepest European drilling was by Shell, operating in 800 m water depth in the Porcupine Seabight area. Other major companies operating in deep waters of Eire include ELF at 400 m, and Phillips at 430 m. In 1976 EXXON completed an exploration study off Thailand in 1061 m, but the deepest commercial exploration well to date is that by the drillship *Discoverer Seven Seas* which completed operations off the Congo, West Africa, in 1334 m water depth. Shell and Texaco are now planning for operations in 1530 m off Laborador with a dynamic-positioned drillship now under construction. A record of drilling progress in deep water is shown in Figure 2.

At present, deep water drilling is conducted from either moored semi-submersible platforms such as the Zapata Ugland (Fig. 3) or from drillships such as the Glomar Challenger (Fig. 4) which employ dynamic positioning systems to maintain a vertical drill string even in high sea states. New platform designs are in the planning stages for deep water operations. These include guyed towers (Finn, 1978), and a variety of mooring configurations utilizing single and multi-line systems (see McClure, 1978, for descriptions). The largest fixed platform now in place is the Shell Oil Cognac structure standing in 314 m of water some 27 km south of the Mississippi River delta. All of these designs depend, in one way or another, upon structures implanted in slope sediments and thus the geotechnical aspects of these materials, and the larger scale factors such as slumps and other unstable conditions, become of prime importance in both exploration and production operations.

Of equal concern are the socio-economic aspects of slope, and perhaps rise, operations. Here

Fig. 3.—One of the newer versions of the semi-submersible drill rig, the *Zapata Ugland* is shown here enroute to a drill site. Unlike most "semi's" which are towed to a location, the Ugland is self-propelled and as such is more like a drill ship (Fig. 4). When operating, the twin hulls are flooded, and the waterline lies at about the level shown at the top of the black legs (photo courtesy of Zapata Offshore Company).

Fig. 4.—The drill ship *Glomar Challenger* provides drilling capabilities for any water depth. Although this vessel has been committed to the Deep Sea Drilling Program, similar ships are active throughout the world. They do not require anchor cables, relying solely on thrusters to maintain position above the drill string. Exploration drilling in the slope environment will utilize similar vessels (photo courtesy of Global Marine, Inc.).

the price of crude oil becomes the critical factor, for production of these deeper deposits will require massive and costly design, construction, operational and maintenance procedures. According to Johnson (1978) it is doubtful that fields with proven reserves of less than 700 million bbls will be considered commercial deposits.

NON-PETROLEUM MINERALS

Although the shelves of the world contain numerous placer deposits of economic minerals left at a time of lower sea level, the slope offers little promise for concentrations of such deposits. In part, this results from the latest low stand of the sea during the Wisconsin sea level minimum when the present slope received a far greater influx of sediment than at present. These recent deposits mask and/or dilute any accumulations of heavy minerals or other materials which reached the present slope surface at that time.

Perhaps the greatest promise for any non-hydrocarbon materials lies in the study of dredging authigenic minerals from the outer shelf and upper slope. These include phosphorite and glauconite, both of which could be recovered for use as fertilizer material and other uses. There is a demonstrated propensity for uranium enrichment in both minerals (Kolodny and Kaplan, 1970) and it is conceivable that at some time a sufficient concentration of these minerals might be found and exploited. For the present we must consider the possibility of economical deposits of primary, hydrothermal or other origin as remote from the standpoint of recoverable slope deposits. A thorough treatment of the potential for deepwater mineral deposits is provided by Emery and Skinner (1977).

SLOPE HAZARDS

Agents and processes described in the previous chapters may be re-assessed in light of their potential hazard to man's activities on the continental slopes of the world. Consideration of those aspects which could cause severe or catastrophic damage include turbidity currents, slumps, and other gravity-induced processes which displace masses of slope sediment at various rates. A description of such processes, including those occurring in submarine canyon margins and outer shelf domains is included in Kelling and Stanley (1976). A more complete treatment of the slope and its various aspects is given in Bouma et al (1978) in their volume dealing with the slope and rise. As noted earlier, Walker (1978) addresses the question of mass movements and their impact on structures.

There are numerous examples in the recent literature which describe large, indeed enormous, slump structures on continental slopes. The largest yet documented is that of the continental margin off southeast Africa in which a slump mass 750 km in length and 106 km wide, having a total volume of over 20,000 km^3 has been identified by Dingle (1977). This feature, and many similar large gravity structures, are related to shelf-edge processes acting during one or more of the eustatic low stands of sea level during the Pleistocene Epoch. Further documentation of large-scale slumping may be found in Embley and Jacobi (1976), Stanley and Silverberg (1969), McGregor and Bennett (1977), Knebel and Carson (in press), and other papers cited herein which result from recent studies of slope structure and stratigraphy. The latter paper cited (Knebel and Carson) has, as it's title, "*Small-scale* slump deposits . . . etc." (italics mine). As is appropriate to workers surveying and mapping the continental slopes, scales of tens to a few thousands of meters are considered "small-scale," and rightly so in the context of geological perspectives. This particular paper treats the results of high-resolution surveys which show ubiquitous "*small-scale*" slumps in the intercanyon areas off the U.S. east coast. They consist of masses 10–90 m thick which extend downslope for 1.8–7.2 km in water depths from 545 to 1500 m. To the geologist these are indeed small scale features, and the age and rate of displacement has been, until now, of secondary importance. However, the engineers and designers responsible for safe and efficient structures and operations on the slope must think in much smaller terms, since displacement of only a few meters may prove disastrous. The challenge of marine scientists is to provide detailed information on very localized sea floor tracts.

As this volume has shown, many geologists are re-evaluating the role of gravity transport and the physical (geotechnical) aspects of marine sediments. Recent work by Hanor and Gorsline (1978) relates such factors as rate of sediment supply, slope, and submarine canyon proximity to slumps in southern California basin margins. Here, mass wasting takes place on slopes as low as 0.8°, and in other areas, large slumps 10–35 m thick have travelled 7–10 km over slopes ranging from 3–5°.

Gravity transport of coherent masses of sediment over low slopes appears to be quite common, especially in tectonically active areas where seismic shocks may induce slumping. Lewis (1971) has demonstrated this in studies off New Zealand, where slumps as large as 250 km^2 and 10–50 m thick have moved over slopes from 1–4°. However, even on passive aseismic margins slumping appears quite common. Bennett et al (1977) have studied the slope off Delaware Bay and on the basis of slope gradient and seismic reflection profiling have published a comprehensive map (Bennett et al, 1978) of this region. Current exploration activity by numerous oil companies active in this area includes the assessment of environmental hazards posed by these features.

CONCLUSION

From the standpoint of the *user* of the continental slope, there are two key factors in the design of any structure or operation, or the deployment of equipment. These are the *rates* and *scales* of environmental processes active in the slope environment. The *rate* aspect includes the frequency of recurrence as well as the duration of the event, while the *scale* encompasses not only areal extent but volmetric (depth) factors as well. At present, our understanding of *rates* and *scales* of slope processes is quite limited, and engineers are justifiably concerned about the magnitude of mass movements on even gentle slopes. Until the dynamics of cause and effect can be determined, extreme caution must be approached in designing facilities which rely upon the sediments and rocks of the slope for their support. The design must obviously cope with the problem of mass movement rather than attempt to withstand it. As exploitation of slope resources proceeds, the marine geologist will play a vital role in the design, installation, operation and maintenance of structures in this dynamic area.

ACKNOWLEDGMENTS

The author wishes to thank Charles Sprankle and Leo Overhizer of the American Telephone and Telegraph Company for information related to submarine cables. Michael Rainbow, Dames & Moore London, kindly provided data on European ocean waste disposal. Jack Musick, Virginia Institute of Marine Sciences, provided information and photographs on the deep sea crab fishery, while Global Marine and Zapata Off-Shore furnished photographs of offshore drilling vessels.

REFERENCES

Bennett, R. H., Lambert, D. N., McGregor, Bonnie A., Forde, E. B., and Merrill, G. F., 1978, Slope Map, a major submarine slide on the U.S. Atlantic continental slope east of Cape May (Map reference) A-5787 USCOMM-NOAA-DC

―――, ―――, and Hulbert, M. H., 1977, Marine Slope Stability: Marine Geotechnology, v. 2, p. 245–261.

Bouma, A. H., Moore, G. T., and Coleman, J. M., eds., 1978, Framework, facies and oil trapping characteristics of the Upper Continental Margin: Assoc. Petroleum Geologists Studies in Geology No. 7, 326 p.

Didyk, B. M., Simoneit, B. R. T., Brassell, S. C., and Eglinton, G., 1978, Organic geochemical indicators of palaeoenvironmental conditions of sedimentation: Nature, v. 272, p. 216–222.

Dingle, R. V., 1977, The anatomy of a large submarine on a sheared continental margin (SE Africa): Jour. Geol. Society, v. 134, p. 293–310.

Embley, R. W., and Jacobi, R. D., 1977, Distribution and morphology of large submarine sediments slides and slumps on Atlantic continental margins: Marine Geotechnical, v. 2, p. 205–228.

Emery, K. O., and Skinner, B. J., 1977, Mineral deposits of the deep-ocean floor: Marine Mining, v. 1, p. 1–72.

――― and Uchupi, E., 1972, Western North Atlantic Ocean: Topography, rocks, structure, water, life and sediments: Am. Assoc. Petroleum Geologists Memoir 17, 532 p.

Fairbridge, R. F., 1966, The Encyclopedia of Oceanography: New York, Reinhold Pub. Co., 1021 p.

Finn, L. D., 1978, A new deep-water platform—the guyed tower: Jour. Petroleum Technology, April, p. 537–544.

Gulland, J. A., 1976, Production and catches of fish in the sea, *in* Cushing, D. H., and Walsh, J. J., ed., The Ecology of the Seas: Philadelphia, W. B. Saunders Co., p. 283–314.

Haefner, P. A., Jr., and Musick, J. A., 1974, Observations on distribution and abundance of red crabs in Norfold Canyon and adjacent continental slope: Marine Fish. Rev., v. 36, no. 1, p. 31–34.

Haner, B. E., and Gorsline, D. S., 1978, Processes and morphology of continental slope between Santa Monica and Dume submarine canyons, southern California: Marine Geology, v. 28, p. 77–87.

Heezen, B. C., 1956, Corrientes de turbidez del Rio Magdalena: Bol. Soc. Geog. Columbia, v. 51–52, p. 135–143.

―――, and Ewing, M., 1955, Orleansville earthquake and turbidity currents: Am. Assoc. Petroleum Geologists, Bull., v. 39, p. 2505–2514.

―――, Menzies, R. J., Schneider, E. D., Ewing, W. M., and Granelli, N. C. L., 1964, Congo submarine canyon: Am. Assoc. Petroleum Geologists, v. 48, p. 1126–1149.

Johnson, A., 1978, International house-hold: Noroil, v. 6, May, p. 60–62.

Kelling, G., and Stanley, D. J., 1976, Sedimentation in canyon, slope and base-of-slope environments, *in* Stanley, D. J., and Swift, D. J. P., eds., Marine Sediment Transport and Environmental Management: New York, Wiley, p. 379–438.

Knebel, H. J., and Carson, B., 1978, Small scale slump deposits, Middle Atlantic continental slope, off eastern United States: Marine Geology, (in press).

Kolodny, Y., and Kaplan, I. R., 1970, Uranium isotopes in sea floor phosphorites: Geochim. Cosmochim. Acta, v. 34, p. 3–24.

Krause, D. C., White, W. C., Piper, D. J. W., and Heezen, B. C., 1970, Turbidity currents and cable breaks in the western New Britain trench: Geol. Soc. America Bull., v. 81, p. 2153–2160.

KRUGER, W. C., JR., 1978, Sediment thickness and percentage estimate of offshore petroleum reserves: Oil and Gas Jour., 16 Jan., p. 88–90.

LEWIS, K. B., 1971, Slumping on a continental slope inclined at 1°–4°: Sedimentology, v. 16, p. 97–110.

MATTICK, R. E., GIRARD, O. W., JR., SCHOLLE, P. A., AND GROW, J. A., 1977, Potential of deep Atlantic targets examined: Oil and Gas Jour., 12 Dec.

MCCLURE, A. C., 1978, Deepwater mooring technology: Petroleum Engineer, v. 50, no. 5, p. 23–26.

MCGREGOR, BONNIE A., 1978, Seismic reflection profiles of the United States East Coast Continental Margin: NOAA Tech. Rept. ERL 398-AOML 28, 17 pp.

———, AND BENNETT, R. H., 1977, Continental slope sediment instability northeast of Wilmington Canyon: Am. Assoc. Petroleum Geologists, v. 61, p. 918–928.

MEADE, T. L., AND GRAY, G. W., JR., 1973, The red crab: University of Rhode Island, Mar. Tech. Rept. Series, no. 11, 21 p.

NOAA, 1976, Report to the Congress on ocean dumping research: January through December, 1975: Washington, D.C., U.S. Dept. Commerce, National Oceanic and Atmos. Admin., 39 p.

———, 1977, Manned undersea and technology fiscal year 1976 report: Washington, D.C., U.S. Dept. Commerce, National Oceanic and Atmos. Admin., 71 p.

PEQUEGNAT, W. E., SMITH, D. D., DARNELL, R. M., PRESLEY, B. J., AND REID, R. O., 1978, An assessment of the potential impact of dredged material disposal in the open ocean: U.S. Army Corps of Engineers, Waterways Experiment Station Tech. Rept. D-78-2, 642 p.

RUSSELL-HUNTER, W. D., 1970, Aquatic Productivity: New York, Macmillan, 306 p.

SHEPARD, F. P., AND DILL, R. F., 1966, Submarine canyons and other sea valleys: Chicago, Rand McNally & Co., 381 p.

STANLEY, D. J., AND SILVERBERG, 1969, Recent slumping on the continental slope off Sable Island Bank, Southeast Canada: Earth Planetary Sci. Letters, v. 6, p. 123–133.

UZMANN, J. R., COOPER, R. A., THEROUX, R. B., AND WIGLEY, R. L., 1977, Synoptic comparison of three sampling techniques for estimating abundance and distribution of selected megafauna; submersible vs. camera sled vs. otter travel: Marine Fish. Rev., v. 39, no. 12, p. 11–19.

WALKER, R. G., 1978, Offshore hazards in deeper water, Section 5, in Offshore Geologic Hazards: Rice University, workshop notes, Office Continuing Studies, 83 p.

WIGLEY, R. L., THEROUX, R. B., AND MURRAY, H. E., Deep-sea red crab, Geryon quinquedens, survey off northeastern United States: NOAA, Marine Fish. Rev., MFR Paper, 1154, v. 37, no. 8, p. 1–21.

SOME GEOTECHNICAL ENGINEERING PROBLEMS OF UPPER SLOPE SITES IN THE NORTHERN GULF OF MEXICO

LELAND M. KRAFT, JR.
McClelland Engineers, Inc.
Houston, Texas
KERRY J. CAMPBELL and MICHAEL R. PLOESSEL
McClelland Engineers, Inc.
Ventura, California

ABSTRACT

During the past 30 years, technology has evolved to allow hydrocarbon production structures to be sited in deeper and deeper water. Structures are now being designed for and installed on upper continental slope areas in water depths greater than 305 m (1000 ft) in the northern Gulf of Mexico where more than 50 blocks have been leased in water depths ranging from about 180 to 610 m (600 to 2000 ft). In addition to pile-supported structures for these deep-water sites, two new concepts are being studied: tension-leg structures and guyed towers. Engineering considerations for these new types of structures include spud-can penetration resistance, anchor drag distance, anchor holding power, breakout resistance, consolidation settlements, and cyclic loading on foundation performance, as well as overall site stability.

INTRODUCTION

Petroleum production structures are now being designed for and installed on the upper continental slope in the northern Gulf of Mexico in water depths greater than 305 m (1000 ft). Figure 1 illustrates the progression of offshore construction of production platforms into deeper waters during the past 30 years. To date, more than 50 blocks have been leased on the upper continental slope in water depths between 180 and 610 m (600 and 2000 ft). Significant advances in technology have been required to keep pace with this advancement to deep water. In addition to large pile-supported structures for these deep-water sites, two new concepts are being studied: guyed towers and tension-leg platforms.

Geologic features or conditions of potential engineering significance in the northern Gulf of Mexico include active faults, relatively weak sediments, creep, mudflows, submarine landslides, relatively steep slopes, irregular seafloor, and local but significant variations in geotechnical properties of the sediments.[1] Historically, selection of the basic design of a structure at a particular site has been determined chiefly by the water depth and the physical properties of the sediments, and not on geologic features and conditions. A thorough understanding of the site and area geology, however, is necessary for safe and cost-effective foundation design, especially when considering the large capital investment required to develop deep-water sites. Deep water, new structural-design concepts, and an awareness of geology have combined to increase the scope and complexity of foundation design and siting considerations for upper slope sites. These factors have also served to emphasize the necessity of integrating subsurface information obtained from in situ testing, borings, and associated laboratory tests and from geophysical data into the design and siting considerations.

The design and siting of structures are influenced by geologic conditions. In the northern Gulf of Mexico, geologic features and conditions that must be considered include the effects of active faults, relatively weak sediments, submarine landslides, mudflows, relatively steep slopes, irregular seafloor, and local but significant variations in sediment properties. Thorough, multidisciplinary geotechnical investigations, commonly on a regional basis, and rigorous assessments are required to design and install safe, environmentally acceptable, and economical structures. These investigations and assessments require that existing engineering, geologic, geophysical and oceanographic techniques and expertise be effectively integrated to provide a multidisciplinary approach, and often require that new equipment and techniques be developed. Most advances will probably be with respect to measuring sediment conditions and properties in situ by both geophysical and mechanical means.

As there is increasing participation of both the

---

[1]The Geotechnical engineer normally uses the term soil, which has a narrower meaning to the geologist, to describe the unconsolidated material composed of discrete solid particles with interstitial gases of liquids; in this paper the term sediments is used.

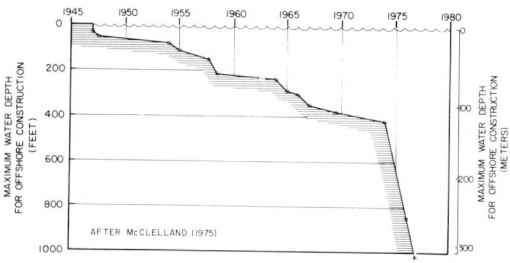

FIG. 1.—Progression of offshore construction into deeper water.

geotechnical engineer and geologist in site evaluation and site characterization for engineering design, the objective of this paper is to provide for a combined audience of engineers and geologists an integrated overview of the geotechnical engineering problems and geologic aspects of designing marine facilities for upper slope sites in the northern Gulf of Mexico. To accomplish this objective, this paper presents past and present concepts of multidisciplinary investigations; briefly describes the basic types of structures being considered for northern Gulf of Mexico sites and the relevant geotechnical design considerations; highlights some common geological features and conditions having potential engineering significance to the design, siting, and maintenance of such structures; and discusses some of the geotechnical problems related to sediment movement. A short discussion of technological improvements that are expected to provide better geotechnical assessment of offshore sites and improved foundation design during the next decade is also included. The information is presented in a language that can be understood and appreciated by both the geologist and engineer.

PAST AND PRESENT CONCEPTS IN GEOTECHNICAL
EVALUATION OF OFFSHORE SITES

General Comment

The primary objective of geotechnical site investigations is to determine the subsurface conditions at the site for foundation design. Pile foundations for the early structures off the Louisiana coast in the late 1940's were designed without benefit of site investigations and were installed in accordance with the practice at the time: drive the piles to practical refusal with the biggest possible hammer. The need for site investigations to select pile penetrations in advance of construction was soon recognized. The scope of these early investigations, conducted using land exploration techniques from small fixed structures, was limited to a single boring with classification and shear strength testing of the sediments. The method of exploration changed in about 1953 to anchored deck barges and in 1962 to anchored boats with wireline percussion sampling.

Soil Borings and Testing

About 1500 geotechnical borings have been made in the northern Gulf of Mexico during the last 30 years. In areas on the continental shelf where conventional pile-supported structures are placed, only one boring per site may be made. In areas where seafloor stability is a potential problem or in frontier areas, several borings are generally made. Borings to 90 m (300 ft) below the seafloor are common, but in some areas such as the Mississippi River delta, borings to 150 m (500 ft) penetration are typical and penetrations have been made to 305 m (1000 ft) depths.

Wire-line sampling has been widely used for offshore site investigations since 1962 (McClelland, 1975) with acceptance of the probability that samples may be partially disturbed and that the quality of shear strength data from recovered samples may be less than the best. This sampling procedure, adopted as a compromise between exploration costs and sample quality, has been a cost-effective method for obtaining invaluable geotechnical data to substantial penetrations below the seafloor in the hostile offshore environment where heavy seas and large boat motions are common. Nevertheless, the uncertainty in sample quality and the emergence of a wider scope of problems at sea, requiring more sophisticated solutions and more comprehensive data, have given impetus to the development of in situ tests to augment test data obtained from recovered samples. In situ testing devices that provide a measure of strength and that are used offshore include the Remote Vane, cone penetrometer, and pressuremeter (McClelland, 1975; Kraft, Ahmad and Focht, 1976; Ferguson, McClelland and Bell, 1977). In addition, piezometers can be used to measure the amount of excess pore pressures for evaluation of consolidation history and for assessment of sample quality.

Present state-of-the-art concepts in investigating and assessing geotechnical problems are relatively new and are far from being universally accepted and applied. Traditional site investigations alone (i.e., sampling and engineering analyses based on the results of sample testing) do not allow for the identification and assessment of "regional" geologic features such as active faults, surficial mudflows, or submarine slides, all of which can impose significant loads on structures. The magnitude and complexity of such features and their potential effects on engineered structures have only been widely appreciated for offshore sites in the past few years. It is now common knowledge that such features are found in many parts of the northern Gulf of Mexico, particularly in the vicinity of the Mississippi delta. As a result, the scope of offshore geotechnical investigations has been expanded to routinely include acquisition of high-resolution geophysical data.

Geophysical Surveys

Although high-resolution geophysical data have been acquired over most leases since about 1973, rigorous engineering geologic interpretation of this information has been made available to the geotechnical engineer only recently on anything approaching a routine basis. Even today, high-resolution geophysical surveys are not always completed until structure design is well underway, thus precluding optimum use of the data.

Geologic considerations, such as rate of mass movement or fault offset, which are least amenable to quantitative analysis in an engineering context, are of particular concern to the geotechnical engineer. Thus, another significant development has been the realization that the geotechnical engineer needs, in addition to the basic identification and mapping of geologic features that could potentially cause geotechnical problems, a critical assessment of the engineering significance of each geologic feature or condition. It is in this regard that the expertise and judgment of the marine engineering geologist are most valuable. To be comprehensive, however, such an assessment requires that the members of a multidisciplinary team, consisting of marine engineering geologists, geophysicists, geochemists, oceanographers, and geotechnical engineers, work closely with one another throughout the project. The importance of a thoroughly integrated multidisciplinary site investigation, however, is still not fully appreciated by industry in general.

The standard survey that is limited in extent by lease block boundaries is not always adequate to fully assess the effects that regional geologic features may impose on a given site. Such increasing awareness has resulted recently in several relatively detailed surveys that included 12 to 25 lease blocks to evaluate conditions for a single site. Most future site studies on the outer shelf and upper slope will probably require some "regional" lines be extended outside the site proper to fully assess the significance of nearby features or conditions, especially those upslope and downslope from the site. Similarly, the potential value of repeated geophysical surveys has been recently recognized. Several detailed geophysical surveys that will be repeated in the future have already been made in the vicinity of the Mississippi delta. Such investigations will help to determine, for example, rates of mudflow advance and the relationship of various types of seafloor failures to hurricanes.

When interpreting high-resolution geophysical data and defining the geologic environment, the difference between geologic hazards and engineering geologic constraints must be recognized and differentiated. A *geologic hazard* is a feature or condition that can cause structural failure, or pollution problems and for which the effects cannot readily or economically be mitigated by engineering design. An engineering *geologic constraint* is a condition that presents difficulties for environmentally sound, engineered construction but whose effects can be mitigated by proper design and construction. Given enough resources, structures can be designed and constructed to safely withstand the effects of most if not all geologic environments, so in practice, the distinction between geologic hazard and engineering geologic constraint must be based on economics, available technology, and adequate data. A given geologic feature may truly be hazardous to a given structural design but merely represents an engineering constraint to another. For example, a guyed tower probably could not be economically constructed to reliably withstand the loads imposed by a 30 m (100-ft) thick mudflow; on the other hand, a pile-supported structure probably could be.

GEOTECHNICAL DESIGN CONSIDERATIONS

General Comments

In the northern Gulf of Mexico, many of the sediments are underconsolidated to normally consolidated clays with very low shear strengths at the seafloor. A typical profile of sediment properties for a site in about 305 m (1000 ft) water is shown in Figure 2. Even though the sediments below about 9 m (30 ft) are of Pleistocene age, they appear to be underconsolidated. A profile of properties at another site whose strength profile exhibits a "crustal zone" is shown for comparison in Figure 3. Strength profiles for sites within a single lease block may vary by 100 percent or more. Foundation design for structures in this environment of weak and variable sediments would be a challenge in itself, even if environmental loads such as winds, waves and seismic shaking and geologic features and conditions such as active faults, mudflows and submarine landslides did not have to be considered. In reality, however, many of these factors must be considered for northern Gulf of Mexico sites, resulting in a truly formidable and complex task for the geotechnical engineer.

The development of technology to address geotechnical design considerations for offshore structures was recently described by Focht and

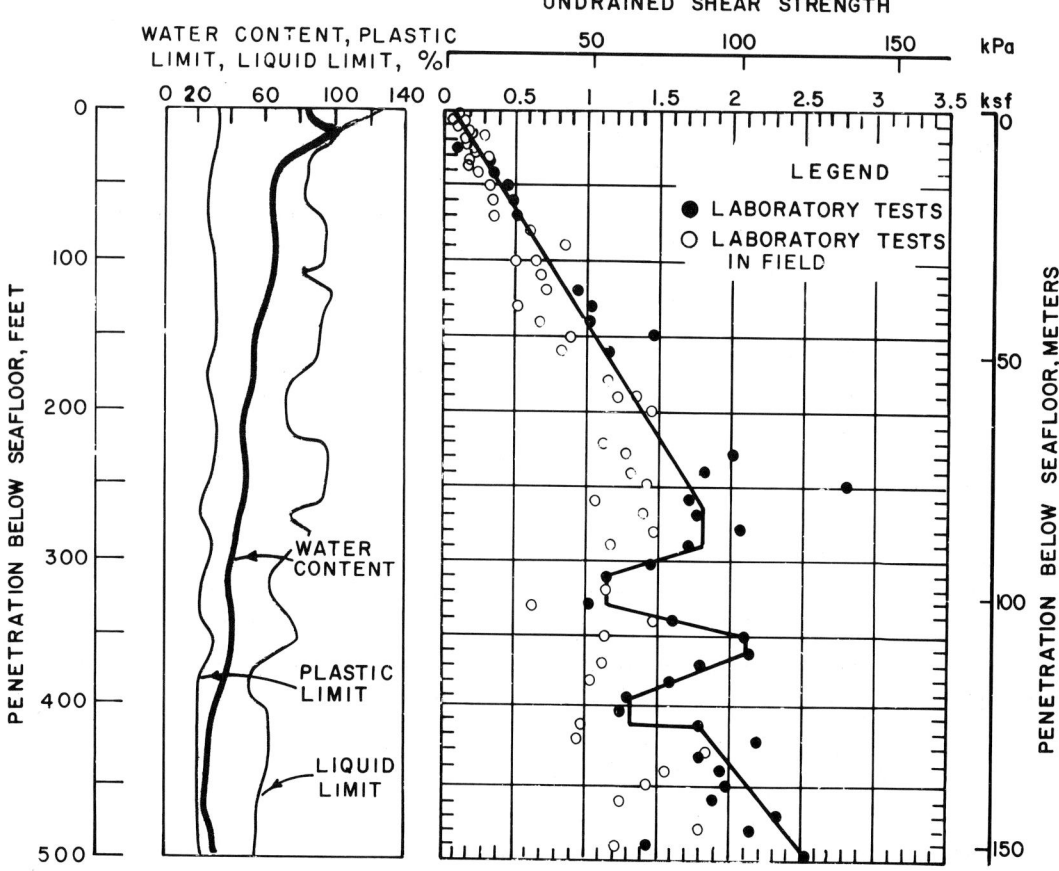

FIG. 2.—Typical properties of sediments, continental slope, norther Gulf of Mexico.

Kraft (1977). Some of the design considerations for each type of structure are useful to illustrate the importance of knowing: (1) the magnitude of and variation in sediment properties, (2) the magnitude of environmental loads, and (3) the nature of any significant geologic features and conditions. Today, the design considerations included in the geotechnical engineer's assignment fall into eight major categories, with a number of items in each category, as shown in Figure 4. Not all categories and every item are involved in each structure design, but the multifold increase in the number of factors, compared to only the consideration of axial pile capacity in the late 1940's, is a striking demonstration of the advances made during the last 30 years.

Platform Design for Upper-Slope Sites

The three basic types of production structures being installed or considered for installation on the upper slope in the northern Gulf of Mexico are pile-supported structures, guyed towers, and tension-leg platforms. The total design, construction, and installation cost of such structures can be as much as several hundred million dollars. Partly because of the ease of recovery and reuse at additional sites, especially in the case of the tension-leg platform, and because the amount of steel is generally less, the guyed tower and tension-leg platform tend to be more economical in deeper waters than pile-supported structures. To provide the reader with a better appreciation for the magnitude of engineering effort and investment required for deep-water production platforms, Figure 5 compares a large pile-supported structure in the northern Gulf of Mexico to some well-known structures.

Pile-Supported Structures

Description.—All production platforms in the Gulf of Mexico today are pile-supported jackets or templates. The base dimensions of some of these structures in the deep waters of the upper slope are greater than 90 m (300 ft). Pile diameters

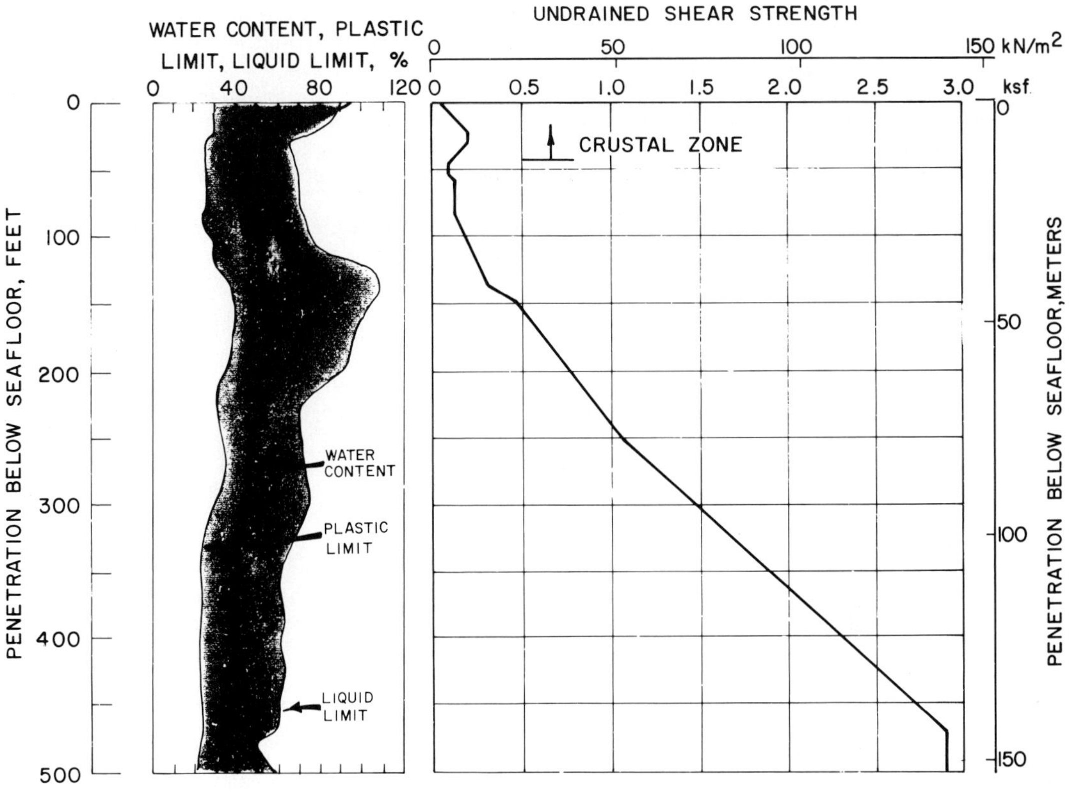

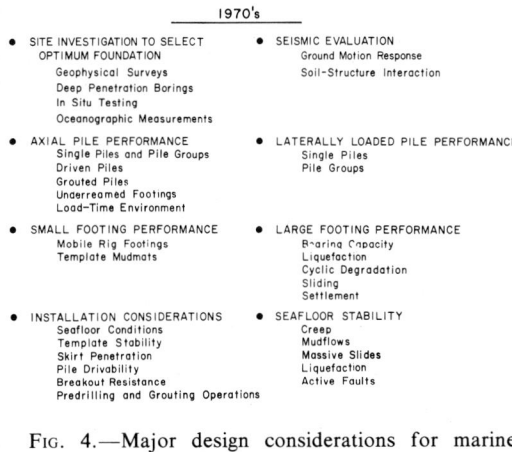

FIG. 3.—Typical properties of sediments west delta area.

1970's

- SITE INVESTIGATION TO SELECT OPTIMUM FOUNDATION
  - Geophysical Surveys
  - Deep Penetration Borings
  - In Situ Testing
  - Oceanographic Measurements
- SEISMIC EVALUATION
  - Ground Motion Response
  - Soil-Structure Interaction
- AXIAL PILE PERFORMANCE
  - Single Piles and Pile Groups
  - Driven Piles
  - Grouted Piles
  - Underreamed Footings
  - Load-Time Environment
- LATERALLY LOADED PILE PERFORMANCE
  - Single Piles
  - Pile Groups
- SMALL FOOTING PERFORMANCE
  - Mobile Rig Footings
  - Template Mudmats
- LARGE FOOTING PERFORMANCE
  - Bearing Capacity
  - Liquefaction
  - Cyclic Degradation
  - Sliding
  - Settlement
- INSTALLATION CONSIDERATIONS
  - Seafloor Conditions
  - Template Stability
  - Skirt Penetration
  - Pile Drivability
  - Breakout Resistance
  - Predrilling and Grouting Operations
- SEAFLOOR STABILITY
  - Creep
  - Mudflows
  - Massive Slides
  - Liquefaction
  - Active Faults

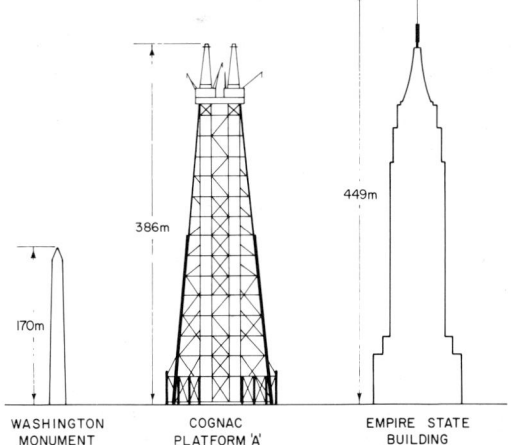

FIG. 4.—Major design considerations for marine foundations.

FIG. 5.—Comparison of structures.

range from 1.22 to 2.14 m (48 to 84 in.) and wall thicknesses are typically 3.8 to 5 cm (1-1/2 to 2 in.). Piles, driven with large pile hammers to 90 to 150 m (300 to 500 ft) below the seafloor, are used to support lateral wind, wave and current loads of 1780 to 2670 kN (400 to 600 kips (1 kip equals 1000 lbs)) per pile and axial working loads of 17,800 to 40,000 kN (4000 to 9000 kips). In areas susceptible to sediment movements, the lateral loads due to the sediment may be tens of thousands of kN (several thousand kips), and pile diameters of up to about 3.66 m (144 in.)

are used. Because large pile-supported structures obtain their stability from deep-penetration piles, they may prove to be the most economical and reliable structures in areas where the upper sediment could experience downslope movement due to creep or mudflow.

Foundation Design.—Foundation design considerations for the installation of pile-supported structures include stability of the structure prior to pile installation and pile drivability. During pile installation the platform must be supported on the seafloor sediments by mud mats or bearing plates made of wood or steel with areas of a few hundred to a few thousand square meters. Calculations of bearing area are critical; if the bearing area is not large enough, for example, the mud mat may penetrate the sediments more than anticipated and place the boat landings below water. Variations in loads on the mud mats or improper assessment of sediment properties at the site, may result in lateral movement of the structure due to wave loading or tilting due to differential penetration.

In the summer of 1977, Shell Oil Company installed the base section of their three-section Cognac Platform A (Fig. 5) in 313 m (1025 ft) of water off the Mississippi River delta. This structure sets a world's record for production structure height and installation water depth. The base section, about 53.4 m (175 ft) high was lowered to the seafloor with two derrick barges, and supported on the seafloor with four 15-m (50-ft)-diameter bearing plates. The elevations of these plates relative to the platform were adjustable to account for the sloping seafloor. This structure is supported by twenty-four 2.14-m-diameter (84-in.) piles that were driven to about 137 m (450 ft) below the seafloor with an underwater hammer having a rated energy of about 814 kN-m (600,000 ft-lbs). The two upper sections were installed during 1978.

The installation of piles is a major cost of the foundation and contributes significantly to the total structure cost. Unexpected coarse-grained layers and variation in the thickness of these layers, for example, can contribute to pile driving difficulties. To minimize the costs, pile wall thicknesses, driving shoe design, energy of pile driving hammers (81.4 to 1356 m-kN or 60,000 to 1 million ft-lbs) and specifications for supplemental procedures must be established in advance based on engineering analyses, boring data, experience, and the results of high-resolution geophysical surveys.

Foundation considerations include axial pile capacity, axial and lateral-load deformation response of the pile under static, cyclic and dynamic loading, and sediment-pile-structure interaction due to combined wind, current and wave loading, to seismic loading and to potential sediment movements. Some typical results of analyses for mud mat penetration and pile capacity are illustrated in Figure 6.

One of the most interesting and challenging aspects of foundation design for pile-supported structures is the sediment-pile-structure interaction. This aspect is of greater significance for structures in deeper water and for structures supported in sediments that experience movements. The resonant period of the structure-foundation system (the time of one cycle of vibration if the system is struck and allowed to vibrate without interference), tends to increase as the water depth and structure height increase. Therefore, the periods of larger waves may coincide with the resonant period where the load amplification factor accounting for the dynamic effect is greatest. For this reason, sediment-structure interaction has received increasing emphasis in recent years in order to determine the load-displacement response of pile foundations. In situations where sediments undergo movements due to wave-induced bottom pressures or gravitational forces, the interaction of sediments and pile is of great importance in evaluating the performance of the structure.

Guyed Towers

Description.—Another structural concept for deep water is the Exxon guyed tower, which reportedly will be an economical structure in deep water to depths of 610 m (2000 ft). The guyed tower consists of a relatively slender steel structure of uniform cross section, rests on a spud can, and is held in position by a set of symmetrically placed bridge strands (Fig. 7). To minimize the danger of collapse, the guy system is designed to provide redundant support.

The guy lines run vertically downward from the deck to fairleads located at some depth below mean water level to avoid being hazardous to navigation and to match approximate elevations of restoring force and wave load. The lines then slope at about a 60-degree angle and connect to anchors through clump weights. Anchor piles or conventional dragtype anchors can be used. The submerged weight of the anchors may be as much as 110 kN (25,000 lbs) and penetration may be as much as a few tens of meters. Since the tower sways between 1 and 2 degrees, the well conductors must be flexible. To comply for tower sway, the spud can base has a paraboloid or bullet shape as shown in Figure 7. The bending moments in the guyed tower are relatively small, and the foundation primarily supports vertical loads.

Foundation Design.—Considerations of foundation design for guyed towers include the penetration resistance that the sediment exerts on the spud can during the installation phase. Creep of

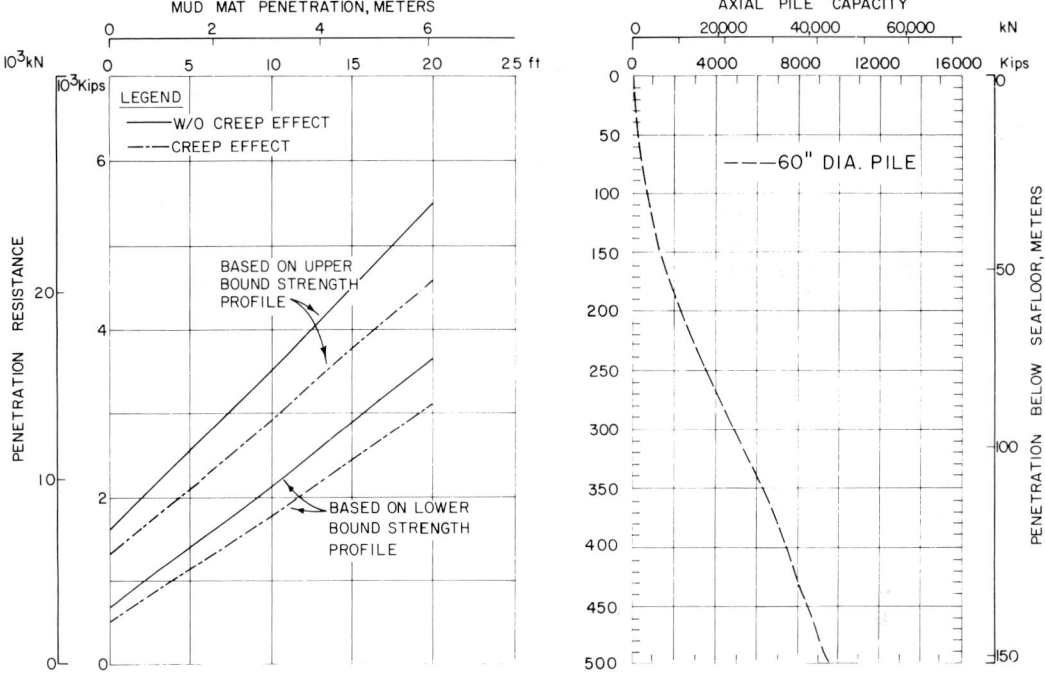

FIG. 6.—Typical results of some foundation analyses for pile-supported structures.

the sediments under load can result in much larger irreversible penetration than expected during the preload operation. Long-term settlement is affected by the preload as well as the net load, and must be considered for conductor and appendage design. The displacements and rotations of the foundation due to lateral and vertical loads on the structure under both static and dynamic loading conditions are needed to evaluate the structure response for the various loading conditions. Other considerations include resistance that the sediments provide to limit the horizontal movement of the spud can during installation and due to environmental loads, bearing capacity for clump weights, and anchor-holding power and drag distance. Some typical design examples of spud-can penetration and anchor performance are illustrated in Figure 8.

Movements of the sediments can affect the performance of a guyed tower. If some of the anchors are founded in sediments that are moving, the structure could be pulled to one side. Depending on the magnitude and rate of movement, this condition might be corrected by adjusting the guys. Moving sediments could also exert sufficiently large lateral forces on the spud can to move the base of the structure, cause problems with well conductors and affect the stability of the structure.

Tension-Leg Structures

Description.—A third structure type that is being considered for deep water is the tension-leg structure shown in Figure 9. This is a floating structure held in place by guys to anchor piles or deadweight anchors. The guys must sustain large tensile loads to minimize lateral movements due to wind, current and wave action. Deadweight anchors of 8900 to 17,800 kN (2 to 4 million pounds), submerged weight, have been considered. The total tensile load for the platform may be 44.5 kN (10,000

FIG. 7.—Guyed tower.

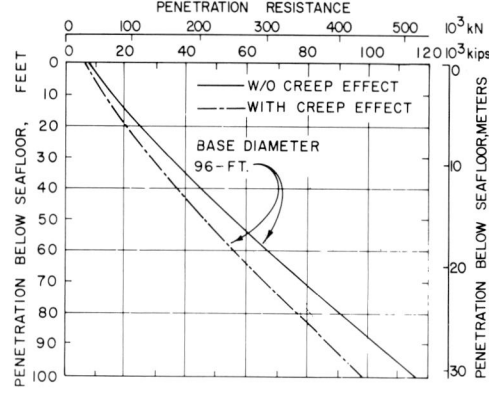

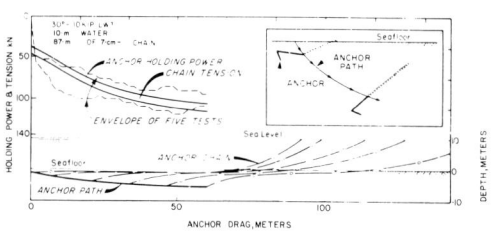

FIG. 8.—Some typical results of some foundation analyses for guyed tower.

kips), and lateral loads may be 4.45 kN (1000 kips). The structure may be held to each anchor by six or more 18 cm-diameter (7-in.) cables. The tension in the cables can be adjusted and maintained by winches. One advantage of the tension-leg structure is that it can be economically reused at additional sites. In some designs using dead-weight anchors, the anchors can be hoisted to the base of the structure, the structure moved to another site, and the anchors lowered again to the seafloor.

Foundation Design.—For tension-leg structures that are held in place with gravity anchors, consideration must be given to the penetration of skirt elements[1] that are used on the base of these anchors to improve the resistance to sliding. In dense sands, where full penetration of skirt elements is difficult, the holding capacity and sliding resistance may be detrimentally affected by less than full penetration. In soft clays, skirt penetration is less of a problem, but the penetration of the whole anchor may be excessive. As scour around the anchors would reduce the holding capacity and sliding resistance, this possibility must be considered. During storms the anchor and sediments holding the anchor are subjected to cyclic loading, which may degrade the strength of the sediments and impair the holding capacity and resistance to sliding.

Slow, creep-like movements of the sediments may have a more significant effect on the wellhead and conductors than on the structure. Rapid movements of the surficial sediments could result in movement of the structure, severing of the conductors between the platform and wellhead.

Although the magnitude of loads and bending stiffness at the tops of anchor piles for a tension-leg structure differ from those at the tops of piles for a template structure, the foundation design considerations are similar. The primary difference between piles for template structures and tension-leg structures is that the major load on the latter is tension rather than compression. The capacity of piles supported in under-to-normally consolidated clays, where the shear strength increases approximately linearly with depth and the pile is subjected to large sustained tensile loads, may degrade with time due to creep effects if these effects are not considered during design. These considerations must include the stress-strain-time (or creep and relaxation) behavior of the sediment and sediment-pile interaction.

Within the scope of this paper, it is not possible to deal effectively with all of the geotechnical design considerations for all three types of structures. This section has provided some background

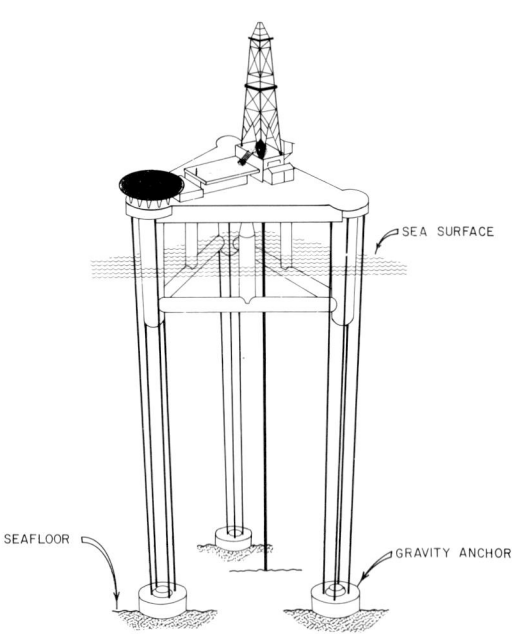

FIG. 9.—Tension-leg Platform.

---

[1]An array of connecting, relatively thin vertical members that extend below the base of the ancher.

by briefly describing a few typical design considerations. In the remaining sections, only the geotechnical design considerations that are associated with sediment movements and pile-structures will be discussed, although many of these concepts are applicable to all structure types.

PRINCIPAL TYPES OF SEDIMENT MOVEMENT

General Comments

Several geologic features and conditions that pose potentially severe problems for siting and foundation design of structures are common in the northern Gulf of Mexico. The most significant of these features and conditions include sediment creep, mudflows, submarine landslides, and active faults. Although not restricted to upper slope sites, such features and conditions are common there, and because of the excessive water depths, the potential geotechnical problems they present are generally greater than on the shelf.

Some of the features and conditions are related to the deformation of the large volume of fine-grained sediment deposited by the present or ancestral Mississippi River or by other large, late-Pleistocene river systems. Others are believed to be related to the movement of large volumes of salt at depth over geologic time (Lehner, 1969). Still others may be closely related to Quaternary fluctuations in sea level. The purpose of this section is to provide a brief introduction to some of the principal types of sediment movement found on the upper slope in the northern Gulf of Mexico that may cause geotechnical problems.

Active Faults

Faults common in upper slope areas of the northern Gulf of Mexico can be divided into two general classes:

1. Deep-seated faults and related subsidiary faults; and
2. Shallow faults affecting only the upper few tens to few hundreds of meters of sediment.

The deep-seated faults are almost invariably growth, (contemporaneous) faults. Most are probably caused by the movement of salt at depth (Lehner, 1969). They typically extend to depths of several thousand meters; the longest of these faults have surface traces several tens of kilometers in length. They are generally subparallel to the coastline and most are down-to-the-basin (Fig. 10); an en-echelon pattern is common in some areas. Similar, but shorter faults with various orientations are formed in the vicinity of salt or shale diapirs. Minor subsidiary faults related to deep seated faults are common.

Many of the deep-seated growth faults on the upper slope are active (Lehner, 1969; Fig. 10). Seafloor expression of these faults ranges from

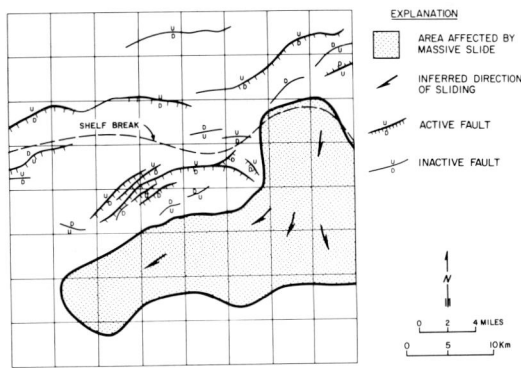

FIG. 10.—Generalized map of area south of Galveston, Texas showing deep-seated growth faults and area affected by large, late-Pleistoncene slide. Group of curvilinear faults between shelf break and northern boundary of slide area is related to growth of a large salt ridge at depth. Each grid square covers an area equivalent to a standard lease block, 23 square km (9 square m).

none to relatively steep scarps up to 17 percent in slope and 30 m (100 ft) in height. Typical throw at subbottom depths of a few hundred meters is on the order of 30 m (100 ft). By definition, throw decreases upwards, and it is generally no more than a few meters in the upper 30 meters of sediment. Limited data suggest the average vertical rate of offset of these faults is on the order of 0.3 to 1 m (1 to 3 ft). per thousand years (derived from data presented by Sidner et al, 1977); the movement is believed to be creep rather than in discrete steps, although the rate of creep may not be constant.

Shallow faults affecting only the upper few tens to few hundred meters of sediment are also found in the northern Gulf of Mexico (Fig. 11). They are most common in the vicinity of the Mississippi Delta in the thick section of rapidly deposited fine-grained sediment. These shallow faults typically have displacements of only a few meters and may range from a few hundred to a few thousand meters in length. Low seafloor scarps, 0.3 to 1.5 m (1 to 5 ft) in height, are commonly associated with them. In detail, such shallow faults undoubtedly have different origins but most probably experience relatively slow movement. These near surface faults are generally related to differential consolidation of shallow sediments, but some are related to local tensional or compressional stresses within the sediment that result from mass movement or diapirism. Most are normal faults although some small reverse faults have been observed (Fig. 11).

An unusual type of shallow fault termed contemporaneous slump-fault known only in the vicinity of the Mississippi delta (Fig. 12, Ploessel and Son, 1978). These faults are characteristically arcuate in plan view. The fault planes flatten into the bedding planes within depths of 300 m (1000 ft). Individual faults or zones of faulting are typically a few kilometers in length. At the seafloor the arcuate shape of the fault trace is typical of the head of large slumps. These features, however, are not known to "toe-out" as slumps or large submarine landslides do. They have continued to move over extended periods of time, but the average rate of movement of the features is slow. The driving force of these faults is gravity; they are not related to the flowage of salt at depth, but flowage of soft clay (shale) at depth may, in some cases, cause them.

Mass Movement Features

Several different mass movement features or phenomena have been identified in upper slope areas of the northern Gulf of Mexico. These include sediment creep, mudflows, and large submarine landslides.

Sediment creep, the gradual and possibly intermittent downslope movement of recently deposited, very soft and underconsolidated sediment, occurs on the outer shelf and upper slope. This phenomena has been observed on slopes as low as 1 percent. Creep typically affects sediments to depths of 1.5 to 15 m (5 to 50 ft) and may affect sediments to depths in excess of 60 m (200 ft). The movement is a result of gravity acting on extremely underconsolidated sediments;

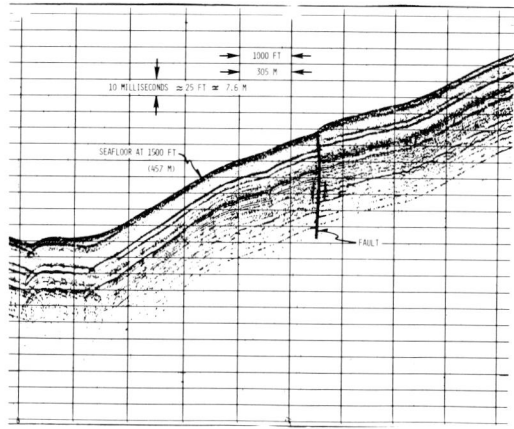

FIG. 11.—Tuned-transducer record of soft, unconsolidated sediments on the upper continental slope offshore the Mississippi Delta. A minor scarp at the seafloor indicates recent activity of this shallow normal fault.

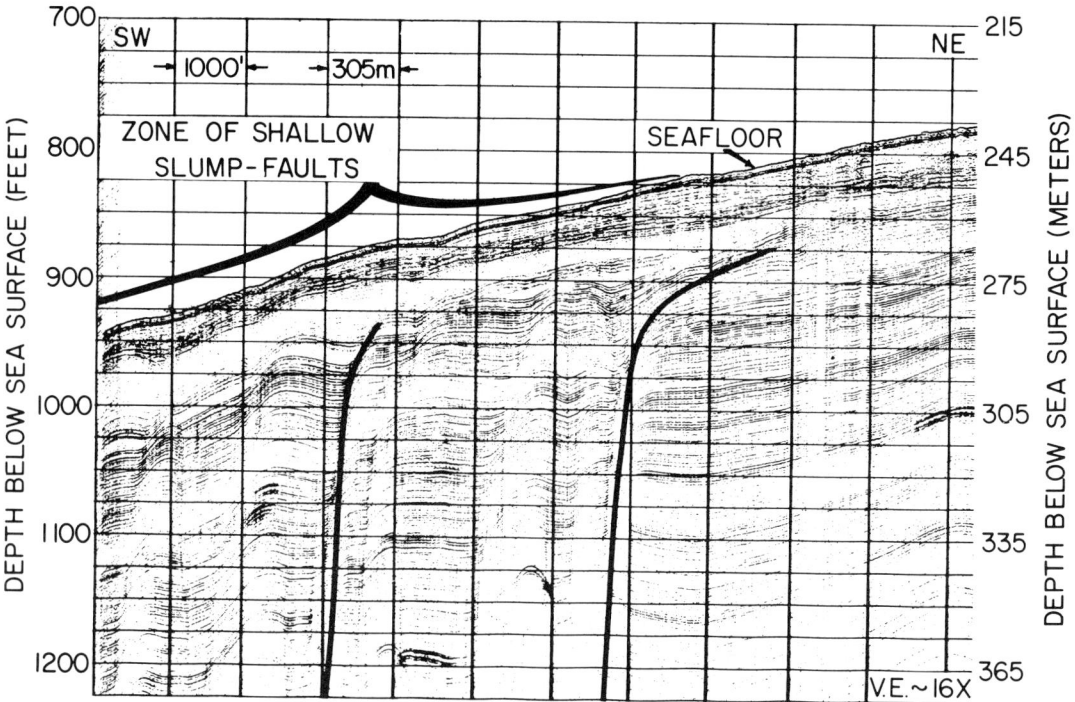

FIG. 12.—Boomer record showing a zone of contemporaneous slump-faults offshore Mississippi Delta. These faults are arcuate in plan and become bedding plane faults within a few hundred meters below the seafloor.

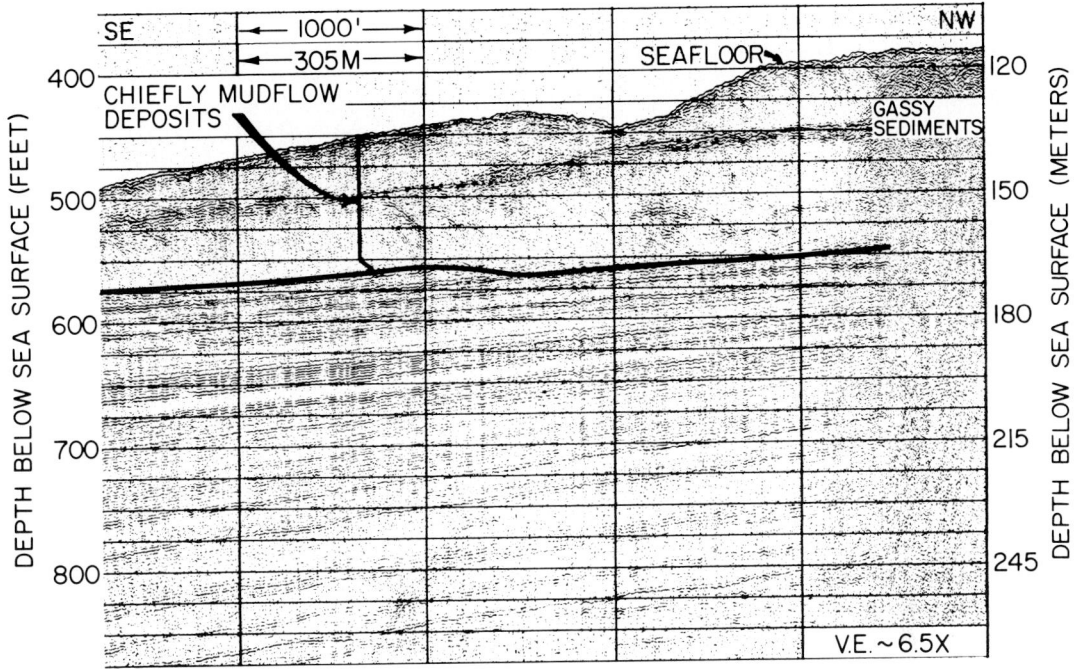

Fig. 13.—Boomer record showing mudflow deposits offshore Mississippi Delta. Note truncation of normal marine deposits at base of mudflow deposits.

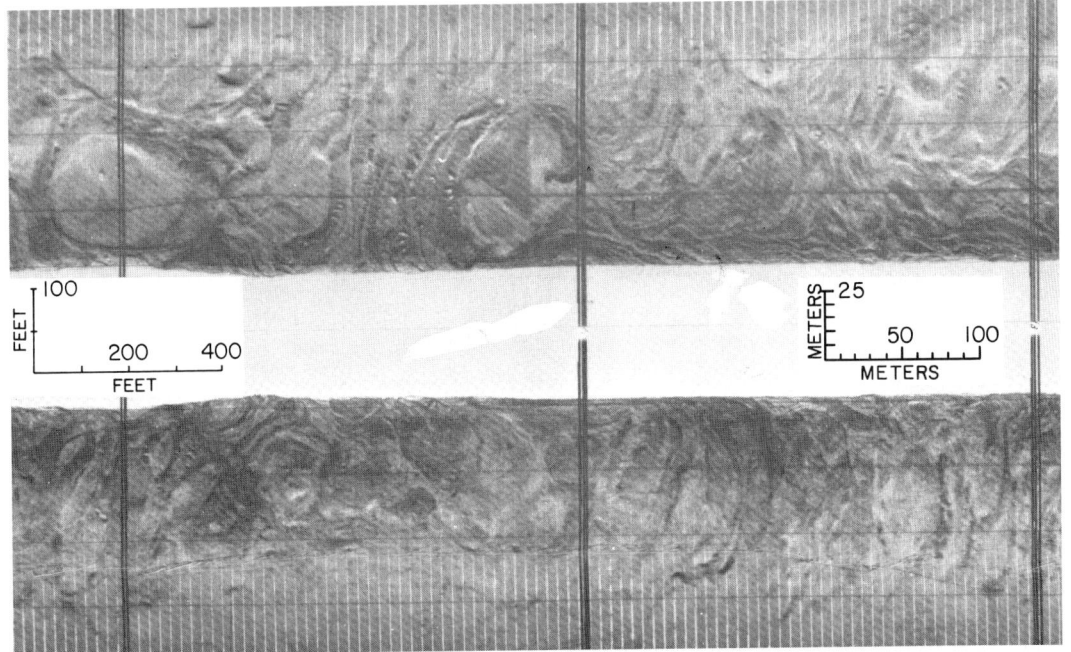

Fig. 14.—Side-Scan Sonar Record showing surface of mudflow deposits offshore Mississippi Delta.

wave-induced loading during storms may play a significant role. Movement presumably decreases with depth and time as the sediments consolidate. Creep is most common in areas of relatively rapid sedimentation of fine-grained material on the upper slope, primarily near the Mississippi delta (Ploessel et al, 1978).

Mudflow deposits are common in the vicinity of the Mississippi delta (Figs. 13, 14). Published data on these features include papers by Coleman (1976); Coleman, Prior and Garrison (1978); and Garrison (1974). In some areas mudflow deposits extend out onto the upper slope. The source of the mudflow material is primarily the sediment deposited rapidly in relatively shallow water around the Mississippi delta. Localized slumping, wave-induced loading, rapid escape of gas or water from sediments or loading from upslope mudflows may trigger the mudflows. Individual mudflows probably flow downslope relatively rapidly as a thick viscous mass. Individual flows are on the order of 1 m (3 ft) to a 20 m (66 ft) in maximum thickness, but repeated mudflows have locally built up over 60 m (200 ft) of mudflow deposits. Actual rates of advance are unknown, but Coleman, Prior and Garrison (1978) have reported that advancement of the leading edge of some flows has been more than 1000 m (3300 ft) in a year. It does not appear that the accumulated mass of mudflow deposits is subject to downslope movement as there are normal, evenly layered reflections on seismic records indicative of normal, undisturbed bedding planes within, or sandwiched between, the mudflow deposits. If the entire mass were moving, these bedding planes would have been disturbed of destroyed.

There are massive submarine landslides at several locations on the outer shelf and upper slope in the northern Gulf of Mexico (Garrison et al, 1977; Robinson, 1977; Sangree et al, 1976; Sidner et al, 1977; Woodbury et al, 1976 and 1977). Commonly, the large slides have each affected

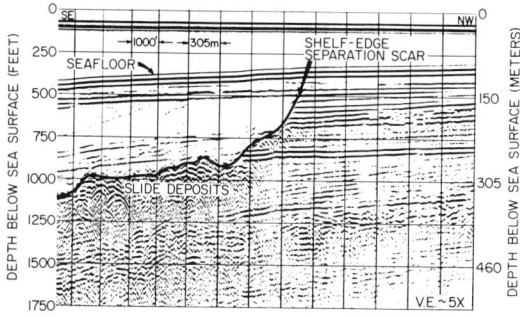

FIG. 15.—Sparker record from offshore Mississippi Delta showing buried head of massive slide ("shelf-edge separation scar") and slide deposits. Sliding is inferred to have taken place during the later Pleistocene.

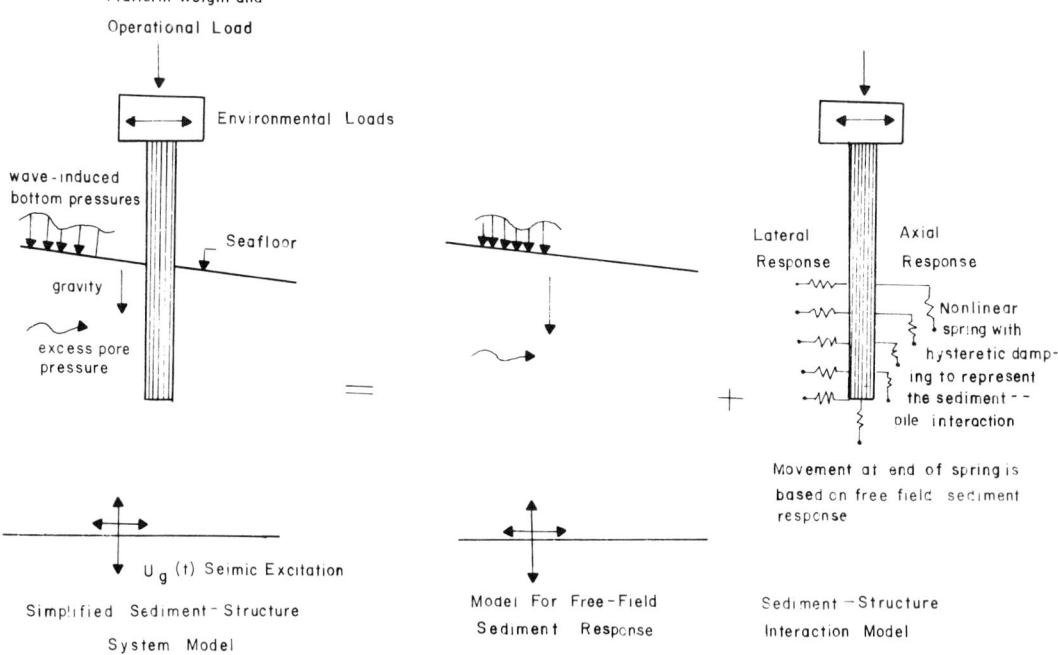

FIG. 16.—Description of analyses for sediment-structure interaction.

areas of several tens of square miles (Fig. 10). Slide deposits, known to be at least 150 m (500 ft) thick in some areas, resulted from catastrophic, rapid sliding. In the vicinity of the Mississippi delta, the truncation at the head of now buried, massive slides that occurred along the ancient shelf-slope break has been termed a "shelf-edge separation scar" (Garrison et al, 1977, Figs. 15, 16). In some areas it appears that there is relatively slow downslope movement of a semicoherent mass along a slide plane (Sidner et al, 1977). Many of the very large slides on the upper slope probably occurred in the late Pleistocene when sea level was lower and sediment was being deposited rapidly in deltas at or near the shelf-slope break. Abnormally high pore pressures, due to the rapid deposition, coupled with the additional effects of wave loading may have triggered some of the large slides.

GEOTECHNICAL PROBLEMS RELATED TO SEDIMENT MOVEMENTS FOR PILE-SUPPORTED STRUCTURES

Problem Description

Soil movement may result from: (1) slow, creep-like movements of the sediment downslope, (2) episodic, large-scale sediment movements due to a catastrophic sediment failure, (3) cyclic sediment movements due to wave-induced differential bottom pressures or seismic loading, (4) diapiric action (mud diapirs), (5) differential consolidation settlements, and (6) differential movement across and along faults. These movements can result in the exertion of a resultant sediment force on structural members of offshore facilities, including well conductors and pipelines, when there is relative movement between the sediment and structural member. The magnitude of the resultant sediment force depends on a multitude of factors that include the magnitude and spatial distribution of sediment movement, the direction of relative movement, the movement rate, and the alterations to the sediment properties due to loading history including cyclic loading. The distribution and rate of soil movements, computed for free-field conditions (i.e., without the presence of the structure) are combined with a sediment-structure interaction analysis that includes a sediment loading criteria based on relative pile-sediment movements to compute the soil loads on the structure. The presence of the structure will affect the sediment movements, but in many cases this effect is probably of second order. Of greater potential importance to the foundation performance is the influence of the sediment-structure interaction on the stress-strain-strength properties of the sediments surrounding the foundation, and in turn the influence of the alteration of sediment properties on the sediment forces that develop. The extent, magnitude and rate of potential sediment movement, amount of new sediment accumulation, if any, and frequency of occurrence must be estimated for the planning and design of safe and economical offshore production facilities in environments where such movements may occur during the service life of the structure. Some potential effects of sediment movements on the foundation performance of pile-supported structures, guyed towers, and tension-leg platforms are described in Table 1.

Figure 16 illustrates the conceptual design process that is used to evaluate the effects of sediment movements on foundation and structure performance. The loads include gravity, wave-in-

TABLE 1.—INFLUENCE OF SEDIMENT MOVEMENT ON FOUNDATION PERFORMANCE

Deformational Mechanisms and Activating Causes	Generalized Description of Foundation Conditions
Faults	Differential movements induce loads on piles and conductors; cause loss of foundation support and differential movement along pipelines.
Wave-Induced Bottom Pressures	Oscillatory and possibly downslope motions of sediments induce complex and dynamic loading on piles, conductors and exposed structural members; results in movements of shallow anchors and spud cans; induce bending and elongation of pipelines.
Mass Movements (Mudflows, Massive Submarine landslides, etc.)	Induces lateral loading on piles, conductors and exposed structural members; results in tilting of platforms and in movement of shallow structural members above the seafloor; loss of foundation support.
Slow Downslope Movement of Sediments (Sediment Creep)	Induces lateral forces on piles, conductors, and spud cans; may cause lateral movements of shallow anchors and pipelines.
Seismic Activity	May induce mass movements, liquefaction and loss of foundation support; results in a complex pile-soil-structure interaction with possible localized liquefaction and degradation of sediments around the foundation.
Diapiric Action	Causes uplift and lateral loading on foundations and pipelines.

duced differential bottom pressures, excess pore pressures, and seismic effects. The pertinent sediment properties are the stress-strain-strength behavior including time dependent effects such as consolidation and creep. The loading and sediment property data are introduced into analytic models with various degrees of sophistication to evaluate the magnitude, rate and distribution of sediment movements and the depth or location of potential failure surfaces in the sediments. Due to uncertainties in sediment properties, loading conditions, and differences between the analytic models and reality, the problem definition must be evaluated with a knowledge of the geologic conditions and tempered with experience and judgment before the results are incorporated in an analysis of sediment-structure interaction.

Data for the sediment-structure interaction model include: (1) sediment and structural properties, (2) wave, current, wind, platform, operational and seismic loads, and (3) free-field sediment movements and sediment loading criteria. The information derived from the interaction model includes displacements and stresses in structural members.

Design Examples

Some typical design conditions are discussed below to illustrate the effects of sediment movements on the structural performance as measured by bending stresses in the pile and by lateral pile movements. Stress is the force per unit area; bending stress is the axial stress due to bending moments and is a measure of the mobilized resistance of the structural member. The first example represents a 24-pile structure with 56 well conductors. The 2.14-m-diameter (84-in.) piles are driven to about 137-m (450-ft) penetration and have a 5.1-cm (2-in.) wall thickness. The computed profile of bending stress along the pile, if the sediment moves at a fast rate (more than 0.3 m (1 ft) per day) down the slope and induces only small downslope movements in the underlying sediment, is shown in Figure 17. The computed maximum bending stresses range from about 48,300 kN/m^2 (7 ksi kips per square inch) for a 9-m-thick (30-ft) slide to 179,400 kN/m^2 (26 ksi) for a 15-m-thick (50-ft) slide zone. If a storm occurs at the same time as the slide, the maximum bending stresses, which are also shown in Figure 17, are increased to about 138 × 10^3 kN/m^2 (20 ksi) for the 9-m-thick (30-ft) slide zone and to 284 × 10^3 kN/m^2 (36 ksi) for the 15-m-thick (50-ft) slide zone. The computed maximum bending stress due only to wave loading is about 104 × 10^3 kN/m^2 (15 ksi). Thus, the effect of the slide may more than double the bending stresses for a 15-m (50-ft) slide zone, requiring thicker walled piles or high yield stress steel. Reducing the allowable pile stresses for the axial pile loads, the allowable bending stress is about 242 × 10^3 kN/m^2 (35 ksi). Comparison of the computed stresses with the allowable stress of 242 × 10^3 kN/m^2 (35 ksi) is a measure of the available resistance that is mobilized.

If the sediment is moving downslope at a creep rate of about 0.4 cm (1 in.) per year, the sediment loading may be less than would be developed at greater rates of movement. Figure 18 compares

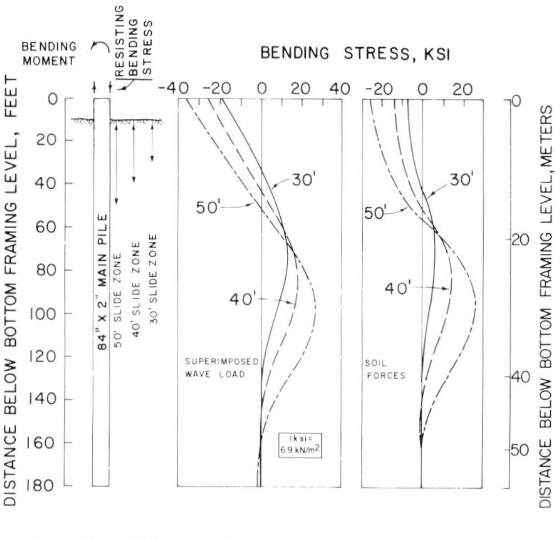

Fig. 17.—Pile bending stresses for 30', 40' & 50' slide zones.

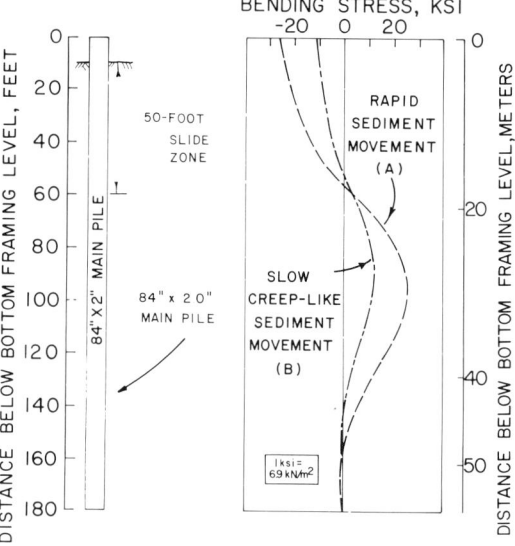

Fig. 18.—Pile bending stresses for 50' slide zone.

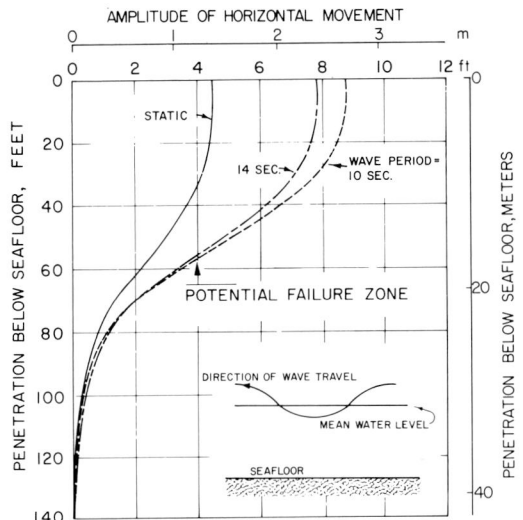

Fig. 19.—Profiles of horizontal soil movements due to wave-induced bottom pressures

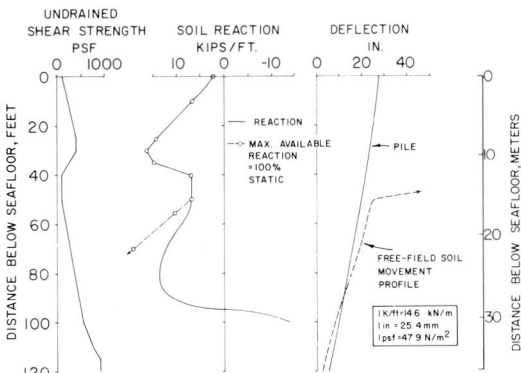

Fig. 21.—Soil-pile interaction for design storm.

the profiles of bending stresses for a 15-m-thick (50-ft) slide zone moving at a "fast" rate (A) and a "creep" rate (B).

Figure 19 shows a profile of computed horizontal displacements in the sediment during passage of a storm. The upper zone of sediment, in some cases to 61-m (200-ft) penetration, may experience large gross downslope movements that are not shown in this figure. Even the sediment below the zone of large movement may exert a loading

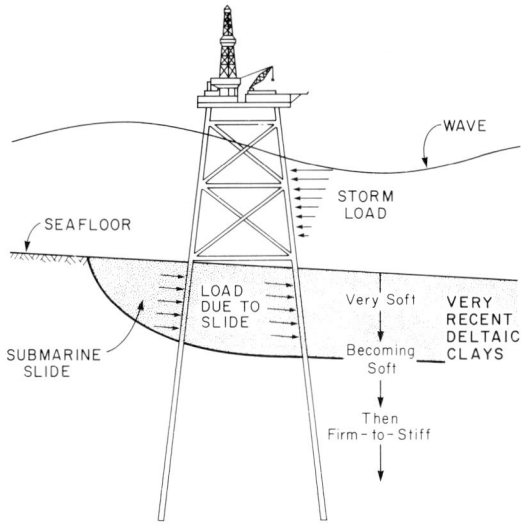

Fig. 20.—Loads due to storm wave and submarine slide.

on the platform. The lateral sediment loads may exceed the total lateral loads due to waves, winds, and currents (see Fig. 20). For example, some designs have shown sediment loads of 35.6×10^3 kN (8000 kips), where the thickness of the slide was about 18 m (60 ft), compared to about 15.6×10^3 kN (3500 kips) for the combined wave, wind and current load.

An example of a sediment-structure interaction analysis is shown in Figure 21 for a four-pile structure located in about 68.7 m (225 ft) of water. The combined wind, wave and current load on this structure is about 8.9×10^3 kN (2000 kips) and the sediment load is about 16.9×10^3 kN (3800 kips). The sediments do not provide any resistance until the pile moves more than the sediments. Thus, for this example the upper 29 m (95 ft) of sediments load the structure rather than provide resistance. The computed lateral movement of the structure at the mudline is about 0.76 m (30 in.). The bending moments for the most severe loading conditions are compared in Figure 22 with the allowable bending moments. The maximum bending moment due to combined wind, wave and current loading without the sediment movements is about 30 percent of the value if the sediment movements are considered.

If the bending stiffness is increased by adding 5.4 cm (2 in.) of steel, the sediment load is increased by about 75 percent and the maximum moment by 30 percent. In this case, the bending stresses decrease since the stiffness increases more than the moment, but if the increased stiffness was due to including knee braces, the bending stresses would increase. In addition, the knee braces would result in an increase in sediment load due to more area of steel being exposed to the moving sediments. Thus, sediment loading may be reduced by making the structure more compliant.

These examples serve to illustrate and quantify

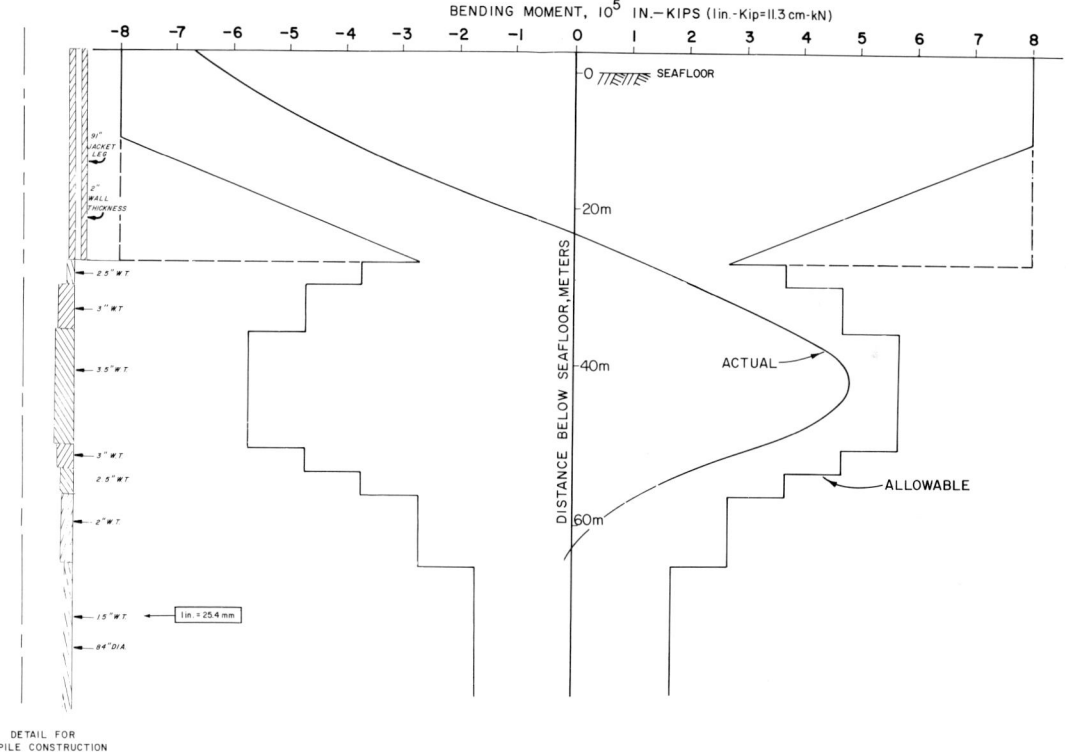

Fig. 22.—Bending moments induced during design storm.

some of the engineering problems for foundations in environments where the sediments experience movement. These examples illustrate the significant lateral loads and bending moments that the moving sediments can generate. These increases result in greater costs to design, build and install platforms in such environments as compared to a case for the same water depth and oceanographical loading conditions where the sediments do not experience movement. Nevertheless, safe and cost-effective structures generally can be designed for areas subject to sediment movements if the stress-strain-strength and time dependent properties of the sediments are known, if the regional and local geology, including the mechanism of the sediment movement, is understood from geophysical data, soil borings and other data, and if the interaction between the water column and subbottom sediments are known.

CONCLUDING COMMENTS

Although very complex and powerful analytic techniques, such as the finite element method, are now available to the geotechnical engineer, confident application of these methods to geotechnical problems only result when the stress-strain-strength properties and their spatial variation are known together with an understanding of the loading characteristics and mechanisms and when these factors are considered in the proper geologic context. As Garrison and Bea (1977) have pointed out, additional field observational data are needed to calibrate existing analytic models. We expect that during the next 5 to 10 years the major trends and developments in marine geotechnology will be in the collection and synthesis of high-resolution geophysical data, in situ testing of sediment properties, sampling and laboratory testing of the sediments, and long-term field instrumentation data. Significant advances will be made in the overall understanding of near-surface geologic processes, particularly of the modes and mechanisms of seafloor instability and rates of movement offshore from the Mississippi delta. Repeat geophysical surveys will be of critical importance to these investigations. Strides will also be made in the development of analytic methods, but these are considered secondary since their applicability will have a low impact without advancement in data acquisition and field instrumentation.

One of the most significant developments in

geotechnical assessments of offshore sites during the past several years has been the integration of geological data derived from high-resolution geophysical records with traditional geotechnical information. This trend will continue as the benefits of this interdisciplinary approach are increasingly realized.

The number of platforms that have been proposed for sites with potentially unstable soils has increased during the last five years. Although foundation design for these structures has been quite challenging, the lack of quantitative field performance data to calibrate the methods of analysis introduces an element of uncertainty in the design and, therefore, in the ultimate cost. The available qualitative data and general experience of those who have worked in the areas of unstable soils indicate that the designs are adequate and catastrophic failure or risk to the public are at levels no greater than those for structures in general. The potential degree of conservatism, however, has large economic implications. Therefore, an interest is developing in industry and government to support field instrumentation programs to monitor (1) the movements of soft sediments during storms, tidal changes, and periods of floods; (2) the interaction between the sediment and water column during storms; and (3) the sediment-structure interaction. The SEAS-WAB project, initiated in 1975 by the U.S. Geological Survey, was the first attempt to acquire data on the response of the seafloor due to wave activity (Garrison, 1977). Additional field performance data are needed to describe and understand the geotechnical environment.

Geologists and geotechnical engineers must not lose their perspective on our ability to build safe structures because we are seeing so many deformational features that suggest seafloor instability. Many of the existing structures may unknowingly have been constructed in areas with deformational features that would have been recognized if high-quality geophysical data had been available. The structures, however, have performed satisfactorily, probably because the rate of the deformation is on a geologic rather than on an historic scale. On the other hand, the foundation failures caused by hurricane Camille illustrate that rates of deformation may vary with time; the few failures to date is no guarantee that there will not be significant failures within the design life of existing structures. Although the effect of deformational features on an ocean structure may not be significant within its normal life of 10 to 50 years, further advances in this technical subarea of geotechnical engineering are needed to achieve confidence in evaluation of seafloor stability where geophysical data reveal deformational features.

REFERENCES

COLEMAN, J. M., 1976, Deltas: Processes of Deposition & Models for Exploration.
―――, PRIOR, D., AND GARRISON, L. E., 1978, Submarine Landslides of the Mississippi River Delta: Preceedings, 10th Annual Offshore Technology Conference, Houston, vol. 2, p. 1067–1074.
FERGUSON, G. H., MCCLELLAND, BRAMLETTE, AND BELL, W. D., 1977, Seafloor Cone Penetrometer for Deep Penetration Measurements of Ocean Sediment Strength: Proceedings 9th Annual Offshore Technology Conference, v. 1, p. 471–478.
FOCHT, J. A., JR., AND KRAFT, L. M., JR., 1977, Progress in Marine Geotechnical Engineering: Journal, Geotechnical Engineering Division, ASCE, v. 103, GT10, p. 1097–1118.
GARRISON, L. E., 1974, The Instability of Surface Sediments on Parts of the Mississippi Delta Front: U.S. Geological Survey open-file report 74-4, 9 p. 3 maps.
―――, 1977, The SEASWAB Experiments: Marine Geotechnology, v. 2, p. 117–122.
―――, AND BEA, R. G., 1977, Bottom Stability as a Factor in Platform Siting and Design: Proceedings 9th Annual Offshore Technology Conference, v. 3, p. 127–133.
―――, TATUM, T. E., BOOTH, J. S., AND CASBY, S. M., 1977, Geologic Hazards of the Upper Continental Slope of the Gulf of Mexico: Proceedings, 9th Annual Offshore Technology Conference, v. 1, p. 51–58.
KRAFT, L. M., JR., AHMAD, NIAZ, AND FOCHT, J. A., JR., 1976, Application of Remote Vane Results to Offshore Geotechnical Problems: Proceedings, 8th Annual Offshore Technology Conference, v. 3, p. 75–96.
LEHNER, P., 1969, Salt Tectonics and Pleistocene Stratigraphy on Continental Slope of Northern Gulf of Mexico: The Amer. Assoc. of Petroleum Geologists Bull. v. 53, p. 2431–2479.
MCCLELLAND, BRAMLETTE, 1975, Trends in Marine Site Investigations: A Perspective, Presented to Offshore Europe '75 Conference, University of Aberdeen, Scotland, September.
PLOESSEL, M. R., AND SON, ROSANNE, 1978, Arcuate Contemporaneous Slump-Faults of the Mississippi Slope, Offshore Lousisiana: Abstract, Geol. Soc. of America Annual Meeting, v. 10, p. 473.
―――, ―――, AND CAMPBELL, K. J., 1978, Geologic Features of Significance for Fiting Structures on the Upper Continental Slope in the Vicinity of the Mississippi Delta, Offshore Louisiana: Assoc. of Engineering Geologists Annual Meeting Abs., Hershey, Pennsylvania.
ROBINSON, N. M., JR., 1977, High Resolution Seismic Reflection Analysis of a Large Submarine Slide, Texas Upper Continental Slope, Geol. Soc. of America Annual Meeting Abs., v 9, p. 1144.
SANGREE, J. B., WAYLETT, D. C., FRAZIER, D. E., AMERY, G. B., AND FENNESSY, W. J., 1976, Recognition of

Continental -Slope Seismic Facies, Offshore Texas-Louisiana: Beyond the Shelf Break, Am. Assoc. Petroleum Geologists Short Course, New Orleans, LA, v. 2, p. Fl–F54.

SIDNER, B. R., GARTNER, STEFAN, AND BRYANT, W. R., 1977, Late Pleistocene Geologic History of the Outer Continental Shelf and Upper Continental Slope, Northwest Gulf of Mexico: Department of Oceanography, College Station, Texas A & M University, 131 p.

WATKINS, D. J., AND KRAFT, L. M., JR., 1976, Stability of Continental Shelf and Slope Off Louisiana and Texas: Geotechnical Aspects, *in* Beyond the Shelf Break: Am. Assoc. Petroleum Geologists Short Course, New Orleans, LA, May, v. 2, p. B1B33.

WOODBURY, H. O., SPOTTS, J. H., AND AKERS, W. H., 1976, Gulf of Mexico Continental Slope Sediments and Sedimentation, *in* Beyond the Shelf Break: Am. Assoc. Petroleum Geologists Short Course, New Orleans, LA, May, v. 2, p. C1C28.

———, ———, AND ———, 1977, Movement of Sediment on the Gulf of Mexico Continental Slope and Upper Continental Shelf: Proc. 9th Annual Offshore Technology Conference, v. 1. p. 59–68.

PETROLEUM POTENTIAL OF PASSIVE MARGIN SLOPES

MAHLON M. BALL
Rosenstiel School of Marine and Atmospheric Sciences
University of Miami
Miami, Florida

ABSTRACT

The Afar Triangle offers an example of a passive continental slope which is undergoing formation at the present time. Its structural style of down-to-the-rift and antithetic faulting has produced a horst and graben topography and an interspersing of blocks of acidic and basic crust. Structural lows are filled with evaporite and clastic deposits while highs are often capped with carbonates.

Basement structure and stratigraphy of the mature analogue, the United States Atlantic passive continental slope, is strikingly similar although it has undergone further considerable evolution in space and time. The structural style and sedimentary patterns provide numerous examples of good potential reservoir rocks and traps in both instances. These have been further enhanced on the Atlantic example by development of major reefal facies and through progradational and erosional patterns.

What is lacking for an accurate assessment of hydrocarbon potential are easily identifiable source beds of sufficient magnitude and quality. The resulting question mark of the presence of economically viable deposits of hydrocarbons on the United States Atlantic margin points out the danger of formulating national policy on estimates of reserves in undrilled frontier areas based primarily upon sediment volumes. These estimates are not taken seriously by professional exploration people. The proper approach is to initiate an agressive drilling policy consistent with reasonable ecological considerations. Only in this way can a rational assessment of our national potential hydrocarbon position be determined.

INTRODUCTION

A major reason for interest in this volume is the slopes' potential for oil and gas. A number of excellent papers have treated various aspects of the petroleum potential of deep ocean areas (Hedberg, 1970; Beck, 1972; Beck and Lehner, 1974; Weeks, 1974; Bally, 1975; Miller and others, 1975; Thompson, 1976; Schlee and others, 1977). I will draw on these works but will not attempt to equal them in scope or depth. Instead, the purpose of this paper is to discuss, in general terms, the oil and gas potential of passive margin continental slopes through two examples, the Afar Triangle which represents a passive margin slope in its primary period of formation, and the United States Atlantic continental slope, a mature trailing edge slope now far removed in both time and space from its spreading center. Perusal of the Afar will identify what structural styles and potential reservoir, seal and source facies are forming there, and therefore may be expected in other passive margin slopes. Then we will see whether or not similar features are found in the mature Atlantic slope. Finally, I will exercise the editorial license explicit in a requested paper on this topic to draw upon the Atlantic example to discuss how estimates of reserves in undrilled frontier areas are made and the inherent dangers in their use in formulation of national energy policy by non-professionals.

Passive or Atlantic type margins are formed along incipient spreading axes as continental blocks break apart and new oceanic crust wells up along the line of separation. The slope generally overlies the transition between light acidic continental basement and thin dense oceanic crust. Passive margins are trailing edges of continental crust moving away from spreading oceanic ridges (Drake and Burk, 1974) except during their initial phase of formation, and, in contrast to active or Pacific type margins, the mature passive margins, remote from spreading ridges, are not characterized by volcanic or earthquake activity.

AFAR TRIANGLE: GENESIS OF A PASSIVE MARGIN SLOPE

The Afar triangle of Ethiopia (Pilger and Rosler, 1975, 1976) lies at the southwest end of the Red Sea and is an excellent analogue for the early rifting phase of passive margin formation. The Afar contains a recently emerged expanse of Neogene seafloor complete with alkaline-rich basaltic pillow lavas and coral capped guyots (Tazieff, 1970; Bonatti and Tazieff, 1970). This embryonic ocean formed along a NNW-SSE spreading center parallel to the axis of the Red Sea. The wedge of new crust lies between the Ethiopian Plateau and a 500 km long block of continental crust, the Danakil Horst, that has rotated some 30 degrees counterclockwise about a point near its north end, opening the triangle and almost closing the southern end of the Red Sea. Rotation at the north end of the horst cut off the marine connection of the Afar depression to the Red Sea during the Pleistocene (Mohr, 1971a, 1972a). The region dessicated leaving its structure and rock distribution laid bare in a setting reminiscent of our own Death Valley. Fumeroles and volcanic activity occur along the axis of the triangle. Varet (1971) reports observations of a lava lake located in a pit crater of Erta'Ale, an active volcano in the northern part of the depression. These phenomena may be related to upwelling of upper mantle material beneath the axial volcanic ranges (Marinelli, 1971). Dakin and others (1971) describe a series of earthquakes that occurred in 1969 at Serdo near the axis of the Afar Triangle.

The Afar is almost completely bound by continental crust. On the west, the Ethiopian Plateau stands as much as 3 km above the triangle floor. Elevations in the Afar range from minus 100 m in the north to several hundreds of meters above sea level in the south. The mountainous terrain on the Danakil Horst on the east contains elevations approaching 2 km. The Somali Plateau on the south rises to heights exceeding 2 km. The southern Afar contains the triple junction of the Danakil spreading axis with the Ethiopian and Gulf of Aden rift trends (Mohr, 1967; Baker and others (1972). However, as pointed out by Barberi and others (1972), most of the oceanic crust has formed along the Afar spreading axis. Mohr (1972b) makes the interesting observation that continental crustal remnants of the Ethiopian and Somalian plates are visible in the Afar as much as 60 km beyond the main plateau escarpments. This broad zone of mixing of continental crustal fragments within new oceanic crust offers some insight regarding the possible dimensions and complexity of transitional crust underlying passive continental margins.

The structural grain of the Afar tends to parallel the orientations of the closest elements of the triple junction. Some crossing trends possibly related to strike-slip faulting have been inferred (Mohr, 1972a). The tectonic style predominantly consists of in echelon normal faults parallel to and down to the rift axes. In some places, antithetic faulting gives rise to horst and graben systems (Mohr, 1971b). Reliefs on the exposed fault scarps are measured in tens to thousands of meters. Tilted fault-bounded blocks rise above salt flats that floor major grabens.

Lowell and Genik (1972) present a very useful synthesis of seismic and well data along two lines of section: one across the northern Afar, the Danakil Horst and Red Sea to Yemen and the other just north of the Afar across the Red Sea. The dominant structural style seen on these sections is one of normal faults down to the axes of the rifts: the Danakil rift and its northern extention, referred to as the western rift by Lowell and Genik (1972), and the axial trough of the Red Sea. The faults bound tilted basement blocks. Antithetic normal faults cut back into the major dislocation planes generating horst and graben topography. The wells confirm existence of Miocene salt deposits greater than 3 km thick. The salt deposits form in structural lows. Over a kilometer thickness of bedded Pleistocene salt occurs in the Danakil depression. Subsequent flowage of evaporites could result in concentrations of pillows and domes in the deeper parts of the structural basins. Counter to affinity of salt deposits for regional structural lows, shallow water carbonates within the Red Sea and Afar region build up on structurally high positions along untilted edges of basement blocks and horsts defined by down-to-the-rift and antithetic normal faults.

The overall conceptual model emerging for passive margins from studies of the Afar Triangle has the following characteristics. The structural style is one of down-to-the-rift normal faults with some antithetic faulting causing tilted basement blocks and horst and graben topography. Elements of acidic crust are interspersed among new basic oceanic materials. Salt and terrigenous clastic sediments thicken into structural lows. Reefal carbonates flank and cap tectonic highs. Given sufficient thickness of evaporties and loading, pillows, domes and diapirs should form as a result of salt flowage.

If source rocks are present, it is obvious that the tectonic style is such that one may expect a variety of structural and stratigraphic traps, the former associated with faulting and salt structure. Both clastic and carbonate reservoirs may be expected with salt seals common. Hydrocarbon accumulation may occur at an early stage of formation as shown by the Mobil-Esso Red Sea C-1 well, near the western end of the northern line of section in the Afar, which blew out of control (Mahdi, 1971) when it encountered gas in the upper nine feet of clastic section beneath massive Miocene salt.

UNITED STATES ATLANTIC SLOPE: A MATURE PASSIVE MARGIN

Recent descriptions of the U. S. Atlantic continental margin are replete with references to basement block faulting during the early Mesozoic rifting phase of margin development (Schlee and others, 1977; Uchupi and others, 1977; Mattick and others, 1978; Dillon and others, in press). The action of down-to-the-basin and antithetic normal faults have created half-grabens and horst and graben topography in basement rocks. Subsequent interaction of erosion and deposition in this structural setting has been responsible for formation of potential trapping configurations (Schlee and others, 1977). Examples of formation of evaporites in structural lows (Schlee and others, 1976; and Mattick and others, 1978) and carbonate upbuilding on basement ridges (Schlee and others, 1976; Mattick and others, 1978; and Dillon and others, in press) are cited in discussions of the regional geology and geophysics of the U. S. Atlantic north, mid and southern margins, respectively. Clearly, the early stages of passive margin slope development as exemplified by the U. S. Atlantic margin share many attributes with the exposed margin in the Afar Triangle.

Now, how has the slope changed with time? A complication in the picture results from the fact that the physiographic slope has migrated. At times of sufficient sediment supply, the slope prograded so that ancient rise deposits are overlain by the slope. Conversely, during the Tertiary, erosion has cut the slope back (Grow and others, in press; Schlee and others, this volume) so that older shelf deposits underlie it. It follows that descriptions of reservoir, seal, and source facies beneath the physiographic slope must include references to both shelf and rise settings. A major carbonate shelf edge reef of Mesozoic age appears to underlie the present slope (Mattick and others, 1978; Grow and others, in press) and extend more-or-less continuously southward from the seaward rim of Georges Bank basin to emerge south of Cape Hatteras as the Blake Plateau escarpment. This carbonate mass accumulated as a result of upbuilding in step with subsidence. What initially began as perpetuation of relief on a basement ridge, eventually resulted in a depositional structure whose relief over the adjacent ocean basin is measured in kms. This reefal mass with its potential for containing porous and permeable reservoir beds sealed by overlying shales or tight carbonates is an attractive exploration target (Mattick and others, 1978). Pinch outs with terrigenous clastic sediments on the reef's shelfward flanks represent possible stratigraphic traps.

Clearly, the history of passive slopes from genesis to maturity, as represented by our two examples, give rise to ample traps and potential reservoir rocks.

SOURCE BEDS

One aspect of passive margin slope's potential that remains problematic is whether souce materials have developed through times which are sufficient for generation of commercial quantities of hydrocarbons. Let us stick with our Atlantic model. Drilling on the shelves has encountered a predominantly terrigenous clastic depositional environment whose source materials are not believed to be rich and are prone toward production of dry gas. The carbonate facies inferred to be present in the Mesozoic shelf edge beneath the present slope could contain prolific producers of organic carbon, but relatively slow deposition rates and attending oxidation of source materials may have prevented creation of oil and gas in this environment. Perhaps the thick sediments underlying the abyssal rise seaward of the present slope may (in places) contain adequate source rocks with permeable migration paths updip to sealed reservoirs. The black shales of Cretaceous age noted in numerous penetrations of the deep sea drilling program (Mattick and others, 1978; Dow, 1978) show that the abyssal setting can, under certain circumstances, contain potential source materials. The question remains whether such materials are preserved deep enough to achieve temperatures capable of generating commercial quantities of hydrocarbons. The recent announcement by the U.S. Geological Survey that the B-3 stratigraphic test on the slope off New Jersey encountered wet gas bearing formations is a significant encouragement for slope exploration.

One must look away from the U.S. Atlantic margin to the west African embayment for an example of a late complication in slope development that can result in greatly enhanced potential for oil and gas production. The structurally controlled embayment is a focal point for drainage off the African continent that has resulted in the formation of the Niger delta. The attending outbuilding of a thick, rapidly deposited accumulation of sandy and shaly sediment has given rise to coexistence of reservoirs, seals, and preserved source beds. Prolific oil and gas fields are the result.

ESTIMATING RESERVES IN FRONTIER AREAS LIKE THE UNITED STATES ATLANTIC MARGIN: METHOD AND MADNESS

Accepting the premise that there are sufficient indications for the existence of structures, reservoirs, seals and source materials beneath passive margin slopes to justify oil exploration in this setting, some thought should be directed toward the nature of the quantitative estimates of recoverable hydrocarbons in these frontier regions. Sediment volumes are invariably a factor in such estimates. Volumes are determined using indirect geophysical measurements, and on the basis of selection of analogue areas, hydrocarbon yields are calculated.

In oil companies and government organizations, these computations are carried out by research personnel on head office staffs or in research laboratories. Every effort is made to consider all aspects of geologic factors such as timing of structural development relative to formation and migration of oil and gas and the history of diagenetic effects that influence reservoir, seal and source bed quality, etc. No matter how arduous the approach, predictions based on these estimates in frontier areas are generally highly inaccurate. The resource assessment researchers are fully aware of this and attempt to allow for these inaccuracies by presenting ranges of possible findings with subjective probabilities assigned to figures within the range of possibilities. Explorationists consider the predictions for frontier areas but are not in the least confused concerning their validity. Non professionals who often use the figures to set national policy are often not aware of their shortcomings. Consider the reactions of the pros. Hedberg (1975), in his usual concise manner, states that the volume of sediment approach is fallacious. Weeks (1974) and Bally (1975) report that regional studies indicate yields from similar settings range through four orders of magnitude. The hard numbers that oil finders deal with are not based on what they hope to find. Instead, like all successful gamblers, oil men base their tactics on what they can afford to lose.

The soundness of this approach is already obvious to most earth scientists. It is clearly born out in recent experiences in the exploration efforts off New Jersey. The top prospect, the Stone Dome (Schlee and others, 1977), is to date unproductive and its potential severely limited by a number of plugged and abandoned wells, including one on its crest. Secondary prospects under the outer shelf have been enhanced by the success of a Texaco test. To everyone's surprise, wet gas shows have been encountered in a stratigraphic test on the slope whose location was designed to avoid any likelihood of penetrating oil or gas bearing strata. New information fed back from the drilling will eventually result in the confirmation or condemnation of this play.

CONCLUSIONS

The Afar Triangle, an analogue of a passive continental slope undergoing formation in the geologic present, and the United States Atlantic slope, a mature passive continental slope, have many stratigraphic and structural features in common. Both have similar down-to-the-basin/down-to-the-rift, antithetic faulting, and similar patterns of sedimentation associated with the basement terrain. Development of traps and reservoir rocks are potentially attractive for hydrocarbon exploration. Development of passive margin reefs and erosion/progradation through geologic time augment the number of potential traps and reservoirs on passive margin slopes making them even more promising targets for exploration of oil and gas. Unfortunately, the development and presence of adequate source beds remains an enigma.

With these factors in mind, estimates of potential reserves based upon sedimentary volumes alone are specious and their use in formulation of national energy policy dangerous. Our policy should be to encourage proving reserves through an active drilling program which takes *reasonable* precautions to protect the environment.

REFERENCES

BAKER, B. H., MOHR, P. A., AND WILLIAMS, L. A. J., 1972, Geology of the East African Rift System: Geol. Soc. America Spec. Paper 136, p. 1–67.
BALLY, A. W., 1975, A geodynamic scenario for hydrocarbon occurrences: Proc. 9th World Petroleum Congress, v. 2, p. 33–44.
BARBERI, F., TAZIEFF, H., AND VARET, J., 1972, Volcanism in the Afar Depression—Its tectonic and magmatic significance: Tectonophysics, v. 15, p. 19–29.
BECK, R. H., 1972, The oceans, the new frontier in exploration: Australian Petroleum Exploration Assoc. Jour., v. 12, pt. 2, p. 7–28.
———, and LEHNER, P., 1974, Oceans, new frontier in exploration: Am. Assoc. Petroleum Geologists Bull., v. 58, p. 376–395.
BONATTI, E., AND TAZIEFF, H., 1970, Exposed Guyot from the Afar Rift, Ethiopia: Science, v. 168, p. 1087–1089.
DAKIN, F. M., GOUIN, P., AND SEARLE, R., 1971, The 1969 earthquakes in Serdo, Ethiopia: Bull. of Geophys. Obs. No. 13, Haile Sellassie I Univ., Addis Ababa, Ethiopia, p. 19–56.
DILLON, W. P., PAULL, C. K., DAHL, A. G., AND PATTERSON, W. C., Structure of the continental margin near the COST GE1 drillsite from a common depth point seismic reflection profile, *in* COST GE1 report: U.S. Geol. Survey circular.
DOW, WALLACE, G., 1978, Petroleum source beds on continental slopes and rises: Am. Assoc. Petroleum Geologists Bull., v. 62, p. 1584–1606.
DRAKE, C. L., AND BURK, C. A., 1974, Geological significance of continental margins, *in* Burk, C. A., and Drake, C. L., The geology of continental margins: New York, Springer-Verlag, p. 1–1009.
GROW, J. A., MATTICK, R. E., AND SCHLEE, J. S., in press, Multichannel seismic depth section and interval velocities over outer continental shelf and upper continental slope between Cape Hatteras and Cape Cod, *in* Geological investigations of continental margins: Am. Assoc. Petroleum Geologists Memoir 29.
HEDBERG, H. D., 1970, Continental margins from viewpoint of the petroleum geologist: Am. Assoc. Petroleum Geologists Bull., v. 54, p. 3–43.

———, 1975, The Volume-of-Sediment Fallacy in Estimating Petroleum Resources, *in* Haun, J. D., 1975, Methods of Estimating the Volume of Undiscovered Oil and Gas Resources: Tulsa, Oklahoma, Am. Assoc. Petroleum Geologists Studies in Geology No 1, p. 1–206.
LOWELL, J. D., AND GENIK, G. J., 1972, Sea-floor spreading and structural evolution of southern Red Sea: Am. Assoc. Petroleum Geologists Bull., v. 56, p. 247–259.
MADHI, M. S., 1971, Geophysical work in oil exploration: Bull. Geophys. Obs. No. 13, Haile Sellasie I Univ., Addis Ababa, Ethiopia, p. 122–123.
MARINELLI, G., 1971, La province geothermique de la depression Danakil: Ann. Mines, p. 123–133.
MATTICK, R. E., GIRARD, O. W., SCHOLLE, P. A., AND GROW, J. A., 1978, Petroleum potential of the U.S. Atlantic slope, rise and abyssal plain: Am. Assoc. Petroleum Geologists Bull., v. 62, p. 492–608.
MILLER, B. M., THOMSEN, H. L., DOLTON, G. L., COURY, A. B., HENDRICKS, T. A., LENNARTZ, F. E., POWERS, R. B., SABLE, E. G., AND VARNES, K. L., 1975, Geological estimates of undiscovered recoverable oil and gas resources in the United States: U.S. Geol. Survey Circular 725, p. 1–78.
MOHR, P. A., 1967, The Ethiopian rift system: Bull. Geophys. Obs. No. 11, Haile Sellessie I. Univ., Addis Ababa, Ethiopia, p. 1–65.
——— 1971a, The Ethiopian triple-rift junction in terms of plate technics: Bull. Geophys. Obs., Haile Sellassia I. Univ., Addis Ababa, Ethiopia, No. 13, p. 1–17.
——— 1971b, Tectonics of the Dobi Graben region, Central Afar, Ethiopia: Bull. Geophys. Obs. No. 13, Haile Sellassie I. Univ., Addis Ababa, Ethiopia, p. 73–89.
——— 1972a, Surface structure and plate tectonics of the Afar: Techtonophysics, v. 15, p. 3–18.
——— 1972b, Regional significance of volcanic geochemistry in the Afar Triple Junction: Geol. Soc. America Bull., v. 83, p. 213–222.
PILGER, A., AND ROSLER, A., eds., 1975, Afar depression of Ethiopia. Vol. I: Inter-Union Commission on Geodynamics Sci. Rept. No. 14, E. Schweizerbart'sche Verlagsbuchhandlung, Stuttgart, p. 1–416.
———, and ———, eds., 1976, Afar between continental and oceanic rifting, Vol. II: Inter-Union Commission on Geodynamics, Sci. Rept. No. 16. E. Schwiezerbart'sche Verlagsbuchhandlung Stuttgart, p. 1–216.
SCHLEE, JOHN, BEHRENDT, JOHN C., GROW, JOHN A., ROBB, JAMES M., MATTICK, ROBERT E., TAYLOR, P. T., AND LAWSON, BARBARA J., 1976, Regional geologic framework off northeastern United States: Am. Assoc. Petroleum Geologists Bull., v. 60, p. 926–951.
———, MARTIN, R. G., MATTICK, R. E., DILLON, W. P., AND BALL, M. M., 1977, Petroleum geology on the United States Atlantic Gulf margins: Proc. Southwestern Legal Foundation, v. 15, p. 47–93.
TAZIEFF, H., 1970, The Afar triangle: Scientific American, v. 222, p. 32–40.
THOMPSON, T. L., 1976, Plate tectonics in oil and gas exploration of continental margins: Am. Assoc. Petroleum Geologists Bull., v. 60, p. 1463–1501.
UCHUPI, ELAZAR, BALLARD, R. D., ELLIS, J. P., 1977, Continental slope and upper rise off western Nova Scotia and Georges Bank: Am. Assoc. Petroleum Geologists Bull., v. 61, p. 1483–1492.
VARET, J., 1971, Recent activity of the Extra'Ale volcano: Bull. Geophys. Obs. No. 13, Haile Sellassie I. Univ., Addis Ababa, Ethiopia, p. 115–119.
WEEKS, L. G., 1974, Petroleum resources potential of continental margins, *in* Burk, C. A., and Drake, C. L., The geology of continental margins: New York, Springer-Verlag, p. 1009.

THE BIOTA AND BIOLOGICAL PROCESSES OF THE CONTINENTAL SLOPE

GILBERT T. ROWE AND RICHARD L. HAEDRICH
Woods Hole Oceanographic Institution
Woods Hole, Massachusetts 02543

ABSTRACT

Life on continental slopes is characteristically zoned with depth. Faunal variation across the slope appears to occur more rapidly than anywhere else in the deep-ocean environment, but the causes of the narrow zonation are not known. A diversity of causes is possible, including competition for a diminishing resource (food), very efficient predation, pressure effects on enzymes, and also conditions such as extremely low dissolved oxygen or variations in the physical properties of the sediments. The composition of the fauna on continental slopes is not entirely unique, sharing species both with the continental shelf and continental rise, but nonetheless the slope can be considered a distinct biological province. The abundance of life and rates of physiological processes are intermediate between those of the shallow continental shelf and the abyss. Rates of change do not follow smooth gradients across the slope. At the base of the slope, where organic-rich sediments from shallow depths have accumulated, there is an important transition zone to the more truly oceanic conditions of the continental rise. Petroleum and phosphorite resources on the continental slope, using biological criteria, would most likely occur in regions historically characterized by upwelling, high productivity and low oxygen concentrations. Like the fauna itself, these resources should occur in zones. Because continental slope depths and beyond are important in the remineralization of organic matter, those involved in the exploration for and utilization of deep-sea resources must keep in mind that alteration of normal biological processes there might adversely affect the natural and vital biochemical cycling.

INTRODUCTION

Continental Slope Explorations

It was once thought that continental slope depths, as well as all of the deep-sea, were devoid of life (Forbes, 1844). This view was short-lived because of discoveries made during extensive regional dredging in the Mediterranean, the west coast of Europe and the eastern United States, summarized by Agassiz (1888), LeDanois (1948), Ekman (1953) and Wolff (1977). More recently, conventional, but detailed studies of the biology of the continental slopes have been conducted in the seas of the U.S.S.R. (Zenkevitch, 1963), off Oregon (Carey, 1965; Griggs, Carey and Kulm, 1968; Barnard, 1971), off southern California (Hartman and Barnard, 1958; Jumars, 1975, 1976), off North Carolina (Rowe and Menzies, 1969; Rowe, 1971a; Menzies, George and Rowe, 1973) and off New England (Bigelow and Schroeder, 1939; Schroeder, 1955; Sanders, Hessler and Hampson, 1965; Wigley and Emery, 1967; Rowe, Polloni and Hornor, 1974; Markle and Musick, 1974; Grassle et al., 1975; Haedrich, Rowe and Polloni, 1975; Rowe, Polloni and Haedrich, 1975; Williams and Wigley, 1977; and others).

Megafauna, defined roughly as those organisms that are big enough to be easily visible, have not been investigated quantitatively as well or as often as the infauna, the small animals that live buried in the sediments. Trawls can catch megafauna, but are not quantitative and grabs, though quantitative, do not sample the sparse megafauna well. Successful approaches have used photography and trawls together, using the photographs to quantify large areas of bottom and trawls to provide actual samples of the species seen in the photographs. Such studies have been completed off North Carolina (Rowe and Menzies, 1969) and off Peru (Menzies et al, 1973). The submersible ALVIN has been used in the Gulf of Maine (Rowe, Haedrich and Polloni, 1975) and off Massachusetts (Grassle et al, 1976; Cohen, 1977) to quantify animals over distances metered by an odometer rolled along the bottom as photographs were being taken.

Zonation of the Fauna: The Typical Slope

The upper biological boundary of continental slopes is more or less equivalent to the depth of the permanent thermocline. The boundary separates the shelf environment, with time-varying temperatures, from greater depths, free of light and free of major seasonal temperature variations. The slope marks the edge of the continent and its gradient is in general steeper than any other major portion of the deep-sea bottom. At its base, near 2000 m in many areas, the slope meets the continental rise. This boundary is marked by a change in the gradient and, proceeding seaward, by a gradual shift to greater and greater pelagic contributions to the sediments.

The slope off New England can be considered a rather typical one, not unlike other continental slopes in many parts of the world. Our data from this region (Fig. 1) show that biologically the slope is not just a transition region from the shelf to the deep ocean, but can be considered a system on its own. Figure 2 shows an example of the sorts of data that contribute to this view. Shown is a cluster based on our data for megafaunal invertebrates, collections being associated on the basis of faunal similarity. While a number of interesting patterns can be seen, the important points here are that the major divisions of the cluster are three, and that each can be identified with a separate physiographic province, that is the shelf, the slope, and the rise.

A universal feature in the distribution patterns of the slope fauna is the tight zonation, the narrow vertical depth ranges suggested, for example, by the relatively large number of subsets within the slope cluster of Figure 2. Despite their restricted vertical range, these populations usually extend great distances along isobaths. This means that slight changes in depth across continental slopes result in radical changes in what lives there, but over extensive distances along a depth contour the composition of a community is altered very little. The greatest change in the kinds of animals living on the sea floor occurs at the continental shelf-continental slope boundary. Down the continental slope early workers believed there was a "gradual replacement of species" with no distinct faunal boundaries all the way down and out across the continental rise (Sanders and Hessler, 1969), a view resulting from insufficient sampling density. More recent work has shown that some animal groups, such as isopod crustaceans (Menzies et al, 1973) and amphipod crustaceans (Mills, 1972), are congregated in such a way that the slope fauna can be subdivided by statistical criteria into groups of species arranged in vertical depth zones. The larger organisms, such as echinoderms, decapod crustaceans, and fishes (together called the megafauna), conform to such zonal patterns in many areas. Across the northeastern United States continental slope major boundaries appear on the upper slope, in the middle of the lower slope and, in a mixed pattern, at the junction between the slope and rise (Menzies et al, 1973; Haedrich et al, 1975).

Analyses of the trawl samples we have collected in the mid-Atlantic Bight illustrate this zonation. In large trawls the addition of species with depth on the slope increases abruptly over narrow depth intervals as newly encountered species are added to the fauna (Fig. 3). The curves shown are for the three major megafaunal groups, the fishes,

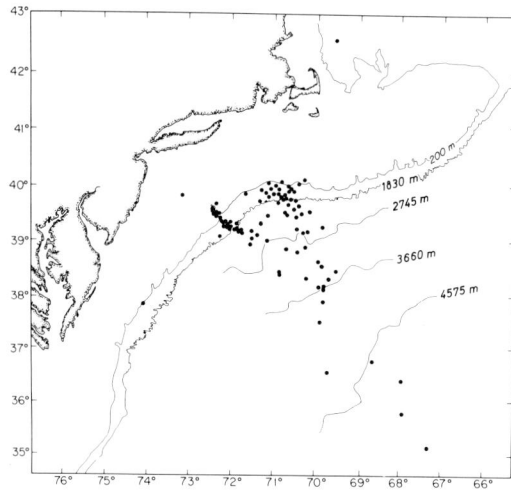

FIG. 1.—Trawl stations off the U.S. continental margin (east coast).

echinoderms and decapod crustaceans. The most marked changes in fishes occur at 800 and 1600 m, separating a homogeneous assemblage, whereas the echinoderm fauna, initially homogeneous, undergoes rapid changes beginning at about 1200 m. Deeper than 800 m, few decapods are added. Comparison of this pattern with those at greater depths, on the continental rise (Fig. 4), indicates that in all these groups greatest faunal variation with depth is to be found on the continental slope. Gardiner and Haedrich (1978), using a statistical test for zonation, found clear evidence for zonation in the total megafauna above 3000 m, but below that depth such could not be distinguished with certainty.

Why animals are narrowly zoned on the continental slope is not known. In shallow water and in the intertidal it seems clear that zonation results from the ability to survive at some set of physical conditions along a gradient of physical extremes or successful competition along a resource gradient. As temperature variations and gradients are slight below the permanent thermocline, it seems unlikely that a gradient of temperature affects the animals' distributions. Other parameters, with the exception of pressure, do not appear to be arranged in a gradient along which the organisms range themselves. Consistent trends over the depth gradient can seldom be seen in sediment particle size, organic matter or sediment physical properties.

FIG. 2.—Cluster diagram based on megafaunal invertebrates from trawl samples, east coast, U.S.A.

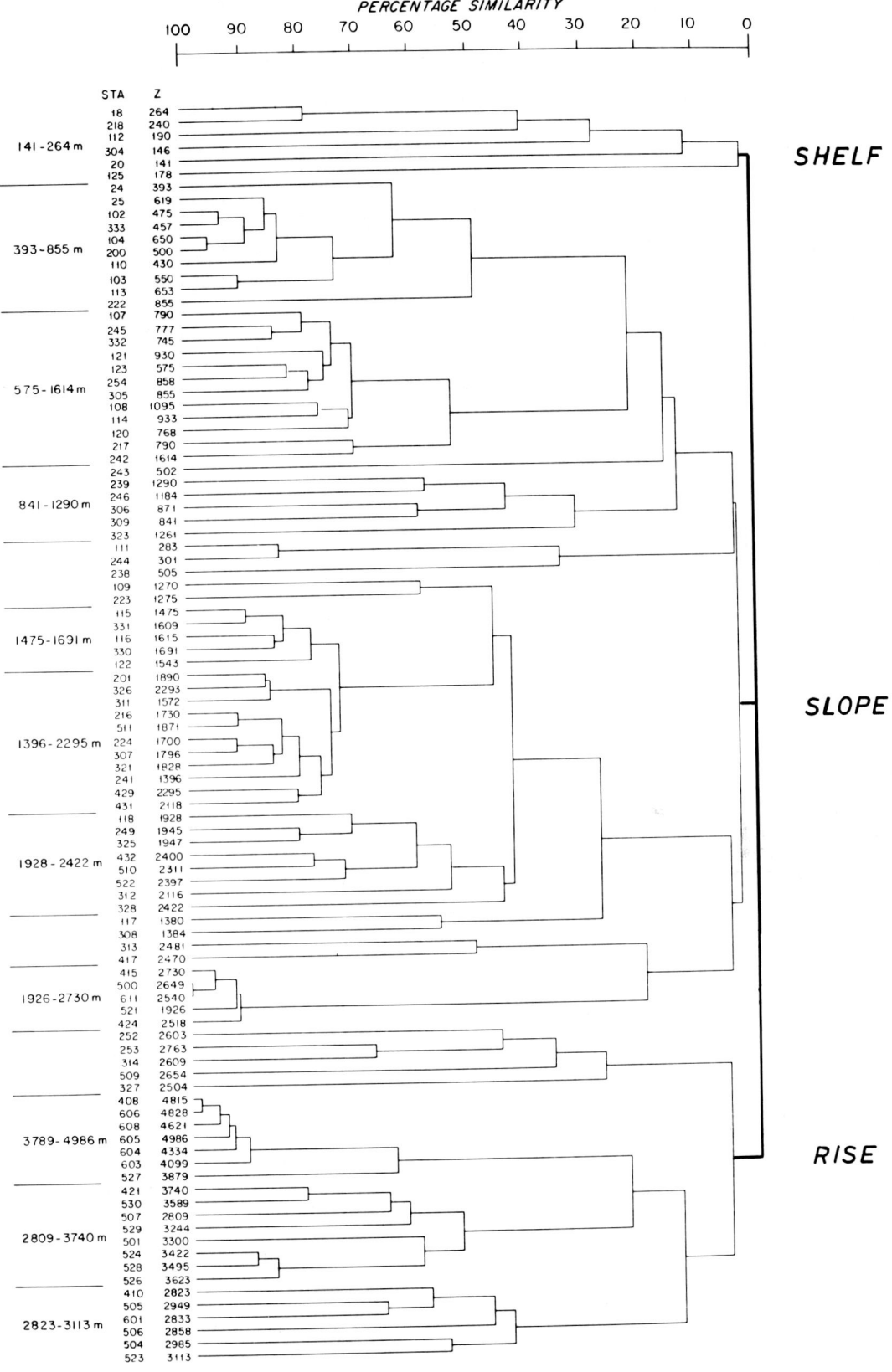

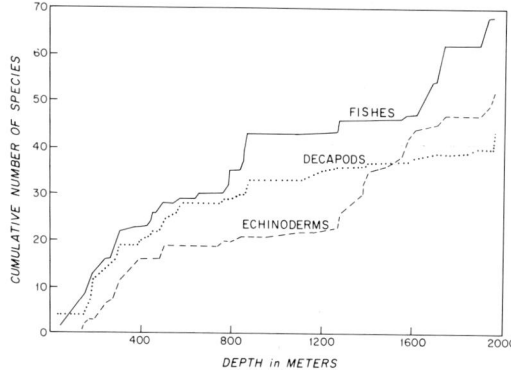

Fig. 3.—Additions of species of fishes, decapods (Crustacea) and echinoderms in trawl samples across continental slope depths off the east coast, U.S.A.

Sediment grain size which could affect zonation can range from well-sorted sand under more active areas such as in the Gulf Stream north of the Blake Plateau (Rowe and Menzies, 1969) to silt off New England in the Mid-Atlantic Bight. A trend with depth usually finds mean grain sizes finer in deeper water (Doyle and Pilkey, this volume). The slope is cut often by steep submarine canyons with somewhat different sediments and faunas compared to outside of the canyons. Those in the western Mid-Atlantic Bight are "inactive," filling with mud, while those south of Georges Bank are "active," with bottoms of rippled sand (Emery and Uchupi, 1974). Despite such known geological differences, biological sampling in the deep ocean has heretofore not been done in adequate density to determine in detail the influence of sediment type on faunal composition. Nonetheless, some generalities can be drawn, as Gray (1974) has done in pointing out that animal diversity is greatest where the substrate is the most heterogeneous. We have gathered some preliminary data on the subject. From samples obtained by DSRV *Alvin* in and adjacent to the extension of the Hudson Submarine Canyon onto the continental rise it appears that grain size, which varied from a mean percent sand of 50% in the canyon axis to 10% outside the canyon, had little effect on infaunal species composition.

There are often important, intimate relationships between the large benthic organisms and the sediments and topography, leading to the assumption that most small-scale features at the sediment water interface are all the products of particular animals. Most erosion, we believe, is catalyzed by animal activity (Rowe et al, 1974), and without animals, erosion on the slope would be rare (Rowe and Boyer, in prep.).

One suggestion advanced to explain zonation is that because bottom currents follow depth contours, the circulation will disperse larvae and eggs of the animals occurring there along isobaths (Rowe and Menzies, 1969). Noting that the abundance of detrital matter corresponded to both the topographic features and currents, the same authors (1968, 1969) suggested the distribution of organic detritus could be reinforcing the influence of currents on where animals live. Differences in feeding strategies in gastropods over much greater depths than just the slope led to the hypothesis that variations in food sources could lead to zonation on a larger scale (Rex, 1977). General changes in feeding strategies have also been noted in polychaete worms (Jumars and Fauchald, 1977), with the presumption of changes in habit based on the morphological characteristics of the species present. Such relationships appear to exist on the continental slope, where numerous species which forage largely on animals of the pelagic deep-scattering layers (DSL) that there impinge on the bottom (Marshall and Merrett, 1977; Lagardère, 1977).

Biological interactions are clearly important in determining the faunal composition of benthic communities (Woodin, 1974, 1976), and it is thus very likely that they play a role too in zonation. Competition is one such interaction. Where animals must compete for a diminishing resource, as for food with increasing depth, some will eventually be eliminated and zonation will result. Predator-prey interactions are another sort, and it can be imagined that, under certain circumstances, a particularly susceptible prey species could be restricted in its range by a very efficient predator. Again, zonation could result.

Finally, a recent provocative discovery is that the kinetics of electrophoretically similar enzymes from animals that live at different depths are affected by pressure (Siebenaller and Somero, 1978). The enzymes show greatest activity at pressures appropriate to the depths at which the animals live. Clearly, these represent adaptations to particular conditions, but since the enzymes do not function everywhere well pressure itself can be imagined to constrain an animal to a certain

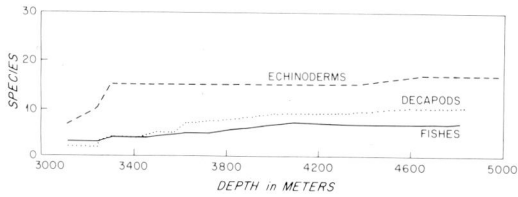

Fig. 4.—Additions of species of decapods (Crustacea), fishes and echinoderms across continental rise depths, east coast, U.S.A.

depth range, producing zonation.

The situation described above, based for the most part on samples from off the northeastern United States, is in its basic elements the situation elsewhere. The European Atlantic slope, the site of most early deep-sea sampling, also has a fauna that is zoned with depth (LeDanois, 1948, summary). As on the upper slope off North Carolina (Menzies et al, 1966) deep-sea corals form banks measuring 10–20 m above bottom and extending for ten's of kilometers in patches along the 400–500 m isobath. About halfway down the slope the fauna also becomes dominated by echinoderms, but LeDanois (1948) recognized no definite patterns below 1000 m. Off Oregon extensive trawling has shown that bottom fishes (Day and Pearcy, 1968) and invertebrates are zoned with depth (Carney and Carey, 1977; Barnard, 1971). The same taxonomic groups are also zoned on slopes in the Arctic (Carey et al, 1974) and Antarctic (McCleave et al, 1977). As yet it is impossible to ascertain how similar the various zonation patterns are, but it is evident in all these studies that animals always line continental margins in ribbons. Even in the horizontal sense, however, marked faunal change may take place over very small distances (Cutler, 1975), a phenomenon little investigated and not well understood.

Thus continental slopes are not the same everywhere. There are, for example, areas off Peru, southwest Africa, and India where a perennial, ubiquitous condition on the upper slope and outer shelf is a well-developed oxygen minimum (<0.5 ml/l). This feature results from biological oxidation of the large amounts of organic matter produced in the surface water due to coastal upwelling. This low oxygen has profound effects on the bottom fauna. At its central depths of about 300 m off Chile, no "macrofaunal" organisms are encountered (Gallardo, 1963). Above and below this core, and in areas where the minimum is not so intense, the communities of benthic animals have very low diversity and are dominated by organisms termed "opportunists," surviving the harsh chemical conditions that eliminate most species. These appear to be primarily foraminiferans (Bandy and Rudolpho, 1964) and nematode worms (Frankenberg and Menzies, 1968). Zonation still occurs as the most prominent feature of the fauna, but the low oxygen causes the zones to be highly modified compared to what we have defined above as typical continental slope patterns.

A recent discovery has been a community of microbes, composed primarily of filamentous sulfur bacteria (*Thioploca*), along the shallow boundary of the anoxic water, extending at depths of about 60–250 m, from 25° S to about 8° S, or over 1500 n.m. (Gallardo, 1977). Some of the organisms oxidize reduced sulfur compounds as a source of energy for the synthesis of new organic matter from CO_2. Other anaerobic microbes used oxidized sulfur compounds as terminal electron acceptors in respiration and produce the reduced sulfur compounds found below the sediment surface. Under certain poorly defined changes in prevailing winds and currents, the anaerobic conditions intensify and denitrification and sulfate reduction also occur up into the water column. The result, called an "aguaje" by Peruvians, is large fish kills and a "red tide" of *Gymnodinium splendens* in surface waters (Dugdale et al, 1978).

It is this same *Thioploca* zone in which extremely high nitrates (up to 2.5 mM/l of interstitial water) in sediment pore waters have been found (Figure 5) (Rowe and Staresinic, in preparation), suggesting the occurrence of chemoautotrophy by microbial nitrification as well as by reduced sulfur compound oxidation. There is an important geochemical implication here. What it means is that a large fraction of the organic matter in the sediments there may be synthesized from CO_2 directly by benthic organisms in addition to phytoplankton from surface waters.

Similar microbial assemblages might be expected on the continental slopes in regions of intense upwelling, including southwest Africa and western India. The occurrence of the sulfur organisms

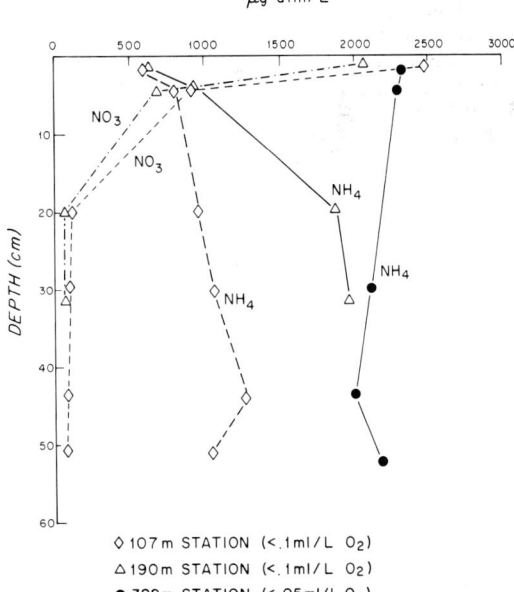

Fig. 5.—Pore water concentrations of NH_4^+ and NO_3^- in surface sediments off Peru. Oxygen concentrations at bottom are bottom water oxygen concentrations at each location.

in permanently anoxic basins, such as the Black Sea and the Cariaco Trench, has yet to be demonstrated.

The *Thioploca* matte assemblage is absent, based on unpublished data from recent cruises, at depths in the center of the O_2 minimum off Peru. It has been sampled at 60–250 m, then it disappeared only to reappear, occasionally, in samples at 800 m. This implies that the organisms are obligate microaerophils and require very low concentrations of oxygen, but cannot survive either without it or with it at high concentrations.

Phosphorite is believed to be depositing presently in sediment pore water on the continental shelves in upwelling zones. Off Peru carbonate sand composed of recent benthic forams was shown to be in the process of changing from carbonate to apatite (Manheim et al, 1974). The required large populations of benthic forams result from the selective zonation referred to above. The importance of similar foram zones to the phosphorite formation in upwellings other than off Peru also remains to be explored.

The Abundance of Life on Continental Slopes

Continental shelf benthos live near the photosynthetic organisms of lighted surface waters. Their great abundance and biomass reflect this proximity, but deep-sea benthos are much lower in abundance and biomass because they live far from the surface and the primary production (Rowe, 1971; Rowe et al, 1974). Continental slope organisms are found in intermediate abundances. The rate of the decrease with increasing depth is erratic across the slope (Fig. 6), presumably because down-slope movements of organic particulate detritus concentrate near the slope base where the continental slope meets the continental rise. Such a pattern of abundance and biomass has been clearly documented off Oregon (Griggs et al, 1969) and a concentration of dead grasses, algae and terrestrial matter at the slope base has been noted by Pérès (1961) in the Bay of Biscaye and Menzies and Rowe (1968) off the eastern United States. This suggests that the slope functions in the transport of food, mostly by gravity, to deeper communities.

Low values of dissolved oxygen render the habitat, as mentioned above, inhospitable to macroinvertebrates. If invertebrates exist under such conditions they are usually very small organisms with little biomass. On the boundaries of these anoxic areas the fauna can reach high biomass and numbers. Off Peru, for example, biomass was highest at 900 m depth, below the oxygen minimum rather than at lesser depths (Rowe, 1971b).

Predicting the abundance of organisms and their biomass across the broadest depth gradient has been attempted (Rowe, 1971b; Rowe et al, 1974) using least squares regressions. Although some confidence can be put in predictions at the greatest depths and nearshore on continental shelves, the variability in processes on the slope make the data from the slope quite variable and therefore less amenable to prediction. The macroinvertebrate infauna ranges from 1000–10,000 animals/m^2, based on a 0.42 or 0.5 mm sieve screen separation from the sediments, and biomass (wet weight) ranges between 0.1 and 10.0 grams per square meter. These values would be an order of magnitude greater on an adjacent continental shelf and somewhat less on the continental rise.

Combining biomass per individual information from trawls with population density estimates from photographs, Haedrich and Rowe (1977) made estimates of the total biomass of slope and rise assemblages, excluding meiofauna (smaller than a 0.42 mm sieve but larger than a 40 μ sieve) and microorganisms (protozoa, fungi and bacteria). They found that among most fishes, the bigger, older individuals live in deeper water (Fig. 7), although total biomass of the community is less in deeper water. No such relationship can be seen in decapods, echinoderms or the infauna as a whole (Polloni et al, 1979). The infauna decrease in biomass with depth much faster than the megafauna, so that the two size groups (macrofauna and megafauna) have about the same biomass at lower slope and upper rise depths. Thus the megafauna cannot depend totally on the macrofauna as a source of food, but must supplement its energy supplies with prey and organic detrital materials from the water column. Food habit studies show that the dominant deep-water benthic fishes do live by scavenging (Haedrich and Henderson, 1974; Pearcy and Ambler, 1974;

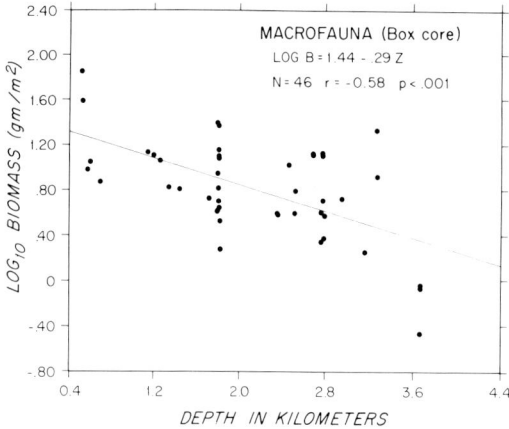

FIG. 6.—Biomass of the benthic infauna (greater than 0.42 mm sieve) off the east coast of the U.S.

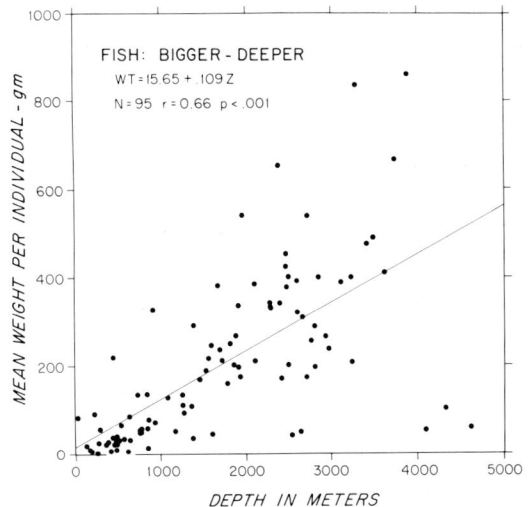

FIG. 7.—Mean weight (wet) per fish in bottom trawl samples, east coast, U.S.A., illustrating bigger-deeper relationships, from Polloni et al (1979).

Isaacs and Schwartzlose, 1975), and detrital plant material derived from shallow water appears important to some deep-sea invertebrates (Schoener and Rowe, 1971; Wolff, 1976).

Rates of Metabolism and Growth

As an index of metabolic rates, the oxygen demand of total assemblages (microbes and infaunal macrobenthos) has been measured using benthic chambers equipped with polarographic oxygen electrodes or by incubating cores at the surface and measuring oxygen depletion in their overlying water. Smith and Teal (1973), using DSRV ALVIN to place chambers at 1850 m depth, found an average O_2 demand of 0.5 ml/m^2-hr, whereas Smith (1974) found demands in the Pacific to be 2.4 ml/m^2-hr. Both measurements are far below similar measurements on the continental shelf or in bays and estuaries. They range from around 10 ml/m^2-hr up to about 100 ml/m^2-hr depending on sediment type, the input of organic matter and temperature, mostly. Both shallow water and slope rates are above those found on the continental rise and abyssal plain (Smith, 1978), which are less than 0.1 ml/m^2-hr.

Although much information exists on the growth and production of shallow water animals, primarily because of their economic importance, little is known about how long deep-sea organisms live or the frequency of their reproduction. A single age estimate on a species of deep-sea bivalve, using a Radium/Thorium method, suggested that the organisms reached maturity after about 50 years (Turekian et al, 1975). This result, based on a somewhat questionable statistical procedure, supports the traditional predisposition that deep-ocean animals should be very old. Studies using traditional seasonal markings, growth rings in otoliths and shells, imply ages in fishes (Rannou, 1976) and brachiopods (Zezina, 1975) comparable to those living in shallow waters. Again, the traditional predisposition is that there are no seasons in the deep sea. Nonetheless, there is evidence of seasonal activity in diverse groups (George and Menzies, 1968; Schoener, 1968; Rokop, 1974; Haedrich and Polloni, 1976), and it is thus not out of the question that annual markers should form. The question is far from settled.

Microbial activity too seems to be very low in the deep-sea, even though plate counts show microbes are very abundant there (ZoBell, 1948). The first clues of low activity were the preservation of a ham sandwich, apple and soup in the submersible ALVIN, unintentionally sunk unoccupied, over a period of 10 months at a depth of 1400 m (Jannasch et al, 1971). Continued work indicates that deep-sea microbes cultured at depth do not use substrate or respire as rapidly as they would near atmospheric pressure (Jannasch and Wirsen, 1973). The implication of this work is that the barophilic (pressure adapted) bacteria suggested by ZoBell (1948) do not exist.

While the microbiology implies rates of biological processes in the deep-sea are low, some organisms appear to grow rapidly given a good food source. Pieces of wood in contact with the bottom are rapidly infested and bored by woodboring bivalve molluscs (Turner, 1973). Experimental blocks were placed on and attached to the bottom in plastic mesh bags and other invertebrates, caught in the bags, grew at rates much faster than would be expected from the small size and old age of the bivalve referred to above. It is presumed these rapidly growing organisms are feeding on the fecal matter and other material excavated from the bivalves' holes in the wood.

Petroleum Resources

Of the four principal requirements for oil fields (Ball, this volume), only the "quantity of . . . hydrocarbons" is related to biological processes on the continental slope. The preservation of hydrocarbons is not favored in most deep-sea sediments, we suspect, because most of the deep-sea biota, being far removed from the photosynthesis in surface waters, are in fact food-limited, as illustrated by our concept based on the U.S. east coast continental margin (Fig. 8). That is, a major fraction of the organic matter sinking into the depths will be eaten by organisms and converted to CO_2. Intuitively, from this reasoning, one would say that most continental slopes that

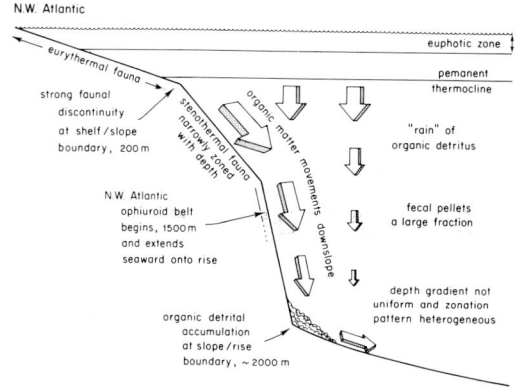

FIG. 8.—Summary of biological conditions and organic cycling, east coast, U.S.A.

have had environments similar to those we describe as typical would not be favorable sites for the formation of petroleum deposits. On the other hand, the upwelling areas, as illustrated in Figure 9, with oxygen minima along the west coasts of continents, we speculate, would be far more amenable to hydrocarbon deposit formation, from a biological viewpoint, because these unique ecosystems are overloaded with organic matter. For the discovery of new resources the biologist, admittedly naive in such ventures, would recommend consideration of areas believed to have been regions of upwelling at acceptable time scales in the past. Furthermore, since zonation is such a ubiquitous feature in the slope faunas, the implication is clear that a useful find (oil, phosphorite) at any depth could successfully be sought 100's of kms away at the same depth. Success might well be less likely only a few 100 m deeper or shallower in the region of the find.

The Future for Biology on the Continental Slope

Just what causes the zonation of populations so characteristic of the continental slope has not been satisfactorily answered, although a number of attractive hypotheses exist. Studies that seek to explain this phenomenon will continue to be of interest. Some current investigations are examining the relationships between narrow zonation and variations in the environment, with special emphasis on the input and cycling of organic matter. Very difficult problems, because of the limited and always expensive opportunities for direct observation and experimentation, involve biological interactions and the spatial and temporal scales on which these occur. Extrapolation from shallow-water studies (e.g., Woodin, 1974) or concentrated work in one small area (e.g., Thistle, 1978) may ultimately be the only recourse. But it is clear now that intra- and interspecific interactions which either restrict or promote animal abundance on very small spatial scales or in narrow depth ranges must be a central concern.

The utilization of fisheries resources, having in most areas affected almost all exploitable shelf species, is spreading onto the continental slope. Species involved in the North Atlantic include red crab (*Geryon quinquedens*), lobster (*Homarus americanus*) and the rattail *Coryphaenoides rupestris*. Knowledge of the reproductive capacity, life histories, secondary production rates, and standing stocks for such species must be gathered in order to set catch quotas and prevent their elimination as useful resources.

Understanding how the biota living in the sediments influence the chemistry of the water column and the early diagenesis of organic matter in sediments is important to ensure a proper, balanced use of continental margin resources. This includes not just fisheries resources, but potential impacts of the exploitation of mineral and petroleum resources on the functioning of the slope ecosystem. While the biological production utilized by fisheries is now and will remain for the most part on the continental shelf, the primary production on the shelves is dependent on limited sources of inorganic nitrogen and phosphorus. A large fraction of these nutrients appears to originate in slope-depth waters offshore (Riley, 1967; Ryther, 1963). Thus, it must be kept in mind that alteration of biological processes on the slope might have large impacts at lesser depths. As we cannot predict what these impacts might be, efforts to investigate the unique qualities of the ecology on continental slopes must continue.

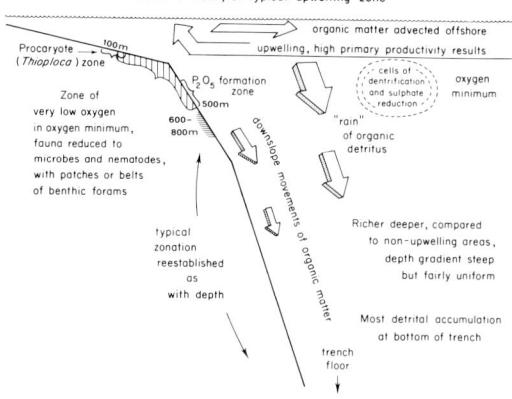

FIG. 9.—Summary of biological and organic geochemical conditions on the west coast of Peru.

REFERENCES

AGASSIZ, A., 1888, Three cruises of the *Blake:* Cambridge, Mass., Harvard Univ., Bull. Museum Comparative Zool., v. 14 and 15.

BALL, MAHLON, 1979, This volume, Petroleum Potential of Passive Margin Slopes.

BANDY, O., AND RUDOLPHO, K. S., 1964, Distribution of foraminifera and sediments, Peru-Chile Trench area: Deep-Sea Res., v. 11, p. 817–837.

BARNARD, J. L., 1971, Gammaridean Amphipoda from a deep-sea transect off Oregon: Smithsonian Contributions to Zoology, v. 61, p. 1–86.

BIGELOW, H. B., AND SCHROEDER, W. C., 1939, Notes on the fauna above mud bottoms in deep water in the Gulf of Maine: Biol. Bull. v. 76, p. 305–324.

CAREY, A. G., JR., 1965, Preliminary studies on animal-sediment interrelationships off the central Oregon coast: Trans. Joint Conf. of Ocean Scientists and Engineers, v. 1, p. 100–110.

CAREY, A. G., JR., RUFF, R. E., CASTILLO, J. G., AND DICKINSON, J. J., 1974, Benthic ecology of the western Beaufort Sea continental margin—preliminary results, *in* REED, J. C., AND SATER, J. E., The Coast and Shelf of the Beaufort Sea: Arlington, Va., Arctic Inst. of North America, p. 665–680.

CARNEY, R., AND CAREY, A. G., JR., 1976, Distribution pattern of holothurians on the northeastern Pacific (Oregon, U.S.A.) continental shelf, slope and abyssal plain: Thallassia, Yugoslavia, v. 12(1), p. 67–74.

COHEN, D. M., 1977, Ten dives of the DSRV *Alvin* in and near the DWD 106 dumpsite, 25 July-3 August 1975, Introduction, station data, general observations and conclusions, *in* Baseline Report of Environmental Conditions in DWD 106: Rockville, Md., NOAA Dumpsite Evaluation Rept. 77–1, v. III, p. 595–609.

CUTLER, E. B., 1975, Zoogeographical barrier on the continental slope off Cape Lookout, North Carolina: Deep Sea Res. v. 22, p. 893–901.

DAY, D. S., AND PEARCY, W. G., 1968, Species associations of benthic fishes on the continental shelf and slope off Oregon: Jour. Fish. Res. Bd. Canada, v. 25, p. 2665–2675.

DOYLE, L., AND PILKEY, O. H., 1979, This volume, Sedimentation on The Eastern United States Continental Slope.

DUGDALE, R. C., GOERING, J. J., BARBER, R. T., SMITH, R. L., AND PACKARD, T. T., 1977, Denitrification and hydrogen sulfide in the Peru upwelling region during 1976: Deep-Sea Res., v. 24, p. 601–608.

EKMAN, S., 1953, Zoogeography of the Sea: London, Sidgwick and Jackson, Ltd., 417 p.

EMERY, K. O., AND UCHUPI, E., 1972, Western North Atlantic Ocean: Tulsa, Am. Assoc. Petroleum Geologists, 532 p.

FRANKENBERG, P., AND MENZIES, R. J., 1968, Some quantitative analyses of deep-sea benthos off Peru: Deep-Sea Res., v. 15: 623–626.

GALLARDO, V. A., 1963, Notas sobre la densidad de la fauna bentonica en el sublitoral del norte de Chile: Gayana v. 8, p. 3–15.

GALLARDO, V. A., 1977, Large benthic microbial communities in sulphide biota under Peru-Chile subsurface counter current: Nature, v. 268, p. 331.

GARDINER, F. P., AND HAEDRICH, R. L., 1978, Zonation in the deep benthic megafauna: Oecologia, v. 31, p. 311–317.

GEORGE, R. Y., AND MENZIES, R. J., 1968, Further evidence for seasonal breeding cycles in deep sea: Nature, v. 220, p. 80–81.

GRASSLE, J. F., SANDERS, H., HESSLER, R., ROWE, G., AND MCLELLAN, T., 1975, Pattern and zonation: a study of the bathyal megafauna using the research submersible *Alvin:* Deep-Sea Res., v. 22, p. 457–481.

GRAY, J. S., 1974, Animal sediment relationships: Oceanography and Marine Biology Annual Review, v. 12, p. 223–261.

GRIGGS, G., CAREY, A., AND KULM, L., 1969, Deep-sea sedimentation and sediment-fauna interaction in Cascadia Channel and on Cascadia Abyssal Plain: Deep-Sea Res., v. 16, p. 157–170.

HAEDRICH, R. L., AND HENDERSON, N., 1974, Pelagic food of *Coryphaenoides armatus*, a deep benthic rattail: Deep-Sea Res., v. 21, p. 739–744.

HAEDRICH, R. L., AND POLLONI, P. T., 1976, A contribution to the life history of a small rattail fish, *Coryphaenoides carapinus:* 75: 203–211.

———, ROWE, G. T., AND POLLONI, P. T., 1975, Zonation and faunal composition of epibenthic populations on the continental slope south of New England: Jour. Mar. Res., v. 33, p. 191–212.

———, AND ———, 1977, Megafaunal biomass in the deep sea: Nature, v. 269, p. 141–142.

HARTMAN, O., AND BARNARD, J., 1958, The benthic fauna of the deep basins off Southern California: Allan Hancock Pacific Exped., v. 22, p. 1–67.

ISAACS, J. D., AND SCHWARTZLOSE, R. A., 1975, Active animals of the deep-sea floor: Sci. Amer. v. 233, p. 85–91.

JANNASCH, H. J., EIMHJELLEN, K., WIRSEN, C. O., AND FARMANFARMAIAN, A., 1971, Microbial degradation of organic matter in the deep-sea: Science, v. 171, p. 672–675.

JANNASCH, H. W., AND WIRSEN, C. O., 1973, Deep-sea microorganisms: In situ response to nutrient enrichment: Science, v. 180, p. 641–643.

JUMARS, P. A., 1975, Environmental grain and polychaete species' diversity in a bathyal benthic community: Mar. Biol., v. 30, p. 253–266.

———, 1976, Deep-sea species diversity: does it have a characteristic scale? Jour. Mar. Res., v. 34, p. 217–246.

———, AND FAUCHALD, K., 1977, Between-community contrasts in successful polychaete feeding strategies, *in* Coull, B. C., ed., Ecology of Marine Benthos: Columbia, S.C., Univ. South Carolina Press, p. 1–20, 467 p.

LAGARDÈRE, J. P., 1977, Recherches sur le régime alimentaire et le comportement prédateur des decapodes benthique de la pente continentale de l'Atlantique Nord oriental (Golfe de Gascogne et Moroc), in Keegan, B. F., Cerdigh, P. O., and Boaden, P. J. S., eds., Biology of Benthic Organisms: New York, Pergamon.

LEDANOIS, E., 1948, Les profoundeurs de la mer, Trente ans de recherches sur la fauna submarine au large des cotes de France: Paris, Payot, 303 p.

MANHEIM, F., ROWE, G., AND JIPA, D., 1975, Marine phosphorite formation off Peru: Jour. Sed. Petrology, v. 45, p. 243–251.

MARKLE, D. F., AND MUSICK, J. A., 1974, Benthic fish associations on the continental slope of the Middle Atlantic Bight: Marine Biology v. 26, p. 225–233.

MARSHALL, N. B., AND MERRETT, N. R., 1977, The existence of a benthopelagic fauna in the deep-sea, in Angel, M., ed., A Voyage of Discovery: George Deacon 70th Anniversary Volume: Oxford, Pergamon Press Ltd.

MCCLEAVE, J. D., DEARBORN, J. H., AND DEWITT, H. H., 1977, Ecology of benthic fishes and echinoderms along the Scotia Arc and the Antarctic Peninsula: Antarctic Jour. U.S., v. 12, p. 19–20.

MENZIES, R. J., PILKEY, O., BLACKWELDER, B., DEXTER, D., HULING, P., AND MCCLOSKEY, L., 1966, A submerged reef off North Carolina: International Review Ges. Hydrobiol, v. 51, p. 393–431.

MENZIES, R. J., AND ROWE, G. T., 1968, The distribution and significance of detrital turtle grass, *Thalassia testudinum*, on the deep-sea floor off North Carolina: International Review Ges. Hydrobiol. v. 54, p. 219–222.

MENZIES, R. J., GEORGE, R. Y., AND ROWE, G. T., 1973, Abyssal environment and ecology of the world's oceans: New York, Wiley, 488 p.

MILLS, E., 1972, T. R. R. Stebbing, the *Challenger* and knowledge of deep-sea amphipoda: Proc. Royal Soc. Edinburgh, Section B (Biology) v. 72, p. 67–87.

PEARCY, W. G., AND AMBLER, J. W., 1974, Food habits of deep-sea macrourid fishes off the Oregon Coast: Deep-Sea Res. v. 21, p. 745–759.

PÉRÈS, J., 1961, Oceanographie biologique et biologie marine I, La vie benthique: Presses Universitaires de France, Paris, 542 p.

POLLONI, P. T., HAEDRICH, R. L., ROWE, G. T., AND CLIFFORD, C. H., 1979, The size-depth relationship in deep ocean animals: International Review ges. Hydrobiol., v. 64, p. 39–46.

RANNOU, M., 1976, Age et croissance d'un poisson bathyal *Nezumia sclerorhynchus* (Macrouride, Gadiformes) en Mer d'Alboran: Cah. Biol. Mar. v. 17, p. 413–421.

REX, M. A., 1977, Zonation in deep-sea gastropods: the importance of biological interactions to rates of zonation, in Keegen, B. F., Ceidigh, P. O., and Boaden, P. J. S., eds., Biology of Benthic Organisms: New York, Pergamon.

RILEY, G. A., 1967, Mathematical model of nutrient conditions in coastal waters: Bull. Bingham Oceanogr. Coll. v. 19, p. 72–80.

ROKOP, F. J., 1974, Reproductive patterns in the deep-sea benthos: Science, v. 186, p. 743–745.

ROWE, G. T., 1971, Observations on bottom currents and epibenthic populations in Hatteras Submarine Canyon: Deep-Sea Res., v. 18, p. 569–581.

———, 1971a, Benthic biomass and surface productivity, in Costlow, J., ed., Fertility of the Sea 2, p. 441–454.

———, 1971b, Benthic biomass in the Pisco, Peru upwelling: Invest. Pesq. v. 35, p. 127–135.

———, 1972, The exploration of submarine canyons and their benthic faunal assemblages: Proc. Royal Soc. Edinburgh (B) 73: p. 159–169.

———, AND BOYER, L. F., The effects of the fauna on the physical properties of deep-sea sediments: (in prep.).

———, AND MENZIES, R. J., 1969, Zonation of large benthic invertebrates in the deep-sea off the Carolinas: Deep-Sea Res. v. 16, p. 531–537.

———, AND STARESINIC, N., A model of nitrification in deep-sea sediments: (in prep.).

———, KELLER, G., EDGERTON, H., STARESINIC, N., AND MACILVAINE, J., 1974, Time-lapse photography of the biological reworking of sediments in Hudson Submarine Canyon: Jour. Sed. Petrology v. 44, p. 549–552.

———, POLLONI, P. T., AND HAEDRICH, R. L., 1975, Quantitative biological assessment of the benthic fauna in deep basins of the Gulf of Maine: Jour. Fish. Res. Bd. Canada v. 32, p. 1805–1812.

———, POLLONI, P. T., AND HORNOR, S. G., 1974, Benthic biomass estimates from the northwestern Atlantic Ocean and the northwestern Gulf of Mexico: Deep-Sea Res. v. 21, p. 641–650.

RYTHER, J. H., 1963, Geographic variations in productivity, in Hill, M. N., ed, The Sea: New York, Interscience, p. 347–380.

SANDERS, H. L., AND HESSLER, R. R., 1969, Ecology of the deep-sea benthos. Science, v. 163, p. 1419–1424.

———, HESSLER, R. R., AND HAMPSON, G., 1965, An introduction to the study of deep-sea benthic faunal assemblages along the Gay Head-Bermuda Transect: Deep-Sea Res., v. 12, p. 845–867.

SCHOENER, A., 1968, Evidence for reproductive periodicity in the deep sea: Ecology, v. 49, p. 81–87.

———, AND ROWE, G., 1970, Pelagic *Sargassum* and its presence among the deep-sea benthos: Deep-Sea Res., v. 17, p. 923–925.

SIEBENALLER, J., AND SOMERO, G. N., 1978, Pressure-adaptive differences in lactate dehydrogenases of congeneric fishes at different depths: Science, v. 201, p. 255–257.

SMITH, K. L., JR., 1974, Oxygen demand of San Diego Trough sediments: an in situ study: Limnol. Oceanogr., v. 19, p. 939–944.

———, 1978. Benthic community respiration in the N.W. Atlantic: in situ measurements from 40 to 5200 meters: Marine Biology, v. 47, p. 337–347.

———, and TEAL, J. M., 1973, Deep-sea benthic community respiration: an in situ study at 1850 meters: Science, v. 179, p. 282–283.
THISTLE, D., 1978, Harpacticoid dispersion patterns: implications for deep-sea diversity maintenance: Jour. Mar. Res., v. 36, p. 377–397.
TUREKIAN, K., COCHRAN, J., KHARKAR, D., CERRATO, R., VAISNYS, J., SANDERS, H., GRASSLE, J., AND ALLEN, J., 1975, The slow growth rate of a deep-sea clam determined by ^{228}Ra chronology: Proc. Nat. Acad. Sci., U.S.A., v. 72, p. 2829–2832.
TURNER, R., 1973, Wood-boring bivalves, opportunistic species in the deep-sea: Science, v. 180, p. 1377–1379.
WIGLEY, R. L., AND EMERY, K. O., 1967, Benthic animals, particularly *Hyalinoecia* (Annelida) and *Ophiomusium* (Echinodermata), in sea-bottom photographs from the continental slope, *in* Hersey, J. B., ed., Deep Sea Photography: Baltimore, The Johns Hopkins Press, p. 235–250.
WILLIAMS, A. B., AND WIGLEY, R. L., 1977, Distribution of decapod crustacea off Northeastern United States based on specimens at the Northeast Fisheries Center, Woods Hole, Mass: NOAA Tech. Rept. NMFS Circular 407, 44 p.
WOLFF, T., 1976, Utilization of seagrass in the deep sea: Aquatic Botany, v. 2, p. 161–174.
———, 1977, Diversity and faunal composition of the deep-sea benthos: Nature, v. 267, p. 780–785.
WOODIN, S. A., 1974, Polychaete abundance patterns in a marine soft-sediment: the importance of biological interactions: Ecol. Monographs, v. 44, p. 171–187.
———, 1976, Adult-larval interactions in dense infaunal assemblages: patterns of abundance: Jour. Mar. Res., v. 34, p. 25–41.
ZENKEVITCH, L. A., 1963, Biology of the Seas of the U.S.S.R.: New York, Wiley, 955 p.
ZEZINA, O. N., 1975, On some deep-sea brachiopods from the Gay Head-Bermuda transect: Deep-Sea Res., v. 22, p. 903–912.
ZOBELL, C., 1946, Marine Microbiology: Waltham, Mass., Chronica Botanica, 240 p.

A REVIEW OF MASS MOVEMENT PROCESSES, SEDIMENT AND ACOUSTIC CHARACTERISTICS, AND CONTRASTS IN SLOPE AND BASE-OF-SLOPE SYSTEMS VERSUS CANYON-FAN-BASIN FLOOR SYSTEMS.

T. R. NARDIN, F. J. HEIN, D. S. GORSLINE, AND B. D. EDWARDS
Department of Geological Sciences
University of Southern California
Los Angeles, California 90007

ABSTRACT

Slopes comprise 10–15% of the Earth's surface and are a major pathway for sediment movement to the deep ocean floor. Mass movement processes are the dominant modes of mobilization, transport and deposition. A review of the literature shows considerable confusion in application of a variety of terms used to describe mass movement phenomena. A summary of the various approaches has been synthesized into a three-part classification in which the major transport and grain support processes can be grouped according to mechanical behavior. Rockfalls and slides (including glides and slumps) represent elastic transport mechanisms. Mass flows move plastically and include debris flows, mud-flows and inertial grain flows. Viscous fluid behavior is represented by viscous grain flows, liquified flows, fluidized flows, and turbidity currents. There are specific assemblages of sedimentary features which have been assigned to each of these processes. A single deposit may contain imprints of more than one process, however. On a less discriminatory scale, particular seismic-reflection patterns can also be used to infer major gravity processes or process groups.

It is evident that much information is needed on the geotechnical properties of slope sediments and the delineation of the variables which define the critical conditions at which incipient failure can occur. Effects of particular triggering forces need examination.

Because of the limitations inherent in field geology and in contemporary marine geological approaches it is clear that cooperative studies by both groups in conjunction with experimental data will be required to yield a complete description of the slope process-deposit-facies problem and to develop a slope-apron-ocean floor model comparable in sophistication to the current canyon-fan-ocean floor model.

INTRODUCTION

Continental slopes, which account for 10–15% of the Earth's surface, may act either as sites of permanent deposition comprising part of the accretionary wedges flanking stable cratons (Moore and Curray, 1963a, b), or as temporary storage areas for sediment in transit to deeper water. Together with submarine canyons, they are the main conduits for transport of terrigenous material into the deep sea.

This symposium volume offers a cross section of current research on slope processes. Because individual research papers by their nature tend to focus on highly specific questions, it seemed appropriate to the authors and the editors that an attempt should be made to provide a framework into which the individual specialized papers could be set. Rather than cover the full spectrum of slope processes discussed in this volume we chose to limit our discussion to mass movement since this topic appears to have received the greatest attention. As is true of any attempt to synthesize a rapidly expanding research field, this short review will certainly be incomplete.

Mass movement, defined as the movement of sediment driven directly by gravity rather than by interstitial fluid motion (Sharpe, 1938; also see Middleton and Hampton, 1976), is the principal transport process operating on slopes and in canyons. Considerable confusion and disagreement exists regarding the terminology used to describe mass movement processes and their depositional products. This problem has been addressed by a number of investigators (Dott, 1963; Sanders, 1965; Walker, 1970, 1978; Fisher, 1971; Middleton and Hampton, 1976; Moore, 1977; Lowe, this volume) and has been reviewed by Carter (1975). One of the principal difficulties with the present nomenclature is that many terms are not defined uniquely. A few examples serve to illustrate this point.

Rockfall deposits associated with deeper water carbonate accumulations have been called *reef talus, marine talus* and *subaqueous talus* (Wilson, 1975; Friedman and Sanders, 1978; Johns, 1978). The terms *olistolith* and *olistostrome* have been applied to a wide range of gravity deposits, including those which show complete homogenization (e.g., Abbate et al, 1969) in addition to those transported without rotation or internal deformation (Johns, 1978). *Debris flow* has been used to describe bouldery and pebbly mudstones as well as deposits which consist of sand and finer sediment (Hampton, 1972; Rodine and Johnson, 1976). The terms *mudflow, incoherent slump* and *debris flow* are sometimes used interchangeably (Lindsay, 1966; Lowe, 1972; this volume; Friedman and Sanders, 1978). The nomenclature used to describe grain flows has been reviewed by Carter (1975) who recommends that *slurry flow* be used if grain flow occurs in the viscous regime. *Modified grain flows* are defined by Lowe (1976b) as flows in which grain support is aided significantly by an overlying current, excess pore-fluid pressure, or a dense interstitial fluid. Historically, the deposits from turbidity currents have been called *turbidites* and they generally conform to the Bouma model (Bouma, 1962). Although the formulation of Bouma's model provided a breakthrough in our understanding of the origins of many deep-sea coarse clastic deposits, there were other deposits that did not follow the model. This was particularly true for very coarse proximal sediments originally called *fluxoturbidites* (Kuenen, 1958; Dzulynski et al, 1959). Fluxoturbidites were thought to be the deposits from currents that were intermediate in nature between slides and true turbidity currents. However, it has become apparent that the term actually includes a wide range of gravity deposits including those transported by mass flows and thus most workers have abandoned its use. Finally, the term *slump*, perhaps the most commonly if not wisely employed word in the mass movement vocabulary (D. G. Moore, pers. comm.), has been used to describe a rotational slide of an essentially undeformed mass along discrete shear surfaces as well as any gravity deposit regardless of the mechanics of motion or degree of internal deformation.

An additional problem has been that a single term such as *slump* is often used to describe the different stages of mass movement, that is, mobilization, transport and deposition. Yet, most terminologies are based only on the processes of mass transport or grain support deduced largely from sediment type, fabric, and deformational and primary sedimentary structures. It is evident that in many cases our understanding of the mechanics of mass movement has not progressed sufficiently to consistently infer specific transport processes from these features alone. Sediment characteristics may reflect a combination of transport mechanisms or only the last in a series of evolving mechanisms (Middleton and Hampton, 1976; Walker, 1978). Furthermore, they may not reflect transport at all but rather the mode of deposition.

In the past five years a number of mass movement classifications have been devised which attempt to resolve these problems. Because the traditional and most commonly used approach in the study of marine sedimentation is to couple seismic-reflection profiling with bottom coring, our objective is to pull previous syntheses together into working classifications which relate mass movement processes to the sedimentary features and characteristic seismic-reflection patterns which potentially could result from them. There are admittedly limitations to this approach. W. R. Normark (pers. comm.) has observed that marine geologists have been confined to relatively narrow windows in the dimensional spectrum of marine observation. They can directly examine only very narrow vertical samples of a sediment body to depths of a few meters. By acoustic profiling, however, they can delimit the three-dimensional geometry of a depositional system such as a contemporary basin. In contrast, outcrop geologists can examine and sample two-dimensional exposures many hundreds of meters in areal extent. They have great difficulty, though, in measuring the third dimension and cannot observe processes in action. Therefore, it is obvious that both approaches are required. Experimentation, which is necessary to objectively examine process-result relationships, is the final ingredient required to more completely describe marine sedimentary phenomena.

In this paper, classifications of mass transport processes and the sedimentary and seismic characteristics of gravity deposits will be reviewed. Other transport mechanisms, such as low density turbidity currents or lutite plumes (Moore, 1969; Drake, 1972; Piper, 1978), nepheloid flows (Ewing and Thorndike, 1965), pelagic sediment fallout, and contour currents (Heezen et al, 1966; Scott, 1966; Embley and Langseth, 1977; Damuth and Hayes, 1977), which may modify or in some cases dominate slope deposition are not discussed. Finally, the characteristics of slope and base-of-slope deposits will be distinguished from those of submarine fan deposits.

CLASSIFICATION OF MASS TRANSPORT PROCESSES

Both descriptive and genetic classifications of deep-sea mass transported sediments have been proposed. Descriptive classifications are based chiefly on the sedimentary features of coarse to fine clastics. These include sorting and size distribution, presence or absence of stratification or cross-stratification, types of grading, and to a lesser extent, the sedimentary fabric. Several descriptive schemes have been integrated into sedimentary models. Perhaps the best known is the Bouma model for sandy and finer sediments (Bouma, 1962). Another is Walker's model for resedimented conglomerates (Walker, 1975; 1978). Crowell (1957) noted the pebbly mudstone facies and discussed its origin. Genetic classifications are not sediment classifications, but rather classifications of transport or grain support mechanisms from which the characteristics of the sedimentary deposits can be inferred (Dott, 1963; Elliott, 1965; Cook et al, 1972; Carter, 1975; Lowe, 1976a, 1976b, this volume; Middleton and Hampton, 1976; Moore, 1977). Most genetic schemes are not restricted to the transport of coarse clastics.

Perhaps the most straightforward and useful terminology is found in Dott's (1963) tripartite classification, a modification of Varnes' (1958) widely accepted systemization of subareal mass movement. Dott (1963) recognized from subareal analogs that a potential continuum of mechanical behaviour, ranging from elastic through plastic to viscous fluid, exists during submarine slope failure. Within the realm of elastic processes are *rockfalls* and *slides*. Slides can be defined as movements of essentially rigid, internally undeformed masses along discrete shear surfaces. Slides can be subdivided into *slumps* where rotary movement accompanies displacement and *glides* where the mass moves along a planar surface. Because permanent internal deformation usually occurs at the base and leading edge of slide masses, a certain amount of plastic behaviour must be accomodated in the definition. Transport processes dominated by plastic behaviour, where shear stress is distributed throughout the mass, are referred to as *mass flows*. Sediments transported by mass flow can exhibit a wide range of rheologic properties including those of Bingham and quasi-plastic materials (Johnson, 1970; Fisher, 1971; Hampton, 1972, 1975). Mass flows are generally thought to move in laminar fashion although turbulent behavior has been suggested (Lindsay, 1966; Enos, 1977). Terms which are used synonomously with mass flow are *high concentration dispersion* (Fisher, 1971), *sediment gravity flow* (Middleton and Hampton, 1976) and *inertia flow* (Sanders, 1965; Carter, 1975). The last terms, however, also encompass transport processes which are modeled after viscous fluids and thus do not strictly describe plastic phenomena. Once the liquid limit of a sediment mass is exceeded the mass no longer behaves as a solid but moves as a viscous fluid. Examples of *fluidal flow* include *turbidity currents* (Kuenen, 1937, 1951), some *grain flows* (Bagnold, 1954, 1956; Lowe, 1976b), *liquified flow* and *fluidized flow* (Lowe, 1976a). Fluidal flows are not to be confused with fluidal density currents which are due to gravitational instabilities arising from density differences between fluid layers. Because the fluid moves the sediment, if any sediment is involved in the flow (Middleton and Southard, 1978), fluidal density currents are not mass movement phenomena.

Middleton and Hampton (1976) and Middleton (1978) have proposed a classification based on the dominant sediment support mechanism operative during transport. The four end members in this classification are: *debris flows* which are supported by their matrix strength, *grain flows* supported by dispersive pressure due to grain interaction, *liquified sediment flows* partially supported by upward motion of intergranular fluid, and *turbidity currents* supported by fluid turbulence. These flow types, grouped under the general heading of *sediment gravity flows*, represent both plastic and fluid behaviour and therefore cannot be equated collectively to mass flows as defined by Dott (1963). However, in practice it is often difficult to determine whether a gravity deposit was transported predominantly in a plastic or fluid mode because both mechanisms may have been operative together or at different times during transport. Thus, sediment gravity flow is a useful term which can be applied generally to anelastic transport mechanisms.

Lowe (this volume), like Dott (1963), has classified flows by their rheology. Fluid and plastic behavior are represented by *fluidal flows* and *debris flows*, respectively. These flow types are further divided, based on a modification of Middleton and Hampton's (1976) grain support scheme. Fluidal flows include turbidity currents, fluidized flows and liquefied flows. Debris flows are described by the Coulomb-viscous equation and can be divided into *cohesive debris flows* (also called *mudflows*) or *grain flows*, depending on whether the cohesion term or internal friction term dominates. Thus, grain flows are considered to behave plastically and no distinction is made between grain flows in the viscous versus the inertial regimes. However, at low volume concentrations (viscous regime), grain flows are probably better described as fluids (Bagnold, 1956).

From this review of the major genetic classifications, it is evident that although appropriate rheologic models are difficult to define (Johnson, 1970), mass movement transport processes can be systemized based upon a combination of mechanical behaviour and transport and sediment support mechanisms. Table 1 is a summary of the major processes in which the important mechanical boundaries follow Dott (1963) in separating elastic, plastic and viscous fluid motion. Transport and sediment support mechanisms are grouped according to mechanical behavior. As is true for all classifications, this scheme is offered as a means of focusing on problems. An obvious problem is the lack of measurements on the geotechnical properties of marine sediments, particularly in situ measurements. Although Table 1 does not attempt to include variables such as grain size distribution, sedimentation rate, and slope gradient, these will obviously influence

TABLE 1.—MAJOR TYPES OF MASS TRANSPORT PROCESSES, THEIR MECHANICAL BEHAVIOR, AND TRANSPORT AND SEDIMENT SUPPORT MECHANISMS (BASED ON DOTT, 1963; MIDDLETON AND HAMPTON, 1976; LOWE, 1976A, B, THIS VOLUME; MOORE, 1977)

Mass Transport Processes			Mechanical Behaviour	Transport Mechanism and Sediment Support
Rock Fall			Elastic	Freefall and subordinate rolling of individual blocks or clasts along steep slopes
Slide		Glide	Elastic	Shear failure along discrete shear planes with little internal deformation or rotation
Slide		Slump	Elastic	Shear failure accompanied by rotation along discrete shear surfaces with little internal deformation
			Plastic Limit	
Sediment Gravity Flow	Mass Flow	Debris Flow– Mud Flow	Plastic	Shear distributed throughout the sediment mass. Strength is principally from cohesion due to clay content. Additional matrix support may come from buoyancy.
Sediment Gravity Flow	Mass Flow	Grain Flows — Inertial / Viscous	Liquid Limit	Cohesionless sediment supported by dispersive pressure. Flow may be in inertial (high concentration) or viscous (low concentration) regime. Usually requires steep slopes.
Sediment Gravity Flow	Fluidal Flow	Liquified Flow	Viscous Fluid	Cohesionless sediment supported by upward displacement of fluid (dilatance) as loosely packed structure collapses, settling into a more tightly packed framework. Requires slopes in excess of 3°.
Sediment Gravity Flow	Fluidal Flow	Fluidized Flow	Viscous Fluid	Cohesionless sediment supported by the forced upward motion of escaping pore fluid. Thin (<10cm) and short-lived.
Sediment Gravity Flow	Fluidal Flow	Turbidity Current	Viscous Fluid	Supported by fluid turbulence

which processes can occur and the boundary conditions for mechanical behaviour.

Slides, mass flows and fluidal flows form a mechanical continuum of mass movement processes (Heezen and Ewing, 1952; Kuenen, 1958; Dott, 1963; Cook et al, 1972; Morgenstern, 1967). The transition from slides to sediment gravity flows involves a change in the physical state of the sediment mass. At the time of mobilization there is a breakdown of metastable grain packing, as well as liquefaction and fluidization in some flows (Moore, 1961; Shepard and Dill, 1966; Andresen and Bjerrum, 1967). Slope failure can be triggered by earthquake shocks (Dott, 1963; Morgenstern, 1967, van der Knapp and Eijpe, 1968) tectonic or erosional oversteepening of the slope (Scholl et al, 1966; Morgenstern, 1967), diapirism (Lehner, 1969), overloading of the sediment with a consequent increase in pore pressure (Dott, 1963), tidal or surface wave action (Dott, 1963; Henkel, 1970; Coleman, 1976), ground water leaching (Rona, 1969; Herzer and Lewis, in press), and gas charging (Coleman, 1976; Doyle et al, 1978). Bioturbation may be important in influencing the stability of the sediment on some slopes (Stanley, 1971; Stanley and Unrug, 1972; Edwards, 1979). Slopes can support a diversity of benthic organisms, which by feeding, nesting and burrowing may change the grain packing and alter the pore water content of the sediment.

During transport the transition from mass flow to fluidal flow (e.g., debris flow to turbidity current) is a consequence of remolding and dilution of the flow in an aqueous medium (Dott, 1963; Allen, 1971). As demonstrated experimentally by Middleton (1967) and from field evidence (Hiscott, 1977; De Long and Middleton, 1978; Krause and Oldershaw, 1978; Hein, 1979), several different flow mechanisms may operate at the same time or in temporal sequence during transport and deposition. Middleton and Hampton (1976), Kelling and Stanley (1976) and Walker (1978) have proposed conceptual models that show the dynamic relationships among flows containing different grain sizes and concentrations of sediment. Among the various sediment gravity flows, the only long distance transport mechanisms are debris flows and turbidity currents (see reviews by Kuenen, 1964; Hesse, 1975; Middleton, 1978). Steady uniform grain flows can only be maintained on high gradient slopes and are generally less than 5 cm thick (Lowe, 1976b). Thus, true grain flows are of small importance in nature except on avalanche slopes (Middleton and Southard, 1977). Modified grain flows may form at the base of a large turbidity current by the downward transmission of shear stress (Middleton, 1970; Lowe, 1976b). Rough calculation shows that if modified grain flow layers exist at the base of large turbidity currents (e.g., 100 m thick) they influence only a very thin laminae of the flow (about 25 cm) (Hein, 1979). In order to accumulate thick grain flow deposits it is necessary that deposition be from successive grain flow layers that continually develop at the base of large turbidity currents.

During mobilization and deposition other gravity flows can predominate producing a deposit that differs significantly from classic turbidites or debris flow deposits (Stauffer, 1967; Aalto, 1972; Davies and Walker, 1972; Hendry, 1972, 1973, 1976; Kruit et al, 1976; Lowe, 1975, 1976a, b; Howell and McLean, 1976; Hiscott, 1977; Hein, 1979). For example, during failure along steep slopes, large liquified or fluidized flows rapidly become turbulent (Lowe, 1976a) and develop into high density turbidity currents. Laminar liquified flows are theoretically effective transport agents only for flows a few centimeters thick and for grains finer than 1 mm. Fluidized flows are unlikely to support debris during the final stages of transport. In addition, immediately prior to deposition, fluidized flows convert to liquified flows. Hence, the final properties of the sediment will most likely reflect the liquified state (Lowe, 1976b; pers. comm.).

CHARACTERISTICS OF GRAVITY DEPOSITS

The only sediment gravity flows that are likely to leave significant imprints on deposits are debris flows, modified grain flows, liquified flows and turbidity currents. The following is a summary of the characteristics (not necessarily definitive) that these flows could potentially leave in a deposit. Turbidites are relatively well-understood, have been extensively reviewed, and will not be discussed further here; however, elastic mass movement phenomena are included. It should be kept in mind that in natural flows several transport and depositional processes may be operating simultaneously or sequentially and thus the observed sedimentary features may reflect a number of them. Table 2 summarizes the characteristics potentially associated with the proposed process types.

Rockfall Deposits

Steep slopes are requisite for the generation of thick rockfall deposits. Rockfall deposits are typically poorly sorted, have angular to subangular clasts or blocks (if the deposits are not resedimented), and are chaotic in appearance. They are free of much fine grained matrix so that the deposit is wholly clast-supported with individual clasts and blocks having a random fabric. Sorting

TABLE 2.—MAJOR TYPES OF MASS MOVEMENT TRANSPORT PROCESSES AND TENTATIVE SEDIMENTARY STRUCTURE ASSEMBLAGES (BASED ON NUMEROUS REFERENCES CITED IN TEXT)

Mass Transport Processes				Mechanical Behavior	Sedimentary Structures
Rock Fall					Grain supported conglomerates, disorganized, open network variable matrix.
Slide		Glide		Elastic	Essentially undeformed, continuous bedding although some plastic deformation may be present particularly at the toe or base. Plow structures, folds, tension faults, joints, slickensides, grooves, rotational blocks if slumped.
		Slump			
				— Plastic Limit —	
Sediment Gravity Flow	Mass Flow	Debris Flow—Mud Flow		Plastic	Matrix supported, random fabric, clast size variable, matrix variable. Rip ups, rafts, inverse grading and flow structures possible.
				— Liquid Limit —	
		Grain Flows	Inertial Viscous		Massive, a-axis parallel to flow and imbricate upstream, inverse grading near base.
	Fluidal Flow	Liquified Flow		Viscous Fluid	dewatering structures, sandstone dikes, flame-load structure, convolute bedding, homogenized sediment.
		Fluidized Flow			
		Turbidity Current			Bouma series

features such as inverse or normal grading are absent and their distinct sheets or bedding are not developed (Cook et al, 1972; Conaghan et al, 1976; Johns, 1978).

Slide Deposits (Glide and Slump)

Steep slopes are not necessary for the generation and propagation of slides (Lewis, 1971; Herzer, 1975; Haner and Gorsline, 1978). The lower surface of a slide may be planar (glide) or curved (slump). Slide deposits are characterized by continuous bedding which has not undergone significant plastic deformation except perhaps at the toe or base. Here, folds and faults may involve only one bed or a considerable wedge of sediment. Large folds are commonly overturned in the direction of movement; however, folds and faults can have any orientation and may even parallel slope-dip (Helwig, 1970; Lajoie, 1972; Thompson, 1973). Pull-apart features may be present, particularly near the head of the slide (Lewis, 1971).

Debris Flow Deposits

Debris flow deposits are matrix supported and show little consistent preferred fabric. Grading and fabric are generally poorly developed. Fabrics that develop can have a-axis patterns that are parallel to flow or random, with imbrications that either dip up or down flow (Lindsay, 1966; Gorler and Reutter, 1968; Harms, 1974). Beds are generally massive and unstratified and a basal zone of inverse grading with imbrication may occur as a consequence of shearing. Internal flow structures can occur at any level within the deposit (Hampton, 1975). Large clasts may project above the bed level (Fisher, 1971; Middleton and Hampton, 1976; Walker, 1975).

Modified Grain Flow Deposits

Deposition from a modified grain flow layer at the base of an overriding turbidity current is en masse. Features of grain flow deposits are not well known. Grain fabric patterns should be a-axis parallel to flow, a-axis imbricate upstream (Davies and Walker, 1974; Walker, 1975). Inverse grading may be present at the base of the beds (Middleton, 1970).

Liquified Flow Deposits

These flows probably are evidenced only by the effects of "quick" sediments that were a product of rapid deposition, and possibly by evidence of syndepositional shearing (Walker, 1978). Characteristic fluid escape structures including dish and pillar structures, and other liquefaction features (Lowe and Lo Piccolo, 1974; Lowe, 1975), form as the result of pore fluid washing out of finest sediment during expulsion from the consolidating sediment. Fluid escape features can also develop due to post-depositional liquefaction of sediment commonly as a result of sudden loading, or shock.

RECOGNITION OF MASS TRANSPORTED DEPOSITS BY SEISMIC PROFILING

Seismic-reflection records provide an assessment of the mode of transport independent of that obtained from smaller scale sedimentary characteristics such as texture, fabric or depositional structures. Traditionally, the approach to analysis of seismic records has been one of seismic-facies analysis. In this approach interpretations are made from the record's acoustic signature or characteristic reflection patterns. The attitude, configuration, continuity, spacing and amplitude of the reflections are combined with geometry and spatial relationships to other patterns to infer sedimentary processes and environmental settings. From these the interpreter predicts facies type and composition (Damuth, 1975; Sangree et al, 1976; Vail and Mitchum, 1977). Because completely different processes can result in reflection patterns that appear similar (Moore, 1977), it is necessary to complement the seismic records with core or well data.

Two types of gravity deposits which can be recognized on seismic-reflection records, particularly those with high resolution, are slides and mass flows. Excellent examples of slump deposits have been reported by Ballard (1966, Fig. 7), Rona and Clay (1967, Fig. 2), Uchupi (1967, Fig. 4), Lewis, (1976, Figs. 3, 4), Emery and Uchupi (1972, Figs. 160, 161), Emery et al (1975, Fig. 42) and Kelling and Stanley (1976, Fig. 6). These records show slope and base-of-slope units with well-defined continous internal reflectors which are only slightly deformed. The reflectors dip back into the slope and evidence rotary motion of the slump block. Occasionally a curved shear surface at the base is visible. A hummocky topography may be present due to pull-aparts, minor folding or faulting or imbricate thrust sheets at the toe. Moore (1977) cautions that some of the seismic characteristics of slides are shared with channel-distributary-levee systems and thus a number of deposits previously interpreted as slides may actually be submarine fans, particularly where trackline density is low.

If hyperbolic reflections are present and internal reflectors are discontinuous, chaotic or absent, an increase in the amount of internal deformation is indicated (e.g., Hyne et al, 1973, Fig. 8; Seibold and Hinz, 1974; Fig. 14; Damuth, 1975, Fig. 7; Jacobi, 1976, Figs. 4 and 5; Sangree et al, 1976, Figs. 14 and 20; Nardin et al, this volume, Fig. 3). It should be noted, however, that hyperbolae are merely the result of an irregular bottom topography and, therefore, may reflect any number of geologic phenomena. For example, in the slope environment they may be due to gullying (Emery and Terry, 1956), small slide blocks (Jacobi, 1976) or tectonic deformation (Piper et al, 1973).

Mass flow deposits appear to be typified by lenses or mound-shaped units which lack internal structure and have low-amplitude bottom echoes. On high resolution records the bottom echo may be prolonged due to surface irregularities too small to be resolved into discrete hyperbolae. Presumably the absence of internal reflectors is due to high water content or deformational homogenization of the sediment mass. However, as is the case with the hyperbolae, these acoustic characteristics may be due to a variety of geologic processes (Damuth, 1975) or to resolution limitations of the seismic tool. Profiles containing examples of mass flow deposits can be found in papers by Wilhelm and Ewing (1972, Figs. 2, 3, 4, 5), Embley (1976, Figs. 3 and 5), Jacobi (1976, Fig. 4), and Nardin et al (this volume, Fig. 3, 4). Table 3 summarizes general acoustic facies characteristics associated with process types.

The recent development of the concept of seismic stratigraphy (Vail and Mitchum, 1977) has lead to a second interpretative approach in which seismic-facies signatures are analyzed within the context of large-scale depositional systems (Brown and Fisher, 1977). A depositional system is a three-dimensional assemblage of lithofacies (or seismic-facies) representing processes which are linked genetically. Slope depositional systems have been summarized by Sangree et al (1976) and Brown and Fisher (1977). Two fundamental slope geometries—offlap and onlap—have been defined. In general, offlap represents a sustained sediment supply such that a basin is filled by progradation of the slope. In large part, deposition is inferred to be from turbidity currents, mass flows and slides, particularly where reflectors are chaotic. In contrast, onlap represents a diminished sediment supply and a recession of the slope by erosion.

Almagor (1976) has used seismic profiling data to develop a model reconstitution of a margin originally formed by offlap to one with characteristic onlap structure.

DISTINCTION BETWEEN CANYON-FAN-SYSTEMS AND SLOPE AND BASE-OF-SLOPE SYSTEMS

Mass transport and deposition of terrigenous material into the deep sea has been discussed most often within the context of the submarine canyon and fan system (Haner, 1971; Nelson and Kulm, 1972; Mutti and Ricci-Lucchi, 1972; Dott and Shaver, 1974; Normark, 1978; Stanley and Kelling, 1978; Walker, 1978). Relatively little research has been done on slope and base-of-slope mass deposition in areas removed from the influ-

TABLE 3.—MAJOR TYPES OF MASS MOVEMENT TRANSPORT PROCESSES AND TENTATIVE ACOUSTIC FACIES ASSOCIATED WITH THESE PROCESSES (BASED ON NUMEROUS SOURCES CITED IN TEXT)

Mass Movement Process				Mechanical Behavior	Acoustic Record Characteristics
		Rock Falls		Elastic	Strong bottom return, poor internal return, hyperbolae.
Sediment Gravity Flow	Slide	Glide		Elastic	Internal reflectors essentially continuous and undeformed but may show some contortion. Hummocky surface locally, hyperbolae if very irregular. Low-angle faults may be visible. Strong basal shear surface reflector. Reflectors dip into slope if slumped.
		Slump			
				— Plastic Limit —	
	Mass Flow	Debris Flow		Plastic	Sea floor reflectors may be hyperbolic, irregular, or smooth; may also be prolonged. Few, if any, internal reflectors (transparent). Mounded or lense shaped. Blunt distal terminations.
		Mud Flow			
		Grain Flows	Inertial		
			Viscous		
				— Liquid Limit —	
	Fluidal Flow	Liquified Flow		Viscous Fluid	Individual flow deposits generally too thin to be resolved acoustically. Generally, as with turbidites, sequences of deposits may produce reflectors but pattern is unknown.
		Fluidized Flow			
		Turbidity Current			Even layered, continuous, onlapping reflectors. Discontinuous or hyperbolic reflectors if channels present.

ence of channelized flow through canyons. The textbook model of the development of the continental rise by turbidity currents originating from point sources to form coalescing submarine fans reflects this bias. The mass transport processes, however, including slides, mass flows and turbidity currents, are identical for both canyon and slope systems. It is the relative volumetric importance of particular types of gravity deposits that is different in each. This in part is due to the difference between sediment input along a line source (slope) versus input from a point source (canyon) (Gorsline, 1978). Because gravity transport processes are a function of the size, composition and rate of sediment supply, in addition to slope gradient, other expectable differences which may aid in field interpretations include the geometry of individual gravity deposits, spatial relationships among facies, dominant grain size, and vertical and lateral grain size trends.

The deeply incised canyons are traps that intercept very coarse sediment delivered to the shore by rivers and redistributed by longshore drift processes. In the canyon, sliding and mass flow are important mechanisms for transporting sediment, eroding canyon walls, and depositing sediment along the canyon's axis and feeder channel (Bouma and Shepard, 1964; Shepard and Dill, 1966; Stanley, 1967; Lewis, 1976; Haner and Gorsline, 1978); however, turbidites rank first volumetrically in submarine fan deposits. Submarine fans are divided into 1) a fan head and upper fan characterized by incised feeder channels with levees, 2) a middle fan built up by the progradation and lateral migration of suprafan lobes and 3) a topographically smooth lower fan. In the upper fan and fan head, clast-supported graded conglomerates and clast-supported pebbly sandstones and massive sandstones are deposited in the channels. Channel systems are relatively free of fine-grained turbidites and shales except during times of diminished sediment supply (Hendry, 1978; Hein, 1979). Similarly, debris flow and slide deposits (*proximal exotic facies* of Nelson and Kulm, 1973) are usually not common in channel fills (Hein, 1979) although they are in the canyon (Stanley, 1967, 1974; Cossey and Ehrlich, 1978). The suprafan lobes commonly have shallow, braided channel sequences which also may be filled with massive sandstones and pebbly sandstones. The smooth outer fan is characterized by even-bedded turbidites which probably represent sheet flow. Thus, in a progradational sequence, the overall trend is coarsening and thickening upward from lower to proximal upper fan settings. Bedding style changes from even-bedded, parallel deposits to channel fills, and facies grade from classic turbidites to massive and pebbly sandstones and coarse conglomerates. Although mud turbidites may predominate in some systems, particularly those off large deltas (Kelling and Stanley, 1976; Piper, 1978; Edwards, 1979) it is clear that the presence of coarse gravity deposits associated with large channels is the diagnostic characteristic of the inner submarine fan environment. Without evidence of this facies in the field, the fan model may not be appropriate to use (Walker, 1978). Again, sea level changes, canyon abandonment and diminished gradient can all alter the basic model.

In the slope setting, sediment input occurs along line sources and turbidites appear to add relatively little to the volume of base-of-slope deposits—slides and mass flows dominate (Gorsline and Emery, 1959; Stanely, 1964; Heezen and Drake, 1964; Morris, 1971; Klein et al, 1972; Lowe, 1972; Stanley and Unrug, 1972). This conclusion is supported by seismic reflection cross-sections which reveal that in many places, particularly where sediment supply is large, the continental rise is built by large allochthonous wedges of sediment moved from the adjacent slope acting as a line source (Ballard, 1966; Rona and Clay, 1967; Uchupi, 1967; Moore et al, 1970; Wilhelm and Ewing, 1972; Emery and Uchupi, 1972; Emery et al, 1975; Moore et al, 1976; Herzer, in press). The extent to which liqufed or grain flows are operative is not known but is unlikely to be large. Studies of ancient and contemporary slopes suggest a general model which is characterized by a basinward transition from slide to mass flow to fluidal flow deposition (Morris, 1971; Klein et al, 1972; Jacobi, 1976; Nardin et al, this volume). Because slopes are more removed than canyons from shoreline sources of coarse sediment except during periods of lowered sea level, the average grain size of slope gravity deposits is likely to be finer. Mass flow deposits will contain smaller clasts and occur as sheets in addition to being channelized. Where large clasts are present they are commonly locally derived from the slope or outer shelf. On heavily dissected slopes, channelized deposits will appear similar to those deposited in upper and middle fan environments and thus, discrimination between base-of-slope and fan environments may be difficult in these cases (McCabe, 1978). Unlike fan models, few overall progradational trends such as fining/coarsening or thinning/thickening are expected.

CONCLUSION

The foregoing discussion and literature review has been aimed at providing a basis for planning research programs and as a background against which many of the following pages can be viewed.

It is evident that we need many more measurements on the geotechnical properties of slope sediments and the critical conditions that cause the failure of these sediments and downslope movement and deposition by particular gravity processes. Experimental data is also required to better understand these processes and the characteristics of the resulting deposit. Most marine geologists agree that the primary controlling factors are sediment characteristics, sedimentation rates and slope gradient. Thus the boundaries of geotechnical characteristics for the various gravity transport processes must be mathematical surfaces defined by these three dimensions. Where critical conditions are present for mass movement a variety of triggering forces can generate a transport process. This then leads to the question of the influence of the triggering force. Does each force trigger the same process for given critical conditions of sediment stability?

Just as the Bouma facies gave us tools for defining particular zones of the canyon-fan-basin floor system, mass movement facies undoubtedly can provide clues about the environment of transport and deposition on slopes. Definition of these associations is an obvious research problem. A slope-apron-basin floor model should be developed paralleling the sophistication of our current models of canyon-fan-basin floor systems.

Have we identified the major acoustic facies associated with mass movement? What conditions inhibit the full development of the spectrum of mass movement morphologies in a slope-deep ocean floor system? Refinement of acoustic profiling methods and field surveys to delineate these sequences can improve our understanding of the mass budgets of sediment delivery to basin floor or deep-ocean floor.

The papers in this volume provide a variety of insights into these problems and also indicate gaps in our present knowledge that should be filled. It is already clear that the 1970's are the decade of emphasis on mass transport comparable to the decades of the 1950's and 1960's when burgeoning discoveries about turbidity currents and the canyon-fan-basin floor system were made. There is much yet to be known about the complete deep ocean sediment deposition system.

ACKNOWLEDGMENTS

Preparation and typing of this paper and support for the research upon which our own contributions have been based comes from National Science Foundation Grant OCE76-00156. Dr. D. W. Lewis read the manuscript and provided additional references.

REFERENCES

Aalto, K. R., 1972, Flysch pebble conglomerate of the Cap-des-Rosiers Formation (Ordovician), Gaspe Peninsula, Quebec: Jour. Sed. Petrology, v. 42, p. 922–926.

Abbote, E., Bortolotti, V., Passerini, P., 1970, Olistostromes and olistoliths: Sedimentology, v. 4, p. 421–557.

Allen, J. R. L., 1971, Mixing at turbidity current heads and its geological implications: Jour. Sed. Petrology, v. 41, p. 97–113.

Almagor, G., 1976, Physical properties, consolidation processes and slumping in Recent marine sediments in the Mediterranean Continental Slope off Israel: Geol. Survey of Israel Rept. MG/4/76, 132 p.

Andresen, A., and Bjerrum, L., 1967, Slides in subaqueous slopes in loose sand and silt in Richards, A. F., ed., Marine Geotechnique: Urbana, Univ. Of Illinois Press, p. 221–239.

Bagnold, R. A., 1954, Experiments on a gravity-free dispersion of large solid spheres in a Newtonian fluid under shear: Proc. Royal Soc. London, A225, p. 49–63.

———, 1956, The flow of cohesionless grains in fluids: Philosoph. Trans. Royal Soc. London, A249, p. 235–297.

Ballard, J. A., 1966, Structure of the lower continental rise hills of the western north Atlantic: Geophysics, v. 31, p. 506–523.

Bouma, A. H., 1962, Sedimentology of some flysch deposits, Amsterdam, Elsevier, 168 p.

———, and Shepard, F. P., 1964, Large rectangular cores from submarine canyons and fan valleys: Am. Assoc. Petroleum Geologists Bull., v. 48, p. 225–231.

Brown, L. F., and Fisher, W. L., 1977, Seismic stratigraphic interpretation of depositional systems: examples from Brazilian rift and pull-apart basins, in Payton, C. E., ed., Seismic Stratigraphy-applications to hydrocarbon exploration: Am. Assoc. Petroleum Geologists Memoir 26, p. 213–248.

Carter, R. M., 1975, A discussion and classification of subaqueous mass-transport with particular application to grain-flow, slurry flow and fluxoturbidites: Earth Science Reviews, v. 11, p. 145–177.

Coleman, J. M., 1976, Deltas: Processes of deposition and models for exploration: Champaign, Ill., Continuing Education Publ. Co., 102 p.

Conoghan, P. J., Mountjoy, E. W., Edgecombe, D. R., Talent, J. A., and Owen, D. E., 1976, Nubrigyn algal reefs (Devonian), eastern Australia, Allocthonous blocks and megabreccias: Geol. Soc. America Bull., v. 87, p. 515–530.

Cook, H. E., McDaniel, P. N., Mountjoy, E. W., and Pray, L. C., 1972, Allocthonous carbonate debris flows at Devobia Bank (reef) margins, Alberta, Canada: Canadian Petroleum Geologists Bull., v. 20, p. 439–497.

Cossey, S. P., and Ehrlich, R., 1978, Growth fault-controlled submarine carbonate debris flows and turbidite deposits from the Jurassic of northern Tunisia: possible canyon fill sequences, in Stanley, D. J., and Kelling, G., eds., Sedimentation in submarine canyons, fans and trenches: Stroudsburg, Pa., Dowden, Hutchinson and Ross, Inc., p. 127–137.

CROWELL, J. C., 1957, The origin of pebbly mudstones: Geol. Soc. America Bull., v. 60, p. 993–1010.
DAMUTH, J. E., 1975, Echo character of the western equatorial Atlantic floor and its relationship to the dispersal and distribution of terrigenous sediments: Marine Geology, v. 18, p. 17–45.
———, AND HAYES, D. E., 1977, Echo characteristics of the east Brazilian continental margin and its relationship to sedimentary processes: Marine Geology, v. 24, p. 73–95.
DAVIES, I. C., AND WALKER, R. G., 1974, Transport and deposition of resedimented conglomerates: the Cap Enrage Formation, Gaspe Peninsula, Quebec: Jour. Sed. Petrology, v. 44, p. 1200–1216.
DE LONG, R. C., AND MIDDLETON, G. V., 1978, Ordovician carbonate megabreccias at Cape Cormorant, western Newfoundland: submarine debris flow: Geol. Assoc. America Program with Abs., v. 10, p. 326.
DOTT, R. H. JR., 1963, Dynamics of subaqueous gravity depositional processes: Am. Assoc. Petroleum Geologists Bull., v. 47, p. 104–128.
———, AND SHAVER, R. H., eds. 1974, Modern and ancient geosynclinal sedimentation: Soc. Econ. Paleontologists Mineralogists Spec. Pub. No. 19, 380 p.
DOYLE, L. J., PILKEY, O. H., WOO, C. C., 1978, Sedimentation on eastern United States continental slope: Am. Assoc. Petroleum Geologists-Soc. Econ. Paleontologists Mineralogists Ann. Mtg., Oklahoma City, p. 61.
DRAKE, D. E., 1972, Distribution and transport of suspended matter, Santa Barbara Channel, California [unpub. Ph.D. thesis]: Los Angeles, Univ. Southern California, 357 p.
DZULYNSKI, S., KSIAZKIEWICZ, M., AND KUENEN, PH. H., 1959, Turbidites in flysch of the Polish Carpathians: Geol. Soc. America Bull., v. 70, p. 1089–1118.
EDWARDS, B. D., 1979, Animal-sediment relationships in bathyl environments, California Continental Borderland [unpub. Ph.D. thesis]: Los Angeles, Univ. Southern California, 193 p.
ELLIOTT, R. E., 1965, A classification of subaqueous sedimentary structures based on rheological and kinematic parameters: Sedimentology, v. 5, p. 193–209.
EMBLEY, R. W., 1976, New evidence for occurrence of debris flow deposits in the deep sea: Geology, v. 4, p. 371–374.
———, AND LANGSETH, M. G., 1977, Sedimentation processes on the continental rise of northeastern South America: Marine Geology, v. 25, p. 279–297.
EMERY, K. O., AND TERRY, R. D., 1956, A submarine slope of southern California: Jour. Geology, v. 64, p. 271–280.
———, AND UCHUPI, E., 1972, Western north Atlantic Ocean: Am. Assoc. Petroleum Geologists Memoir 17, 532 p.
———, UCHUPI, E., BOWIN, C. O., PHILLIPS, J., AND SIMPSON, E. S. W., 1975, Continental margin off western Africa, Cape St. Francis (South Africa) to Walvis Ridge (Southwest Africa): Am. Assoc. Petroleum Geologists Bull., v. 59, p. 3–59.
ENOS, P., 1977, Flow regimes in debris flows: Sedimentology, v. 24, p. 133–142.
EWING, M., AND THORNDIKE, E. M., 1965, Suspended sediment in deep-ocean water: Science, v. 147, p. 1291–1294.
FISHER, R. V., 1971, Features of coarse-grained high-concentration fluids and their deposits: Jour. Sed. Petrology, v. 41, p. 916–927.
FRIEDMAN, G. M., AND SANDERS, J. E., 1978, Principles of Sedimentology, New York, Wiley, 792 p.
GORLER, K., AND REUTTER, K. J., 1968, Entsrelung und merkmale der Olistostrome: Geol. Rund., v. 57, p. 484–514.
GORSLINE, D. S., 1978, Anatomy of margin basins: Jour. Sed. Petrology, v. 48 p. 1055–1068.
———, AND EMERY, K. O., 1959, Turbidity current deposits in San Pedro and Santa Monica Basins off southern California: Geol. Soc. America Bull., v. 70, p. 279–290.
HAMPTON, M. A., 1972, The role of subaqueous debris flow in generating turbidity currents: Jour. Sed. Petrology, v. 42, p. 775–793.
———, 1975, Competence of fine-grained debris flows: Jour. Sed. Petrology, v. 45, p. 834–844.
HANER, B. E., 1971, Morphology and sediments of Redondo submarine Canyon: Geol. Soc. America Bull., v. 82, p. 2413–2432.
———, AND GORSLINE, D. S., 1978, Processes and morphology of continental slope between Santa Monica and Dume submarine canyons, southern California: Marine Geology, v. 28, p. 77–87.
HARMS, J. C., 1974, Brusby Canyon Formation, Texas: a deep-water density current deposit: Geol. Soc. America Bull., v. 85, p. 1763–1784.
———, SOUTHARD, J. B., SPEARING, D. R., AND WALKER, R. G., 1975, Depositional environments as interpreted from primary sedimentary structures and stratification sequence: Soc. Econ. Paleontologists and Mineralogists Short Course Notes No. 2.
HEEZEN, B. C., AND DRAKE, C. L., 1964, Grand Banks slump: Am. Assoc. Petroleum Geologists Bull., v. 48, p. 221–225.
———, AND EWING, M., 1952, Turbidity currents and sub-marine slumps and the 1929 Grand Banks earthquake: Am. Jour. Sci., v. 250, p. 849–873.
———, HOLLISTER, C. D., AND RUDDIMAN, W. F., 1966, Shaping of the continental rise by deep geostrophic contour currents: Science, v. 152, p. 502–508.
HEIN, F. J., 1979, Deep-sea valley fill sediments: Cap Enrage Formation, Quebec [unpub. Ph.D. thesis]: Hamilton, MacMaster University, 570 p.
HELWIG, J., 1970, Slump folds and early structures, northeastern Newfoundland Applachians: Jour. Geology, v. 78, p. 172–187.

HENDRY, H. E., 1972, Breccias deposited by mass flow in the breccia nappe of the French pre-Alps: Sedimentology, v. 18, p. 277–292.
———, 1978, Cap des Rosiers Formation at Grosses Roches, Quebec—deposits of the mid-fan region on an Ordovician submarine fan: Canadian Jour. Earth Sci., v. 15, p. 1472–1488.
HENKEL, D. J., 1970, The role of waves in causing submarine landslides: Geotechnique, v. 20, p. 75–80.
HERZER, R. H., 1975, Uneven submarine topography south of Mernoo Gap: the result of volcanism and submarine sliding: Jour. Geol. Geophys., v. 18, p. 182–188.
———, in press, Submarine slides and submarine canyons on the continental slope off Canterbury, New Zealand: N. A. Jour. Geol. Geophys.,
———, AND LEWIS, D. W., in press, Growth and burial of a submarine canyon off Motunau, North Canterbury, New Zealand: North American Jour. Geol. Geophys.
HESSE, R., 1975, Turbiditic and non-turbiditic mudstones of Cretaceous flysch sections of the eastern Alps and other basins: Sedimentology, v. 22, p. 387–416.
HISCOTT, R. N., 1977, Sedimentology and regional implications of deep-water sandstones of the Tourelle Formation, Ordovician, Quebec [Unpub. Ph.D. thesis]: Hamilton, McMaster Univ., 542 p.
HOSKINS, H., 1967, Seismic reflection observations on the Atlantic continental shelf, slope and rise southeast of New England: Jour. Geology, v. 75, p. 598–611.
HOWELL, D. G. AND MCLEAN, H., 1976, Middle Miocene paleogeography, Santa Cruz and Santa Rosa Islands, in Howell, D. G., ed., Aspects of the geologic history of the California Continental Borderland: Am. Assoc. Petroleum Geologists Pacific Sect., Misc. Pub. 24, p. 266–293.
HYNE, N. J., GOLDMAN, C. R., AND COURT, J. E., 1973, Mounds in Lake Tahoe, California-Nevada: a model for landslide topography in the subaqueous environment: Jour. Geology, v. 81, p. 176–188.
JACOBI, R. D., 1976, Sediment slides on the northwestern continental margin of Africa: Marine Geology, v. 22, p. 157–173.
JOHNS, D. R., 1978, Mesozoic carbonate rudites, magabreccias and associated deposits from central Greece: Sedimentology, v. 25, p. 561–574.
JOHNSON, A. M., 1970, Physical processes in geology: San Francisco, Freeman Pub. Co., 571 p.
KEITH, B. D., AND FRIEDMAN, G. M., 1977, A slope-fan-basin plain model, Taconic sequence, New York and Vermont: Jour. Sed. Petrology, v. 47, p. 1220–1241.
KEELING, G., AND STANLEY, D. J., 1976, Sedimentation in canyon, slope and base-of-slope environments, in Stanley, D. J., and Swift, D. J. P., eds, Marine sediment transport and environmental management: New York, Wiley, p. 378–435.
KLEIN, G. D., DE MELO, U., AND FAVERA, J. C. D., 1972, Subaqueous gravity processes on the front of Cretaceous deltas, Reconcavo Basin, Brazil: Geol. Soc. America Bull., v. 83, p. 1469–1492.
KRAUSE, F. F., AND OLDERSHAW, A. E., 1978, Continental slope-rise deposits from Lower Cambrian Sekwi Formation, Mackenzie Mountains, Northwest Territories, Canada: Am. Assoc. Petroleum Geologists-Soc. Econ. Paleontologists and Mineralogists Ann. Mtg. Abs., p. 83.
KUENEN, PH.H., 1958, Experiments in geology: Trans. Geol. Soc. Glasgow, v. 23, p. 1–28.
———, 1964, Deep-sea sands and ancient turbidites, in Bouma, A. H., and Brouwer, eds., Turbidites: Amsterdam, Elsevier, p. 3–33.
LA JOIE, J., 1972, Slump fold axis orientations: an indication of paleoslope?: Jour. Sed. Petrology, v. 42, p. 584–586.
LEHNER, P., 1969, Salt tectonics and Pleistocene stratigraphy on continental slope of northern Gulf of Mexico: Am. Assoc. Petroleum Geologists Bull., v. 53, p. 2431–2479.
LEWIS, D. W., 1976, Subaqueous debris flows of early Pleistocene age at Motunau, North Canterbury, New Zealand: New Zealand Jour. Geol. & Geophys., v. 19, p. 535–567.
LEWIS, K. B., 1971, Slumping on a continental slope inclined at 1°–4°: Sedimentology, v. 16, p. 97–110.
LINDSAY, J. F., 1966, Carboniferous subaqueous mass-movement in the Manning-Macleay Basin, Kempsey, New South Wales: Jour. Sed. Petrology, v. 36, p. 719–732.
LOWE, D. R., 1972, Implications of three submarine mass-movement deposits, Cretaceous, Sacramento Valley, California: Jour. Sed. Petrology, v. 42, p. 89–101.
———, 1975, Water escape structures in coarse-grained sediments: Sedimentology, v. 22, p. 157–204.
———, 1976a, Subaqueous liquified and fluidized sediment flows and their deposits: Sedimentology, v. 23, p. 285–308.
———, 1976b, Grain flow and grain flow deposits: Jour. Sed. Petrology, v. 46, p. 188–190.
———, 1979, Sediment Gravity flows: Their classification and some problems of its application to natural flows and deposits: this volume.
———, AND LO PICCOLO, R. D., 1974, The characteristics and origins of dish and pillar structures: Jour. Sed. Petrology, v. 44, p. 484–501.
MCCABE, P. J., 1978, The Kinderscoutian delta (Carboniferous) of North England: a slope influenced by density currents, in Stanley, D. J., and Kelling, G., eds, Sedimentation in submarine canyons, fans and trenches: p. 116–126.
MIDDLETON, G. V., 1967, Experiments on density and turbidity currents, III, deposition of sediment: Canadian Jour. Earth Sci. v. 4, p. 475–504.
———, 1970, Experimental studies related to problems of flysch sedimentation, in Lajoie, J., ed., Flysch sedimentation in North America: Geol. Soc. Canada Spec. Paper No. 7, p. 253–272.

———, 1978, Coarse sediments deposited at base of continental slope by sediment-gravity flows (examples from northern Appalachians): Am. Assoc. Petroleum Geologists-SEPM Abs. Ann. Mtg, p. 94–95.

———, AND HAMPTON, M. A., 1976, Subaqueous sediment transport and deposition by sediment gravity flows, in Stanley, D. J., and Swift, D. J. P., eds, Marine sediment transport and environmental management: New York, Wiley, p. 197–218.

———, AND SOUTHARD, J. B., 1977, Mechanics of sediment movement, Soc. Econ. Paleontologists and Mineralogists Eastern Section Short Course Notes No. 3.

MOORE, D. G., 1961, Submarine slumps: Jour. Sed. Petrology, v. 31, p. 343–357.

———, 1969, Reflection profiling studies of the California continental borderland: structure and Quaternary turbidite basins: Geol. Soc. America Spec. Paper 107, 142 p.

———, 1977, Submarine slides, in Vaight, B., ed., Rockslides and avalanches, v. 1, Natural phenomena: Developments in Geotechnical Engineering, 14A, p. 563–604.

———, CURRAY, J. R., AND EMMEL, F. J., 1976, Large submarine slide (olistostrome) associated with Sunda Arc Subduction Zone, northeast Indian Ocean: Marine Geology, v. 21, p. 211–226.

MOORE, T. C., VAN ANDEL, T. H., BLOW, W. H., HEATH, G. R., 1970, Large submarine slide off northeastern continental margin of Brazil: Am. Assoc. Petroleum Geologists Bull., v. 54, p. 125–128.

MORGENSTERN, N. R., 1967, Submarine slumping and the initiation of turbidity currents, in Richards, A. F., ed., Marine Geotechnique: p. 189–220.

MORRIS, R. C., 1971, Classification and interpretation of disturbed bedding types in Jackfork flysch rocks (Upper Mississippian), Ouachita Mountains, Arkansas: Jour. Sed. Petrology, v. 41, p. 410–424.

MUTTI, E., AND RICCI-LUCCHI, R., 1972; Turbidites of the northern Apennines: Introduction to facies analysis: International Geol. Rev. v. 20, p. 125–166.

NARDIN, T. R., EDWARDS, B. D., AND GORSLINE, D. S., 1979, Santa Cruz Basin, California Borderland: dominance of slope processes in basin sedimentation: this volume.

NELSON, H., AND KULM, L. D., 1973, Submarine fans and channels, in Middleton, G. V., and Bouma, A. H., eds, Turbidites and deep-sea sedimentation: Soc. Econ. Paleontologists and Mineralogists Short Course Notes, p. 39–78.

NORMARK, W. R., 1978, Fan valleys, channels and depositional lobes on modern submarine fans: characteristics for recognition of sandy turbidite environments: Am. Assoc. Petroleum Geologists Bull., v. 62, p. 912–931.

PIPER, D. J. W., 1978, Turbidite muds and silts on deep-sea fans and abyssal plains, in Stanley, D. J., and Kelling, G., eds, Sedimentation in submarine canyons, fans and trenches: p. 163–176.

———, VON HUENE, R. AND DUNCAN, J. R., 1973, Late Quaternary sedimentation in the active eastern Aleutian Trench: Geology, v. 1, p. 19–22.

RODINE, J. D. AND JOHNSON, A. M., 1976, The ability of debris heavily freighted with coarse clastic material, to flow on gentle slopes: Sedimentology, v. 23, p. 213–234.

RONA, P. A., 1969, Middle Atlantic continental slope of United States: deposition and erosion: Am. Assoc. Petroleum Geologists Bull., v. 53, p. 1453–1465.

———, AND CLAY, C. S., 1967, Stratigraphy and structure along a continuous seismic reflection profile from Cape Hatteras, North Carolina to the Bermuda Rise: Jour. Geophys. Res., v. 72, p. 2107–2130.

SANDERS, J. E., 1965, Primary sedimentary structures formed by turbidity currents and related resedimentation mechanisms, in Middleton, G. V., ed., Primary sedimentary structures and their hydrodynamic interpretation: Soc. Econ. Paleontologists and Mineralogists Spec. Pub. p. 192–219.

SANGREE, J. B., WAYLETT, D. C., FRAZIER, D. F., AMERY, G. B., AND FENNESSY, W. J., 1976, Recognition of continental slope seismic facies offshore Texas-Louisiana, in Bouma, A. H., Moore, G. T., and Coleman, J. M., eds, Beyond the shelf break: p. F1–F54.

SCOTT, K. M., 1966, Sedimentology and dispersal pattern of Cretaceous flysch sequence Patagonian Andes, southern Chile: Am. Assoc. Petroleum Geologists, Bull., v. 50, p. 72–107.

SHARPE, C. F. T., 1938, Landslides and related phenomena: New York, Columbia University Press, 137 p.

SHEPARD, F. P., AND DILL, R. F., 1966,, Submarine canyons and other sea valleys: Chicago, Rand-McNally and Co., 381 p.

SCHLEE, J. S., DILLON, W. P., AND GROW, J. A., 1978, Structure of Atlantic slope of eastern North America: Am. Assoc. Petroleum Geologists-Soc. Econ. Paleontologists and Mineralogists Ann. Mtg. Abs., p. 111–112.

SCHOLL, D. W., BUFFINGTON, E. C., AND HOSKINS, D. M., 1966, Exposure of basement rock on the continental slope of the Bering Sea: Science, v. 153, p. 992–994.

SEIBOLD, E., AND HINZ, K., 1974, Continental slope construction and destruction, west Africa, in Burk, C. A., and Drake, C. L., eds., Geology of the Continental Margins: Berlin, Springer-Verlag, p. 179–198.

STANLEY, D. J., 1964, Nonturbidites in flysch type sequences; their significance in basin studies: Geol. Soc. America Spec. Paper 76, p. 155–156.

———, 1967, Company patterns of sedimentation in some modern and ancient submarine canyons: Earth and Planetary Sci. Letters, v. 3, p. 371–380.

———, 1971, Bioturbation and sediment failure in some submarine canyons: Vie et Milieu, p. 531–555.

———, 1974, Pebbly mud transport in the head of Wilmington Canyon: Marine Geology, v. 16, M1–M8.

———, AND UNRUG, R., 1972, Submarine channel deposits fluxoturbidites and other indicators of slope and base-of-slope environments, in Rigby, J. K., and Hamblin, W. K., eds., Recognition of ancient sedimentary environments: Soc. Econ. Paleontologists Mineralogists Spec. Pub. 16, p. 287–340.

———, AND KELLING, G., 1978, Sedimentation in submarine canyons, fans and trenches: Stroudsburg, Pa., Dowden, Hutchinson and Ross, Inc., 395 p.

STAUFFER, P. H., 1967, Grain flow deposits and their implications, Santa Ynez Mountains, California: Jour. Sed. Petrology, v. 37, p. 487–508.

THOMSON, A., 1973, Soft sediment faults and the Tesnus Formation and their relationship to paleoslope: Jour. Sed. Petrology, v. 43, p. 525–528.

UCHUPI, E., 1967, Slumping on the continental margin southeast of Long Island, New York: Deep-sea Research, v. 14, p. 635–639.

VAIL, P. R., AND MITCHUM, R. M. JR., 1977, Seismic-stratigraphy and global sea level changes, Part I: overview, in Payton, C. E., ed., Seismic stratigraphy—applications to hydrocarbon exploration: Am. Assoc. Petroleum Geologists Memoir 26, p. 51–52.

VAN DER KNAPP, W., AND EIJPE, R., 1968, Some experiments on the genesis of turbidity currents: Sedimentology, v. 11, p. 115–124.

VARNES, D. J., 1958, Landslide types and processes, in Eckel, E. B., ed, Landslides and engineering practice: Highway Res. Bd. Spec. Rept. 29, p. 20–47.

WALKER, R. G., 1970, Review of the geometry and facies organization of turbidites and turbidite bearing basins, in Lajoie, J., ed., Flysch sedimentology in North America: p. 219–251.

———, 1975, Generalized facies models for resedimented conglomerates of turbidite association: Geol. Soc. America Bull., v. 86, p. 737–748.

———, 1978, Deep-water sandstone facies and ancient sub-marine fans: models for stratigraphic traps: Am. Assoc. Petroleum Geologists Bull., v. 62, p. 932–966.

WILHELM, O., AND EWING, M., 1972, Geology and history of the Gulf of Mexico: Geol. Soc. America Bull., v. 83, p. 575–600.

WILSON, J. L., 1975, Carbonate facies in geological history: Berlin, Springer-Verlag, 263 p.

SEDIMENT GRAVITY FLOWS: THEIR CLASSIFICATION AND SOME PROBLEMS OF APPLICATION TO NATURAL FLOWS AND DEPOSITS

DONALD R. LOWE
Department of Geology
Louisiana State University
Baton Rouge, Louisiana 70803 USA

ABSTRACT

A system of classification and nomenclature for sediment gravity flows is proposed based on that developed by Middleton and Hampton (1973, 1976). Flows are classified first by their rheology, with fluidal flows and debris flows distinguished by their fluid and plastic behaviour, respectively. These two basic flow types are further subdivided into five types on the basis of the dominant coarse-particle support mechanism: *turbidity currents* (turbulence), *fluidized flows* (fluidization), *liquefied flows* (escaping pore fluid), *grain flows* (dispersive pressure) and *mudflows* (matrix cohesiveness).

Several considerations influence the application of this terminology to natural flows and deposits. Buoyancy, although an important lift force in natural flows, depends on the presence of other suspended sedimentary particles and is not included as an end-member support mechanism. Multiple support mechanisms characterize most flows, sometimes acting simultaneously on a single grain population or differentially on separate populations, and/or sometimes acting serially over the life of a flow. Many sediment gravity flows evolve from completely laminar to fully turbulent systems. The proposed terminology is not well adapted to the entire range of behaviour and the term *high-density turbidity currents* should be included to described turbulent flows originating as laminar fluidized, liquefied, or grain flows. Many natural flows also include both suspended and traction loads.

The correct interpretation of natural systems must include consideration of all of the above factors and be based on a firm understanding of how each influences the texture and structuring of flows and their deposits.

INTRODUCTION

Recent studies of submarine sedimentation have shown that a spectrum of processes moves and distributes coarse-grained sediment in the deep sea. The most effective of these processes are driven by gravity acting on the excess weight of entrained solids and have been termed sediment gravity flows, sediment flows, or mass flows. These flows are distinct from those in which moving fluids entrain and transport passive sediment particles. The latter include fluid gravity flows, wave- and tide-driven currents, and contour currents. They also stand in contrast to slumps and slides, where gravity pulls large coherent or semi-coherent blocks downslope with shear concentrated along one or a limited number of glide planes. Although the evolution of subaerial sediment flows from slumps is well documented and is probably common in subaqueous environments as well, transitions between sediment gravity flows and fluid gravity flows appear to be limited. The only documented examples appear to be sediment-laden rivers which, upon entering lakes or oceans, persist beneath the standing water as turbidity currents (Smith and others, 1960).

The evolution of sediment flow classification and nomenclature has been summarized by Dott (1963), Sanders (1965), Middleton (1967), Middleton and Hampton (1973, 1976), and Carter (1975), and will not be reviewed here. Modern terminologies have been offered by Carter (1975); Middleton and Hampton (1973, 1976), the latter modified slightly by Lowe (1976b); and Nardin et al (this volume). Carter's classification is based mainly on the laminar vs. turbulent state of flow and the dominance of viscous vs. inertial grain interactions. It has not been widely adopted because of its complexity, because it lumps together flows of differing rheological properties, because it is difficult to apply to flow deposits where the nature of particle interactions during movement may be indeterminate, and because our understanding of grain interactions within both sediment and fluid gravity flows is rather slight and cannot yet serve as a rational basis for flow definition and discrimination.

The present discussion takes as its basis the classification of Middleton and Hampton (1973, 1976) in which flow types are defined by the nature of the dominant sediment-support mechanisms (Fig. 1). This approach provides for a limited number of end-member models based on flow mechanics. These models should lend themselves to mathematical and empirical testing more easily than less precisely and more complexly defined flow types. Analyses of turbidity currents (Bagnold, 1962; Middleton, 1966, 1967; and many others), grain flows (Bagnold, 1954, 1956; Lowe, 1976a) debris flows (Johnson, 1965, 1970), and liquefied flows (Lowe, 1976b) substantiate the

FLOW TYPE	SEDIMENT SUPPORT MECHANISM
TURBIDITY CURRENT	FLUID TURBULENCE
FLUIDIZED FLOW	ESCAPING PORE FLUID
GRAIN FLOW	DISPERSIVE PRESSURE
DEBRIS FLOW	MATRIX STRENGTH

(MIDDLETON AND HAMPTON, 1973, 1976)

FIG. 1.—Sediment gravity flow nomenclature of Middleton and Hampton (1973, 1976).

FLOW TYPE	SEDIMENT SUPPORT MECHANISM
TURBIDITY CURRENT	FLUID TURBULENCE
FLUIDIZED FLOW	(FULL SUPPORT) ESCAPING PORE FLUID
LIQUEFIED FLOW	(PARTIAL SUPPORT) ESCAPING PORE FLUID
GRAIN FLOW	DISPERSIVE PRESSURE
DEBRIS FLOW	MATRIX STRENGTH

(LOWE, 1976)

FIG. 2.—Sediment gravity flow nomenclature of Middleton and Hampton (1973, 1976) as modified by Lowe (1976b).

usefulness of end-member support mechanisms in classifying and studying sediment flows.

As the nomenclature of Middleton and Hampton has been applied to real flows and flow deposits, a number of problems have arisen. These reflect imprecision in the application of certain terms, such as debris flow and grain flow; the interaction within individual flows of more than one support mechanism; difficulties in relating conceptual flow models to actual flows and flow evolution; our present inability to characterize quantitatively many flow types; and the paucity of experimental studies. The present discussion will focus on aspects of the first three of these problems.

PROPOSED CLASSIFICATION

Middleton and Hampton (1973, 1976) defined four types of sediment gravity flows (Fig. 1):

turbidity currents in which the sediment is supported by turbulence within the flow;

"*fluidized sediment flows* in which the sediment is supported by the upward flow of fluid escaping from between the grains as the grains are settled out by gravity";

grain flows in which the grains are supported by grain dispersive pressure, and

debris flows in which the larger grains are supported by a cohesive matrix.

Lowe (1976b) subsequently pointed out that sediment grains cannot simultaneously be fully supported and settling out and suggested that fluidized flow be redefined and the term "liquefied flow" added (Fig. 2):

fluidized flows in which the sediment is fully supported (fluidized) by upward moving fluid; and

liquefied flows in which the sediment is not fully supported but is settling through its pore fluid, which is displaced upward.

Perhaps the major problem in standardizing sediment flow terminology has been in conceptually distinguishing debris flows, grain flows, and liquefied flows. The present author would suggest that the problem has arisen mainly because of imprecise application of the term debris flow. Middleton and Hampton (1976, p 209) indicate that the matrix strength of "true debris flows" originates with the cohesive properties of entrained clay: "Support of grains by cohesion of the fluid phase distinguishes true debris flow from grain flow and turbidity current flow" and the "competence of a true debris flow is controlled by the strength and density of the clay-water fluid." The Coulomb-viscous rheological model of debris flows used by Middleton and Hampton (1973, 1976) was developed by Johnson (1965, 1970):

$$T = C + \sigma_\mu \tan \phi + \mu \frac{du}{dy}, \quad (1)$$

where

T = shear stress
C = cohesion,
σ_μ = internal normal stress,
ϕ = angle of internal friction,
μ = viscosity of flow, and
$\frac{du}{dy}$ = velocity gradient.

Middleton and Hampton (1976, p. 209) observe that "within a steadily moving debris flow, the driving stress is equal to the resistance supplied by cohesion C plus friction $\sigma_\mu \tan \phi$ plus viscosity."

It is clear from the above discussion that two types of flows are being discussed: "true debris flows" in which larger grains are supported by the cohesion of a clay-water matrix slurry and rheological debris flows which are described by equation (1). The former represents the type of debris flow included within the family of sediment gravity flows distinguished and defined on the basis of the sediment-support mechanism. The latter includes all sediment flows which exhibit plastic behaviour.

Allusions and direct references to both types of debris flows are present in the literature. Descriptions of pebbly mudstones and muddy, matrix-supported conglomerate and breccia interbedded within submarine flysch or base-of-slope deposits (Crowell, 1957; Dott, 1963; Peterson, 1965; Stanley, 1969; Abbate, Bortolotti, and Passerini, 1970; Lowe, 1972) leave little doubt that the larger clasts were truly supported, at least during the last stages of movement, by cohesive matrix strength. In other cases, however, the presence of fine-grained matrix material in a flow or flow deposit is taken as sufficient evidence of debris flow transport (Curry, 1966, as cited in Middleton and Hampton, 1973, 1976; Enos, 1977; Johnson, 1970; Montjoy and Playford, 1972). Although it seems probable that many of the described flows and deposits represent plastic or Bingham substances, there is commonly little evidence that the larger grains were actually supported by the cohesive matrix.

The author would suggest that some of the nomenclatorial problems can be relieved by incorporating this dual nomenclature into our terminology. Figure 3 presents the proposed nomenclature in which flow behavior, either plastic or fluid, serves the distinguish between *debris flows* and *fluidal sediment gravity flows*, respectively (see also Nardin et al, this volume). These two general types of flow are divided into specific flow types based on the particle support mechanisms. Instead of applying the term *debris flow* to both Coulomb-viscous flows and those within which larger grains are supported by matrix cohesiveness, the latter are termed *mudflows, cohesive debris flows*, or *cohesive flows*. Although resurrecting the term *mudflow* may meet with some objections, the name has been applied historically to flows "in which mud, although not necessarily quantitatively predominant, endows the mass with specific properties and modes of behavior which distinguish it from flows of debris devoid of mud" (Sharp and Nobles, 1953, p. 550). The distinguishing property of mud, a mixture of clay, silt, and sand, is its cohesiveness. Hence, the definition of mudflow of Sharp and Nobles is coincident with the range of flow types discussed here as cohesive debris flows. It is noteworthy that Sharp and Nobles (1953) considered mudflows to be one variety of debris flow, although they defined debris flows differently than has been done here.

The proposed terminology merely formalizes the common tendency to apply the term *debris flow* to flows of Bingham substances, whether their strength is frictional or cohesive in origin. To this extent, it better reflects current usage than does the support-mechanism-based definition of debris flow. It is also compatible with current sedimentological usage of the term *mudflow*.

The proposed terminology also offers a clear distinction between cohesive debris flow and grain flow that has been lacking in previous discussions. Most mudflows show plasticity because of the cohesive properties of suspended clay particles, although the yield strength and shear behaviour of clay-water suspensions vary greatly with the composition of the clay, and, once the yield strength is surpassed and deformation begins, clay-water suspensions behave essentially as viscous fluids. If the applied shear stress drops below the yield strength, the flows freeze. Grain flows are characterized by frictional strength because of the role of grain interactions in maintaining the dispersion against gravity. This strength is reflected in the relatively high slopes, generally greater than 20°, required for the maintenance of steady grain flows of uniformly sized particles (Bagnold, 1956; Lowe, 1976a). On moving onto lower slopes, the frictional strength of the colliding grains exceeds the applied tangential component of gravity, and the flows freeze. Thus, mudflows are characterized by cohesive strength and grain flows by frictional strength. In exhibiting a finite yield strength, and as a result the tendency to freeze from the top down during resedimentation, both are here included as debris flows.

In terms of equation (1), most mudflows can be described by the relationship

$$T = C + \mu \frac{du}{dy}$$

FLOW BEHAVIOR	FLOW TYPE		SEDIMENT SUPPORT MECHANISM
FLUID	FLUIDAL FLOW	TURBIDITY CURRENT	FLUID TURBULENCE
		FLUIDIZED FLOW	ESCAPING PORE FLUID (FULL SUPPORT)
		LIQUEFIED FLOW	ESCAPING PORE FLUID (PARTIAL SUPPORT)
PLASTIC (BINGHAM)	DEBRIS FLOW	GRAIN FLOW	DISPERSIVE PRESSURE
		MUDFLOW OR COHESIVE DEBRIS FLOW	MATRIX STRENGTH MATRIX DENSITY

FIG. 3.—Proposed nomenclature for sediment gravity flows.

whereas the general grain-flow model is

$$T = \sigma_\mu \tan \phi + \mu \frac{du}{dy}.$$

The application of grain flow theory to natural transport problems is probably less well understood than that of any other sediment flow mechanism. As pointed out by Lowe (1976a), the results of experiments by Bagnold (1954) are not directly applicable to natural flows of cohesionless grains. Especially problematic is the role of inertial versus viscous grain interactions. Fully inertial grain flows are commonplace, represented by avalanches on the slipfaces of ripples and dunes. Viscous grain interactions undoubtedly occur, but, as suggested by Lowe (1976a, p. 191) and implied by the experimental results of Bagnold (1954), true viscous grain flows, in which the grains are fully supported by viscous dispersive pressure, require slopes in excess of those required for similar inertial flows. Such flows, if ever developed, could occur only on slopes greater than the angle of repose of the material. Viscous interactions undoubtedly characterize some mudflows, liquefied flows, and the bed load layers of traction or turbidity currents, but, in the absence of additional support mechanisms, viscous dispersive pressure will not maintain flowing particle dispersions over most natural subaqueous slopes. Except on subaqueous slopes in excess of 25° to 30°, therefore, unmodified viscous grain flows probably cannot exist. Suggestions that viscous grain flows can be regarded as fluidal flows because of their fluid-like behavior (Nardin et al, this volume) overlook the fact that, in the absence of external stress or additional support mechanisms, this behavior is manifest only on steep slopes where the downslope gravity component exceeds the internal resistance of the interacting grains: the sediment-water mixtures are behaving effectively as plastic substances.

The dynamics of natural liquefied and fluidized dispersions is not well understood. At relatively low shear rates, concentrated suspensions of rigid solids apparently behave as Newtonian fluids (Roscoe, 1953; Metzner and Whitlock, 1958) but at high shear rates or if confined, can show considerable frictional resistance and dilatancy (Freundlich and Jones, 1936; Seed and Lee, 1966). Hence, although frictional resistance may develop near the bases of thick flows of liquefied sediment, it is unlikely to characterize the flows as a whole (Lowe, 1976b). Liquefied and fluidized flows are most likely to behave as fluidal flows but may, near their bases, show strength and overlap into the area of debris flow if shear rates or flow thickness is great. They also differ from mudflows and most grain flows in resedimenting from the base up.

Relationship Among Flow Types

The sediment-flow types defined in Figure 3 do not represent a parallel series in which the end-members differ from one another only in the dominant process of particle support (Fig. 4). Turbidity currents, for instance, can maintain themselves as steady flows only if turbulent whereas grain flows, fluidized flows, and mudflows can exist and transport sediment as steady, laminar flows. Liquefied flows are, by definition, unsteady because sediment is continuously settling out. They are thus hydraulically similar to the waning, laminar resedimenting tails of turbidity currents in that the prevailing conditions do not permit sediment transport. The main importance of liquefied flows thus lies in the relatively long distances fine-grained sediment may travel before fully resedimenting (Morganstern, 1969; van der Knaap and Eijpe, 1969; Middleton, 1970; Lowe, 1976b) and in their evolution into turbidity currents through the development of turbulence (van der Knaap and Eijpe, 1969; Lowe, 1976b).

When considering turbulent flows, the support-mechanism-based nomenclature is not applicable (Fig. 4). A liquefied flow in which turbulence becomes fully developed is no longer liquefied; it is a *high-density turbidity current*. The same is true for grain flows. Because cohesionless fine-grained sand, silt, and clay is the only natural material likely to be fluidized (Lowe, 1976b), accelerating fluidized flows are likely to evolve into turbidity currents ranging from relatively dilute suspensions to high-density flows. Cohesive debris flows may commonly become turbulent (Enos, 1977), but apparently the effects of sediment cohesion are important in preventing particle size-segregation and in aiding in clast support even

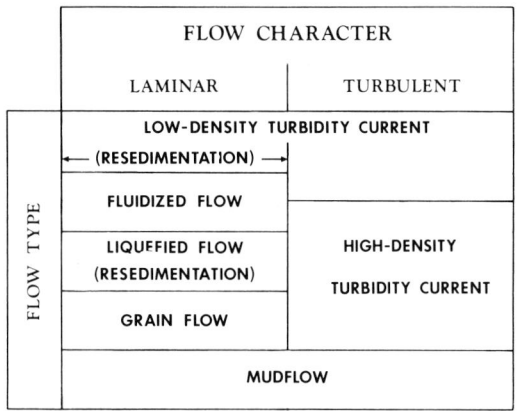

FIG. 4.—Sediment gravity flow nomenclature for both laminar and turbulent flows.

in fully turbulent flows. Thus, the term *turbulent mudflow* seems appropriate.

DISCUSSION

The classification of sediment flows based on rheology and particle-support mechanisms is useful because the support mechanisms represent discrete end members, and hence, are conceptually distinct, and because both rheology and support mechanism control the mechanics of transport and deposition and, thus, are likely to be reflected in the textures and structures of the final deposits. In fact, a number of problems complicate both the theoretical and practical application of the proposed terminology (see also Nardin et al, this volume).

Buoyancy

Buoyancy is perhaps the single most important coarse-particle support mechanism in mudflows, but is not included as an end-member in terms of flow terminology. Rodine and Johnson (1976) have emphasized the role of buoyancy in supporting large blocks within debris flows. Because the specific gravity of debris-flows is commonly between 1.5 and 2.0 (40% to 60% quartz-density solids), from 30% to 70% of the excess weight of larger quartz-density particles may be supported by the buoyancy of the surrounding solid-liquid mixture. In the case of still-plastic, semi-lithified mudstone chunks, their entire mass may be suspended by buoyant lift.

Rodine and Johnson (1976) discuss several examples of the effectiveness of buoyancy in supporting larger blocks in debris flows. Another example is provided by Lowe (1972). A debris-flow deposit from the Cretaceous Great Valley Sequence of western California is made up largely of pebbly mudstone composed of 15 to 20% rounded, pebble- and cobble-sized exotic clasts suspended within a massive mudstone matrix. Large blocks of bedded mudstone up to 10 m across "float" at the top of the flow. The paucity of mudstone blocks within the flow suggests that they may have been supported largely or entirely by the buoyancy of the pebbly mudstone.

Suspended or floating mudstone clasts are ubiquitous components of proximal turbidites or "fluxoturbidites." Whether these sandstone units were deposited by high-density turbidity currents (Kuenen, 1951; Middleton, 1966, 1967), sandy debris flows (Hampton, 1975), or liquefied flows (Lowe, 1976b), the mud clasts were probably supported largely by the buoyancy of the dense sand-water mixture. Mud clasts with a specific gravity between 1.5 and 2.0 would be entirely, supported, for instance, by suspensions having quartz sand contents between 20% and 40%, well within the probable range of such flows.

Although an additional end-member flow type could be defined based on buoyant support, buoyant lift beyond that provided by water alone occurs only when a significant quantity of other, smaller grains are also entrained. The processes by which these smaller grains are supported is necessarily not a function of buoyant lift and perhaps can be considered to be the dominant support mechanism within the flow: without the smaller grain population, there would be no buoyant support of the larger grains. Consequently, there is no separate end-member flow type characterized only by buoyant support of grains and no need to define one. In grain mixtures, however, buoyancy must be regarded as an important support mechanism for larger particles.

Support Versus Lift

It is essential to recognize the difference between support and lift mechanisms in sediment gravity flows. Actual lifting of the grains against gravity can be accomplished by flow turbulence, dispersive pressure, escaping pore fluids, and buoyancy. All of these mechanisms and flow strength can support the grains, but the latter cannot by itself provide lift. Pore fluids escaping due to liquefaction can only provide partial support for the larger grains. In many sediment flow deposits, such as pebbly mudstones or pebbly sandstones, larger clasts occur evenly dispersed within a structureless fine-grained matrix. The suspension of larger, heavy particles within the matrix suggests that they were, at least during the late stages of flow, supported by matrix strength and that the flows were mudflows. Although the clasts, once suspended, may have been partially supported by matrix strength, they could not have been lifted by matrix strength. Most such flows probably originate through the flowage and mixing of interstratified and gravitationally unstable mud, sand, and gravel. The sinking of gravel and sand into lower density mud, combined with shear during flowage, could produce the observed texture, but, in the absence of lift forces, there would be a tendency for the clasts to work their way to the base of the flow. The common nearly uniform dispersion of clasts is best explained by the existence of dispersive pressure between the colliding gravel-sized particles. Such pressure would maintain a uniform dispersion against the tendency of gravity to cause the particles to settle. Within mud-matrix flows, the bulk of the excess weight of the larger particles is probably supported by the buoyancy of the mud-water matrix and additional support is provided by flow strength, but the actual dispersion is produced and maintained by dispersive pressure. In some flows where dispersive pressure is relatively high, perhaps due to a relatively high concentration of gravel, inverse grading may develop even though the matrix is cohesive mud (Fisher, 1971).

Attempts to name such flows on the basis of particle support mechanism are futile; there is no single or dominant particle support mechanism. The final character of the flow deposit, however, in particular its matrix-supported texture, will reflect the cohesive strength of the mixture during the final stages of movement. The terms *cohesive debris flow* or *mudflow* accurately characterize this type of mass movement even though cohesiveness may not have been either the dominant lift or support mechanism.

Enos (1977) has recently suggested that many debris flows are turbulent. Within turbulent flows, turbulence could provide part or all of the lift force required to support the larger grains. Upon declining to laminar flows, dispersive pressure would replace turbulence as the principal lift mechanism until the flows freeze.

Multiple Support Mechanisms

Middleton and Hampton (1976), Lowe (1976) and others have stressed that debris in real sediment flows is generally kept suspended by more than one support mechanism. At least three different conditions of multiple support may occur. (1) Several processes of lift and/or support may act simultaneously on the same grains. Buoyancy is effective in most sediment flows in addition to the primary support mechanism. In the preceeding discussion of pebbly mudstones, it was suggested that the simultaneous interaction of buoyancy, dispersive pressure or turbulence, and cohesive matrix strength maintains many suspensions of gravel-sized clasts. (2) In some sediment gravity flows, different size or composition populations of grains can be simultaneously supported by different mechanisms. In gravel flows termed *density-modified grain flows* by Lowe (1976a), it was suggested that the gravel-sized clasts were supported by the combined processes of sand-mud-water matrix buoyancy and dispersive pressure and that sand-sized grains were supported by a combination of mud-water matrix buoyancy, turbulence, and possibly cohesiveness. In liquefied flows, sand grains may be resedimenting in a liquefied state, mud clasts are fully supported by flow buoyancy, and finer silt and clay are fluidized and elutriated by the escaping pore water. (3) Many flows probably undergo a secular evolution involving changes in the relative effectiveness of a number of support mechanisms. A mass of sediment may fail as a slump, liquefy, accelerate and become a turbulent high-density turbidity current, and finally slow and resediment as a liquefied flow (Lowe, 1976b).

Each of these varieties of multiple support probably leaves a distinctive imprint on the final flow deposit. The accurate interpretation of sediment flow deposits depends, first, on the identification of textures and structures characteristic of each mechanism of particle support and deposition and, secondly, on the discrimination of compound textures and structures produced through the interaction of several support mechanisms acting simultaneously and/or serially.

Multiple Transport Mechanisms

Within many debris flows, it is probable that many of the larger transported clasts are not fully supported. Motion pictures of the 1968 mudflows at Wrightwood, California, taken by D. M. Morton of the U.S. Geological Survey, show that large cobble- and boulder-sized blocks were actually shoved along as bed load, rolling or, less commonly, sliding but remaining in more-or-less continuous contact with the bottom or with other underlying clasts. In some flows, blocks over 0.5 m in diameter were rolled along as bed load within flows less than 0.4 m thick moving at speeds of about 1 m/sec. In other instances, there was a jostling and collision of blocks within a tightly packed mass of material, accompanied by forward grain rotation but relatively slight net shear. The clasts were shoved and dragged forward by the surrounding, more rapidly moving sediment-water mixture, but were not fully supported or suspended within it. This type of movement, with the periodic formation and breakdown of debris dams, may, in fact, represent the main mode of clast transport in many mudflows.

In the debris flow deposit in the Great Valley Sequence discussed previously (Lowe, 1972), blocks of tonalite and gabbro larger than 0.4 m in diameter occur only at the base of the deposit. Although most probably moved several kilometers, they were apparently not suspended within the flow inasmuch as their tendency to settle should have decreased as flow velocity declined and the yield strength of the mud-gravel mixture was approached. Hence, they probably slid or rolled as bed load at the flow base. Associated larger blocks of lithified sandy sedimentary rock up to 50 m high were certainly pushed or rolled along in continuous contact with the bottom.

These observations suggest that rheological debris flows can include both traction and suspended loads as do turbidity currents. This conclusion is not unexpected inasmuch as, beyond their yield strengths, plastic substances exhibit fluid-like behaviour. Hence, larger grains whose weight exceeds the combined effects of yield strength and buoyant lift force of the mud-water matrix will tend to settle. They may be pushed along the bottom as bed load. Smaller pebbles are entrained as suspended load by the combination of cohesive matrix support and lift provided by flow turbulence (Enos, 1977) or, in laminar flows, dispersive pressure. Still smaller sand and silt grains may be fully supported by matrix strength.

It is unlikely that bed load material can exist in most liquefied flows because the continuous rapid resedimentation of liquefied sediment would result in the absence of a well-defined flow base. However, both suspended load, maintained by one or more particle-support mechanisms, and bed load may exist for the other varieties of sediment-flows just as for fluid gravity flows.

Modeling of sediment flow dynamics and evolution and of sediment entainment, transport, and deposition must, therefore, include consideration of multiple modes of sediment transport and the probability of sediment exchange among the loads. Equally important, the textural properties and internal structuring of all types of sediment flow deposits may reflect the existence of more than one mode of sediment transport within a single flow. Instead of evaluating the textures and structures of a deposit in terms only of support mechanisms for the suspended load, they must be evaluated in terms of a traction load as well. The textural identification of these loads may be possible using techniques similar to those used to study fluid gravity flow deposits (Visher, 1969; Middleton, 1976).

CONCLUSIONS

The present discussion has focused on some of the problems of sediment gravity flow classification and nomenclature. It has not sought to develop any fundamentally new concepts or hypotheses, but to provide a practical taxonomy for sediment flows, compatible with that already in use by sedimentologists, and to explore some of the shortcomings, complexities, and implications of that taxonomy.

A system has been proposed in which flows are named, firstly, on the basis of their rheological behaviour and, secondly, by the dominant coarse-particle support mechanism. Although the taxonomy is conceptually simple, its application to natural flows and deposits must as yet be undertaken with caution. The textures and structures of natural deposits reflect not only the dominant support mechanism, if one exists, but also, (1) the simultaneous or serial interaction of multiple support mechanisms, (2) the character of flow, either laminar or turbulent, before and during deposition; (3) the existence of both suspended and traction loads, and (4) the effects of particle support or lift mechanisms not formally incorporated into the support-mechanism based nomenclature, such as buoyancy. Other processes beyond the scope of the present discussion, most notably water escape, also strongly influence the character of the final deposit.

The greatest problem in the use of sediment flow terminology lies not in its application to relatively well-understood phenomena, such as turbidites or mudflow deposits, but in its application to flow types that are not well understood, as such fluidized flows, liquefied flows, and some debris flows. It is too often convenient to interpret deposits exhibiting textures and structures whose origins are problematic in terms of processes whose dynamics are equally poorly understood. Until objective criteria are developed to identify and evaluate each of the processes active within natural flows and its effect on the composition, texture, and structures of flow deposits, any system of nomenclature based on genetic interpretations must be applied with caution.

REFERENCES

ABBATE, E., BORTOLOTTI, V., AND PASSERINI, P., 1970, Olistostromes and Olistoliths: Sed. Geology, v. 4, p. 521–557.
BAGNOLD, R. A., 1954, Experiments on a gravity-free dispersion of large solid spheres in a Newtonian fluid under shear: Proc. Royal Soc., Ser. A, v. 225, p. 49–63.
———, 1956, The flow of cohesionless grains in fluids: Phil. Trans. Royal Soc., Ser. A, v. 249, p. 235–297.
———, 1962, Auto-suspension of transported sediment; turbidity currents: Proc. Royal Soc., Ser. A, v. 265, p. 315–319.
CARTER, R. M., 1975, A discussion and classification of subaqueous mass-transport with particular application to grain-flow, slurry-flow, and fluxoturbidites: Earth Sci. Rev., v. 11, p. 145–177.
CROWELL, J. C., 1957, The origin of pebbly mudstones: Geol. Soc. America Bull., v. 68, p. 993–1009.
CURRY, R. R., 1966, Observations of alpine mudflows in the Tenmile Range, central Colorado: Geol. Soc. America Bull., v. 77, p. 771–776.
DOTT, R. H., JR., 1963, Dynamics of subaqueous gravity depositional processes: Am. Assoc. Petroleum Geologists Bull., v. 47, p. 104–128.
ENOS, P., 1977, Flow regimes in debris flow: Sedimentology, v. 24, p. 133–142.
FISHER, R. V., 1971, Features of coarse-grained, high-concentration fluids and their deposits: Jour. Sed. Petrology, v. 41, p. 916–927.
FREUNDLICH, H., AND JONES, A. D., 1936, Sedimentation volume, dilatency, thixotropic, and plastic properties of concentrated suspensions: Jour. Physical Chemistry, v. 40, p. 1217–1236.
HAMPTON, M. A., 1975, Competence of fine-grained debris flows: Jour. Sed. Petrology, v. 45, p. 834–844.

JOHNSON, A. M., 1965, A model for debris flow [unpub. Ph.D. thesis]: University Park, Pa., The Pennsylvania State Univ., 305 p.
———, 1970, Physical Processes in Geology: San Francisco, Freeman, Cooper, and Co., 571 p.
KUENEN, PH. H., 1950, Turbidity currents of high density: Rept. 18th Int. Geol. Congr., London, pt. 8, p. 44–52.
LOWE, D. R., 1972, Implications of three submarine mass-movement deposits, Cretaceous, Sacramento Valley, California: Jour. Sed. Petrology, v. 42, p. 89–101.
———, 1976a, Grain flow and grain flow deposits: Jour. Sed. Petrology, v. 46, p. 188–199.
———, 1976b, Subaqueous liquefied and fludized sediment flows and their deposits: Sedimentology, v. 23, p. 285–308.
METZNER, A. B., AND WHITLOCK, M., 1958, Flow behaviour of concentrated (dilatant) suspensions: Trans. Soc. Rheologists, v. 2, p. 239–254.
MIDDLETON, G. V., 1966, Experiments on density and turbidity currents. I. Motion of the head: Canadian Jour. Earth Sci., v. 3, p. 523–546.
———, 1967, Experiments on density and turbidity currents. III. Deposition of sediment: Canadian Jour. Earth Sci., v. 4, p. 475–505.
———, 1969, Grain flows and other mass movements down slopes, in Stanley, D. J., ed., The New Concepts of Continental Margin Sedimentation: American Geol. Inst., Short Course Lecture Note, GM-B-1 to GM-B-14.
———, 1970, Experimental studies related to problems of flysch sedimentation: in Lajoie, J., ed., Flysch Sedimentology in North America. Geol. Assoc. Canada Spec. Paper 7, p. 253–272.
———, 1976, Hydraulic interpretation of sand size distributions. Jour. Geol., v. 84, p. 405–426.
———, AND HAMPTON, M. A., 1973, Sediment gravity flows: mechanics of flow and deposition, in Turbidites and Deep Water Sedimentation: Soc. Econ. Paleontologists Mineralogists, Pacific Sec. Short Course Lecture Notes, p. 1–38.
———, AND ———, 1976, Subaqueous sediment transport and deposition by sediment gravity flows, in Stanley, D. J., and Swift, D. J. P., eds., Marine Sediment Transport and Environmental Management: New York, Wiley, p. 197–218.
MORGENSTERN, N. R., 1967, Submarine slumping and the initiation of turbidity currents, in Richards, A. F., ed., Marine Geotechnique: Urbana, University of Illinois Press, p. 189–220.
MOUNTJOY, E. W., AND PLAYFORD, P. E., 1972, Submarine megabreccia debris flows and slumped blocks of Devonian of Australia and Alberta—a comparison (abs.): Am. Assoc. Petroleum Geologists Bull., v. 56, p. 641.
PETERSON, G. L., 1965, Implications of two Cretaceous mass transport deposits, Sacramento Valley, California: Jour. Sed. Petrology, v. 5, p. 401–407.
RODINE, J. D., AND JOHNSON, A. M., 1976, The ability of debris, heavily freighted with coarse clastic material, to flow on gentle slopes: Sedimentology, v. 23, p. 213–234.
ROSCOE, R., 1953, Suspensions, in Hermans, J. J., ed., Flow Properties of Disperse Systems: New York, North-Holland, p. 1–38.
SANDERS, J. E., 1965, Primary sedimentary structures formed by turbidity currents and related resedimentation mechanisms: Soc. Econ. Paleontologists Mineralogists, Spec. Pub., v. 12, p. 192–219.
SEED, H. B., AND LEE, K. L., 1966, Liquefaction of saturated sands during cyclic loading: Jour. Soil Mech. Fdns. American Soc. Civil Engineers., v. 92, SM6, p. 105–134.
SHARP, R. P., AND NOBLES, L. H., 1953, Mudflow of 1941 at Wrightwood, California: Geol. Soc. America Bull., v. 64, p. 547–560.
SMITH, W. O., VETTER, C. P., CUMMINGS, G. B., AND OTHERS, 1960, Comprehensive survey of sedimentation in Lake Mead (Arizona—Nevada), 1948-49: U.S. Geol. Survey, Professional Paper 295, 254 p.
STANLEY, D. J., 1969, Submarine channel deposits and their fossil analogs ('fluxoturbidites'), in Stanley, D. J., ed., The New Concepts of Continental Margin Sedimentation: American Geol. Inst., Short Course Lecture Notes, DJS-9-1 to DJS-9-17.
VAN DER KNAAP, W., AND EIJPE, R., 1968, Some experiments on the genesis of turbidity currents: Sedimentology, v. 11, p. 115–124.
VISHER, G. S., 1969, Grain size distributions and depositional processes: Jour. Sed. Petrology, v. 39, p. 1074–1106.

PART II

MODERN CONTINENTAL SLOPES

A. ATLANTIC AND GULF OF MEXICO SLOPES
B. PACIFIC AND ANARCTIC SLOPES

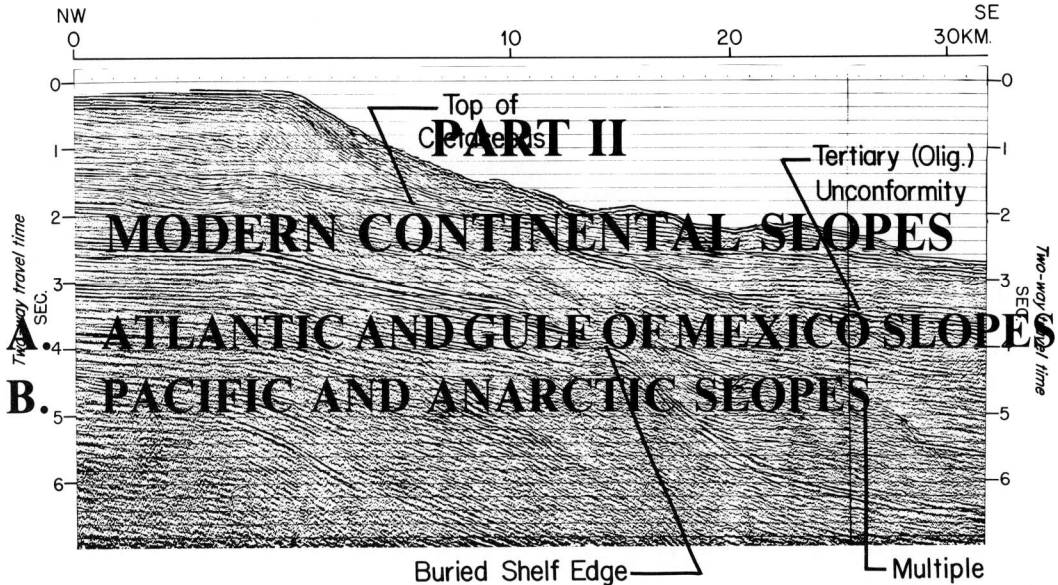

CURRENTS IN SUBMARINE CANYONS AND OTHER TYPES OF SEAVALLEYS

FRANCIS P. SHEPARD
Geological Research Division
Scripps Institution of Oceanography
University of California San Diego
La Jolla, California 92093

ABSTRACT

Currents along the floors of submarine canyons and other types of slope valleys are now rather well documented after obtaining some 25,000 hours of records from various parts of the world and at a great variety of depths. At least two distinct types of currents exist: (1) the ordinary currents that rarely exceed 50 cm/sec alternate, flowing almost continuously up and down the valley axes at periods that are related to the depth and the range of tide, and (2) the occasional surges of strong downcanyon flow that according to our four records do not exceed 100 cm/sec. The latter interpreted as turbidity currents are believed to occur at intervals which may well be as little as a few days wherever rivers are discharging considerable loads of sediment into the sea in close proximity of submarine valley heads. In turn submarine valleys apparently develop wherever rivers have built deltas across the continental shelf so that foreset slopes have sufficient steepness to become unstable and hence suspend sediments into the water column and therefore produce a density flow.

Although high-speed turbidity currents have not been recorded by current meters, the evidence of their existence cannot be disregarded. However, the slower type of turbidity current should alone be capable of transporting great quantities of sediment down the submarine valleys and hence contributing actively to sedimentation in the fans, troughs, and basins on their seaward margins.

INTRODUCTION

The rapid expansion of the field of marine geology has lagged somewhat in the study of bottom currents. This gap in our expanding knowledge of the ocean and its floor was apparently the result of being a borderline field lying between physical oceanography and marine geology. Since only a few theoretical papers had touched on the field in the physical oceanography literature and since oceanographers had developed very efficient current meters that allowed investigators to observe the currents along the sea floor, it seemed desirable for marine geologists to begin to fill this gap. This we started to do in 1968 with our canyon current-meter project at Scripps Institution of Oceanography and have now assembled our results with the hope that they will at least help geologists to understand more clearly what is happening on the sea floor in these valleys along which so much sediment is being transported to basins of deposition.

With the help of Neil Marshall, Gary Sullivan, and Patrick McLoughlin in both field and laboratory, I have now gathered a large amount of information involving some 200 records that include about 25,000 hours of continuous measurement of the currents in 25 different submarine canyons and other types of sea valleys from various parts of the Pacific and Atlantic Oceans. This is by far the most comprehensive project of this type.

This article summarizes ten years of current-meter operations in submarine canyons and hence includes some material that has been published previously, along with considerable updating of most of the information and the ideas that have been derived from it.

Our current meters are the Isaacs-Schick Savonius rotor type of freefall vehicle which is dropped into the valley axes and suspended from the weight at the bottom by floats until an explosive release severs a nylon cord and allows the floats to raise the meter, or meters, to the surface where an attached buoy with radio and flashing light allows recovery (Fig. 1).

Our work, starting as a small-boat operation used in the canyons directly adjacent to Scripps Institution of Oceanography in La Jolla, California, has expanded so that it now includes submarine canyons from various parts of the world as well as from a wide area off the California coast (Fig. 2). Much of this work has become possible through the cooperation of many other oceanographic agencies and institutions as the result of invitations to join their expeditions or their local operations in order to obtain our records at relatively low cost.

ALTERNATING UP- AND DOWNVALLEY CURRENTS.

It is perhaps rather surprising that prior to our investigations it was not known that there are almost continuous currents flowing alternately up and down the axes of the various types of valleys of the continental margins. In 1938 and 1939 a

Copyright © 1979, The Society of Economic Paleontologists and Mineralogists

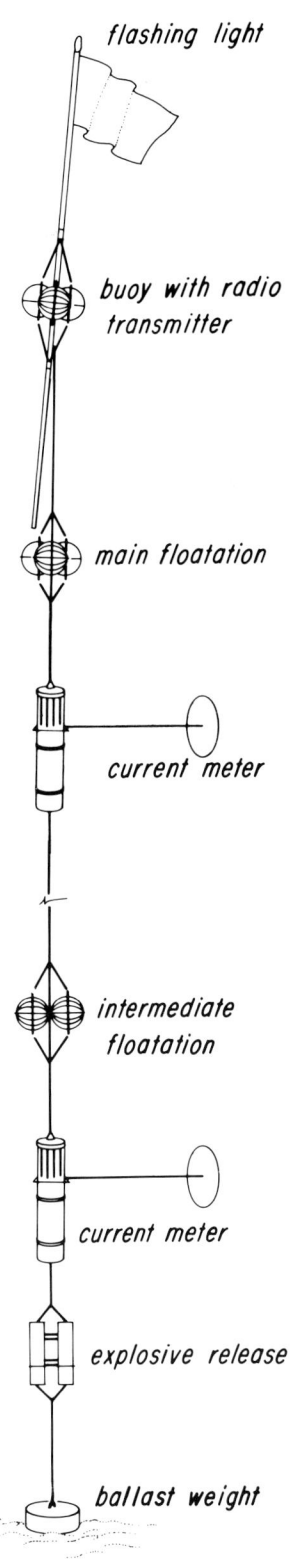

Fig. 1.—Current meters as they are deployed in the axes of valleys. The ballast weight is dropped after a prearranged interval by the explosive release, and the floats return the package to the surface with the buoy and its radio and flashing light that assist in its recovery.

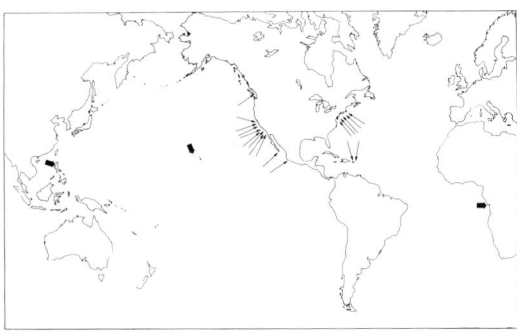

Fig. 2.—Index map showing the localities (arrows) where current-meter observations have been made during the project.

few measurements were made (Shepard et al, 1939), and we did find some reversals in directions apparently not related to tides, but virtually nothing was done until 30 years later when with better instruments we initiated the studies now being reported. As our information grew it became clearly evident that not only do these alternations occur but that they are practically continuous not only in submarine canyons and in their fan valley deep-water continuations but even may be a characteristic of currents in the fault valleys of the continental margins.

Furthermore, the currents which we would like to call *ordinary currents* rarely exceed 50 cm/sec, but they are often of sufficient speed so that from deep diving vehicles they can be seen to stir up the fine sands and silts that are so common on the floors of valleys and are often quite capable of developing ripple marks.

The cycles of reversal are clearly related in most cases to the depth of water in the valley axis and to the range of the tides in the local area (Shepard, 1976). This is demonstrated in a composite diagram which we have put together from our study of currents along the axes of 25 canyons and other types of seavalleys (Fig. 3). In this diagram the lines connecting points along valley axes have depths that are the ordinate and length of reversal cycles are the abscissa. The lines are differentiated according to tidal range. The diagram shows a general increase of cycle

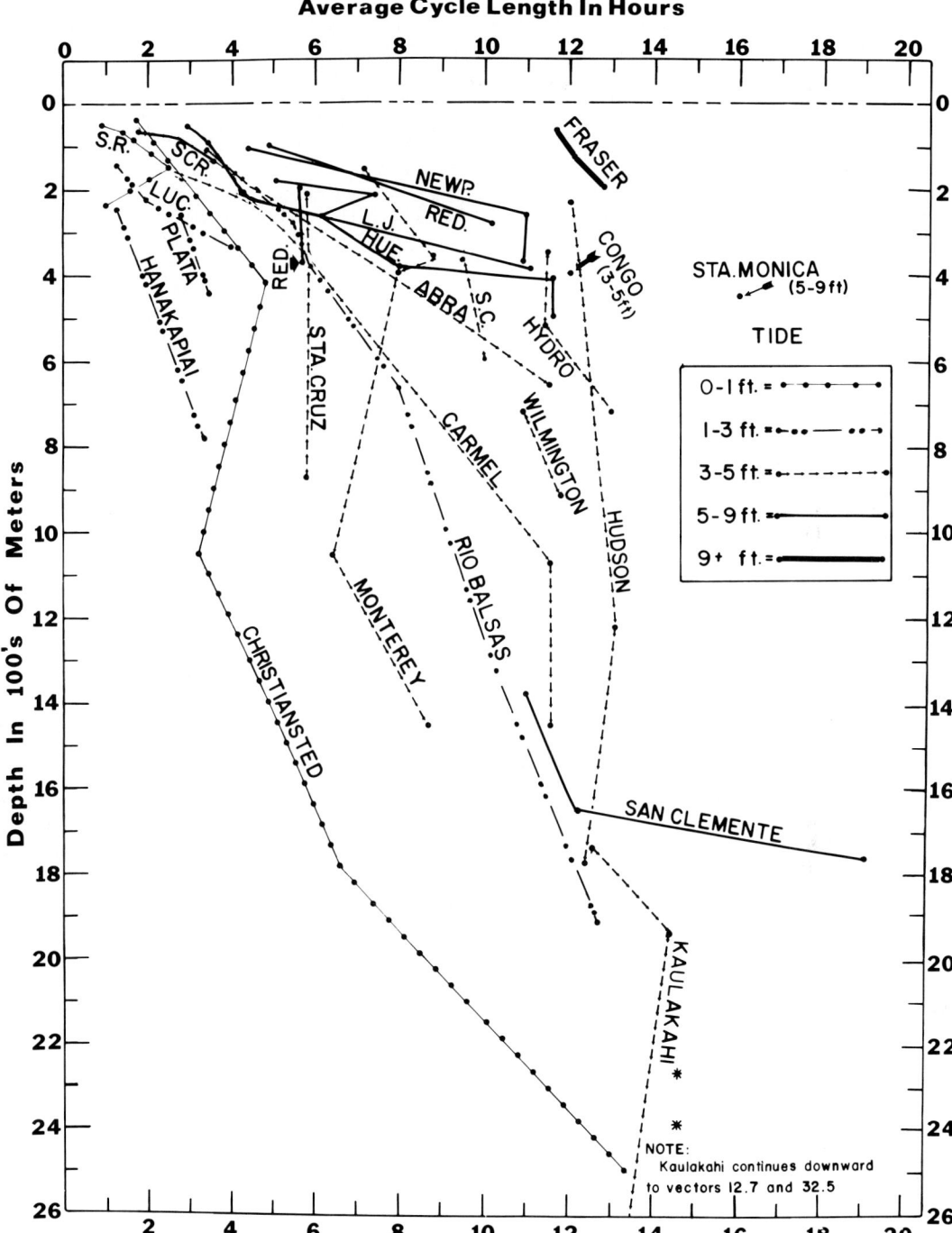

Fig. 3.—Composite diagram with abscissa the average cycle length of up- and downvalley alternation and with ordinate as the depth of the axis. The stations of each canyon or seavalley are connected by lines, the nature of which indicate the tidal range of the area (see inset upper right). Note that with few exceptions the cycle length increases with depth until it attains close to that of the semidiurnal tide (12.4 hours), but it is also dependent on the tidal range.

length with axial depth until the cycles reach about 10 to 12.4 hours and then have little if any change in length at greater depths. However, where the tides have small range (dotted lines), the depth at which this tidal relation is indicated is very great, as for example in Christiansted Canyon in the West Indies where tides are only about 0.3 m. On the other hand in Fraser Seavalley, the depth is very small (solid line) where the cycles are essentially tidal at 60 m and here the tidal range is about 5 m.

"INTERNAL WAVES" OR "INTERNAL TIDES" MOVING ALONG VALLEY AXES

Another discovery which also applies to the ordinary valley currents is that it is often possible to trace a sequence of alternations of up- and downvalley currents along a canyon axis by using a time shift as in Figure 4A where we have shifted the time-velocity curve from an axial depth of 1,445 m in Carmel Canyon until it fits rather well with that of a record at 1,070 m taken at the same period of time. In this case we seem to have a sequence that occurs some 90 minutes later at the station farther up Carmel Canyon. Similarly in San Clemente Rift Valley some 90 km off the coast at San Diego we had two stations separated by 12 km and rather different currents relative to net flow occurred at the two stations, but although the records lasted for an entire month it is quite possible to find a duplication of the general character during this entire period if we shift the time so that the arrivals are about 2 hours and 55 minutes later at the shallower station than at the deep station (Fig. 4B).

These repetitions are not always as good as the ones shown, but they strongly suggest that they are due to the advance of internal waves (or perhaps tides) along the canyon axes (Shepard, 1973); such waves are known to advance across

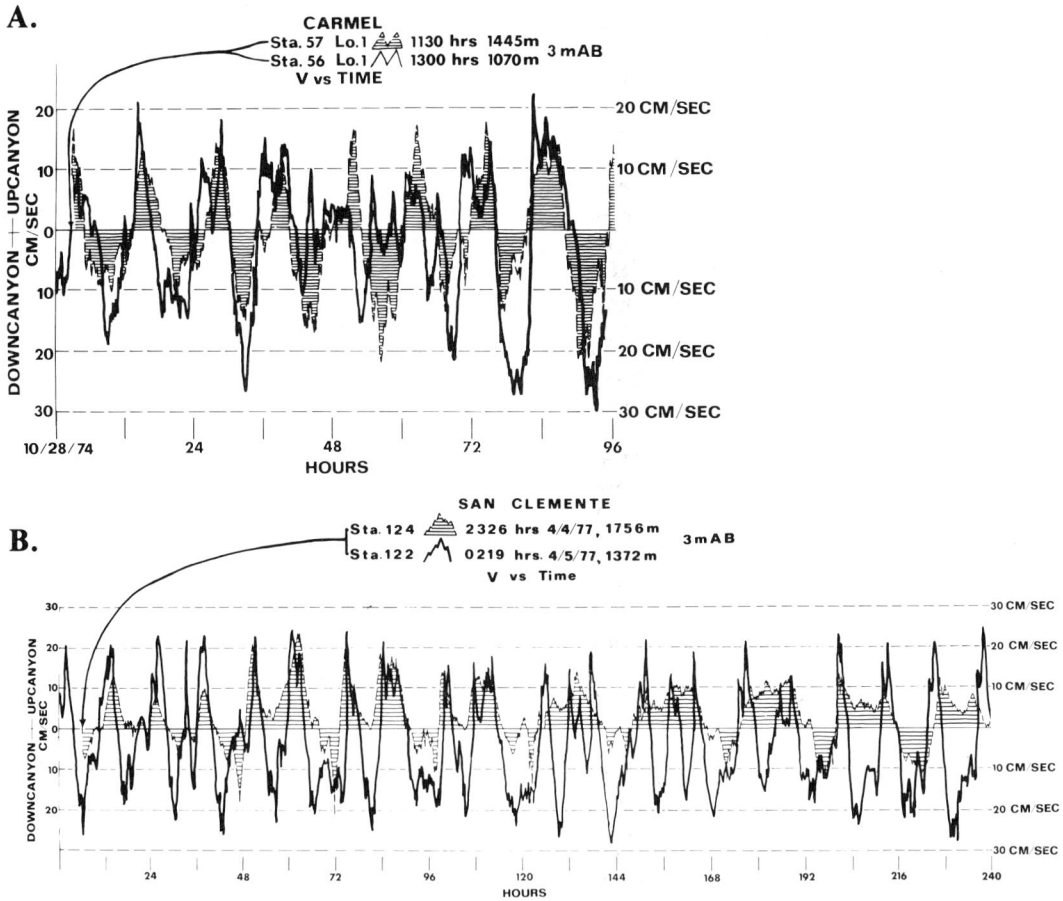

FIG. 4.—Two examples of how internal waves or internal tides advance up the axis of valleys so that the time velocity of curves of two contemporaneous records may be repeated at a later time in an upvalley station.

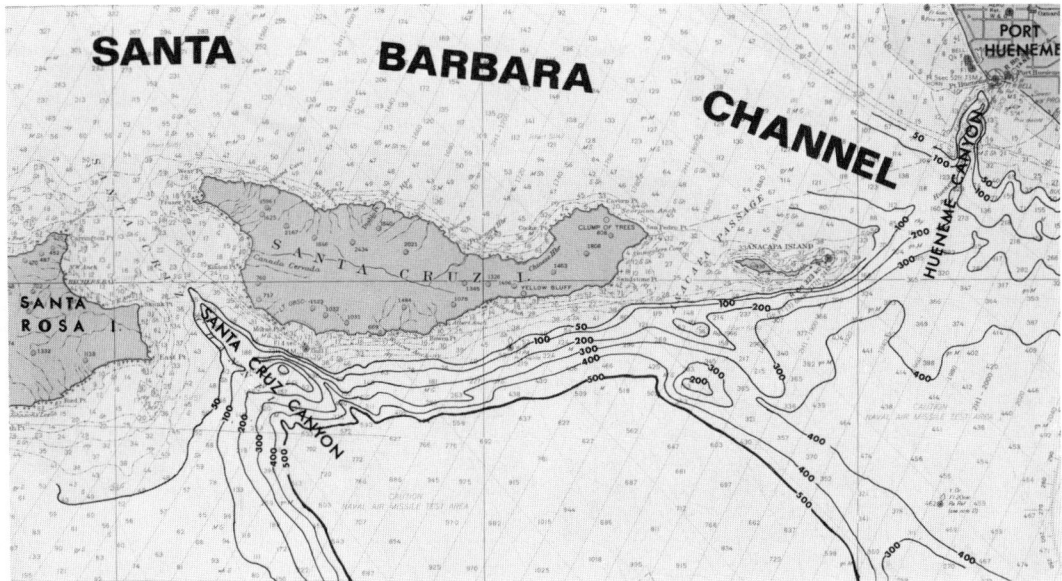

Fig. 5.—Map showing how the head of Santa Cruz Canyon is located in a strait between the two islands where large outflow of water from Santa Barbara Channel occurs during the usual strong northwest winds. This may explain the advance of internal waves down rather than up the axis of Santa Cruz Canyon which has been observed.

the continental shelves (LaFond, 1962).

One of our puzzling features in connection with these migrating internal waves is that some of them rather clearly move down the valleys instead of up as might be expected from information relative to the continental shelf. As possible explanations for this anomaly, we have the introduction of water masses into the system at the canyon or valley head that may reverse the internal wave motion. Thus the head of Santa Cruz Canyon is in the strait between Santa Rosa and Santa Cruz Islands seaward of Santa Barbara Channel along the California coast. Here the ordinary wind is from the northwest, and this funnels water through the straits to the south from Santa Barbara Channel (Fig. 5). Here we might expect internal waves to move downcanyon, and our two comparative records seem to indicate that this is actually the case. Another example is off the mouth of the Fraser River where flood waters introduce an enormous quantity of water into the head of Fraser Seavalley that extends down the delta front where again we have internal waves apparently moving down the valley axis. The effect of the mass of fresh or brackish water moving into the system may be responsible for this reversal of the ordinary upvalley advance of internal waves. It would be advisable, however, to test other cases before too much support is given to this explanation.

TURBIDITY CURRENTS.

The discoveries of greatest importance to geology coming from the current-meter studies are certainly those related to relatively fast-flowing downvalley currents which we interpret as turbidity currents. During the first few years of our studies we were of the opinion that we would obtain very little information regarding turbidity currents unless we were sucessful in starting such a flow by artificial methods. We knew that they occurred from time to time in Scripps Canyon because of loss of instruments both by Inman (1970) and ourselves during storm conditions. We supposed that the turbidity currents were so strong that they would carry away our current meters, and we knew that great masses of kelp found on the bottom would aid this process at least in the La Jolla canyons. Finally in 1972 we had two current meters in La Jolla Canyon when a storm developed with wind gusts up to 33 knots. The current meters failed to surface at the allotted time so we imagined they had been swept away. However, we had the use of a *Nekton* submarine a few days after they were activated to come up and after diving on the site we moved down La Jolla Submarine Canyon looking for the meters. In half a kilometer, we observed the floats that lift the current meters attached to a nylon cord and extending out of a mass of tangled kelp. Pulling on one of the floats, fortunately we were able to extract the current meters from the kelp so we could bring them to the surface. We discovered that the instruments were not damaged by the journey and the records were intact. We

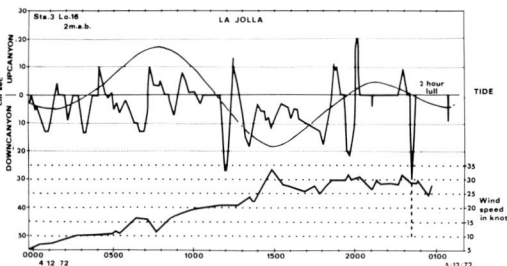

FIG. 6.—The time-velocity curve in La Jolla Canyon that preceded the transport of current meters down the canyon by a turbidity current. Lower curve shows the wind speed as a storm developed. The rather strong downcanyon surge occurring about midnight of the second day is thought to represent a small turbidity current with speeds measured up to 50 cm/sec and followed by no current for more than an hour. The record was apparently stopped when kelp entangled it, and the current meter was carried down the canyon for 0.5 km where it was later found.

found from the record that during the heart of the storm a small turbidity current had occurred with downcanyon velocities of 50 cm/sec (Fig. 6) and that this was followed by a gap with no currents and then a short period with renewed currents that terminated the record, clearly representing the time when kelp had stopped the turning of the rotors and this had been followed by the transportation of the current meters downcanyon. Here was our first evidence that some turbidity currents might not be of a violent nature since the current meters were not damaged (Shepard and Marshall, 1973).

Beginning in 1975 we have developed a new realization of the possibility of measuring turbidity currents. In April we visited the first of 4 areas where submarine canyons or seavalleys head directly off the mouths of rivers and where kelp or other entangling sea grasses are of no importance. In three of these areas our short time measurements (just a few days) produced downcanyon currents of from 70 to 100 cm/sec during the relatively short periods that the current meters were emplaced. All of them are quite distinct from ordinary valley currents, and we are convinced that they represent a type of turbidity current (Shepard et al, 1977).

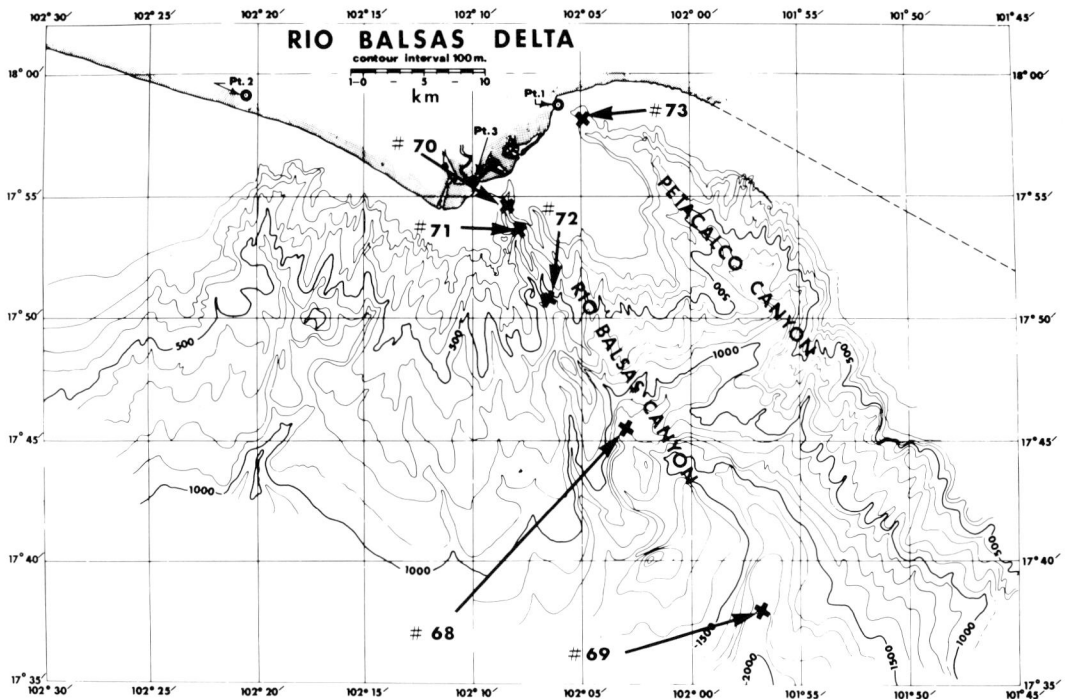

FIG. 7.—Sea floor off the delta of the Rio Balsas along the west coast of Mexico showing the various canyons and seavalleys which cut the advancing prodelta slope. The locations of the current meters are shown. The turbidity current was recorded at Station 70 in Rio Balsas Canyon.

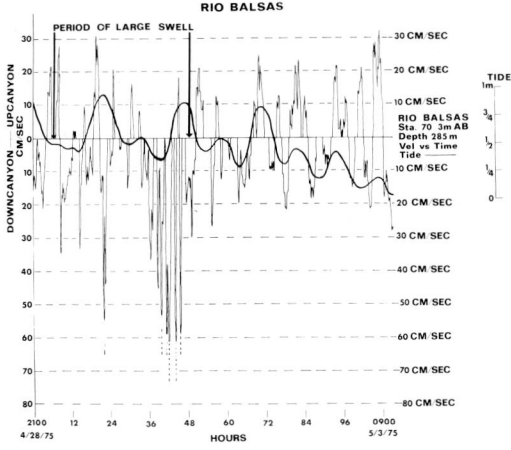

Fig. 8.—The time-velocity curve at 285 m in Rio Balsas Canyon during and after a period of large swells when turbidity currents developed as a single surge followed by a series of large surges. Normal currents occurred both before and after the time of the large swells. The tide gauge record was damped by being covered with sediment towards the end of this period, but earlier the surges seem to be related to times of high tide, which may have occurred earlier at the gauge in the river mouth than in the canyon.

First we used a Scripps ship for work off the Rio Balsas Delta where we found that canyons and seavalleys crease the slope of the forward-building delta (Fig. 7). These resemble the valleys off the passes of the Mississippi Delta (Shepard, 1955) and off the Magdalena Delta of Colombia (Heezen, 1956; Shepard, 1973). Apparently these canyons are found wherever deltas have built beyond the continental shelf. In Rio Balsas Canyon we had a current meter in 285 m of water at a time when the swell from the open ocean showed a decided increase in size. Our records showed a series of downcanyon surges confined entirely to the period of these large swells (Fig. 8). These surges appear to be associated also with times of high tides. The first pulse was isolated and was followed by a lull and then by normal currents for 12 hours, but the others came at rather irregular intervals after short gaps, the longest 69 minutes. Each surge built up rapidly, and most of them had a much slower decay. The current meter had a coating of sand although the floats should have prevented it from touching bottom.

The most interesting feature of these turbidity-current pulses is that they very likely occur whenever large waves of any type come into the area. This is indicated by the fact that similar currents were observed by Reimnitz (1971) in a tributary to Rio Balsas Canyon during another

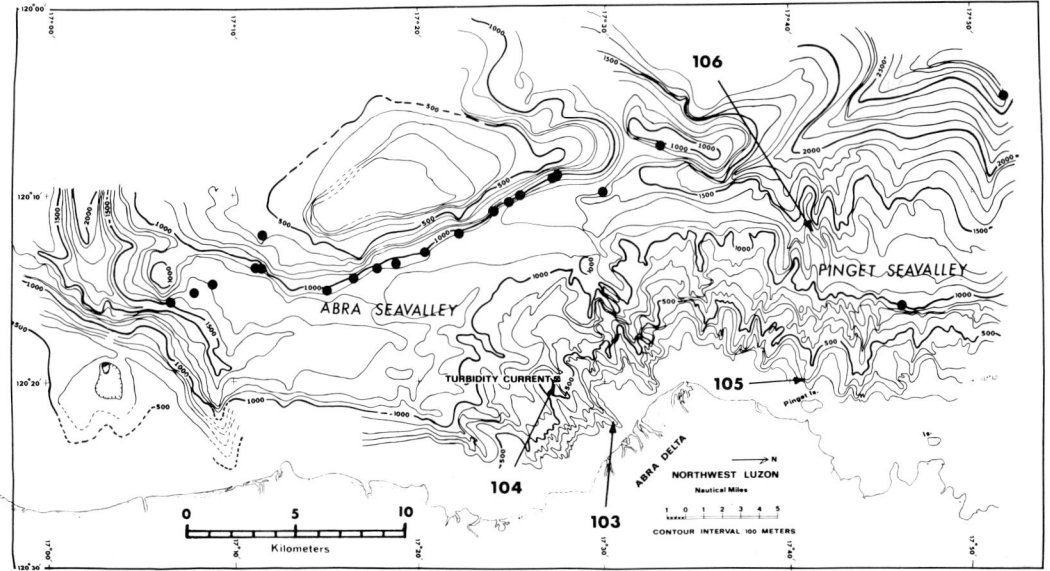

Fig. 9.—The sea floor off the Abra Delta of northwest Luzon showing the seavalleys in the foreset delta slope. Current-meter stations are shown. In Abra Seavalley a turbidity current occurred at Station 104. Beyond the delta slope a fault valley is shown, and the circles indicate positions where seismic profiles indicated underlying faults mostly at the steep inner margin of a bank.

period when the normal swell had shown a considerable increase. Thus in the only two episodes when it was possible to observe such turbidity currents they occurred suggesting this is a usual result of the production of relatively large turbulence in the shallow water at the canyon head. This frequent occurrence could be a very important factor in the transportation of sediment down the submarine front of the Rio Balsas Delta.

Our next investigation of the valleys bordering deltas was off the Abra Delta of northwest Luzon. Here again we have a delta building out the continental slope and again there are several valleys that cut the slope (Fig. 9). Our investigation of the area on the Scripps ship R/V *Thomas Washington* only lasted about 10 days, and we had current meters in the canyon off the main river mouth for only 4 days. However, this was during a time of high rainfall in the neighboring mountains and the Abra River was pouring quantities of sediment into the ocean as we could see from the clouds of muddy water extending well out into the Philippine Sea.

The record from our current meters at 622 m in Abra Seavalley again gave us a considerable surge that can only be interpreted as a turbidity current (Fig. 10). This took place 80 hours after the start of the record and was indicated both at 3 m above bottom and at 30 m, with maximum speeds of 73 cm/sec at 3 m and 55 cm/sec at 30 m. Both instruments came up well coated with sandy sediment although the floats must have kept them both well clear of the bottom. The strong surge was preceded again by a somewhat unusually strong upvalley flow, and after the surge there were long periods with no measurable current, lasting 14.5 hours at 30 m and lasting 3.5 hours at 3 m. This is a remarkably long time at 30 m compared to our many other records at this height where there are almost no lengthy gaps in the current flow. There seems little doubt that the turbidity current completely threw out of phase the current-producing causes. But the most significant discovery is perhaps that the currents must have had a thickness in excess of 30 m in order to effect the higher current meter.

In January, 1977, we had still another opportunity to measure currents off a river mouth although in this case with no appreciable delta. We planted a current meter in a small canyon at 238 m and 2 km off the mouth of the Rio de la Plata (Fig. 11). Here we encountered unusually strong currents of the ordinary type flowing alternately up- and downcanyon at the short alternation period characteristic of an area of small tides. After the

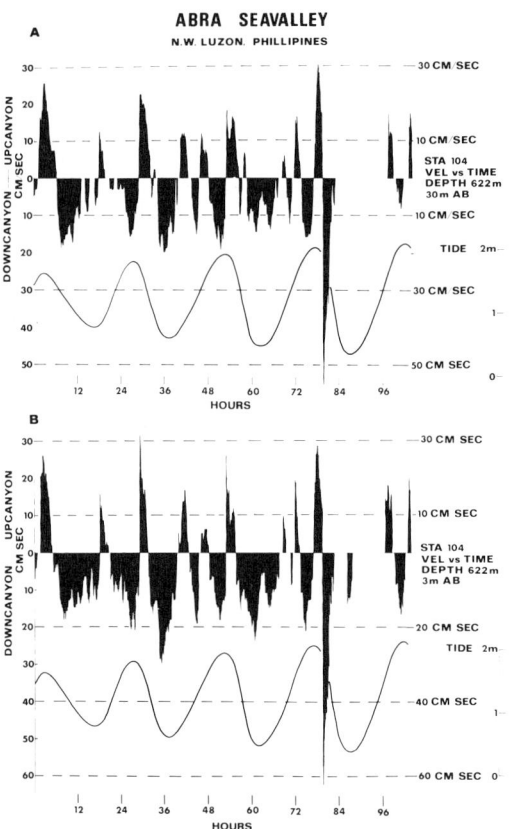

FIG. 10.—Time-velocity curves of the record in Abra Seavalley showing clear example of a turbidity current that interrupted the ordinary up- and downvalley alternations at hour 80 both at 3 m and 30 m above bottom. Note the large upvalley current that preceded the turbidity current and the long interval with no measurable current that followed. The underlying curves show the tide.

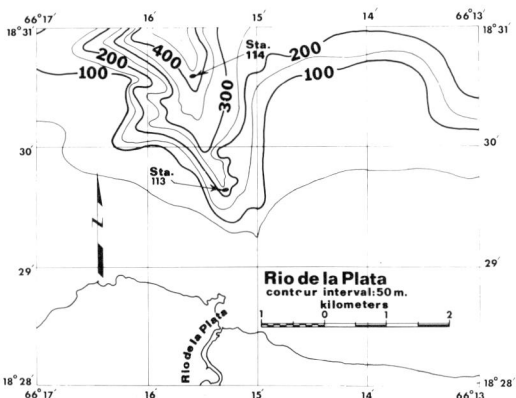

FIG. 11.—The canyon off Rio de la Plata on the north coast of Puerto Rico where a turbidity current occurred during a flood period. Location of the current-meter station is indicated.

current meter had been operating about 58 hours there was another example of a strong downcanyon surge, this one attaining a speed of about 100 cm/sec (Fig. 12). Preceding this fourth example of a turbidity current we had a fairly large upcanyon current as in the other cases. What happened after the turbidity current decayed we do not know because unfortunately the record was only a short one and the release allowed the floats to bring the instrument back to the surface as soon as the surge had died down.

The most interesting feature of this Rio de la Plata turbidity current is that it occurred just two days after the river gauge showed the crest of a flood, strongly suggesting that here an unusual amount of sediment carried into the canyon head had started a slide on the steep slope which initiated the turbidity current.

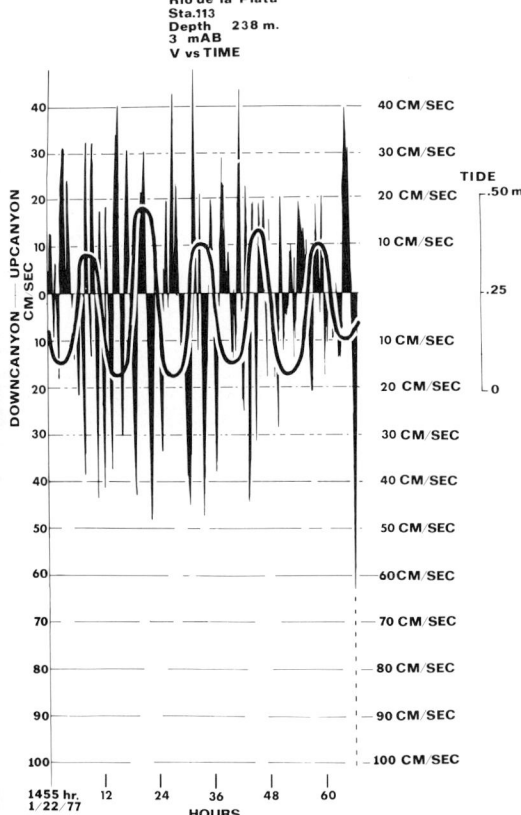

Fig. 12.—Time-velocity curve in Rio de la Plata Canyon at 238 m, showing unusually large ordinary currents in both up- and downcanyon directions that are followed by a turbidity current at hour 66. The latter followed a relatively large upcanyon flow. Unfortunately the current meter was set to return to the surface just after the turbidity current had subsided.

Among our recent operations we have included measurements off the Fraser River mouth where again great quantities of sediment are being introduced and again there is no kelp. The valleys that exist off the mouth of the Fraser must also be receiving great quantities of sediment. Here, however, we failed to obtain records of turbidity currents in spite of measurements made during two periods of relatively high river flow. A possible explanation for absence of the strong surges here is that the sea floor at the mouth of the Fraser is undergoing extensive dredging which may remove the sediment before it builds up unstable slopes that result in slides such as would cause turbidity currents.

Summarizing our information on turbidity currents it would appear that they may be of frequent occurrence wherever rivers are dumping sediment into canyon heads, but our results also seem to indicate that the typical turbidity current is not the high-speed flow such as has been often considered but that it is a relatively slow current with, however, sufficient speed to transport great quantities of sediment.

ANCIENT TURBIDITES AND SLOW TURBIDITY CURRENTS

In view of this evidence that we have obtained relative to the slow turbidity currents as against the great torrents that are often pictured as typical turbidity currents, the vast literature about turbidites may throw some light on this subject. One gets the impression that the most typical turbidites are sand layers, often fine-grained sands, that alternate with mudstones and grade upward from sand to mud. These are certainly what we usually find in our cores of sea-floor sediments. Are not these just what one would expect to be transported by currents that rarely exceed speeds of one or two knots? Is it not quite possible that the typical turbidity currents of the past were to a great extent the result of such slow currents set off by slides on the unstable slopes that extended down into the troughs and basins of the past?

On the other hand, there can be little doubt that occasionally torrential turbidity currents do occur. How, otherwise could we find gravel and even boulders in deposits even in the outer fan valleys at very great depths or the unsorted conglomerates among the ancient turbidites which have been carried well out into sedimentary basins. Great earthquakes such as that of the Grand Banks in 1929 probably produce such major flows. Large submarine landslides may also set up large-scale turbidity currents with great transporting power. Unfortunately our studies at Scripps Institution of Oceanography have not thrown any light on events of this sort, and we would have little chance of recording them with current meters since the latter would be carried away. However, it should be possible to obtain records of at least moderately large turbidity currents, as suggested by Neil Marshall, by suspending transducers from wires anchored to the canyon walls and suspended across the canyons at two or more points so that the transducers could record the passage of turbidity currents down the canyon floors. Transducers are quite capable of obtaining echoes from the top of the advancing front of the turbidity currents and two or more along the axis can provide a good estimate of their speed.

ACKNOWLEDGMENTS

The work was made possible through National Science Foundation Grants GA 826, GA 11463, GA 19492, OCE74-22089, and OCE77-17570 and through Office of Naval Research Contracts USN N00014-69-A-0200-6006 and USN N00014-75-C-0152. The current meters were operated by Neil F. Marshall, Gary G. Sullivan, and Patrick A. McLoughlin, who were also responsible for a large part of the laboratory study of the records, which led to the construction of the various graphs depicting the currents.

REFERENCES

HEEZEN, B. C., 1956, Corrientes de turbidez del Rio Magdalena: Bol. Soc. Geog. Colombia, nos. 51, 52, p. 135-142.
INMAN, D. L., 1970, Strong currents in submarine canyons: Trans. Am. Geophysical Union Abs., v. 51, p. 319.
LaFOND, E. C., 1962, Internal waves, in Hill, M. N., ed., The Sea: Interscience, v. 1, p. 731-763.
REIMNITZ, E., 1971, Surf-beat origin for pulsating bottom currents in the Rio Balsas Submarine Canyon, Mexico: Geol. Soc. America Bull., v. 82, p. 81-90.
SHEPARD, F. P., 1955, Delta-front valleys bordering the Mississippi distributaries: Geol. Soc. America Bull., v. 66, p. 1489-1498.
———, 1973, Sea floor off Magdalena Delta and Santa Marta area, Columbia: Geol. Soc. America Bull., v. 84, p. 1955-1972.
———, 1976, Tidal components of currents in submarine canyons: Jour. Geology, v. 84, p. 343-350.
———, AND MARSHALL, N. F., 1973, Storm-generated current in a La Jolla submarine canyon, California: Marine Geology, v. 15, p. M19-M24.
———, McLOUGHLIN, P. A., MARSHALL, N. F., AND SULLIVAN, G. G., 1977, Current-meter recordings of low-speed turbidity currents: Geology, v. 5, p. 297-301.
———, REVELLE, R. R., AND DIETZ, R. S., 1939, Ocean-bottom currents off the California coast: Science, v. 89, p. 488-489.

STRUCTURE OF THE CONTINENTAL SLOPE OFF THE EASTERN UNITED STATES

JOHN S. SCHLEE, WILLIAM P. DILLON, AND JOHN A. GROW
U.S. Geological Survey
Woods Hole, Massachusetts 02543

ABSTRACT

The continental slope off eastern United States is built across a zone of deeply buried (6–12 km) fault blocks that formed during the rifting of continents. Associated with these blocks are intrusives, salt diapirs, and reef complexes; all are underlain by a zone of thinned continental, transitional, and oceanic crust beneath the shelf, slope, and upper continental rise. Four basic sedimentary units appear to underlie much of the slope: (1) a probable terrigenous clastic-evaporite-volcanic sequence associated with fault blocks; (2) a carbonate platform reef sequence, which probably formed along the ancestral shelf-slope break; (3) a marine and non-marine sequence built over the carbonate rocks north of the Blake Plateau as the shelf in some areas prograded seaward during the Cretaceous; and (4) a thick rise prism of sediments of Tertiary and younger age that laps up on the base of the slope north of the Blake Spur. Much of the present slope is an erosional surface, cut into lower Tertiary and Upper Cretaceous strata during Cenozoic marine regressions as sediment moved downslope to build the rise prism; the shelf-slope break has retreated 5–30 km from the break that existed during the Late Cretaceous. Over the Blake Plateau, periodic scour by currents associated with the ancestral Gulf Stream caused a 300-km landward jump of the shelf break at about the end of Paleocene as the plateau continued to subside.

Laterally, the slope varies in the distance that it retreated (maximum to the south) and in the degree to which reeflike masses are developed. Continuous well-developed carbonate banks or reefs controlled location of the shelf edge of the southern Blake Plateau; they were terminated both there and to the north near the end of the Early Cretaceous, although reef formation may have shifted to locations beneath the present continental shelf.

Local progradation of the outer shelf and slope occurred during parts of the Jurassic and Cretaceous, depending on the sediment supply and the effects of transgression and regression of sea level. During the Tertiary, however, erosion and retreat of the shelf edge was dominant.

INTRODUCTION

Continental slopes bridge the relatively steep (4° world average, Ross, 1977) area between the continental shelves and the ocean basins, and their origin has always been somewhat of an enigma. The presence of such a marked declivity makes one wonder if this is a recent feature or is inherited from earlier stages of margin development. Figure 1, a multichannel reflection profile taken east of Virginia, shows the present slope and also indicates that an earlier formed slope, now buried, existed seaward of the present slope. The purpose of this paper is to explore the structure and stratigraphy of the Continental Slope along the entire U.S. Atlantic margin through the medium of 12 multichannel seismic reflection profiles. We wish to infer the basic structure and stratigraphy beneath the slope on the basis of analysis of the character and arrangement of the reflectors and their interval velocity, and data from shelf and deep-ocean drill holes. Using the patterns of gravity and magnetic anomalies, we shall infer what the deep crustal structure may be below the slope sedimentary sequences.

Along the Atlantic margin, profiles like the one shown in Figure 1 have been used to work out the structural framework of marginal basins, infer ages of prominent groups of reflectors, and interpret the types of rocks present and their geologic history (Schlee and others, 1976, 1977; Jansa and Wade, 1975; Given 1977; King 1975; Mattick and others, 1978; Shipley and others, 1978; Dillon and Paull, 1978; Dillon and others, in press; Dillon and others, 1979; Grow and Markl, 1977; Grow and Schlee, 1976; Grow and others, 1979; Buffler and others, 1979; Klitgord and Behrendt, 1979). In the mid-Atlantic and Georges Bank area, Schlee and others (1976) inferred the ancient shelf-slope are to be a probable carbonate reef-platform complex on the basis of the high interval velocity shown by the structure, the loss of reflectors in it, the development of a reef "flank" on the seaward side, and the bowed cross-section outline it displays on the profile.

Buried ancient carbonate buildup under the outer Continental Shelf was inferred by Emery and Uchupi (1972) from a profile off Cape Hatteras. Early refraction studies by Drake and others (1959) revealed velocities greater than 4.5 km/second at about 2–5 km beneath the outer shelf; though they interpreted rocks at this depth to be basement, the layered appearance on seismic reflection profiles led to the interpretation that the rocks are limestone (Sheridan, 1974, 1976; Behrendt and others, 1974; Mayhew, 1974; Schlee and others, 1976). Thin high velocity horizons

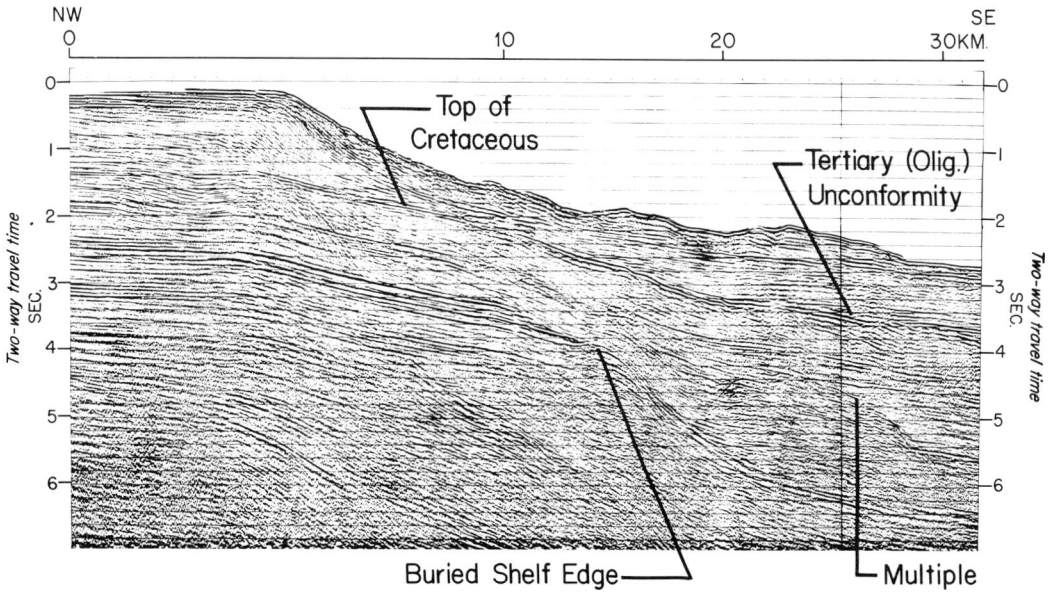

Fig. 1.—Multichannel seismic profile of line 11 across the outer shelf, slope, and upper rise, (southern Baltimore Canyon trough). Vertical exaggeration is 4:1 for topography. Location shown on Figure 2.

(5.0 km/sec) 2.6 km beneath the shelf edge off New Jersey have been shown to correlate with a sedimentary horizon that is the top of the Lower Cretaceous.

Ancient reef limestones have been dredged on the Blake Escarpment (Heezen and Sheridan, 1966; Sheridan and others, 1969). They are Lower Cretaceous oolitic calcarenite and chalky calcilutite. From Deep Sea Drilling Project (DSDP) holes 390 and 392 (Benson and others, 1976) samples were taken of shallow-water reefal or platform limestone of Early Cretaceous age on the Blake Nose (2600 m water depth). On the southeast side of Georges Bank, Ryan and others (1978), using a submersible, sampled Lower Cretacous reef limestones in "Heezen" Canyon (near Corsair Canyon) at 1200 m water depth.

Some areas of the Continental Slope show evidence of erosional truncation of reflectors, major unconformities, thinning of sequences, and slumping downslope. Such evidence of slope retreat is concentrated mainly in the youngest (Cenozoic) phase of slope development (Grow and Markl, 1977; Grow and others, 1979),

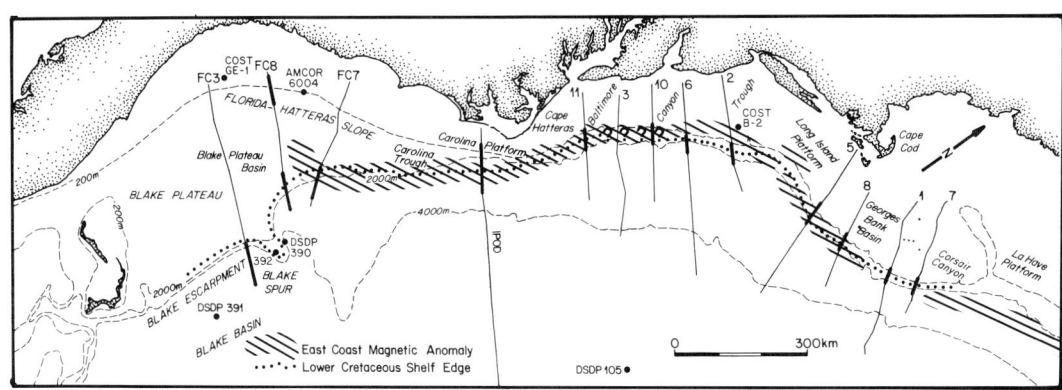

Fig. 2.—Index map of U.S. Atlantic continental margin showing multichannel lines discussed in this paper and the main structural features. The dark heavy line segments indicate the parts of the lines that are illustrated in the report.

where the retreat resulted in the beveling of more extensive Cretaceous and Tertiary strata beneath the slope. Parsons (1975) showed similar evidence of major erosional truncation beneath the Laurentian Fan off eastern Canada, where paleosubmarine canyons cut rocks as old as Eocene. Under the slope off northwest Africa, Seibold and Hinz (1974) delineated two major disconformities, which they inferred as middle Cretaceous and Oligocene in age. DSDP drilling in the area at hole 397 (Ryan and others, 1976) showed major hiatuses beneath the West African slope (DSDP hole 369) in the Eocene, Paleocene, and Middle Cretaceous (von Rad and Einsele, in press). Because of the contrast of slope erosion and slope aggradation, King and Young (1977) classified paleocontinental slopes of the Canadian-Atlantic margin into destructional and constructional types, mainly on the basis of arrangement of reflectors beneath the slope.

PROFILE DESCRIPTION AND INTERPRETATION

From 1973 to the present, the U.S. Geological Survey has contracted to collect 20,500 km of common depth point (CDP) multichannel profiles over the Atlantic margin, a few of which are used here (Fig. 2). They were collected mainly by geophysical service companies, but some were collected by the French Petroleum Institute and the University of Texas. Most of the profiles (Fig. 1, 4, 5, 11, 14, 15, 16) are plotted in distance (km) and two-way travel time (seconds). This type of display tends to distort the record in that the deeper older part of the section is shortened and some of the shallow reflectors near the slope surface appear to bend subparallel to the slope although they are probably nearly horizontal (see Tucker and Yorston, 1973, p. 23-26). The problem can be rectified by converting the time sections into depth sections (Grow and others, 1979); only 4 of the 12 profiles are exclusively time sections.

We have attempted to show the relationship between structure of the slope and the main tectonic features of the margin; therefore, discussion of the profiles will be related to the main basins and platforms (Fig. 2) under the shelf. For the slope part of each profile, we shall summarize the arrangement of reflectors at shallow and intermediate depth, the change beneath shelf and rise, the inferred ages of key reflectors, and the presence of major hiatuses. Much of the mapping of key horizons on the shelf part of these and other lines has been based on placing the horizons on major unconformities. The technique is described by Vail and other's (1977) and is of significance because some of the unconformities are dated as the same age regardless of the margin on which they occur. Thus, utilizing key holes on the margin, the stratigrapher can infer the age of different intervals between major unconformities over wide areas of the margin (Fig. 3).

In the deep sea, DSDP holes drilled on Leg 11 (99, 100, 101, 105, 106) and Leg 44 (389, 390, 391) and the early Joint Oceanographic Institutions Deep-Earth Sampling (JOIDES) program (Bunce and others, 1965) provided ages on key horizons of Late Jurassic age and younger. With the exception of a conspicuous unconformity in the lower Tertiary (Horizon A unconf.), the deep-sea horizons are related to lithologic units (top of Neocomian limestone; lower Eocene chert bed) and traced on profiles over wide areas across the continental rise.

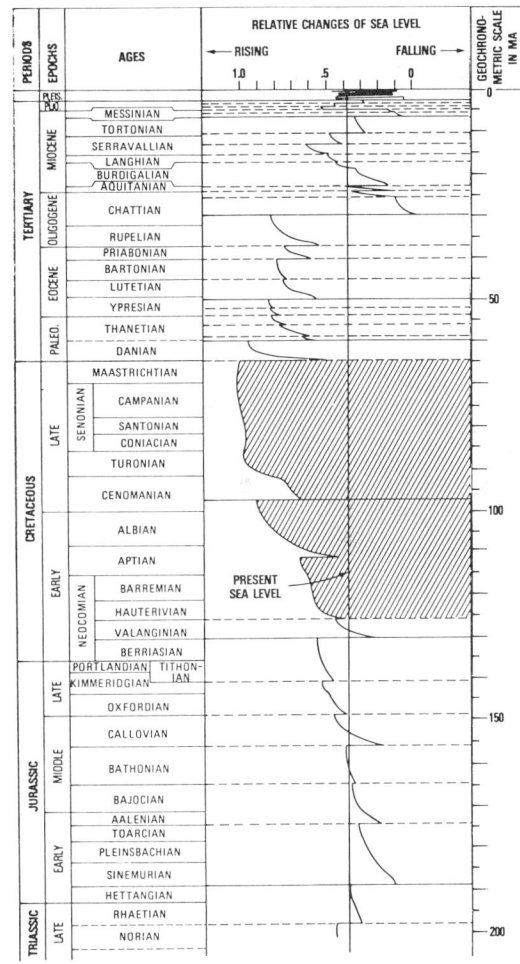

FIG. 3.—A curve showing global cycles of relative change in sea level during the Jurassic, Cretaceous, and Tertiary (from Vail and others, 1977, Fig. 2). Detailed relative changes in sea level are not shown in the cross-hatched part because that data is not released for publication.

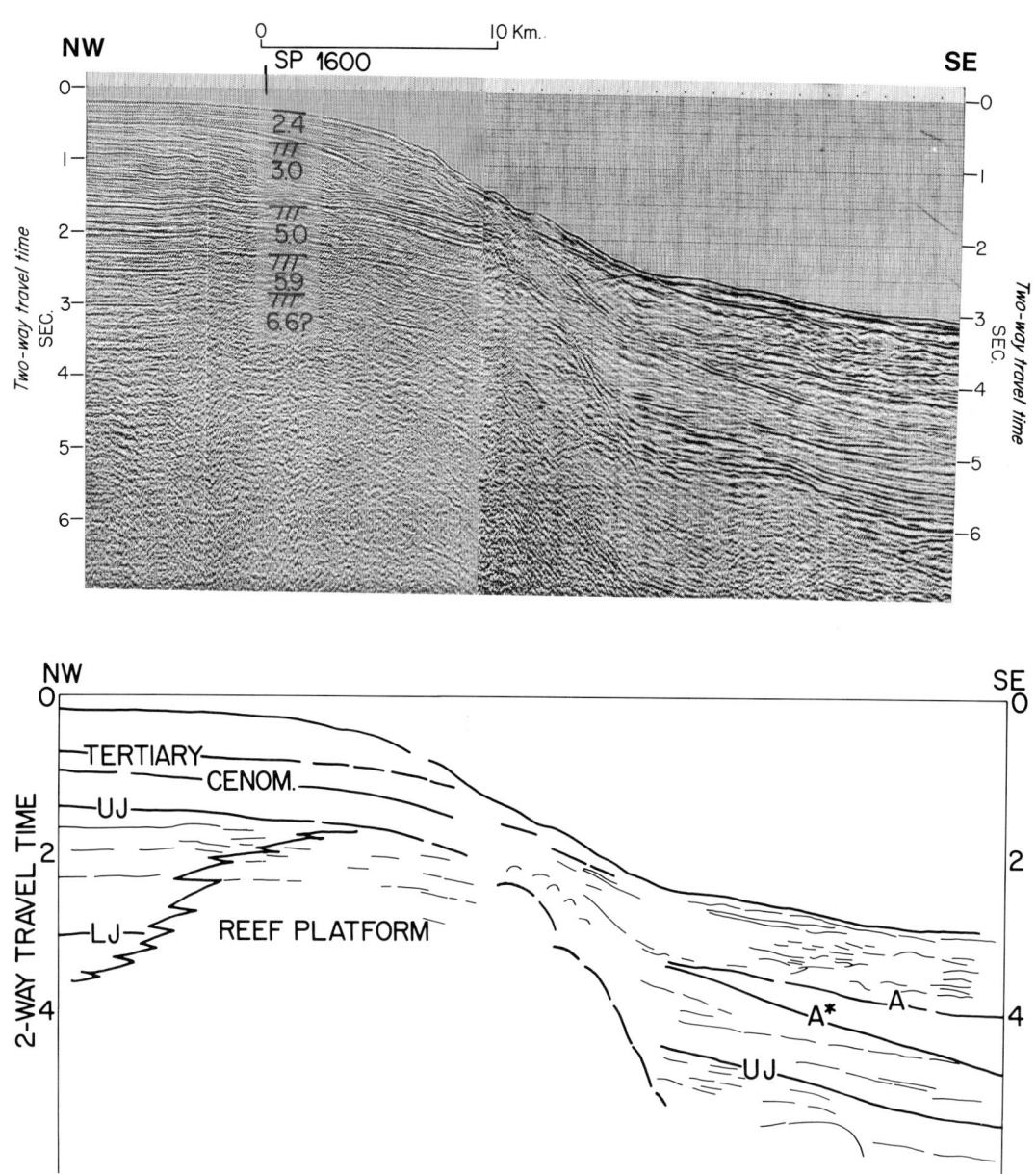

Fig. 4.—Multichannel profile of line 1, across the slope and outer shelf part of Georges Bank. Interpretation is given in 4B. Tertiary marks reflectors inferred to be near the base of the Tertiary section, Cenom, marks a hiatus within the section of Cenomanian Age, UJ delineates reflectors thought to be near top of the Upper Jurassic section and LJ marks the top of a strong group of reflectors thought to be within the section near the base of the Jurassic System. Beneath the lower slope and rise, Horizon A is inferred to approximate a chert of Eocene age and A* is inferred to be near the top of a prominent group of black shale of middle Cretaceous age. See figure 2 for location of profile. High velocity refraction values (5.0 km/sec) in sedimentary rocks, presumably limestone, occur at 1800 m depth beneath the outer shelf edge in Upper Jurassic and/or Lower Cretaceous units (Grow and others, 1978).

Georges Bank Basin

Georges Bank is built over a northeast-trending basin (Schlee 1978; Uchupi and others, 1977) that gives way to the block-faulted La Have platform to the northeast and the block-faulted Long Island platform to the southwest. The basin contains at least 10 km of Mesozoic and Cenozoic sedimentary rock under the south-central part of the bank and is inferred by a slight rise in the basement rocks (Klitgord and Behrendt, 1979) and a carbonate platform above.

Reflectors can be traced from the basin under the outer shelf (Figs. 4, 5) to the vicinity of the inferred carbonate platform (outer shelf-upper slope), where they become faint and disappear (Schlee and others, 1976). Line 1 (Fig. 4) taken across south-central Georges Bank shows a tectonic style intermediate to styles shown on profiles

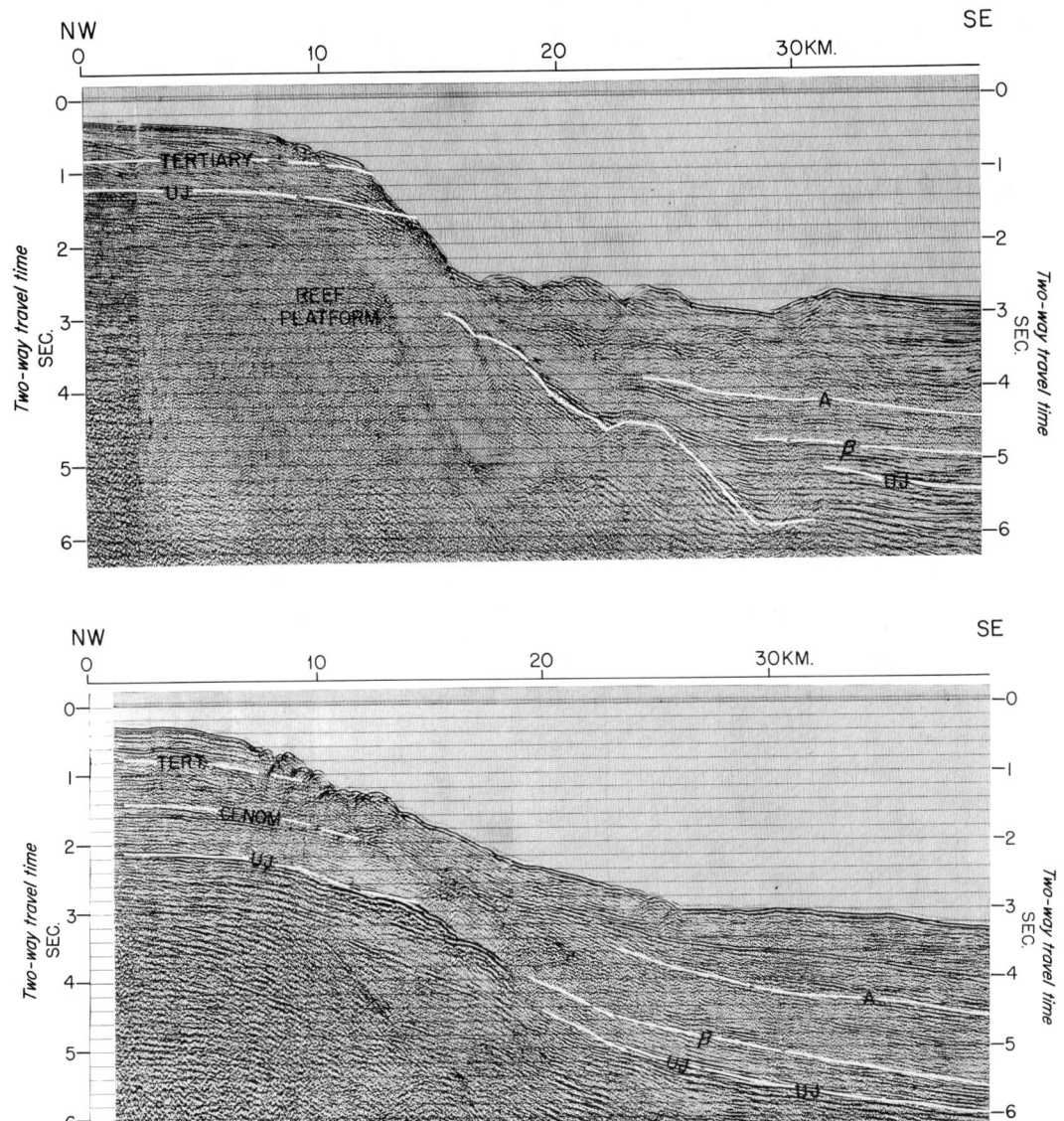

FIG. 5.—Multichannel profiles of lines 7 and 8, across the slope south of Georges Bank. A) Line 7; notice that the reef-platform complex may crop on the shelf. Horizon β is inferred to approximate the top of a group of Neocomian limestones. B) Line 8; notice that the probable reef-platform complex is buried beneath several hundred meters (1.5 seconds) of younger sediment. Location of both profiles is shown in Figure 2.

7 and 8 to the east and west (Fig. 5). Under the outer part of the shelf and slope is a poorly defined structureless mass, the upper part of which we infer to be a Jurassic-Lower Cretaceous limestone platform. The top of it is about 1.5 seconds (1.8 km) below the shelf edge. On its seaward side, the mass has a hummocky inclined front that has an apparent minimum relief of 2.5 seconds, against which the deepest reflectors (4–5 seconds) terminate abruptly. Shallower reflectors beneath the rise bend upward and lose their amplitude and continuity near the inferred platform front. Acoustic basement beneath the upper rise is an irregular surface at about 5–6 seconds (7–8 km below sea level) and is probably the top of oceanic basement (Schlee and others, 1976).

Over the top of the mass, reflectors are hummocky and discontinuous. They change laterally into the evenly layered reflectors beneath the upper part of the shelf. Beneath the lower slope reflectors show a channeled anastomosing pattern.

The inferred ages of the reflectors (Fig. 4b) show that much of the reef(?) and adjacent shelf section is Jurassic. A group of strong continuous reflectors at approximately 1.75 seconds under the slope may be Neocomian reef limestone sampled by Ryan and others (1978) near Corsair Canyon (eastern end of Georges Bank, Fig. 2) Using a research submersible, they sampled a 140 m thick sequence of limestone and calcareous sandstone (1200–1300 m water depth) which they dated by Calpionellids as Neocomian and interpreted as having formed in a open carbonate platform. The oldest strong deep-sea horizons are inferred to be reef-flank limestones of Late Jurassic age; all the deep-sea acoustic stratigraphy is extended from a broad survey by Kim D. Klitgord and J. A. Grow (unpub. data) from DSDP holes, particularly hole 105 (Fig. 2). A prominent unconformity (marked by obvious onlap) occurs 3–4 seconds below sea level (Fig. 4a), which we infer to represent a hiatus in the middle Cretaceous; horizon A* (Fig. 4b) approximates this hiatus, and, in deep-sea drill holes, it has been correlated with the top of a sequence of black claystone of Middle Cretaceous age (Ewing and Hollister, 1972). Horizon A (lower Eocene chart) is thought to be adjacent to an unconformity approximately 1 second below the rise; it loses continuity below the lower slope in the same place where major hiatuses are masked by slump deposits.

The other two profiles (Fig. 5) show many of the same features seen on line 1. Seaward of the slope on line 7, a block-faulted acoustic basement is buried by as much as 2.5 seconds of channeled and slumped rise sediment (Fig. 5a). Under the slope, deep coherent reflectors are absent so that only the upper 1.5 seconds of section show subhorizontal reflectors, which appear to represent beds that probably crop out on the slope in water depths shallower than 1200 m (1.6 seconds two-way travel time). Below them, the almost structureless mass which we infer to be a carbonate platform should crop out or be thinly veneered by slump deposits. On the other profile (Fig. 5b), the same structureless mass (marked by diffractions) is present, but it is covered by 1.5 seconds of sediment. If the correlations are correct (Fig. 5b), the platform existed into the Early Cretaceous as a feature that had as much as 2 seconds of relief (minimum 1500 m) facing a narrowed Atlantic ocean. It was a positive feature against which ancestral rises were built. Furthermore, it acted as a gradually accreting platform barrier flanking a subsiding basin to the northwest. The net shift landward in the present slope has been small, only a few kilometers.

Long Island Platform

To the southwest, the margin is built over the Long Island platform—a structural high between the Georges Bank basin and the Baltimore Canyon trough (Fig. 2). The Long Island platform is a complexly faulted area cut by northeast trending grabens. Line 5 (Fig. 6) crosses the northeastern part of the platform and is similar to the Georges Bank basin profiles except that acoustic basement beneath the shelf is relatively shallow, about 7 km (left side of profile). Reflectors above acoustic basement are fairly continuous and extend beneath the slope except for the deeper reflectors, where continuity ceases and a carbonate buildup is inferred, on the basis, in part, of high interval velocities (Fig. 6) (4.7 km/sec, Grow and others, 1979) there. Above the inferred carbonate platform is a zone of weak returns suggestive of slump deposits; beneath the upper rise, these deposits are broadly onlapped by the rise wedge of sediment. Also present just above the carbonate platform is an unconformity (Fig. 6), which bevels strata inferred to be Early Cretaceous in age. At a depth of 9 km, oceanic basement can be traced landward as far as the upper rise, where a complexly faulted transition zone underlies the outer shelf-upper slope sedimentary wedge (diffractions at a depth of 7 km, Fig. 6).

The ages of key horizons (Fig. 6) are similar to those already shown and indicate that the present slope has moved landward about 20 km from the carbonate platform front. Only the upper section in this area has been sampled (Valentine, 1978; Weed and others, 1974) in Veatch and Atlantis Canyons. The canyons flank line 5; in them, siltstone and shale of Late Cretaceous age (Santonian) and chalk of Eocene age have been obtained. Fauna suggests that an upper slope-

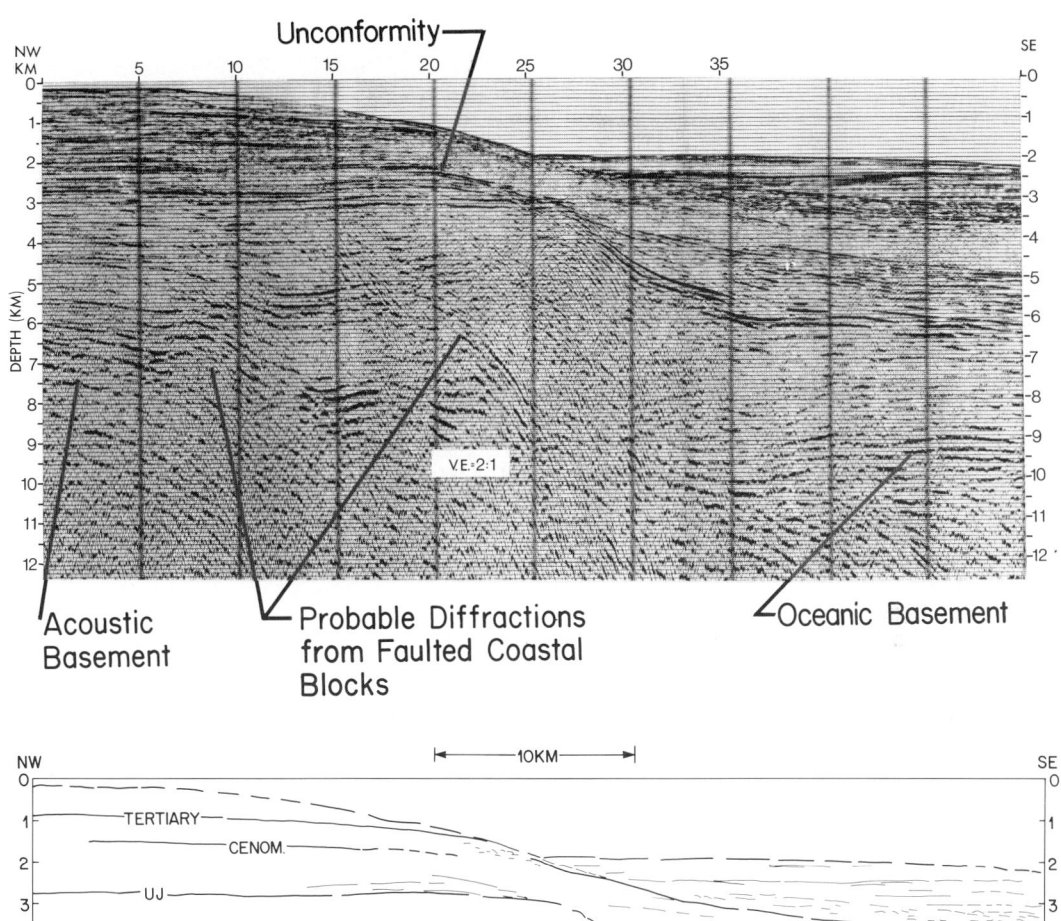

FIG. 6.—Depth section of line 5 over the slope-rise part of the profile. Interpretation is given in 6B below the section. Location of profile shown on Figure 2.

outer shelf milieu existed in the area during deposition of these rocks (Page Valentine, pers. comm., January 1978).

Baltimore Canyon Trough

In the next three profiles across the seaward side of the Baltimore Canyon trough, two of the three sections (lines 2 and 6) show the presence of an ancient carbonate platform; the third (line 10) shows an upbuilding of the sedimentary prism.

However, the platform has been complexly downfaulted, and, hence, it does not form a marked seaward slope similar to slopes elsewhere (Figs. 7–9).

Off the northern Baltimore Canyon trough (line 2, Fig. 7), the slope transition is much wider than it is to the north (40 km versus 15 km), and the record indicates that a prominent reef-like buildup is present 25 km seaward of the shelf break. The main part of the carbonate platform can be seen

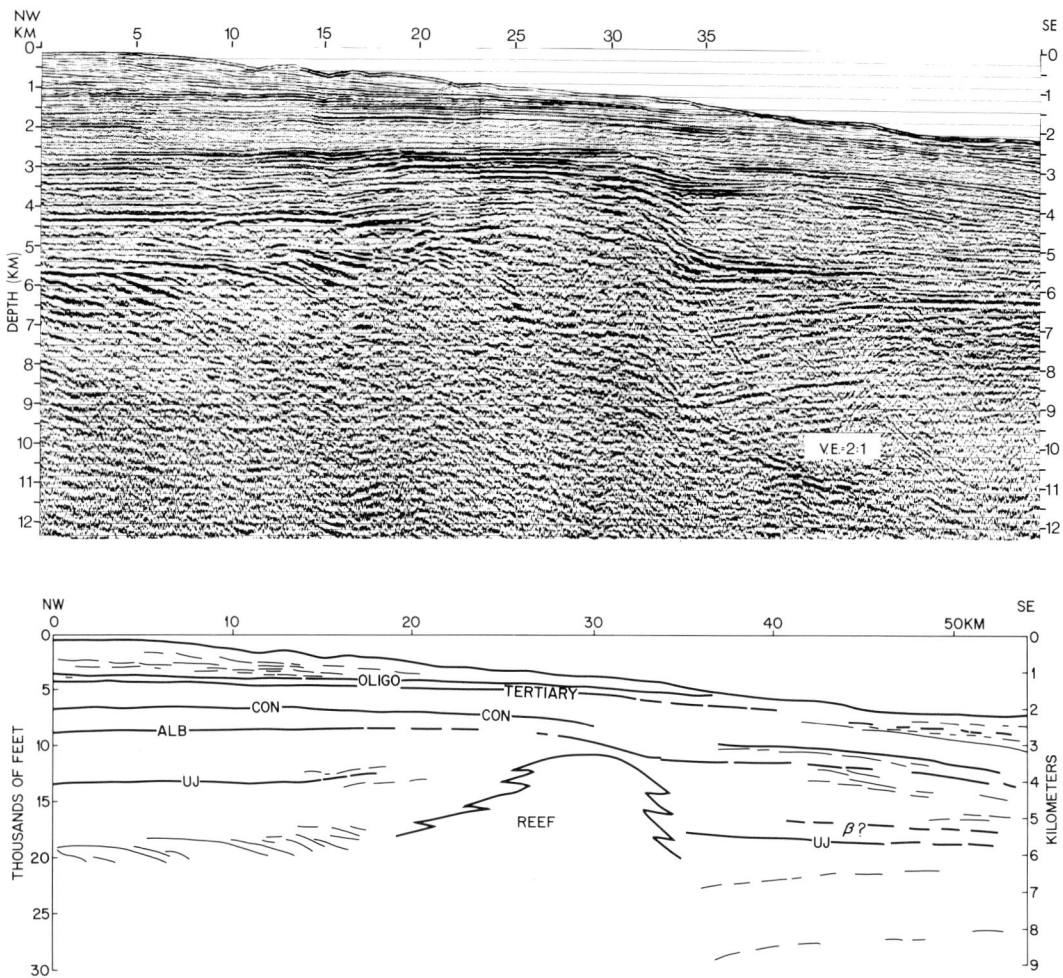

Fig. 7.—Depth section of line 2 over the slope and upper rise. On 7a notice the probable reef buildup 2 km beneath the central part of the slope. Reflectors marked A1b, Con, and Oligo on 7b are thought to be within rocks of Albian Coniacian, and Oligocene age, respectively. Location of profile shown on Figure 2.

at a depth of 5–6 km (left side of Fig. 7), where strong reflectors are arranged in a clinoform pattern dipping at a low angle to the right. Reflectors appear at shallower depths to the southeast and finally lose their identity in a reef-like mass 4 km across buried 2.5 km below the slope (Fig. 7). Again, deeper coherent returns are missing below this mass and below the inferred carbonate platform, landward of it. Scattered discontinuous reflectors are seen below 6 km, but they are under the shelf, well landward of the shelf break (Schlee and others, 1976). Immediately seaward of the reefold mass, deep reflectors appear to terminate abruptly. The reef and platform appear to be built over a deeper discontinuous basement ridge (Klitgord and Behrendt, 1979) inferred by magnetic depth estimates to be 7–8 km below sea level.

The younger part of the sedimentary section has built over the reef and platform complex to form a gently inclined sequence of slope sediments. The formation of the present slope has resulted from the incision of older sediments.

The ages of the sequences (Fig. 7) are inferred from extension of reflectors adjacent to the COST (Continental Offshore Stratigraphic Test) No. B-2 well (Scholle, 1977) drilled on the shelf 45 km to the northwest (Fig. 2). Deep-sea reflectors are extensions from DSDP stratigraphy. (Ewing and Hollister, 1972). The inferred ages show that reef complex was actively up building until the Albian, after which it was buried.

Farther to the south, on line 6 (Fig. 8) where the slope is narrow, the record shows that the seaward edge of the buried platform is marked

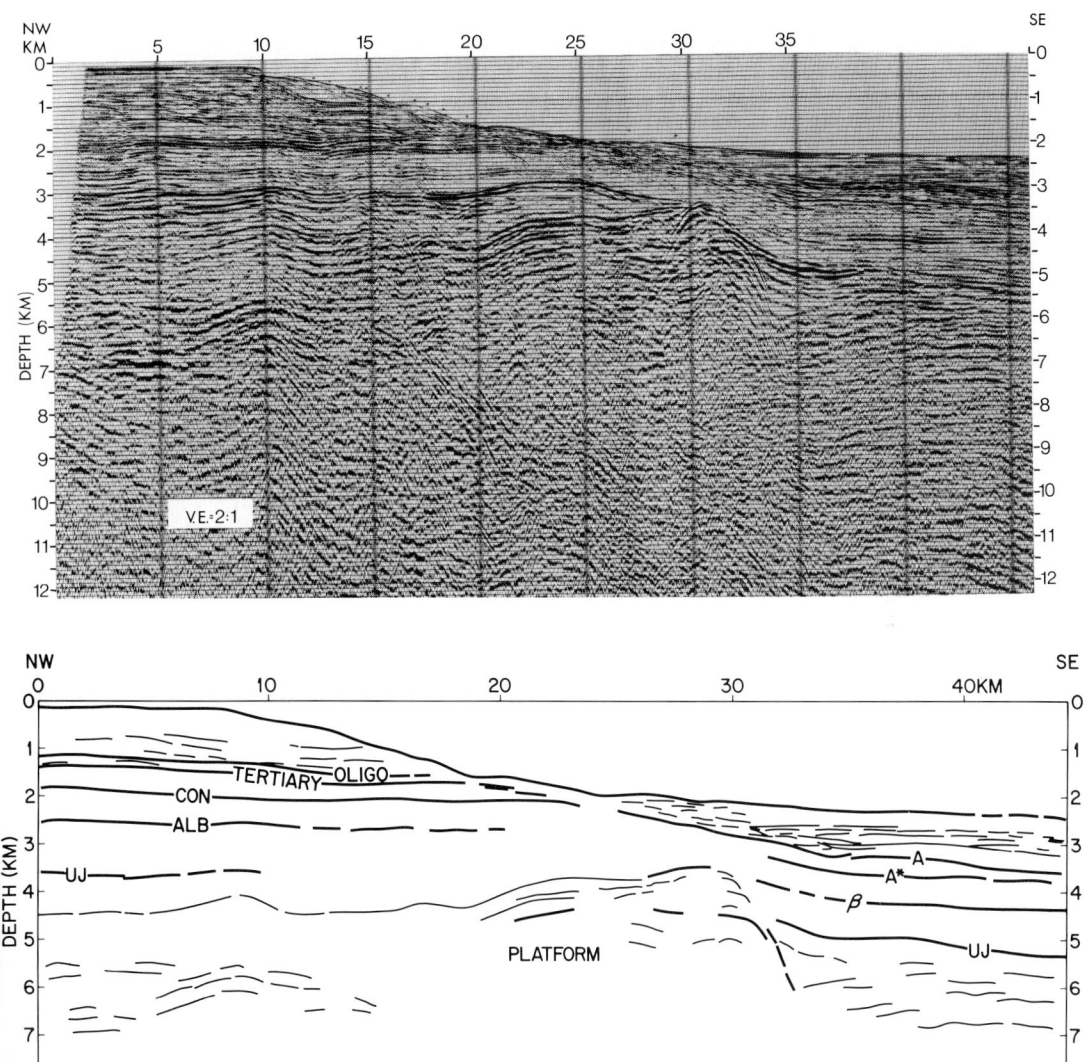

FIG. 8.—Depth section of part of line 6, over slope and upper rise. On 8a notice the diffraction at the edge of the platform block and landward tilting of deeper reflector beneath the lower slope.

by a diffraction. Landward of the platform edge, reflectors either are horizontal or dip gently back toward the shelf basin. Faint discontinuous reflectors, inferred to be part of an Upper Cretaceous sequence, overlie the platform. Reflectors representing the uppermost sedimentary sequences under the shelf show a broad out-building pattern; under the slope, reflectors indicate that the section is complexly channeled and eroded.

Under the rise, a broad sedimentary apron is present as characterized by a broad onlapping of subhorizontal, moderately continuous reflectors. The major changes in the continental slope on line 6 (Fig. 8) as compared with line 2 (Fig.

7) are: (1) a steepening of the slope (20 km versus 41 km on Line 2); and (2) a thinning of the Cretaceous and Tertiary sequence covering the reef-platform complex (1.5 km versus 2.3 km on line 2).

On the central part of the Baltimore Canyon trough (line 10), the cross section is somewhat changed from line 6 in that an obvious declivity, or edge, of the carbonate platform is not present (Fig. 9), and the Cretaceous paleocontinental slope deposits appear to be missing. Line 3 to the south shows much the same arrangement of reflectors as on line 10 (Schlee and others, 1976). As on line 6, the deep reflectors on line 10 suggest that

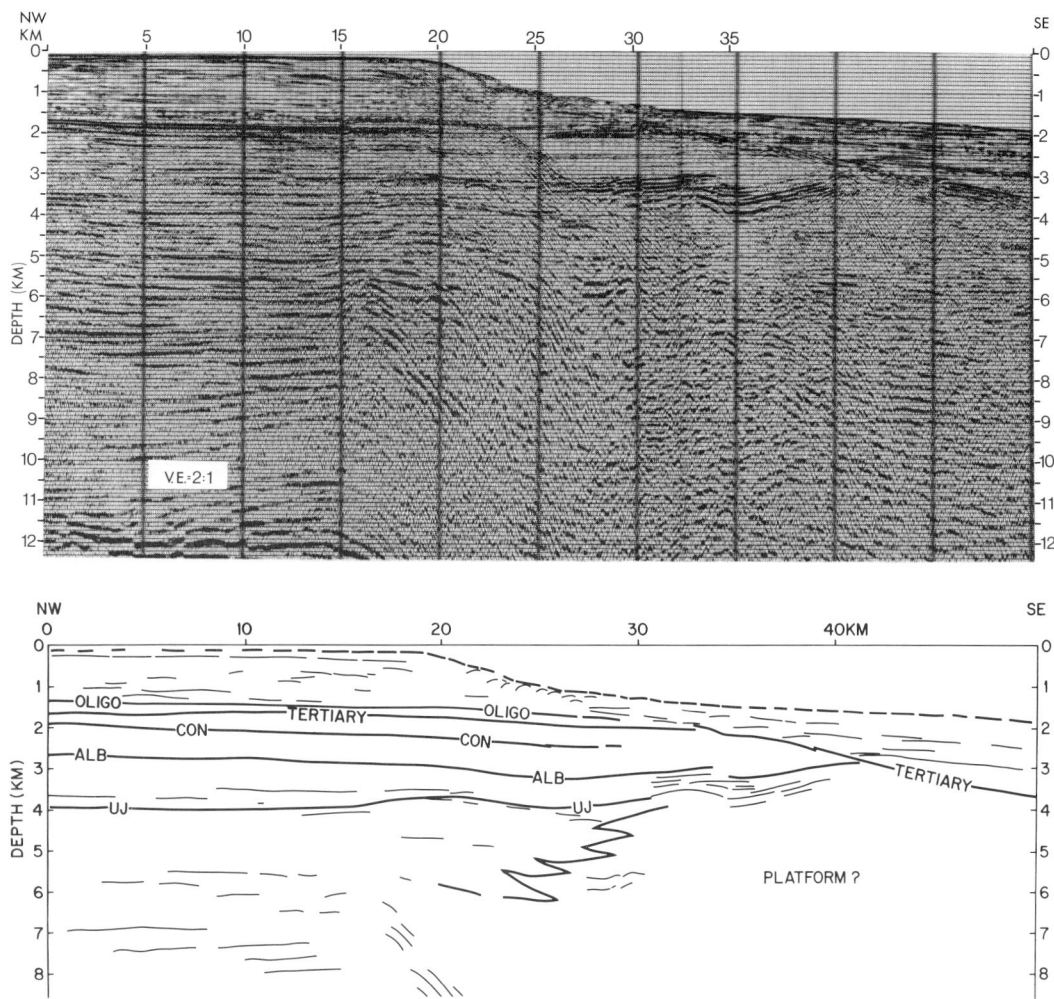

Fig. 9.—Depth section of a part of line 10, over slope and upper rise. On 9a notice the onlapping of the rise wedge and the irregular landward tilting of deeply buried blocks beneath the upper continental rise. Vertical exaggeration is 2:1.

the strata are level or backtilted at about the same depth (2 km below the sea floor). In the vicinity of Line 10, the Lower Cretaceous and Jurassic platform appears to be broken up by block faulting.

Again, the thickest part of the basin (more than 12 km) is under the outer edge of the shelf. Seaward of the shelf break, reflectors below 7 km lose acoustic character. Though faint and discontinuous, a few shallow reflectors continue out under the slope where they are cut out by a broad unconformity, which forms the boundary of the rise wedge. Beneath the slope, the shallow reflectors change acoustic character to hyperbolic returns, which suggest a change to slump deposits there.

The rise wedge in line 10 is thicker and extends farther landward than in the sections to the north. The slope is only about 13 km wide, and the upper part of the wedge extends up to 1400 m below sea level as compared with 2000 m below sea level on line 6. Along line 10 farther seaward than is shown in Figure 9, a few strong, discontinuous reflectors can be traced landward beneath the rise wedge subparallel to the unconformity; they are inferred to be Early Cretaceous (B) and Late Jurassic in age (UJ).

We interpret the arrangement of reflectors under the slope and rise to indicate (1) a probable series of deeply buried fault blocks (to explain sudden changes in reflector dips) and (2) a poorly defined block-faulted carbonate platform. Exact locations of the faults are difficult to infer (Fig. 9) because of a lack of continuous deep reflectors to show offsets; the bowed arrangement of reflectors at 3.5 km depth below 35 in the kilometer scale may mark the edge of one block. The buried platform front so obvious on the profiles across the northern part of the Baltimore Canyon trough is not apparent on this profile (no strong horizontal reflectors, no diffraction pattern of reflectors, and no indications of an abrupt onlapping by seaward-dipping reflectors). The similarity in depths of key chronostratigraphic factors indicates that the subsidence history for this sector of the shelf basin is similar to that farther north, although UJ on line 10 is somewhat deeper than on lines 2 and 6. The zone of carbonate accumulation (loss of deep coherent reflectors from kilometer 20 to right-hand edge of Fig. 9) broadly coincides with the zone of inferred fault blocks, which suggests that the foundation of the basin may be down faulted along this sector of the margin. The low-angle approach of deep-sea reflectors inferred to be of Late Jurassic and Early Cretaceous age to the area of the back-tilted blocks suggests that a gentle slope probably existed here.

In the southern part of the Baltimore Canyon trough (Line 11, Fig. 1), the section is similar to Line 6 in that the Lower Cretaceous and Upper Jurassic platform buildup is obvious (at 4 seconds) beneath the middle part of the Continental Slope, and a broad apron of inferred Cretaceous slump deposits has covered it (1.5 seconds); Cenozoic pelagic and slump deposits add 0.7 of a second covering above Cretaceous deposits. Landward, a thick section (5 seconds −8 km) of Mesozoic and Cenozoic sedimentary rocks underlie the main part of the shelf.

Cape Hatteras Area

The slope here is largely an erosional feature cut into Cretaceous and Tertiary shelf strata (Grow and Markl, 1977). The thickest part of a sedimentary basin is inferred to underlie the Continental Slope and upper Continental Rise (Fig. 10), and again a carbonate bank edge(?) has been detected at a depth of 5 km below sea level 30 km seaward of the present shelf edge (SPN 850) and under the upper continental rise (Grow & Markl, 1977). The sedimentary wedge thickens abruptly beneath a narrow shelf to more than

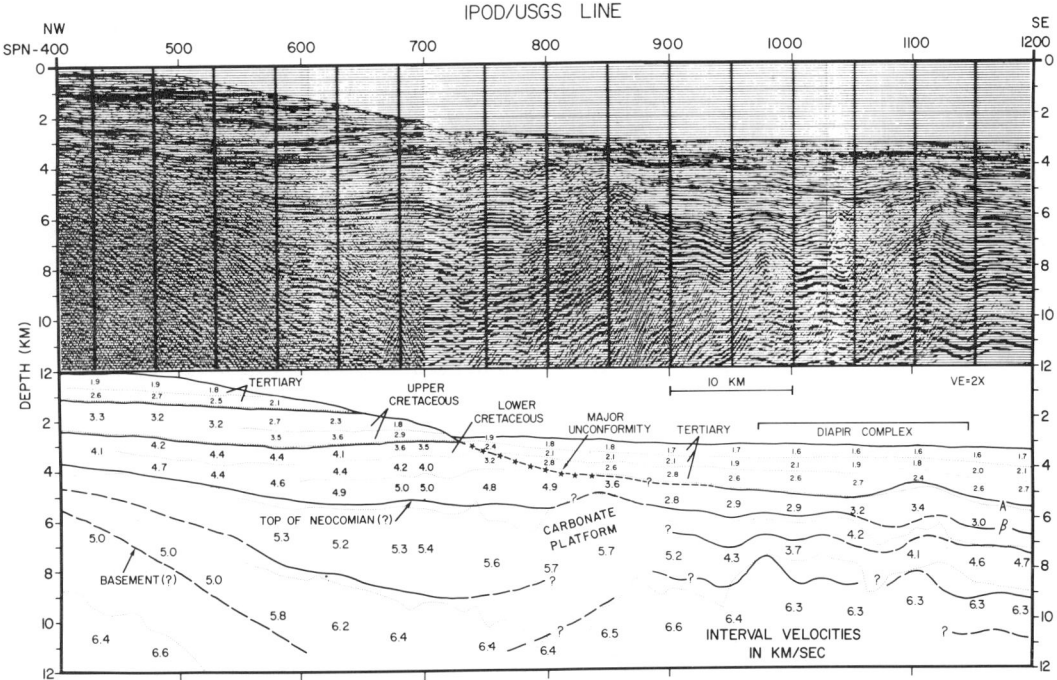

FIG. 10.—Depth section (10a) and line-drawing interpretation (10b) of IPOD-USGS line (from Grow and Markl, 1977). Section between Shotpoint (SPN) 700 and 1200 has also been migrated. Interval velocities used for depth conversion were averaged every 5 km along horizontal bands separated by dotted lines. Velocity values down to 4 to 5 km between SPN-400 and SPN-700 and 6 to 7 km between SPN-700 and SPN-900 has a complex topography and a major unconformity resulting in less reliable velocities and some artificial undulations of horizons.

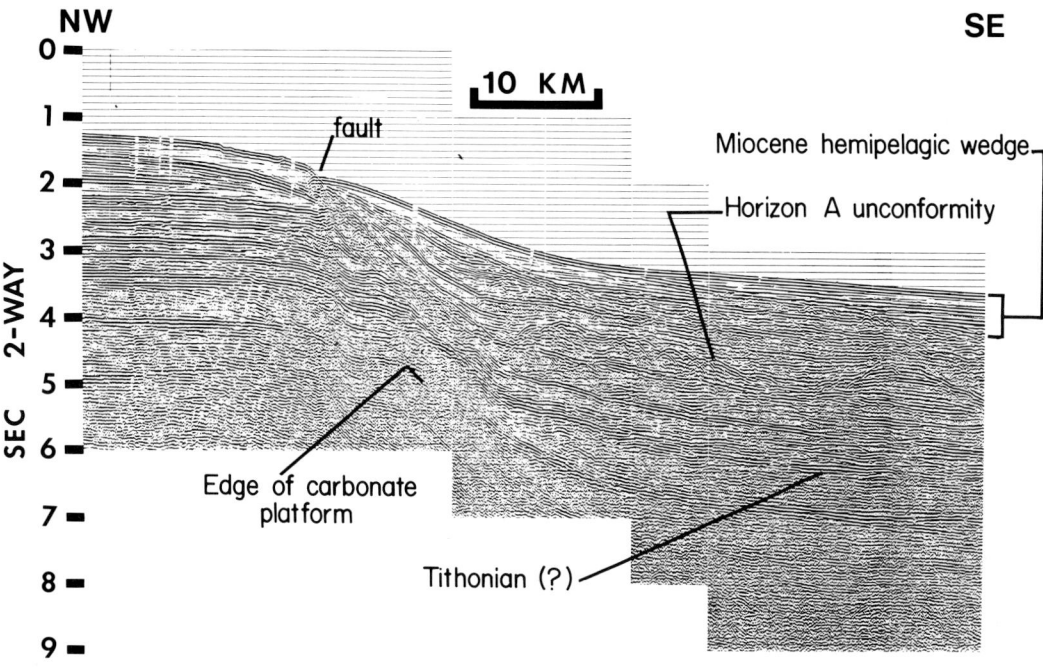

Fig. 11.—A part of multichannel seismic reflection profile FC7 over the outer Blake Plateau, slope, and rise. Notice the probable gravity induced normal faults near the upper part of the slope; and the loss of reflectors at a depth of 4–6 seconds beneath the slope. Location of profile shown on Figure 2.

7 km and possibly 9 km under the slope. The deep seismic reflectors are again discontinuous in the vicinity of the ancient shelf edge. Strata younger than Early Cretaceous are cut either by the present slope or by the erosional interval which preceded deposition of the continental rise wedge. *At least* two major erosional breaks are indicated—an early Tertiary break, which existed prior to formation of the continental rise prism, and a later Tertiary break, which resulted in the present surface. The slope moved landward nearly 30 km from the Early Cretaceous platform edge, and probably much of the erosional debris from the Tertiary retreat of the shelf lies in the rise wedge as broadly overlapping fans separated by erosional hiatuses.

The Cape Hatteras area shows more erosion than any other area shown in the profiles; the slope off Cape Hatteras is essentially cut into an older shelf sedimentary prism. Seaward of the platform edge, older horizons are deformed by a diapir complex that probably originated from Jurassic or Lower Cretaceous sedimentary rocks (Grow and Markl, 1977). These intrusions are under the rise and appear similar in their setting to the deep-water diapir complexes off the Nova Scotia shelf (Grow and Markl, 1977; Jansa and Wade, 1975). The Hatteras diapir complex appears to be centered under the East Coast Magnetic Anomaly, and the belt continues to the southwest at least to the latitude of Charleston.

Blake Plateau

The continental margin south of Cape Hatteras is dominated by the broad, flat Blake Plateau at depths of approximately 600–1000 m. This plateau breaks the continental slope into two slopes—the Florida-Hatteras slope descending from shelf depths to about 600 m and the slope seaward of the plateau. The seaward part of this divided slope is also very different in its northern and southern regions; the change occurs at the Blake Spur (Fig. 2). South of the spur, the Blake Escarpment descends very steeply to oceanic depths (4800 m) and no continental rise is present, whereas north of the spur a more typical slope and rise are present.

Estimated ages of tops of the units shown in Figures 12, 13, and 17 have been made by tracing reflectors from known ages at drill sites and dredge locations, by identification based on acoustic appearance in the deep-sea basin compared to locations where horizons have been drilled, and by structural relationships (Dillon and Paull, 1978; Dillon and others, 1979). Because only the top of the Cretaceous has been drilled at only three

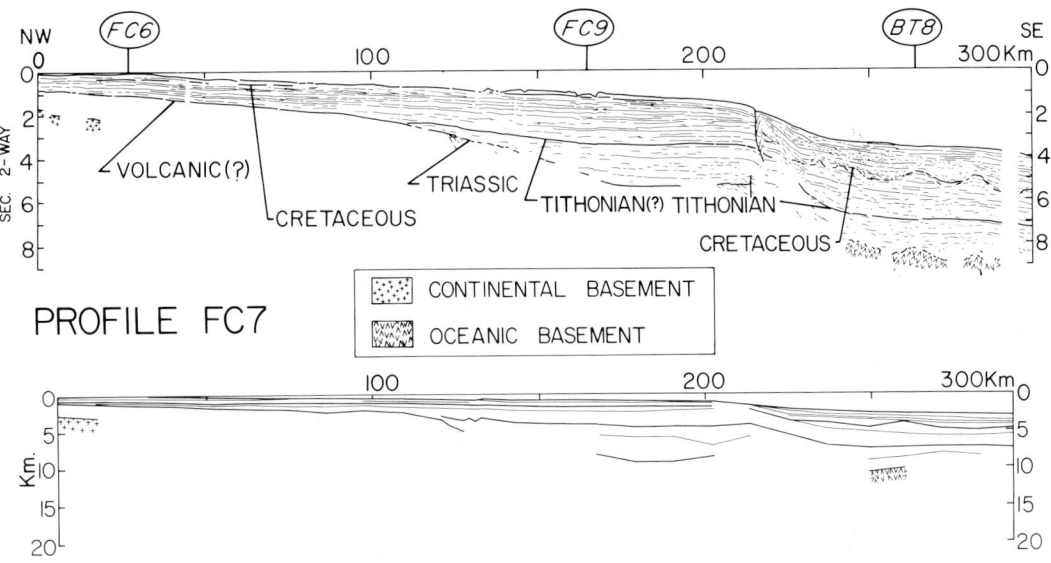

Fig. 12.—Interpretation of multichannel seismic reflection profile FC7. Vertical exaggeration is about 8:1 for the time section (upper profile) and 2:1 for the depth section (lower profile). Profile collected by Institut Francais du Pétrole.

locations on the Blake Plateau and on the Florida-Hatteras slope and shelf (AMCOR 6004 on Fig. 2, COST GE-1, DSDP 390–392) (Shipley and others, 1978; Dillon and others in press), the inference of stratigraphic ages is hazardous and should be considered preliminary.

South of Cape Hatteras the early formation of the continental margin was dominated by subsidence of two troughs floored by transitional crust, which probably consisted of mantle-derived mafic material extensively mixed with continent-derived sediments—essentially a modified ocean-

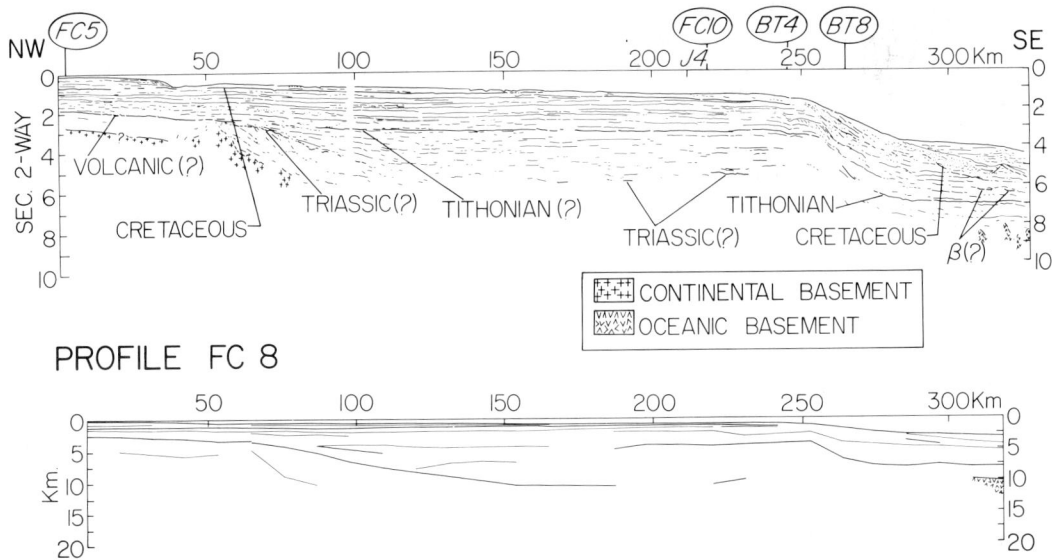

Fig. 13.—Interpretation of multichannel seismic reflection profile FC8. Vertical exaggeration is about 8:1 for the time section (upper profile) and 2:1 for the depth section (lower profile). Profile collected by Institut Francais du Pétrole.

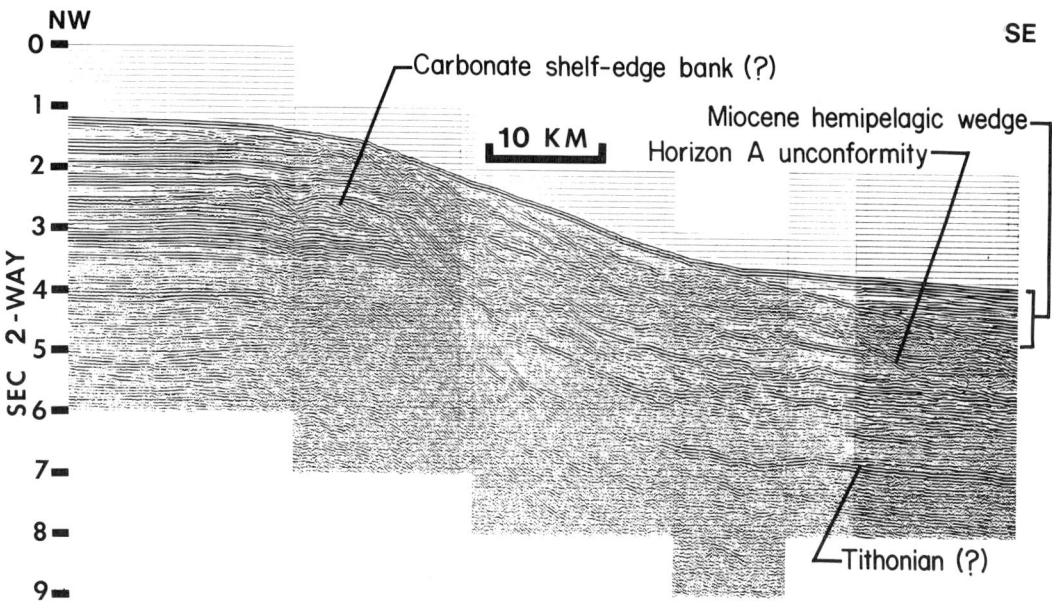

FIG. 14.—A part of multichannel seismic reflection profile FC8 showing a marked unconformity rising toward the left (4.9 seconds at right hand side of photo) and an irregular high area containing prograded reflectors (left hand side of photo beneath outer Blake Plateau) that probably represent a Cretaceous shelf-edge carbonate bank or reef.

ic crust (Klitgord and Behrendt, 1979). The narrow Carolina trough (profile FC7—Fig. 12) parallels the continental margin and underlies the continental slope east of the Carolinas. To the south, the Blake Plateau underlies the main part of the plateau. Reflectors in the photograph of profile FC7 (Fig. 11) show no interruption near the seaward edge of the Blake Plateau. That edge was the shelf edge through most of the continental margin development until late Paleocene time (Dillon and others, in press). Reflectors seem to extend continuously from the deep sea past the shelf break and, therefore, do not indicate significant reefs. At a greater depth below the slope (4–6 seconds, Fig. 11), a probable carbonate platform makes tracing difficult through older deep-sea reflectors to the plateau. The line drawing (Fig. 12) shows the relatively narrow zone of subsidence during the Jurassic represented by pre-Tithonian-post-Triassic deposits. This narrow zone, from approximately km 140 to km 220, represents deposition controlled by subsidence of the southern end of the Carolina Trough. A characteristic of the region is that early subsidence (Jurassic) was irregular and resulted in large variation in sediment thickness, whereas later subsidence (Cretaceous and Cenozoic) covered a broader area and resulted in a more uniform sediment thickness. The shelf break shows a minor retreat of less than 10 km from the Early(?)

Jurassic [reflectors directly above the unconformity at the top of the Triassic] to the top of Cretaceous strata. A prominent unconformity containing deep channels intersects the right side of the profile at a depth of 5.2 seconds. Cores from DSDP drilling show that the hiatus probably formed during the early Eocene; hence, it is termed the Horizon A unconformity (DSDP 99, 100, 101, 105, and 391; Ewing and Hollister, 1972; Benson, Sheridan, and others, 1976). A listric normal fault that has a surface offset appears at the seaward edge of the Blake Plateau at a water depth of 1.8 seconds. A set of at least four back-rotated antithetic fault blocks is also present. Though common in areas of rapid sedimentation, such a large gravity fault is unusual on the U.S. Atlantic margin.

The broad, flat Blake Plateau results from continued subsidence accompanied by much reduced rates of deposition during Late Cretaceous and Cenozoic time. Actually, erosion is common on the Blake Plateau, as shown on line drawing of profile FC7 (Fig. 12); scour features at the sea floor are particularly noticeable between kms 130–170.

South of profile FC7, the line drawing of profile FC8 (Fig. 13) shows the much broader but shallower pattern of Jurassic subsidence (Triassic top to Tithonian top) which is characteristic of the sill area between the Carolina trough and Blake

Plateau basin. Again, the profile seems to indicate a shelf break that maintained its position as a carbonate platform from Early Jurassic time until the end of the Cretaceous by a balance of accretion and subsidence. No well-developed reef is evident, but an irregular, bank-like accumulation where reflectors dip steeply seaward (Fig. 14), appears at the edge of the Blake Plateau in the middle of the Cretaceous section (depth about 2.5 seconds). Relatively high-interval velocities of 4.4 to 4.7 km/sec and a velocity inversion associated with this feature in an adjacent profile, as well as its structure, suggest that it was formed as a shelf-edge carbonate bank. Beneath the outer Blake Plateau, interval velocities in the upper 2.5 km of section of profile FC8 average 200–400 m/sec higher than the corresponding section of FC7. This change suggests a general increase in carbonate rock southward.

Again, the channeled unconformity at the top of Cretaceous deposits (Horizon A unconformity) appears on FC8, intersecting the right side of the photograph at 5.3 sec (Fig. 14). Above the channeled unconformity, is a smoother angular wedge of strongly acoustically layered strata. It intersects the area shown on the right side of the photograph at 4.9 seconds. This hiatus is probably of late Miocene age, separating carbonate turbidites below from hemipelagic deposits above (Sheridan and others, 1974).

As is typical throughout the region, the continental margin continued to subside through Cenozoic time. Subsidence rate exceeded sedimentation rate so that the Blake Plateau was formed and is now at 600 to 1000 m depth. The present shelf, as seen on profile FC8 (Fig. 15), is a progradational feature built against the flank of the Gulf Stream. The photograph shows the scoured nature of the sea floor seaward of the accreted Florida-Hatteras slope.

Our southernmost profile, FC3 (Fig. 16), shows the steep Blake Escarpment, which drops abruptly from about 1000 to about 5000 meters (over 15 km—13° slope). Two zones of no reflectors beneath the outer Blake Plateau shown in the line drawing interpretation are considered to be reefs (Fig. 17). Results of dredging information (Heezen and Sheridan, 1966) and drilling (Benson, Sheridan and others, 1976), as well as our seismic correlations (Dillon and others, in press) suggest that reef growth was terminated at about the end of Early Cretaceous time in this area. South of the Blake Spur, the reefs acted as a sediment dam, largely preventing sediment transport to the deep sea. Thus, the deposition in the southern Blake Basin is more comparable to an abyssal plain than to an ordinary continental rise. As the basement of the continental margin (Blake Plateau basin) subsided, it dragged down the oceanic basement to the east. A group of nearly vertical faults that

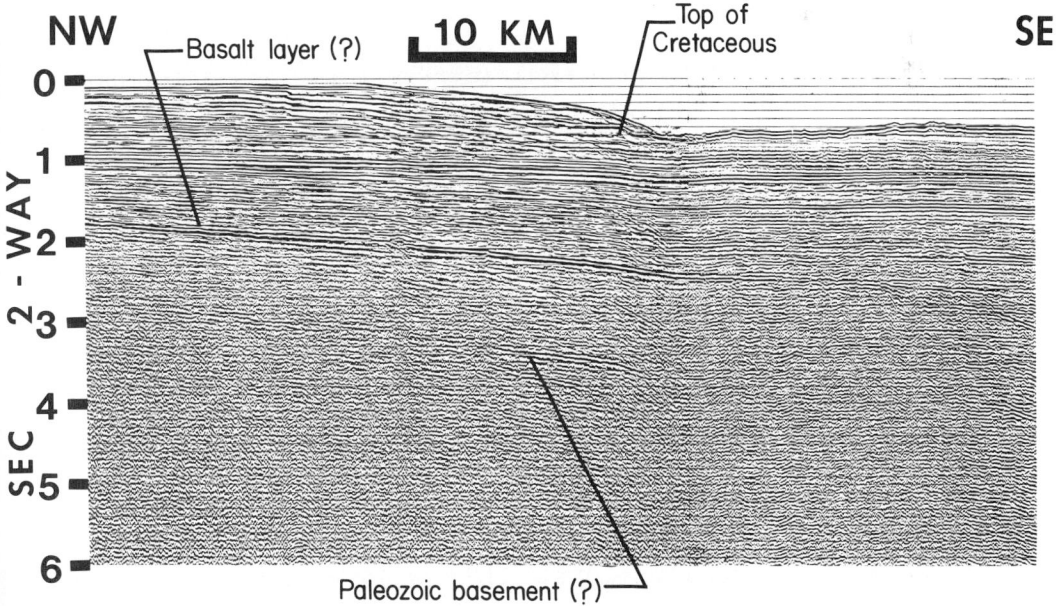

Fig. 15.—A part of multichannel seimic reflection profile FC8 showing the progradation of shelf deposits beneath the Florida-Hatteras slope.

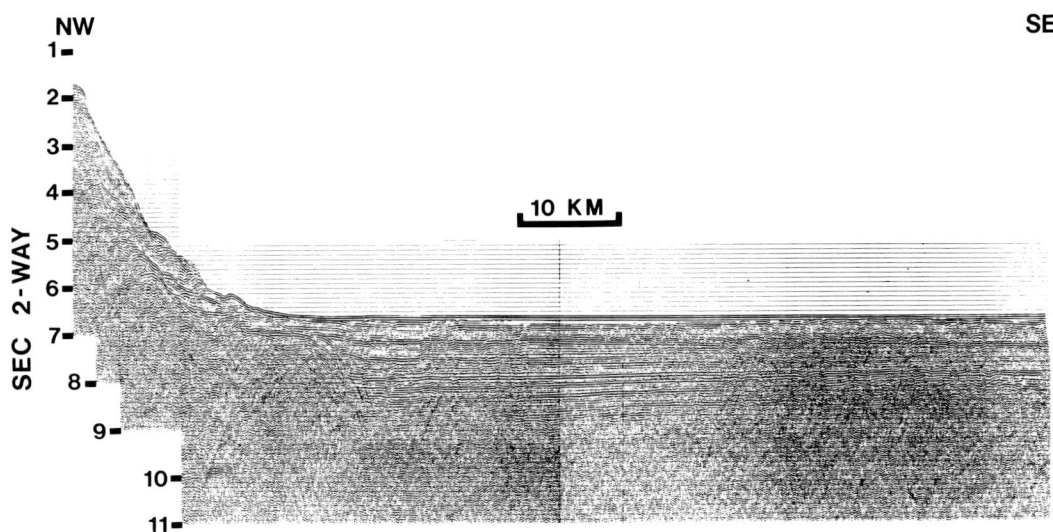

Fig. 16.—A part of multichannel seismic reflection profile FC3 across the outer edge of the Blake Plateau and escarpment. Notice the absence of reflectors adjacent to escarpment where reef limestone have been dredged.

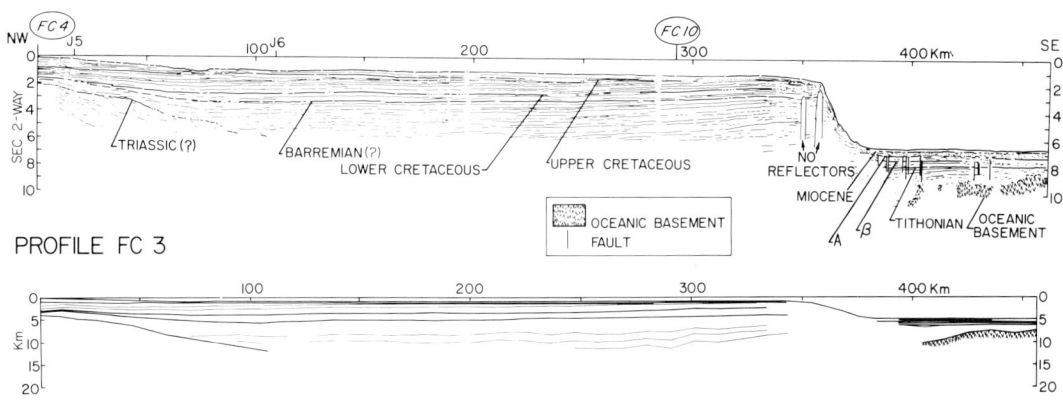

Fig. 17.—Interpretation of multichannel seismic reflection profile FC3. Vertical exaggeration is about 8:1 on the time section (upper profile) and 2:1 on the depth section (lower profile). Location of profile shown on Figure 2.

have very small displacement just seaward of the Blake Plateau probably indicate a partial decoupling of the Blake Plateau basement from oceanic basement.

DISCUSSION

The profiles indicate that slope development has been a complex process involving both constructional and destructional phases similar to the processes observed off northwest Africa by Seibold and Hinz (1974). Constructional phases include both prograding by carbonate banks and clastic sequences; destructional phases appear as submarine unconformities resulting in a retreat of the slope edge. In the north where a more typical shelf-slope-rise transition is present, an ancient carbonate platform appears to have formed the shelf edge during Late Jurassic and Early Cretaceous time. In the south, a carbonate buildup is also evident, probably as a reef, that existed in the Late Cretaceous along the seaward side of the Blake Plateau basin. Along the southeastern edge of the Carolina trough, a gradual buildup of the margin has occurred, but at diminished rates on the ancestral slope (FC7, Fig. 11, 12). An intermediate format occurs between the

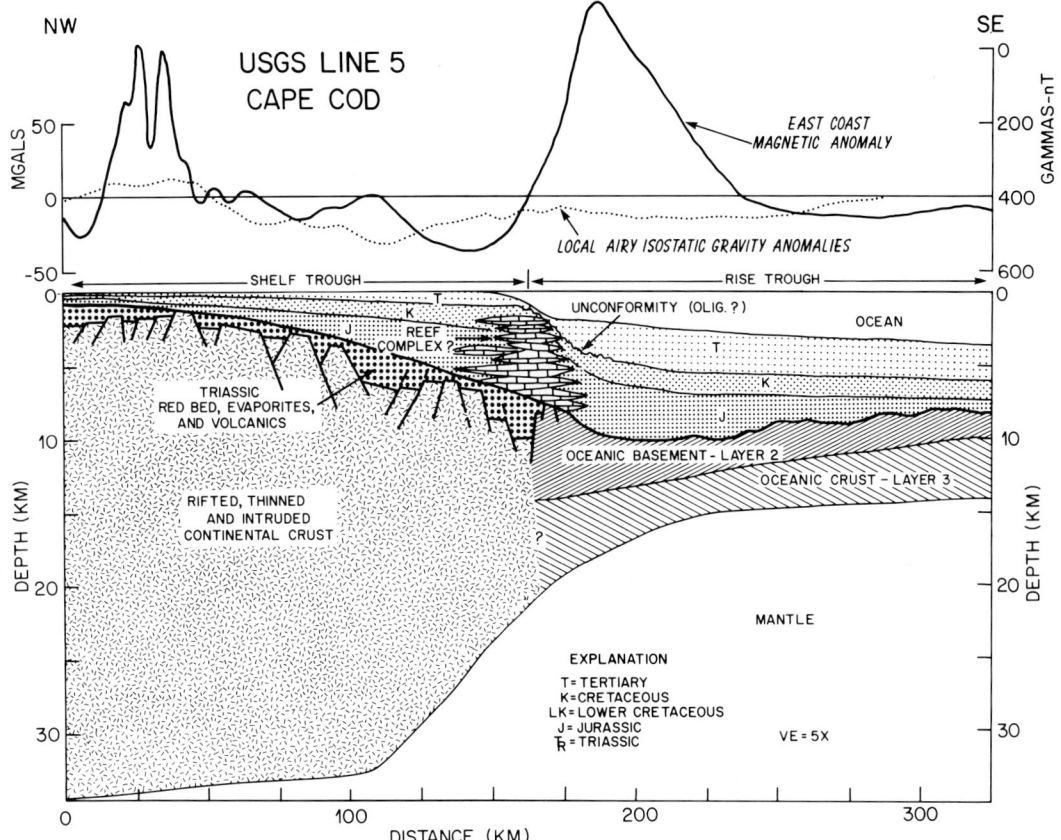

FIG. 18.—Schematic section of line 5 summarizing reflection data and magentic and isostatic gravity anomalies. On line 5, the top of oceanic basement can be traced to the axis of the East Coast Magnetic Anomaly. The line is typical of a narrow transition zone between continental and oceanic crusts.

Carolina trough and Blake Plateau basin where probable carbonate platform is evident under the seaward edge of the plateau. The southern area has the added complexity of periodic scour by the ancestral Gulf Stream during the Cenozoic that, in effect, helped to create an additional slope (Florida-Hatteras slope). Before discussing the sedimentologic processes that shaped the slope, we need to ask what underlies the Jurassic and Lower Cretaceous sedimentary rocks.

Deep Structure

A study of crustal structure beneath the slope shows that in many areas, the carbonate platforms sit above a zone of transitional crust and thickened oceanic crust (Klitgord and Behrendt, 1979; Grow and others, 1979), marked by the East Coast Magnetic Anomaly (ECMA). Though the carbonate rocks are adjacent to different structural settings (basins, platforms), there is a similarity in the deep crustal structure (block-faulted continental crust-thickened oceanic layer 2), though the width of thinned-rifted continental crust varies (Figs. 18, 19). Most of our inference on the crustal structure beneath the sedimentary section is based on refraction, magnetic, and gravity data. Klitgord and Behrendt, 1979, present preliminary depth estimates for magnetic sources over the ECMA (Figs. 18, 19) from a recent aeromagnetic survey. Their results indicate weak to moderate strength sources at depths of 7 to 8 km (Fig. 19) in the middle of the disturbed zone and with stronger sources on either side of the anomaly at depths of 10 km or more. Although their results could be explained by weakly magnetized basement blocks, Lower Cretaceous dikes and sills intruding sedimentary units might also explain the weak and moderate sources at 7 to 8 km depth for the areas off the Baltimore Canyon trough and Georges Bank.

Although little is presently known about the velocity structure of the crust beneath the shelf,

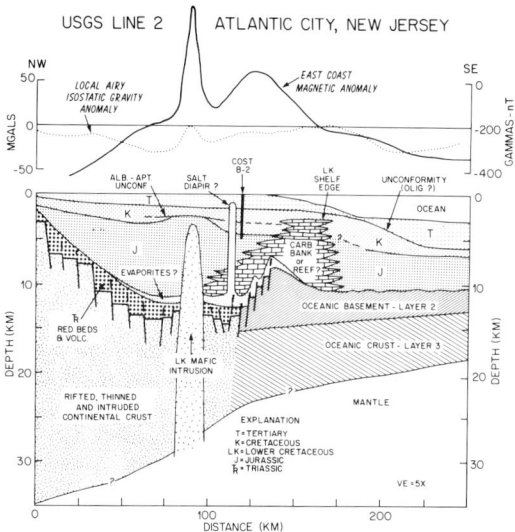

Fig. 19.—Schematic section along line 2, through a wide ocean-continent transition zone associated with the Baltimore Canyon trough off New Jersey. The diapir is projected from 10 km south of line 2, and the COST No. B-2 well is projected from 10 km north of the line.

seismic refraction data suggest that normal oceanic crust and mantle can be traced to within 50 to 75 km of the ECMA (Sheridan, 1976) and that oceanic crustal layer 3B can be traced laterally to the Anomaly itself. Following Keen and Keen's (1974) hypothesis, we have assumed that the ECMA marks the fundamental boundary between continental and oceanic crust (Figs. 18, 19). However, the continental crust beneath the 12 km of shelf sediments must be fragmented and attenuated. Finally, we have assumed that the disturbed zone is probably underlain by somewhat thickened oceanic crust.

The composition of the zone where coherent reflectors are absent (depths of 6–12 km in the zone 50–100 km seaward of the ECMA) remains a matter for continuing debate. The observation of probable back-tilted sedimentary blocks (Fig. 8, 9) and diapirs (IPOD, Fig. 10) in this zone suggests that some of it is sedimentary rock. The diapirs evident on the IPOD line (Grow and Markl, 1977) have been found to extend 40 km north and 200 km south of the IPOD line in a more recent survey (Grow, Dillon, and Sheridan, 1977). None of the diapirs have a detectable magnetic or gravity signatures and are therefore are probably composed of sediment or salt. Manheim and Hall (1976) reported hypersaline brines of 55 parts per thousand at a depth of 1700 m in the Caldrill 15 slope hole off New Jersey. They project that saturated brines should exist at 3 to 4 km depth (i.e., 1 to 2 km) beneath the Lower Cretaceous shelf edge (see lines 2, 6, 10, Figs. 7, 8, 9). Although these concentrations could be explained by local evaporite basins, they could also be explained by diapirs.

Our inferred crustal structure (Figs. 18, 19) is schematic and intended only to summarize only the major features and give a likely hypothesis for the disturbed zone. Additional seismic studies plus careful modeling of magnetic and gravity data will be needed to discriminate among these hypotheses.

Carbonate Platform

Mesozoic carbonate platforms are a widespread phenomena bordering part of the western North Atlantic and the Gulf of Mexico (Emery and Uchupi, 1972). They are also present off northwest Africa (Lehner and de Ruiter, 1977) as buried shelf edges associated with salt diapirs in some areas (Aymé, 1965).

The ancient platform edge off the eastern United States (Fig. 2) generally coincides with the present continental slope. Between Georges Bank and Cape Hatteras, it is under the lower part of the continental slope (1000–2000 m). At Cape Hatteras (Fig. 10), it is under the rise, probably because of the larger amount of slope retreat since the Cretaceous. Between Cape Hatteras and the southern Blake Plateau, the ancient platform edge is under the upper part of the Blake Escarpment, although along part of this slope, the reef has been eroded back (Ryan 1978). The present shelf edge (approximated by the 200 m isobath—Fig. 2) is consistently landward of the ancient carbonate platform edge over the entire margin. The amount varies from a few kilometers off Georges Bank to 300 km off Florida. Off eastern Nova Scotia, the area of carbonate deposition is mainly under the outer shelf, landward of the 200 m isobath (Given, 1977).

The positions of the platform edge and the ECMA (Fig. 2) generally coincide for the margin north of Cape Hatteras. South of the Cape, the zone of the ECMA swings to the west and ends, but the platform edge continues southward along the Blake Escarpment. If the ECMA marks the boundary between oceanic and continental or transitional crust (Grow and others, 1979; Klitgord and Behrendt, 1979), then the carbonate platform has tended to center over this boundary.

Constructional Processes

Slope development following the carbonate phase (Fig. 20) has been complex. The carbonate platform accreted over the zone of crustal separation, as indicated by the ECMA. Along some parts of the margin (Figs. 1, 4, 6, 7, 8, 11), the relief

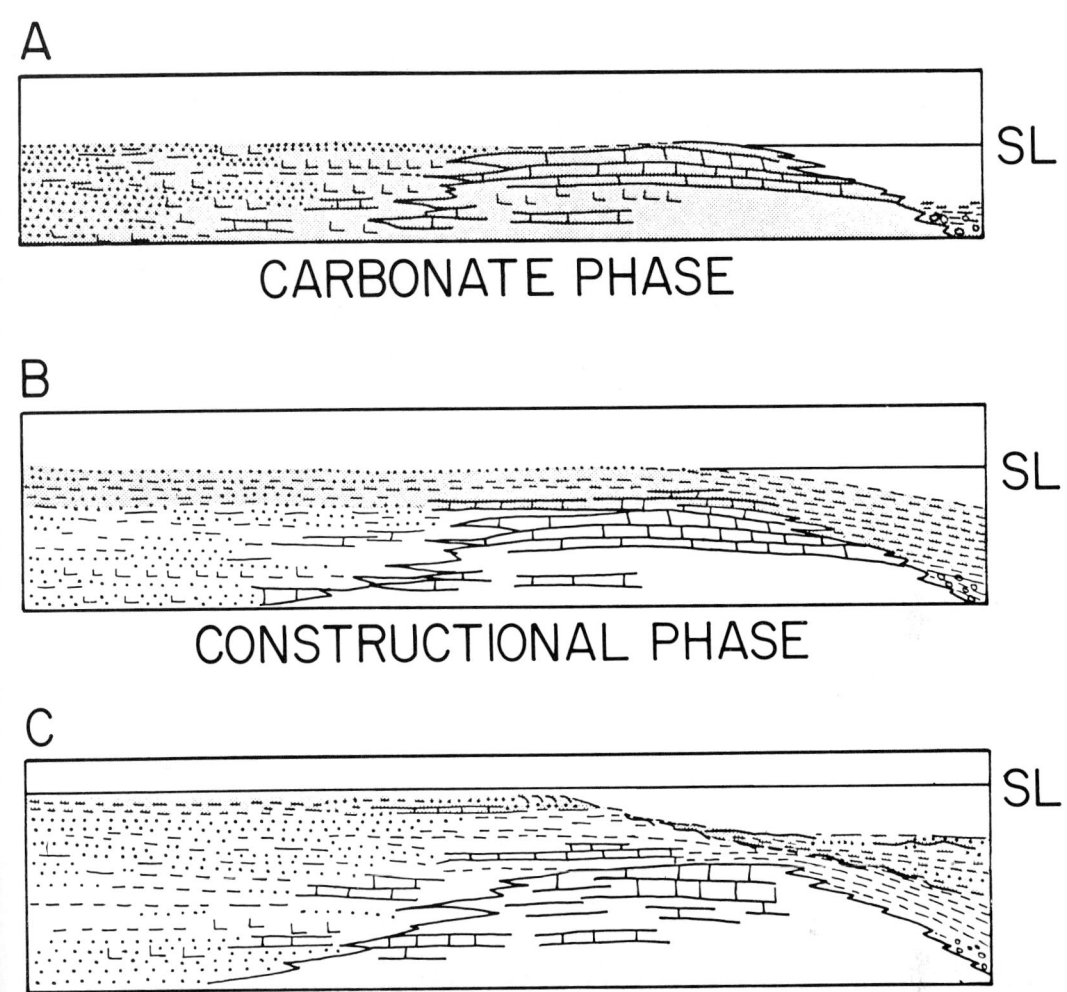

FIG. 20.—Sequence of schematic sections showing the main phases of slope development. In the south the sequence was modified by a step of the shelf edge to the west during the Cretaceous and by extensive erosion of the escarpment, to create a relatively steep slope.

caused by the buildup of limestone was lessened through the subsequent deposition of probable hemipelagic muds during the late Early and Late Cretaceous. The broad extension of shelf reflectors seaward (Fig. 7) suggests that the shelf prograded seaward over the carbonate platform and reef to build a gentle constructional slope (Grow and others, 1979); the lack of continuity and the low intensity of reflectors between Horizons B and A (Figs. 6, 7, 8), leads us to suspect that a clayrich sequence accumulated on the slope during the late Early and Late Cretaceous. In the deep ocean, widespread organic-rich clays were deposited in the North Atlantic basin during the Early Cretaceous (Arthur and Schlanger, 1977), and these were followed by gray mud deposits during the late Cretaceous (Ewing and Hollister, 1972). Along other parts of the margin (Figs. 5, 16), bottom currents scoured the slope and eroded part of the reef (Ryan, 1978) or prevented the accumulation of much sediment on the slope; it was deposited mainly on the shelf and rise.

During the late Oligocene and early Miocene, a broad deltaic outbuilding of the shelf occurred off New Jersey and New York (Garrison, 1970; Schlee and others, 1976). The sequence is as much as 1 km thick and formed subsequent to a major regression in the middle Oligocene (Figs. 3) postulated by Vail and others, 1977. The shelf broadly prograded seaward as a series of deltas deposited sand and mud over the area in response to a slow marine transgression. In the south (Fig. 15), a broad progradation of the shelf can be seen in the upper 0.5 second on the left of the section. Timing of the progradation was during the Tertiary when the shelf was built toward the inner Blake Plateau.

The other main area of sediment accumulation during the Tertiary was the upper continental rise, where as much as 3 km of sediment accumulated above a broad unconformity that cuts out some of the Cretaceous section (Figs. 9, 10). The rise sedimentary wedge is complexly channeled off Georges Bank (Figs. 4, 5) and the northern Blake Plateau (Figs. 11, 12). Off the mid-Atlantic area and of the Blake Plateau (Figs. 6, 7, 10, 14), it broadly onlaps older horizons. The wedge probably formed in response to a periodic retreat of the continental slope through erosion.

Destructional Processes

In one of several stages of slope retreat, the continental slope was cut back (10 to 300 km—Fig. 1). Unconformities formed during the Eocene, late Oligocene, and Pleistocence along parts of the continental slope and Blake Plateau. In essence, the slope became an erosional feature as older Cretaceous and Tertiary deposits were exposed. Nor was the erosion restricted to the slope, for conspicuous unconformities are evident on the rise (Fig. 14). The preservation *and* exposure of Tertiary deposits on the slope suggests that intervals of sediment accumulation (probably during stages of marine transgression) alternated with intervals of slope erosion (marine regression—relative sea level drop). Vail and others (1977) have provided a relative sea-level curve (Fig. 3), which shows that the level was a worldwide transgression in the Cretaceous and that it began to fall during the Tertiary; major short intervals of marine regression (sea level drop) took place during the Danian (early Paleocene), Ypresian (early Eocene), Chattian (middle Oligocene), and Messinian (late Miocene). The regression of Oligocene age was particularly strong and has been correlated by Grow and others (1979) with a major rise-lower slope unconformity seen on six lines off the middle Atlantic area (lines IPOD, 11, 10, 6, 5, 2). Along the southern edge of Georges Bank, some canyon cutting and slope erosion is indicated in post-Miocene time; Stetson (1949) dredged outcrops of Miocene sandstone on the sides of several canyons there. Ryan and others (1978) also found evidence of slope erosion in the same area during the pre-Campanian to Maestrichtian, the middle Eocene, and the Pleistocene, that is on sampling traverses from the submersible ALVIN. Off the Long Island platform (Fig. 6), a beveling of Lower Cretaceous strata which probably occurred at middle Cretaceous time beneath the present lower slope, indicates that some erosion of the upper slope then. In Hudson Canyon, erosion during the post-Miocene is indicated by exposures of tan sandstone of Miocene age recovered from there and in "Double" Canyon, 55 km to the northeast of Hudson Canyon (Weed and others, 1974). In the south, major erosional events took place during the early Tertiary when the broadly channeled unconformity formed (Figs. 12, 13) seaward of the Blake Plateau and in the post-Miocene, when sediments of the Blake basin onlapped upper Miocene turbidite deposits that had covered the older hiatus (Sheridan and others, 1974; Benson and others, 1976). On the Blake Plateau, major hiatuses are in the middle Cretaceous, late Paleocene, late Eocene, Miocene, and post-Miocene times (Schlee, 1977; Benson and others, 1976) when scour by the ancestral Gulf Stream eroded the plateau and helped to decrease the rate of sedimentation over a long period of time.

CONCLUSIONS

The present Continental Slope was formed over two earlier-formed slopes (Fig. 20). Although little is known about the slope during the Early and Middle Jurassic, the Upper Jurassic and Lower Cretaceous shelf edge existed 10–300 km seaward of the present shelf edge and appears to have marked the beginning of a platform composed of carbonate rocks having a high velocity. The discontinuous carbonate platform and slope in front of it were present at least from the Canadian Maritimes to the Bahamas. It existed during the Jurassic off Canada but continued into the Early Cretaceous off the United States. Strong reef growth typified the platform in the vicinity of the northern Baltimore Canyon trough and the southern Blake Escarpment.

The platform formed a partial barrier to the dispersal of sediment to the ocean, so that terrestrial debris collected in the rapidly subsiding basins as a sequence of red beds, probably interlayered with carbonate and evaporite units (Given, 1977; Jansa and Wade, 1975; Schlee, unpub. data). The abrupt onlapping of deeply buried seaward-dipping reflectors (during inferred Late Jurassic and Early Cretaceous time) suggests that the platform existed as a major topographic feature. The carbonate platform formed adjacent to structural platforms or rapidly subsiding basins.

Off the southeastern part of the Baltimore Canyon trough where magnetic basement deepens, the carbonate buildup appears to be broken up as a zone of tilted fault blocks and to be covered by a broad sedimentary apron; the same degree of subsidence and sediment accumulation appears to have taken place here as farther to the north.

Assuming that the East Coast Magnetic Anomaly outlines the zone of initial separation in the Late Triassic, then by the Late Jurassic and Early Cretaceous the margin had built as much as 50 km seaward of the initial break in the area of the northern Baltimore Canyon (Grow and others, 1979). Little or no progradation of the margin from to the zone of separation seems to have occurred south and east of Cape Cod, east of Cape Hatteras, and the northern Blake Plateau by the beginning of the Cretaceous (Fig. 2). On Georges Bank, the platform edge is about 30 km. from the anomaly.

During the remainder of the Late Cretaceous, a gentle slope formed off the middle Atlantic and Georges Bank margins. Where still preserved (Figs. 1, 4, 5, 7, 11), the slope built over the carbonate platform to create a gentler one seaward of it. Prograding of the shelf was most pronounced seaward of the northern Baltimore Canyon trough during the middle Cretaceous. The acoustic signature of these sediments suggests that they are probably fine grained because the returns are poor and continuity of reflectors is low. Contributing factors in the formation of a gentler slope may have been an influx of terrigenous detritus during the Early and Late Cretaceous. Perhaps incursions of anoxic water terminated the carbonate buildup (Arthur and Schlanger, 1977) as broad Cretaceous transgressions (Vail and others, pt. 4, 1977); the position of coarse debris was shifted landward so that hemipelagic muds could mantle the slope area to create a more subdued topography.

Most of the slope erosion and retreat occurred during the Tertiary, when a series of regressions helped to build up clastic deposits on the outer shelf, erode Cretaceous slope deposits (Parsons, 1975; Grow and Markl, 1977), and spread slump and fan sediments on the rise. For much of the U.S. Atlantic margin, what King and Young (1977) classified as a series of destructional paleoslopes formed as sea level fluctuated (Fig. 3) during the Eocene, late Oligocene and Pleistocene. A margin interval in slope retreat occurred during the Oligocene when a drop in sea level exposed parts of the shelf and Atlantic Coastal Plain and resulted in a deposition of a thin sequence on the outer shelf (Grow and others, 1979); sediments bypassed the mid-Atlantic shelf to be directed through submarine canyons to the Continental Rise and Abyssal Plain.

Compared to the Upper Jurassic and Lower Cretaceous shelf edge, the present shelf-slope break is landward 10 to 300 km. (Fig. 2). In some areas (Fig. 6), the amount is minor so that older Cretaceous slope deposits are preserved, whereas off Cape Hatteras and in particular farther south off the Blake Plateau, the erosional surfaces have cut back into the upper part of the older sedimentary basin, and the Upper Cretaceous deposits are partly removed. Obviously, the history of the Atlantic slope is a complex one involving the interplay of subsidence, sea-level shifts, terrigenous sedimentation, and biogenic input during the early stages of carbonate platform construction. Later stages still involved the balancing of subsidence, terrigenous contribution, and shifts in sea level. The last factor became critical in affecting the Cenozoic development of the Continental Slope, in part because the rate of margin subsidence has slowed (thickness of the Tertiary section on the shelf is much less than the Cretaceous section for much of the shelf), and in part because the Cenozoic was a time of slow drop in sea level punctuated by rapid relative drops in sea level. The overall trend of sea level during the Cretaceous was either slowly rising (Early) or static (Late), as shown by Figure 3.

A last important question that the study answers is what reflectors can be tied from the shelf to the deep sea. Very few reflectors found beneath the continental rise carry through to the Continental Shelf, and these are mainly Cretaceous rocks. For older horizons, the carbonate platform was a physiographic and lithologic barrier for the tying of correlative shelf and rise reflectors. In younger horizons, hiatuses associated with the continental rise wedge cut out much of the section inferred to be Late Cretaceous and younger.

ACKNOWLEDGMENTS

The paper has benefited from the comments of many of our colleagues, in particular Kim D. Klitgord, D. W. Folger, and M. M. Ball, J. C. Behrendt, and R. E. Sheridan. Drafting was done by Linda Sylwester, and the manuscript was typed by Janet Burke, Barbara Van Dyk, Terry Dunne, and Peggy Mons-Wengler.

REFERENCES

ARTHUR, M. A., AND SCHLANGER, S. O., 1977, Middle Cretaceous "Oceanic anoxic events" as causal factors in development of giant oil fields: Society Exploration Geophysicists (abs.): Program 47th Ann. mtg., Society Exploration Geophysicists, Calgary, p. 40.

AYMÉ, J. M., 1965, The Senegal salt basin, in Salt basins around Africa: London, Institute of Petroleum, p. 83–90.

BEHRENDT, J. C., SCHLEE, JOHN, AND FOOTE, R. Q., 1974, Seismic evidence for a thick section of sedimentary rocks on the Atlantic outer continental shelf and slope of the United States (abs.): EOS, American Geophysical Union Transactions, v. 55, p. 278.

BENSON, W. E., SHERIDAN, R. E., AND OTHERS, 1976, In the North Atlantic deep sea drilling: Geotimes, v. 21, p. 23–26.

BUFFLER, R. T., WATKINS, J. S., AND DILLON, W. P., 1979, Geology of the offshore Southeast Georgia Embayment, U.S. Atlantic continental margin, based on multichannel seismic reflection profiles, *in* Watkins, J., Montadert, L., and Dickerson, P., eds., Geological and geophysical investigations of continental margins: Am. Assoc. Petroleum Geologists Memoir 29, p. 11–25.

BUNCE, E. T., EMERY, K. O., GERARD, R. D., KNOTT, S. T., LIDZ, LOUIS, SAITO, TSUNEMASA, AND SCHLEE, JOHN, 1965, Ocean drilling on the continental margin: Science, v. 150, no. 3697, p. 709–716.

DILLON, W. P., AND PAULL, C. K., 1978, Interpretation of multichannel seismic reflection profiles of the Atlantic continental margin off the coast of South Carolina and Georgia: U.S. Geol. Survey Miscellaneous Field Studies Map MF 936, 1 sheet.

———, PAULL, C. K., BUFFLER, R. T., AND FAIL, J. P., 1979, Structure and development of the Southeast Georgia Embayment and northern Blake Plateau: preliminary analysis, *in* Watkins, J., Montadert, L., and Dickerson, P., eds., Geological and geophysical investigations of continental margins: Am. Assoc. Petroleum Geologists Memoir 29, p. 27–41.

———, PAULL, C. K., DAHL, A. G., AND PATTERSON, W. C., 197, Structure of the continental margin near the COST GE-1 drillsite from a common depth point seismic reflector profile (in press).

DRAKE, C. L., EWING, MAURICE, AND SUTTON, G. H., 1959, Continental margins and geosynclines: the east coast of North America north of Cape Hatteras, *in* Ahrens, L. H., and others, eds., Physics and chemistry of the earth: London, Pergamon Press, v. 3, p. 110–198.

EMERY, K. O., AND UCHUPI, ELAZAR, 1972, Western North Atlantic Ocean: topography, rocks, structure, water, life, and sediments: Am. Assoc. Petroleum Geologists Memoir 17, 532 p.

EWING, J. I., AND HOLLISTER, C. H., 1972, Regional aspects of deep sea drilling in the western North Atlantic: Deep Sea Drilling Project Initial Rept., v. 11, p. 951–973.

GARRISON, L. E., 1970, Development of continental shelf south of New England: Am. Assoc. Petroleum Geologists Bull., v. 54, p. 109–124.

GIVEN, M. M., 1977, Mesozoic and Cenozoic geology of offshore Nova Scotia: Bull. Canadian Petroleum Geol., v. 25, no. 1, p. 63–91.

GROW, J. A., DILLON, W. P., AND SHERIDAN, R. F., 1977, Diapirs along the Continental Slope off Cape Hatteras (abs.): Program 47th Ann. Mtg., Society Exploration Geophysists, Calgary, p. 51.

GROW, J. A., JAWORSKI, B. L., AND MEEDER, C. A., 1978, Sedimentary rock velocity trends across Georges Bank (abs.): Geol. Soc. America Abs. with Programs, v. 10, no. 2, p. 45–46.

———, AND MARKL, R. G., 1977, IPOD-USGS multichannel seismic reflection profile from Cape Hatteras to the mid-Atlantic ridge: Geology, v. 5, p. 625–630.

———, MATTICK, R. E., AND SCHLEE, J. S., 1979, Multichannel seismic depth sections and interval velocities over outer Continental Shelf and upper Continental Slope between Cape Hatteras and Cape Cod, *in* Watkins, J., Montadert, L., and Dickerson, P., eds., Geological and geophysical investigations of continental margins: Am. Assoc. of Petroleum Geol. Memoir 29, p. 65–83.

———, AND SCHLEE, JOHN, 1976, Interpretation and velocity analysis of U.S. Geological Survey multichannel reflection profiles 4, 5, and 6. Atlantic continental margin: U.S. Geol. Survey Miscellaneous Field Studies Map MF 808.

HEEZEN, B. C., AND SHERIDAN, R. E., 1966, Lower Cretaceous rocks (Neocomian-Albian) dredged from the Blake Escarpment: Science, v. 154, p. 1644–1647.

JANSA, L. F., AND WADE, J. A., 1975, Paleogrography and sedimentation in the Mesozoic and Cenozoic, southeastern Canada, *in* Yorath, D. J., Parker, E. R., and Glass, D. J., eds., Canada's continental margins and offshore petroleum exploration: Canadian Soc. Petroleum Geol. Memoir 4, p. 79–102.

KEEN, C. E., AND KEEN, M. J., 1974, Continental margins of eastern Canada and Baffin Bay, *in* Burk, C. A., and Drake, C. L., eds., The geology of continental margins: New York, Springer-Verlag, p. 381–390.

KING, L. H., 1975, Geosynclinal development of the continental margin south of Nova Scotia and Newfoundland, *in* Van der Linden, W. J. M. and Wade, J. A., eds., Offshore Geology of eastern Canada: Canada Geology Survey Paper 74-30, v. 2, p. 199–206.

KING, L. H., AND YOUNG, I. F., 1977, Paleocontinental slopes of east coast geosyncline (Canadian Atlantic margin): Canadian Jour. of Earth Science, v. 14, p. 2553–2564.

KLITGORD, K. D., AND BEHRENDT, J. C., 1979, Basin structure of the U.S. Atlantic continental margin, *in* Watkins, J., Montadert, L., and Dickerson, P., eds., Geological and geophysical investigations of continental margins: Am. Assoc. Petroleum Geologists Memoir 29, p. 85–112.

LEHNER, P., AND DERUITER, P. A. D., 1977, Structural history of Atlantic margin of Africa: Am. Assoc. Petroleum Geologists Bull., v. 61, no. 7, p. 961–981.

MANHEIM, F. T., AND HALL, R. E., 1976, Deep evaporitic strata off New York and New Jersey—evidence from interstitial water chemistry of drill holes: Journal of Research of the U.S. Geological Survey, v. 4, no. 6, p. 697–702.

MATTICK, R. E., GIRARD, O. W. JR., SCHOLLE, P. A., AND GROW, J. A., 1978, Petroleum Potential of the U.S. Atlantic Slope, Rise and Abyssal Plain: Am. Assoc. Petroleum Geologists Bull., v. 62, no. 4, p. 592–608.

MAYHEW, M. A., 1974, Geophysics of Atlantic North America, in Burk, C. A., and Drake, C. L., eds., The geology of continental margins: New York, Springer-Verlag, p. 409–427.
PARSONS, M. G., 1975, The geology of the Laurentian fan and Scotian Rise, in Yorath, C. J., Parker, E. R., and Glass, D. J., eds., Canada's continental margins and offshore petroleum exploration: Canadian Soc. Petroleum Geol. Memoir 4, p. 155–167.
ROSS, D. A., 1977, Introduction to oceanography: Englewood Cliffs, Prentice-Hall, 438 p.
RYAN, W. B. F., 1978, Upbuilding, outbuilding and defacement of the continental slope (abs.), in The Second Maurice Ewing Memorial Symposium: Implications of deep drilling results in the Atlantic Ocean, p. 29–30.
———, VON RAD, ULRICH, AND OTHERS, 1976, Passive continental margin: Geotimes, v. 21, no. 10, p. 21–24.
———, CITA, M. B., MILLER, E. L., HANSELMAN, D., NESTEROFF, W. D., HECKER, B., AND NIBBELINK, M., 1978, Bedrock geology in New England submarine canyons: Oceanologia Acta, v. 1, no. 2, p. 233–254.
SCHLEE, JOHN, 1977, Stratigraphy and Tertiary development of the continental margin east of Florida: U.S. Geol. Survey Professional Paper 581-F, 25 p.
———, 1978, Geology of Georges Bank: Proc. New England Section of National Association of Geology Teachers, Kingston R. I., p. 88–92.
———, BEHRENDT, J. C., GROW, J. A., ROBB, J. M., MATTICK, R. E., TAYLOR, P. T., AND LAWSON, B. J., 1976, Regional geologic framework off northeastern United States: Am. Assoc. Petroleum Geologists Bull., v. 60, no. 6, p. 926–951.
———, MARTIN, R. G., MATTICK, R. E., DILLON, W. P., AND BALL, M. M., 1977, Petroleum geology on the United States Atlantic-Gulf of Mexico margins, in Cameron, V. S., ed., Exploration and economics of the petroleum industry—new ideas, new methods, new developments: Southwestern Legal Foundation: New York, Matthew Bender and Co., v. 15, p. 47–93.
SCHOLLE, P. A., ed., 1977, Geological studies on the COST No. B-2 well, U.S. mid-Atlantic outer continental shelf area: U.S. Geol. Survey Circular 750, 71 p.
SEIBOLD, EUGEN, AND HINZ, KARL, 1974, Continental slope construction and destruction, West Africa, in Burk, C. A., and Drake, C. L., eds., The geology of continental margins: New York, Springer-Verlag, p. 179–196.
SHERIDAN, R. E., 1974, Atlantic continental margin of North America, in Burk, C. A., Drake, C. L., eds., The geology of continental margins: New York, Springer-Verlag, p. 391–407.
———, 1976, Sedimentary basins of the Atlantic margin of North America, in Bott, M. P. H., ed., Sedimentary basins of the continental margins and cratons: Tectonophysics, v. 36, no. 1–3, p. 113–132.
———, GOLOVCHENKO, XENIA, AND EWING, J. I., 1974, Late Miocene turbidite horizon in Blake-Bahama basin: Am. Assoc. Petroleum Geologists Bull., v. 58, no. 9, p. 1797–1805.
———, SMITH, J. D., AND GARDNER, J., 1969, Rock dredges from Blake Escarpment near Great Abaco Canyon: Am. Assoc. Petroleum Geologists Bull., v. 53, no. 12, p. 2551–2558.
SHIPLEY, T. H., BUFFLER, R. T., AND WATKINS, J. S., 1978, Seismic stratigraphy and geologic history of Blake Plateau and adjacent western Atlantic continental margin: Am. Assoc. Petroleum Geologists Bull., v. 62, no. 5, p. 792–812.
STETSON, H. C., 1949, The sediments and stratigraphy of the east coast continental margin—Georges Bank to Norfolk Canyon: Massachusetts Institute of Technology and Woods Hole Oceanographic Institution, Papers in Physical Oceanography and Meteorology, v. 11, no. 2, p. 1–60.
TUCKER, P. M., AND YORSTON, H. J., 1973, Pit falls in seismic interpretation: Society Exploration Geophysics Monograph, no. 2, 50 p.
UCHUPI, ELAZAR, BALLARD, R. D., AND ELLIS, J. P., 1977, Continental slope and upper rise off western Nova Scotia and Georges Bank: Am. Assoc. Petroleum Geologists Bull., v. 61, no. 9, p. 1483–1492.
VAIL, P. R., MITCHUM, R. M. JR., AND THOMPSON, S. III, 1977, Seismic stratigraphy and global changes of sea level, part 4: global cycles of relative changes of sea level, in Payton, C. E., ed., Seismic stratigraphy—applications to hydrocarbon exploration: Am. Assoc. Petroleum Geologists Memoir 26, p. 83–97.
VALENTINE, P. C., 1978, Shallow subsurface stratigraphy of the continental margin off southeastern United States (abs.): Geol. Soc. of America Abs. with Programs, v. 10, no. 2, p. 90.
VON RAD, ULRICH, AND EINSELE, GERHARD, in press, Mesozoic-Cenozoic subsidence history and paleobathymetry of the northwest African continental margin (Aaiun Basin to DSDP Site 397): Philosophical Trans. Royal Soc. London.
WEED, E. G. A., MINARD, J. P., PERRY, W. J. JR., RHODEHAMEL, E. C., AND ROBBIN, E. I., 1974, Generalized pre-Pleistocene geologic map of the northern United States Atlantic continental margin: U.S. Geol. Survey Miscellaneous Inventory Series Map I-861, Scale 1:1,000,000.

SEDIMENTATION ON THE EASTERN UNITED STATES CONTINENTAL SLOPE

LARRY J. DOYLE
Department of Marine Science
University of South Florida
St. Petersburg, Florida 33701

ORRIN H. PILKEY
Department of Geology
Duke University
Durham, North Carolina 27708

C. C. WOO
U.S.G.S.
Quisett Campus
Woods Hole, Massachusetts 02543

ABSTRACT

Along the continental margin of the eastern United States, a major break in intercanyon continental slope sedimentation occurs at Cape Hatteras. North of Cape Hatteras, slope sediments are dominantly silts which exhibit little change in grain size from upper slope to lower slope. Because of the Florida Current, south of Cape Hatteras the upper slope has a greatly increased sand fraction, compared to the north, and grain size decreases downslope. All slope sediments are high in mica relative to the inshore portions of the margin, even the estuaries. Slope sediments have an exotic authigenic heavy mineral suite containing iron carbonates and iron sulfides. Detrital heavy minerals are depauperate in the densest fraction but otherwise mirror adjacent shelf suites. Slope benthic foraminiferal suites are mixtures of shelf and slope forms.

Intercanyon portions of the slope are active depocenters. North of Hatteras, hemipelagic sedimentation is the dominant process with shelf spillover a secondary contributor. South of the Cape, spillover is much more important. Although mass wasting phenomena have been shown in the literature on the grand scale, there is little evidence for them in the upper six meters of sediment.

Clay mineralogy and mica distribution on the continental margin indicate that some sediment being carried by rivers is getting through the estuaries, by-passing the shelf, and accumulating on the slope. They further show that winnowing on the adjacent shelf, both now and during the Holocene transgression, has provided additional sediment to the slope system.

INTRODUCTION

Objectives

By far the least studied portions of the continental margins of the eastern United States are the intercanyon areas of the continental slope. Emery and Uchupi (1972) summarized the status of United States Atlantic slope studies as follows:

"Surprisingly few studies have been made on the segments of continental slopes that lie between submarine canyons because of greater depths of the slopes. The number of slope studies has also been less than those of the deep-sea floor because of the smaller area of the slopes. Another reason the slopes are less studied is that many of them are dissected by submarine canyons and many marine geologists have found the canyons to be of greater interest than the relatively featureless slopes."

As the search for hydrocarbons intensifies in ever deeper waters, the slope province, its sediments, and sedimentary processes become of greater interest. The continental slope is transitional between the sedimentary environments of the continental shelf and the continental rise-ocean basin floor. It is a zone of flux which can be viewed as a critical link in the big picture of continental margin sedimentation. The nature of this link is poorly understood.

Since the continental slope is a frontier area about which little is known, a broad-scale investigation was designed to establish an overall sedimentological framework within which to attack specific problems.

Our objectives in this investigation are as follows:

1. What are the diagnostic characteristics of slope sediments of an Atlantic United States type margin; how can they be distinguished from other major marine sedimentary environments?

2. What are the principle sedimentary proc-

esses which operate on the slope? At what rate do they act?

3. What is the role of the adjacent continental shelf in furnishing sediment to the continental slope? What do sedimentary events on the slope reveal concerning models of shelf sedimentation?

4. What are the respective roles of submarine canyons and intercanyon areas of continental slopes in furnishing sediment to the continental rise and deep-sea during high versus low stands of sea level?

Geologic Setting

The continental slope of eastern North America is part of a tectonically stable trailing edge margin of the Atlantic type (Heezen, 1974). Over most of its length, the shelf break is well defined and lies at a depth of about 100 meters north of Cape Hatteras and 75 meters south of the Cape. North of Cape Hatteras, the slope is dissected by numerous submarine canyons (Emery and Uchupi, 1972) and terminates at about 2000 to 2500 meters, at the top of the continental rise. Bathymetry of the slope is rugged, with local relief in many places on the order of 10's to 100's meters, apparently as the result of numerous submarine rock outcrops and erosion by mass wasting phenomena such as sliding, slumping, and creep (McGregor, 1977; Embley and Jacob, 1977).

From Cape Lookout to Cape Canaveral, the continental slope is divided into two parts separated by the Blake Plateau. Between the Plateau and the shelf lies the upper continental slope (Florida to Hatteras) sampled for this investigation, which extends from the shelf break to the beginning of the plateau at 600 to 700 meters. Seaward of the Blake Plateau lies the Blake Escarpment, not sampled for this investigation.

Gorsline (1963), Emery and Uchupi (1972), Milliman and others (1972), Hathaway (1972), and Stanley (1978) have described some surface sediments from grab samples or short cores from portions of the uppermost continental slope. However, most of the data available on sediments of the intercanyon slope are based upon seismic reflection profiling. Moore and Curray (1963), Krause and others (1966), Hoskins (1967), Rona (1969), Uchupi (1970), Emery and Uchupi (1972), and Schlee and others (1978) have outlined the geologic framework of the province. McGregor (1977) has examined the upper few hundred meters of sediment with seismic lines along the strike of the slope. These profiles were taken in conjunction with the coring program of the present investigation. In addition, McGregor and Bennett (1977) have geophysically examined in detail what they interpret to be a massive submarine slide northeast of Wilmington Canyon.

Several surveys have been conducted on a number of the submarine canyons of the northeastern United States continental slope and a number of direct observations have been made from submersibles (see for example Trumbull and McCamis, 1967). Fenner and others (1971), Keller and others (1973), and Keller (1975) have reported that south of Hydrographer Canyon surface sediments in the canyons are chiefly silts, clayey silts, and silty clays. Predominance of these fine sediments suggests that most eastern margin canyons are inactive and may be filling up. Although some fine grained sediments may be moving down canyons, they are apparently not conduits of sands to the deep-sea during the present high stand of sea level. In contrast, Hydrographer Canyon and other canyons adjacent to Georges Bank are floored with sand and may be active transport sites (Keller, 1975).

Methods

One hundred thirty piston cores (Fig. 1) averaging between five and six meters in length were acquired in thirty-four transects perpendicular to the strike of the slope between Welker Canyon to the north and Cape Canaveral to the south. It should be born in mind that piston cores often miss the topmost sediments. All cores were x-radiographed whole, split, described, and sampled at 50 cm intervals for analysis of grain size, percent $CaCO_3$, heavy mineralogy, and mica content. Selected cores south of Cape Hatteras were analyzed for clay mineralogy and several north of Cape Hatteras for organic carbon.

Standard methodology was used throughout as described in Carver (1971). Grain size analysis was by sieve and pipette. Heavy minerals in the 125–250 micron fraction were separated using bromoform or a mixture of tetrabromethane and diethyl formanide of 2.85 density. Whole grain mounts were examined optically, by x-ray diffraction, and by SEM with a Kevex nondispersive x-ray analyzer. Clays were examined by x-ray diffraction using the techniques of Griffin (1962) and Grimm (1968). $CaCO_3$ was determined by gas displacement and organic carbon by combustion

---

Fig. 1.—Locations of the piston cores analyzed for this study and the discrete sand layers present within them. Transects made up of several cores are numbered. Individual cores are designated with the transect number and a letter beginning closest to the coast. For instance, 14A, 14B, 14C, 14D, 14E, and 14F are cores in transect 14 with 14A nearest the coast and 14F out on the rise.

EASTERN U.S. SLOPE SEDIMENTATION

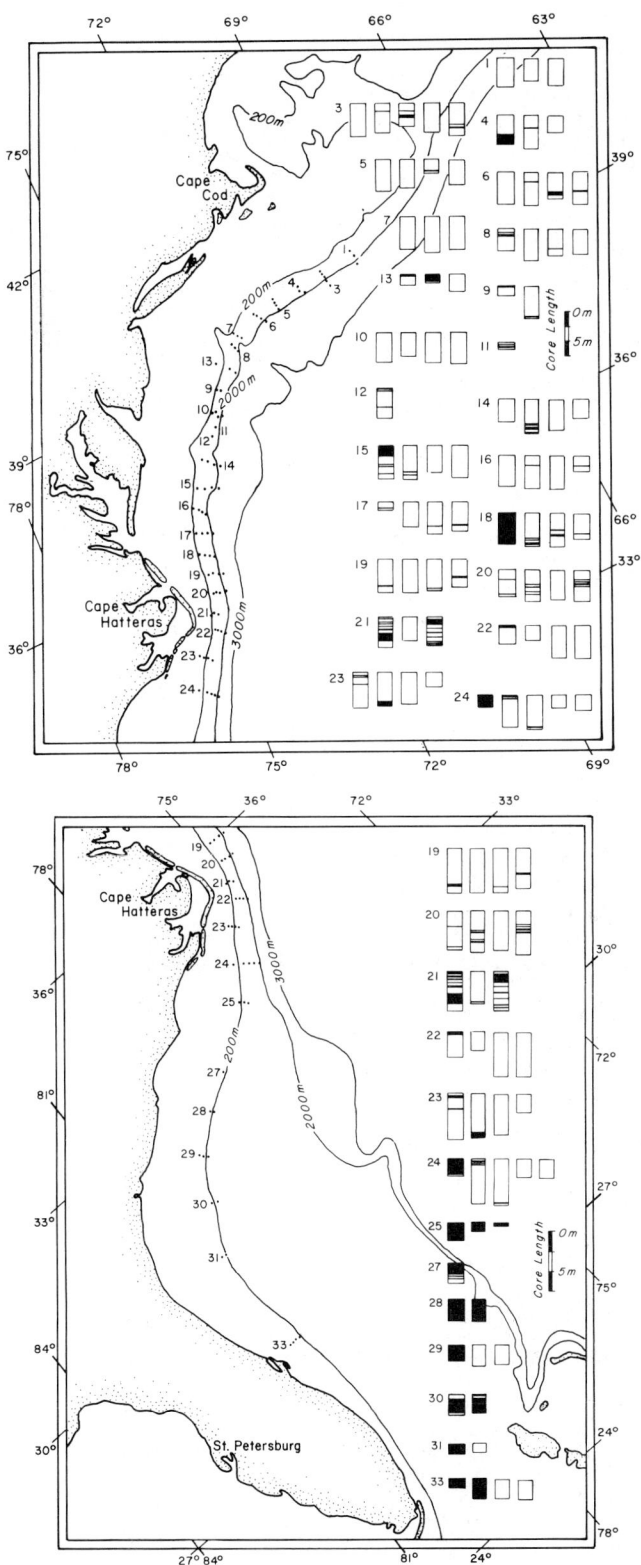

in an Angstrom carbon analyzer. Samples for organic carbon determination were not frozen. Four samples were dated by radiocarbon analysis at the Radiocarbon Dating Laboratory, University of Miami.

CHARACTERISTICS OF EASTERN UNITED STATES SLOPE SEDIMENTS

Texture

North of Cape Hatteras, the continental slope of the eastern United States is a depo-center for silt-sized sediments. Silt contents of cores from this area vary from about 50% to over 85%, and average over 60%. Sediments classed as silts, clayey silts, and sandy silts dominate with silty clays being locally important. Figure 2 shows the sand, silt and clay content of a typical core north of Cape Hatteras. The sand size fraction generally comprises less than 15% of the sediment. Once beyond the shelf break, no distinct trend toward finer grain sizes is apparent downslope.

A major break in grain size of slope sediments occurs at Cape Hatteras. South of the Cape, sand becomes a significant component, often comprising over 80% of the sediments. Percent sand decreases downslope.

The distribution of discrete sand layers in the continental slope cores shown in Figure 1 illustrates the general textural pattern. More numerous and thicker sand layers are present south of the Cape. Some of the small number of sand layers north of Cape Hatteras are graded, suggesting a possible turbidity current origin. South of the Cape, grading is not apparent in the sands.

The concept of a texturally graded continental shelf was widely held prior to extensive sedimentologic investigation of continental shelves following World War II. By this theory, sand should be found in the near-shore zone associated with beach systems and grade progressively to mud offshore. Shepard (1948) showed that the requisite gradation did not hold for continental shelves and the theory was discarded. The fact that the continental slope is the province of silt sized sediments leads us to suggest that the concept should be revived on a grander scale and applied to the whole of the continental margin, with sands being typical of continental shelves, silts and silty clays of continental slopes, grading to progressively finer sediments on the rise and in the abyssal plain (with the obvious exception of the turbidite sands).

Calcium Carbonate

Carbonate content of slope sediments north of Cape Hatteras is generally low (<10%) with a few layers up to 25%. South of Cape Hatteras, carbonate content is significantly higher, usually over 30%. This relative increase parallels the occurrence of higher carbonate percentages on the adjacent shelf documented by Milliman and others (1972), and Pilkey and others (1969). The calcium carbonate increase is about the same for the sand fraction as for the fine fraction. A large percentage of the sand fraction is composed of quartz grains swept over the slope break. The dominant components of the sand sized carbonate fraction are the tests of planktonic foraminifera. Occasionally, especially on the upper slope, molluscan shell hash similar to that found on the shelf makes up the carbonate sand fraction.

Tests of benthic foraminifera typical of the inner and central continental shelf are found within the slope cores further showing the loss of shelf sand sized sediment to the slope. Tests of shelf species identified in cores of transects 6 and 10 include *Islandiella norcrossi*, *Cribrononion* sp., *Elphidium* sp., and *Cassiculina laevigata*. Foraminifera displaced from the shelf are common and even dominant in most cores.

Heavy Minerals

The heavy mineral suite on the continental slope differs markedly from that reported on the adjacent shelf (Pilkey, 1963; Ross, 1970; and Milliman and others, 1972) in that it contains an important authigenic fraction. Authigenic heavy minerals are dominated by pyrite which is found as casts of foraminifera, worm tubes, radiolarians, and other

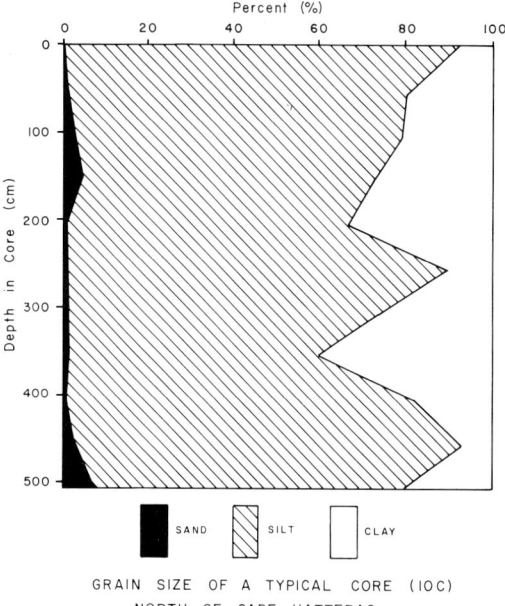

FIG. 2.—Sand, silt, and clay content of a typical core north of Cape Hatteras.

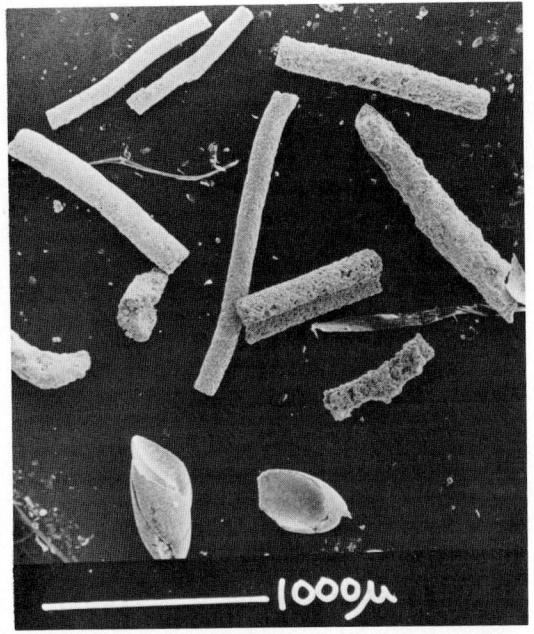

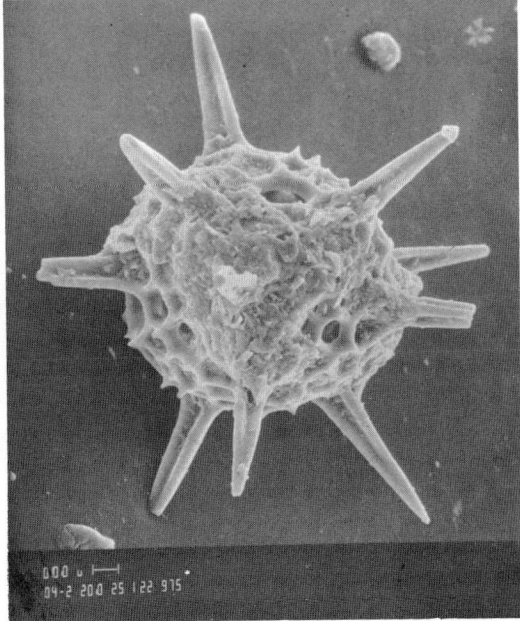

Fig. 3.—A) SEM micrograph showing some typical forms in which pyrite is found in slope sediments. B) SEM micrograph showing a radiolarian test filled with siderite.

organisms and free as globular masses within the sediments. Figure 3A shows some typical pyrite forms found in the sediments. Another authigenic mineral found less commonly in slope sediments is shown in Figure 3B, a radiolarian test filled with siderite. Other minerals of this exotic authigenic suite include ankerite, limonite, and greigite. The proportion of authigenics decreases in the heavy mineral fraction of slope sediments south of Cape Hatteras. Some authigenic minerals are relict or residual. Glauconite and phosphorite are abundant in core transects 25 and 27 (Fig. 1). Phosphorite is probably derived from Tertiary outcrops on the slope or near the shelf break.

The detrital heavy mineral suite in its dominant aspects parallels the suite of the adjacent continental shelf and beach zones. That is, where the shelf-beach system is dominated by hornblende and characterized by epidote, as it is south of Cape Hatteras, so is the slope. Cape Hatteras appears to be a major boundary between nothern and southern shelf provinces. The impoverishment of opaque minerals and garnet south of Cape Hatteras is the most striking difference. In addition, while similar in dominant species, the slope detrital heavy mineral suite appears to differ from that of the shelf by being depauperate in the densest fractions. Opaques, zircon, and rutile are less common in slope sediments, relative to adjacent continental shelf areas.

Mica

In the fine sand fractions, mica is the hydraulic equivalent of even finer fractions. This property has been used by Doyle and others (1968) to indicate the intensity of winnowing processes in various areas of the continental shelf.

Table 1 shows the distribution of mica grains per 10,000 grains of the 125–250 micron size fraction in both sand layers and lutite sections from the continental margin and deep-sea off the eastern United States. Little mica is present in continental shelf sands and only relatively small amounts in estaurine muddy sediments. A dramatic increase in mica content occurs in both the sand layers and lutite layers of continental slope sediments. Clearly, mica is diagnostic of slope sands and lutites when compared to estaurine muds and shelf sands.

Organic Carbon

Figure 4 shows the distribution of organic carbon content within piston cores on transects 15 and 16 (see Fig. 1). Curves of organic carbon content versus depth are typical of cores north of Cape Hatteras. Levels approach two percent at the top of most cores and decrease gradually to about 0.5 percent about 200 cm within the cores. Below that depth, organic carbon levels fluctuate between 0.1 percent and 0.5 percent to the bottom of the cores. South of Cape Hatteras in the more sandy slope sediments swept by the

TABLE 1.—EASTERN U.S. CONTINENTAL MARGIN MICA
(MICA EXPRESSED IN NO. GRAINS PER 10,000 GRAINS OF INSOLUBLE RESIDUE IN THE 125–250μ SIZE FRACTION)

Estuaries Lutite Layers	Southern Continental Shelf	Northern Continental Slope		Southern Continental Slope	
		Sand Layers	Lutite Layers	Sand Layers	Lutite Layers
New River	0	90	1490	70	1580
(Coastal Plain)	5	330	4390	110	2240
16	5	20	4890	280	2550
20	21	320	1750	50	1730
0	0	60	2340	50	
0	11	340	1340		
4	4	210	2710		
49	15	280	1280		
9	0	320	3480		
21	32	290	2030		
	2	540	3200		
Neuse River	0	180	8740		
(Piedmont)	75	600	1190		
0	9	80	1160		
20	1	140			
5	7	320			
0	3				

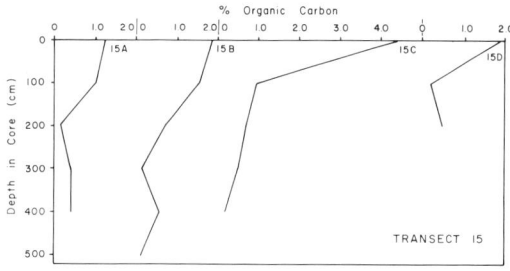

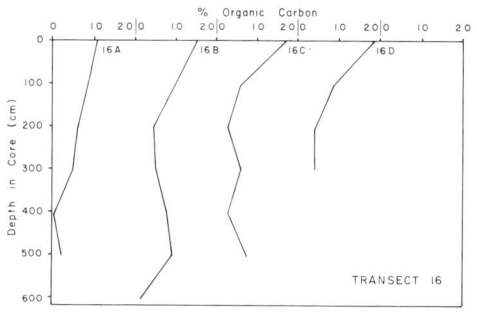

FIG. 4.—Organic carbon content within piston cores from transects 15 and 16. Curves are representative of organic carbon values north of Cape Hatteras.

Florida Current, surface organic carbon values may be expected to be much lower.

Clay Mineralogy

Cores south of Cape Hatteras show a mixed suite of clay minerals with Kaolinite, smectite, and illite all being present in appreciable quantity. Figure 5 shows a profile of clay mineralogy in a typical core.

SEDIMENTARY PROCESSES OF THE CONTINENTAL SLOPE

Textural parameters, composition, sedimentary structures as determined by x-radiography, plus the character of various sand grains as determined by scanning electron microscopy enables us to infer the sedimentary processes responsible for the upper few meters of slope sediments and how slope sedimentation interacts with the processes of the rest of the continental margin. North of Cape Hatteras where sediments are dominated by the silt and clay size fraction, x-radiographs show that the upper six meters of sediment are lightly to heavily burrowed. In the upper six meters, sedimentary structures related to current activity or mass wasting phenomena are rare. Core 14B on the middle slope (see Fig. 1) has been dated top and bottom by whole sample radio-carbon techniques. It shows an average age of 4,655 ± 110 years B.P. for the top 50 cm and an average of 27,000 ± $\frac{880}{990}$ years for the bottom 50 cm. This core is adjacent to the southwestern scarp of the massive slide scar east of Delaware Bay described by Embley and Jacob (1977). The upper few meters of slope deposits represented in this core are the result of hemi-pelagic sedimentation rather than slumping, turbidity currents, and other down-slope types of transport. On the upper rise, Core 14E, within what Embley and Jacob (1977)

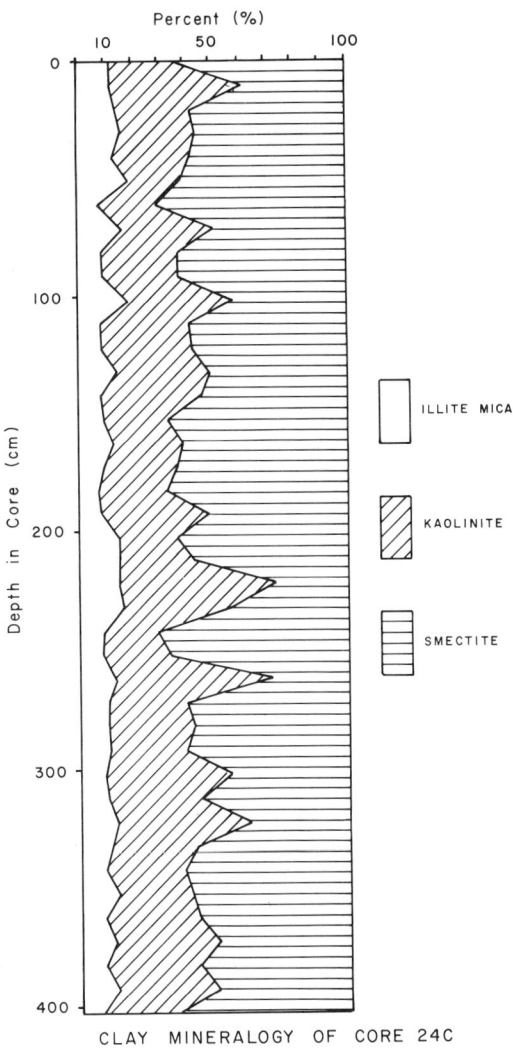

FIG. 5.—Clay mineralogy of a typical core south of Cape Hatteras showing relative distribution of smectite, kaolinite, and illite.

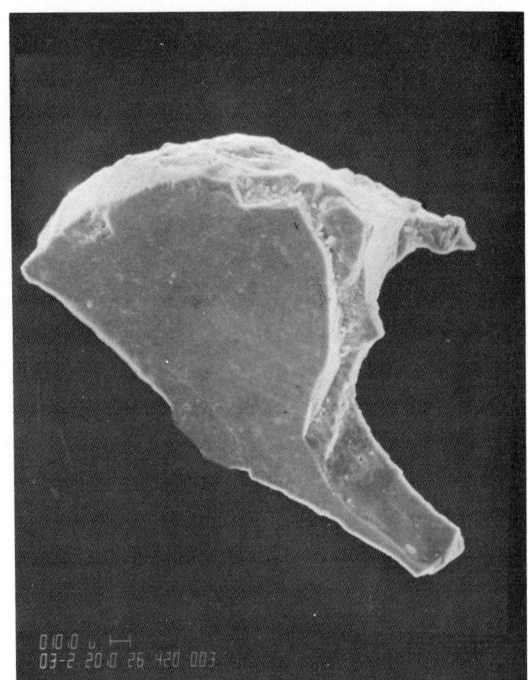

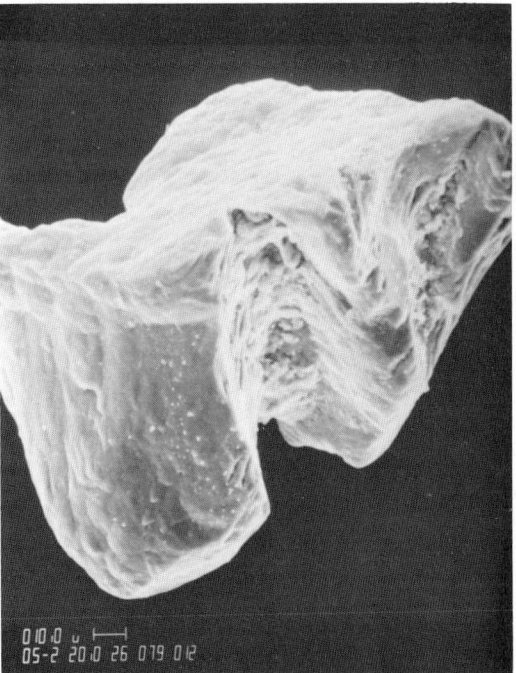

FIG. 6.—A and B) SEM micrograph of quartz grains which are interpreted as being glacially transported, based upon texture, angularity, and edge characteristics.

interpret as the slide deposit, gives an age of $25,515 \pm {}^{735}_{810}$ for the 10 to 30 cm interval and $26,465 \pm {}^{685}_{750}$ for the 500 to 540 cm interval. Similarity of top and bottom ages show that the sediment on the upper rise has been thoroughly mixed as expected within such a deposit.

Displaced middle and outer shelf benthic foraminifera present throughout the slope cores, quartz sand, and a detrital heavy mineral suite similar to that of the adjacent shelf (with the diminution of the densest minerals) suggest that shelf spill-over is a constant though muted process north of Cape Hatteras.

As expected, north of Baltimore Canyon, continental slope sediments show evidence of contribution by ice rafting. Figures 6A and 6B are scanning electron micrographs of sharp angular quartz grains with the typical conchoidal fracture pattern indicative of ice rafting (Krinsley and Takahashi, 1962; and Blackwelder and Pilkey, 1972). Scattered pockets of granules and gravels a few centimeters in diameter in a lutite matrix in northern slope cores further suggest ice rafting. While present, gravel pockets are rare, even in northern cores, indicating that ice rafting may not be as important in late Pleistocene slope deposition in mid-latitudes as could be expected.

South of Cape Hatteras, a major increase in quartz sand content, decreasing rapidly downslope, demonstrates that shelf spill-over is an important depositional process. Relative increase in sand must be due at least in part to the effect of the Florida Current winnowing fines from slope deposits. Incursions of the Florida Current onto the outer continental shelf may aide in sweeping sands over the shelf break. Although shelf spill-over is more important south of the Cape, presence of a considerable fine fraction in many slope cores and decreased sand content on the lower portion of the slope argue that hemipelagic sedimentation is still an important factor.

The Shelf / Slope System

In preceeding sections, we have discussed the presence of quartz sand and a heavy mineral suite, which is by and large similar to that of the adjacent shelf, in slope sediments. These factors suggest that there is some cascading of bottom sediments over the shelf break and down the slope, a process which seems to be especially important south of Cape Hatteras. This concept is further reinforced by the presence and often the dominance of shelf benthic foraminifera throughout many of the cores. In addition, we noted that the densest fraction of the heavy mineral suite was absent. One way to accout for this distribution of heavy minerals may be that the shelf acts as a giant Willfley table, a shaker table on which large storms act as a vibrator. Densest minerals and relatively coarse grained sediments are caught on the baffle of the shelf break and less dense fractions are passed over the edge, down the slope.

If the continental slope heavy mineral assemblages are a product of a Willfley table-like process, the sand size sediments must have arrived when the sea was at least partially covering the shelf. Thus the slope sediment cover must have been derived during the Holocene transgression and perhaps is still arriving at the present time.

Several lines of evidence also lead to the conclusion that some fine grained sediments are escaping the estuaries, by-passing the shelf, and are being deposited, at least temporarily on the slope. Distribution of mica in continental margin sediments is shown in Table 1. From an overall standpoint, the greater abundance of mica in continental slope sediments as compared to the shelf reflects the lower energy conditions prevalent in the deeper water environment. However, the fact that marsh muds are less micaceous than slope muds is a measure of the high degree of effectiveness of the various nearshore processes in trapping mud (fecal pellet formation, coagulation, etc.). Apparently the sum effect of the processes which are reviewed by Drake (1976) is to create mud particles with less winnowing potential than mica flakes. The large slope-shelf difference in mica content is an indication of present day by-passing of material from the nearshore zone to deeper water.

Clay mineralogy of the continental margin south of Cape Hatteras reinforces the notion of by-passing by fines of estuary and shelf and deposition on the slope which has been suggested by the heavy mineral and mica distribution. Pevear (1972) has shown that rivers south of Cape Hatteras, which rise in the Piedmont, carry a clay mineral suite dominated by kaolinite, reflecting the dominance of that mineral in Piedmont rocks. The Coastal Plain and Coastal Plain river load is dominated by smectite. Pevear (1968) found that smectite also dominated the clay fraction of sediments of the continental shelf of the southeastern United States. This is not unexpected since shelf strata are continuations of Coastal Plain formations with a reworked veneer of sediments a few meters thick overlying them (Milliman and others, 1972; Pilkey and others, 1976).

Doyle and others (1978) have shown that the clay mineral suite in the suspended sediments of the southeastern United States continental margin is dominated by kaolinite. Analysis of clay minerals from cores on the continental slope south of Cape Hatteras shows a mixed suite of kaolinite, illite, micas and smectite. The fact that kaolinite being born by Piedmont Rivers apparently largely by-passes the estuaries and the fact that kaolinite is important in the slope clay mineral assemblage strongly argues that fines are by-passing the entire shelf at the present time. If this is not the case, we are left with the problem of what happened to Piedmont river kaolinite. In addition, the smectite in slope cores further suggests that shelf sediments are being winnowed of fines which are at least partially redeposited and are residing, at least for a time, on the slope.

Doyle and others (1978) have also shown that, as expected, highest concentrations of suspended particulates are nearshore and load generally de-

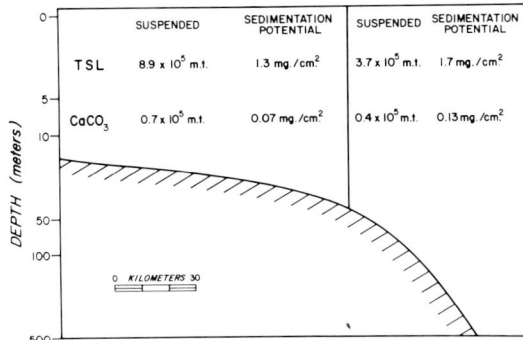

Fig. 7.—The average distribution of total suspended load and calcium carbonate suspended load over the continental shelf and slope of the southeastern United States based upon four quarterly cruises, after Doyle and others (1978). Also shown is the sediment potential for the shelf and slope (see text).

creases offshore. Slope values per unit volume are significantly lower than those of the shelf. Figure 7 shows the average total amount of suspended particulate matter and of $CaCO_3$ in the water column of the southeastern shelf and slope based upon data from four seasonal cruises. Again, as expected, the total amounts in suspension over the shelf are greater than those over the slope. Figure 7 also shows what we call the sedimentation potential of the water column. It is a hypothetical case never achieved in nature in which all particulates in suspension in the water column settle to the bottom immediately below. Its value is in allowing comparison of potential sedimentation in areas which have varying amounts of material in suspension and water columns of unequal depth, while ignoring water movement of all kinds. Perusal of this parameter shows that the slope has a significantly higher sedimentation potential than the shelf even though concentrations per unit volume within the water column are much lower, because the water column over the slope is much thicker than that over the shelf.

Our discussion of data to this point suggests that the intercanyon portions of the continental slope are active depo-centeres. Actual rates of sedimentation remain to be determined. Up to this time, we have been able to obtain only four whole sample radio-carbon dates which were discussed previously. Core 14B from the upper slope showed an age of 4,655 years B.P. near the top and 27,000 near the bottom, which translate to a sedimentation rate of about 22 cm/1000 years, roughly corresponding to the rate in some of the active basins of the southern California Borderland. These data, while suggestive of relatively high rates of sedimentation, need to be augmented with many more.

SUMMARY AND CONCLUSIONS

This investigation is best summarized by examining each of our originally stated objectives:

1. What are the diagnostic characteristics of slope sediments of an Eastern United States type margin; how can they be distinguished from the other sedimentary environments?

Slope sediments are the realm of silt sized particles. They are very high in mica relative to adjacent shelf sediments and even relative to estuarine muds. In addition, they have an exotic authigenic heavy mineral suite (ankerite, siderite, pyrite, greigite), while the detrital heavy mineral suite is lacking or is poor in the densest minerals (the opaques, zircon, and rutile). Finally, slope benthic foraminiferal assemblages may be mixtures of shelf and slope forms or shelf forms may completely mask endemic slope species. A mixed foraminiferal assemblage or a shelf assemblage in sediments with other slope characteristics might then be diagnostic. South of Cape Hatteras, grain size on the upper slope increases dramatically.

2. What are the principal sedimentary processes which operate on the slope? At what rates do they act?

The principal processes apparent in the uppermost sediment cover of the continental slope are hemipelagic sedimentation combined with shelf spill-over. South of Cape Hatteras, the Florida Current plays an important role by winnowing the sediment cover; while, north of Cape Hatteras, there is little evidence of current activity playing an important role in sedimentation. Seismic studies indicate that mass wasting is an important process, but this is not evident in the upper sediment cover. Turbidites are infrequent and generally small in areal extent and thickness.

We envision the slope sedimentation big picture to be as follows. Sediment accumulates relatively rapidly, largely by hemipelagic process. Most evidence concerning rates of sedimentation are indirect. High organic content, shelf sediments in slope cores, evidence of fine sediment escaping the estuaries, by-passing the shelf and settling on the slope, and mass wasting all suggest relatively high rates during the present high stand of sea level. These inferences are supported by our single pair of radio-carbon dates on the upper slope which give a depostion rate of 22 cm/1000 years. Most turbidity currents are generated at the heads of canyons which received large sediment supplies from rivers during times of lowered sea level. The canyons act to funnel material across the slope, adding little material to the slope. Occasionally, large scale mass movement must transport intercanyon area material to the base of the slope and beyond.

3. What is and has been (during the Holocene Transgression) the role of the adjacent continental shelf in furnishing sediment to the slope? What do sedimentary events on the slope reveal concerning models of shelf sedimentation?

Clay mineralogy and mica distribution suggest that fine sediments brought down rivers appear to be getting through the estuaries and by-passing the shelf. At least part of them are being deposited on the continental slope. Clay mineralogy indicates that fines have also been winnowed from the shelf and deposited on the continental slope. Detrital heavy mineralogy, shelf foraminifera and shell hash in slope sediments, and phosphorite and glauconite from the shelf in slope cores show that shelf sand size material as well as fines are being transported over the shelf edge into the slope system.

Thus the slope, particularly the intercanyon area, is a regime of very active sedimentation at this time of high sea level. It would appear that there is validity in both onshore (Meade, 1969) and offshore (McCave, 1972) shelf sediment transport models for the U. S. Atlantic Continental Margin. Evidence from previous studies indicates onshore transportation of both fines (Meade, 1969) and sand sizes (Pevear and Pilkey, 1968) is occurring at present. This investigation indicates that newly arrived fluvial sediment plus shelf produced material is also traveling across the shelf to the slope.

During the Holocene Transgression, as the shoreline crossed the continental shelf, it is now apparent that barrier islands migrated concurrently (Swift, 1975; Field and Duane, 1976). As preexisting back-barrier deposits were destroyed by the rising sea level, the fines thus contributed to the shoreface must have furnished a continuous supply of material to the slope, along with some of fluvial contribution at any given time. Since the sea level rise has slowed during the past 5–7000 years, the contribution from overrun backbarrier deposits has been temporarily reduced and it would be expected that the present day rate of slope sedimentation is somewhat lower than that of a few thousand years ago.

4. What are the respective roles of submarine canyons and intercanyon areas of continental slopes in furnishing sediment to the continental rise and deep-sea?

During the present high stand of sea level, intercanyon portions of the slope are active depocenters. Keller (1975) and Keller and others (1973) have shown that many canyons on the eastern margin of North America are filling with sediment, rather than acting as active conduits of material to the deep-sea. Geometrically, intercanyon areas must have steeper gradients than submarine canyons so it is quite possible that they are contributing more material to the rise and deep-sea than are the canyons. During low stands of sea level, rivers directly flow into canyon heads and so canyons become the major conduits of sediment to the rise and adjacent plains.

Whatever the case, intercanyon areas are where the action is today.

ACKNOWLEDGMENTS

The authors wish to acknowledge the NOAA/AOML Laboratory in Miami which provided shiptime aboard the R/V RESEARCHER and logistic support for collection of the cores. U.S.G.S. Contracts 14-08-0001-1625 and 14-08-0001-15681 funded portions of this work. Mark Ayers, Gary Hayward, Linda Lehman, and Rick Wall carried out much of the laboratory analysis. Dr. Robert Fleisher of Exxon Production Research identified the foraminifera.

REFERENCES

Blackwelder, P. L., and Pilkey, O. H., 1972, Electron microscopy of quartz grain surface textures: The U.S. Eastern Atlantic Continental Margin: Jour. Sed. Petrology, v. 42, p. 520–526.

Carver, R. E., 1971, Procedures in Sedimentary Petrology: New York, Wiley Interscience, 635 p.

Doyle, L. J., Betzer, P. R., Peacock, M. A., and Wall, F., 1978, Seston of the southeast Georgia Embayment: Final Report to U.S. Geol. Survey Contract No. 14-08-0001-15681, 72 p.

———, Cleary, W. J., and Pilkey, O. H., 1968, Mica: its use in determining shelf-depositional regimes; Marine Geology, v. 6, p. 381–389.

Drake, D. E., 1976, Suspended sediment transport and mud deposition on continental shelves, in Stanley, D. J., and Swift, D. J. P., eds., Marine Sediment Transport and Environmental Management: Stroudsburg, Dowden, Hutchinson, and Ross, p. 127–158.

Embley, R. W., and Jacob, R., 1977, Distribution and morphology of large submarine sediment slides and slumps on Atlantic continental margins: Marine Geotechnology, v. 2, Marine Slope Stability, p. 205–228.

Emery, K. O., and Uchupi, E., 1972, Western North Atlantic Ocean: Topography, rocks, structure, water, life, and sediments; Am. Assoc. Petroleum Geologists Memoir 17, 532 p.

Fenner, P., Kellinn, G., and Stanley, D. J., 1971, Bottom currents in Wilmington submarine canyon: Nature Phys. Science, v. 229, p. 52–54.

Field, M., and Duane, D., 1976, Post-Pleistocene history of the United States inner continental shelf: Significance to origin of barrier islands: Geol. Soc. America Bull., v. 87, p. 691–702.

GORSLINE, D. S., 1963, Bottom sediments of the Atlantic shelf and slope off the southern United States: Jour. Geology, v. 71, p. 422–440.

GRIFFIN, G. M., 1962, Regional clay-mineral facies-products of weathering intensity and current distribution in the northeastern Gulf of Mexico: Geol. Soc. America Bull., v. 73, p. 737–768.

GRIMM, R. E., 1968, Clay Mineralogy: New York, McGraw-Hill, 596 p.

HATHAWAY, J. C., 1972, Regional clay mineral facies in estuaries and continental margins of the United States east coast, in Nelson, B. W., Environmental framework of coastal plain estuaries: Geol. Soc. America Memoir 133, p. 293–316.

HEEZEN, B. C., 1974, Atlantic-type continental margins, in Burke, C. A. and Drake, C. L., eds., The Geology of Continental Margins: New York, Springer-Verlag, p. 13–24.

HOSKINS, H., 1967, Seismic reflection observations on the Atlantic shelf, slope and rise southeast of New England: Jour. Geology, v. 75, p. 598–611.

KELLER, G. H., 1975, Sedimentary processes in submarine canyons off northeastern United States, in Proceedings IXme Congres International de Sedimentologie Theme 6, p. 77–81.

———, LAMBERT, D., RONA, G., AND STARESINIC, N., 1973, Bottom currents in the Hudson Canyon: Science, v. 180, p. 181–183.

KRAUSE, D. C., CHARAMEIC, M. A., WALSH, G. M., AND WISOTSKY, S., 1966, Seismic profile showing Cenozoic development of the New England continental margin; Jour. Geophys. Research, v. 71, p. 4327–4332.

KRINSLEY, D. H., AND TAKAHASI, TARO, 1962, Applications of electron microscopy to geology: New York Acad. Sci. Trans., Ser. 2, v. 25, p. 3–22.

MCCAVE, I. N., 1972, Transport and escape of fine-grained sediment from shelf areas, in Swift, D. J. P., Duane, D. B., and Pilkey, O. H.,eds., Shelf Sediment Transport: Stroudsburg, Dowden, Hutchinson and Ross, Inc., p. 225–248.

MCGREGOR, B., 1977, Geophysical assessment of submarine slide northeast of Wilmington Canyon: Marine Geotechnology, v. 2, Marine Slope Stability, p. 229.

MCGREGOR, B. A., AND BENNETT, R. H., 1977, Continental slope sediment instability northeast of Wilmington Canyon: Am. Assoc. Petroleum Geologists Bull., v. 61, p. 918–928.

MEADE, R. H., 1969, Landward transport of bottom sediments in estuaries of the Atlantic coastal plain: Jour. Sed. Petrology, v., 39, p. 222–234.

MILLIMAN, J. D., PILKEY, O. H., AND ROSS, D. A., 1972, Sediments of the continental margin off the eastern United States: Geol. Soc. America Bull., v. 83, p. 1315–1334.

MIXON, R., AND PILKEY, O. H., 1976, Reconnaissance geology of the submerged and emerged Coastal Plain province, Cape Lookout area, North Carolina: U.S. Geol. Survey Professional Paper 859, 45 p.

MOORE, D. G., AND CURRAY, J. R., 1963, Sedimentary framework of the continental terrace off Norfolk, Virginia and Newport, Rhode Island: Am. Assoc. Petroleum Geologists Bull., v. 47, p. 2051–2054.

PEVEAR, D. R., 1968, Clay mineral relationships in recent river nearshore marine, and continental slope sediments of the southeastern United States [Ph.D. Thesis]: University of Montana, 164 p.

———, 1972, Source of recent nearshore marine clays, southeastern United States, in Nelson B. W.,ed., Environmental Framework of Coastal Plain Estuaries: Geol. Soc. America Memoir 133, p. 317–335.

———, AND PILKEY, O. H., 1966, Phosphorite in Georgia shelf sediments: Geol. Soc. America Bull., v. 77, p. 849–858.

PILKEY, O. H., 1963, Heavy minerals of the U.S. South Atlantic Continental shelf and slope: Geol. Soc. America Bull., v. 74, p. 641–648.

———, BLACKWELDER, B. W., DOYLE, L. J., ESTES, E., AND TERLECKY, P. M., 1969, Aspects of carbonate sedimentation on the Atlantic continental shelf off the southeastern United States: Jour. Sed. Petrology, v. 39, p. 744–768.

RONA, P. A., 1969, Middle Atlantic continental slope of the United States: deposition and erosion: Am. Assoc. Petroleum Geologists Bull, v. 53, p. 1435–1465.

ROSS, D. A., 1970, Atlantic continental shelf and slope: Heavy minerals of the continental margin from southern Nova Scotia to northern New Jersey: U.S. Geol. Survey Professional Paper 529-G, 40 p.

SCHLEE, J. S., DILLON, W. P., AND GROW, J. A., 1979, Structure of the Atlantic Slope of Eastern North America, in Doyle, L. J., and Pilkey, O. H., eds., Continental Slopes; Soc. Econ. Paleontologists Mineralogists Spec. Pub. (in press).

SHEPARD, F. P., 1948, Submarine Geology: New York, Harper and Brothers, 348 p.

STANLEY, D. J., AND WEAR, C. M., 1978, The "mud-line": An erosion-deposition boundary on the upper continental slope: Marine Geology, v. 28, p. M19–M29.

SWIFT, D. J. P., 1975, Barrier island genesis: Evidence from the middle Atlantic shelf of North America: Sedimentary Geology, v. 14, p. 1–43.

TRUMBULL, J. V. A., AND MCCAMIS, M. J., 1967, Geological exploration in an east coast submarine canyon from a research submersible: Science, v. 158, p. 370–372.

UCHUPI, E., 1970, Atlantic continental shelf and slope of the United States: Shallow Structure; U.S. Geol. Survey Professional Paper 529-I, 44 p.

GEOTECHNICAL PROPERTIES OF CONTINENTAL SLOPE DEPOSITS—CAPE HATTERAS TO HYDROGRAPHER CANYON

GEORGE H. KELLER
School of Oceanography
Oregon State University
Corvallis, OR 97331

DOUGLAS N. LAMBERT AND RICHARD H. BENNETT
NOAA, Atlantic Oceanographic and Meteorological Labs
Miami, FL 33149

ABSTRACT

The continental slope off the northeastern United States commonly displays gradients ranging from 3 to 10°, is heavily dissected by submarine canyons and valleys, and is an area of considerable slumping activity. A study of the geotechinal properties (sediment texture, shear strength, water content, wet bulk density, porosity, and Atterberg Limits) of 73 sediment cores from 21 transects across the continental slope from Cape Hatteras to Hydrographer Canyon provides insight into the general distribution and variation of these properties within the near-surface deposits of this province of the seafloor.

Although a general gradation in sediment texture from coarse to fine prevails in a down-slope direction all along the continental slope, fine-grained sediments (silty clay) appear to comprise the predominant sediment type along the slope particularly within the central portion of the Middle Atlantic Bight. This depositional pattern appears to account for the occurrence of generally higher water contents and porosities as well as the lower wet bulk densities found in the slope deposits of the Middle Atlantic Bight. Relatively coarse-grained sediments of low water content and porosity and high bulk density make up the slope deposits to the north of Block Canyon as well as in the general area of Cape Hatteras. Higher values of shear strength [7 to 14 kPa (1–2 psi)] are commonly found in the lower mid- to lower slope deposits except in the vicinity of submarine canyons where lower values [2 to 4 kPa (0.3–0.6 psi)] appear to be related to a combination of increased concentrations of organic matter and fine-grained sediments. Sediment sensitivities range from 1 to 12 with a mean of 3, giving the indication that these deposits may be "slightly quick" in places, but they are predominantly classed as "medium sensitive." Porosities vary from 44–82% with the higher values occurring along the lower slope. The mean value of 71% for these sediments is slightly higher than that reported for the hemipelagic sediments of the North Atlantic. An analysis of the plasticity characteristics of the mid and lower slope sediments indicates that they vary little from those of abyssal plain deposits which are classed as inorganic clays of low, medium, and high plasticity. An exception is found in the central part of the Middle Atlantic Bight where a large proportion of the sediments are classified as organic clays of medium to high plasticity and micaceous.

Distribution of the mean values of the various geotechnical properties in the near-surface (1–290 cm) deposits of the slope indicates that there is a general increase in water content, liquid limit, plastic limit, clay content, and porosity from off New England towards Cape Hatteras. Analyses of slope deposits, using the infinite slope analysis method, to determine the effects of overburden pressure on their stability revealed no indication of instability. Factors other than overburden, however, undoubtedly play a role in the slumping of slope deposits.

INTRODUCTION

The continental slope is the most consistent and significant topographic discontinuity in the ocean basins, yet it has escaped study to any degree approaching that devoted to continental shelves and abyssal plains. The submarine slope off the east coast of the United States has received probably more attention than any other continental slope in the world, with investigations dating back to the 1930's (Shepard, 1934; Stetson, 1938; Veatch and Smith, 1939). Studies of the slope have usually been part of broad regional investigations of the continental margin (Stetson, 1949; Drake et al, 1959; Uchupi and Emery, 1967; Knott and Hoskins, 1968; Pratt, 1968; Uchupi, 1968a, 1970; Emery et al, 1970, Emery and Uchupi, 1972; Milliman et al, 1972; McGregor 1978) or have dealt with stratigraphic, morphological or sedimentary features on a very local scale (Moore and Curray, 1963; Roberson, 1964; Krauss et al, 1966; Hoskins, 1967; Rona and Clay, 1967; Uchupi, 1967; Emery and Ross, 1968; Trumbull and Hathaway, 1968; Uchupi, 1968b; Oser, 1969; Kelling and Stanley, 1970; MacIlvaine, 1973). More recently, as part of a program to assess offshore petroleum resources, the U.S. Geological Survey has conducted a study of the continental margin's structural and stratigraphic characteristics off the eastern United States (Schlee et al, 1976).

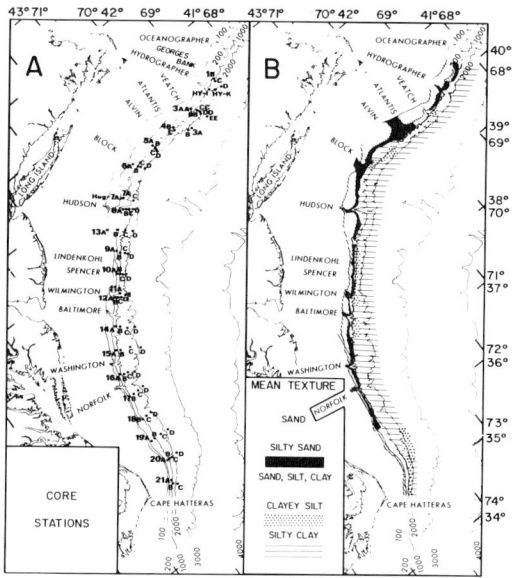

FIG. 1.—A) Core Stations; submarine canyons identified along inshore margin; isobaths in meters; B) Sediment texture distribution, mean composition for the respective cored intervals. Classification after Shepard (1954).

In 1974 the NOAA Atlantic Oceanographic and Meteorological Laboratories in cooperation with a number of university investigators initiated a geological study of the continental slope from Cape Hatteras to Hydrographer Canyon (Fig. 1A). The primary purpose of this effort was to improve the definition of the geological history, shallow structure, morphology, and sedimentary processes associated with the slope province. The basic field data collected during four crusies in the area consisted of five continuous seismic (air gun) reflection and narrow-beam echo sounder profiles paralleling the shelf break and approximately equally spaced along the slope from the shelf break to the lower slope-upper rise (McGregor et al, 1975; McGregor, 1978). In addition, 76 piston and 73 gravity cores were collected along 21 traverses; each one of which extended from the upper to the lower slope with two (traverses) continuing out onto the continental rise (Doyle et al, 1975) (Fig. 1A). The discussion presented here reports on the geotechnical properties of the slope deposits and the stability of these sediments in regard to both gravitational and bottom transport (currents) processes. In another section of this volume, Doyle, Pilkey, and Woo discuss the textural and mineralogical aspects of the piston cores collected during this study.

Location, Morphology, and Geological Setting

The continental slope between Cape Hatteras and Hydrographer Canyon is a classic example of a continental slope, bordered on the landward side by a relatively wide continental shelf and by a typical continental rise province seaward. The continental slope proper is a relatively steep incline, clearly delimited by the 120–140 and 2000 m isobaths (Fig. 1A). Considerable diversity in both bottom morphology and gradient is characteristic of the continental slope off the northeastern United States. The most pronounced morphological features on the slope are the submarine canyons which occur in great abundance. At least 190 canyons as well as numerous valleys entrench the slope from Labrador to Cape Hatteras (Oser, 1969; Emery and Uchupi, 1972; McGregor et al, 1975; McGregor and Bennett, 1977). The large number of canyons north of the Hudson Canyon undoubtedly owe their occurrence to the position of the former Wisconsin ice sheet and its associated drainage. To the south of Hudson Canyon the major canyons such as Wilmington, Baltimore, Washington, and Norfolk appear to have been linked to drainage through various estuarine systems carrying glacial meltwater from the north (Shideler and Swift, 1972; Milliman, 1973; Stout and McClennen, 1977; Twichell et al, 1977). Apparently the Pleistocene lowering of sea level resulted in the formation and/or re-excavation of a large number of the submarine canyons presently found along this margin of the Atlantic.

Morphologically, the continental slope from Georges Bank to Hudson Canyon consists of an upper and lower slope (Fig. 1A). Off Georges Bank the upper slope extends from the shelf break (120–140 m) to about 400 m with a gradient of 2°, and the lower slope extends to a depth of 2000 m with a declivity of about 7°. Between Georges Bank and the Hudson Canyon the upper slope extends to a depth of 1000 m with a gradient of 2°, and the lower slope to about 2200 m and has a declivity of 4°. From Hudson Canyon to Wilmington Canyon the continental slope narrows and increases in gradient to about 10°. The slope from Wilmington Canyon to lat. 36°30′ N consists of a relatively steep (11°) upper slope between the depths of 120 to 1200 m and a much more gentle (1°) lower slope extending to a depth of 2200 m. Here the lower slope takes on an appearance similar to that of an area that has undergone extensive erosion due to currents and/or massive slumping. Between lat. 36°30′ N and Cape Hatteras the continental slope displays a more uniform gradient than that to the north and is not cut by any submarine canyons, although numerous gullies are present (Fig. 1A). Off Cape Hatteras gullies and a number of canyons appear to form part of a dendritic drainage system which comprises the Hatteras Canyon system (Newton and Pilkey, 1969). From 36°30′N, with a gradient of about 3°, the slope steepens toward Cape Hatteras where it has a declivity of about 9°. Locally, along the eastern margin, gradients are often considerably more complex and variable, with slopes in some areas reaching angles of 20 to 25°.

The sedimentary framework of the overall continental slope appears to be one comprised of a thick Tertiary sequence which has buried Cretaceous strata as sediments were prograded seaward and deposited conformably to the present slope. Along much of the slope these Tertiary sediments have been since truncated as the slope has steepened due to erosion (Uchupi and Emery, 1967). Mass movements affecting Quarternary sediments reveal that slumping has been a continuing process along many parts of the slope (MacIlvaine, 1973; McGregor and Bennett, 1977).

Methods

In order to obtain the least disturbed samples as was practical, the gravity-type hydroplastic corer (Richards and Keller, 1961) was used to collect cores 8.2 cm in diameter and up to 3 m in length. The presence of open worm burrows in a number of the cores attests to the relatively low degree of sample disturbance.

All cores were x-radiographed aboard ship shortly after being collected and then stored vertically at a temperature of 4°C until the various geotechnical tests were performed.

Measurements of shear strength were made with a miniature laboratory vane-shear apparatus, a technique which is simple and particularly well suited to the testing of relatively low-strength submarine sediments. The tests were made on the center-most portion of the core (inner core) by inserting a small, four-bladed vane (2.50 × 1.25 cm) into the sample at right angles to the bedding and applying an increasing torque until shear occurred (Evans and Sherratt, 1948; Richards, 1961). In order to insure an undrained test condition, which is required for a valid vane-shear measurement, a vane rotation rate of $60°/min$ (1.7×10^{-2} rad/s) was used. Remolded shear strength was determined immediately after the natural strength measurement on the same material which had been thoroughly mixed with a spatula.

Considering the short core lengths (20–290 cm), relative to the scale of the study area, the respective values for each property along the entire length of each core were averaged. For example, if eight water content determinations were made on a core, the mean of these values along with the mean values of all the other cores, was used to define the areal distribution of this property as discussed in a later section.

AREAL VARIATIONS OF GEOTECHNICAL PROPERTIES

Sediment Texture

Complicated by the transport of sediments from the continental shelf, deposition of biogenic material, the presence of submarine canyons, the slumping of slope deposits themselves, and the effects of bottom currents, the sedimentation pattern along the continental slope off the northeastern United States is quite varied. Textural data from the cores collected during this program complemented by information from studies by Oser (1969, 1973), Hathaway (1971), and MacIlvaine (1973) serve as the basis for the sediment distribution pattern shown in Figure 1B. As seen from these data, clayey silts and silty clays are the predominant near-surface (0–3 m) sediment types found on the mid and lower portions of the continental slope, whereas sands and silts comprise the upper slope deposits. A marked change in sedimentation is noticed in the vicinity of Alvin Canyon (70°30′W) where to the east only the lowermost slope deposits are silty clays. Surface and near-surface deposits between Alvin and Hydrographer Canyons are considerably coarser-grained than those found elsewhere in the area north of the Hudson Canyon. The uppermost portions of the submarine canyons off New England commonly consist of reworked Pleistocene sand and gravels as the canyons erode farther into the shelf. Such headward erosion was observed in Veatch and Hudson Canyons from the submersible ALVIN by the senior author. The combination of strong bottom currents and a source of coarse-grained sediments on the adjacent shelf undoubtedly are prominent factors contributing to the greater concentration of coarse sediments along this segment of the continental slope. Current measurements and time-lapse camera observations in Hydrographer Canyon have recorded current velocities of up to 53 cm/sec and the actual migration of sand ripples seaward through the canyon down to depths of 710 m (Keller and Shepard, 1978). In contrast to these active canyons off New England, Hudson Canyon and those to the south are relatively inactive and presently are primarily depositional areas for fine-grained sediments.

In the vicinity of Cape Hatteras both the continental shelf and slope narrow considerably with the result that more of the coarse-grained shelf deposits appear to be spilling farther onto the slope. Relatively strong and complex currents in this area also undoubtedly contribute to the greater extension of coarse sediments down the slope. The extent of downslope transport of coarse-grained sediment along the entire study area is clearly seen from an analysis of the sand content of these deposits (Fig. 2A).

Shear Strength

Shear strength of a cohesive sediment is a function of the internal friction of the material, the cohesion, and the effective stress normal to the shear plane, which can be expressed as

$$\tau_f = c + \bar{\sigma} \tan \phi$$

where c is cohesion, $\bar{\sigma}$ is the effective stress, and ϕ the angle of internal friction. In theory, saturated, fine-grained sediments which are stressed without loss of pore water behave with

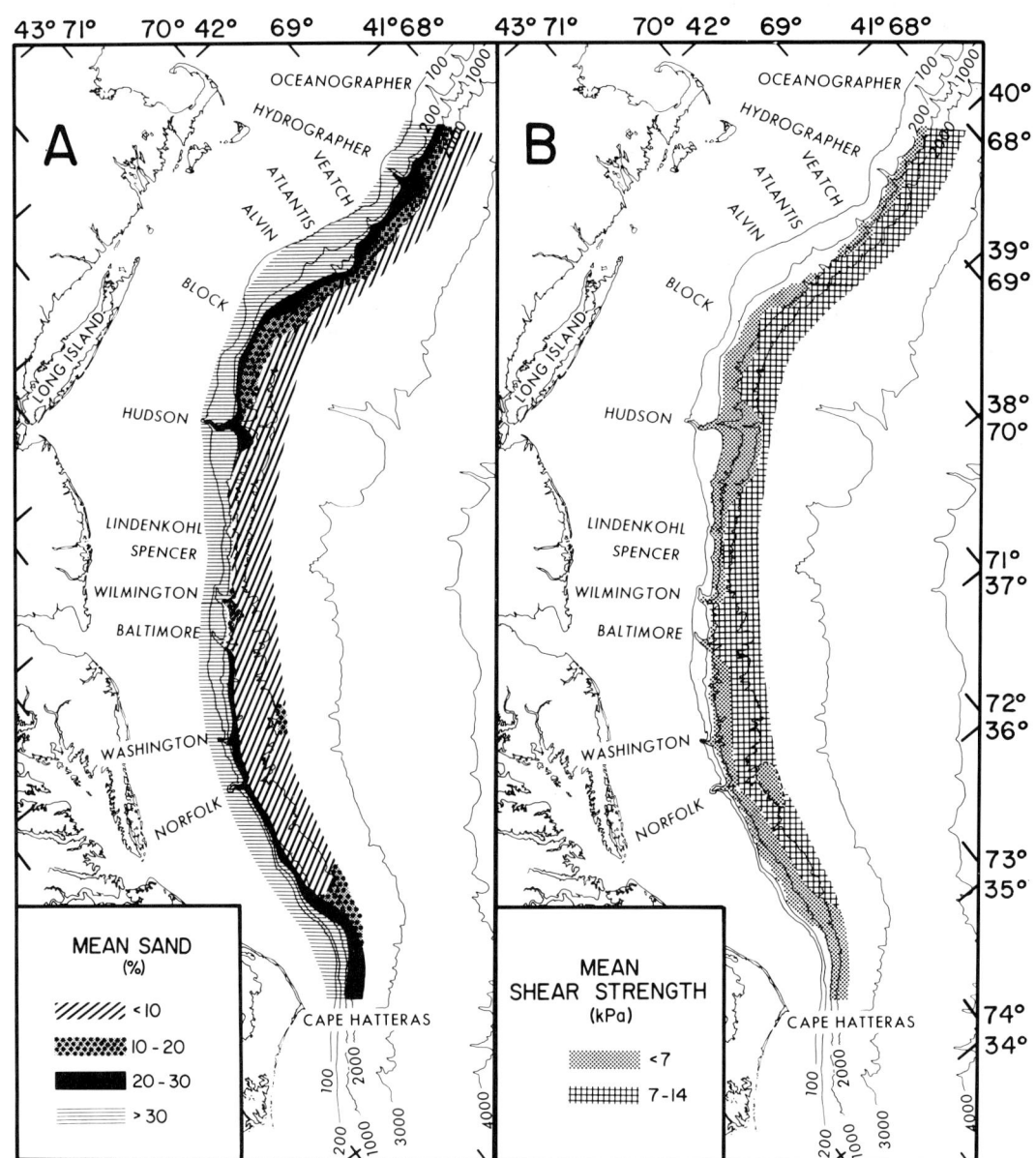

FIG. 2.—A) Sand distribution, mean concentrations for the respective cored intervals; B) Shear strength distribution, mean values for the respective cored intervals.

respect to an applied load as if they were cohesive materials without any internal friction ($\phi = 0$). In such a case, shear strength then is equal to cohesion ($\tau_f = c$). Care must be used, however, in assuming that shear strength and cohesion are synonymous since in reality some degree of internal friction does exist in these samples. Shear strength determinations were made on sub-samples comprised primarily of fine-grained cohesive sediments; intervals of high percentages of sand were not tested.

Mean shear strength values for the near-surface slope deposits range from 1.5 to 1.48 kPa (0.2–2.1 psi). An examination of these values for the continental slope as a whole reveals the presence of a generalized distribution pattern (Fig. 2B). Relatively low values [<7 kPa (1 psi)] comprise the upper and middle slope, and higher values [7–14 kPa (1–2 psi)] typify the lower slope as well as the upper rise. Overall, the near-surface slope deposits display a mean shear strength of 8.3 kPa (1.2 psi) (Table 1). Owing to the high concentrations (40% or more) of sand along the uppermost slope and at the shelf break, mean shear strength values based on vane shear tests are considered to be questionable and therefore have not been reported for this portion of the slope (Fig. 2B).

Lower shear strengths [2–4 kPa (0.3–0.6 psi)] are also generally found in association with a number of the submarine canyons. Indirectly, this may be an effect of the relatively high biological productivity commonly found in the canyons, which especially in the less active canyons results in higher concentrations of organic matter being incorporated into the fine-grained sediments. The combination of high organic carbon and fine-grained sediments may be a contributing factor to lower shear strengths particularly if any degree of remolding takes place (Pusch, 1973). An exception to the general shear strength distribution pattern is seen just south of the Hudson Canyon where relatively low values [<7 kPa (1 psi)] extend across the entire slope. An explanation for this occurrence is not readily apparent, but it is possible the very slight increase in sand content found in these sediments may be a contributing factor. The combination of textural components (sand, silt, clay) in the lower slope deposits in this area is somewhat similar to those of the lower upper slope sediments to the south, but is distinctly different from the mid and lower slope deposits farther south.

Off New England higher strength deposits [7–14 kPa (1–2 psi)] appear to blanket a considerable portion of the slope, in some areas extending from the mid slope on to the upper rise. The presence of strong bottom currents and the general setting of an erosional or nondepositional environment leads to the exposure of more cohesive sediment at the seafloor and the removal or nondeposition of soft, relatively low-shear strength material. A similar effect by currents on fine-grained sediments blanketing topographic highs has been found in other areas (Keller, 1970).

The lower and portions of the mid slope of the Middle Atlantic Bight are characterized by relatively high shear strengths. An explanation for this is not obvious from the available data, but a number of inferences may be made. The distribution pattern appears to correlate to some extent with the higher concentrations of clay-size material in the same area. In a number of areas the higher strengths are possibly due to the exposure of more cohesive sediments at the seafloor as a result of the extensive erosion that has affected the area in the past (Uchupi and Emery, 1967). Although adequate data are lacking, there is a good possibility that compositional characteristics such as unusual concentrations of various mineral assemblages may also contribute to the higher shear strengths displayed by these deposits. Any combination of these factors might possibly contribute to the higher shear strengths found along this portion of the slope.

Shear strengths are notably lower on the mid and lower slope in the vicinity of Cape Hatteras. This may be a reflection of the decrease in the

TABLE 1.—VARIATION OF GEOTECHNICAL PROPERTIES IN SLOPE DEPOSITS AND MEAN VALUES FOR THE ATLANTIC BASIN

	Maximum	Minimum	Mean	Altantic Mean[1]
Water Content (%)	165	33	88	86
Specific Gravity	2.84	2.60	2.72	2.73
Wet Bulk Density (Mg/m^3)	1.97	1.31	1.52	1.52
Porosity (%)	82	44	71	66
Shear Strength (kPa)	24.1	0.9	8.3	5.2
Sensitivity	12	1	3	4
Liquid Limit	124	23	79	65
Plastic Limit	56	16	36	77
Plasticity Index	74	3	42	34
Liquidity Index (%)	478	71	132	170

[1]from Keller and Bennett (1970)

proportions of fine-grained sediment in this area.

Sensitivity, the ratio of natural to remolded shear strength, is a measure of the strength lost as the result of disturbance or shock to the sediment. Sensitivities for the slope deposits off the northeastern United States vary from 1 to 12 thereby classifying them as insensitive to slightly quick (Rosenquist, 1953). Mean sensitivities for the cored intervals, however, range from 2 to 8 with values of 2–4 predominating for most of the slope deposits (Fig. 3A). Isolated areas of sediment with mean sensitivities of from 4 to 8 occur on the upper, middle, and lower slope, as well as on the upper rise. No clear pattern for the distribution of these more sensitive sediments can be defined from our data. Using the range of mean sensitivities reported here, slope deposits can be expected to lose from 50 to 87%

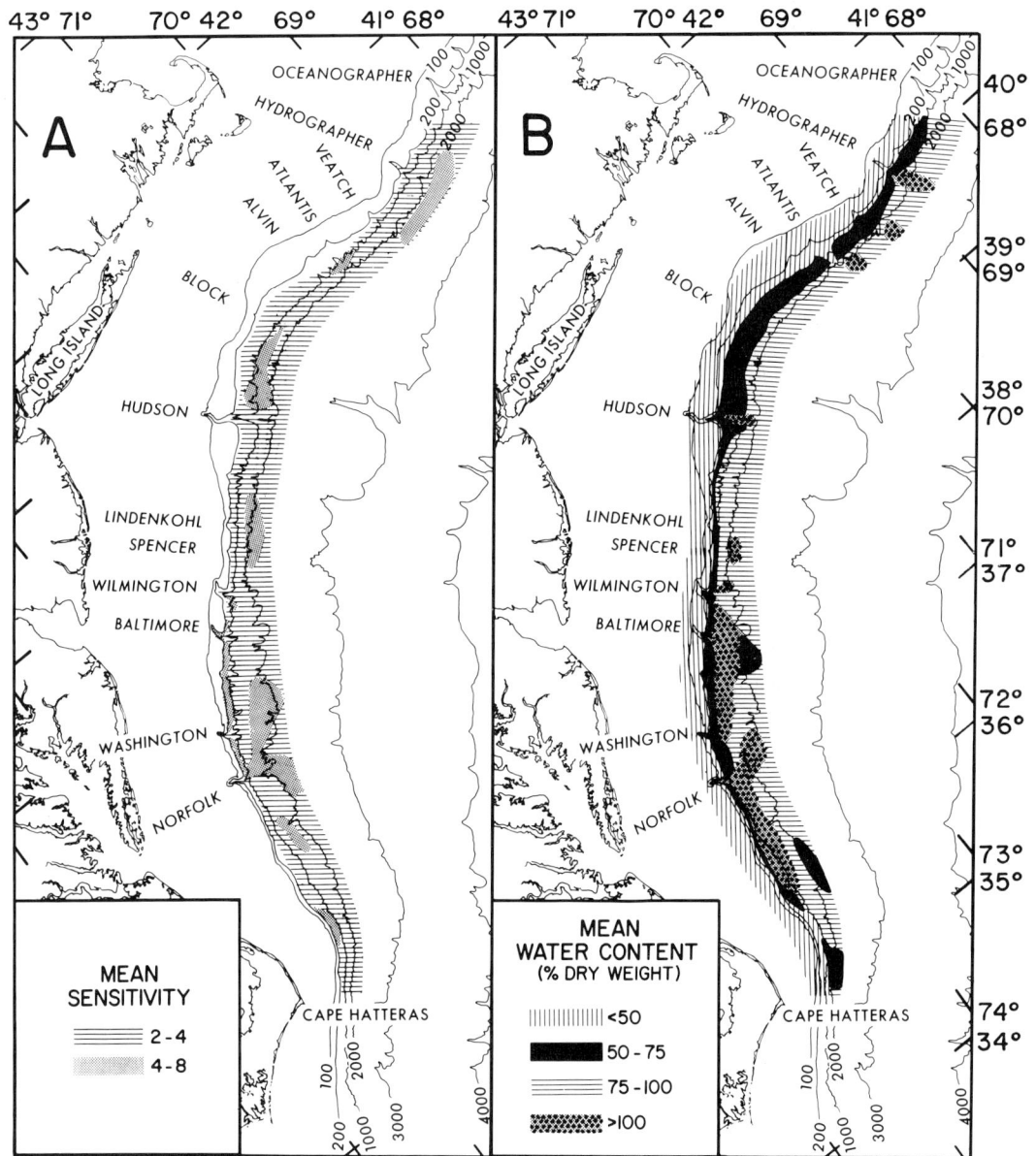

FIG. 3.—A) Sensitivity distribution, mean values for the respective cored intervals; B) Water content distribution, mean values for the respective cored intervals.

of their strength as a result of remolding (Richards, 1962). In comparison to the near-surface deposits of the Atlantic basin, which have a mean sensitivity of 4 (Keller and Bennett, 1970), the slope sediments display comparable characteristics.

Water Content

Water content as used here is the ratio, given as a percent, of the weight of water to the weight of oven-dried (110° C) solids in a given sediment mass. Correction for salt content has not been made in any of the data presented.

Mean water contents for the slope deposits range from 37 to 140% with an overall mean of 85%. The highest water contents are found in association with a number of submarine canyons and along the slope from Wilmington Canyon to about 36°N (Fig. 3B). Distribution of these relatively high values appears to be correlated with those areas in which sediments are fine-grained and contain relatively high concentrations of organic matter. It appears that the Hudson Canyon denotes a boundary between sediments possessing distinctly different water contents (Fig. 3B). To the north, water contents generally range from 45–75% except on the uppermost part of the slope where values of 37 to 45% are common. Water contents which are exceptional to the ranges noted above are associated with the lower portions of submarine canyons where values of 95 to 140% occur. South of Hudson Canyon, water contents are considerably higher for the most part, ranging from 75–130% over the mid and lower slope. This distribution pattern clearly reflects changes in sediment texture and the decree of induration in some cases, with the coarser-grained deposits north of Hudson Canyon having relatively lower water contents. This same area, to the north of Hudson Canyon, is also primarily one of erosion and non-deposition where exposed surface sediments are slightly indurated and possess lower water contents than would be found in recently deposited material. The coarser slope deposits off Cape Hatteras are also characterized by lower water contents. As might be expected, the lowest water contents along the entire slope occur in the vicinity of the shelf break and in the heads of submarine canyons where the sediments are the coarsest. In comparison to the adjacent rise and abyssal sediments, where water contents in the near-surface sediments range from 50–100% (Keller, 1968), slope deposits in the Middle Atlantic Bight display considerably higher water contents. An overall mean value of 85% for the slope deposits, however, corresponds remarkably well with the mean water content value (86%) reported for north Atlantic sediments as a whole (Keller and Bennett, 1970).

Wet Bulk Density

Wet bulk density, or wet unit weight, is the weight per unit volume of a sediment mass. The samples studied here are considered to be at 100% saturation and thus the term "saturated unit weight" applies to the values reported here.

Mean wet bulk densities for sediment cores collected along the continental slope range from 1.32 to 1.87 Mg/m^3 (82.4–116.7 pcf) with an overall mean of 1.52 Mg/m^3 (94.8 pcf) (Table 1). Relatively low bulk densities of 1.32 to 1.50 Mg/m^3 (82.4–93.6 pcf) predominate for the mid and lower slope deposits in the Middle Atlantic Bight (Lindenkohl Canyon to 35°45'N) as well as along most of the adjacent upper continental rise (Fig. 4A). Lower density sediments are commonly found also in the outer portions of a number of the submarine canyons. This occurrence is particularly pronounced in Hudson Canyon where fine-grained, organic-rich sediments predominate (Keller, 1975).

A distinct contrast in wet bulk densities occurs in the vicinity of Lindenkohl Canyon where slope deposits to the north possess somewhat higher densities than those to the south. The general distribution pattern of bulk density values for slope deposits off the northeastern United States appears to correlate reasonably well with variations in sediment texture; the higher densities are commonly associated with coarser-grained deposits. This relationship is particularly evident in the vicinity of Atlantis Canyon where high densities are associated with relatively coarse deposits blanketing the entire slope. Mean bulk densities for the mid and lower slope deposits south of Lindenkohl Canyon as well as for the upper rise are generally lower than the 1.50 to 1.75 Mg/m^3 (93.6–109.2 pcf) values reported for the adjacent North Altantic (Keller and Bennett, 1970). These lower values from the Middle Atlantic Bight correlate with the increase in clay-size material and may possibly also reflect the slightly higher concentration of organic matter reported in these margin sediments (Emery and Uchupi, 1972).

Porosity

Porosity is the ratio, expressed as a percentage, of the volume of voids to the total volume of a given sediment mass and is calculated based on the measured water content, wet unit weight, and grain specific gravity. A detailed discussion of porosity determination as well as that of any of the other geotechnical properties discussed above can be found in most texts dealing with soil mechanics.

The overall porosity distribution pattern as shown in Figure 4B takes on much the same appearance as that displayed by bulk density (Fig.

4A). Mean porosity values for the slope as a whole range from 50 to 81%, but more commonly vary from 67 to 73%. The highest values (70 to 80%) predominate in the mid and lower slope deposits of the southern part of the study area as well as in the upper rise sediments. Porosities are considerably lower along the slope off New England where values commonly range from 62 to 65% and rarely exceed 67%. In the coarser, uppermost slope deposits of this area, porosities generally vary from 53 to 58%. It is clear from the distribution pattern that areas of coarser-grained or slightly indurated sediment are relfected by lower porosities.

Plasticity Characteristics

A means of classifying the plastic characteristics of cohesive sediments was worked out some three decades ago when Casagrande (1948) developed the plasticity chart. By means of this

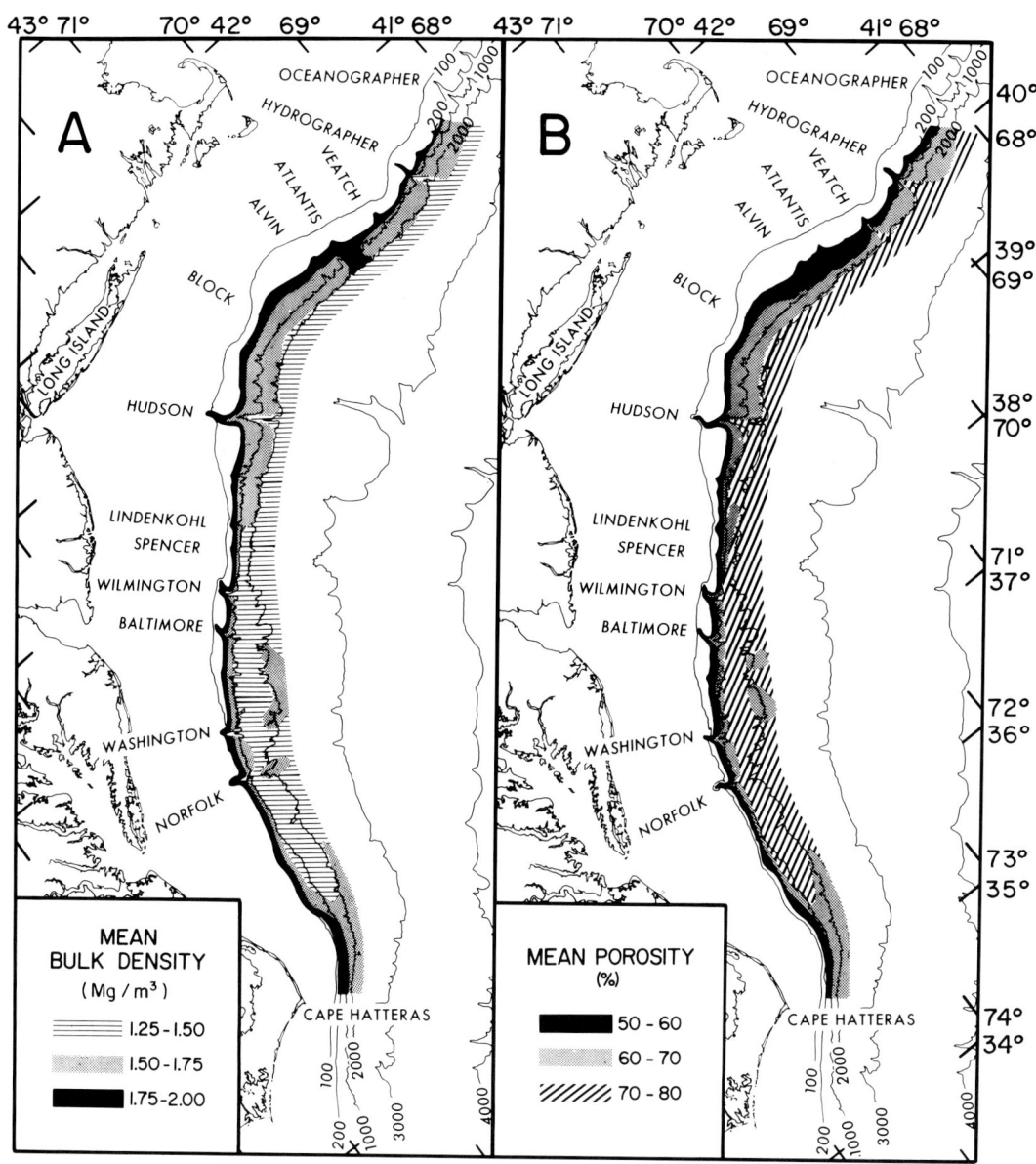

FIG. 4.—A) Bulk wet density distribution, mean values for the respective cored intervals; B) Porosity distribution, mean values for the respective cored intervals.

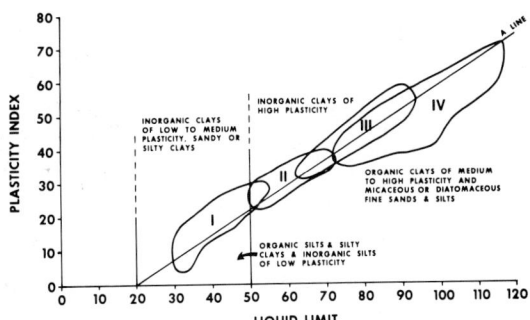

Fig. 5.—Plasticity chart.

chart, which relates plasticity index (range of water content between the liquid and plastic limit of the soil) to the liquid limit, cohesive sediments can be classed into eight categories (Terzaghi and Peck, 1948). An "A"-line is shown on the chart to represent an empirical boundary between inorganic clays, above the line, and organic clays along with inorganic silts below (Fig. 5).

Using Casagrande's classification, the continental slope deposits off the northeastern United States fall into four classes. The coarser-grained, upper-slope sediments are classed as inorganic clays of low to medium plasticity with sandy or silty clays and are designated as Group I (Fig. 5). This classification also applies to those sediments found in the heads of most submarine canyons. Sediments with these characteristics are found to extend down to mid-slope depths off New England (Fig. 6). Group II sediments which are inorganic clays of medium to high plasticity typify mid-slope deposits from Hudson Canyon to Cape Hatteras. In the Middle Atlantic Bight, south of Wilmington Canyon, the lower slope sediments are primarily classed as organic clays of medium to high plasticity containing micaceous sands and silts (Group IV). It is interesting to note that except for the outer portions of Hudson and Block Canyons, these highly plastic clays are limited essentially to this area of the continental slope. Lower slope deposits between Hudson and Wilmington Canyons and those of the upper rise along the entire northeastern margin of the United States are classed as inorganic clays of high plasticity (Group III) and are somewhat similar to the Group II material, but have higher liquid limits.

In summary, it can be seen from Figures 5 and 6 that considerable variation in the plasticity characteristics exists among the slope deposits off the northeastern United States. There is, however, a clear indication that the degree of plasticity increases toward the lower slope. As might be expected, the sediments as a whole off

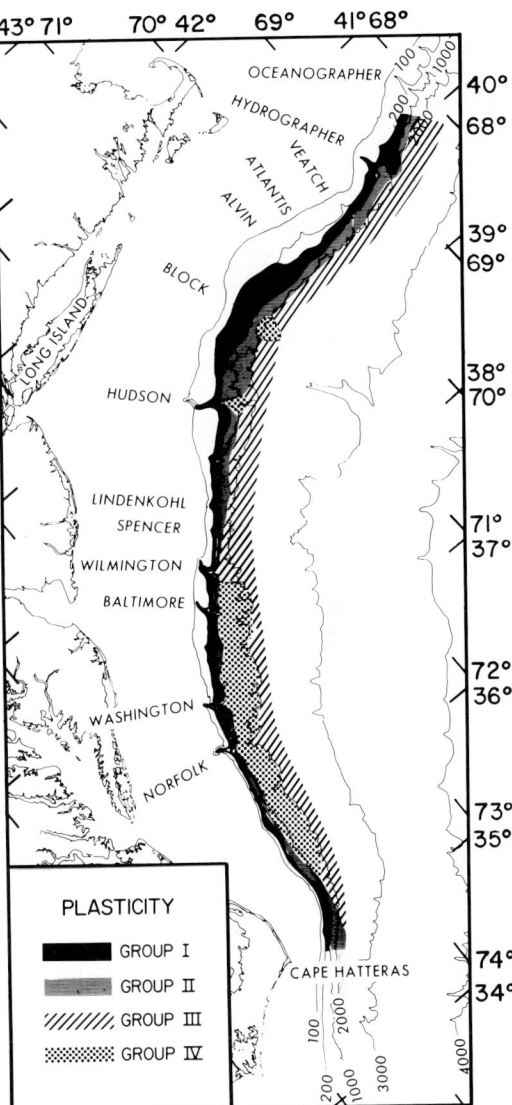

Fig. 6.—Distribution of plasticity characteristics. See Figure 5 for the basis of the four groups and the classification used.

new England and in the vicinity of Cape Hatteras display considerably lower plasticity indices than those blanketing the remainder of the slope off the northeastern United States. The highest plasticity indicies are primarily restricted to those sediments mantling the lower slope between Wilmington Canyon and 36°N. It is interesting to note that the occurrence of these high plasticity sediments coincides to a large degree with the area of anomalous bottom topography which is probably the result of a combination of extensive current erosion and slumping. Although the plasticity chart is a rather simple method of sediment classification, it provides a useful means of identifying similar and dissimilar depositional environments.

VERTICAL VARIATIONS OF GEOTECHNICAL PROPERTIES

The areal distribution of geotechnical properties has shown that considerable diversity exists among these properties along the continental slope off the northeastern United States. In order to better understand these variations on a local scale, it is appropriate to examine the vertical distribution and interrelationships of these properties from different areas of the slope. A series of vertical profiles of the geotechnical properties from four core transects is presented for this purpose (see Fig. 1A for core locations).

Georges Bank Transect

Continental slope deposits off New England exhibit large concentrations of sand extending across much of the slope, and only at its lower boundary are the near-surface sediments relatively free of sand (core 3EE) (Fig. 7). As might be anticipated in an environment influenced by oscillations of high energy, with erosion and deposition of coarse and fine sediments at varying intervals, the geotechnical properties vary considerably with depth. On the lowermost slope (core 3EE) the degree of these variations is markedly reduced. Progressing along the transect in a down-slope direction it is readily apparent that water contents and porosities increase as bulk densities decrease. It can also be seen that because of the more homogeneous nature of the sediments farther down on the slope, the geotechnical properties tend to vary with depth in a more uniform or defined manner. On the upper slope (cores 4B and 4C) the overall variation of any one property, except shear strength in core 4C, does not reveal any particular pattern of change with depth. This reflects the variations in the textural properties, particularly the sand content, which are somewhat variable throughout the sam-

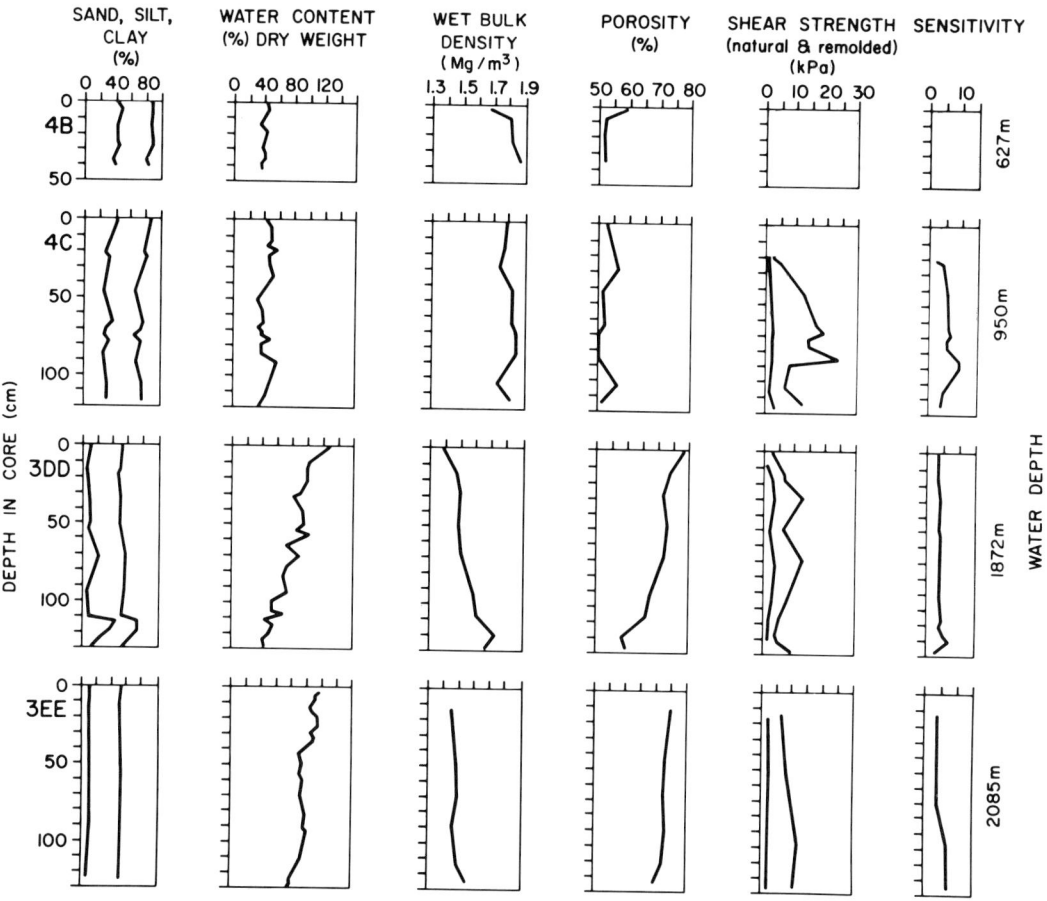

FIG. 7.—Vertical distribution of geotechnical properties, Georges Bank slope transect.

pled interval. Lower-slope sediments (cores 3DD & 3EE) display the more or less expected trends with depth of decreasing water contents and porosities and increasing bulk densities. Shear strengths vary only moderately along this core transect, except for the interval (45-85 cm) in core 4C of unusually high strength [25 kPa (3.6 psi)] which appears to reflect a zone of greater induration.

New Jersey Transect

As seen from the distribution of mean values of the geotechnical properties, slope sediments between Hudson and Wilmington Canyons appear to have somewhat distinct characteristics. Comparison of data profiles from the New Jersey slope reveals that the ranges of values for the various parameters vary relatively little from core to core (Fig. 8). Shear strength is an exception to this statement, in that the uppermost core (9B) in the transect is characterized by much lower shear strengths than are found in the deposits farther down on the slope.

At mid-slope (core 9B), variations in the geotechnical properties with depth are gradual and reflect the normal effects of burial. Further down slope (cores 9C & 9D), changes of these properties are also relatively uniform with depth, but with some minor variations being attributed to differences in sand and clay content. The marked

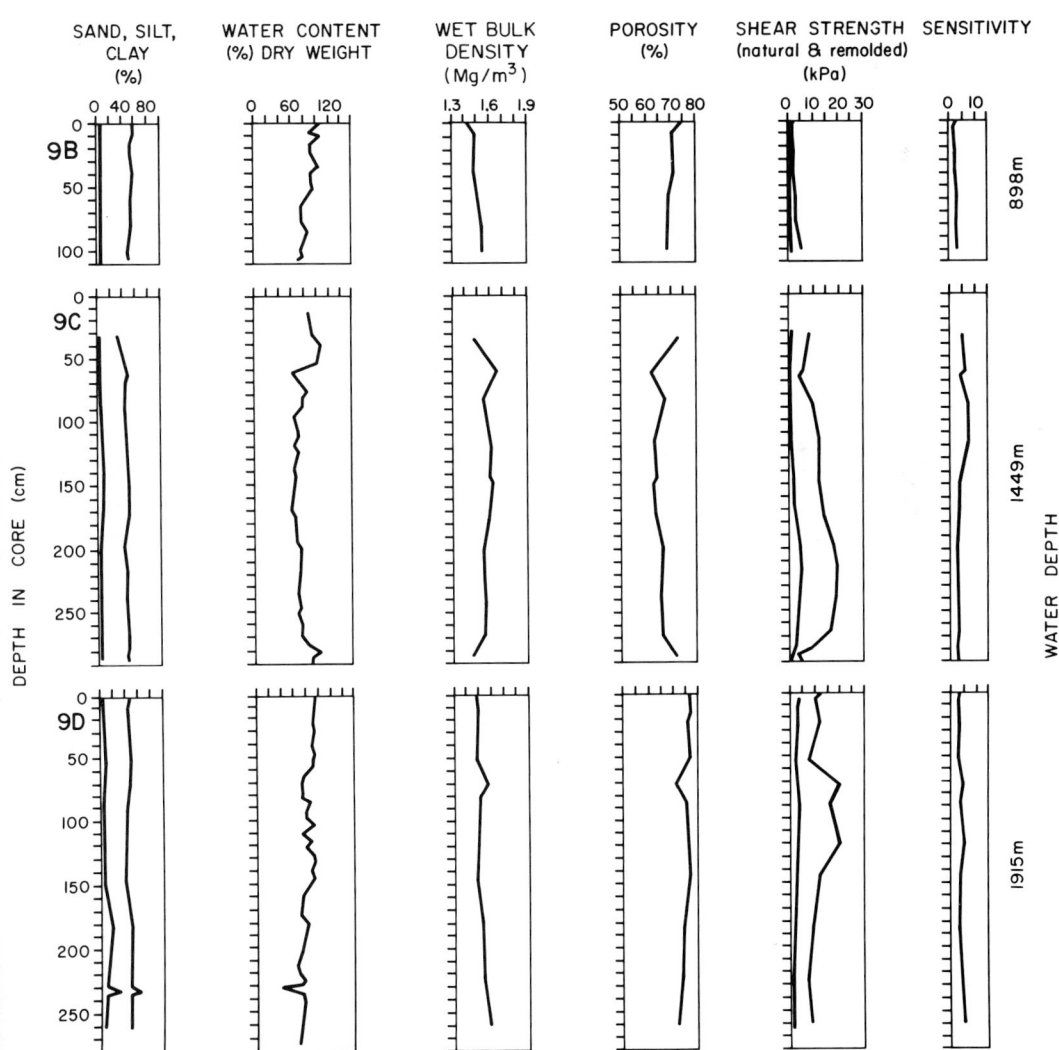

FIG. 8.—Vertical distribution of geotechnical properties, New Jersey slope transect.

changes such as higher water content and lower shear strength in the lower part of core 9C are attributed to burrowing activity.

Middle Atlantic Bight Transect

As noted earlier, the continental slope of the Middle Atlantic Bight, particularly from Wilmington Canyon south to about 36°N, displays sediment characteristics differing from those of other parts of the slope off the northeastern United States. Examination of profiles of geotechnical data from a series of cores across the slope off Virginia reveals pronounced variations from core to core (Fig. 9).

The uppermost slope sediments (core 15A) are characterized by high sand contents with relatively little change in water content (45–55%), wet bulk density [1.67–1.77 Mg/m^3 (104.2–110.4 pcf)], and porosity (55–60%) within the upper 50 cm. To a depth of 100 cm, the mid and lower-slope deposits (cores 15B & 15C) are almost free of sand and are relatively homogeneous. Although the rates of change of the geotechnical properties with depth are quite different for the two areas, the relatively uniform rate of change is attributed to the normal effects of overburden. The slightly anomalous values in core 15C at a depth of 70–77 cm are due to the presence of a worm burrow. On the lower most slope (core 15D) increased concentrations of sand result in lower water contents and porosities and higher wet bulk densities in the near-surface (20–30 cm) sediments. Below about 40 cm the geotechnical properties become quite uniform except for shear strength which decreases markedly through the remainder of the cored interval. This anomalous condition is attributed to a high degree of bioturbation found in the lower portion of this core.

Cape Hatteras Transect

The influence of the increased concentration of sand on the slope in the vicinity of Cape Hatteras in quite apparent from the profiles of geotechnical properties along a transect approxi-

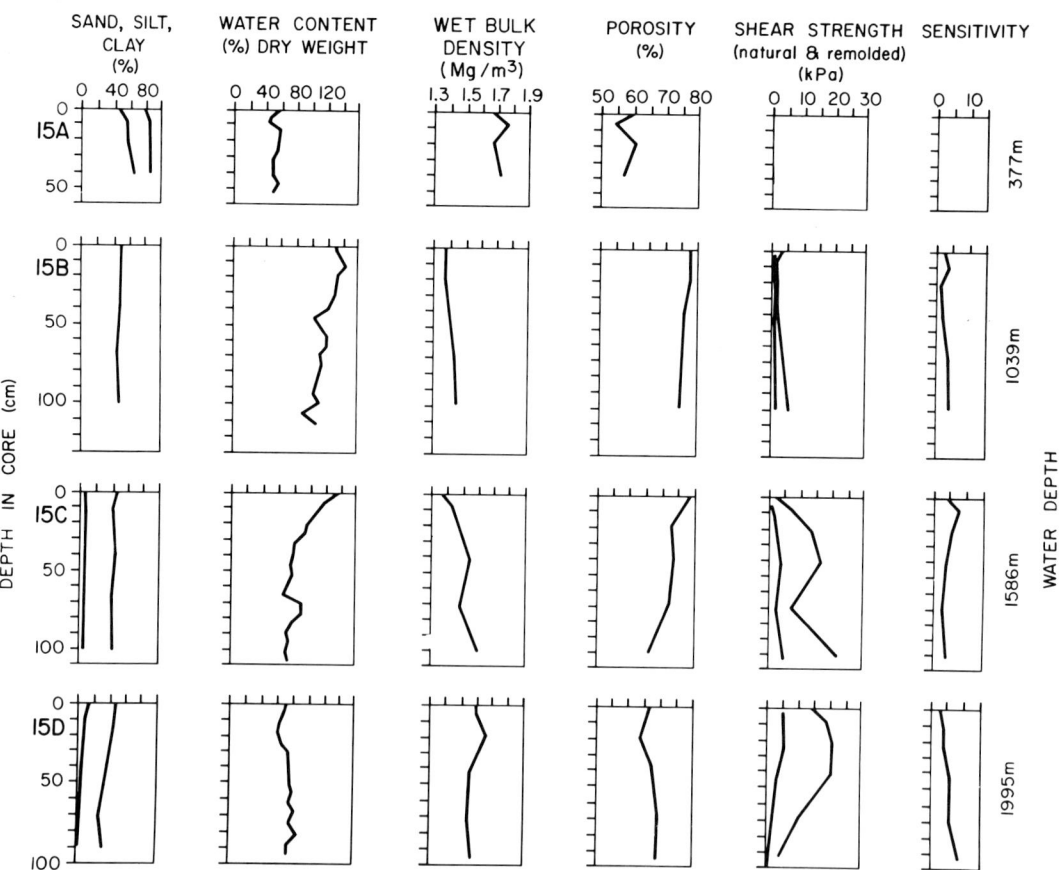

Fig. 9.—Vertical distribution of geotechnical properties, Middle Atlantic Bight slope transect.

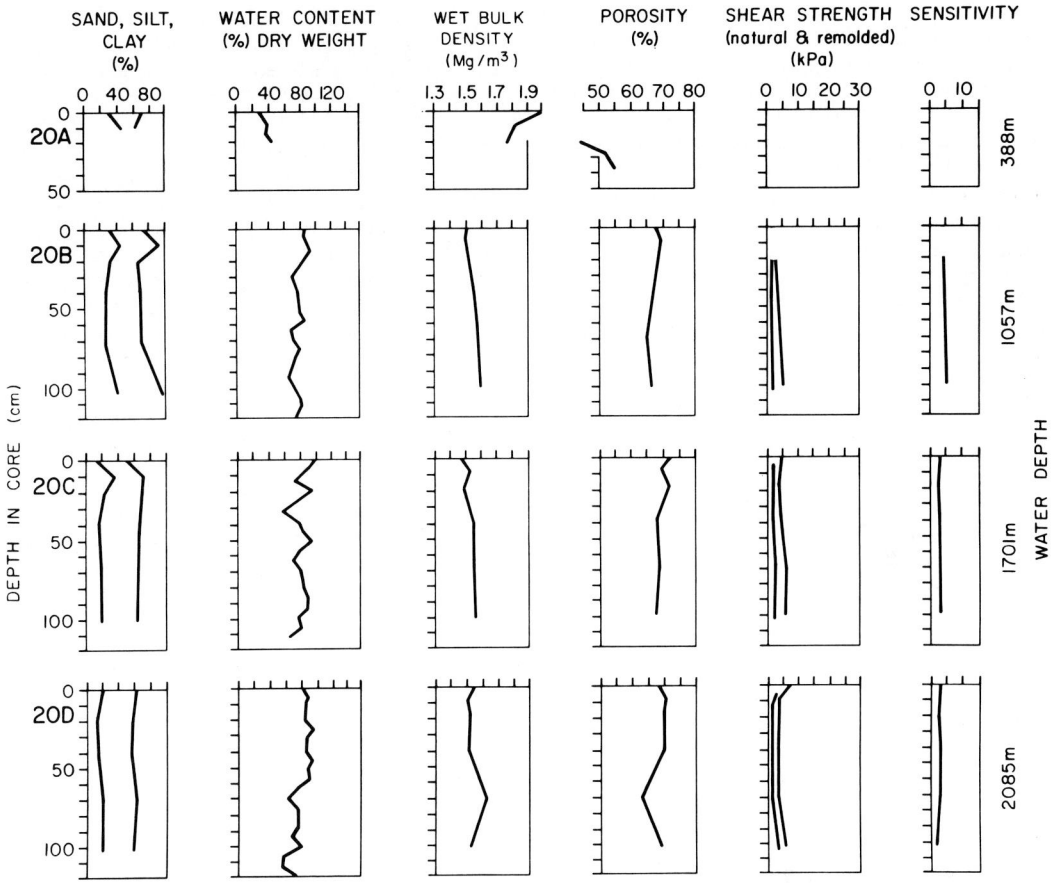

FIG. 10.—Vertical distribution of geotechnical properties, Cape Hatteras slope transect.

mately 65 km north of Cape Hatteras (Fig. 10). Sand is a major constituent to a depth of at least 100 cm even in the lowermost slope deposits of this area. A progressive change in the geotechnical properties in a downslope direction is clearly evident from the profiles shown in Figure 10. Water contents of 30 to 40% along with porosities as low as 44% are characteristic of the coarse-grained deposits on the uppermost slope (core 20A). Bulk densities of 1.97 Mg/m^3 (122.9 pcf) from the upper slope are the highest found along the entire northeastern slope. At mid-slope (Core 20B), sand contents are noticeably lower, whereas water contents and porosities have increased considerably, and bulk densities are reduced drastically from those found at the upper slope. Although there is a continued decrease in the coarse-grained fraction from mid-slope to the base of the slope, the other geotechnical properties remain rather uniform and do not vary markedly within the upper meter. Those vertical variations that do occur are attributed to minor changes in sand content along with the effect of overburden.

SEDIMENT STABILITY

General Setting

Seismic reflection and bathymetric surveys of the continental slope reveal that the margin off the northeastern United States has undergone extensive deposition, erosion due to currents, and mass movement of sediment. The incised nature of the slope with literally hundreds of submarine canyons, valleys, and gullies readily points to the history of intensive erosion. The lower stands of sea level during the Pleistocene probably accounted for most of the erosional features revealed by the present bathymetry. There is, however, evidence from off Georges Bank that in this area canyon erosion and infilling may have been an alternating process dating back to at least the Oligocene (Uchupi et al, 1977). Evidence such as turbidite sequences clearly indicates that a number of the major canyons have served as conduits for the transport of coarse-grained shelf deposits out onto the rise and abyssal plain (Horn et al, 1971; Cleary et al, 1977). Although erosion due to currents is still a prominent process in a few of the canyons, particularly in those off New England, indications are that the majority of the East Coast canyons are inactive at this time (Keller and Shepard, 1978).

Slumping is a process which continues strongly to alter the morphology of the continental slope between Cape Hatteras and Hydrographer Canyon. Slumping along with the extensive erosion mentioned above has been, and in some areas may continue to be, the primary processes contributing to the morphological characteristics of the slope today. The prominence of slumping along the entire continental slope off the northeastern United States is clearly documented by numerous reports of slump structures and scars (Moore and Currey, 1963; Uchupi, 1967, 1968b; Rona, 1969; Kelling and Stanley, 1970; Emery and Uchupi, 1972; MacIlvaine, 1973; Wilkniss, 1973; McGregor et al, 1977; Embley and Jacobi, 1977; McGregor and Bennett, 1977; Uchupi et al, 1977). On-site submersible observations of slump features along the slope just south of Baltimore Canyon (Malahoff et al, 1977) and in the vicinity of Norfolk Canyon (J. Musick, pers. comm., 1977) have more recently been made.

An examination of the bathymetry between Wilmington and Norfolk Canyons reveals a broad flattening of the mid and lower slope relative to the East Coast slope as a whole (Fig. 1A). This flattening at the foot of the slope takes on a morphological character similar to that ascribed to slumping off Nova Scotia by Stanley and Silverberg (1969) and may well reflect massive as well as small-scale slumping in this area. A seismic reflection profile across this portion of the slope shows jumbled subsurface reflectors in the vicinity of the lower slope which Schlee et al (1975) interpret to represent large-scale slumping. A similar conclusion was drawn by Embely and Jacobi (1977) based on their seismic reflection data from across the slope in an area just south of Baltimore Canyon. Another explanation for the broadening of the lower slope may be the extensive erosion believed to have been caused by turbidity currents along with slumping (Uchupi and Emery, 1967). Recent seismic reflection studies of this area tend to support the role turbidity or strong bottom currents may have played in reshaping the bottom topography (McGregor, 1978). The strong evidence for both current erosion and slumping suggests that this part of the slope probably has been affected significantly by both processes. The anomalous character of these deposits is also reflected in their geotechnical properties which differ considerably from those of other areas of the continental slope.

Slumping

A number of methods using strength data are available for making slope stability analyses. In the case of submarine deposits where sediment strength, slope gradients, and slumping mechanics are not sufficiently known, a relatively simple method such as the infinite slope stability analysis is appropriate. This method as described by Taylor (1948) and applied to submarine deposits by Moore (1961), Morelock (1969), Ross (1971), and Almagor and Wiseman (1977) assumes an infinite slope (meaning a constant slope of unlimited extent relative to the thickness of the sediment mass) and provides a means of using vane shear or other shear strength data to assess the stability characteristics of the near-surface slope deposits. Although these authors have discussed the method of analysis to varying degrees it is advisable to outline briefly the procedure used to arrive at the results presented here.

The forces acting on a saturated sedimentary slope of inclination i and sediment thickness Z along a shear plane, AB, are represented as in Figure 11A. The weight on plane AB is $\bar{\gamma}Zb \cos i$, where $\bar{\gamma}$ is the submerged sediment density and b the unit area. Vertical stress, P_v, is the effective weight of the overlying sediment divided by the unit area, b, or $P_v = \bar{\gamma}Z \cos i$. The diagram also

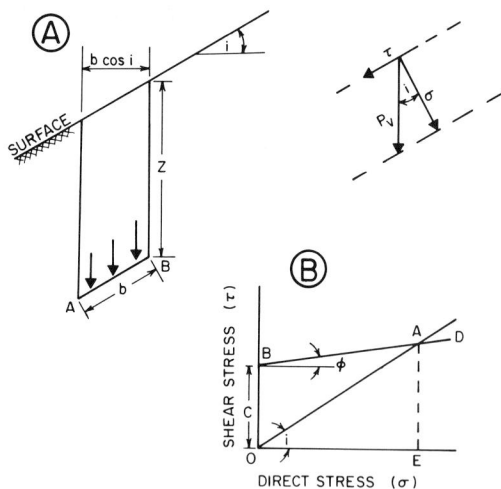

FIG. 11.—A) Schematic of elements of vertical pressure on a saturated sedimentary slope; B) illustration of the limited sediment accumulation that can occur on a slope when the gradient, i, is greater than the angle of internal friction, ϕ, of the sediment.

shows that the vertical stress has two components, shear stress τ directed downslope, and direct stress σ acting normal to the shear plane.

An illustration of these forces in relation to the strength envelope of a sediment is shown in Figure 11B. The vertical stress, P_v, for a given sediment is shown as OA; for the slope angle, i, to the direct stress, OE or σ, and has the corresponding shear stress, AE or τ. Also shown is the strength envelope, BAD, for the given sediment. It can be seen that a vertical stress of less than OA results in a shear stress lower than the shear strength of a given sediment. As long as the shear stress, τ, developed by the overburden pressure is less than the shear strength of the sediment, a stable condition is assumed to exist. If the vertical stress exceeds OA, the sediment mass may become unstable (Fig. 11B).

It is clear from Figure 11B that if vertical homogeneity of the sediment mass properties is assumed and the angle of internal friction φ of the sediment is greater than the slope i, the two lines will never intersect, and infinite sediment accumulation is theoretically possible.

Referring to Figure 11A, shear stress, τ, is determined using the equations

$$\tau = P_v \sin i$$

and

$$P_v = \bar{\gamma} Z \cos i,$$

therefore,

$$\tau = \bar{\gamma} Z \sin i \cos i.$$

Submerged density $\bar{\gamma}$, as used here, is equal to the wet bulk density minus the unit weight or density of sea water (1.028 Mg/m³).

Using the method just described and by plotting the vane shear data as least-squares lines against burial depth, a comparison of shear strength and the shear stress generated by the overburden can be made. In all cases the data indicate that the sediments are presently stable. Of the 68 cores examined, only two indicated a potential for instability by extrapolation of existing data. Using an approximate slope of the seafloor at the respective core station as determined from small-scale bathymetric maps and the mean bulk density for the particular core, it can be seen from Figure 12 that extensions of the two lines in each case would result in their intersection. This analysis indicates that for cores 12C and 8D, given the respective slopes, at depths equal to 9 and 16 m failure might be expected to take place.

Comparing least-squares plots of shear strength data from a representative number of cores against

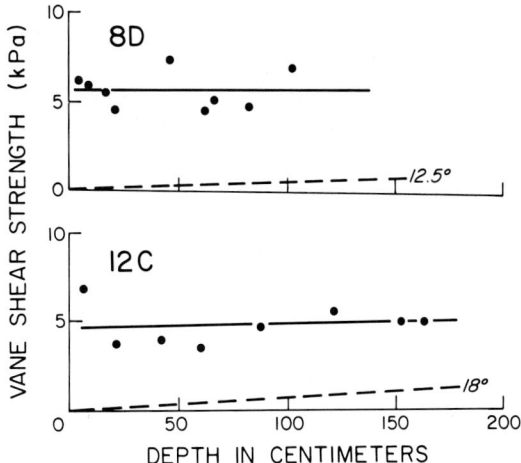

FIG. 12.—Least squares lines of vane shear strength versus depth for cores 8D and 12C. Dashed lines are the shear stresses due to overburden pressure with increasing sediment thickness on the respective slopes (12.5 and 18°) at sites 8D and 12C.

shear stresses developed at slope gradients of 5, 10, 20, and 30° reveals that these deposits would be stable at relatively high slopes (Fig. 13). A bulk density of 1.52 Mg/m³ (94.8 pcf) was used for these shear stress determinations. It can be seen from Figure 13 that at a mean gradient for the continental slope off northeastern United States of 7° (Emery and Uchupi, 1972), the near-surface sediments are stable insofar as overburden pressures are concerned using the infinite slope

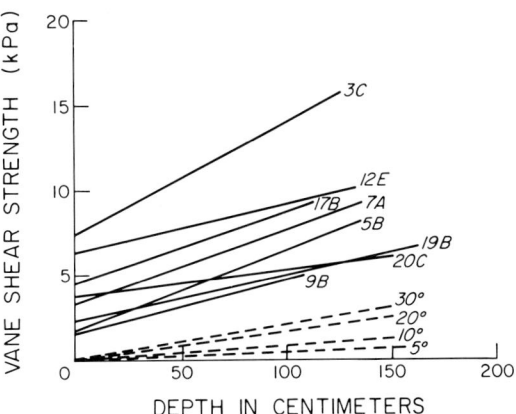

FIG. 13.—Least squares lines of vane shear strengths versus depth for a representative number of slope cores. Dashed lines are the shear stresses generated by overburden pressures with increasing sediment accumulation on slopes of 5 to 30°.

method of analysis. Even at gradients of up to 30°, a considerable thickness of sediments would have to be deposited before failure would occur.

Although the infinite slope analysis is a commonly used method for slope stability determinations, a word of caution is appropriate in light of at least three problems inherent in our analysis. First, it is assumed that the linear regressions determined from these short cores (1 to 2.9 m) are representative of the shear strength profiles which are extrapolated to depth; second, it is also assumed that the mass physical properties, specifically the wet unit weights used, are representative over the extrapolated depths; and finally, the actual slope gradient at the core sites are only approximately known. Additionally the infinite slope analysis method is limited in that it relates only to failure resulting from excess overburden on an inclined slope. It should be noted also that the shear strengths used in this study were undrained vane tests, thus, theoretically, the undrained strength is equal to the cohesion of the sediment if stressed without loss of pore water. Therefore, the stability analyses presented above should be considered as conservative estimates if the sediment were to fail under drained conditions.

With the above considerations in mind, it is interesting to note that two areas exhibiting possible incipient instability (cores 8D & 12C) are in submarine canyons where fine-grained, organic-rich sediments are being deposited. Evidence of slumping in Wilmington Canyon (core 12C) has been reported by Killing and Stanley (1970) and observed by the senior author in Hudson Canyon (core 9D) during submersible dives.

As pointed out earlier, slumping is a prominent process along portions of the continental slope, but solutions from our limited analyses indicate that slopes are generally stable. Mechanisms other than excess overburden such as the effect of internal waves on upper slope deposits or infrequent seismic shocks, which were not considered in the infinite slope analysis, are undoubtedly important factors contributing to the slumping process. Obviously, much longer cores are required and actual slopes at the respective core locations must be known for a rigorous analysis of slope stability.

Hydrodynamic Processes

Bottom currents play a significant role in the sedimentary processes along major portions of the continental slope off the northeastern United States. Although relatively few near-bottom (1 to 3 m above the bottom) current measurements have been made along the slope, there is sufficient evidence from bottom photographs to substantiate the fact that sediment transport occurs frequently along many portions of the slope. The few current observations that do exist indicate that velocities are generally higher along the slope off New England than off the Middle Atlantic Bight. Velocities as high as 70 cm/sec have been reported in Hydrographer Canyon (B. Heezen, pers. comm, 1973) and along the slope between Alvin and Block Canyons (Emery and Ross, 1968). Peak velocities reported from the area between Hudson Canyon and Cape Hatteras rarely exceed 20 to 30 cm/sec (Lyall et al, 1971; Rowe, 1971; Keller and Shepard, 1978; McGregor, 1979). It is obvious from bottom photographs, however, that the hydrodynamic processes resulting in the transport of bottom sediments are highly irregular even within a local area.

In order to relate current velocities to sediment transport it is appropriate to consider the threshold velocities (the velocity slightly less than that required to initiate motion for a certain sediment grain size) of the respective sediment types occurring on the continental slope. Since most of the current measurements referred to in the following discussion were made at or near a height of 300 cm above the bottom, the threshold velocity curve for $\bar{\mu}_{300}$ (current velocity 300 cm above the bottom) presented by Keller and Shepard (1978) is used as the basis for the threshold velocities mentioned herein. In using such a diagram, certain limitations must be kept in mind. Variations in conditions (fluid viscosity and grain shape and density) from those upon which the diagram is based will result in different threshold velocities. It must also be pointed out that the theory of threshold velocity and sediment transport is poorly developed for very fine-grained sediments. With this in mind, the threshold velocities presented here are considered to be approximations yet suitable for the overall discussion of sediment transport on the continental slope.

Although bottom currents off New England are relatively strong, their effect on the slope deposits varies markedly. In their study of East Coast canyons, Keller and Shepard (1978) recorded bottom currents of up to 53 cm/sec and the seaward migration of sand ripples along the axis of Hydrographer Canyon down to depths of 710 m. The study also found that threshold velocities of the canyon sediments were exceeded 26% of the time during a 11.5 day observation period. In the same general area, between Block and Alvin Canyons, Emery and Ross (1968) reported current velocities of 70 cm/sec, but found no expression of current activity in the microtopography of the sea floor. The reported median diameters (.009 to .017 mm) of the surface sediments in that area (Emery and Ross, 1968) should have threshold velocities of 18 to 20 cm/sec using the $\bar{\mu}_{300}$ curve, yet the 70 cm/sec velocities did not appear to be eroding or transporting these sediments. It is apparent that in dealing with very fine-grained sediments, grain-size in itself may not necessarily provide the basis for determing threshold velocity. Southard (1977), in studying fine-grained cohesive sediments, found that in addition to grain size such factors as water content, organic content, mineral composition, and amount of biogenic reworking may significantly influence the threshold velocity of a particular sediment.

The strong currents off New England are effective in transporting sand and silt-size material across the slope, but in those areas of fine-grained cohesive sediment little erosion appears to take place. At most, the currents tend to smooth out sharp irregularities in the microtopography and keep additional sediments from being deposited (McIlvaine, 1973). The resulting fine-grained sediments are found to have relatively high shear strengths and bulk densities and low water contents, all of which contribute to higher threshold velocities.

Although Keller and Shepard (1978) report currents in Hudson and Wilmington Canyons that exceed theoretical threshold velocities required to transport the bottom sediments, currents along the slope in the Middle Atlantic Bight do not at present appear to be a prominent factor in transporting bottom sediments. Bottom photographs and current measurements indicate that threshold velocities in the Bight are seldom exceeded except on the uppermost slope where coarser sediments commonly display streaming features and ripple marks.

Velocities of up to 30 cm/sec have been measured along the midslope between Baltimore and Washington Canyons (McGregor, 1979), but it is questionable as to whether they exceed the threshold velocity of the fine-grained (.008 mm median diameter) deposits found there. Bottom photographs from a mid-slope site with similar sediments, approximately 56 km to the southwest of McGregor's current meter station, showed no evidence of current activity.

In an examination of numerous bottom photographs from the slope and continental rise off Cape Hatteras, Rowe (1971) found evidence for a patchy distribution of current activity (sediment streamers, ripples) down to depths of 4400 m. Current measurements on the slope in the same area revealed near-bottom (3 to 25 m above the bottom) currents with velocities up to 20 cm/sec (Rowe, 1971). At the lower slope off Cape Hatteras, Barrett (1965) reported near-bottom velocities of 10 to 15 cm/sec based on geostrophic calculations. Although these velocities do not exceed the threshold velocities of the slope deposits, based on the grain-sizes (.008 to .025 mm) reported to Hathaway (1971), it is obvious from bottom photographs that currents are causing bottom sediments to be transported along this portion of the East Coast continental slope.

Hydrodynamic processes along the continental slope off the northeastern United States are influential in transporting bottom sediments along the uppermost slope and in those areas where coarse-grained sediments extend farther onto the slope. Although rather strong currents persist over certain parts of the slope, they appear to have little or no effect in eroding and transporting fine-grained cohesive sediments.

In contrast to the non-depositional and erosional environments of the slope off New England, the slope off the Middle Atlantic Bight, with its weaker currents and large sources of fine-grained sediment, is presently a zone of deposition for fine-grained material. A similar conclusion was reached by Doyle et al (1975) based on their study of mica distribution on the continental slope and by Keller and Shepard (1978) as a result of their investigation of sedimentary processes in the major East Coast canyons. It is, however, clear from an examination of the sedimentary structures in the cores collected during this study that considerable variation in the hydrodynamic processes has occurred over long periods of time. The presence of coarse-grained laminae, some of which display cross bedding, within thick sequences of fine-grained sediment attests to the drastic changes in the energy level of bottom currents from time-to-time.

SUMMARY

The continental slope off the northeastern United States is a province of the seafloor which has been the site of extensive deposition (marginal outbuilding) in the past. More recently (late Tertiary to present) it has become an area where erosion as well as large and small scale slumping are the predominant processes affecting the slope. As might be expected these deposits display considerable variability as is apparent from an examination of the areal and vertical distribution of their geotechnical properties (Table 1). Based on these properties, however, it appears that the near-surface slope deposits off the northeastern United States can be divided into distinct sedimentological provinces.

Factors, such as the existing strong bottom currents and previous glacial activity, combine to give the slope deposits off New England mass physical properties which strongly contrast with those of most other slope sediments off the northeast margin of the United States (Table 2). From just north of Hudson Canyon to Hydrographer Canyon, the slope is blanketed by coarse-grained sediments from the adjacent shelf and/or eroded from the uppermost slope, or is composed of fine-grained, relatively cohesive material which is exposed at the seafloor as a result of erosion. Textural characteristics appear to contribute strongly to the display of distinctly different geotechnical properties for this portion of the continental slope (Table 2). On the whole, the slope deposits off New England are characterized as having relatively low water contents, porosities and plasticity indices, and high wet bulk densities and shear strengths.

The area from the vicinity of Hudson Canyon to Lindenkohl Canyon is unique in that it is delineated by a certain degree of uniformity among the mass physical properties across the entire slope. Relatively low shear strengths along with intermediate values of water content and bulk density are typical from the upper to the lower slope. Normally, some degree of gradation in these properties would be expected across the slope. Gross textural characteristics are not distinctly different between this area and that to the south, but there appear to be subtle changes in the sand content of the lower slope sediments that result in their possessing geotechnical properties similar to those of the lower upper-slope deposits of the Middle Atlantic Bight.

As the result of extensive current erosion of a prograded margin coupled with large and small scale slumping, the mid and lower slope of the Middle Atlantic Bight is considerably wider and not as steep as found elsewhere. Bulk densities are notably low for the slope as a whole just as shear strengths and water contents are relatively high. Plasticity characteristics of the mid and lower-slope deposits are also significantly different from those found elsewhere along the entire slope. Plasticity indices of up to 70, the highest measured during this study, are unique to the slope deposits off the Middle Atlantic Bight. Relatively high concentrations of clay-size material plus the dynamic effects of currents and slumping appear to be major factors contributing to the observed distribution of geotechnical properties.

In the vicinity of Cape Hatteras, the effects of a narrowing shelf and slope plus the large concentration of sand on the adjacent shelf contribute to a distinct sedimentary environment. Varying concentrations of sand which has been transported down the slope appear to be the primary factor influencing the geotechnical properties of these slope deposits. As might be expected, bulk densities are relatively high, whereas water contents, porosities, and shear strengths are low.

Sediments in the major submarine canyons along the continental slope usually display mass physical properties considerably different from those of the adjacent slope. Axial deposits of canyons off New England commonly are relatively coarse-grained and thus possess geotechnical properties normally associated with coarse sediments. In the outermost part of these canyons, finer sediments prevail with usually higher water contents and porosities, and lower bulk densities. Those canyons from Hudson Canyon south generally are blanketed with fine-grained, organic-rich sediments and display higher water contents and porosities, but lower wet bulk densities and shear strengths than the nearby slope deposits.

Vertical variations of geotechnical properties in the near-surface (50 to 290 cm) slope deposits are relatively minor and are generally attributed to the normal effects of overburden. Variations of these properties from the normal gradation with depth are, however, observed at a number of sites which are influenced by irregularities in sand content or by what appear to be zones of increased induration.

It is readily apparent from reviewing the data that textural characteristics are a primary factor influencing the observed variations in the geotechnical properties (Table 2). Commonly, increases in water content and porosity along with decreases in bulk densities with distance downslope tend to correlate with increasing percentages of clay-size material and decreasing amounts of sand. Silt content varies remarkably little in rela-

TABLE 2.—Comparison of mean values of geotechnical properties for cored intervals of slope sediments

Region	Water Content (%)	Specific Gravity	Wet Bulk Density (Mg/m^3)	Porosity (%)	Shear Strength (kPa)	Sensitivity	Liquid Limit	Plastic Limit	Sand (%) >62 μm	Silt (%) 2–62 μm	Clay (%) <2 μm
New England (cores 1–6)	78	2.71	1.57	66	7.8	3	62	30	18	44	38
Mid-Atlantic (cores 8–19)	91	2.72	1.50	70	7.4	3	78	36	9	43	48

tion to sand and clay content in the slope deposits as a whole, normally ranging from 37 to 45%.

A comparison of mean values of the various geotechnical properties for the slope with those of the adjacent Atlantic basin indicates that there is considerable similarity between such properties as water content, wet bulk density, grain specific gravity, and sensitivity (Table 1). Mean values for porosity, shear strength, and plasticity index are, however, notably higher in the slope deposits.

An assessment of the effects of overburden pressure on the stability of the slope sediments, using the infinite slope analysis method, indicates that the areas sampled are stable. Analyses of cores from Hudson and Wilmington Canyons showed that unstable conditions might exist if additional sediment accumulations on the order of 9 and 16 m, respectively, should occur. Owing to the short lengths of core and thus the need to extrapolate geotechnical values to greater depths as well as being able only to approximate the slope of the sea floor at the core site results in severe limitations being placed on the stability analyses we were able to perform. Slumping is clearly a significant process along the northeastern margin of the United States and therefore factors other than overburden pressure, such as the impact of internal waves on the upper slope, undercutting by bottom currents, or infrequent seismic shocks, undoubtedly contribute to the slumping of the slope deposits.

Hydrodynamic processes acting on the continental slope off the northeast coast of the United States are effective in transporting bottom sediments in a number of areas. Of the areas considered in this study, bottom currents appear to be strongest along the New England slope. Here current velocities of up to 70 cm/sec have been recorded both on the slope and in a submarine canyon. Despite these strong currents, it is apparent that they are effective in transporting only relatively coarse-grained material. In areas of fine-grained sediment, these strong currents appear to do little more than serve to keep the area free of new sediments or at most to smooth out minor irregularities in the microtopography.

In contrast to the high energy conditions off New England, bottom currents along the slope of the Middle Altantic Bight are reported to have considerably lower velocities. Although a number of measurements have been made, velocities seldom have been found to exceed 30 cm/sec and more commonly are less than 15 cm/sec. Indications based primarily on bottom photographs are that transport of bottom sediments along this portion of continental slope does occur, but is rather limited. In contrast to the erosional and/or non-depositional environment of the New England slope, the slope off the Middle Atlantic Bight appears to be a depositional zone for fine-grained material.

ACKNOWLEDGMENTS

The authors acknowledge with great appreciation the outstanding efforts of George Lapiene during the field phase of this study. A successful field program was made possible by the fine support and cooperation received from the officers and crew of the NOAA ship RESEARCHER for which we are most grateful. To our colleagues, Bonnie McGregor, who reviewed the manuscript and assisted us at sea, as well as William Sawyer and Pamela Bates who carried out much of the laboratory work, we express our many thanks. This study was primarily funded by NOAA, with additional support to the senior author from NSF Grant OCE76-21869.

REFERENCES

ALMAGOR, G., AND WISEMAN, G., 1977, Analysis of submarine slumping in the continental slope off the southern coast of Israel: Marine Geotechnology., v. 2, p. 349–388.

BARRETT, J. R., JR., 1965, Subsurface currents off Cape Hatteras: Deep-Sea Res., v. 12, p. 173–184.

CASAGRANDE, A., 1948, Classification and identification of soils: Amer. Soc. Civil Engineers Trans., v. 113, p. 901–931.

CLEARY, W. S., PILKEY, O. H., AND AYERS, M. W., 1977, Morphology and sediments of three ocean basin entry points, Hatteras abyssal plain: Jour. Sed. Petrology, v. 47, p. 1157–1170.

DOYLE, L. J., PILKEY, O. H., HAYWARD, G. L., AND ARBOGAST, J. S., 1975, Sedimentation on the northeastern continental slope of the United States: proc. IX Congress International de Sedimentologie, Theme 6, p. 51–56.

———, ———, AND WOO, C. C., Sedimentation on the eastern United States continental slope: This volume.

DRAKE, C. L., EWING, M., AND SUTTON, G. H., 1959, Continental margins and geosynclines: the east coast of North America north of Cape Hatteras, in Physics and Chemistry of the Earth: London, Pergamon Press, v. 3, p. 110–198.

EMBLEY, R. W., AND JACOBI, R. D., 1977, Distribution and morphology of large submarine sediment slides and slumps on Atlantic continental margins: Marine Geotechnology, v. 2, p. 205–228.

EMERY, K. O., AND ROSS, D. A., 1968, Topography and sediments of a small area of the continental slope south of Martha's Vineyard: Deep-Sea Res., v. 15, p. 415–422.

———, UCHUPI, E., PHILLIPS, J. D., BOWIN, C. O., BUNCE, E. T., AND KNOTT, S. T., 1970, Continental rise off eastern North America: Am. Assoc. Petroleum Geologists Bull., v. 54, p. 44–108.

——, AND UCHUPI, E., 1972, Western north Atlantic ocean: topography, rocks, structure, water, life, and sediments: Am. Assoc. Petroleum Geologists Memoir 17, 532 p.

EVANS, I. AND SHERRATT, G. G., 1948, A simple and convenient instrument for measuring the shear resistance of clay soils: Jour. Scientific Inst. and Phys, in Industry, v. 25, p. 411–414.

HATHAWAY, J. C., 1971, Data file, continental margin program, Atlantic coast of the United States, v. 2, Sample collection and analytical data: Woods Hole Oceanog. Inst. Ref. No. 71-15, 489 p.

HORN, D. R., EWING, M., HORN, B. M., AND DELACH, M. N., 1971, Turbidities of the Hatteras and Sohm abyssal plains, western north Atlantic: Marine Geology, v. 11, p. 287–323.

HOSKINS, H., 1967, Seismic reflection observations on the Atlantic shelf, slope, and rise southeast of New England: Jour. Geology, v. 75, p. 598–611.

KELLER, G. H., 1968, Shear strength and other physical properties of sediments from some ocean basins, in Proc. Civil Eng. in the Oceans: Amer. Soc. Civil Eng., p. 391–417.

——, 1970, Mass properties of the sea floor in a selected depositional environment: Proc. Civil Eng. in the Oceans II, Amer. Soc. Civil Eng., p. 857–877.

——, 1975, Sedimentary processes in submarine canyons off northeastern United States: Proc. IX Congress International de Sedimentologie, Theme 6, p. 77–86.

——, AND BENNETT, R. H., 1970, Variations in the mass physical properties of selected submarine sediments: Marine Geology, v. 9, p. 215–223.

——, AND SHEPARD, F. P., 1978, Currents and sedimentary processes in submarine canyons off the northeastern United States, in Stanley, D. J., and Kelling, G., eds., Sedimentation in submarine canyons, fans, and trenches: Stroudsburg, PA., Dowden, Hutchinson and Ross, Inc., p. 15–32.

KELLING, G., AND STANLEY, D. J., 1970, Morphology and structure of Wilmington and Baltimore submarine canyons, eastern United States: Jour. Geology, v. 78, p. 637–660.

KNOTT, S. T., AND HOSKINS, H., 1968, Evidence of Pleistocene events in the structure of the continental shelf off northeastern United States: Marine Geology, v. 6, p. 5–43.

KRAUSS, D. C., CHRAMIEC, M. A., WALSH, G. M., AND WISOTSKY, S., 1966, Seismic profile showing Cenozoic development of the New England continental margin: Jour. Geophys. Res., v. 71, p. 4327–4332.

LYALL, A. K., STANLEY, D. J., GILES, H. N., AND FISHER, A., 1971, Suspended sediment and transport at the shelf-break and on the slope, Wilmington Canyon area eastern U.S.A.: Jour. Mar. Tech. Soc., v. 5, p. 15–27.

MACILVAINE, J., 1973, Sedimentary processes on the continental slope off New England: Woods Hole Oceanog. Inst. Ref. No. 73-58, 211 p.

MALAHOFF, A., EMBLEY, R. W., PERRY, R. B., AND FEFE, C., 1977, Sedimentation processes on the continental shelf, slope and upper rise near Baltimore Canyon: Trans. Am. Geoph. Union, v. 58, p. 1160.

MCGREGOR, B. A., 1979, Current meter observations on the U.S. Atlantic continental slope, variation in time and space: Marine Geology, v. 29, p. 209–219.

——, 1978, Seismic reflection profiles of the United States east coast continental margin: NOAA Tech. Rept. ERL 398-AOML28, 17 p.

——, KELLER, G. H., AND BENNETT, R. H., 1975, Seismic profiles along the U.S. northeast coast continental margin: Trans. Am. Geoph. Union, v. 56, No. 6, p. 382.

——, AND BENNETT, R. H., 1977, Continental slope sediment instability northeast of Wilmington Canyon: Am. Assoc. Petroleum Geologists Bull., v. 61, p. 918–928.

——, BENNETT, R. H., AND MERRILL, G. F., 1977, Continental slope south of Baltimore Canyon, U.S. East Coast: Proc. Geol. Soc. America, Ann. Mtg., p. 1089.

MILLIMAN, J. D. 1973, Marine Geology, in Coastal and offshore environmental inventory. Cape Hatteras to Nantucket Shoals: Providence, RI, Univ. of Rhode Island, Marine pub. series No. 3, 91 p.

——, PILKEY, O. H., AND ROSS, D. A., 1972, Sediments of the continental margin off the eastern United States: Geol. Soc. America Bull., v. 83, p. 1315–1334.

MOORE, D. G., 1961, Submarine slumps: Jour. Sed. Petrology, v. 31, p. 343–357.

——, AND CURRAY, J. R., 1963, Sedimentary framework of continental terrace off Norfolk, Virginia, and Newport, Rhode Island: Am. Assoc. Petroleum Geologists Bull., v. 47, no. 12, p. 2051–2054.

MORELOCK, J., 1969, Shear strength and stability of continental slope deposits, western Gulf of Mexico: Jour. Geophys. Res., v. 74, p. 465–482.

NEWTON, J. G., AND PILKEY, O. H., 1969, Topography of the continental margin off the Carolinas: Southeastern Geology, v. 10, No. 2, p. 87–92.

OSER, R. K., 1969, Bottom environmental oceanographic data report—Hudson Canyon area, 1967: Naval Oceanog. Office Informal Rept. No. 69-8, 43 p.

——, 1973, Seafloor bathymetry and sediment properties in deep water dump area A, in Wilkniss, P. I., Environmental condition report for deep water dump area A: Washington, D.C., Naval Research Lab., NRL Report 7553, p. 103–112.

PRATT, R. M., 1968, Atlantic continental shelf and slope of the United States—physiography and sediments of the deep-sea basin: U.S. Geol. Survey Professional Paper 529-B, 44 p.

PUSCH, R., 1973, Influence of organic matter on the geotechnical properties of clays: Nat. Swedish Building Res. Document 11, 63 p.

RICHARDS, A. F., 1961, Investigations of deep-sea sediment cores, I. Shear strength, bearing capacity, and consolidation: U.S. Navy Hydrographic Office, Tech. Rept. 63, 60 p.

———, 1962, Investigation of deep-sea sediment cores, II, Mass physical properties: U.S. Navy Hydrographic Office Tech. Rept. 106, 146 p.

———, AND KELLER, G. H., 1961, A plastic-barrel sediment corer: Deep-Sea Res., v. 8, p. 306–312.

ROBERSON, M. L., 1964, Continuous seismic profiler survey of Oceanographer, Gilbert and Lydonia Submarine Canyons, Georges Bank: Jour. Geophys. Res., v. 69, p. 4779–4789.

RONA, P. A., 1969, Middle Atlantic continental slope of United States: deposition and erosion: Am. Assoc. Petroleum Geologists Bull., v. 53, p. 1453–1465.

———, AND CLAY, C. S., 1967, Stratigraphy and structure along a continuous seismic reflection profile from Cape Hatteras, North Carolina to the Bermuda Rise: Jour. Geophys. Res., v. 72, p. 2107–2130.

ROSENQUIST, I., 1953, Consideration on the sensitivity of Norwegian quick-clays: Géotechnique, v. 3, p. 195–200.

ROSS, D. A., 1971, Mass physical properties and slope stability of sediments of the northern Middle America Trench: Jour. Geophys. Res., v. 76, p. 704–712.

ROWE, G. T., 1971, Observations on bottom currents and epibenthic populations in Hatteras submarine canyon: Deep-Sea Res., v. 18, p. 569–581.

SCHLEE, J., BEHRENDT, J. C., MATTICK, R. E., AND TAYLOR, P. T., 1975, Structure of continental margin of Mid-Atlantic states (Baltimore Canyon Trough): U.S. Geol. Survey Open File Rept. 75-60, 59 p.

———, BEHRENDT, C., GROW, J. A., ROBB, J. M., MATTICK, R. E., TAYLOR, P. T., AND LAWSON, B. J., 1976, Regional geological framework off northeastern United States: Am. Assoc. Petroleum Geologists Bull., v. 60, p. 926–951.

SHEPARD, F. P., 1934, Canyons off the New England coast: Am. Jour. Sci., 5th Ser., v. 27, p. 24–36.

———, 1954, Nomenclautre based on sand-silt-clay ratios: Jour. Sed. Petrology, v. 24, p. 151–158.

SHIDELER, G., AND SWIFT, D. J. P., 1972, Seismic reconnaissance of Post-Miocene deposits; middle Atlantic continental shelf-Cape Henry, Va. to Cape Hatteras, North Carolina: Marine Geology, v. 12, p. 165–185.

SOUTHARD, J. B., 1977, Erosion and transport of fine cohesive marine sediments: Trans. Am. Geoph. Union, v. 58, p. 1161.

STANLEY, D. J., AND SILVERBERG, N., 1969, Recent slumping on the continental slope off Sable Island Bank, southeast Canada: Earth and Planetary Sci. Letters, v. 6, p. 123–133.

STETSON, H. C., 1938, The sediments of the continental shelf off the eastern coast of the United States: Mass. Inst. Technology and Woods Hole Oceanographic Inst. Papers in Phys. Oceanog. and Meterol., v. 5, No. 4, p. 5–48.

———, 1949, The sediments and stratigraphy of the east coast continental margin—Georges Bank to Norfolk Canyon: Mass. Inst. Technology and Woods Hole Oceanog. Inst., Papers in Phys. Oceanog. and Meteorol., v. 11, No. 2, p. 1–60.

STOUT, P. M., AND MCCLENNEN, C. E., 1977, Buried valley segments on the continental shelf off Delaware Bay and New Jersey: Geol. Soc. America Northeast Reg. Mtg. (Abs.), March, p. 322.

TAYLOR, D. W., 1948, Fundamentals of soil mechanics: New York, Wiley, 700 p.

TERZAGHI, K., AND PECK, R. B., 1948, Soil mechanics in engineering practice: New York, Wiley, 566 p.

TRUMBULL, J. V. A., AND HATHAWAY, J. C., 1968, Further exploration of Oceanographer Canyon: Woods Hole Oceanog. Inst. Ref. No. 68-32, 57 p.

TWICHELL, D., KNEBEL, H., AND FOLGER, D. W., 1977, Delaware River evidence for its former extension to Wilmington submarine canyon: Science, v. 195, p. 483–485.

UCHUPI, E., 1967, Slumping of the continental margin southeast of Long Island, New York: Deep-Sea Res., v. 14, p. 635–639.

———, 1968a, Atlantic continental shelf and slope of the United States physiography: U.S. Geol. Survey Professional Paper 529-C, 30 p.

———, 1968b, Seismic profiling survey of the east coast submarine canyons, Part 1, Wilmington, Baltimore, Washington, and Norfolk Canyons: Deep-Sea Res., v. 15, p. 613–616.

———, 1970, Atlantic continental shelf and slope of the United States—shallow structure: U.S. Geol. Survey Professional Paper 529-I, 44 p.

———, EMERY, K. O., 1967, Structure of continental margin off Atlantic coast of United States: Am. Assoc. Petroleum Geologists Bull., v. 52, p. 1162–1193.

———, BALLARD, P. D., AND ELLIS, J. P., 1977, Continental slope and upper rise off western Nova Scotia and Georges Bank: Am. Assoc. Petroleum Geologists Bull., v. 61, p. 1483–1492.

VEATCH, A. C., AND SMITH, P. A., 1939, Atlantic submarine valleys of the United States and the Congo Submarine Valley: Geol. Soc. America Spec. Paper 7, 101 p.

WILKNISS, P. E., 1973, Environmental condition report for deep water dump area A: Washington, D.C., Navy Research Lab. Rept. 7753, 116 p.

RECENT HISTORY OF MASS-WASTING ON THE UPPER CONTINENTAL SLOPE, NORTHERN GULF OF MEXICO, AS INTERPRETED FROM THE CONSOLIDATION STATES OF THE SEDIMENT

JAMES S. BOOTH
U.S. Geological Survey
Office of Marine Geology
P.O. Box 6732
Corpus Christi, Texas 78411

ABSTRACT

As a marine sediment undergoes progressive burial, its response to the effective overburden pressure is predictable and measurable. It is possible, therefore, to detect a departure from the normal process of consolidation. Further, the type and extent of the departure can often be used to interpret the cause of that departure. Dynamic events, such as slumping, for example, are especially amenable to this type of interpretation. Thus, the occurrence of mass-wasting in an area may be interpreted from a detailed knowledge of the consolidation states of the sediment. One objective of this research was to demonstrate the usefulness of this principle as a tool in marine geology research. The second objective was to determine the history of mass-wasting for several localities on the upper Continental Slope of the northern Gulf of Mexico.

Interpretation was based on the geotechnical and geologic properties of 13 piston cores. Consolidation states were documented by 1) using shear strength and plasticity index data to predict the overburden that would be present if the sediment was normally consolidated, 2) comparing this overburden value to the amount of overburden determined by summing bulk densities (the "actual" overburden), and 3) analyzing the qualitative and quantitative differences between the two. This last step provides the insight into consolidation by revealing an apparent condition of underconsolidation, overconsolidation, or normal consolidation. Interpretation of the cause of these states is the link to the history of mass-wasting of an area.

The upper Continental Slope of the northern Gulf of Mexico has been a dynamic environment during at least the last 10,000 years. Off South Texas evidence related to the repeated occurrence of overconsolidated sediment indicates frequent mass movement involving sections tens of meters thick. The Mississippi Depocenter can be characterized by rapid and variable deposition and frequent small-scale mass-wasting events. In contrast, the Mississippi Trough, located on the southwestern periphery of the active depocenter, shows little evidence of recent mass-wasting. It is typified by highly underconsolidated sediment near the head, however, and thus may be in a condition of incipient instability.

The cause of this omnipresent condition of instability is the rapid deposition of fine-grained sediment. High excess pore pressure resulting from rapid deposition render the effective overburden pressure inadequate as an agent to increase shear strength. Yet, although there is little change in shear strength at a given level within an ever-thickening sediment column, the shearing stress continues to increase. Finally, the shear stress surpasses the abnormally low shear strength and failure occurs. This sequence does not develop unless rates of deposition are high, and unless these high rates are present for a sufficient length of time. Sediments deposited on the Gulf of Mexico upper slope since the last sea level rise, for example, are not inherently unstable despite a high rate of deposition (3 mm/y). However, this rate is rapid enough to cause an unstable section to develop, and because equal or greater rates probably existed in the past during a lowered sea level, then unstable sections have been developing for at least 25,000 years. It is this circumstance that has promoted failure in the present. Thus, although mass-wasting has been frequent and widespread in recent times, and will probably continue to be so, the conditions for the instability were initiated much earlier.

INTRODUCTION

As a marine sediment undergoes progressive burial, pore water is expelled and the porosity is reduced. This response to the increasing overburden is continuous and reflects a system trying to maintain an equilibrium condition. At equilibrium, no excess pore pressure exists, and the degree of consolidation reflects the amount of overburden. The sediment is then normally consolidated. If the degree of consolidation does not reflect the true amount of overburden, then disequilibrium exists and the sediment is either overconsolidated (insufficient overburden to account for the degree of consolidation) or underconsolidated (excess overburden relative to the degree of consolidation).

In the typical case, where deposition is continuous and the sediment is delivered to the sea floor at rates less than 1 mm/y, the developing column will be normally consolidated. Episodes of rapid deposition, or interruptions or changes in the established pattern of accumulation, however, can result in a complex consolidation history—one that has combinations of the possible consolidation states. The recognition of these different consolidation states and the detailed interpretation of their causes can provide a bridge between the consolidation history and the history of mass-wasting of an area.

Demonstrating the method by which consolidation data may be used to interpret the history of mass-wasting was one of the two primary objectives of this research. Although this method has not been applied in a comprehensive manner before, glimpses of its potential were furnished by Fisk and McClelland (1959), and McClelland, (1967). They discussed some of the geologic causes of different consolidation states and applied, in a general manner, their interpretations to selected data derived from sediment borings taken on the Louisiana Continental Shelf and the Mississippi Delta. Morgenstern (1967) linked consolidation to instability for these same sediments and showed that the degree of underconsolidation could be correlated to rate of deposition by using consolidation theory. The geologic implications of consolidation states were further discussed by Bryant et al (1967), who suggested that cementation may explain the apparent overconsolidation of many southwestern Gulf of Mexico slope sediments.

The other primary objective of this research was to determine the recent history of mass-wasting on the upper Continental Slope in the northern Gulf of Mexico. Establishing the frequency, magnitude, and causes of sediment mass movement in this area is of paramount interest because of the extensive commercial development within the region.

That the Continental Slope of the northern Gulf of Mexico has been a dynamic area in the past has been established by Lehner (1969), Woodbury, et al (1973), and Woodbury (1977). Through the use of seismic records, faunal studies, and coarse-fraction analysis, they showed the existence of large-scale slumps and the ubiquitous presence of displaced shelf sediments on the slope. These investigations, however, were directed toward Pleistocene sediments and did not emphasize the more recent history of mass-wasting of the northern Gulf of Mexico region.

METHODS

Samples used in this study were selected from a sample pool created by U.S. Geological Survey piston coring operations on the upper Continental Slope of the northern Gulf of Mexico. Figure 1 shows the general study areas and station locations of the cores used in this research.

Because excessive sample disturbance can so effectively disable a planned geotechnical investigation, precautions were taken during the cruise and in the laboratory to assure that the samples used were of the highest possible quality. The piston corer was rigged to house a liner of unconventionally large inside diameter (89 mm) to reduce disturbance during the act of sampling. On board the vessel the liner was extruded, cut with a tube cutter, and the core parted with a wire saw. After shear strength measurements were made and samples for water content were taken, the sections were capped, sealed, and stored securely in a refrigerated van in the in situ (i.e., vertical) orientation.

In the laboratory, shear strength and water content were again determined at points as close as possible (~ 5 cm) to the original measurement positions, with major differences in the equivalent data indicating potential disturbance. Careful visual and X-ray examination for evidence of disturbance gave further indication of sample quality. In general, moderate to severe disturbance was rare and confined to the upper meter or so.

"Undisturbed" shear strength was measured by a four-bladed, 12.7-mm-square laboratory vane which was inserted normal to the laminae and buried at least 20 mm into the sample. The rotation rate was 0.0262 rad/s (90°/min). Remolded strength (strength of thoroughly kneaded sample) was also determined. Other analyses included moisture content, liquid and plastic limit, and grain specific gravity. Specific gravity was determined by use of air comparison pycnometer; all other data were obtained in accordance with the procedures recommended by the ASTM (1977). Precisions were as follows: vane shear, ± 0.30 kPa; moisture content, $\pm 3\%$ (relative); liquid limit, $\pm 3\%$ (absolute); plastic limit, $\pm 2\%$ (absolute); grain specific gravity, $\pm 1\%$ (relative). Plasticity index and bulk density values were derived from the basic data set. These and the other variables were determined at intervals of 0.5 to 1 m and at lithologic changes.

Consolidation tests were performed on selected samples by L. E. Shephard and W. R. Bryant at Texas A&M University. They used backpressure consolidometers and determined preconsolidation pressures (maximum past effective overburden experienced by the sample) by the Casagrande method. Error associated with preconsolidation pressure values is $\pm 10\%$ (L. E. Shephard, 1978, oral comm.). Sedimentation rates (also $\pm 10\%$ error) were determined by C. W. Holmes and E. A. Martin of the U.S. Geological Survey, Corpus Christi, by the ^{210}Pb method.

INTERPRETIVE PROCEDURE

Approach

The essence of the interpretive procedure was 1) to predict, assuming normal consolidation, the

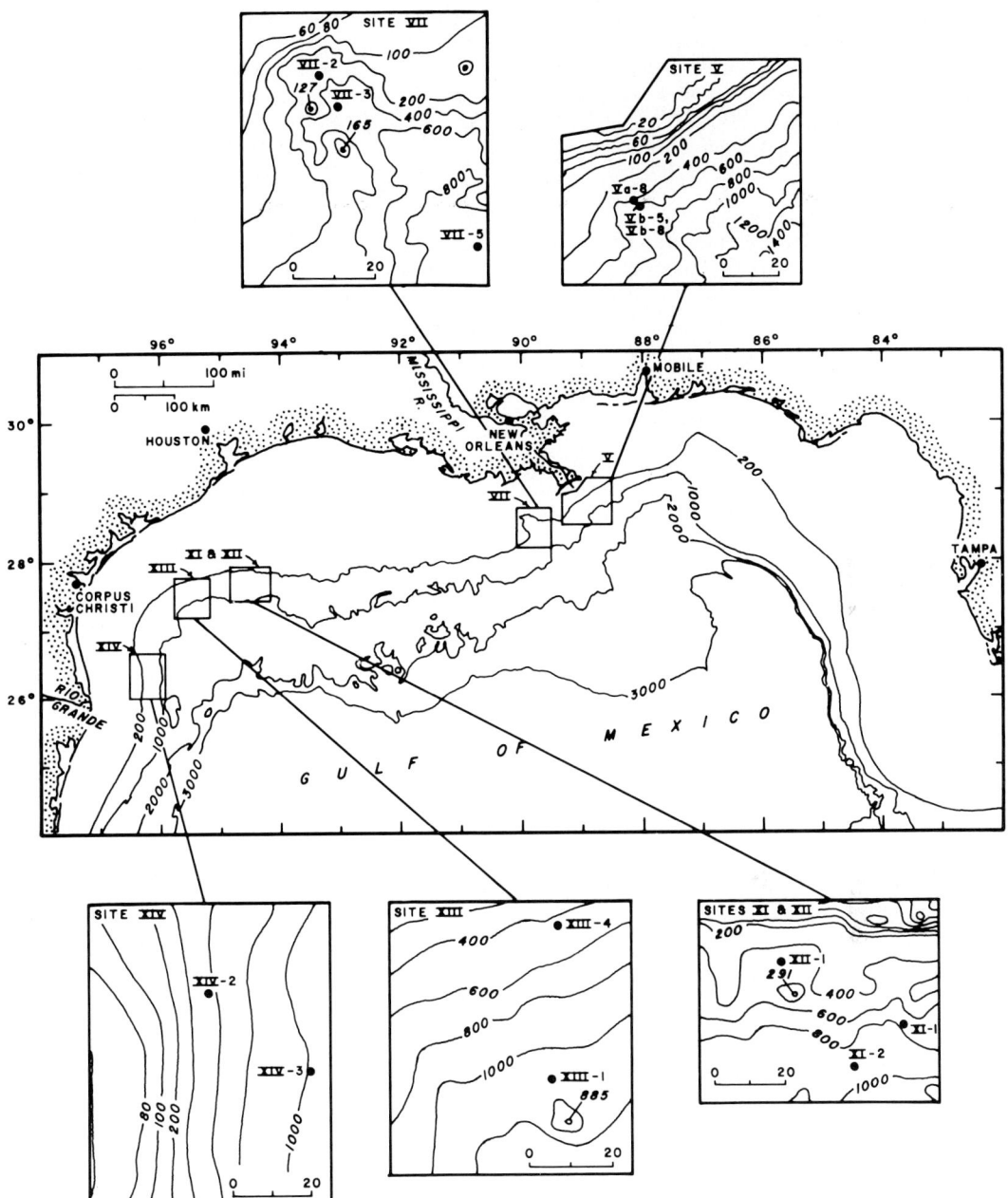

FIG. 1.—Locations of general study sites and stations. Scales are in kilometers, contours are in meters.

amount of effective overburden above a given level within the sediment column, 2) to compare this value with the amount of effective overburden determined by summing bulk densities above that level, and 3) to analyze the qualitative and quantitative difference between the two. This last stage provides the key to interpreting the mass-wasting history and the potential for mass-wasting of the site.

Method

The first stage, predicting effective overburden, is the critical element in the procedure. Both direct and indirect methods are available for this task, but direct methods (i.e., using a consolidometer) are time consuming and were judged inappropriate because of the large quantity of samples. Therefore, indirect methods were used and the amount of effective overburden was inferred from measurements of other mechanical properties. These methods are generally simple and easily applied, and therefore detailed documentation of consolidation states is readily obtainable. The method selected is based on an empirical relationship established by Skempton (1954) for predicting the strength/effective overburden ratio for normally consolidated sediments. The formula was primarily derived from marine and estuarine clays. The formula is:

$$S_u/\bar{\sigma}_v = 0.11 + 0.0037\,Ip$$

Effective overburden pressure was the key unknown, thus the formula was rearranged accordingly:

$$\bar{\sigma}_v = S_u/(0.11 + 0.0037\,Ip)$$

Thus, the effective overburden ($\bar{\sigma}_v$) may be predicted knowing shear strength (S_u) and plasticity index (Ip) (the difference between liquid and plastic limit). This predicted value represents the overburden anticipated for conditions of normal consolidation. A departure of this value from the determined amount of overburden signals the presence of an anomaly; that is, a departure from a state of normal consolidation. The direction (overconsolidation or underconsolidation) and magnitude of this anomaly may then be ascertained.

Evaluation of Method

Before this empirical formula could be applied, its accuracy had to be assessed. Both Morgenstern (1967) and Sangrey (1972) concluded that the formula was of sufficient accuracy for application under appropriate circumstances and that it tended to yield slightly conservative strength values (which means slightly liberal overburden values in this study). Karlsson and Viberg (1967) showed that, for a variety of normally consolidated soils, predictions based on the formula resulted in considerable scatter about the true values. However, their data indicated that on marine sediment of moderate to low sensitivity (ratio of "undisturbed" to remolded shear strength less than 8), such as the sediment used in this investigation, the Skempton relationship provided good agreement with the true values. Further, normally consolidated clays on the Louisiana shelf were shown by Fisk and McClelland (1959) to have a $S_u/\bar{\sigma}_v$ ratio that showed strong correlation with the ratio predicted by the Skempton formula. The consensus is, therfore, that the formula is sound and is appropriate for use in this study.

One of the keys to applying the formula successfully, however, is the accuracy of the data used with it. Although reasonable precautions were taken to prevent or reduce disturbance, sampling itself releases in situ stresses that can cause a significant reduction in shear strength. Subsequent handling, storage, and processing of the sample, no matter how careful, cause additional disturbance with resulting increased or decreased shear strength. In this study, the change in shear strength due to these various factors was estimated by comparing the values of overburden predicted by the empirical formula to those determined by direct consolidation testing at Texas A&M University. The scatter diagram (Fig. 2) shows a good correlation between the two independent sets of data (correlation coefficient $(r) = 0.80$). The average departure of the predicted values from the preconsolidation pressure is 30%. That is, on the average, the strength values used in the prediction need to be increased by 30% to equal the average of the determined preconsolidation pressure. This comparison also reflects on the integrity of the formula itself, of course, as well as on sources or error inherent to each of the two methods. However, the high correlation coefficient coupled with the use of error bands, which represent the 30% disparity, establish an

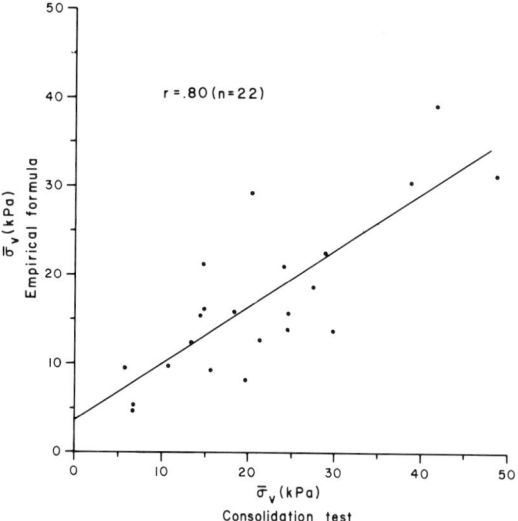

FIG. 2.—Relationship between effective overburden values predicted by the empirical formula ($\bar{\sigma}_v = S_u.0.11 + .0037\,Ip$) and the effective overburden values (preconsolidation pressures) determined as a result of consolidation testing by L. E. Shephard and W. R. Bryant of Texas A&M University.

adequate frame of reference on which to base the interpretations. I emphasize, however, that although the amount of error appears to be within certain bounds, insertion of the strength data into the formula results in predictions of overburden that are essentially approximations.

Application

Although the method provides a means of evaluating sediment characteristics in terms of consolidation state, the inferred consolidation state may not be the result of the physical process of compaction. Outside agents, such as gas or cement, for example, can create a condition within the sediment such that the true consolidation state is masked and the indicated consolidation state is actually unrelated to the settlement process. Thus, the method only provides an indication of *apparent* consolidation state. In order to verify the validity of an inferred consolidation state, therefore, criteria were established for the recognition and causes of the different consolidation states (Table 1). The table is incomplete in that only common explanations appropriate to this investigation are included. As shown in the table, many of the recognition criteria are related to the consolidation data alone; that is, they do not require other data. Thus, the consolidation data not only provide the consolidation state, but often, by showing trends and magnitudes, provide insight into cause as well. Because the focus of the research was on mass-wasting, the criteria for recognizing the effects of outside agents were used only in the process of verifying true consolidation state. Examples of the occurrence of these outside agents will not be discussed separately in this paper.

Identification of the anomalies and their trends was made by plots (e.g., Fig. 3) that showed the difference between the measured column height at a given sample level (summed bulk densities) and the height that the column would have been if it were normally consolidated (prediction with empirical formula). Conversion of overburden to

TABLE 1.—INTERPRETATION OF APPARENT CONSOLIDATION STATE—COMMON CAUSES AND THEIR RECOGNITION

Underconsolidation		Overconsolidation	
Cause	Recognition	Cause	Recognition
rapid deposition	—down core trend of increasing degree of underconsolidation	origin cohesion (basic interparticle attraction)	—slight overconsolidation in upper 2 m
gas	—abrupt to gradual transition toward underconsolidation; extreme underconsolidation in upper few meters; gas analysis	mass movement (e.g., slump, slide)	—generally abrupt and dramatic appearance of overconsolidation of sediment with little or no change in other properties; seismic evidence of scarp and/or anomalous deposit downslope
change to finer texture	—same or increase in water content; textural analysis		
disturbance	—visual or X-ray analysis showing mechanical disruption of sample integrity; abnormally low strength; low sensitivity, etc.	erosion (non mass movement type, e.g., scour)	—Same as for recognition of mass movement except absence of evidence of anomalous deposit down-slope; also, evidence of current action (e.g., lag deposit)
		cementation	—sensitivity; detection of high levels of nonparticulate $CaCO_3$, Fe_2O_3, or other potential cementing agents
		flow deposits (e.g., debris, turbidity)	—gradual to abrupt appearance of apparently overconsolidated sediment often with the presence of large quantities of sand; particularly if in conjunction with lower water contents; usually a return to a normal or underconsolidated section below the interval
		disturbance	—compaction in core liner; post-sampling dessication

column height was made by assuming a bulk density of 1.5 g/cm^3 (submerged bulk density of 0.47 g/cm^3, which is the average value for surficial, fine-grained marine sediment). At each level down the sediment column (core), therefore, the amount to be added or subtracted to achieve normal consolidation was approximately known (vertical axis). The horizontal line at mid-graph is the "normal" consolidation line. Negative values indicate overconsolidation (sediment apparently missing), and positive values indicate underconsolidation (sediment apparently thicker than expected). Hachured areas in figures 3–8 represent the range of values between zero and an average of 30% error.

RESULTS AND INTERPRETATIONS

The upper slope off the Mississippi Delta is site V (Fig. 1). Figure 3 shows that a common trend toward underconsolidation is shared by three cores. Thus, rapid deposition is indicated. Rates of just under 4 mm/y up to almost 6 mm/y were determined for this region by ^{210}Pb dating. core Va-8 (Fig. 3A) was recovered from the nose of a mudflow in the area. A spike toward the overconsolidation field at about 5 m represents a coarse layer of sediment and not actual overconsolidation. This is believed to be deposition by a mass-wasting event or current. Inclined laminae within this layer in conjunction with underconsolidated sediment beneath the zone (i.e., indicating material has probably not been removed) form the basis for this interpretation. An overconsolidated zone is also present in core Vb-5 (Fig. 3B), which was taken on a ridge seaward of the mudflow. Instantaneous deposition is again indicated. In this core, a lithologic change to a coarser sediment is found in the interval of apparent overconsolidation, which is sandwiched between two zones trending toward underconsolidation. The third core, Vb-8 (Fig. 3C), was taken adjacent to Vb-5 but does not show the minor depositional event in the consolidation data or the core descriptions. This testifies to the local nature of the mass-wasting in the area. It is concluded that this site off the delta experiences

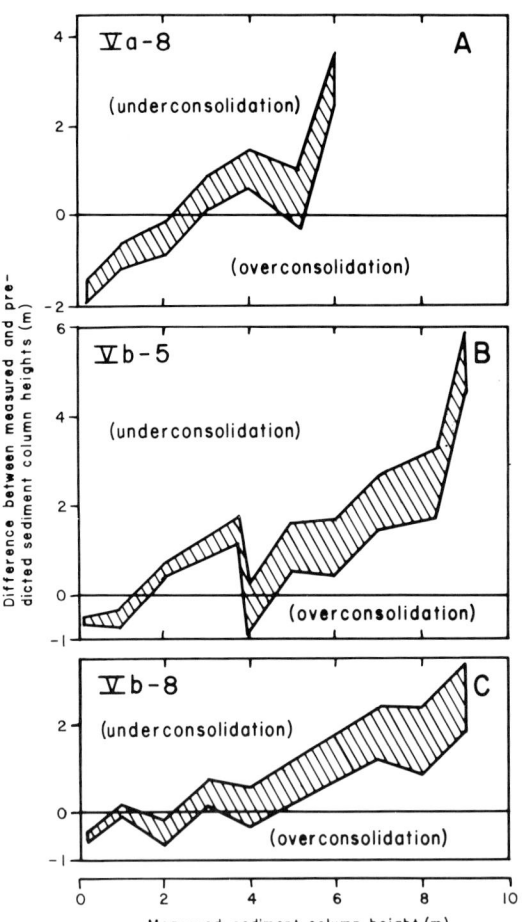

FIG. 3.—Progressive consolidation state of sediments in site V: A) core Va-8, B) core Vb-5, C) core Vb-8. Hachured areas are error bands. Detailed explanation of figure format is in text.

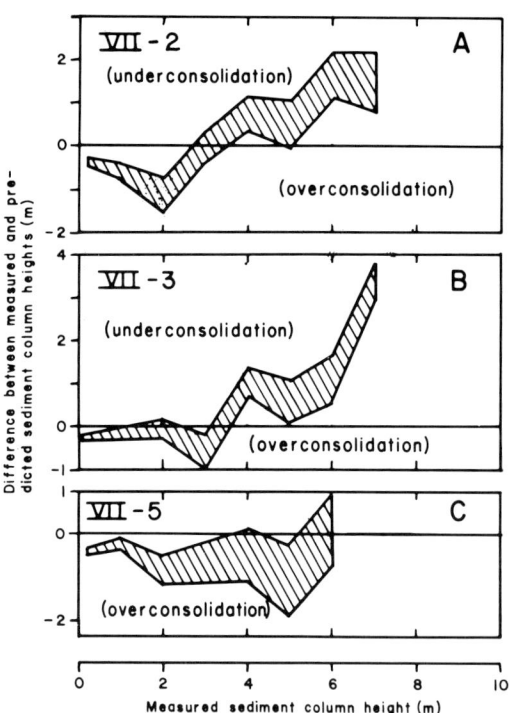

FIG. 4.—Progressive consolidation state of sediments in site VII: A) core VII-2, B) core VII-3, C) core VII-5. Hachured areas are error bands. Detailed explanation of figure format is in text.

frequent small-scale episodes of mass-wasting or a type of deposition under a relatively high flow regime within a general climate of rapid deposition. Previous work (Booth and Garrison, 1978) has shown that these small-scale events may be related to the advancing Balize (modern) lobe of the Delta, which has created a climate of rapid deposition in an area that was once subjected to much lower rates of deposition. No evidence of gas, large-scale incipient movement, or erosion was found in these surficial sediments.

The adjacent Mississippi Trough (Site VII, Fig. 1) stands in contrast to the Mississippi depocenter. Plots of cores VII-2, VII-3, and VII-5 indicate a general absence of dynamic events, even small-scale ones, within the trough (Fig. 4). All cores are either normally consolidated or trending toward underconsolidation. Rates of deposition range from 1 mm/y (VII-5) to more than 7 mm/y (VII-3), the latter being the highest rate determined in this study. The difference in the consolidation-state trends of these cores (Figs. 4B, 4C), which have similar textures throughout, emphasizes the effect produced by two different sedimentation rates.

Sites XI and XII (Fig. 1) show dramatic evidence of Holocene mass movement and suggest current conditions of fairly rapid deposition. Core XI-1 (Fig. 5A) shows the familiar trend toward underconsolidation, although the upper 4 m are normally consolidated. Figure 5B (Core XI-2) shows a similar pattern in its upper 4 m, but then the consolidation state changes abruptly from being normal to overconsolidated. (Note that the vertical scales of Figs. 5A and 5B are different.) At this depth within the sediment column (about 4 m) the equivalent of approximately 10 m of normally consolidated overburden in excess of the current (4 m) was once present. Without evidence of the influence of outside agents, it is concluded that a section approximately 14 m thick was involved in a slump (or other related event) on this location during the Holocene. (The height of the column would vary, of course, depending on the consolidation state of the material that was removed.) The consolidation-state trend becomes complex below this level, but the general state of overconsolidation is sustained. This fact, coupled with the absence of change in grain size at or below the appearance of overconsolidated sediment (the section is fine-grained throughout), rules out a depositional event.

A slide of approximately the same thickness (about 15 m) took place in the area represented by Core XII-1 (Fig. 6). A clear trend toward

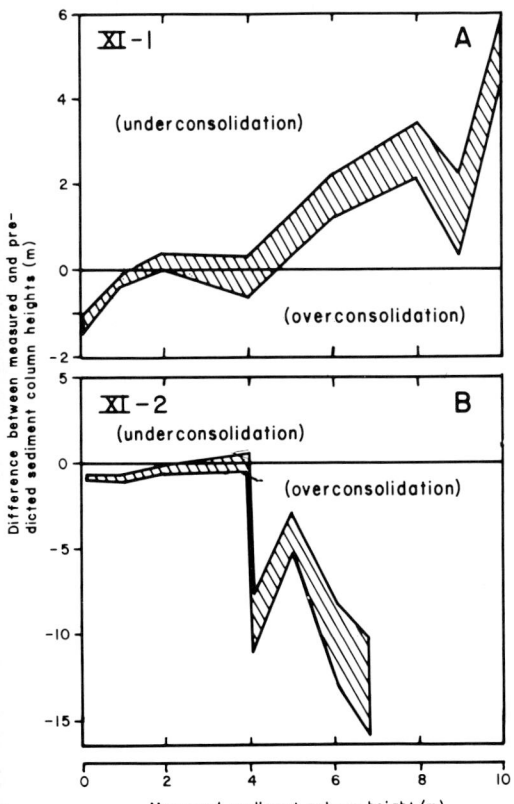

FIG. 5.—Progressive consolidation state of sediments in site XI: A) core XI-1, B) XI-2 (Note change in vertical scale). Hachured areas are error bands. Detailed explanation of figure format is in text.

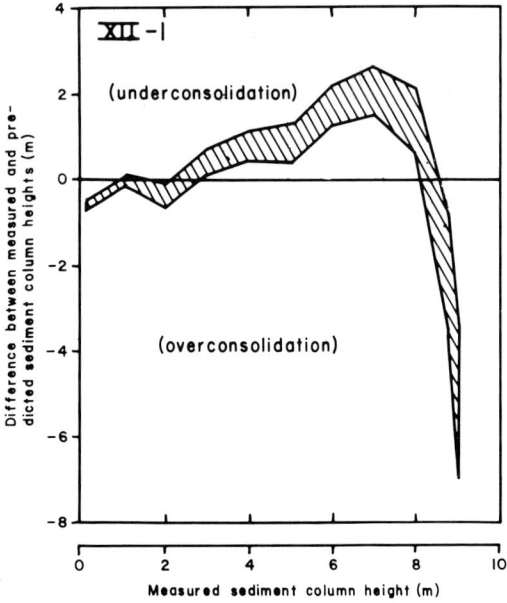

FIG. 6.—Progressive consolidation state down core XII-1. Hachured area is error band. Detailed explanation of figure format is in text.

increasing underconsolidation is seen in the upper 8 m of the column, which indicates fairly rapid deposition, but then an abrupt change to overconsolidation occurs in the lower part of the core. About 5 or 6 m of overburden is missing, which suggests that as the depth in core is almost 9 m, at one time a mass movement event of about 14 or 15 m took place. If the determined rate of deposition of 2.44 mm/y can be extrapolated back to the event, then it occurred about 3000–4000 years BP (before present). If something close to this rate applies to the proximal Core XI-2, the date of the slide (4 m of modern sediment over scar) would be about 1500–2000 years BP. Further, the modern sedimentation rate may be conservative, that is, less than it was in the past. If so, these dates represent a maximum age. It can be concluded that this general area, sites XI and XII, has experienced frequent mass movement and has current rates of deposition that likely will lead to more mass-wasting events.

The most spectacular evidence for mass movement was found on the upper slope due east of Corpus Christi (site XIII, Fig. 1). Core XIII-1 (Fig. 7A) penetrated an overconsolidated zone about 6 m below the mudline. The particular cored section had more than 40 m of overburden above it at one time.

Although the data in general suggest a prominent mass movement event, the interpretation is more complex. Within the overconsolidated zone, the core is distorted and has blocky subsections. Further, the laminae within these subsections have a spectrum of orientations. Thus, this core may have penetrated the deposit rather than the scar of a mass movement event. The suggestion that 40 m of overburden is missing above the top of the overconsolidated zone may therefore only apply to that uppermost subsection, which was in an unknown position within the original (pre-slump) sediment column. The erratic nature of the consolidation-state (Fig. 7A) beneath the initial level of overconsolidation underscores the possibility that several different subsections of the slumped material were penetrated. No sedimentation rates were determined in the overlying 6 m of more recent sediment. However, it probably exceeds 1 mm/y which makes the time of mass movement at or within the last 6000 years BP. The other core, XIII-4 (Fig. 7B), maintains a consolidation state trend that stays fairly close to the line of normal consolidation with a minor trend toward underconsolidation. However, because the trend is basically flat, the deposition rate at this location is probably less than at most locations heretofore discussed. On the basis of evidence within these cores, site XIII is experiencing low to moderate rates of deposition during recent times, and has been a region sub-

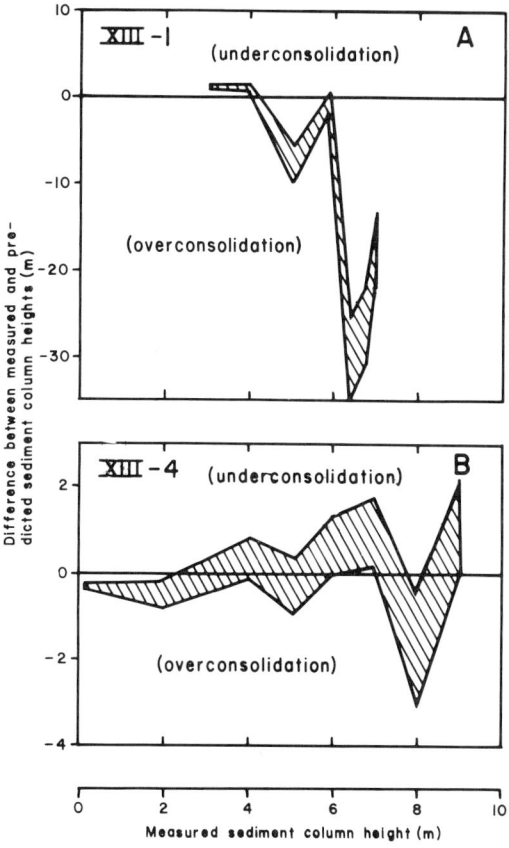

Fig. 7.—Progressive consolidation state of sediments in site XIII: A) core XIII-1 (note vertical scale), B) core XIII-4. Hachured areas are error bands. Detailed explanation of figure format is in text.

jected to dramatic mass movement.

Mass movement has also apparently occurred on the slope off the Rio Grande Delta (Site XIV, Fig. 1). Figure 8B shows that about 18 m of overburden is missing from core XIV-3 at the 5 m level. Thus, mass movement involving 23 m of sediment took place at this site. The date of the event is uncertain. Analysis of core XIV-2, (Fig. 8B) which is upslope from XIV-3, yielded a rate of deposition of 0.63 mm/y—the lowest rate found in the study. If a rate of deposition having this order of magnitude is applied to Core XIV-3, then the mass movement event took place about 10,000 years BP. Core XIV-2 itself does not show evidence of mass wasting, but contains a marked change in consolidation state at 6 m. This change from a normal consolidation trend to one of underconsolidation may reflect a change in sedimentation rate at that location. That is, the core contains sediment deposited during a

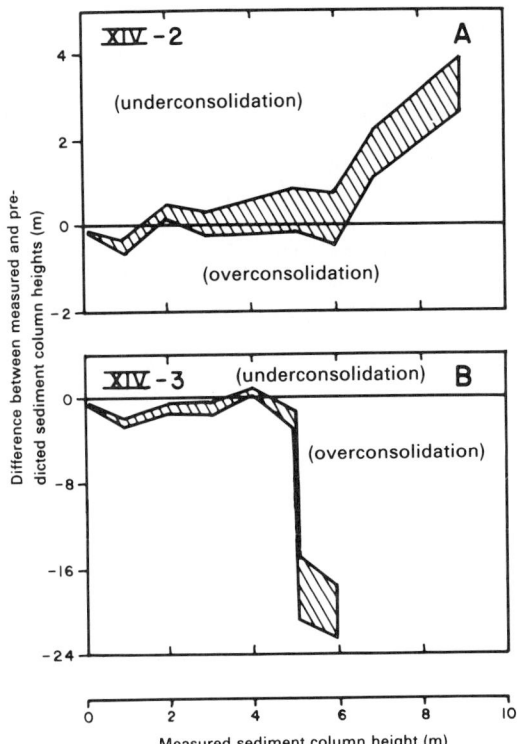

Fig. 8.—Progressive consolidation state of sediments in site XIV: A) core XIV-2, B) core XIV-3 (note change in vertical scale). Hachured areas are error bands. Detailed explanation of figure format in text.

determining their cause. Additionally, these events have been taking place throughout the time interval represented by the cores: more than 10,000 years. Depending on the accuracy of extrapolated rates of deposition, the data show a spread of 1500 to 10,000 years for the times of occurrence for these events. And these dates are assumed to be fortuitous—the actual range may extend from the present to before the time represented by this particular collection of cores. This time factor must also be considered in determining the cause of the mass-wasting events. Clearly, the conditions responsible for the events are omnipresent. The cause of the mass-wasting must also be: 1) capable of initiating movement in sediment sections as much as several tens of meters thick, 2) capable of operating on the near-horizontal (1°–2°) surfaces that are common on this continental slope, and 3) capable of initiating movement in the absence of a prominent, direct triggering mechanism, such as major storms (most sites are below the effective wave base even for hurricanes) and earthquakes (Gulf of Mexico tectonism is essentially shock free), although movement could also be caused by such mechanisms.

Rapid deposition fulfills these criteria and is considered the most likely cause of the conditions responsible for the mass-wasting. The chain of events is as follows:

rapid deposition
↓
excess pore pressure
(above hydrostatic)
↓
abnormally low effective overburden pressure
↓
inadequate shear strength
↓
failure

This sequence may be viewed in a more detailed fashion by using slope stability analysis and consolidation theory.

Current rates of deposition on the slope average about 3 mm/y. According to the theory developed by Gibson (1958), an underconsolidated, and hence overpressured, section would form even at these rates. In the past, that is, when the upper slope was closer to sediment sources just prior to, and in association with, the last low stand of sea level, sedimentation rates may have been equal to or greater than present rates. If so, then a section characterized by excess pore-water pressures has been accreting for at least 25,000 years. The actual length of time is an important factor, for, although consolidation theory suggests that rates of deposition on the order of 3 mm/y are not significant in terms of causing short-term

time interval when sea level rose from a lower level to its present position. The assumed greater rates of deposition in the past, when sediment sources were closer, may have produced an underconsolidated section which, because of the reduced rates at present (0.63 mm/y), was subsequently topped by a more normally consolidated section. An alternative explanation of the abrupt change in consolidation-state trend is that gas, probably biogenic methane, is present in the sediments below 6 m in this location.

All conclusions regarding the occurrence of mass movement (Sites XI, XII, XIII, and XIV) are supported by seismic evidence developed by Tatum (1978), in a study done concurrently with this investigation.

CAUSE OF MASS-WASTING

Evidence of mass wasting was found at several sites on the upper continental slope of the northern Gulf of Mexico. The fact that these events are frequent and widespread is not only an important aspect of the recent geologic history of the area, but establishes one of the primary criteria for failure, they are significant when great amounts of time are involved. Specifically, the time parameter devised by Gibson is a function of time (t) as well as rate of deposition (m). In addition, the coefficient of consolidation (C_v), which we consider a constant in this discussion, is also involved. Gibson's time parameter has the form

$$\frac{m^2 t}{c_v}$$

and is directly related to excess pore pressure. As this time parameter increases in value, any given level within the sediment column has more of its overburden supported by the pore pressure and less by interparticle contact. When the buoyant effect of excess pore pressure becomes significant relative to the weight of the overlying material (i.e., their ratio approaches one), the shear strength becomes inadequate with respect to the shear stress, and failure occurs (Gibson, 1958, Fig. 3).

The relationship of overburden and shear strength can be examined more closely by introducing slope stability analysis. The model of the infinite slope (thickness of sediment insignificant compared to lateral dimensions) is appropriate in this investigation: the formula for evaluation is:

$$F = 1 - \left[\frac{U_e}{\bar{\gamma} Z \cos^2_\alpha} \right] \frac{\tan \bar{\phi}}{\tan \alpha}$$

where F is the factor of safety, U_e is excess pore water pressure (i.e., above hydrostatic, $\bar{\gamma}$ is buoyant unit weight (submerged bulk density) of the sediment, Z is the sediment thickness, $\bar{\phi}$ is the angle of internal friction with respect to the effective stresses, and α is the slope angle. For the purposes of this analysis, $F = 1$, and, because the slopes are generally 1°–2°, $\cos^2 \alpha \sim 1$. Thus the equation may be rewritten in the following form:

$$\tan \alpha = \left[1 - \frac{U_e}{\bar{\gamma} z} \right] \tan \bar{\phi}$$

If no excess pore-water pressure exists, then

$$\tan \alpha = \tan \bar{\phi}$$

at failure. This criterion is derived directly from the Mohr-Coulomb failure theory. Because $\bar{\phi}$ is typically between 20° and 30°, failure is not likely to occur on anything but extreme slopes. However, where excess pore-water pressure approaches $\bar{\gamma}$ (which represents the overburden pressure), the product of normal stress and tan $\bar{\phi}$ is reduced until failure results. In short, the shear strength is in large part dependent on the effective overburden pressure, and the lower the effective overburden pressure exerted by a given column height, the lower the strength. This effect may be responsible for the basic unstable condition on the upper slope.

The intensity of this unstable condition, however, is not present equally throughout a given sediment column. Failure is most likely to occur in the upper part. Gibson (1958, Fig. 3) has shown that the pore pressure effect is enhanced (relative to overburden) as the top of the underconsolidated section is approached. Thus, although the upper slope sediments may be underconsolidated for 100 m or more (≥ 3 mm/y for $\geq 25,000$ yr), the part of the column that is the most susceptible to failure may be an order of magnitude less in thickness and may be perched on top of the column. This, in fact, was the circumstance found in this study. Slumps or slides initially 10–20 m thick were suggested by the data. A quantitative treatment of this type of consolidation-slope stability problem has been presented by Hampton et al, (1978).

In addition to this basic instability, small-scale phenomena, such as the emplacement of a turbidite a few meters thick, can act as a triggering mechanism for deeper seated phenomena such as a 10 to 20 m slide (Dwight A. Sangrey, 1978, oral comm.). The instantaneous deposition of a sediment layer can act to increase pore pressures further by the amount equivalent to the new overburden in an already underconsolidated section. This effect may cause failure in the deposit well below the level of the turbidite cap. Shelf-edge events, therefore, which conceivably may be induced by major storms, internal waves, or other forces, may in turn be responsible for initiating some of the movements associated with the underconsolidated sediment. These types of small-scale events have been operating on the slope off the Mississippi Delta (Booth and Garrison, 1978), and probably most shelf-edge sites are vulnerable to this effect.

The fundamental instability of the upper slope sediments also makes them vulnerable to failure triggered by earthquakes. Although shocks are rare in the Gulf of Mexico, the occasional seismic event that does occur could initiate movement directly or, similar to the storm effect, bring down sediments from the shelf edge to act as a further contributor toward failure on the underconsolidated column.

SUMMARY AND CONCLUSIONS

The recent history of mass-wasting on the upper slope of the northern Gulf of Mexico has been marked by frequent mass movement and other types of mass-wasting; it has been and is a dynamic environment. Consolidation data also show that variable and rather high rates of deposition are prevalent and that for the past several thousand years there have been no major changes in the basic fine-grained texture of the sediment delivered. This rapid deposition of fine-grained sediment has led to an underconsolidated sediment column. Occasionally, coarser material is deposited, probably in consequence to shelf-edge phenomena.

The upper slope off the Mississippi Delta is currently the site of frequent small-scale depositional and mass movement events. Silty sediments are being deposited at fairly rapid rates (about 5 mm/y) and the general depositional environment changed in conjunction with the outbuilding of the Balize lobe of the delta (Booth and Garrison, 1978). This environment will become even more active as the delta continues to build toward it.

The least active environment studied is the Mississippi Trough, which, paradoxically, is located adjacent to the Mississippi Depocenter. Fine-grained material has been accumulating at varying rates in the trough throughout recent time. No evidence of mass-wasting was found in the cores. The sediment does tend to be underconsolidated, however, and thus the conditions for mass movement exist.

The upper slope on the continental margin off Texas has also been receiving fine-grained sediment during the past several thousand years. Episodes of mass movement have taken place throughout the region. In the southern part of the Texas upper slope (off the Rio Grande), rates of deposition have slowed considerably since sea level rose; the current rate of deposition is less than 1 mm/y in an area where an underconsolidated section was once forming. Thus, in recent time, a steady, but gradually decreasing rain of silt and clay has been blanketing the region. Slumps, slides, or related events periodically interrupt this steady accumulation of sediment.

During the past several thousand years, the upper continental slope of the northern Gulf of Mexico has been experiencing rates of deposition that lead to an underconsolidated section. The instability that has led to frequent and widespread mass movement, however, was established by long term rapid deposition related to the last low stand of the sea. In other words, the mass-wasting events have occurred in response to conditions established prior to recent time.

The present study has led to the following conclusions:

1. Consolidation states may be used qualitatively to detect evidence of mass movement and other phenomena related to mass-wasting in marine sediments. In addition, the thickness of sediment involved in a mass movement event can often be estimated and, under certain conditions, relative rates of deposition can be inferred.

2. The upper Continental Slope of the northern Gulf of Mexico has been the site of frequent and widespread mass-wasting during the past 10,000 years, and probably even long before then: it is a dynamic environment.

3. The slumps, slides, and kindred phenomena range in thickness from a few meters to more than 40 m (at a location east of Corpus Christi).

4. High rates of deposition are the basic cause of the instability. Rates of less than 1 mm/y to more than 7 mm/y were determined; the average rate is 3 mm/y. Deposition rates of this magnitude (3 mm/y) ultimately lead to an underconsolidated sediment, which, because of abnormally high excess pore pressures, and thus abnormally low strength, is prone to fail. These high rates of deposition have probably been present since prior to the last low stand of the sea (about 25,000 years ago at least).

ACKNOWLEDGMENTS

I am grateful to Louis E. Garrison for his support and to Dwight A. Sangrey for his useful comments and discussion. Charles W. Holmes and Ann Martin provided the ^{210}Pb dates, and William R. Bryant and Les E. Shephard of Texas A&M University provided the results of their consolidation tests.

REFERENCES

American Society for Testing and Materials, 1977, 1977 annual book of ASTM standards, part 19, Natural building stones; soil and rock; peats, mosses, and humus: Philadelphia, 494 p.

Booth, J. S., and Garrison, L. E., 1978, A geologic and geotechnical analysis of the upper Continental Slope adjacent to the Mississippi Delta: Proc. 10th Annual Offshore Technology Conference, Houston, Texas, OTC #3165, p. 1019–1028.

Bryant, W. R., Cernock, Paul, and Morelock, Jack, 1967, Shear strength and consolidation characteristics of marine sediments, in Richards, A. F., ed., Marine Geotechnique: Urbana, Ill., Univ. Illinois Press, p. 41–62.

Fisk, H. N., and McClelland, B., 1959, Geology of Continental Shelf off Louisiana: its influence on offshore foundation design: Geol. Soc. America Bull., v. 70, p. 1369–1394.

Gibson, R. E., 1958, The progress of consolidation in a clay layer increasing in thickness with time: Geotechnique, v. 8, p. 171–182.

HAMPTON, M. A., BOUMA, A. H., CARLSON, P. R., MOLNIA, B. F., CLUKEY, E. C., AND SANGREY, D. A., 1978, Quantitative study of slope instability in the Gulf of Alaska: Proc. 10th Annual Offshore Technology Conference, Houston, Texas, OTC #3314, p. 2307–2318.

KARLSSON, R., AND VIBERG, L., 1967, Ratio C/P' in relation to liquid limit and plasticity index, with special reference to Swedish clays: Proceedings, Geotechnical Conference, Oslo, 1967, v. 1, p. 43–47.

LEHNER, PETER, 1969, Salt tectonics and Pleistocene stratigraphy on Continental Slope of northern Gulf of Mexico: Am. Assoc. Petroleum Geologists Bull., v. 53, p. 2431–2479.

McCLELLAND, BRAMLETTE, 1967, Progress of consolidation in delta front and prodelta clays of the Mississippi River, *in* Richards, A. F., ed., Marine Geotechnique: Urbana, Ill., Univ. Illinois Press, p. 22–40.

MORGENSTERN, N. R., 1967, Submarine slumping and the initiation of turbidity currents, *in* Richards, A. F., ed., Marine Geotechnique: Urbana, Ill., Univ. Illinois Press, p. 189–220.

SANGREY, D. A., 1972, Obtaining strength profiles with depth for marine soil deposits using disturbed samples: Special Technical Publication 501, American Society for Testing and Materials, p. 106–121.

SKEMPTON, A. W., 1954, Discussion of the structure of inorganic soil: Journal of Soil Mechanics and Foundation Division: Proc. ASCE, v. 80; Sept. no. 478, p. 19–22.

TATUM, T., 1977, Shallow geologic features of the upper continental slope, northern Gulf of Mexico [Masters thesis]: Texas A&M University.

WOODBURY, H. O., 1977, Movement of sediment of the Gulf of Mexico Continental Slope and Upper Continental Shelf: Marine Geotechnology, v. 2, p. 263–274.

WOODBURY, H. O., MURRAY, I. B., JR., PICKFORD, P. J., AND AKERS, W. H., 1973, Pliocene and Pleistocene depocenters, Outer Continental Shelf, Louisiana and Texas: Am. Assoc. Petroleum Geologists Bull., v. 57, p. 2428–2439.

DEEP CARBONATE BANK MARGIN STRUCTURE AND SEDIMENTATION IN THE NORTHERN BAHAMAS

HENRY T. MULLINS AND A. CONRAD NEUMANN
Moss Landing Marine Labs and San Jose State University, P.O. Box 223, Moss Landing, California 95039, and Curriculum in Marine Sciences, 12-5 Venable Hall, University of North Carolina, Chapel Hill, North Carolina 27514

ABSTRACT

The deep (>200 m) carbonate bank margins in the northern Bahamas display complex and diverse seismic facies, structure, morphology and sedimentary facies. On the basis of analysis of more than 1200 km of high-frequency, high-resolution seismic reflection profiles and 149 bottom samples seven deep carbonate bank margin types have been recognized in the northern Bahamas: (1) Type 'A'—Open Ocean Windward margin east of Little Bahama Bank (LBB); (2) Type 'B'—Open Ocean Windward margin north of LBB; (3) Type 'C'—Extended deep bank margins off the northwest corners of both LBB and Great Bahama Bank (GBB); (4) Type 'D'—Open Seaway Leeward margin west of LBB and GBB; (5) Type 'E'—Open Seaway Leeward margin south of LBB; (6) Type 'F'—Open Seaway Windward margin north of GBB; and (7) Type 'G'—Eroded margins along the mouth of the Northwest Providence Channel (NWPC).

In general, leeward deep bank margins tend to be steeper, narrower, more dissected, and contain significantly greater amounts of coarse grained sediment than windward margins. The processes most responsible for the development of deep carbonate bank margins include: (1) basement faulting: (2) direction and magnitude of off-bank sediment transport; (3) oceanic circulation; (4) gravity and pelagic sedimentation; (5) submarine cementation; and (6) biological buildups. Of these, basement faulting is primarily responsible for the initiation of carbonate platform edges. Off-bank sediment transport is controlled by the physical energy flux (winds, waves, storms) at the sea surface and controls the availability of shallow-water sediment for transport to the deep flanks. Oceanic circulation (bottom currents) winnows and redistributes sediment and may aid in submarine cementation which appears responsible for the stabilization of carbonate slopes through hardground development. Pelagic carbonate sediments are ubiquitous in deep-water carbonate environments but are volumetrically important only where they are not winnowed by bottom currents, diluted by sediment gravity flow deposits or deposited below the CCD. Much of the material deposited on carbonate slopes is allochthonous, transported via a variety of sediment gravity flows including turbidity currents, grain flows and debris flows. Deep-water biological buildups are locally important and may add to flank accretion.

Previous models of deep carbonate bank margins derived solely from analysis of the ancient appear to be oversimplifications. A new knowledge of the diverse types of modern deep carbonate bank margins and an understanding of the processes that control their formation should aid in the interpretation of ancient deep-water carbonate sequences as well as in the exploration for and exploitation of hydrocarbons from off-platform carbonates.

INTRODUCTION

Modern carbonate platforms, such as the Bahamas, have been extensively studied in order to better understand the nature of these carbonate deposits and to construct modern analogs for ancient limestone sequences. Much of this research has been aimed at determining the sedimentary facies relationships and early diagenesis of shallow-water carbonate sediments of the platform interiors. Until very recently, however, little was known about the structural or sedimentological framework and facies relationships of modern carbonate bank margins.

A research program has been undertaken at the University of North Carolina to study both shallow (0–200 m water depth) and deep (>200 m water depth) modern carbonate bank margins along open seas and seaways in the northern Bahamas. The initial thrust of this program was aimed at the shallow bank edges. High resolution seismic reflection (uniboom) profiling and bottom sampling of the shallow margins around Little Bahama Bank (LBB) and northern Great Bahama Bank (GBB) indicate that there is considerable complexity and diversity of structure, sedimentary facies and growth history along the shallow bank margins in the northern Bahamas (Hine and Neumann, 1977).

The present study is an extension of the work on the shallow bank margins in the northern Bahamas in order to expand our study of modern carbonate bank margins to deeper slope and basin environments. The purpose of this initial research is to determine the gross structural and sedimentological framework of modern deep carbonate bank margins along open seas and seaways in the northern Bahamas. The major objectives are to: (1) Determine the variability of deep bank margin types in the northern Bahamas; (2) Determine the structural and seismic facies relationships of each deep bank margin type from the analysis of high frequency-high resolution air-gun and uniboom seismic reflection profiles; (3) Determine the major processes of deep bank margin sedimentation and sedimentary facies relationships from the analysis of samples recovered in piston and gravity cores, rock dredges and surface grabs as well as bottom photographs; (4) Determine the major controls and processes of deep carbonate bank margin development; and (5) Establish a set of recognition criteria and generalized models for each deep margin type.

Economic Significance

Nearly half of the world's proven petroleum reserves are in carbonate reservoirs (Sheridan, 1977). Ancient carbonate platforms constitute important, prolific hydrocarbon reservoirs in many parts of the world, particularly in the United States, Canada, North Africa, Mexico, Southeast Asia and the Middle East (Bubb and Hatfield, 1977). Although shallow-water reservoirs (reefs, oolite shoals, etc.) have provided the bulk of the recovered hydrocarbons to date the search for hydrocarbons is expanding to include deeper water environments (Scholle, 1977a,b; Cook and Enos, 1977). In fact, since 1970, more than 2.5 billion barrels of oil and 8 trillion cubic feet of gas have been recovered from deep-water carbonate facies of the North Sea (Scholle, 1977b) and significant production has been obtained from deep-water carbonates in Texas, California and Mexico (Scholle, 1977a; Enos, 1977). Thus our need to better understand these deep-water carbonate environments and the transition from shallow to deep has become increasingly important.

Regional Setting

Structure of the Bahama Platform.—The Bahama platform is presently tectonically stable (Barazangi and Dorman, 1968; Molnar and Sykes, 1969) and consists of a series of carbonate platforms built along the subsiding continental margin of North America which extends more than 1400 km from southern Florida to Puerto Rico. The northwestern Bahamas consist of large shallow-water carbonate platforms that are separated by linear, deep-water channels (Fig. 1). Drill hole data on the platforms indicate that the thickness of Jurassic to Recent shallow-water limestones beneath the banks is at least 5.4 km and geophysical data suggest that in some places it is as great as 7–14 km (Uchupi et al, 1971; Meyerhoff and Hatten, 1974; Mullins and Lynts, 1975, 1977; Sheridan, 1974, 1976). On the southwest the Bahama Platform is bounded tectonically by the Cuban orogen and to the south by the Caribbean lithospheric plate. To the east, the platform drops precipitously along the Bahama Escarpment to

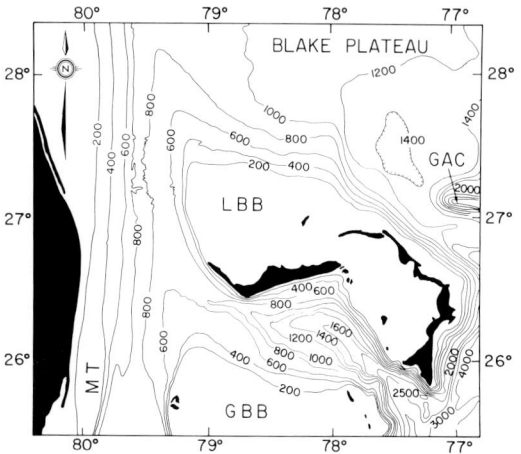

Fig. 1.—Generalized bathymetric map of the study area. GAC = Great Abaco Canyon: LBB = Little Bahama Bank; GBB = Great Bahama Bank; MT = Miami Terrace. Large land mass to west is Florida. The Florida Straits separate Florida and the Bahamas. NWPC separates LBB and GBB. Contour interval is 200 m. Based on Belding and Holland (1970).

abyssal depths of the Atlantic. To the north and northwest the Bahamas are contiguous with the Blake Plateau and the South Florida Platform, constituting one of the largest carbonate provinces in the world.

The origin of this vast carbonate province is still controversial. It is at present uncertain whether the Bahamian basement is oceanic (Dietz et al, 1970; Sheridan, 1974) or continental (Meyerhoff and Hatten, 1974; Mullins and Lynts, 1977). There also is not complete agreement as to the timing and control of platform segmentation. Many workers believe that basement structure controls the location of the platform edges (Talwani, 1960; Ball, 1967; Sheridan, 1974; Mullins and Lynts, 1977). However, Dietz et al (1970) and Paulus (1972) suggest that the location of the platform margins has been controlled solely by reef growth.

Bathymetry.—Immediately beyond the shelf break (~50 m) of the Bahama Banks a nearly vertical marginal escarpment extends to approximately 150–200 m of water depth. Seaward of this marginal escarpment the gradient decreases and extends along variable down-to-basin slopes (Table 1). Maximum basin depths are also highly variable ranging from 800 m in the Florida Straits west of LBB to 4500 m in the open Atlantic east of LBB (Fig. 1). A generalized bathymetric map of the study area is given in Figure 1.

Physical Energy Flux.—The Bahama Platform is strongly influenced by the easterly trade wind

TABLE 1—AVERAGE DOWN-TO-BASIN SLOPES[1]

	Average Slope		
	m/km	Ratio	Degrees
North of LBB (Profile #1)	60	1:16	4
N.W. of LBB (Profile #3)	5	1:200	1
**West of LBB	81	1:12	6
*South of LBB			
a) Off Grand Bahama	90–180	1:11–1:5	6.2–12.5
b) Off Mores Island	135	1:7	9.5
c) Off Gorda Cay	300	1:3	20
ˣEast of LBB	—	—	28–40
**N.W. of GBB	6	1:166	1
*North of GBB	15	1:66	1.5

[1]These are average slopes from the bank tops to the toe of the slope. All areas exhibit much steeper slopes close to the bank tops.
ˣBased on data from Emiliani (1965)
*Based on bathymetric chart of Andrews et al (1970)
**Based on bathymetric chart of Malloy and Hurley (1970)

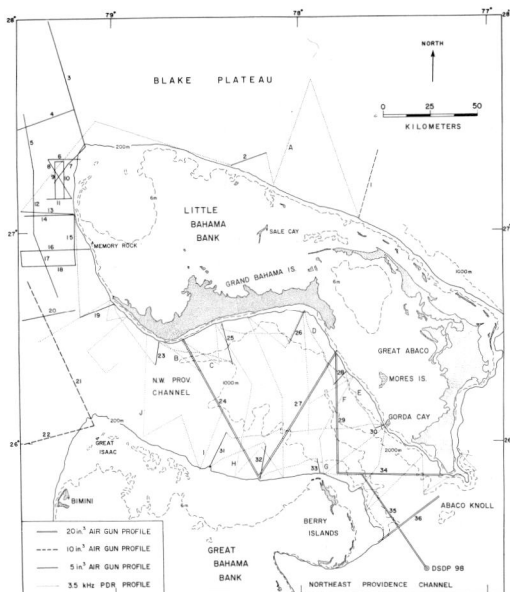

FIG. 2.—Location map of seismic reflection profiles referred to by number in text.

belt. To quantify the effects of the physical energy flux (winds, storms, and waves) on platform sedimentation, available data on the wind regime, frequency and direction of passage of tropical storms and hurricanes, and wave energy was compiled by Hine (1975) and presented in Hine and Neumann (1977). The results of this compilation indicate that LBB is dominated by easterly to northeasterly winds and waves. As a result of this energy flux there is a dominance of *on-bank* transport of shallow-water sediment along *windward* margins (north and east of LBB, and north of GBB) and *off-bank* transport of shallow-water sediment along *leeward* margins (west and south of LBB; Hine and Neumann, 1977).

Ocean Currents.—Surface water flow in the northern Straits of Florida, west of LBB and GBB, is dominated by the swift, northerly flow of the Florida Current (Duing, 1975), having maximum velocities greater than 160 cm/sec. Bottom water flow in the northern Straits of Florida along the eastern or Bahamian side of the Straits is also to the north, at velocities of up to 50 cm/sec and greater (Neumann and Ball, 1970; Neumann et al, 1977). In the NWPC water movement is in opposite directions on either side of the channel. South of LBB surface and bottom water flow is to the west, whereas north of GBB water transport is to the east (Richardson and Finlen, 1967).

METHODS

For this study, more than 1200 km of high frequency-high resolution single channel air-gun seismic reflection profiles were obtained from the deep bank margins in the northern Bahamas (Fig. 2). Most of the seismic data were obtained by the firing of a single Bolt air-gun equipped with a 5 in.^{3} (82 cm^3) chamber. The gun was fired at a two or three second interval at pressures of up to 2000 lbs./in^2 (13.78 × 10^7 dynes/cm^2). Reflections were received by a single array of

TABLE 2—INTERPRETATION OF BASIC SEISMIC FACIES PATTERNS*

Seismic Facies Pattern	Interpretation
Mounded Chaotic or Contorted/Disrupted	Mass movements; slumps or slides
Discontinuous	High-energy turbidity current transport
Sigmoid Terrace	Low-energy turbidity current transport
Oblique Even Layered	Bottom-flowing current transport
Parallel Layered or Conformable Drape	Pelagic suspension transport; uniform rates of deposition
Chaotic	Highly variable deposits, relatively high-energy environment
Reflection Free	Homogenous, non-stratified units such as salt, shale or igneous masses

*Based on Sangree et al (1976) and Mitchum et al (1977).

50 double Teledyne hydrophones equipped with a pre-amplifier, towed approximately 137 m astern of the ship. Signals were then filtered at 63 or 80 to 205 Hz, amplified and then recorded graphically. Ship speed was approximately 5 knots (9 km/hr.) and a combination of Loran C and satellite determinations were used for navigation with fixes taken at 15 minute intervals and at major course changes.

Resolution on the 5 in.^{3} air-gun profiles is about 5 m. Seismic stratigraphic methods outlined by Sheriff (1976), Sangree et al (1976), Mitchum et al. (1977), and in Payton (1977) were used in the interpretation of the reflection profiles. A list of some of the basic seismic reflection patterns and their interpretation is given in Table 2.

Bottom Samples

Forty-four piston cores and twenty-seven gravity cores (Fig. 3) were obtained from the northern Bahamas. Cores were taken in plastic core liners, sectioned, measured and brought back to the laboratory where they were X-radiographed to determine sedimentary structures, split and described. The bottom of each core was qualitatively analyzed for its content of planktonic Foraminifera to determine the age of each core. Subsamples from selected sand layers were quantitatively analyzed for their constituent components in order to determine the source of the >63 um fraction. Cores were interpreted in terms of major processes of deposition based on sedimentary structures revealed in X-radiographs and/or by visual observations of split cores, as outlined by Middleton and Hampton (1973). Core data were supplemented by grab samples, dredge samples and bottom photographs (Fig. 3).

DEEP CARBONATE BANK MARGIN TYPES

On the basis of seismic facies, structure, morphology and sedimentary facies, seven deep carbonate bank margin types have been recognized along the open seas and seaways in the northern Bahamas. From this, a general classification scheme for modern deep carbonate bank margin types is proposed (Table 3). This classification scheme is based upon: (1) whether the deep margin is presently accreting or eroding; (2) whether the accretion is uniformly lateral with respect to the platform edge or extensional from the margin along a single axis; (3) whether the lateral accretion is out into an open ocean, an open seaway or a closed seaway; and (4) along a windward or leeward margin (with respect to the physical energy flux at the sea surface).

The seven types of deep bank margins identified in the northern Bahamas (Fig. 4) include: (1) Type A—*Open Ocean Windward* margin seaward of the steep, high relief Bahama Escarpment east of LBB; (2) Type B—*Open Ocean Windward* margin facing the Blake marginal plateau, located north of LBB; (3) Type C—*Extended* deep margins off the northwest corners of both LBB and GBB; (4) Type D—*Open Seaway Leeward* margin with downslope lithoherms west of both LBB and GBB; (5) Type E—*Open Seaway Leeward* margin south of LBB; (6) Type F—*Open Seaway Windward* margin north of GBB; and (7) Type G—*Eroded* margins along the mouth of NWPC.

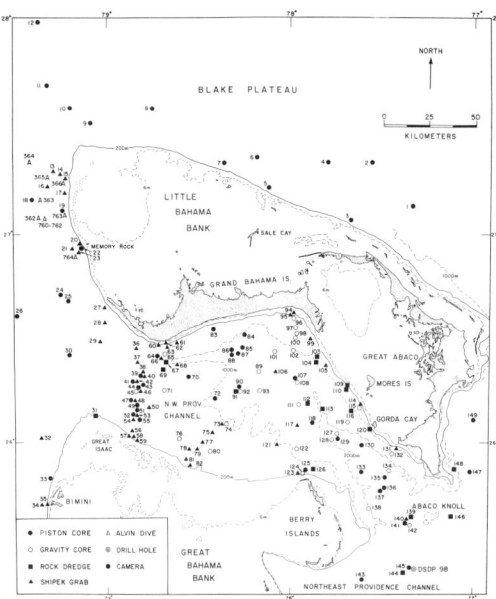

FIG. 3.—Location map of bottom samples referred to by number in text.

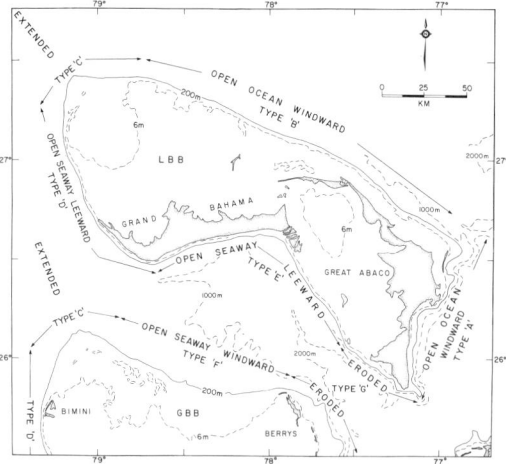

FIG. 4.—Location map of deep carbonate bank margin types in the northern Bahamas.

TABLE 3—CLASSIFICATION OF DEEP CARBONATE BANK MARGINS IN THE NORTHERN BAHAMAS

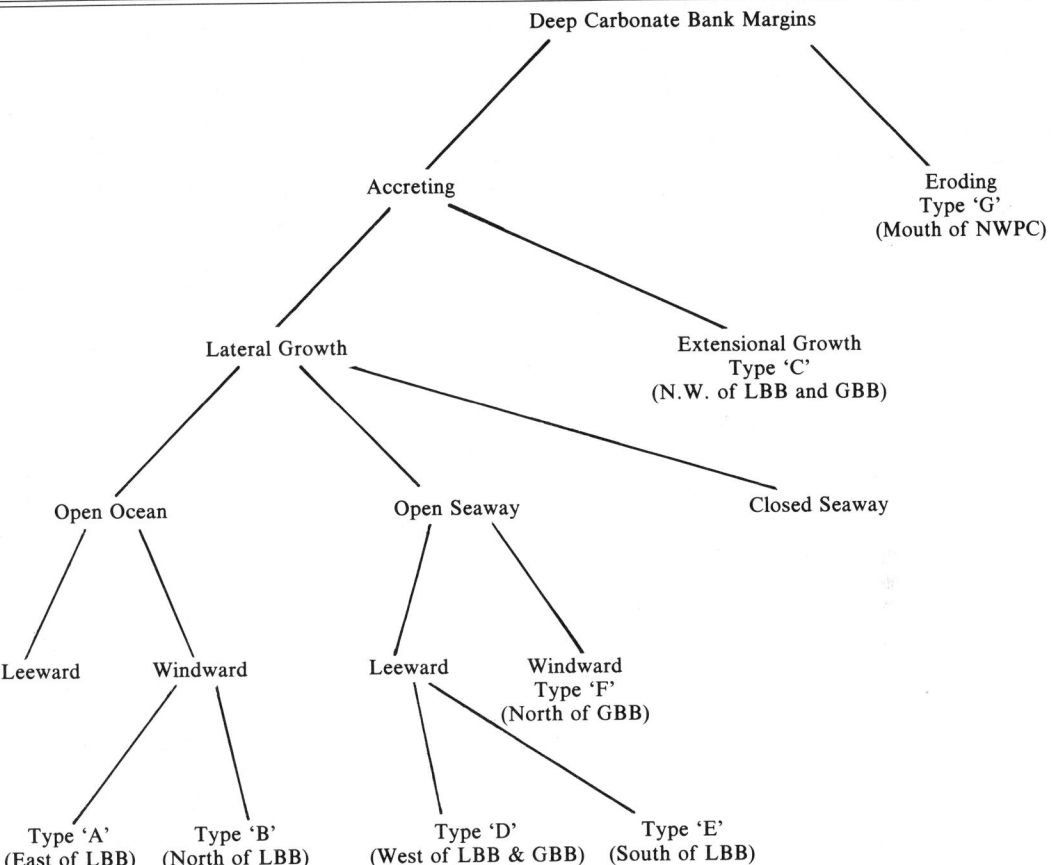

A description and discussion of the results of the geophysical and geologic data from each of these deep bank margin types beginning with Type A east of LBB and proceeding counterclockwise around LBB is presented below, preceeded by a brief description of the general setting of each area.

Type A—Open Ocean Windward—East of LBB

Setting.—The shallow bank margin east of LBB faces the open Atlantic and is a high energy windward margin with well developed coral reefs along its shallow edge (Storrs, 1964). Seaward of the shelf edge is the Bahama Escarpment which drops precipitously along slopes of 28–40 degrees (Emiliani, 1965) to the abyssal depths (4,500 m) of the Atlantic. Slopes here are too steep to record any useable seismic data, but the linear nature and great relief of the Bahama Escarpment makes it a unique type of deep bank margin based on morphology alone.

Core Data.—Cores recovered east of LBB contained very different sediment (Fig. 5). Core 147 from 4,000 m of water contained an upper 1.5 m of white-gray pelagic ooze overlying 2.4 m of coarse shallow-water sands. This sand layer consisted of coarse, inversely graded shallow-water sand becoming massive upwards and capped by a 30 cm thick section of shallow-water sands exhibiting normal grading (Fig. 5). This sand layer is interpreted to have been deposited by modified grain flow. This deposit contains the sedimentary structures of the "hypothetical" grain flow deposit of Middleton and Hampton (1973) such as injection structures at the base, inverse grading near its base becoming massive as well as larger "floating" clasts. Middleton and Hampton (1973) also point out that theoretically large slopes (18–37 degrees) are required for true grain flows. Slopes along the Bahama Escarpment east of LBB average 28–40 degrees which would tend to enhance the probability of downslope movement of coarse sand by grain flow processes.

The second core recovered east of LBB (core

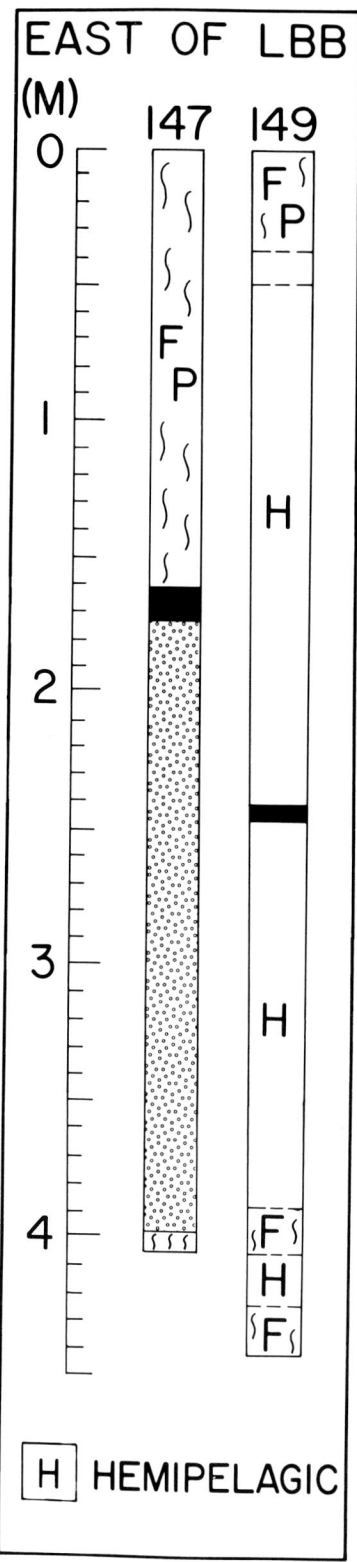

FIG. 5.—Schematic illustration of core data from east of LBB. For location of cores see Figure 3. See Figure 7 for legend. Depth in core in meters marked on left.

149; 4,695 m) contained very little sand. Instead it consisted primarily of dark brown hemipelagic muds with lesser amounts of white-gray pelagic ooze (Fig. 5). X-ray diffraction analysis of the hemipelagic units indicate the presence of low-Mg calcite, quartz, illite and kaolinite.

Structure.—Submersible observations along the Bahama Escarpment indicate that flanking strata are thin or absent and that shallow-water lagoonal limestones outcrop along the escarpment (Ryan et al, 1977). Seismic refraction data across this escarpment indicate drastic differences of crustal thickness on either side. Total crustal thickness beneath the eastern edge of the Bahama Platform is 23.5 km of which about 10 km is carbonate rock, whereas total crustal thickness, in the Atlantic is only 12 km, including 5 km of water (Ball et al, 1971). These data prompted Mullins and Lynts (1977) to suggest that the Bahama Escarpment may mark the boundary between continental and oceanic crust and that the linearity of this escarpment was initially controlled by basement faults produced by the transcurrent rifting of the Bahamas from Africa in Late Triassic time.

Depositional Model.—Although we do not have an abundance of data from the deep bank margin east of LBB, there is enough information to make some generalizations. The major characteristic of this deep margin type is the great relief (4,500 m) and declevity (28–40 degrees) of its down-to-basin slope. Because of the steep slopes grain flow deposits and large talus blocks have accumulated in a narrow zone at the base of the slope. Seaward of these deposits pelagic and hemipelagic sediments are found. This deep bank margin should be recognizable in the ancient by the occurrence of coarse grain flow and talus deposits at the base of a nearly vertical escarpment. A schematic depositional model of this deep bank margin type is presented as Figure 6A.

Type B—Open Ocean Windward—North of LBB

Setting.—The northern windward margin of LBB faces the Blake Plateau and is exposed to the open Atlantic. The shallow bank edge here is a reef/oolite shoal complex containing well developed reefs, relict and active sand shoals and a line of small rocky islands (Hine and Neumann, 1977). Beyond the marginal escarpment at depths greater than 150–200 m the slope north of LBB is broad and gentle having an average down-to-basin slope of only 4 degrees (Table 1).

Cores.—Piston cores recovered from the slope

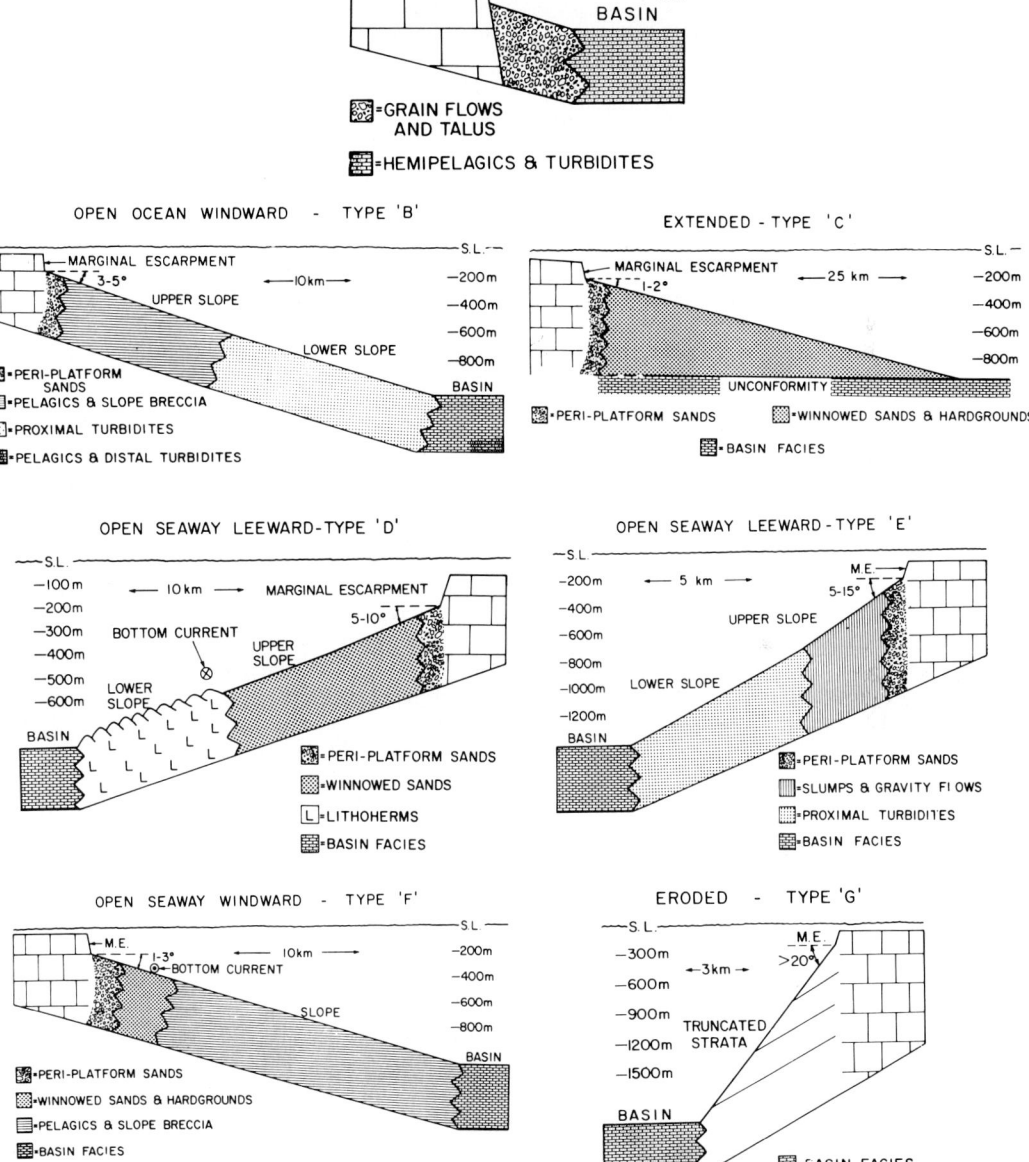

Fig. 6.—Schematic illustrations of depositional models of deep carbonate bank margins in the northern Bahamas.

north of LBB (Fig. 3) contain highly variable sediment types (Fig. 7). Those cores (Nos. 3, 5, 6, 8) taken from the upper slope in close proximity to active coral reefs of the shallow bank margin contain only fine grained pelagic sediments and muddy slope "breccias." These slope breccias consist of poorly sorted, subrounded to subangular clasts (up to 5 cm across) of well lithified foram pteropod biomicrites floating in a fine-grained muddy matrix. The poor sorting, floating clasts, and muddy matrix suggest deposition by debris flow (Middleton and Hampton, 1973). However, it is uncertain at this time as to whether these deposits are true debris flows or whether they represent some type of intraformational conglomerate.

In contrast to the muddy sediments found on the upper slope, cores recovered from the lower slope (Nos. 2, 4) contain thick biolithoclastic turbidites (Fig. 7). These well graded deposits are up to 2 m thick and contain both shallow- and deep-water allochems. Graded throughout, these turbidites also posess a coarse (up to gravel size) lower unit which probably is equivalent to Bouma sequence A (Bouma, 1962), and thus would be considered proximal in nature (Walker and Mutti, 1973).

A plot of percent pelagic sediment/core versus water depth for the cores recovered north of LBB (Fig. 8) reveals that there are very high percentages of muddy pelagic sediments on the upper slope and in the basin whereas sands and gravel dominate on the lower slope. Because the turbidites found on the lower slope contain significant percentages of shallow-water material (such as *Halimeda*) it appears as though coarse carbonate material is by-passing the upper slope, possibly via submarine canyons, to be deposited on the lower slope as a proximal turbidite facies.

Depositional Model.—The outstanding characteristics of this open ocean windward deep bank margin type are its gentle (3–5°), broad (up to 50 km) down-to-basin slope and facies relationships. Core data suggest an upper slope facies of fine-grained pelagic sediments and muddy slope breccia and a lower slope proximal turbidite facies (Fig. 6B). A peri-platform sand facies is inferred along this deep bank margin just seaward of the marginal escarpment (Fig. 6B). This facies will be discussed in more detail under Type D deep bank margins.

Type C—Extended—Northwest of LBB and GBB

Setting.—Large sediment "noses" extend up to 100 km to the north into the Florida Straits and onto the Blake Plateau from the northwest corners of GBB and LBB (Fig. 1). Present day bottom water flow in the northeastern Florida Straits is to the north at velocities of 50 cm/sec and greater west of LBB and GBB (Neumann and Ball, 1970). The shallow bank margins adjacent to these noses are deep, sediment barren, rocky shoulders at depths of 20–30 m with minor leeside sand accumulations behind rock ridges (Hine and Neumann, 1977). Down-to-basin slopes along these extended wedges of sediment are very broad and gentle averaging less than 1 degree (Table 1).

Air-Gun Seismic Profiles.—Air-gun profile '3' along the crest of the nose off LBB (Fig. 9) clearly indicates that this nose has prograded out onto the Blake Plateau over a horizontal unconformity. Based on correlation with JOIDES (1965) drill hole data on the Blake Plateau, this unconformity is middle Miocene in age. Progradation of the nose apparently began during the middle Miocene in response to an intensification of the Florida Cur-

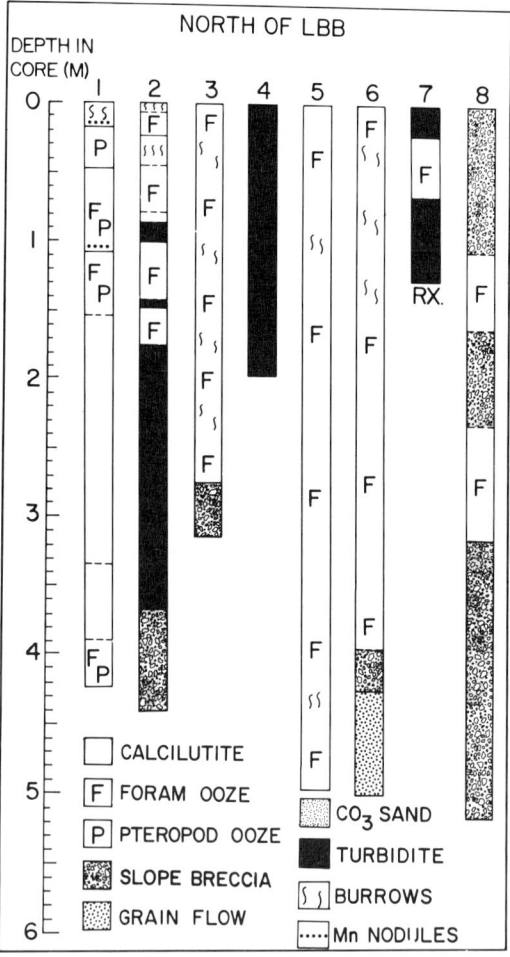

FIG. 7.—Schematic illustration of core data from north of LBB—Type B—deep bank margin. For location of cores, see Figure 3.

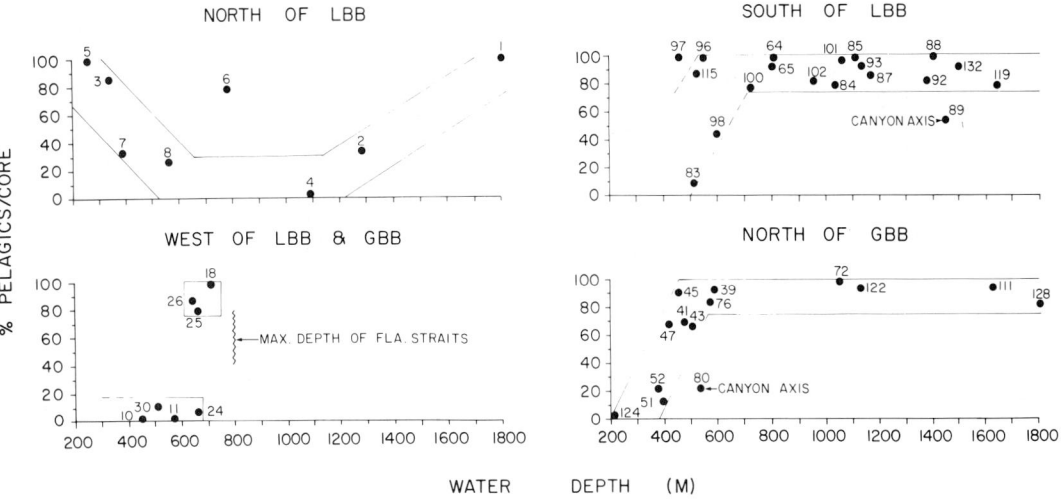

FIG. 8.—Graphs plotting percent pelagics per core versus water depth for various deep bank margins in the northern Bahamas.

rent at this time (Mullins and Neumann, 1977, 1979).

This wedge of sediment thickens appreciably toward the Bahama Banks (to the south) and the internal reflectors are oblique progradational to divergent, pinching out to the north indicating a southerly source for the sediments. The oblique even layered seismic facies of this sediment wedge (Fig. 9) suggests deposition by bottom flowing currents (Table 2). At the south end of profile '3' (Fig. 9), which is 12 nautical miles (22 km) from the bank top, this wedge of sediment is more than 400 m thick.

A similar but more complex nose extends to the north from the northwest corner of GBB (Fig. 10). Profile '21' (Fig. 10) along the crest of the GBB nose reveals that a large mass of contorted chaotic strata is present at its northern terminus (Fig. 10). This contorted zone may represent a large slump block that has moved down slope over a zone of irregular bottom topography. Oblique even layered to slightly diverging layers pinch out against an unconformity near the south end of profile '21' (Fig. 10) and overlies what appears to be a slump or erosional scar.

Profile '20' taken perpendicular to the strike of the GBB nose on its distal end (Fig. 2) shows in cross section a large (22 km wide and up to 200 m thick) lens shaped mound of sediment (Fig. 11). Internal reflectors are oblique progradational suggesting that this nose is building out laterally perpendicular as well as parallel to the strike of the nose. A zone of mounded chaotic reflectors is present along the western edge of this profile (Fig. 11), which is interpreted as a lateral progradation of sediment over a band of irregular topography (lithoherms; see Type D margins).

The reflectors beneath the sediment mound in profile '20' (Fig. 11) exhibit a distinct lateral change of seismic facies. To the east adjacent to LBB, the seismic pattern is discontinuous, wavy-subparallel-chaotic whereas to the west the reflectors are much more even and continuous. This change in reflection pattern is interpreted as a sedimentary facies change from east to west of coarse grained sediment gravity flows and winnowed sands to pelagic oozes (Table 2).

Cores.—Four piston cores were attempted on the LBB nose (Fig. 3). Of these, two (cores Nos. 10, 11) recovered less than a meter of sediment consisting of winnowed carbonate sand (Fig. 12) whereas the core cutters on cores Nos. 9 and 12 were crumpled on hard rock surfaces (Fig. 12). These data suggest that large portions of the LBB nose are well lithified. This is not surprising as a large area west of LBB is known to consist of well lithified hardgrounds and coral mounds (Neumann and Ball, 1970; Neumann et al, 1977). One complete core (No. 30) was recovered from

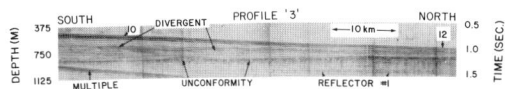

FIG. 9.—Photograph of 5 in.^{3} seismic reflection profile '3' from Extended—Type C—deep bank margin northwest of LBB. For location of profile see Figure 2.

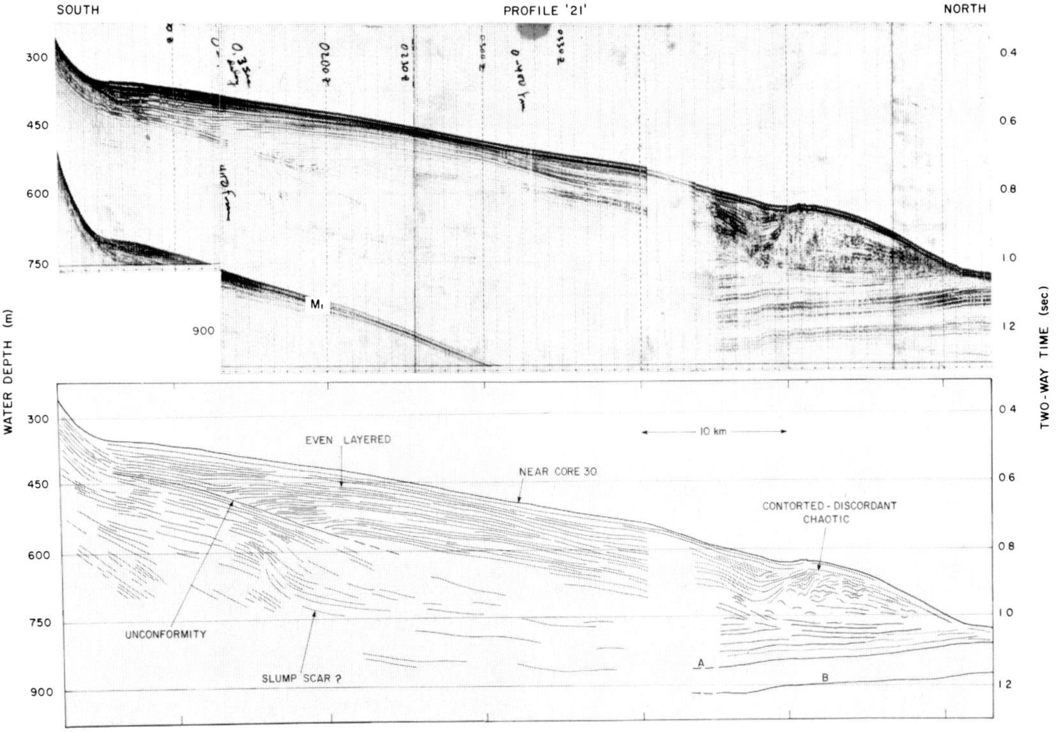

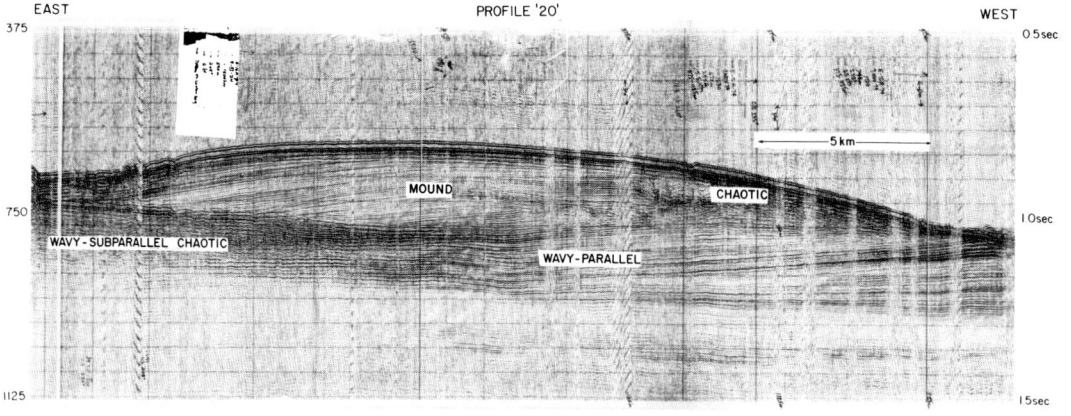

FIG. 10.—Photograph and accompanying line drawing interpretation of 10 in.³ air-gun seismic reflection profile '21' from Extended—Type C—deep bank margin northwest of GBB. For location of profile see Figure 2.

FIG. 11.—Photograph of 5 in.³ air-gun seismic reflection profile '20' from Extended—Type C—deep bank margin northwest of GBB. Note oblique progradational internal reflectors within the mound. For location of profile see Figure 2.

the GBB nose. This core consisted almost entirely of winnowed, silty sand (Fig. 12) made up of planktonic forams, pteropods, deep-water rock fragments and lesser amounts of bank derived material.

The seismic facies, position and shape of the sediment noses which extend many kilometers from the Bahama Banks and the nature of the sediment of which they are made indicate that these very large sand bodies are the product of

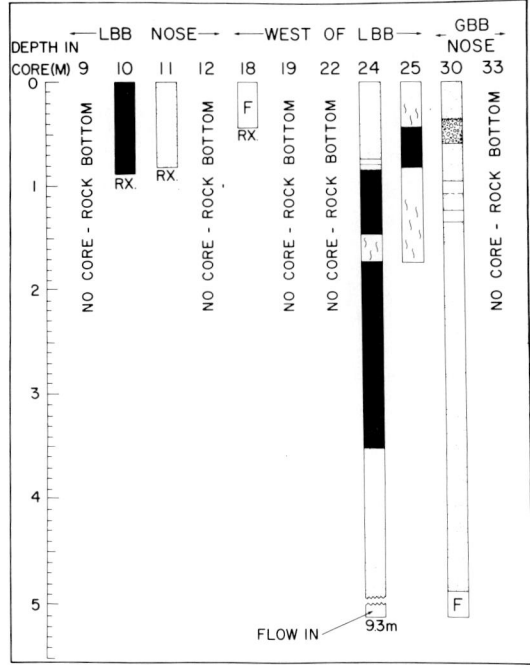

FIG. 12.—Schematic illustration of core data from west of LBB and GBB. For location of cores see Figure 3. See Figure 7 for legend.

sediment accumulation and winnowing due to the bottom water flow of the Florida Current since the middle Miocene.

Depositional Model.—The resultant depositional model for Type C—Extended—deep bank margins is a simple one. These deep margins are large (up to 100 km), thick (up to 600 m), wedge shaped accumulations of winnowed carbonate sands, and widespread hardgrounds (Fig. 6C). Extensive progradation is occurring along gentle (1-2° or less) slopes over a basinal facies of pelagic oozes and thin distal turbidites. Based on the seismic profiler data, bedding is gently inclined seaward with numerous down-slope pinch outs and the entire sedimentary package thickens appreciably toward the shallow-water platforms (Fig. 6C).

Type D—Open Seaway Leeward Margin—West of LBB and GBB

Setting.—The shallow bank edge along the western margin of LBB is a leeward margin characterized by a broad zone of carbonate sand bodies. Small active reefs are present along this margin where they are protected by rock ridges from burial by sand. Where unprotected extensive off-bank sediment transport has buried early Holocene reefs (Hine and Neumann, 1977). Large volumes of shallow-water sands are being transported off LBB along this margin due to the physical energy flux from the east-northeast (Hine and Neumann, 1977). Slopes along this deep margin are concave upward and of moderate steepness (average of 6 degrees). At the base of the slope in depths of 600-700 m (Fig. 1) is a zone of very irregular bottom topography which is now known to consist of north-south oriented mounds (Neumann et al, 1977). The deep bank margin here is directly affected by northward bottom water flow of the Florida Current (Neumann and Ball, 1970).

Air-Gun Seismic Profiles.—The outstanding feature along this deep bank margin is the 10-15 km wide bank of deep-water lithified bioherms (lithoherms) at depths of 600-700 m. This band of lithoherms extends for over 200 km from the Blake Plateau along the western margin of LBB to Bimini (Neumann et al, 1977). Profile '14' (Fig. 13) oriented perpendicular to the bank edge typifies the position and lateral extend of these features on the lower slope. Beneath the lithoherms and upslope from them the internal reflectors (Fig. 13) are very discontinuous to chaotic. One possible interpretation (Table 2) of this seismic facies is that it represents high-energy sediment gravity flows. Graded sand layers have been cored off this deep bank margin (Figs. 3, 12; cores 24 and 25). However, submersible observations (Neumann et al, 1977) indicate widespread occurrences of lithoherms, hardgrounds and sand waves, and petrographic studies (Wilber, 1976) indicate that there is little shallow-water sediment in these rocks at depths greater than 350 m. Most of the sands transported off the western edge of LBB and down the adjacent marginal escarpment are remobilized into north facing ripple marks and sand waves by bottom currents and are transported to the north parallel to the slope (A. C. Hine, pers. comm.). This may be a source of sediment for the large sediment noses northwest of LBB and GBB. Thus the discontinuous-chaotic seismic facies in Figures 11 and 13 may in part be due to complex vertical and lateral variations of lithoherms, hardgrounds, winnowed sands and turbidites.

Seaward of the lithoherms, internal reflectors become more continuous (Fig. 13) suggesting increasing proportions of uniformly deposited pelagic sediments. The lithoherm area itself takes on a mounded-chaotic appearance on the seismic record (Fig. 13) which would classically be interpreted as slump blocks. However, bottom observations from submersibles (Fig. 14) strongly suggest that these mounds are the result of in situ growth of individual mounds via sedimentological and biological buildup intermittent with submarine cementation (Neumann, et al, 1977).

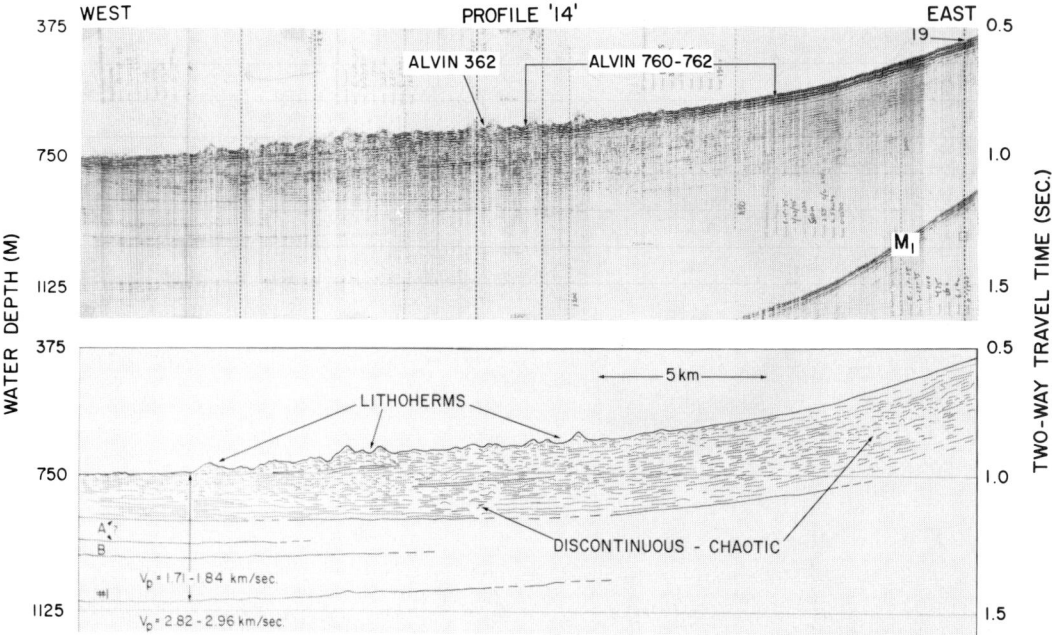

Fig. 13.—Photograph and accompanying line drawing interpretation of 5 in.³ air-gun seismic reflection profile '14' from Open Seaway Leeward—Type D—deep bank margin west of LBB. Note downslope position of lithoherms. Location of ALVIN dives shown on upper part of figure. For location of profile see Figure 2. Compressional (P-wave) wave velocity data from Sheridan et al. (1966).

Fig. 14.—Bottom photographs taken from DSRV ALVIN in lithoherm area from Open Seaway Leeward—Type D—deep bank margin west of LBB. (Left) Dive 761; water depth = 650 m. Crest of a lithoherm with sponges and ahermatypic coral (*Lophelia* sp.). Note steep sides of lithoherm. (Right) Dive 762; water depth = 660 m. Undercut submarine cemented crust along crest of a lithoherm. Crinoids and alcyonarians face into northward current (lower left to upper right). For additional photographs of lithoherms see Neumann et al. (1977).

Individual lithoherm mounds are very numerous and are 30–50 m in relief and 100's of meters long (Fig. 15A). Detailed examination of air-gun profiles from the lithoherm area indicates that individual mounds (Fib. 15B) are up to 70 m thick and appear to have built up from a hard flat horizontal reflector and are separated by hardgrounds.

Cores.—It was difficult to core in this area west of LBB because of extensive submarine

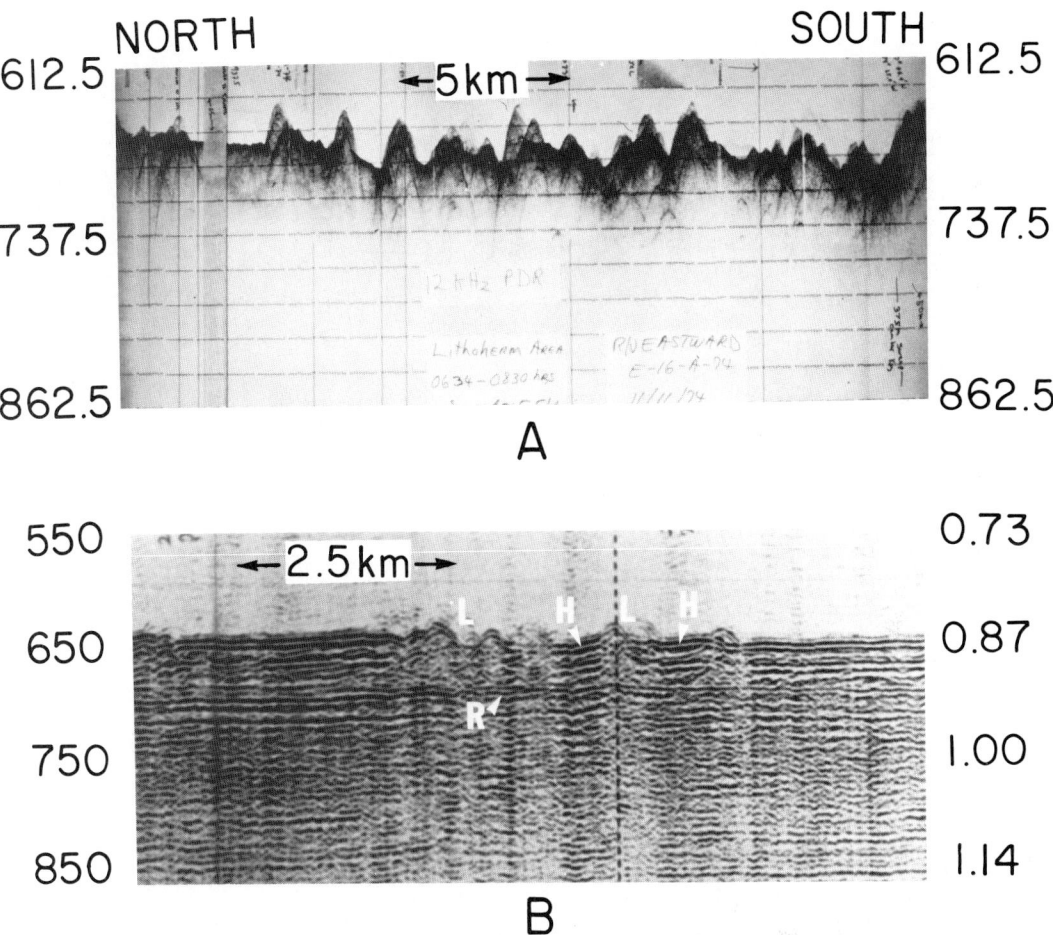

FIG. 15.—(A) Photograph of 12 kHz PDR profile '12' across lithoherms from Open Seaway Leeward—Type D—deep bank margin west of LBB. Note numerous mounds from which hyperbolic echoes originate. Water depth in meters on both right and left. (B) Enlarged portion of photograph of 5 in.^{3} air-gun seismic reflection profile '12' across lithoherms. Lithoherms (L) appear to have built up from hard, horizontal reflector (R) and are separated by hardgrounds (H). Using P-wave velocity data from Sheridan et al (1966) the lithoherms here are up to 70 m thick. Water depth in meters marked on left; seconds of two-way travel time on right. For location of profile see Figure 2.

cementation. Cores that were recovered consisted mainly of thick bioclastic turbidites and winnowed sand layers with minor amounts of pelagic oozes (Fig. 12). A plot of percent pelagics per core versus water depth for the area west of LBB and GBB (Fig. 8) indicates that at depths down to about 650–700 m there is a dominance of sand sized material and then in deeper water there is an abrupt increase ih the percent of pelagic sediments. This abrupt change from sands to muds is probably controlled by the ability of bottom currents to winnow the finer size fractions.

Grab Samples.—Grab samples taken close to the bank edge in water depths between 155 m and 350 m seaward of the marginal escarpment indicate the presence of muddy sand containing both shallow- and deep-water allochems and submersible observations west of GBB indicate the presence of large talus blocks (Neumann and Ball, 1970). Platform coralgal sands such as *Halimeda* are very abundant. Apparently these shallow-water sands have moved down the nearly vertical marginal escarpment by grain flow and creep and have admixed with pelagic sediments. The resultant facies is a narrow (1–3 km) band of muddy sands and talus blocks adjacent to the shallow-water platform. This facies is termed here the "peri-platform sand facies" because of its proximity to the shallow-water platform and the contribution of platform sediments to it. Hubbard et al (1976) have documented via observations and sediment traps the downslope transport of *Halimeda* sand by grain flow and creep processes south of Grand Bahama Island, and similar peri-platform deposits have been observed in the fore-reef areas of Jamaica (Burne, 1974; Goreau and Land, 1974; Moore et al, 1976) and Belize (Ginsburg and James, 1973).

Rock Samples.—Numerous rock samples have previously been collected from the deep-bank margin west of LBB by rock dredging and submersible vehicles (Wilber, 1976). Detailed petrographic examination of lithoherm and hardground samples from this area (Wilber, 1976; Wilber and Neumann, 1976, 1977) indicate that these rocks are mud-to grain-supported and were cemented in the submarine environment by micritic high-Mg calcite of about 14 mole percent Mg. Ahermatypic coral, planktonic Foraminifera, pteropods and minor amounts of shallow-water debris are the principle allochems of these well cemented biomicrites and biomicrudites (Fig. 16) with porosities of up to 50 percent and more (Wilber and Neumann, 1977).

Depositional Model.—The outstanding characteristic of Type D deep bank margins is the presence of lower slope lithoherms and extensive hardgrounds (Fig. 6D). The entire slope here is swept by strong bottom currents, and where lithoherms are not present, winnowed sands dominate (Fig. 6D). Sediment gravity flow deposits (turbidites) are also found admixed with winnowed sands and pelagic layers. Because this deep bank margin type is a leeward margin there also is a well developed peri-platform sand facies adjacent to the shallow-water platform. These slope facies grade seaward into a basinal facies of pelagic oozes and thin distal turbidites.

Type E—Open Seaway Leeward Margin—South of LBB

Setting.—The southern bank margin of LBB is a leeward margin (Fig. 1). Where the shallow bank margin is protected by Grand Bahama Island active coral reefs have developed along Pleistocene rock ridges at the edge of a very narrow shelf. The bank edge that is not protected by islands consists of a broad band of skeletal and oolitic sand bodies. (A. C. Hine, pers. comm., 1978). Grand Bahama Island consists of Pleistocene-Holocene oolite shoals, lagoonal deposits and low sand dunes, and is anomalous in that it is the only large island in the Bahamas that has developed along a leeward margin. The slope south of LBB is steep and narrow (see 1000 m isobath in Figs. 2 or 3). Average down-to basin slopes seaward of Grand Bahama are variable ranging from 6.2 to 12.5 degrees, slopes seaward of Mores Island average 9.5 degrees, and slopes

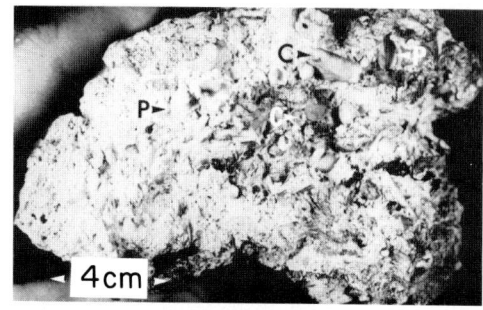

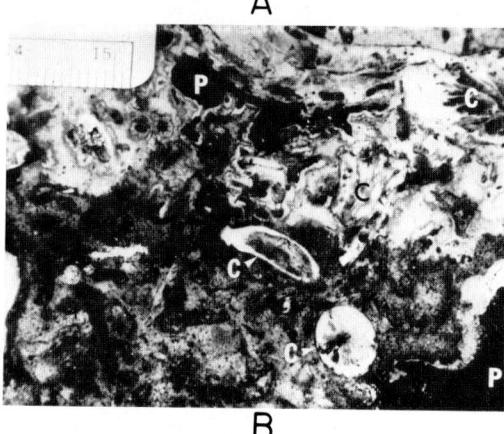

FIG. 16.—Photographs of lithoherm rocks from Open Seaway Leeward—Type D—deep bank margin west of LBB inwater depths of 600–700 m. (A) Fresh broken surface of coarse, grain-supported coral, pteropod biomicrudite. Principal allochems are deep-water ahermatypic coral (C) and pteropods (P) with lesser amounts of planktonic foraminifera and minor shallow-water debris. Cement is micritic high-Mg calcite of about 14 mole percent Mg. (B) Polished slab section of framework-supported deep-water coral biomicrudite. Note ahermatypic coral (C) and macroporosity (P) produced by endolithic sponges. Scale is in centimeters. For more detail on the petrology of lithoherm and hardground rocks see Wilber (1976). Photographs supplied by R. Jude Wilber.

become very steep (greater than 20 degrees) off of Gorda Cay (Table 1).

Air-Gun Seismic Reflection Profiles.—The shallow seismic stratigraphy of the NWPC is well illustrated on profile '23' (Fig. 17). Four major reflecting horizons are present: A, B, No. 1, No. 4. These horizons are identifiable based on stratigraphic position and reflection amplitude throughout the NWPC. Horizons A and B are identified for the first time in the present study and the identification of No. 1 and No. 4 is based on comparative analysis with Ewing et al (1966). These horizons have not been sampled directly

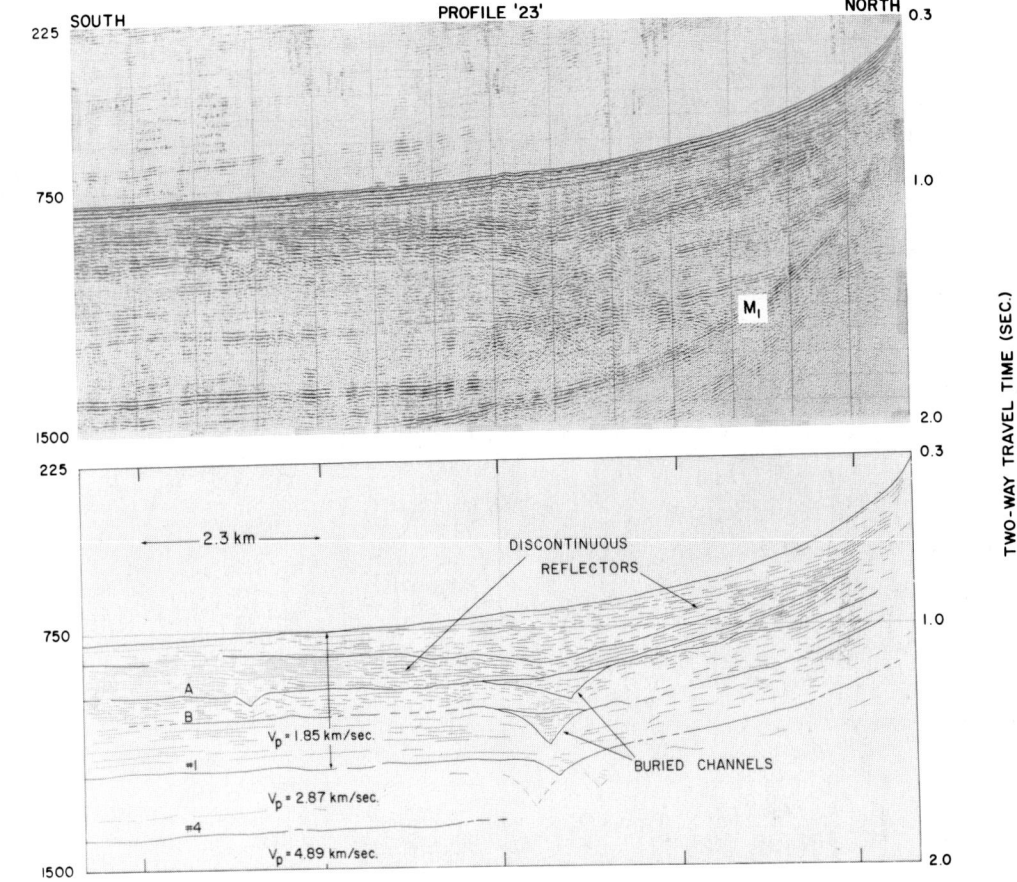

FIG. 17.—Photograph and accompanying line drawing interpretation of 5 in.³ air-gun seismic reflection profile '23' illustrating the shallow seismic stratigraphy of Northwest Providence Channel. P-wave velocity data from Sheridan et al (1966). For location of profile see Figure 2.

and thus their ages and lithologies are uncertain. Ewing et al (1966) based on correlation with drill hole data on land suggest that horizons No. 1 and No. 4 correlate with the top of the Paleocene and the Late Cretaceous respectively. However, correlation with the seismic stratigraphy worked out for the Northeast Providence Channel (NEPC) based on DSDP drill site 98 (Hollister et al, 1972) and air-gun seismic profiles (Mullins and Lynts, 1976) suggests that horizons No. 1 and No. 4 in the NWPC may represent a middle Eocene chert layer and a Paleocene-Late Cretaceous hard chalk layer respectively. Above horizon No. 4 numerous diffraction patterns outline a stacked series of buried channels (Fig. 17) suggesting intermittent episodes of erosion and sedimentation (Schlager et al, 1976).

Profile '25' (Fig. 18) taken perpendicular to the bank edge seaward of the center of Grand Bahama Island (Fig. 2) exhibits a complete seismic facies transition from chaotic reflectors beneath the upper slope to discontinuous reflectors beneath the broad lower slope and finally into even parallel continuous reflectors beneath the flat basin floor (Fig. 18). This transition of seismic facies basinward is interpreted as a transition from a mixed and variable sequence of high energy gravity flow deposits and slumps beneath the upper slope to a proximal turbidite facies beneath the lower slope to basinal pelagics and thin distal turbidites in the basin. Above the chaotic seismic facies of the upper slope on profile '25' (Fig. 18) is a thin layer of transparent to even parallel layered sediment (about 75 m thick) which probably represents a late stage of pelagic deposition. This upper unit pinches-out downslope and is absent at depths greater than about 825 m (Fig. 18).

Cores.—Cores retrieved from the slope south of LBB (Fig. 19) do not agree well with the gross interpretation of the seismic profiles. A possible

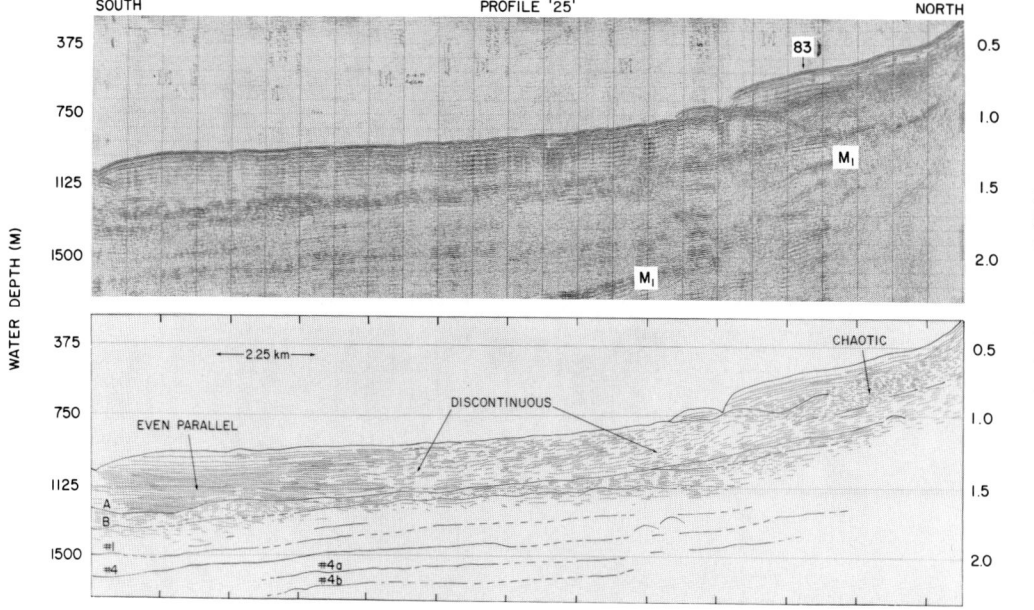

Fig. 18.—Photograph and accompanying line drawing interpretation of 5 in.³ air-gun seismic reflection profile '25' from Open Seaway Leeward—Type E—deep bank margin south of LBB. For location of profile see Figure 2.

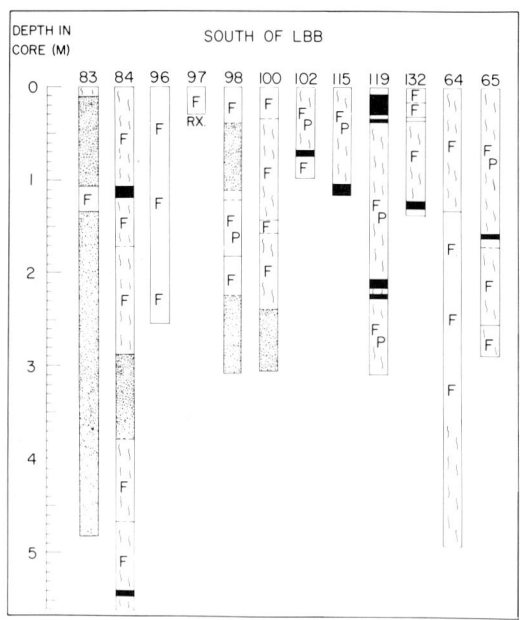

Fig. 19.—Schematic illustration of core data from Open Seaway Leeward—Type E—deep bank margin south of LBB. For location of cores see Figure 3. For legend see Figure 7.

explanation for this is that ten of the twelve cores from this slope were collected south of Grand Bahama Island where seismic profile '25' (Fig 18) indicates there is a thin surficial layer of transparent to even parallel layered sediment. Cores taken from this layer indicate that it consists of Pleistocene-Holocene pelagic oozes and muddy slope breccia with only minor thin turbidite sands (Fig. 19). The anomalous presence in the Pleistocene of Grand Bahama Island along this leeward bank margin may have blocked off-bank sediment transport thus explaining the lack of near surface sands.

Cores taken from basinal areas of NWPC (Fig. 20) consist of pelagic oozes, thin turbidites and an occasional thin layer of winnowed foram-pteropod sand. Conspicuously absent from these cores are debris flow and grain flow deposits (Fig. 20). Pelagic oozes constitute 60 percent of these cores retrieved from the basin-environment of NWPC whereas 35 percent is distal turbidites and 5 percent winnowed carbonate sand layers.

Turbidites found in basin cores are thin having an average thickness of 17 cm, and typically do not contain a complete Bouma Sequence (A-B-C-D-E; Bouma, 1962). Rather, based on X-radiographs most turbidites consist of Bouma units C-D-E or B-C-D-E and would thus be considered distal in nature. (Walker and Mutti, 1973). Seismic reflection patterns from basinal areas of NWPC (Fig. 18) are characterized by continuous even

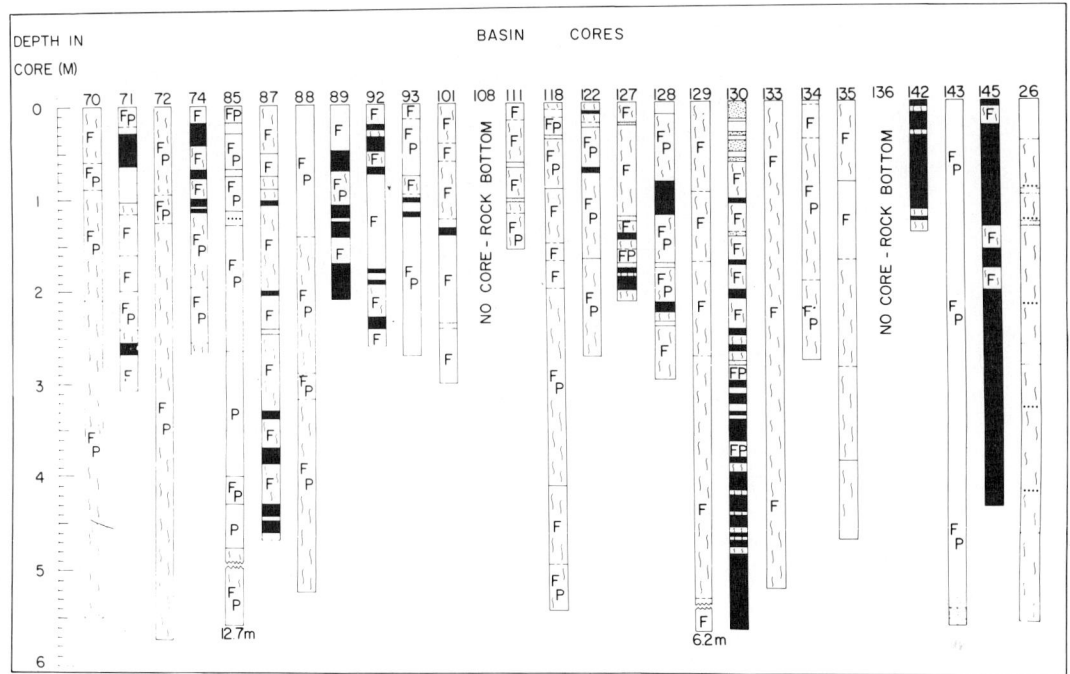

FIG. 20.—Schematic illustration of core data from basinal areas in Florida Straits and Northwest Providence Channel. For location of cores see Figure 3. For legend see Figure 7.

parallel reflections, which agrees well with the core data that indicate widespread uniform deposition of basin sediments.

Depositional Model.—The depositional model for Type E deep bank margins (Fig. 6E) is based principally on its seismic character. An upper slope chaotic to mounded chaotic seismic facies is interpreted as a mixed and variable sediment gravity flow and slump facies whereas the discontinuous lower slope seismic facies is interpreted as a proximal turbidite facies (Fig. 6E). Seaward, this lower slope turbidite facies grades into a basinal facies of even layered pelagic oozes and thin distal turbidites. Because this bank margin south of LBB is a leeward margin, there is also a well developed peri-platform sand facies adjacent to the shallow-water platform to a depth of about 300–400 m.

Type F—Open Seaway Windward Margin—North of GBB

Setting.—The northern margin of GBB is a windward margin that faces an open seaway (Fig. 1). The shallow bank margin here is a broad, deep, sediment-barren rocky shelf with no coral reefs. There is very little sediment along this shallow bank margin and that which is present is in the form of leeside sand accumulations behind patch reefs and rock ridges as a result of the physical energy flux from the northeast (A. C. Hine, pers. comm., 1978). The slope north of GBB is very broad and gentle having an average down-to-basin slope of only 1.5 degrees (Table 1).

Air-Gun and Uniboom Seismic Profiles.—Profile '32' (Fig. 21) typifies the deep bank margin north of GBB. Reflecting horizons A, B, No. 1, and No. 4 are not as obvious on this profile as elsewhere in the NWPC (for example profiles '23' and '25'; Figs. 17, 18) but are present (Fig. 21). Profile '32' is further complicated by structural and erosional complexities but the overall seismic facies patterns appear to be much simpler. The upper part of the slope contains an irregular erosional surface in the shallow subsurface made up of broad U-shaped and one narrow V-shaped channels (Fig. 21). A wavy-subparallel chaotic seismic facies beneath the broad shallow canyons (Fig. 21) suggest the presence of locally thick sediment gravity flows, possibly debris flow deposits. Under the central part of the slope north of GBB on profile '32' (Fig. 21) the internal reflectors are warped and upturned beneath a broad anomalous rise of the sea floor.

On the lower part of the slope on profile '32' (Fig. 21) at depths greater than 525 m the subbottom reflectors are remarkably even continuous and parallel suggesting widespread, uniform de-

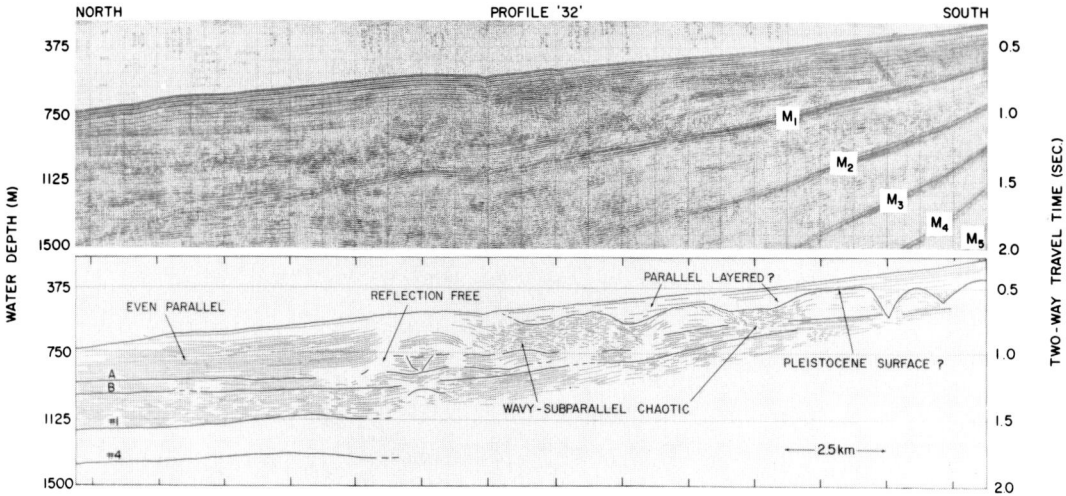

FIG. 21.—Photograph and accompanying line drawing interpretation of 5 in.³ air-gun seismic reflection profile '32' from Open Seaway Windward—Type F—deep bank margin north of GBB. M_1—M_5 represent first through fifth multiple reflections. For location of profile see Figure 2.

position. The shallow sub-surface layers beneath the upper slope also appear to be even-parallel layered (Fig. 21) but this is difficult to see because of bubble-pulse interference. A uniboom profile (Fig. 22) taken along profile '31' (Fig. 2) seaward of the marginal escarpment demonstrates that the upper 20 m or so of sediment is even-parallel layered with only an occasional small slump structure (Fig. 22).

Core Data.—Cores retrieved from the slope north of GBB (Fig. 23) consist predominatly of muddy pelagic sediments and slope breccia. Very little sand was found on this deep bank margin. Turbidite sands were recovered in core 80 from the axis of a submarine canyons (Fig. 23) and bottom current winnowed sand and hardgrounds were recovered from a narrow zone between 200–300 m of water depth (Mullins et al, in press).

Depositional Model.—This open seaway windward deep bank margin appears to consist predominantly of muddy deposits. With the exception of a narrow band of current winnowed peri-platform sands high on the upper slope, very little sand was recovered. The resultant depositional model for Type F deep bank margins (Fig. 6F) consists of a narrow poorly developed current winnowed peri-platform facies with hardgrounds high on the upper slope. This facies grades into a broad lower slope deposit of pelagic oozes and muddy slope breccia which grade seaward along a broad, gentle slope into a basinal facies of pelagic oozes and thin distal turbidites (Fig. 6F).

Type G—Eroded Margins—North NWPC

Setting.—Down-to-basin slopes along the NWPC southeast of Gorda Cay and east of the Berry Islands (Figs. 2 or 3) are very steep averaging more than 20 degrees (Table 1), presumably

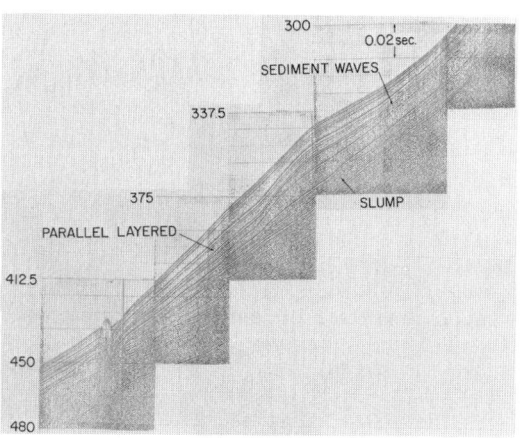

FIG. 22.—Photograph of uniboom seismic reflection profile from Open Seaway Windward—Type F—deep bank margin north of GBB. Note overall sigmoid Terrace form and continuous, even parallel layered reflectors, a small slump structure as well as possible talus block at left end of profile. Water depth in meters marked on left. Horizontal distance across figure is approximately 5 km. Data taken along profile '31' in Figure 2. Data supplied by A. C. Hine.

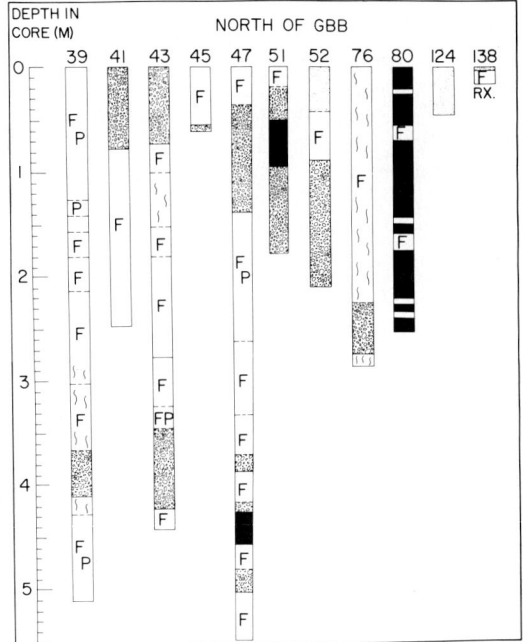

FIG. 23.—Schematic illustration of core data from north of GBB. For location of cores see Figure 3. See Figure 7 for legend.

due to erosion and oversteepening (Andrews et al, 1970).

Air-Gun Seismic Profiles and Bottom Samples.—Profile '30' off Gorda Cay (Fig. 2) illustrates the steepness of the slope here and the inability of the seismic system to obtain useful data along such slopes (Fig. 24). Numerous side echoes can be seen near the base of profile '30' and one sub-bottom reflector is present here between 1.5 and 2.0 seconds of two-way travel time (Fig. 24). This horizon which truncates near the base of the slope may be horizon No. 4 but this is speculative. Rock dredging along the upper portions of this steep slope recovered a mixed, well preserved coral and mollusk assemblage of Recent age containing the hermatypic corals *Montastrea sp.* and *Manicina sp.* suggesting a shelfal, shallow-water (less than 200 m) origin. The recovery of this Recent predominantly shallow-water assemblage from 375–925 m indicates that this margin is being undercut and oversteepened to allow downslope displacement of shallow-water material.

Profile '33' north of the Berry Islands (Fig. 2) also demonstrates the erosional nature of the deep bank margins in this region of the NWPC. Even parallel layered reflectors beneath the upper slope can be seen to truncate along the escarpment between 525 and 750 m of water depth (Fig. 24).

Dredge 126 (Fig. 3) recovered well cemented planktonic foram-pteropod biomicrites capped by thin brown phosphatic (?) rinds. Seismic horizons A, B, No. 1, and No. 4 appear to be present along this profile (Fig. 24) but their identification as such is tenuous due to the chaotic nature of the overlying strata.

Depositional Model.—Slope sediments apparently were once deposited along the deep bank margins at the mouth of the NWPC between the Berry Islands and Great Abaco Island but have subsequently been modified by erosion. These margins are characterized by very steep (> 20 degrees) down-to-basin slopes with outcrops of truncated deep-water limestones (Fig. 6G). The basin seaward of these steep slopes is heavily incised by basin canyons floored by thick turbidites, that strike *parallel* to the bank margin, and pelagic sediments have accumulated on intercanyon highs.

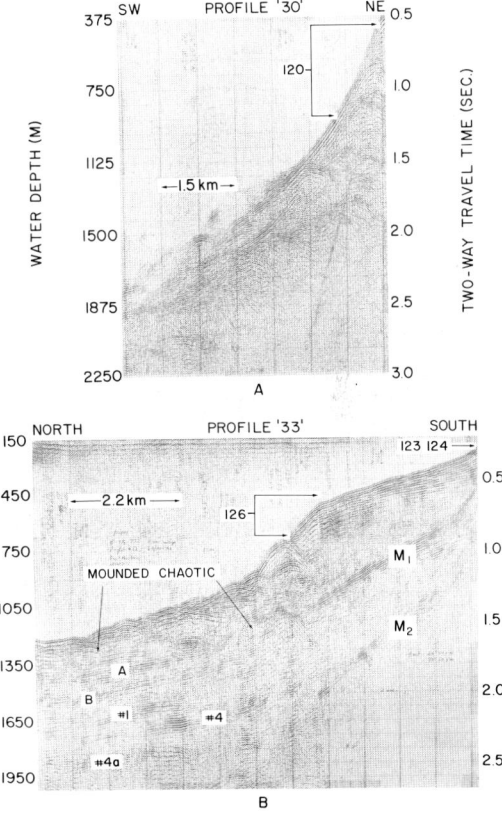

FIG. 24.—Photographs of 5 in.^{3} air-gun seismic reflection profiles from Eroded—Type G—deep bank margins along the mouth of NWPC. M_1 and M_2 denote first and second multiple reflections respectively. Numbers 120 and 126 indicate locations of rock dredge hauls. For location of profiles see Figure 2.

DISCUSSION AND SUMMARY

Processes and Controls of Deep Carbonate Bank Margin Development

On the basis of seismic facies, structure, morphology, and sedimentary facies, seven deep carbonate bank margin types have been recognized in the northern Bahamas (Figs. 4, 6, 25; Table 3). A map illustrating the sedimentary facies relationships of each area and a summary of the recognition criteria for each deep bank margin type are presented as Figure 25 and Table 4. Each deep margin type is distinctly different from any other as a result of a unique set of processes which act on and within each area. The major controls of deep bank margin development are (Fig. 26): (1) basement faulting, (2) direction and magnitude of off-bank sediment transport, (3) gravity and pelagic sedimentation, (4) oceanic circulation, (5) submarine cementation, and (6) biological build-ups.

Structure appears to play an imporatnt role in the initial development of platform edges. In the Bahamas, the initial location of bank margins appears to be controlled at depth by basement faults (Talwani, 1960; Ball, 1967; Ball et al, 1969; Sheridan, 1974; Mullins and Lynts, 1977) possibly related to continental rifting processes during the Late Triassic. Although faulting may be a major control in the initial development of carbonate platform edges, it appears to be of little if any significance to the subsequent development of the deep bank margins in the Bahamas. However, faulting may play a more important role along other, more tectonically active carbonate areas.

Of the six contolling factors listed, the direction and magnitude of off-bank sediment transport and oceanic circulation appear to have the greatest effect on the type of deep bank margin that develops. Shallow-water bank margin types are controlled to a large degree by the direction and magnitude of off-bank sediment transport which in turn is controlled by the physical energy flux at the surface (Hine and Neumann, 1977). These factors in turn control the amount of shallow-water sediment available for down-slope transport by gravity processes. Along shallow windward margins the dominate shallow-water sediment transport is onto the bank, whereas along leeward margins it is off-bank (Hine and Neumann, 1977). Thus, carbonate sand produced on the bank tops along shallow windward margins may not be an important source of sediment to the adjacent deep margin. Sediment produced in deeper fore-reef zones along windward margins would not be affected by the physical energy flux at the sea surface, and may contribute significant amounts of sand to the deep bank margin. Conversely, along shallow leeward margins, where not blocked by islands, there is a net off-bank transport of sediment and the resultant deep bank margin should contain significantly more sand sized material.

Strong bottom water flow along deep bank margins can remobilize sediment and significantly affect the resultant sediment types. West of LBB, bottom water flow associated with the Florida Current is actively transporting shallow-water sands to the north, which may supply sediment to the large sediment noses which extend to the north off the northwest corners of LBB and GBB.

The type of windward or leeward shallow bank margin which develops also exerts controls on the adjacent deep bank margin. The shallow northern margins of LBB and GBB are both windward margins but differ in that north of LBB the shallow bank edge is rimmed by coral reefs and sand shoals whereas the northern margin of GBB is a deep, sediment barren rocky shelf (Hine and Neumann, 1977; Enos, 1974a). As a result there is much less sand sized material available for downslope transport north of GBB than LBB. This is reflected in the type of sediment found on these deep bank margins. North of LBB thick biolithoclastic turbidites containing peri-reef debris have been cored whereas primarily pelagic oozes and muddy slope breccias have been cored on the slope north of GBB.

Pelagic and gravity flow processes control the

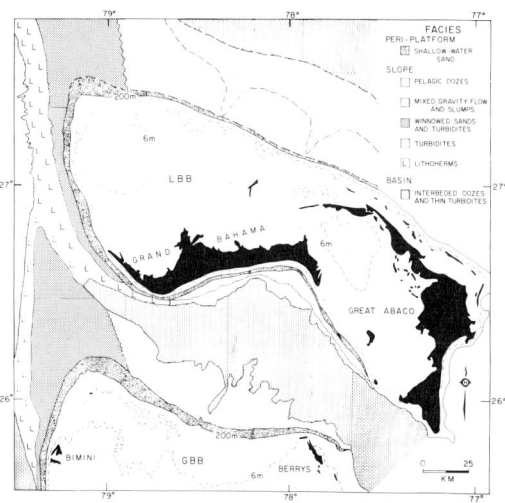

FIG. 25.—Generalized near surface sedimentary facies map of the deep carbonate bank margins in the northern Bahamas. Map is based on all available bottom samples and seismic reflection profiles. Extrapolation between data points was based on bathymetric data of Malloy and Hurley (1970) and Andrews et al (1970). Based line indicate that facies boundaries are poorly known.

TABLE 4—SUMMARY OF RECOGNITION CRITERIA FOR DEEP CARBONATE BANK MARGIN TYPES IN THE NORTHERN BAHAMAS

Bank Margin Type	Down-to-Basin Slopes	Seismic Facies	Processes and Controls	Sedimentary Facies
Type 'A'—Open Ocean Windward—East of LBB	Very narrow; steep (30–60 degrees).	—	Basement faulting; carbonate buildup; submarine erosion; grain flows; rock falls; pelagic sedimentation; turbidity currents.	Outcrops of shallow-water limestones along slope; wedge of thick massive to inversely graded bioclastic grain flows and talus blocks at base of slope which grade rapidly seaward into pelagic and hemipelagic muds.
Type 'B'—Open Ocean Windward—North of LBB	Broad (up to 50 km wide); gentle (2–5 degrees).	Even parallel continuous to transparent on upper slope; discontinuous to chaotic on lower slope.	On-bank transport of back reef sediments; downslope transport of fore-reef material; pelagic sedimentation; debris flows; turbidity currents.	Peri-platform facies adjacent to bank; upper slope pelagic oozes and slope breccia; lower slope peri-reef biolithoclastic turbidites.
Type 'C'—Extended—Northwest of LBB and GBB	Very broad (up to 100 km and more); very gentle (< 1 degree).	Oblique progradational; divergent to even parallel continuous.	Pelagic sedimentation; sediment winnowing, reworking and deposition by bottom currents; submarine cementation.	Peri-platform facies adjacent to bank; massive current winnowed sands; strata form a thick wedge which thickens toward the bank.
Type 'D'—Open Seaway Leeward—West of LBB and GBB	Moderate width (< 25 km); moderate steepness (5–10 degrees).	Discontinuous-chaotic on upper slope; mounded chaotic on lower slope.	Off-bank transport; bottom current flow; biological buildups; pelagic sedimentation; turbidity currents; submarine cementation.	Peri-platform facies adjacent to bank; massive submarine cemented current winnowed sands and turbidites on upper slope; linear band(s) of current oriented submarine lithified bioherms on lower slope; hardgrounds.
Type 'E'—Open Seaway Leeward—South of LBB	Narrow (5–15 km wide); steep (6–20 degrees)	Chaotic to mounded chaotic on upper slope; discontinuous on lower slope.	Off-bank transport; slumps and gravity flows; turbidity currents; pelagic sedimentation.	Peri-platform facies adjacent to bank; mixed gravity flow deposits and slumps on upper slope; biolithoclastic turbidites on lower slope.
Type 'F'—Open Seaway Windward—North of GBB	Broad (25 km wide or more); Gentle (< 2 degrees).	Even parallel continuous; minor wavy subparallel chaotic.	On-bank transport; pelagic sedimentation; debris flows; submarine cementation; bottom currents.	Peri-platform facies adjacent to bank; slope consists of pelagic oozes and muddy slope breccia with hardgrounds and current winnowed sands on uppermost slope minor turbidites.
Type 'G'—Eroding Mouth of NWPC	Very narrow (< 5 km wide); very steep (> 20 degrees).	—	Submarine erosion.	Slope facies thin or absent; truncated strata.
Basin	—	Even parallel continuous	Pelagic sedimentation; turbidity currents; reworking by bottom currents.	Pelagic oozes interbedded with thin distal turbidites.

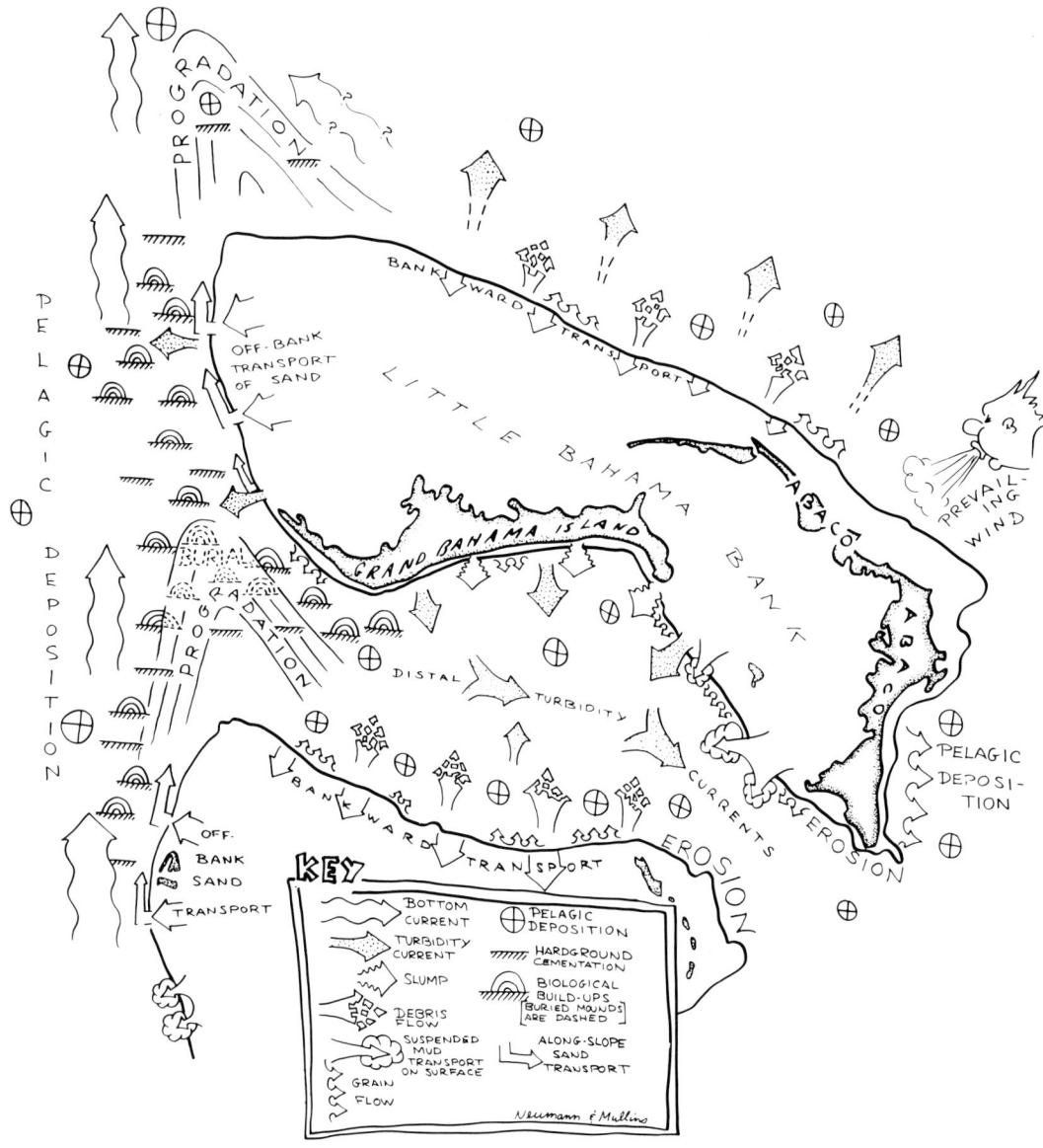

FIG. 26.—Schematic cartoon illustrating the major processes which control deep carbonate bank margin development in the northern Bahamas.

deposition of sediment on deep bank margins. Pelagic carbonate sedimentation is ubiquitous along the deep bank margins in the Bahamas. However, fine grained pelagic carbonates are volumetrically important only where they are not winnowed by bottom currents, diluted by gravity flow deposits or deposited below the calcite compensation depth. Areas in the Bahamas where pelagics constitute a major type of sediment are the windward upper slope north of LBB, the windward slope north of GBB, and in the basins where they make up about 60 percent of the deposits.

Much of the sediment on the carbonate slopes in the Bahamas has been deposited by gravity flow processes. However, the nature and distribution of these deposits varies significantly, with the primary controls being availability of sediment and steepness of the slope. Thick (average of 55 cm), coarse, graded biolithoclastic turbidites are common on the lower slopes north, west and south of LBB. Turbidites are also common in the basins interbedded with pelagic oozes, but due to increased distance from their source and low gradients, are much finer in grain size and thinner (average of 17 cm) than slope turbidites. Slope breccia (debris flows ?) characterized by the poor sorting of clasts in a muddy matrix, are dominant on the gentle, muddy slopes north of LBB and GBB. Grain flows, characterized by their inversely graded to massive bedding are found in significant proportions only at the base of very steep (>18 degrees) slopes. Significant grain flow deposits are only found in narrow zones seaward of the marginal escarpment which rims the Bahama Banks to a depth of about 150 m as part of the peri-platform facies, and at the base of the Bahama Escarpment east of LBB.

Submarine cementation is now thought to be a major process in the maintenance of carbonate slopes (Ginsburg and James, 1973; Neumann, 1974; Neumann and Hine, 1974 Ginsburg et al, 1974; Neumann et al, 1977). The steep slopes in the Bahamas which allow extensive downslope movement of sediment by gravity flow processes, were classically thought to be the result of coral reef upgrowth (Newell, 1955; Hess, 1960; Dietz et al, 1970; Paulus, 1972). However the lack of coral reefs in drill holes in the Bahamas (Paulus, 1972; Meyerhoff and Hatten, 1974; Tator and Hatfield, 1975) and the widespread occurrences of submarine lithified limestones along the margins of the Bahama Platform (Gibson and Schlee, 1967; Neumann, 1974; Schlager et al, 1976; Wilber, 1976; Neumann et al, 1977) favor the concept of slope maintenance by submarine cementation (Mullins and Lynts, 1977; Hine and Neumann, 1977). The actual mechanism of submarine cementation is presently unknown but the association of extensive hardgrounds and litho herms with strong bottom water flow of the Florida Current west of LBB suggests a possible cause and effect relationship. Wilber (1976) has proposed that winnowing by bottom currents west of LBB removes the finer sized material and results in a porous open framework sediment through which interstital waters can percolate and cementation (by high-Mg calcite) occurs. This hypothesis is supported by the fact that many of the samples of lithified carbonate rock recovered from the deep bank margin west of LBB are grain-supported biomicrites and biomicrudites (Wilber, 1976).

Deep-water biological buildups (lithoherms) west of LBB and GBB appear to play a significant role in the growth of this deep bank margin (Neumann, 1974; Neumann et al, 1977). These lithoherms are covered by a dense, diverse benthic fauna which actively baffles and traps sediment and contributes skeletal material. Lithoherms appear to develop in situ as a consequence of biological and sedimentological build-up intermittent with submarine cementation.

Possible Ancient Analogs

Ancient carbonate platforms and off platform sequences similar to the modern Bahamas appear to be common in the rock record. Examples of ancient shallow-water platforms separated by deeper-water basins include the Devonian of western Canada (Bassett and Stout, 1967), the Permian Basin of west Texas (Oriel et al, 1967), the Triassic Dolomites of northern Italy (Bosellini and Rossi, 1974), the Jurassic—Cretaceous carbonates of the Mediterranian Alps (Bernoulli and Jenkyns, 1974; D'Argenio et al, 1975), the middle Cretaceous of eastern Mexico (Enos, 1974b) and the Miocene of St. Croix (Multer et al, 1977).

A survey of the literature on ancient deep-water carbonates ranging in age from Cambrian to Miocene indicates that specific examples with similar facies relationships as six of the seven deep carbonate bank margin types identified in this study are recognizable in the ancient. The only deep bank margin type that we could not find an ancient analog for is the Extend—Type C deep bank margins. A list of *possible* ancient analogs for our bank margin types is given in Table 5. For more details on ancient carbonate facies the reader is referred to Wilson (1975).

Hydrocarbon Potential of Deep Carbonate Bank Margins

Deep carbonate bank margins and basins appear to offer new objectives for hydrocarbon exploration (Cook et al, 1972; Scholle, 1977a,b; Cook and Enos, 1977; Enos 1977). The question which arises however is what type of deep bank margins should be explored? For hydrocarbons to accumulate in any setting there must be source rocks, sufficient heat and pressure for the transformation of kerogens to liquid hydrocarbons, porous/permeable reservoir rocks, traps and impermeable seals. Mullins et al (1978) have discussed the potential for these requirements in the deeper water environments of the Bahama Platform.

Enos (1977) has suggested that those deep margins formed by gravity flow processes would have the greatest potential for good reservoirs and consequently for the accumulation of hydrocarbons. Leeward deep bank margins, because of off-bank sediment transport and steep slopes along these margins would most likely contain the greatest quantities of coarse detrital carbonates which under favorable diagenetic conditions would have the greatest potential as good reservoirs. These would include Types 'D' and 'E' deep bank margins. Deep-water bioherms (lithoherms) may also be potentially attractive reservoir

TABLE 5—POSSIBLE ANCIENT ANALOGS FOR BAHAMIAN DEEP CARBONATE BANK MARGIN TYPES

Bank Margin Type	Ancient Analog	Reference
Type A—Open Ocean Windward	Boundary limestone, Middle Cambrian of western Canada	McIlreath (1977)
Type B—Open Ocean Windward	Cambrian Taconic Sequence of eastern New York and western Vermont	Keith and Friedman (1977)
Type C—Extended	None	None
Type D—Open Seaway Leeward	Jurassic in the High Atlas Mountains of Morocco	Evans and Kendall (1977)
Type E—Open Seaway Leeward	Middle Pennsylvanian to Lower Permian Hare Fiord Formation of the Sverdrup Basin. Lower Pennsylvanian Dimple Limestone of West Texas. Cretaceous Tamabra Limestone of Mexico west of the Golden Lane	Davies (1977) Thomson and Thomasson (1969) Enos (1977)
Type F—Open Seaway Windward	Cretaceous Tamabra Limestone of Mexico east of the Valles Platform	Carrasco—V (1977)
Type G—Eroded	Triassic Dolomites of Italy. Miocene Miami Terrace	Leonardi (1967) Mullins and Neumann (1979)

possibilities. The band of lithoherms west of LBB is a heavily bored, porous coral-framework-supported unit over 200 km long, 10–15 km wide and 70 m thick! Type 'C' (Extended) deep bank margins may also be potential reservoirs. These deep bank margins are thick (>400 m) wedge shaped sand bodies of sorted current winnowed carbonate sand that have prograded great distances (up to 100 km) from the platform edges and are surrounded by pelagic sediments.

Windward deep bank margins, because of the bankward flux of sediment, would appear to have less potential for good reservoirs. North of GBB (Type 'F') slope deposits consist of pelagic oozes and muddy slope breccia which give way downslope to basinal pelagic muds and thin distal turbidites. Along those deep bank margins that are rimmed by shallow-water reefs there is a source of detrital carbonates and the potential for favorable reservoirs would increase. North of LBB (Type 'B') there is a broad, lower slope biolithoclastic turbidite facies. Similar facies in the rock record could be potential reservoirs.

Major quantities of hydrocarbons have been recovered from deep bank margin carbonates (Cretaceous Tamabra Limestone) of the Poza Rica trend in eastern Mexico (Enos, 1977). The Tamabra Limestone located west of the Golden Lane Platform is dominated by bioclastic grainstone/packstone facies deposited by sediment gravity flows, with lesser amounts of slope breccia and lime mudstones, similar to the leeward bank margins in the northern Bahamas. To date, oil production from the Poza Rica trend has been in excess of 2 billion barrels (Enos, 1977). Although only one Poza Rica has thus far been discovered, when one considers the large volume of off-platform carbonates in the rock record it is unlikely that only one Poza Rica exists (Enos, 1977).

CONCLUSIONS

A marine geological/geophysical survey of the deep carbonate bank margins that face open seas and seaways in the northern Bahamas has defined a diversity of deep bank margin types and complex facies relationships. These findings demonstrate that all deep carbonate bank margins are not simply areas of accumulation of "fore-reef talus" and pelagic sediments.

Seven modern deep carbonate bank margin types have been identified on the basis of seismic facies, structure, morphology and sedimentary facies. These seven types are: (1) Type A—Open Ocean Windward margin east of LBB; (2) Type B—Open Ocean Windward margin north of LBB; (3) Type C—Extended margins off the northwest corners of LBB and GBB; (4) Type D—Open Seaway Leeward margins west of LBB and GBB; (5) Type E—Open Seaway Leeward margin south of LBB; (6) Type F—Open Seaway Windward margin north of GBB; and (7) Type G—Eroded margins along the mouth of NWPC.

Each deep bank margin type is distinct from any other because of a unique set of processes which act on and within each margin. The processes most responsible for the development of deep carbonate bank margins include: (1) basement faulting, (2) direction and magnitude of off-bank sediment transport; (3) gravity and pelagic sedimentation; (4) oceanic circulation; (5) submarine cementation; and (6) deep-water biological buildups.

Although the Bahama Platform is a unique physiographic feature on the surface of the Earth today, similar carbonate platforms appear to have existed in the geologic past. Models of modern deep carbonate bank margin sedimentation derived from the northern Bahamas also appear to be directly applicable to the rock record.

In ancient deep-water carbonate environments, coarse gravity flow deposits (particularily along leeward deep bank margins) deep-water bioherms and winnowed carbonate sand bodies have potential as hydrocarbon reservoirs whereas fine-grained pelagic limestones are potential source and/or sealing beds. Ancient deep carbonate bank margins (such as the Poza Rica) thus appear to offer new objectives in the exploration for hydrocarbons because of the primary depositional juxtaposition of potential source, reservoir and sealing rocks.

ACKNOWLEDGMENTS

This research was supported by National Science Foundation Grants GA-36674 and OCE-76-04330 to A. C. Neumann and A. C. Hine at the University of North Carolina as part of a larger program on Bahamian Carbonate Bank Margin Structure and Sedimentation. Use of the R/V *Eastward,* R/V *Gillis,* and DSRV *Alvin* all funded by the UNOLS program of NSF, was provided by the Cooperative Oceanography Program of the Duke University Marine Laboratory, the University of Miami, and Woods Hole Oceanographic Institution respectively. Some piston core descriptions were provided by the core lab of the Lamont-Doherty Geological Observatory supported by NSF Grant OCE-76-18049 and ONR Contract N00014-75-C-0210. O H. Pilkey and the geology department at Duke University provided use of their X-radiography unit.

J. J. W. Rogers, R. E. Sheridan, A. C. Hine, J. M. Bane, D. A. Textoris, and C. S. Martens critically reviewed portions of this manuscript. We would also like to acknowledge valuable discussions with G. W. Lynts, P. Enos, R. E. Sheridan, A. C. Hine, O. H. Pilkey, M. Boardman, R. J. Wilber, R. Lighty, B. R. Rosendahl, and C. R. Newton.

REFERENCES

ANDREWS, J. E., SHEPARD, F. P., AND HURLEY, R. J., 1970, Great Bahama canyon: Geol. Soc. America Bull., v. 81, p. 1061–1078.
BALL, M. M., 1967, Tectonic control of the configuration of the Bahama Banks: Gulf Coast Assoc. Geol. Socs. Trans., v. 17, p. 265–267.
———, HARRISON, C. G. A., HURLEY, R. J., AND LEIST, C. E., 1969, Bathymetry in the vicinity of the northeastern scarp of the Great Bahama Bank and Exuma Sound: Mar. Sci. Bull., v. 19, p. 243–252.
———, DASH, B. P., HARRISON, C. G. A., AND AHMED, K. O., 1971, Refraction seismic measurements in the northeastern Bahamas: EOS, Amer. Geophys. Union Trans., v. 52, p. 252.
BARAZANGI, M., AND DORMAN, J., 1969, World seismicity maps compiled from ESSA, Coast and Geodetic Survey, Epicenter data, 1961–1967: Seismol. Soc. America Bull., v. 59, p. 369–380.
BASSETT, H. G., AND STOUT, J. G., 1967, Devonian of western Canada: Int. Symp. Devonian System, Alberta Soc. Petroleum Geologists, p. 717–752.
BELDING, H. F., AND HOLLAND, W. C., 1970, Bathymetric maps of eastern continental margin, U.S.A.: Am. Assoc. Petroleum Geologists, Sheet 2 of 3, Atlantic Ocean South of Cape Hatteras.
BERNOULLI, D., AND JENKYNS, H. C. 1974, Alpin Mediteranean and central Atlantic Mesozoic facies in relation to the early evolution of the Tethys: Soc. Econ. Paleontologists and Mineralogists Spec. Pub. No. 19, p. 129–160.
BOSELLINI, A., AND ROSSI, D., 1974, Triassic carbonate buildups of the Dolomites, northern Italy: Soc. Econ. Paleontologists and Mineralogists Spec. Pub. No. 18, p. 209–233.
BOUMA, A. H., 1962, Sedimentology of some flysch deposits: Amsterdam, Elsevier, 168 p.
BUBB, J. N., AND HATFIELD, W. G., 1977, Seismic recognition of carbonate buildups: Am. Assoc. Petroleum Geologists Memoir 26, p. 185–204.
BURNE, R. V., 1973, The deposition of reef-derived sediment upon a bathyal slope: The deep off-reef environment, north of Discovery Bay, Jamaica: Marine Geology, v. 16. p. 1–19.
CARRASCO, V. B., 1977, Albian sedimentation of submarine autochthonous and allochthonous carbonates, east edge of the Valles-San Luis Potosi Platform, Mexico: Soc. Econ. Paleontologists and Mineralogists Spec. Pub. No. 25, p. 263–273.
COOK, H. E., MCDANIEL, P. N., MOUNTJOY, E. W., AND PRAY, L. C., 1972, Allochthonous carbonate debris flows at Devonian bank ('reef') margins Alberta, Canada: Canadian Petroleum Geologists Bull., v. 20, p. 439–497.
———, AND ENOS, P., eds., 1977, Deep-water carbonate environments: Soc. Econ. Paleontologists and Mineralogists Spec. Pub. No. 25, 366 p.
D'ARGENIO, B., CASTRO, P. D., EMILIANI, C., AND SIMONE, L., 1975, Bahamian and Apenninic limestones of identical lithofacies and age: Am. Assoc. Petroleum Geologists Bull., v. 59, p. 524–533.
DAVIES, G. R., 1977, Turbidites, debris sheets, and truncation structures in Upper Paleozoic deep-water carbonates of the Sverdrup Basin, Artic archipelago: Soc. Econ. Paleontologists Mineralogists Spec. Pub. No. 25, p. 221–248.
DIETZ, R. S., HOLDEN, J. C., AND SPROLL, W. P., 1970, Geotectonic evolution and subsidence of Bahama platform: Geol. Soc. America Bull., v. 81, p. 1915–1927.

DUING, W., 1975, Synoptic studies of transients in the Florida Current: Jour. Mar. Res., v. 33, p. 53–73.
EMILIANI, C., 1965, Precipitous continental slopes and considerations on the transitional crust: Science, v. 147, p. 145–148.
ENOS, P., 1974a, Surface sediment facies of the Florida-Bahamas plateau: Geol. Soc. America Map and Chart MC-5, scale 1:3, 168,000, 1 sheet.
———, 1974b, Reefs, platforms and basins in middle Cretaceous in northeast Mexico: Am. Assoc. Petroleum Geologists Bull., v. 58, p. 800–809.
———, 1977, Tamabra Limestone of the Poza Rica trend, Cretaceous, Mexico: Soc. Econ. Paleontologists Mineralogists Spec. Pub. No. 25, p. 273–314.
EVANS, I., AND KENDALL, C. G. ST. C., 1977, An interpretation of the depositional setting of some deep-water Jurassic carbonates of the central High Atlas Mountains, Morocco: Soc. Econ. Paleontologists and Mineralogists Spec. Pub. No. 25, p. 249–262.
EWING, J. I., EWING, M., AND LEYDEN, R., 1966, Seismic profiler survey of Bake Plateau: Am. Assoc. Petroleum Geologists Bull., v. 50, p. 1948–1971.
GIBSON, T. B., AND SCHLEE, J., 1967, Sediments and fossiliferous rocks from the eastern side of the Tongue of the Ocean, Bahamas: Deep Sea Res., v. 14, p. 691–702.
GINSBURG, R. N., AND JAMES, N. P., 1973, British Honduras by submarine: Geotimes, v. 18, p. 23–26.
———, JAMES, N. P., LAND, L. S., MOORE, C. H., AND NEUMANN, A. C., 1974, Exploration of modern reef and carbonate platform margins by submersible: Geol. Soc. America Abs. with Programs, v. 6, no. 7, p. 754–755.
GOREAU, T. F., AND LAND, L. S., 1974, Fore-reef morphology and depositional processes, north Jamaica: Soc. Econ. Paleontologists and Mineralogists Spec. Pub. No. 18, p. 77–89.
HESS, H. H., 1960, The origin of Tongue of the Ocean and other great valleys of the Bahama Banks: Caribbean Geol. Conf., 2nd, Mayaguez, Puerto Rico 1959, Trans., p. 160–161.
HINE, A. C., 1975, Shallow carbonate bank margin structure and depositional processes: northwestern Little Bahama Bank, Bahamas [Ph.D. thesis]: Columbia, Univ. South Carolina, 215 p.
———, 1977, Lily Bank, Bahamas: History of an active oolite sand shoal: Jour. Sed. Petrology, v. 47, p. 1554–1582.
———, AND NEUMANN, A. C., 1977, Shallow-carbonate-bank margin growth and structure, Little Bahama Bank, Bahamas: Am. Assoc. Petroleum Geologists Bull., v. 61, p. 376–406.
HOLLISTER, C. D., et al, 1972, Site 98-Northeast Providence Channel: Initial Rept. Deep Sea Drilling Project, v. 11, p. 9–50.
HUBBARD, D. K., WARD, L. G., FITZGERALD, D. M., AND HINE, A. C., 1976, Bank margin morphology and sedimentation, Lucaya, Grand Bahama Island: Columbia, Univ. of South Carolina Dep't. Geol. Tech. Rept. 7-CRD, 36 p.
JOIDES, 1965, Ocean drilling on the continental margin: Science, v. 150, p. 709–716.
KEITH, B. D., AND FRIEDMAN, G. M., 1977, A slope-fan-basin-plain model, Taconic Sequence, New York and Vermont: Jour. Sed. Petrology, v. 47, p. 1220–1241.
LEONARDI, P., 1967, LE DOLOMITI: Geologic dei monti tra Isarco e Piave, v. 1 and 2, 1019 pp Rome, Nat. Res. Council.
MALLOY, R. J., AND HURLEY, R. J., 1970, Geomorphology and geologic structure: Straits of Florida: Geol. Soc. America Bull., v. 83, p. 251–272.
MCILREATH, I. A., 1977, Accumulation of a Middle Cambrian, deep-water limestone debris apron adjacent to a vertical submarine carbonate escarpment, southern Rocky Mountains, Canada: Soc. Econ. Paleontologists and Mineralogists Spec. Pub. No. 25, p. 113–124.
MEYERHOFF, A. A., AND HATTEN, C. W., 1974, Bahamas salient of North America: Tectonic framework, stratigraphy and petroleum potential: Am. Assoc. Petroleum Geologists Bull., v. 58, p. 1201–1239.
MIDDLETON, G. V., AND HAMPTON, M. A., 1973, Sediment gravity flows: Mechanics of flow and deposition, in Middleton, G. V., and Bouma, A. H., eds., Turbidites and Deep-water sedimentation: Soc. Econ. Paleontologists and Mineralogists Pacific Sec., Los Angeles, Calif., p. 1–38.
MITCHUM, R. M., JR., VAIL, P. R., AND SANGREE, J. B., 1977, Stratigraphic interpretation of seismic reflection patterns in depositional sequences: Am. Assoc. Petroleum Geologists Mem. 26, p. 117–134.
MOLNAR, P., AND SYKES, L. R., 1969, Tectonics of the Caribbean and Middle America regions from focal mechanisms and seismicity: Geol. Soc. America Bull., v. 80, p. 1639–1684.
MOORE, C. H., JR., GRAHAM, E. A., AND LAND, L. S., 1976, Sediment transport and dispersal across the deep fore-reef and island slope (−55 m to 305 m), Discovery Bay, Jamaica: Jour. Sed. Petrology, v. 46, p. 174–187.
MULLINS, H. T., 1978, Deep carbonate bank margin structure and sedimentation in the northern Bahamas [Ph.D. thesis]: Chapel Hill, N.C., Univ. North Carolina, 166 p.
———, BOARDMAN, M. R., AND NEUMANN, A. C., in press, Echo character of off-platform carbonates: Marine Geology.
———, AND LYNTS, G. W., 1975, Plate tectonic origin for northwestern Bahama Platform: Geol. Soc. America Abs. with Programs, v. 7, no. 7, p. 1207.
———, AND ———, 1976, Stratigraphy and structure of Northeast Providence Channel, Bahamas: Am. Assoc. Petroleum Geology Bull., v. 60, p. 1037–1053.
———, AND ———, 1977, Origin of the northwestern Bahama Platform: Review and reinterpretation: Geol. Soc. America Bull., v. 88, p. 1447–1461.

———, AND NEUMANN, A. C., 1977, Miocene submarine erosion in the northeastern Straits of Florida: Geol. Soc. America Abs. with Programs, v. 9, no. 2, p. 170–171.

———AND———, 1979, Geology of the Miami Terrace and its paleo-oceanographic implications: Marine Geology, v. 30, p. 205–232.

———, LYNTS, G. W., NEUMANN, A. C., AND BALL, M. M., 1978, Characteristics of the deep Bahama channels in relation to hydrocarbon potential:Am. Assoc. Petroleum Geologists, Bull., v. 62, p.

MULTER, H. G., FROST, S. H., AND GERHARD, L. C., 1977, Miocene "Kingshill-Seaway"—A dynamic carbonate basin and shelf model, St. Croix, U.S. Virgin Islands: Am. Assoc. Petroleum Geologists Studies in Geology No. 4, p. 329–352.

NEUMANN, A. C., 1974, Cementation, sedimentation and structure on the flanks of a carbonate platform, northwestern Bahamas, *in* Recent advances in carbonate studies, Abs. V: Fairleigh Dickinson Univ., West Indies Lab Spec. Pub. 6, p. 26–30.

NEUMANN, A. C., AND BALL, M. M., 1970, Submersible observations in the Straits of Florida: Geology and bottom currents: Geol. Soc. America Bull., v. 81, p. 2861–2874.

NEUMANN, A. C., AND HINE, A. C., 1974, Shallow and deep bank margin structure and sedimentation: Little Bahama Bank: Geol. Soc. America Abs. with Programs, v. 6, no. 7, p. 888.

NEUMANN, A. C., KOFOED, J. W., AND KELLER, G. H., 1977, Lithoherms in the Straits of Florida: Geology, v. 5, p. 4–10.

NEWELL, N. D., 1955, Bahamian platforms: Geol. Soc. America Spec. Paper 62, p. 303–315.

ORIEL, S. S., MYERS, D. A., AND CROSBY, E. J., 1967, West Texas Permian Basin region: U.S. Geol. Survey Professional Paper 515, p. 21–64.

PAULUS, F. J., 1972, The geology of Site 98 and the Bahama Platform: Initial Rept. Deep Sea Drilling Project, v. 11, p. 877–900.

PAYTON, C. E., ed., 1977, Seismic stratigraphy-applications to hydrocarbon exploration: Amer. Assoc. Petroleum Geologists Memoir 26, 516 p.

RICHARDSON, W. S., AND FINLEN, J. R., 1967, The transport of Northwest Providence Channel: Deep Sea Res., v. 7, p. 361–367.

RYAN, W. B. F., LYNDE, R., AND HEEZEN, B. C., 1977, The post Neocomian succession on the Bahama Escarpment: EOS, Amer. Geophys. Union Trans., v. 58, p. 417.

SANGREE, J. B., WAYLETT, D. C., FRAZIER, D. E., AMERY, G. B., AND FENNESSY, W. J., 1976, Recognition of continental-slope seismic facies offshore Texas-Louisiana, *in* Bouma, A. H., Moore, G. T., and Coleman, J. M., eds., Beyond the Shelf Break: Am. Assoc. Petroleum Geologists Marine Geology Short Course, New Orleans, La., p. F1–F54.

SCHLAGER, W., HOOKE, R. L., AND JAMES, N. P., 1976, Episodic erosion and deposition in the Tongue of the Ocean (Bahamas): Geol. Soc. America Bull., v. 87, p. 1115–1118.

SCHOLLE, P. A., 1977a, Chalk diagenesis and its relation to petroleum exploration: Oil from chalks, a modern miracle?: Am. Assoc. Petroleum Geologists Bull., v. 61, p. 982–1009.

———, 1977b, Deposition, diagensis, and hydrocarbon potential of "deeper-water" limestones: Am. Assoc. Petroleum Geologists Short Courses in Exploration Geology, 25 p.

SHERIDAN, R. E., 1974, Atlantic continental margin of North America, *in* Burk, C. A., and Drake, C. L., eds., Geology of Continental Margins: New York, Springer-Verlag, p. 391–407.

———, 1976, Sedimentary basins of the Atlantic margin of North America: Tectonophysics, v. 36, p. 113–132.

———, 1977, Blake plateau offers good potential for large petroleum reserves: Oil and Gas Jour., v. 75, p. 218–229.

———, DRAKE, C. L., NAFE, J. E., AND HENNION, JR., 1966, Seismic refraction study of continental margin east of Florida: Am. Assoc. Petroleum Geologists Bull, v. 50, p. 1972–1991.

SHERIFF, R. E., 1976, Inferring stratigraphy from seismic data: Am. Assoc. Petroleum Geologists Bull, v. 60, p. 528–542.

STORRS, J. F., 1964, Ecology and oceanography of the coral-reef tract, Abaco Island, Bahamas: Geol. Soc. America Spec. Paper 79, 98 p.

TALWANI, M., 1960, Gravity anomalies in the Bahamas and their interpretation [Ph.D. thesis]: New York, Columbia Univ., 89 p.

TATOR, B. A., AND HATFIELD, L. E., 1975, Bahamas present complex geology: Oil and Gas Jour., v. 73, no. 43, p. 172–176; no. 44, p. 120–122.

THOMSON, A. F., AND THOMASSON, M. R., 1969, Shallow to deep water facies development in the Dimple Limestone (Lower Pennsylvanian), Marathon region, Texas: Soc. Econ. Paleontologists and Mineralogists Spec. Pub. No. 14, p. 57–77.

UCHUPI, E., 1969, Morphology of the continental margin of southeastern Florida: Southeastern Geology, v. 11, p. 129–134.

———, MILLIMAN, J. D., LUYENDYK, B. P., BROWN, C. O., AND EMERY, K. O., 1971, Structure and origin of southeastern Bahamas: Am. Assoc. Petroleum Geologists Bull, v. 55, p. 687–704.

WALKER, R. G., AND MUTTI, E., 1973, Turbidite facies and facies associations: Soc. Econ. Paleontologists and Mineralogists Pacific Sec. Short Course, Los Angeles, Calif., p. 119–158.

WILBER, R. J., 1976, Petrology of submarine-lithified hardgrounds and lithoherms from the deep flank environment

of Little Bahama Bank (northeastern Straits of Florida) [M.S. thesis]: Durham, N.C., Duke Univ., 241 p.
——, AND NEUMANN, A. C., 1976, Petrology of subsea cemented carbonate mounds, lithoherms, in northern Straits of Florida: Am. Assoc. Petroleum Geologists Bull., v. 60, p. 733.
——, AND NEUMANN, A. C., 1977, Porosity controls in subsea cemented rocks from deep-flank environment of Little Bahama Bank: Am. Assoc. Petroleum Geologists Bull., v. 61, p. 841.
WILSON, J. L., 1975, Carbonate facies in geologic history: New York, Springer-Verlag, 471 p.

SEDIMENT FACIES OF PLATFORM-BASIN TRANSITION, TONGUE OF THE OCEAN, BAHAMAS

WOLFGANG SCHLAGER AND ANDREW CHERMAK
Comparative Sedimentology Laboratory
Rosenstiel School of Marine and Atmospheric Science
University of Miami at Fisher Island Station, Miami, FL 33139
Chevron USA Exploration, 1111 Tulane Avenue, New Orleans, LA 70124

ABSTRACT

Inter-platform troughs in the Bahamas are either U-shaped basins or V-shaped canyons. Basins are thought to have formed where parts of the originally continuous Florida-Bahama platform were drowned, canyons originate where erosion by turbidity currents cuts headward into the flat basin floors.

Southern Tongue of the Ocean (TOTO) is a U-shaped basin, in shape and setting comparable to many ancient carbonate basins. Carbonate platforms with high-sediment production rim the basin and fill it largely with neritic material. Lime mud, winnowed from the platform, simply augments the rain of planktonic material. Lime sand and rubble are carried to the basin floor by turbidity currents, possibly also by debris flows and grain flows. The platform sheds debris all along its margin and delivers it to the basin floor through numerous small gullies rather than few large canyons. Therefore, it acts as a line source, not as a point source of sediment and the familiar pattern of large canyons and deep-sea fans is not developed in TOTO. Rather, the facies pattern is concentric around the imaginary mid-point of the basin, with boundaries running parallel to the bank edge. Three major facies belts could be mapped in closely spaced cores and 3.5 kHz seismic profiles: (1) Basin interior with medium to fine sand turbidites and layered reflection character; (2) basin margin with closely spaced and sometimes coalescing small fans of coarser, more proximal turbidites and opaque reflection character (prolonged seabed echo without subbottom reflections); (3) gullied bypass slope with mainly mud and only strings of clastic material in gullies; seismic echo consists of large hyperbolae with parallel subbottom reflections where sediment is mud, of hyperbolae without subbottom reflections where gullies are floored with sand and rubble. Layers of coarse clean sand and rubble seem to directly cause the change from layered to opaque echo character. Neither grain size of intervening ooze nor microtopography of sea floor correlate with seismic reflection character.

Compared to ancient carbonate basins, TOTO and other Bahamian basins are rather small but their flanks are high. Consequently, the slope environment as a trap of mud and a bypass-zone of sand and rubble may be more clearly developed than in many ancient examples; on the other hand, the distal part of the platform-basin transition, notably the gradation into fully pelagic facies, is missing.

INTRODUCTION

This report describes the facies pattern of a platform-basin transition in the Bahamas as a modern example of a situation frequently encountered in the geologic record. The juxtaposition of shallow platforms and deep, narrow basins is one of the most characteristic features of carbonate continental margins and, with gentler relief, also found along the margins of intracratonic basins. Numerous platform-basin transitions in both continental margins as well as intracratonic basins have been described from the geologic record (e.g., Mojsisovics, 1879; Newell and others, 1953; Playford and Lowry, 1966; Thomson and Thomasson, 1969; Cook and Taylor, 1977). In a careful analysis and synthesis, Wilson (1975, p. 24) pointed out the basic pattern underlying them, namely the gradual decrease of the neritic component and the increase of pelagic features with increasing distance from the platform margin.

The inherent difficulty with ancient platform-basin transitions is the correlation of sedimentary facies with morphologic features such as slope and basin floor, and water depth. Even in tectonically undisturbed examples, this correlation remains hypothetical because of incomplete exposure, insufficient stratigraphic control and distortion from diagenetic compaction. Recent examples can provide the required calibration of the sediment facies. Mullins and Neumann (in press) described different types of platform slopes from the northern Bahamas, notably those facing the open ocean, current swept sea ways or major canyons. This report focuses on flat-floored Bahamian basins with sluggish circulation; this setting—we feel—is more comparable to the numerous intracratonic basins of the geologic record. We specifically addressed the question: is there a persistent succession of facies beyond the platform edge and if so, to what extent does the facies pattern reflect major topographic features, such as slope and flat basin floor, and the proximity to the platform?

It turned out that intercalations of sand and

Copyright © 1979, The Society of Economic Paleontologists and Mineralogists

rubble in the carbonate ooze are sensitive indicators of these changes and have several advantages over the fine pelagic sediments: (1) they are easy to recognize macroscopically in cores; (2) they will be preserved in spite of diagenetic alteration because of the large grain size and the large scale of their depositional structures; (3) they profoundly influence the echo character of high-frequency seismic records and can thus be mapped partly with the seismic profiler.

METHODS

Data for this study were acquired by piston and gravity coring, dredging, seismic reflection profiling from the sea surface and direct examination of the sea floor by submersible (Fig. 1).

All surface work was done with the University of Miami's ships *James M. Gilliss* and *Columbus Iselin* in 1975–1977. Cores were taken in plastic liners, and described immediately aboard the ship. Grain size was estimated with a visual grain size comparator developed by Shell Research, Holland. Description of depositional structures is partly based on 30 cores from the University of Miami archives that had been impregnated with resin, slabbed and X-rayed.

Ten dives with the submersible *Alvin* in 1975 and 1977 allowed us to directly examine the sea floor of the gullied slopes and the basin floor and sample the rocky cliffs. Additional rock samples were collected with the dredge.

Observations on the talus slope and marginal escarpment are based on 34 dives with the submersible *Nekton* in 1976.

Seismic surveys of the sea floor were carried out with a 12 kHz precision depth recorder, and a 3.5 kHz profiler, both with ship-mounted transducers.

BAHAMIAN TROUGHS: BASINS VS. CANYONS

A glance over contour maps (Andrews et al, 1970; Athearn 1962a; Belding and Holland, 1970; Hurley and others, 1962; King, 1969) shows that some Bahamian troughs are flat-floored, U-shaped basins, e.g., Exuma Sound, Southern TOTO, Columbus Basin, others are deeply incised, V-shaped canyons, such as the Providence Channels (Fig. 2). Both, basins and canyons are flanked by steep slopes (overall dips of 5-27°) that are riddled with gullies. The platform edge above the gullied slopes is an almost vertical escarpment, footed by ungullied talus slope. Several authors (e.g., Hess, 1960; Gibson and Schlee, 1967; Andrews and others, 1970; Schlager and others, 1976) pointed to the essentially erosional character of gullied slopes and canyons and concluded that besides upbuilding of the banks, erosion by gravity flows is the most important factor in shaping the troughs. For the Tongue of the Ocean and the Providence Channels, Hooke and Schlager (in prep.) proposed that long term headward erosion by gravity flows has gradually transformed the flat floored basins into V-shaped canyons. The

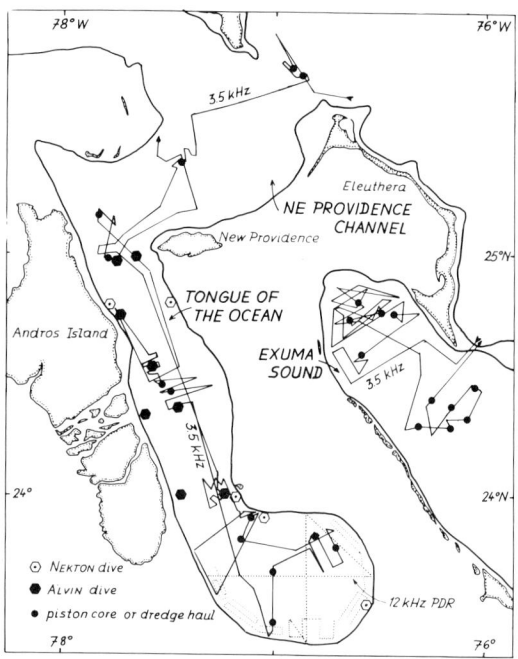

FIG. 1.—Northeastern Bahamas with Tongue of the Ocean (TOTO), Northeast Providence Channel and Exuma Sound; track lines and sample locations.

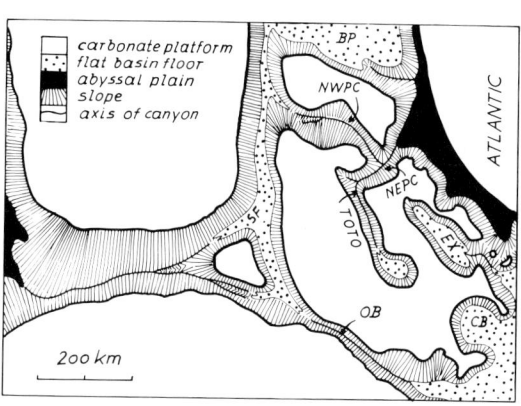

FIG. 2.—Carbonate platforms, U-shaped flat-floored basins, V-shaped canyons and abyssal plains in the Florida-Bahama region. Basins are Straits of Florida (SF), southern TOTO, Exuma Sound (EX), Columbus Basin (CB); also drowned but not surrounded by growing platforms is Blake Plateau (BP). Major canyons are Northwest Providence Channel (NWPC), Northeast Providence Channel (NEPC), Old Bahama Channel (OB) and northern TOTO.

pattern in other troughs (Fig. 2) is consistent with this hypothesis: All flat basin floors seem to be connected with the abyssal plain of either the Atlantic or the Gulf of Mexico through a V-shaped canyon that cuts headward into the flat floor like a rejuvenated river into an alluvial valley. The system of gullied platform slopes and V-shaped canyons can be looked upon as a drainage system that disposes of the excess sediment of the platforms by carrying it to the flat basin floors and the abyssal plains as the ultimate repositories.

The flat basin floors are probably the surface expression of drowned carbonate platforms now buried under basin deposits. Such a situation clearly exists in the Blake Plateau, north of the Bahamas, where a deep-water sequence overlying drowned shallow-water carbonate platforms has been recovered by the drill (Benson, Sheridan and others, 1976) and a reflector interpreted as the top of a Cenomanian (?) carbonate platform has bern traced seismically across the plateau (Ewing and others, 1966; Shipley and others, 1978). A similar reflector marked by a velocity jump from 2.8 to 4.8 km/sec. has been mapped underneath the basin fill of Northwest Providence Channel by Mullins and Neumann (in press). Water depth of the flat basin floors is the result of the subsidence of the platform since its drowning minus the thickness of the basin fill. It may provide a rough indication of the time of drowning. We speculate that the basins in TOTO and Exuma Sound are underlain by early Cretaceous platforms like the outer Blake Plateau, whereas the deep basins SE of Great Bahama Bank (e.g., Columbus Basin) may have been drowned earlier.

The distinction between canyons and basins among the Bahamian troughs is of some importance for a sedimentologic analysis. Gravity-driven sediment flows tend to deposit their load on the flat basin floors and low angle slopes. They generally bypass the steep flanks and narrow axes of the canyons. Thus, the V-shaped troughs tend to record mainly the perennial rain of fine sediment, (see Hollister, Ewing and others, 1972 on DSDP hole 98 for an example), while the flat basin floors display the full spectrum of autokinetic (=gravity-displaced) sediments delivered by the platforms. In search for modern analogs of most ancient interplatform carbonate basins we will restrict outselves to the U-shaped parts of the Bahamian troughs.

SEDIMENTATION

Sediment in the Bahamian troughs arrives there in two modes. A more or less continuous rain of sediment is provided by planktonic coccoliths, globigerinids, pteropods, as well as by carbonate mud from the banks. Sand and gravel from the banks and the upper slopes, on the other hand, are delivered by episodic pulses of gravity flows. The result are graded layers of rubble, sand and silt with sharp bases and gradational tops. They were described in detail by Rusnak and Nesteroff (1964) and interpreted as turbidites. All our observations support this view.

In the semi-enclosed troughs with sluggish bottom circulation, the perennial rain of fines is spread evenly, the input from gravity flows varies laterally in response to relief and distance from source. The lateral variation of these clastic layers was thus used to subdivide the deep-water depositional environment in TOTO. Contrasting types are illustrated in Figure 3. Beds of medium sand to silt, 5-20 cm thick, with parallel or occasional ripple cross-lamination are found throughout the basin and are most common in the basin interior. Beds of coarse sand and rubble, up to 50 cm thick, occur only near the toe-of-slope. They, too, are graded but show a more rapid decrease in grain size, occasionally even sharp upper contacts. Convolute bedding has not been observed in any of the graded beds. However, the fine sand in the middle of a graded cycle is frequently a quicksand that flows out of the core barrel. We believe that under burial many of these quicksands will develop the convolute laminations so common in ancient turbidites. In a few cores from the basin margin, we recovered other types of clastic layers: Poorly sorted rubble with mud or sand matrix below a layer of graded sand is tentatively interpreted as a deposit of debris flow (Middleton and Hampton, 1976); clean sand and rubble, over 1 m thick, with sharp top and no apparent size-grading was found near the mouth of a major gully and was possibly transported partly by grain flow (Middleton and Hampton, 1976). Slumps are represented by pebbly muds, often several meters thick and intensively folded.

In other Bahamian basins, slumps and coarse debris-flow (?) deposits seem to be more common than in TOTO. Crevello (1978) traced a 2–3 m thick debris sheet across the whole basin in Exuma Sound and Mullins et al (in press) described similar clastic beds from Northwest Providence Channel.

We believe that all these clastic sediments are autokinetic (Kuenen in Sanders, 1965), i.e., "driven downslope by their own excess density and not via a medium with a movement of its own." In some cases, these sediment movements were almost certainly triggered by sediment sliding and slope failure of the upper part of the platform flanks; in other cases, excessively cold or hypersaline bank water cascading down the slope may have stirred up mud and set off a turbidity current. Hypersaline bank water has been found to intrude into the water column of TOTO down to ca. 300 m (Koczy and others, 1958) and recently,

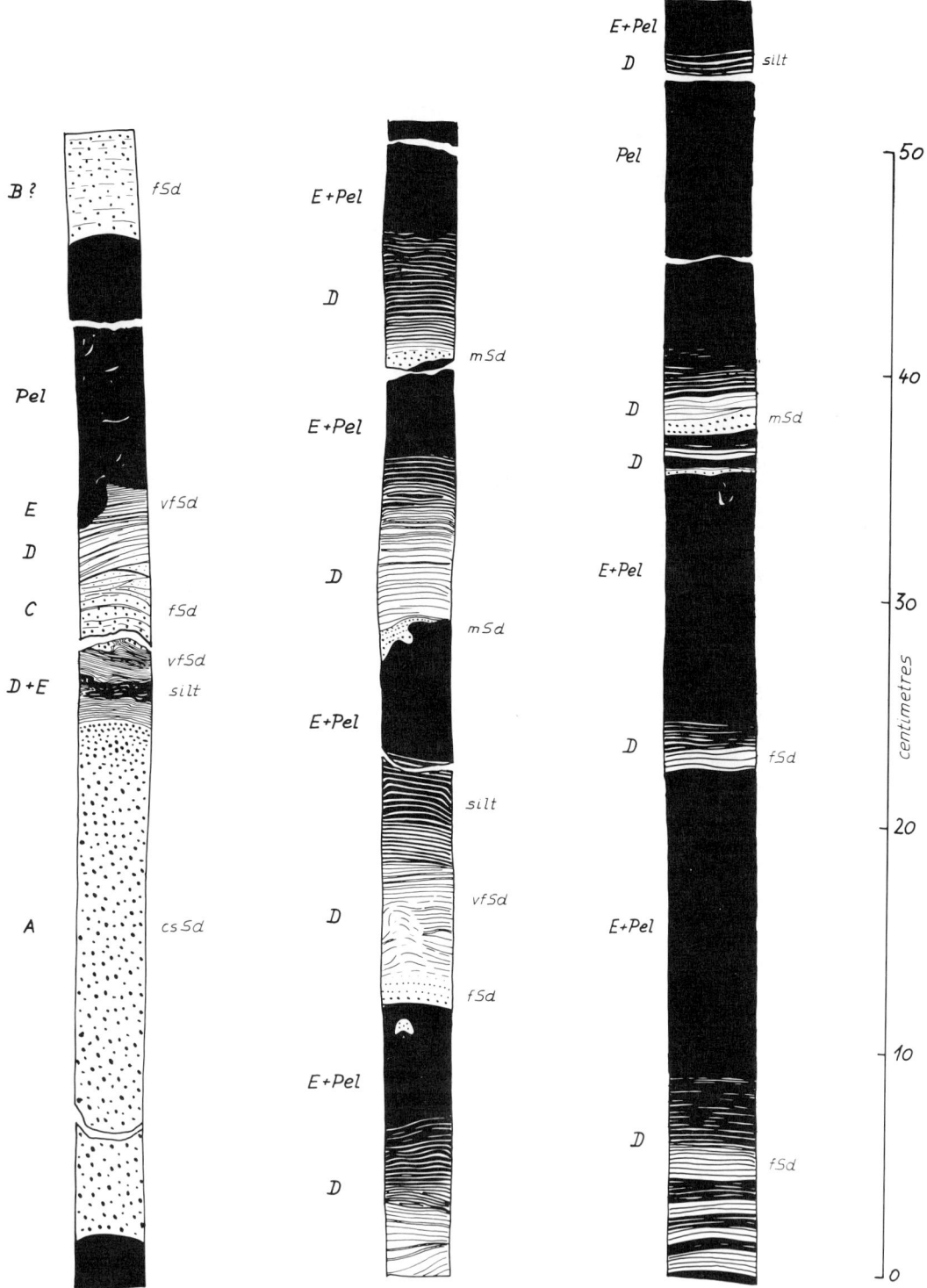

Robertson and others (in prep.) have pointed to the general importance of this phenomenon for off-bank transport. It seems, however, that from intermediate depth on down, turbidity currents must take over and carry the sediment all the way down in order to create beds with traction current lamination at the basin floor.

Sediment input into TOTO varied considerably during the Quaternary, mainly in response to rises and falls of sea level. Aragonite-rich and aragonite-poor intervals alternate in the cores, and have been related to flooding and emergence of the mud-producing platforms (Supko, 1963; Kier and Pilkey, 1971). We found a similar variation in content of organic matter and abundance of shallow water sand. On a meter-scale, "green" intervals with mainly greenish muds, rich in organic matter and with abundant layers of shallow-water sand, alternate with "white" intervals that are low in organic content, have fewer shallow-water sands but abundant pteropods and globigerinids. The topmost pair is a green interval of 1-2 m thickness underlain by a white interval of 1.5-2.5 m. Biostratigraphy of Lynts and others, (1973) (Core 6278), as well as radiocarbon dates by G. A. Rusnak and W. D. Nesteroff (University of Miami archives, unpub.) suggest a postglacial age (less than ca. 8000 years) for the top green interval. It is thus probable that the green intervals represent the high stands of sea level when the banks were flooded and shed abundant carbonate sand and sea grass detritus. Conversely, the white intervals might represent the low-stands when the banks were exposed and the pelagic component was dominant in the deep-water sediments.

The terminology of the Bahamian deep-water sediments deserves a special note. The clastic layers are classified as turbidites, debris-flow and grain-flow deposits as explained above. The intercalations of fine perennial sediments between the graded beds seem to have settled as gentle rain from the water column, a widely accepted criterion for pelagic sediments (Arrhenius, 1963, p. 655; Davies and Gorsline, 1976, p. 21). Pelagic carbonates are generally subdivided into nannofossil ooze, foraminiferal ooze, pteropod ooze or a combination of these types (e.g., Berger, 1974, p. 214). Because of the high admixture of neritic aragonite and magnesian calcite, the perennial sediment in the troughs does not easily fit into any of these categories and was thus called "peri-platform ooze" by Schlager and James (1978). Because of the limited source of neritic fines as well as their high solubility at great depth, peri-platform ooze can be expected to form only narrow belts around carbonate platforms and not extend far out into the abyssal plains. Occurrences elsewhere in modern oceans support this assumption (Friedman, 1965; Berner and others, 1976; Garrison and others, 1975).

The whole sequence of turbidites and peri-platform ooze is classified as "hemipelagic." Kuenen (1950, p. 316-321) introduced this term for "pelagic deposits with admixtures of terrestrial mineral grains of silt or fine sand grade sizes," and Berger and v. Rad (1972, p. 843-854) and Berger (1974, p. 213-214) expanded this definition to explicitly include turbidites from terrigenous or neritic sources.

CORRELATION OF SEISMIC DATA WITH SEDIMENT FACIES

In the course of this study, 1600 km of 3.5 kHz seismic lines were run in TOTO, Exuma Sound, and NE Providence Channel (see Fig. 1). Ship speed was 7 knots, length of outgoing pulse 5 msec, broadcast power 5.4 kW. The recordings were made with a 1 second sweep.

Three major types of bottom/sub-bottom reflection character can be distinguished in these records: opaque, layered, and hyperbolic. The opaque type (Fig. 4a) is characterized by strong, extended echo from sea floor having relatively low relief. Sub-bottom reflections are weak or absent. This category corresponds approximately to echo types IIA and IIB of Damuth and Hayes (1977). The layered echo (Fig. 4b) displays a short, relatively weak sea floor reflection, followed by parallel to gently divergent sub-bottom reflections. Maximum sub-bottom penetration (up to 60 msec) is obtained in areas with this reflection character. It corresponds to Damuth's and Hayes' (1977) type IB. The hyperbolic echo type is characterized by high density of point-source diffraction hyperbolae, with or without conformable sub-bottom reflections (types IIIA and IIIB of Damuth and Hayes, 1977). Similar echo types have been reported from the northern Bahamas by Mullins et al (in press).

In all these reports, the hyperbolic reflection character has been interpreted as the result of hummocky sea floor topography (areas with slumps, erosional gullies or large sediment dunes).

FIG. 3.—Turbidites from basin margin (left) and basin interior (center and right). Capital letters to left of column indicate Bouma interval, symbols to right grain size (visual estimate of mode). Lime mud of Bouma interval E and pelagic sediment shown in black. Highly variable thickness, coarse grain size and top truncation of basin margin turbidites contrast sharply with more uniform thickness and fine grain size of basin interior turbidites, which commonly lack basal intervals of Bouma cycle. Copied directly from impregnated and slabbed gravity cores.

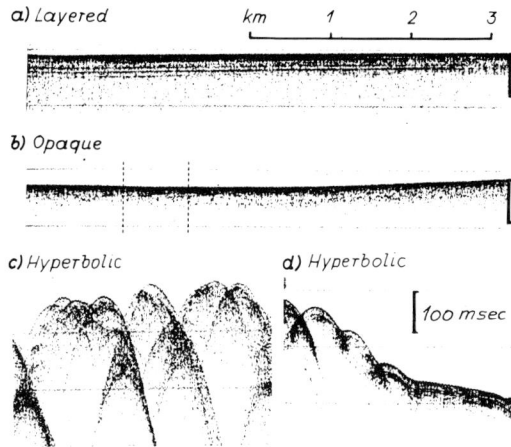

FIG. 4.—Reflection character of 3.5 kHz seismic records from TOTO. a) *Layered:* short sea bed return, 4–6 parallel sub-bottom reflections (basin interior with medium to fine sand turbidites) b) *Opaque:* prolonged sea bed return, very faint and discontinuous sub-bottom reflections (basin margin with coarse turbidites) c) *Hyperbolic:* closely spaced hyperbolae due to gully-and-ridge topography; few conformable sub-bottom reflections (gullied slopes) d) *Hyperbolic:* wide hyperbolae with conformable sub-bottom reflections (gullied slopes with more gentle relief than in 4c, muddy slope sediment).

This interpretation also holds good for TOTO and other Bahamian basins where the hyperbolic echo is exclusively found on the gullied slopes.

The interpretation of the layered and opaque echo is more difficult. A well-layered return is typical for most deep-sea sediments, including all types of carbonate ooze. Mayer (1978) shows that individual reflections do not necessarily indicate individual layers in the sediment. Rather, they may be caused by interference of the signal with small scale impedance variations in the subsurface. The change from layered to opaque echo has been related to changes of at least three different attributes of deep-water sediments: 1) increase in grain size, thickness, and abundance of silt-sand layers in the otherwise muddy sediment (Normark and Dickson, 1976); 2) increase in grain size of carbonate ooze by increasing content of skeletal fragments (Berger and Mayer, 1978), and 3) change from smooth sea floor to a rough microtopography with sand ripples or scours (Bryan and Markl, 1966; Ewing and others, 1973; Damuth and Hayes, 1977, p. 89). In the Northwest Providence Channel (Bahamas) Mullins et al (in press) observed that the change from layered to opaque echo correlates with a change in abundance of sand layers as well as a change in sea floor topography. However, abundance of sand seems to be the more important parameter.

The data from TOTO show very good correlation between reflection character and the grain size and distribution of sand layers in the uppermost part of the sediment column. Figure 5 shows a statistical break down for the entire suite of available core data. Clastic layers in areas of opaque reflection character are coarser, and somewhat more abundant than in areas of layered echo. In addition, we find a small but distinct group of layers over 30 cm thick in opaque areas and none in layered areas.

Figures 6–8 illustrate the important relationship between reflection character and position of clastic layers within the core. A few coarse layers in the upper two meters of sediment are sufficient to cause an opaque return. This is illustrated by Figure 6. Two cores from the slope were taken close to one another but in zones of contrasting reflection character. Both cores consist largely of carbonate ooze, but the core from the opaque zone contains several layers of coarse skeletal sand.

Figures 7 and 8 show that the coarse layers do not have to be more than a few tens of centimeters thick to cause an opaque return as long as they occur in the upper part of the sediment column. A few, but thin, coarse layers in Figure 7 and a thick sand body in Figure 8 generate very similar opaque echos. Thus, the 3.5 kHz tool can be used to map the distribution of coarse, more proximal turbidites, but it will not necessarily lead to the discovery of very thick sand bodies.

All our data suggest that the reflection character changes from layered to opaque when layers of coarse sand or gravel, at least several tens of centimeters thick, appear in the top 2 m of the sediment column.

The alternative explanations for the layered and opaque character, grain size distribution of ooze and sediment surface microtopography, are not satisfactory in the case of TOTO. The histogram

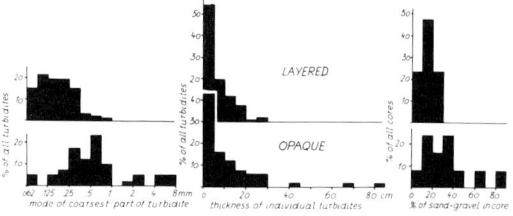

FIG. 5.—Coarseness, thickness and abundance of sand-gravel portions of turbidites in areas with layered and with opaque seismic echo. Note marked difference in grains size and in abundance of clastic layers but only slight difference in bed thickness (due to rapid upward decrease of grain in size in basin-margin turbidites)

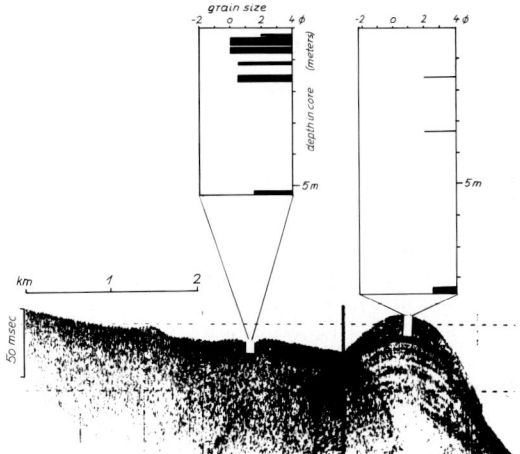

FIG. 6.—Layered and opaque echo on slope of Northeast Providence Channel. Inserts indicate position, thickness and maximum grain size of sand layers in cores. Thickness by height of bar, grain size by length of bar. Note that cores from opaque and layered zone differ mainly in the upper two meters.

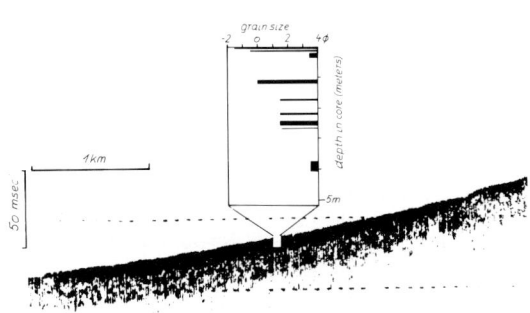

FIG. 7.—Opaque record caused by few and thin coarse turbidites in uppermost part of sediment column. Symbols as in Figure 6.

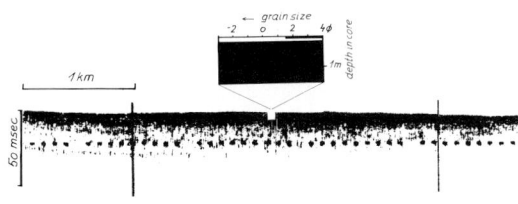

FIG. 8.—Opaque record caused by thick accumulation of sand and rubble near mouth of major gully. Small fan indicated by bulge in middle part of topographic profile and gradation to slightly layered echo to left.

plot of mud fraction grain size distribution in areas of opaque and layered reflection character show no correlation between the two (Fig. 9). The mean grain sizes of basin margin and basin interior oozes with contrasting echos are practically identical, whereas slope sediments, which are also frequently layered in reflection character, are distinctly coarser than the sediments from the layered basin interior.

Similarly, the effect of sea floor topography appears insignificant in the case of TOTO. Bryan and Markl (1973), and Ewing and others, (1973) have suggested that opaque reflection character can be caused by multiple side echoes from local relief too small to be resolved into discrete point-source hyperbolae. In the case of a sedimentary basin floor, this relief takes the form of sand waves and erosional furrows, and is most commonly associated with areas of strong bottom water circulation. Unlike the nearby and open Straits of Florida, and Providence Channels where bottom currents are strong and permanent, the semi-enclosed TOTO basin has no well developed longform bottom circulation (Koczy and others, 1958, Busby, 1962). Our observations from the submersible *Alvin* and camera surveys by Busby (1962) and Athearn (1962b) show the basin floor of central and southern TOTO to be uniformly flat and muddy. No significant differences in the appearance of the bottom in areas of opaque and layered reflection character are noted. This situation differs from that in NW Providence Channel, where the opaque character of the current-swept upper slopes can be attributed to a surface blanket of coarse sand (Mullins et al, in press).

In summary, we believe that reflection character of the flat basin floor seaward of the high relief slopes and canyons is due to a combination of the grain size, abundance, and stratigraphic position of turbidite sand layers whose ultimate source is the adjacent shallow platform.

The reflection events observed in both layered and opaque areas are due fundamentally to the same physical process, reflection of seismic energy from one or more interfaces separating sediments with different acoustic impedances. The acoustic impedance of a material is defined by the product of its bulk density and P-wave velocity. In a layered elastic medium, such as the water column and sediment fill in TOTO, the amplitude of the seismic wavelet reflected from the interface between two layers with different acoustic impedances is given by:

$$A = A_o \frac{\rho_2 V_2 - \rho_1 V_1}{\rho_2 V_2 + \rho_1 V_1} \quad (1)$$

where A and A_o are the amplitudes of the reflected

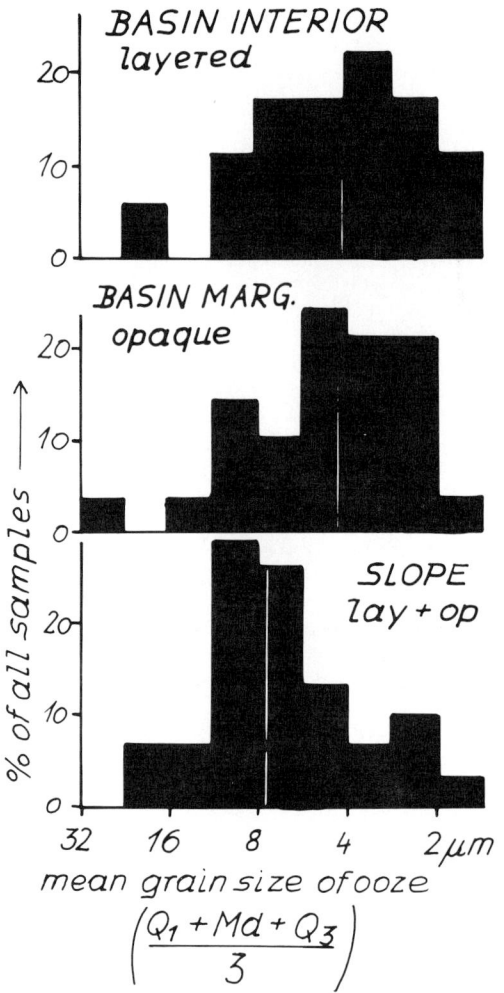

FIG. 9.—Grain size distribution of carbonate mud in three facies with different reflection characters. White line is arithmetic mean. No difference in grain size between basin margin with opaque echo and basin interior with layered echo. Slope, although mostly layered, has coarser grained mud than opaque basin margin. Thus, grain size variation in mud intervals does not seem to explain changes in reflection character. Grain size data from Busby, 1962.

and incident waveforms, and ρ_n, V_n are the bulk densities and sonic velocities of the two layers.

In the case where impedance contrasts are relatively small, individual reflections are correspondingly lower in amplitude since the transmission of seismic energy is given by:

$$T = A_o \left(1 - \frac{\rho_2 V_2 - \rho_1 V_1}{\rho_2 V_2 + \rho_1 V_1} \right) \qquad (2)$$

Sub-bottom penetration is great in sediments whose internal impedance contrasts are small or few. The result of this situation is a layered echo observed in many of the TOTO profiles. On the other hand, penetration is reduced in sediments with large or more numerous impedance contrasts. An opaque record appears where the impedance contrast across near-surface interface is very high or where a sufficient number of interfaces are present in a near surface interval whose thickness is comparable to, or less than the length of the incident seismic wavelet. These situations are illustrated diagrammatically in Figure 10.

Hyperbolic reflection character is an attribute of high relief and is not fundamentally due to the internal constitution of the sediments. Both layered and opaque sediments, as well as completely lithified material may assume this character which is illustrated by Figures 4c and 4d.

FACIES BELTS

A clear facies pattern emerged when the sea floor of southern TOTO was mapped by closely spaced cores and high-frequency seismic profiles. Three major facies belts, slope, basin margin and basin interior, form a concentric pattern around the basin center, with boundaries running approximately parallel to the bank edge, (Table 1). This pattern is the response of gravity driven sediment transport to sea floor topography. The facies belts will be discussed in terms of seafloor topography, seismic character and sedimentary record as revealed by bottom samples.

Topography

Athearn's 1962a contour map as well as our surveys show a clear differentiation in southern TOTO into slope and basin floor (Fig. 11). The upper slope with marginal escarpment and talus slope resembles the profiles described from Belize and Jamaica (James and Ginsburg, in press; Goreau and Land, 1976) and is almost certainly the result of the up and down of sea level during the Pleistocene. The upper slope will not be dealt with here any further. The lower slope from ca. 300–500 m on down, is riddled with gullies a few tens to hundreds of meters wide and 20–100 m deep, often flanked by erosional cliffs of chalk and limestone. The gullies in these slopes are thought to be carved by turbidity currents (Andrews and others, 1970; Hooke and Schlager, in prep.). The basin floor consists of a marginal portion with a slight basinward dip (slope 6–8m/km) and occasional small hummocks. The basin interior is smooth and essentially flat. Athearn's (1962a) map shows a very gentle swell and swale topography and an axial dip to W and NW of less than 1m/km.

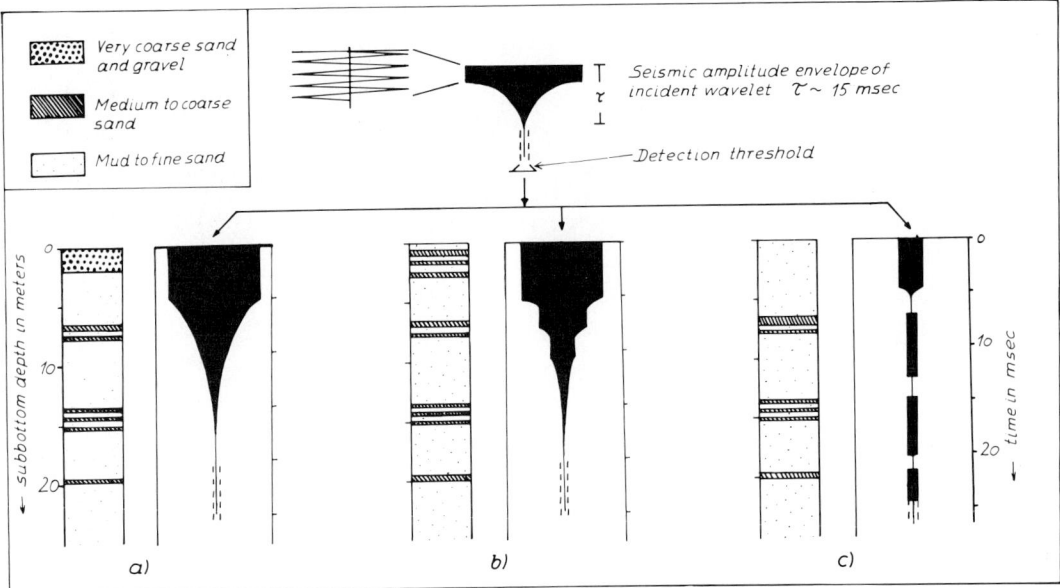

Fig. 10.—Model for the lithology dependent origin of seismic reflection character of TOTO basin sediments: (a) opaque reflection character caused by single interface of high impedance at or near sea floor; (b) opaque reflection character caused by closely-spaced, moderate impedance contrast interfaces near sea floor; (c) layered reflection character due to widely-spaced, low-impedance contrast interfaces throughout sediment column.

TABLE 1.—CHARACTERISTICS OF FACIES BELTS IN SOUTHERN TOTO

Name/Water Depth	Seafloor Topography	Seismic Record	Sedimentary Record
Gullied Slope 400–1300 m	Slope dips 6–9°, dissected by parallel gullies, (20–100 m, often flanked by erosional cliffs of chalk and limestone).	Large overlapping hyperbolae, with or without conformable subbottom reflections (maximum penetration 60 millisecs, 9–10 reflections).	Bioturbated, silty carbonate ooze; sand and gravel less than 4%, occurring in thin lenses or burrow fills; gullies floored with coarse sand and boulders, erosional cliffs harden to oxide-coated, bored hard-grounds.
Basin Margin 1150–1370 m	Sea floor flat with occasional hummocks (large boulders?) dipping basinward at less than 0.5°	Opaque; prolonged sea bed return; none or very few indistinct subbottom reflections.	Alternation of graded coarse sand-silt layers (turbidites) and carbonate ooze; occasional coarse ungraded clean sands over 1 m thick (fluxoturbidites) and muddy gravel beds (debris-flow deposits); slumped beds; bioturbated.
Basin Interior 1300–1400 m	Flat and smooth, very gentle swells and swales; axial dip W and NW at less than 1 m/km	Layered; semi-prolonged sea bed return; 4–6 parallel subbottom reflections, penetration 20–35 millisecs.	Alternation of graded, medium sand-silt layers (turbidites) and carbonate ooze; bioturbated.

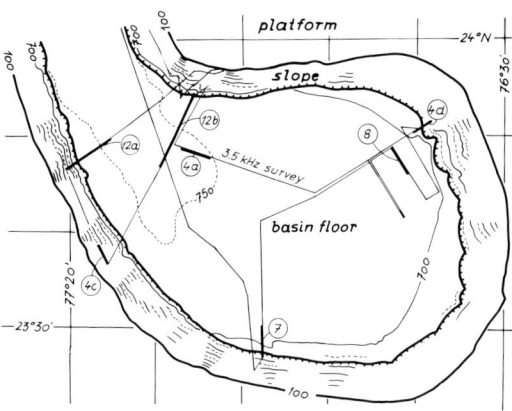

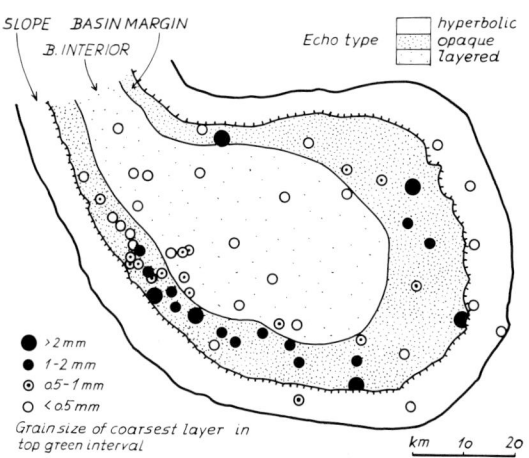

FIG. 11.—Topography of southern TOTO based on Athearn, 1962 and our own surveys. Track lines of 3.5 kHz survey. Bold lines with encircled numbers refer to figures of this report. Contours in fathoms; 100-fathom contour closely follows the marginal escarpment and is good approximation of platform margin.

FIG. 13.—Echo type of 3.5 kHz seismic record and grain size of turbidites in top green stratigraphic interval (upper 1.2–2.0 m of sediment, Latest Pleistocene-Holocene). Grain size plotted is visual estimate of mode of coarsest part of coarsest turbidite at a particular location. Note good correlation between opaque reflection character and distribution of very coarse sand and rubble along basin margin.

Seismic Record

In a cross section from the gullied slopes to the basin interior, the 3.5 kHz seismic profiles show a constant succession of three echo types (Figs. 12, 13): The gullied slopes appear as a series of large, overlapping hyperbolae, many of them with conformable sub-bottom reflections. The basin margin shows a prolonged sea bed echo with occasional discontinuous and indistinct sub-bottom reflections. Towards the basin interior, the sea bed return becomes shorter and the opaque echo gives way to a layered echo with 20–35 millisec penetration and 4–6 parallel sub-bottom reflections.

Sediment Record

Samples from dredging, coring and submersible dives show the topographic boundary between slope and basin floor to be also a first order facies boundary. Because of their steepness, the gullied slopes are bypassed by most gravity flows and the sedimentary record consists of silty muds with thin strings of sand. The axes of gullies are filled by laminated sand and silt with layers of gravel and boulders and flanked by erosional cliffs of chalk and limestones that hardened at the sea floor (Schlager and James, 1978). On the flat basin floor accumulates a rhythmic sequence of graded turbidites and periplatform ooze. The facies of the clastic sediments gradually changes from the basin margin to the basin interior. Near the slope, the graded beds are coarser, thicker and occasionally mixed with very coarse, ungraded sands (fluxoturbidites), muddy layers of sand and gravel (debris-flow deposits) and slumped beds. Towards the basin interior, the beds become thinner, finer and are almost exclusively C–E or D–E turbidites with well developed lamination.

Relationship of Seismic and Sediment Record

The facies change from basin margin to basin interior is gradual, not nearly as abrupt as the change from slope to basin floor. However, it is accompanied by a rapid change from opaque to layered echo which corresponds closely to the basinward end of coarse sand in the Holocene turbidites (Fig. 13). We have used this change to define the boundary between basin margin and basin interior, because all available evidence suggests that the change in echo character is a direct response to the change of turbidite facies observed in the cores (see section on seismic record). The slope province owes its characteristic hyperbolic

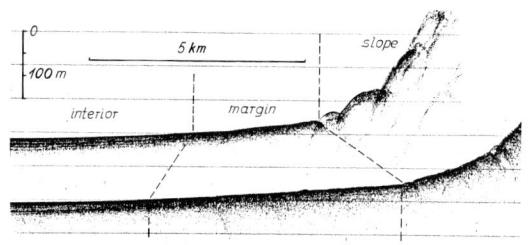

FIG. 12.—Slope-to-basin transition in 3.5 kHz profiles of southern TOTO with characteristic succession of hyperbolic, opaque and layered echos.

echo to the gully-and-ridge topography mentioned above. The muddy character of the slope sediment (only 3–5% sand layers as opposed to 18–20% in the basin) expresses itself in frequent conformable subbottom reflections. Zones with opaque return on the slope are gully floors with coarse sediment and ridges with lithified material.

GEOMETRY OF INDIVIDUAL TURBIDITES

We have used closely spaced gravity cores to trace individual sand layers in the basin. This has been done before in other Bahamian basins by Bornhold and Pilkey (1971); Sieglie and others (1976); Bennetts and Pilkey (1976); Ditty and others (1977). In TOTO, however previous attempts to correlate individual sand layers were not very successful (Rusnak and Nesteroff, 1964, p. 496; Pilkey and others, in press). We started our correlation program in the SW corner of the Cul de Sac of TOTO, in an area with only one or two sand layers in the top two meters of sediment. With a core spacing of ±1 km, the sequences in the cores were almost identical and the correlation obvious. Later, the spacing of the cores was increased to 2 km, then 4 and finally 10 km, and areas with more abundant sand layers were studied.

The following parameters were used for correlation:

1) sub-bottom depth of the turbidite, grain size, composition, and number of other turbidites above a particular layer;

2) characteristics of the mud intervals between the sand layers, in particular color bands produced by variations of organic content. The fine tails of turbidites display typical greenish colors because of a high content of organic matter, probably sea grass detritus from the platforms; in this way, the silt-mud tail of a turbidite could be traced beyond the range of the sand lobe.

3) the lithostratigraphy described above was finally checked by coccolith biostratigraphy. The top green interval and the upper part of the underlying white interval were consistently found to belong to the *Emiliania huxleyi* acme-zone of Gartner and Emiliani (1976) (0–65,000 years).

The sand turbidites correlated and traced in this way (Fig. 14) turned out to be tongue-shaped and vary in size from 150 to over 500 km^2, in volume from 10^6 to 10^8 m^3. The coarse proximal parts of the sand lobes are confined to the basin margin, medium to fine sand, silt and mud of the larger turbidites may be swept all the way across the basin.

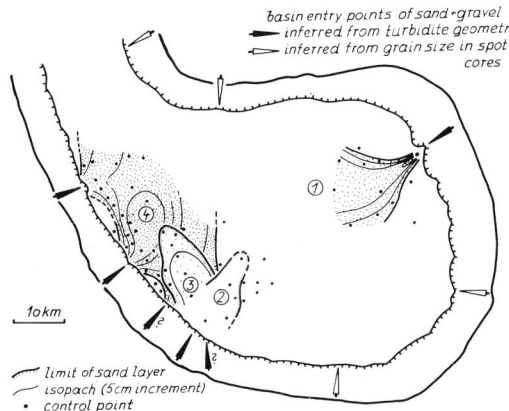

FIG. 14.—Geometry and grain size of some individual turbidites in top 2 m of sediment. Map shows extent of sand only; fine tails of turbidites reach much farther out into the basin. Layer 1 has clear point source with over 1 m of rubble at mouth of major gully; layers 2 and 3 are typical small sand lobes; layer 4 is probably a lateral amalgamation of several lobes built out simultaneously from different gullies. Notice that material for layers 1–4 was shed from at least 4, probably 6 small point sources. Several other point sources are indicated by local occurrence of coarse sediment near mouths of gullies. The basin margin facies is thought to consist of the coarse proximal parts of numerous overlapping and coalescing small sand lobes like the ones shown in this figure.

DISCUSSION

Concentric Facies Belts—Line Source vs. Point Source

The facies belts defined by autokinetic sediments form a remarkably regular concentric pattern, with facies boundaries running parallel to the margin of the platform. This concentric pattern including the basin-margin sand belt appears to illustrate a principle: A carbonate platform tends to act as a *line source* of sediment, because it produces and sheds debris all along its margin. The classical delta-canyon-fan system of siliciclastic sediments, on the other hand, is structured by major point sources. The slopes of a carbonate platform reinforce the line source effect because of their immature morphology of small but numerous, parallel gullies. Major canyons that would capture large volumes of sediment and funnel it to the basin are lacking. The system is illustrated schematically in Figure 15. The "line-source effect" can be noticed in other Bahamian basins as well: In Exuma Sound, gravity flows reached the basin through numerous entry points from all sides (Crevello, 1978); in Columbus and Hispaniola Basins of southern Bahamas, turbidites of about equal size have been shown to prograde from opposite flanks of the basin (Bornhold and Pilkey, 1971; Ditty and others, 1976).

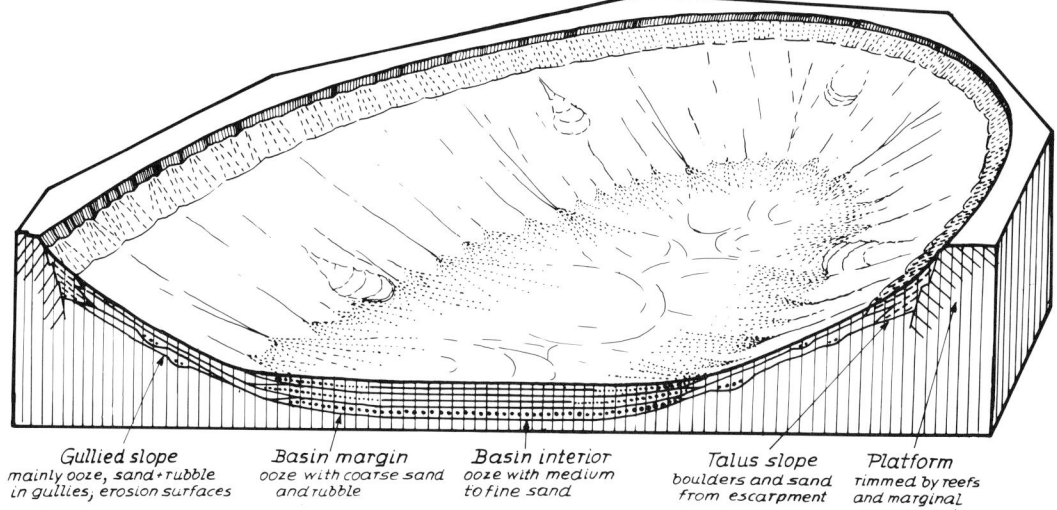

Fig. 15.—Schematic diagram of southern TOTO. Shallow water platforms on both sides provide sediment all along their margin ("line source"); turbidity currents originate on steep upper slope, bypass and erode gullied slopes and deposit coarse load in small, coalescing fans at basin margin, medium to fine sand and silt in overlapping lobes in basin interior; slumps move across toe-of-slope; large scale failure of slopes may create gravity flows that cover whole basin with debris sheets (front profile only; inferred from seismic records in TOTO and core studies in Exuma Sound by Crevello, 1978).

Gullied Slopes: Bypassing of Coarse Sediment

One of the most remarkable features in the platform-basin transition of southern TOTO is the zone of predominantly muddy sediment between the coarse reef talus and the turbidite sands of the basin floor (Fig. 15). This zone, the gullied slopes, extending from 400 to ca. 1300 m in depth, is characterized by erosion in the gullies and deposition of fine perennial sediment on ridges and swells. The change from erosion to deposition at the base of the gullied slope is obviously related to the change in slope gradient from 2–7° to 0.5° on the basin floor. The change from talus accumulation to erosion at the upper end of the gullied zone is probably due to a change in transport mechanism from rockfall at the foot of the reef wall to mass-flow and turbulent suspensions on the gullied slopes.

Bypassing of the gullied slopes is not complete. Sand and gravel accumulate locally in gullies, but these deposits are typically channeled, not sheetlike as the turbidites on the basin-floor and they consitute less than 1% of the total sediment cover as indicated by cores and submersible dives. Bypass slopes are typical for many platform flanks in the Bahamas, e.g., the flanks of Exuma Sound (Crevello, 1978), the Northeast Providence Channel and the Bahama escarpment ("erosional platform margins" of Mullins and Neumann (in press): they are also common amongst recent and ancient siliciclastic environments (e.g., Walker, 1971; Kelling and Stanley, 1976, p. 389) but they have not been described from the flanks of ancient carbonate platforms. Usually fore reef talus or shallow-water sands of ancient carbonate platforms pass gradually into basin deposits (e.g., Permian of the Delaware Basin, Newell and others, 1953; Late Paleozoic Dimple Limestone, Thomson and Thomasson, 1969; Late Paleozoic of Sverdrup Basin, Davies 1977; Triassic of the Dolomites, Bosellini and Rossi 1974; Cretaceous Golden Lane Platform, Enos, 1977.) An example for a fossil bypass slope is the Cambrian Cathedral Escarpment (McIlreath, 1977); however, this almost vertical cliff resembles more the reef wall of the Bahamas than the much gentler gullied slopes.

We can only speculate about the lack of good examples of ancient carbonate bypass slopes. Obviously, the height of a slope is important. If the Bahamian slopes were 300 m rather than over 1000 m high, tongues of fore reef talus would reach the floor in most basins. Furthermore, slope angle may be critical. Thus, the absence of a bypass-slope facies may often be the result of gentle slopes and low relief. However, certain fossil carbonate build-ups rose 1000 m or more above the basin floor, with slopes as steep or steeper than those in the Bahamas. There, we feel, the search for bypass slopes should be intensified. As criteria for this environment we suggest: Bioturbated muddy sediments, erosion surfaces including steep cliff-like hardgrounds, channel fills with lenses of coarse sand and gravel.

Vertical Variation of Facies Pattern

Numerous gravity cores and the opaque seismic echo that probably originates in the near-surface sediment, provide excellent control of the facies belts in the top 1.5–2 meters of sediment. This interval corresponds approximately to the latest Pleistocene—Holocene rise of sea level. Long piston cores indicate that the pattern of a basin-margin belt of more proximal turbidites and an interior zone of distal turbidites perisists to a sub-bottom depth of at least 10 m, that is Mid-Pleistocene (Gephyrocapsa oceanica-zone). The composition of the clastic layers varies slightly with pteropods and globigerinids being most abundant in what seem to be low-stand intervals.

Little is known about the frequency and distribution of thick debris sheets like those found in Exuma Sound (Crevello, 1978). In TOTO, no such deposit is within the reach of the piston corer. Likewise, Columbus Basin, Navidad Basin and Hispaniola-Caicos Basin in the southern Bahamas show only turbidites in the upper 10 m of the basin fill (Bornhold and Pilkey 1971; Bennetts and Pilkey, 1976; Sieglie and others, 1976).

Coarse debris-flow deposits, although less common than turbidites, may nonetheless be a normal part of the basin fill. We picture the typical sequence of Bahamian basin sediments to consist of perennial mud with several intercalations of graded sands (turbidites) per meter and beds of muddy rubble, several meters thick, only in intervals of tens or hundreds of meters. A similar rhythm can be observed in many ancient carbonate basins (e.g., Devonian of Alberta Basin, Cook et al, 1972; Permian of Delaware Basin, especially Rader and Lamar members of Bell Canyon formation, Newell et al, 1953, 48–94 and pers. obser.; Pennsylvanian to Permian of Sverdrup Basin, Davies 1977; Cretaceous adjacent to El Doctor platform, Mexico, Enos 1974, p. 806; Triassic Cassian Formation, Dolomites, Italy, Bosellini and Rossi 1974, 216 and pers. obser.). While the large mass-flows deserve attention as a potential reservoir facies in the basin fill, the small turbidites may be used to subdivide the basin facies.

Influence of Basin Size on Facies Pattern

Any attempt to use the facies pattern in modern Bahamian basins as a guide for the geologic record will have to consider size and relief of modern and ancient examples. The relief between platform and basin floor in the Bahamas is larger than

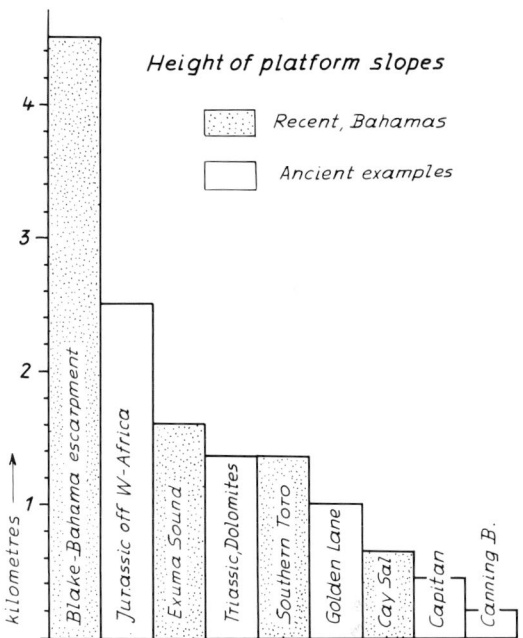

FIG. 16.—Height of Bahamian platform slopes compared with platform slopes in the geologic record (only particularly high ancient platforms have been considered, many ancient carbonate build-ups were surrounded by water less than 200 m deep). Sources of ancient examples: Jurassic off West Africa—Todd and Mitchum 1977; Triassic, Dolomites—Wilson, 1975, 238; Golden Lane (Cretaceous)—Enos, 1977, 286–305; Capitan (Permian)—Newell et al 1953; Canning Basin (Devonian)—Playford and Lowry 1966.

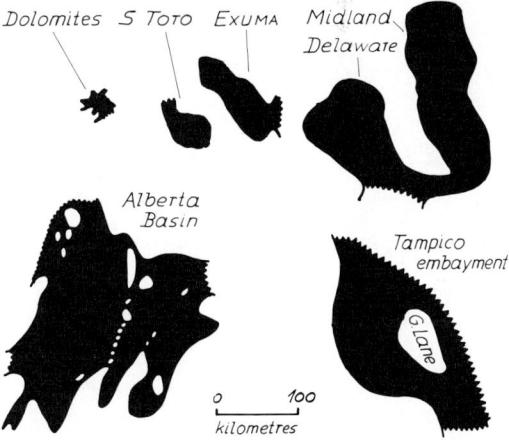

FIG. 17.—Size of Bahamian basins compared with ancient carbonate basins. Serrated edges—connections with open sea. Sources: Dolomites (Triassic)—Bosellini and Rossi, 1974, 230. Midland and Delaware Basin (Permian)—Wilson, 1975; Alberta Basin (Late Devonian)—Davies, 1975; Tampico Embayment (Cretaceous)—Enos 1977, Wilson 1975, 326.

in most ancient examples (Fig. 16). The surface area however is small compared to many ancient carbonate basins (Fig. 17). We have to keep this in mind, when we use the Bahamian troughs as modern analogs of ancient basins. Because of the narrowness of the Bahamian basins, we can expect to miss the more distal facies in the platform-basin transition, notably the gradation into truly pelagic deposits. Because of the high relief and the width of the basin flanks, on the other hand, the bypass slope is particularly well developed. Autochthonous platform sediments and redeposited neritic material in the basin may thus be more clearly separated than in many ancient examples of intra-cratonic basins.

CONCLUSIONS

Deepwater sediments beyond the bank edge in the Tongue of the Ocean are not uniform but can be subdivided into distinct facies belts. The subdivision is based on the principle that sediment from the banks is fractionated in size on its way into the basin: while mud simply augments the pelagic rain, sand and rubble are carried by gravity flows that vary in response to topography and distance from source. Thus, layers of coarse sediment provide the criteria to define facies belts such as the marginal escarpment with its coarse talus, the gullied bypass slope with only discontinuous layers of sand and rubble, and the basin floor with a rhythmic sequence of graded sands and mud. All these topographic provinces produce a sediment record that is distinct enough to recognize them if the basin had to be reconstructed from its sediment fill alone, without knowledge of topography.

The turbidites in the Bahamian basins are not associated with large deep sea fans. Rather, they form a continous belt of ephemeral sand lobes that coalesce and overlap. This reflects the nature of the source area. The platform delivers sediment all along its margin and channels it through numerous small gullies rather than a few large canyons. We are thus dealing with a line source and not a point source of sediment.

The 3.5 kHz seismic tool responds directly to changes in coarseness and thickness of sand layers in the upper 2 meters of sediment. Microtopography of the sea floor and grain size of carbonate ooze, elsewhere invoked to explain changes in echo character, do not vary to any significant degree and cannot be used to explain the changes of echo character in our case.

ACKNOWLEDGMENTS

We thank P. D. Crevello, H. T. Mullins, A. C. Neumann, and O. H. Pilkey for providing manuscripts in the early stages of preparation. Wally Charm made available to us unpublished data on TOTO sediments. P. A. Rona first drew our attention to the potential of the 3.5 kHz seismic tool in turbidite studies. R. N. Ginsburg reviewed the manuscript and helped greatly to clarify ideas and concepts presented here. We gladly acknowledge the cooperation of the skilled crews of DSRV *Alvin*, DSRV *Nekton*, RV *James M. Gillis* and RV *Columbus Iselin*. The senior author was supported by NSF Granst no. OCE74-22717 and OCE76-21932, the junior author by American Chemical Society PRF Grant 8066-AC2 and a fellowship by Gulf Oil Company.

REFERENCES

ANDREWS, J. E., SHEPARD, F. P., AND HURLEY, R. J., 1970, Great Bahama Canyon: Geol. Soc. America Bull., v. 81, p. 1061–1078.
ARRHENIUS, G. 1963, Pelagic sediments, *in* Hill, M.N., ed., The Sea: New York, Wiley, v. 3, p. 655–727.
ATHEARN, W. D., 1962a, Bathymetry of the Straits of Florida and the Bahama Islands. II. Bathymetry of the Tongue of the Ocean, Bahamas: Mar. Sci. Bull., v. 12: p. 365–377.
———, 1962b, Bathymetric and sediment survey of the Tongue of the Ocean, Bahamas. II. Bottom photographs: Massachusetts, Woods Hole Ocean. Inst., Ref. No. 62-27, p. 1–29.
BELDING, H. F., AND HOLLAND W. C., (compilors), 1970, Bathymetric maps, eastern continental margin, U.S.A. Sheet 2., Am. Assoc. Petroleum Geologists, Tulsa.
BENNETTS, K., AND PILKEY, O. H., 1976, Characteristics of three turbidites, Hispaniola-Caicos Basin: Geol. Soc. America Bull., v. 87: p. 1291–1300.
BENSON, W. E., SHERIDAN R. E., and others, 1978, Initial Reports of the Deep Sea Drilling Project, Washington, D. C., Govt. Print. Off., v. 44, p. 1–1005.
BERGER, W. H., 1974, Deep-sea sedimentation, *in* Burk, C. A., and Drake, C. L., (eds.). The Geology of Continental Margins: p. 213–241. Springer Verlag, New York.
———, AND MAYER, L. A., 1978, Deep-sea carbonates: Acoustic reflectors and lysocline fluctuations, Geol. v. 6: p. 11–15.
———, AND VONRAD, U., 1972, Cretaceous and Cenozoic Sediments from Atlantic Ocean, *in* Initial Reports of the Deep-sea Drilling Project: v. 14: p. 787–954. Govt. Printing Off., Washington.
BERNER, R. A., BERNER, E. K., AND KEIR, R. S., 1976, Aragonite dissolution on the Bermuda pedestal: its depth and geochemical significance: Earth Planet. Sci. Letters, v. 30: p. 169–178.
BORNHOLD, P., AND PILKEY, O. H., 1971, Bioclastic turbidite sedimentation in Columbus Basin, Bahamas: Geol. Soc. America, Bull., v. 82 p. 1314–1354.

BOSELLINI, A., AND ROSSI, D. 1974, Triassic carbonate buildups of the Dolomites, Northern Italy, in Laporte, L.F., ed., Reefs in time and space: Soc. Econ. Paleontologists Mineralogists Spec. Pub. v. 18, p. 209–233.
BRYAN, G., AND MARKL, R., 1966, Microtopography of the Blake-Bahama Region: Lamont-Doherty Geol. Observ., Tech. Rpt., 8, CU-8-66, Contr. Nobsr 85077, p. 1–26.
BURNE, R. V., 1974, The deposition of reef-derived sediment upon a bathyal slope, the deep off-reef environment, north of Discovery Bay, Jamaica: Mar. Geol., v. 16: p. 1–19.
BUSBY, R. F., 1962, Submarine geology of the Tongue of the Ocean, Bahamas: Washington, D.C., U.S. Naval Oceanogr. Office Tech. Rpt., 108: p. 1–84.
COOK, H. E., MCDANIEL, P. N., MOUNTJOY, E. W., AND PRAY, L. C., 1972, Allochthonous carbonate debris flows at Devonian bank ("reef") margins, Alberta, Canada: Canada. Petroleum Geologists Bull., v. 20, p. 439–497.
———, AND TAYLOR, M. E., 1977, Comparison of continental slope and shelf environments in the Upper Cambrian and lowest Ordovician of Nevada. in Cook, H. E., and Enos, P., Deep-water carbonate environments: Soc. Econ. Paleontologist Mineralogist Spec. Pub. No. 25, p. 51–81.
CREVELLO, P., 1978, Debris-flow deposits and turbidites in a modern carbonate basin, Exuma Sound, Bahamas [unpub. Masters thesis]: Univ. Miami, p. 1–133.
DAMUTH, J. E., 1975, Echo character of the western equatorial Atlantic floor and its relationship to the dispersal and distribution of terrigenous sediments: Mar. Geol., v. 18, p. 17–45.
———, AND HAYES, D. E., 1977, Echo character of the east Brazilian continental margin and its relationship to sedimentary processes: Mar. Geol., v. 24, p. 73–95.
DAVIES, G. R., 1975, Introduction, in Davies, G. R., ed., Devonian reef complexes of Canada, III–IX: Canadian Soc. Petroleum Geologists, Calgary.
———, 1977, Turbidites, debris sheets, and truncation structures in Upper Paleozoic deep-water carbonates of the Sverdrup Basin, Arctic Archipelago, in Cook, H. E., and Enos, P., Deep-water carbonate environments: Soc. Econ. Paleontologist Mineralogist Spec. Pub. No. 25, p. 221–247.
DAVIES, T. A., AND GORSLINE, D. S., 1976, Oceanic sediments and sedimentary processes, in Riley, J. P., and Chester, R., Chemical Oceanography, (2nd Ed.): London, Academic Press, v. 5, p. 1–80.
DITTY, P. S., HARMON, C. J., PILKEY, O. H., BALL, M. M., AND RICHARDSON, E. S., 1977, Mixed terrigenous-carbonate sedimentation in the Hispaniola-Caicos turbidite basin: Mar. Geol., v. 24, p. 1–20.
ENOS, P., 1974, Reefs, platforms, and basins of Middle Cretaceous in northeast Mexico: Am. Assoc. Petroleum Geologists Bull., v. 58, p. 800–809.
———, 1977, Tamabra limestone of the Poza Rica trend, Cretaceous, Mexico, in Cook, H. E., and Enos, P., Deep-water carbonate environments: Soc. Econ. Paleontologist Mineralogist Spec. Pub. No 25, p. 273–314.
EWING, M., EMBLEY, R. W., AND SHIPLEY, T. H., 1973, Observations of shallow layering utilizing the Pinger Probe Echo Sounding System: Mar. Geol., v. 14, p. 55–63.
EWING, J. I., EWING, M., AND LEYDEN, R., 1966, Seismic profiler survey of Blake Plateau: Am. Assoc. Petroleum Geologists Bull., v. 50, p. 1948–1971.
FRIEDMAN, G. M., 1965, Occurrence and stability relationships of aragonite, high-magnesian calcite and low-magnesian calcite under deep-sea conditions: Geol. Soc. America Bull., v. 76, p. 1191–1196.
GARRISON, R. E., SCHLANGER, S. O., AND WACHS, D., 1975, Petrology and paleogeographic significance of Tertiary nannoplankton-foraminiferal limestones, Guam: Palaeogeogr. Palaeoclim., Palaeoecol., v. 17, p. 49–64.
GARTNER, S., AND EMILIANI, E., 1976, Nannofossil biostratigraphy and climate stages of Pleistocene Brunhes epoch: Am. Assoc. Petroleum Geologists Bull., v. 60, p. 1562–1564.
GIBSON, T. G., AND SCHLEE, J., 1967, Sediments and fossiliferous rocks from the eastern side of the Tongue of the Ocean, Bahamas: Deep-Sea Research, v. 14, p. 691–702.
GOREAU, T. F., AND LAND, L. S., 1974, Fore-reef morphology and depositional processes, North Jamaica, in L.F. Laporte, ed., Reefs in time and space: Soc. Econ. Paleontologist Mineralogists Spec. Pub. No. 18, p. 77–89.
HAMILTON, E. L., 1977, Sound velocity-density relations in sea-floor sediments and rocks: Jour. Acoust. Soc. America v. 63, p. 366–377.
HESS, H. H., 1960, The Origin of the Tougue of the Ocean and other great valleys of the Bahama Banks: 2nd Carribbean Geol. Conference Trans. Puerto Rico, p. 160–161.
HOLLISTER, C. D., EWING, J. I., AND OTHERS. 1972, Initial reports of the Deep-Sea Drilling Project: Washington, D.C., v. XI, p. 1–1077.
HOOKE, R. L., AND SCHLAGER, W., Geomorphic evolution of the Tongue of the Ocean and the Providence Channels, Bahamas: Mar. Geol., (in prep.).
HORN, D. R., HORN, B. M., AND DELACH, M. N., 1968, Correlation between acoustical and other physical properties of deep-sea cores: Jour. Geophysical Res., v. 73, p. 1939–1956.
HURLEY, R. J., SIEGLER, V. B., AND FINK, K. L., 1962, Bathymetry of the Straits of Florida and the Bahama Islands. I. Northern Straits of Florida: Marine Sci. Bull., v. 12, p. 313–321.
JAMES, N. P., AND GINSBURG, R. N., The deep seaward margin of Belize barrier and atoll reefs: Int. Assoc. Sedimentologists Spec. Pub. No. 3, (in press).
KELLING, G., AND STANLEY, D. J., 1976, Sedimentation on canyon slope and base-of-slope environments, in Stanley, J. D., and Swift, D. J. P., Marine sediment transport and environmental management: New York, Wiley, p. 379–635.
KIER, J. S., AND PILKEY, O. H., 1971, The influence of sea level changes on sediment carbonate mineralogy,

Tongue of the Ocean, Bahamas: Mar. Geol., v. 11, p. 189–200.
KING, P. B. (compilor), 1969, Tectonic map of North America: Washington, D.C., U.S. Geol. Survey
KOCZY, F. F., CHEW, F., FEINSTEIN, A., RHIAN, E., RICHARD, J. D., SIEGLER, V. B., AND WENNEKENS, M. P., 1958, Oceanographic survey of the Tongue of the Ocean: Univ. Miami Marine Lab. Tech. Rpt., p. 1–161.
LYNTS, G. W., JUDD, J. B., AND STEHMAN, C. F., 1973, Late Pleistocene history of Tongue of the Ocean, Bahamas: Geol. Soc. America Bull., v. 84, p. 2665–2684.
MAYER, L., 1978, High-resolution acoustic stratigraphy in carbonates: EOS Trans. American Geophysical Union Abs., v. 59, p. 299.
MCILREATH, I. A., 1977, Accumulation of a Middle Cambrian, deep-water limestone debris apron adjacent to a vertical, sub-marine carbonate escarpment, southern Rocky Mountains, Canada: *in* Cook, H. E., and Enos, P., Deep-water carbonate environemnts: Soc. Econ. Paleontology Mineralogy Spec. Pub. No. 25, p. 113–124.
MIDDLETON, G. V., AND HAMPTON, M. A., 1976, Subaqueous sediment transport and deposition by sediment gravity flows, *in* Stanley, D. J., and Swift, D. J. P., Marine Sediment Transport and Environmental Management: New York, Wiley, p. 197–218.
MOJSISOVICS, E., 1879, Die Dolomit-Riffe von Suedtirol und Venetien: Vienna, A. Hoelder, p. 1–552.
MULLINS, H. T., AND NEUMANN, A. C., in press, Deep carbonate bank margin structure and sedimentation in the northern Bahamas; *in* Doyle, L. and Pilkey, O. H., eds, Geology of continental slopes: Soc. Econ. Paleontologists Mineralogists Spec. Pub. No. 27.
———, BOARDMAN, M. R., AND NEUMANN, A. C., In press, Echo character of off-platform carbonates, Northwest Providence Channel, Bahamas: Mar. Geol.
NEWELL, N. E., RIGBY, J. K., FISCHER, A. G., WHITEMAN, A. J., HICKOX, J. E., AND BRADLEY, J. S. 1953, The Permian Reef Complex of the Guadalupe Mountains region, Texas and New Mexico: San Francisco, Freeman and Company, p. 1–236.
NORMARK, W. R., AND DICKSON, F. H., 1976, Sublacustrine fan morphology in Lake Superior: Am. Assoc. Petroleum Geologists Bull., v. 60, p. 1021–1036.
PILKEY, O. H., LOCKER, S. D., AND CLEARY, W. J., in prep., Comparison of sand layer geometry on the flat floors of ten modern depositional basins.
PLAYFORD, P. E., AND LOWRY, D. C., 1966, Devonian reef complexes of the Canning Basin, Western Australia: Geological Survey Western Australia Bull., v. 118, p. 1–50.
ROBERTS, H. H., HUH, O. K., AND ROUSE, L. J., In prep., Generation of hyperpycnal conditions, density cascading, and sediment transport off shallow Banks and shelves.
RUSNAK, G. A., AND NESTOROFF, W. D., 1964, Modern Turbidites, terrigenous abyssal plains versus bioclastic basin, *in* Miller, R. L. Papers in marine geology: New York, London, p. 488–507.
SANDERS, J. E., 1965, Primary sedimentary structures formed by turbidity currents and related resedimentation mechanisms, *in* Middleton, G. V., ed., Primary sedimentary structures and their hydrodynamic interpretation: Soc. Econ. Paleontology Mineralogy Spec. Pub. No. 12, p. 192–219.
SCHLAGER, W., HOOKE, R. L., AND JAMES, N. P., 1976, Episodic erosion and deposition in the Tongue of the Ocean. Bahamas: Geol. Soc. America Bull., v. 87, p. 1115–1118.
———, AND JAMES, N. P., 1978, Low-magnesian calcite limestones forming at the deep-sea floor, Tongue of the Ocean, Bahamas: Sedimentology, v. 25, p. 675–702.
SCHREIBER, B. C., 1968, Sound velocity in deep sea sediments: Jour. Geophysical Res., v. 73, p. 1259–1268.
SHIPLEY, T. H., BUFFLER, R. T., AND WATKINS, J. S., 1978, Seismic stratigraphy and geologic history of Blake Plateau and adjacent Western Atlantic continental margin: Am. Assoc. Petroleum Geologists Bull., v. 62, p. 792–812.
SIEGLIE, G. A., FROELICH, P. N., AND PILKEY, O. H., 1976, Deep-sea sediments of Navidad Basin, correlation of sand layers: Deep-Sea Res., v. 23, p. 89–101.
SUPKO, P. R., 1963, A quantitative X-ray diffraction method for the mineralogical analysis of carbonate sediments from Tongue of the Ocean [Thesis]: Miami, Univ. Miami, p. 1–158.
TODD, R. G., AND MITCHUM, R. M., 1977, Seismic stratigraphy and global changes of sea level, part 8: Identification of Upper Triassic, Jurassic and Lower Cretaceous seismic sequences in Gulf of Mexico and offshore West Africa, *in* Payton, C. E., ed. Seismic stratigraphy—applications to hydrocarbon exploration: Am. Assoc. Petroleum Geologists Mem. 26, p. 145–163.
WALKER, R., 1971, Nondeltaic depositional environments in the Catskill clastic wedge (Upper Devonian) of Central Pennsylvania: Geol. Soc. America Bull., v. 82, p. 1305–1326.
WILSON, J. L., 1975, Carbonate Facies in Geologic History: New York, Springer-Verlag, p. 1–471.

SANTA CRUZ BASIN, CALIFORNIA BORDERLAND: DOMINANCE OF SLOPE PROCESSES IN BASIN SEDIMENTATION

T. R. NARDIN, B. D. EDWARDS, AND D. S. GORSLINE
Department of Geological Sciences
University of Southern California
Los Angeles, California 90007

ABSTRACT

Quaternary sedimentation in Santa Cruz Basin, California Borderland, has been dominated by gravity processes which are locally exemplified by a downslope progression from slides through a sequence of mass flows. Transport processes have been inferred from (1) the morphology and acoustic characteristics of these deposits as interpreted from seismic-reflection records, (2) from sedimentary characteristics observed in box and piston cores, and (3) from bottom photography. Four types of gravity deposits have been recognized acoustically. In the first type, slides, failure took place elastically and only minor internal deformation occurred. In the remaining three gravity deposits, to which we have assigned the designations types "A," "B," and "C" until the details of the processes are better known, the sediment masses were emplaced plastically albeit with different viscosities and strengths. The A, B, and C types are characterized respectively by hyperbolic reflections, an acoustically transparent lense-shaped mass, and thin units which have blunt-nosed terminations. It is also suggested that the surficial meter or two of slope sediments which possess high water content are shed as low viscosity flows and perhaps as true turbidity currents at the moment of slide activation. Thus, although gravity processes are gradational, the resulting deposits are relatively sharply defined by sets of acoustic signatures. The presence of all four types may typify slope areas where the full spectrum of mass-transport processes have been operative. Partial sequences would indicate slopes where sediment properties, slope gradient or barriers inhibit the complete development.

INTRODUCTION

History of Mass Movement Studies

Although geologists have long recognized mass movement as an important transport mechanism for coarse sediments in the deep sea (e.g., Twenhofel, 1926), Daly's (1936) hypothesis for the formation of submarine canyons, and the experimental work of Kuenen (1937, 1951) shifted research emphasis to turbidity currents and their deposits. The consequent search for turbidites in the stratigraphic record and for contemporary models in the oceans became a dominant theme in marine geology and rapidly lead to formulations of increasingly more sophisticated turbidite facies models (e.g., Kuenen and Migliorina, 1950; Gorsline and Emery, 1959; Bouma, 1962; Walker, 1967, 1973; Mutti and Ricci-Lucchi, 1972). Mass movement was relegated to the secondary role of a generating mechanism for the turbidity currents (e.g., Heezen and Ewing, 1952; Heezen and Drake, 1964). In this regard it is interesting to note that turbidites are commonly assumed to be volumetrically more important than other gravity deposits. This infers that most of the displaced mass is transformed into a turbulent suspension, a premise that is difficult to accept when the mechanics of mass transport are considered (e.g., Dott, 1963; Morgenstern, 1967; Allen, 1971; Hampton, 1972).

In the past decade significant advances have been made in our appreciation of the role of mass movement and in our understanding of the processes although much needs to be done experimentally. In particular, depositional conditions which lead to slope failure, the behaviour of sediments in transit, and the sediment-ambient fluid mixing have received a great deal of attention (Bagnold, 1954, 1956; Kuenen, 1955; Terzagni, 1956; Moore, 1961, 1977; Dott, 1963; Morgenstern, 1967; van der Knapp and Eijpe, 1967; Allen, 1971; Fisher, 1971; Hampton, 1972; Rodine and Johnson, 1976). Shear strength tests of slope sediments (Moore, 1961; Morelock, 1964) have shown that sedimentation rate, slope gradient and sediment bulk properties are interrelated and determine the conditions of stability, metastability and instability on contemporary slopes (Lewis, 1971; Haner and Gorsline, 1978). For example, fine sediments deposited at slow rates on relatively steep slopes may be stable but the same sediment in areas of high accumulation on gentle slopes may fail (Coleman, 1976; Coleman and Prior, 1978).

It is generally recognized that a spectrum of emplacement processes exists ranging from slides through mass flows to turbidity currents. Ideally, these evolve from each other as sequences having depositional characteristics that may be stratigraphically useful in identifying the various stages (Dott, 1963; Morgenstern, 1967; Hampton, 1972;

Carter, 1975; Jacobi, 1976; Middleton and Hampton, 1973; Kelling and Stanley, 1976; Embley, 1976; Stanley et al, 1978; Walker, 1978; Lowe, 1976a, 1976b, and in review). The problems concerning sedimentary and deformational structures and their rheologic inferences, as well as facies relations, have been addressed by Crowell (1957), Sanders (1965), Cook et al (1972), and Carter (1975) among others. Despite these prior studies and the advantages accruing from the study of subaerial analogs (Sharpe, 1938; Eckel, 1958; Zaruba and Mencl, 1969; Johnson, 1970; Hsü, 1975; Coates, 1977) modern examples of submarine slope complexes which illustrate the spectrum of gravity deposits and their evolution during mobilization are rare (e.g., Hyne et al, 1975; Jacobi, 1976; Embley, 1978; Haner and Gorsline, 1978). Ancient examples, however, have been documented by numerous workers, including Morris (1971), Cook et al (1972) Klein et al (1972) Lewis (1976), Keith and Friedman (1977), Shearman (1976), and Stanley et al (1978). Thus our knowledge of slope processes and deposits is still a fraction of those describing models for turbidite systems. A number of workers as noted are turning to examination of mass movement and we are on the threshold of a period of renewed emphasis on these processes (see Emery and Uchupi, 1972; Stanley and Kelling, 1978).

In this paper we will draw from field studies in Santa Cruz Basin, California Borderland, for an example of sequences of mass movement deposits reflecting the operation of an ordered spectrum of gravity processes. The data forming the basis of the paper come from seismic-reflection profiling (3.5 KHz and air gun), radiographs of piston cores collected by Barnes (1970), box coring and bottom photography.

BORDERLAND

The general dimensions, characteristics and geologic history of the California Continental Borderland have been described by Shepard and Emery (1941), Emery (1960), Gorsline and Emery (1959), Moore (1969), Vedder et al (1974) and Howell (1976). Our area of interest is Santa Cruz Basin (Barnes, 1970) which lies along the east flank of one of several large northwest-trending *en echelon* anticlinoria which are reflected in the bathymetry of the Santa Rosa-Cortes Ridge. Its northern limit is formed by the Santa Monica-Santa Cruz Fault System, the seismically active east-west boundary between the Peninsular and Transverse Ranges. To the southeast the basin is closed by its deepest sill which is probably of fault origin. Deep water representing the North Pacific Intermediate Water of low oxygen content flows over this sill from the south. Surface circulation is dominated by a major area of upwelling

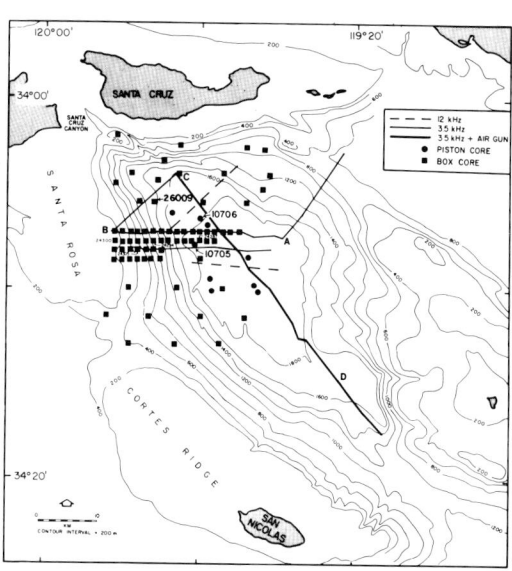

FIG. 1.—The bathymetry of Santa Cruz Basin. Seismic-reflection tracklines and core and bottom photography stations are also shown by symbols explained on the figure.

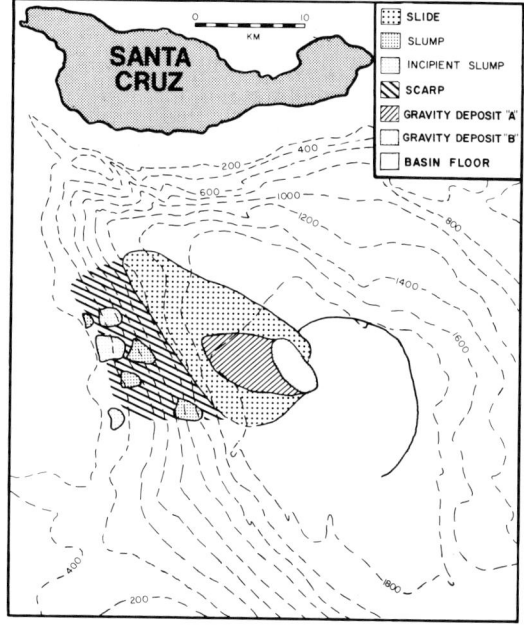

FIG. 2.—Areas covered by different mass movement types in the northwestern quadrant of Santa Cruz Basin. Slumps incipient slumps were identified by Greene et al (1975).

generated by topographic effects and wind stress (Reid, et al, 1958).

Submarine canyons enter the basin from the channel between Santa Rosa and Santa Cruz Islands, the southern sill, and the flanks of the San Nicolas and Santa Barbara Insular shelves. The Santa Cruz Canyon is active (Felsher, 1971) and there is evidence of activity associated with the southern canyons although the major turbidity currents have been rare since the last rise in sea level about 12,000 years B.P. (Barnes and Gorsline, 1972).

GEOPHYSICAL EVIDENCE

Seismic-reflection records show that mass movement has long been an important depositional process along the periphery of Santa Cruz Basin (also see Barnes, 1970). This study focuses on a region of slope instability located in the northwest quadrant of the basin between Santa Cruz Submarine Canyon and the sill separating northern Santa Rosa-Cortes Ridge from the San Nicolas Insular Shelf (Figs. 1, 2).

Seismic-reflection data indicate an average gradient of 9.5° for the upper slope between the 300 to 1300 m isobaths. The upper slope, for the most part, is interpreted as a zone of erosion (area = 100 km^2) produced by multiple episodes of mass movement. High resolution (3.5 kHz) profiles indicate that undisturbed sediments thicker than 1 m the limit of acoustic resolution, form patchy surfaces only (Fig. 3); however, a number of slumps and metastable masses (incipient slumps) have been identified by Greene et al, 1975). In areas of the slope below 800 m where the unconsolidated sediment cover cannot be resolved acoustically, box coring revelas a minimum of 50 cm of soft mud. In contrast, the lower slope (1300 to 1850 m) has an average gradient of about 2° and is an area of mass deposition covering approximately 150 km^2.

Gravity deposits on the lower slope locally can

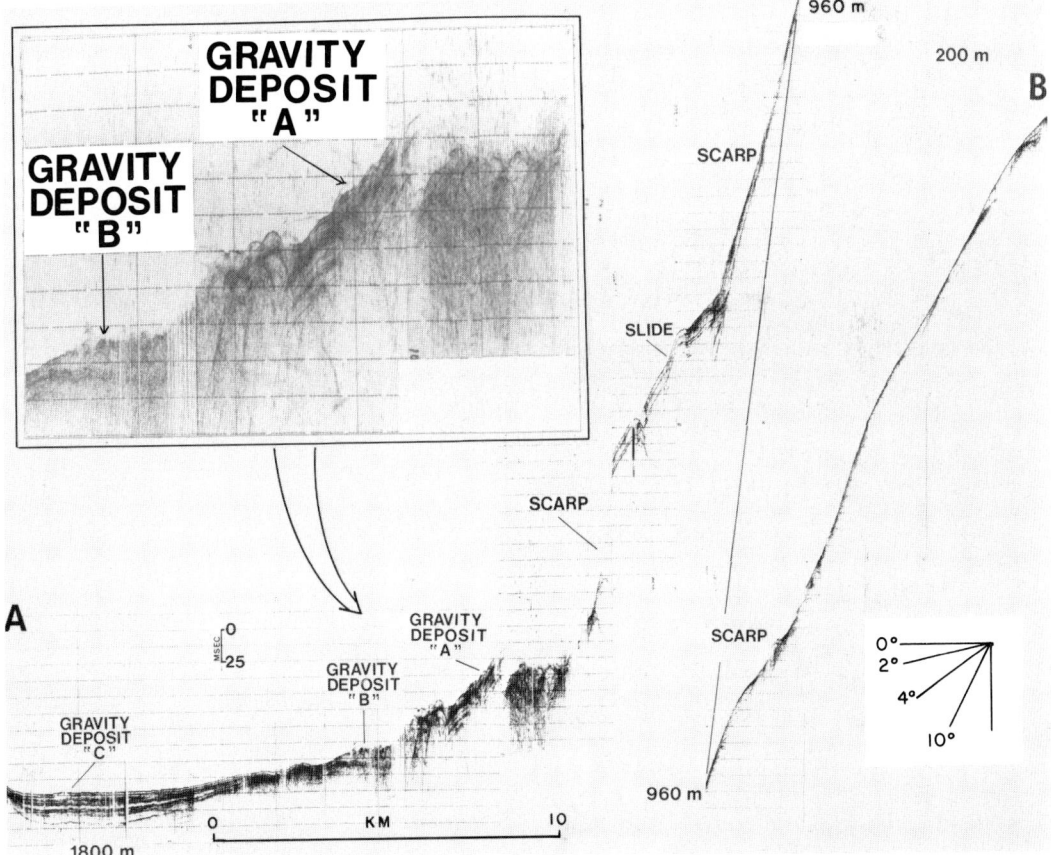

FIG. 3.—High resolution (3.5 kHz) profile A-B illustrating slide scarp and sequence of gravity deposits ranging from slides through various mass flow types of decreasing strength and viscosity. Travel time is one-way.

be classified into four types which are distinguishable on the basis of acoustic character, geometry, gradient, and spatial relationship to adjacent deposits. Although specific rheologic models (other than elastic or plastic) have not been assigned to the transport processes responsible for the deposits, it is thought that the deposits reflect a mechanical continuum of decreasing viscosity and strength in the downslope direction. During fractionation critical values cause abrupt changes in the actual deposit laid down thus producing a limited sequence of discrete sediment bodies.

Beginning at the top of the lower slope, line A-B (Fig. 3) depicts the first of these types, the slide, in the 1300–1460 m depth interval where the average slope is 3.7°. Slide deposits are characterized by a surface which is produced either by gentle compressional features or tension fractures where individual slide blocks have pulled apart. Although the slip planes are difficult to identify, the presence of well-defined continuous internal reflectors, only mildly deformed, shows that the displaced material failed elastically.

The second type in order of increasing depth (1460–1640 m) is distinguished acoustically by diffraction hyperbolae which reflect a highly irregular seafloor topography (gravity deposit A). The zone of hyperbolae is bounded upslope by a scarp (1460–1640 m) with a gradient of about 8°. The scarp cuts slide deposits and provides evidence for an episode of remobilization of the lower part of the original slide mass. Similar acoustic signatures are evident in seismic-reflection records of mass movements in Lake Tahoe, on the Atlantic slope near Senegal, on the continental rise off Brazil, and in the Aleutian Trench (Hyne et al, 1973; Piper et al, 1973; Damuth, 1975; Jacobi, 1976). Jacobi (1976) suggests that the hyperbolae in records from the Senegal slope represent deposits sufficiently strong to form either blocky terrane or sediment piles which have undergone more mixing than slide deposits but less than lower viscosity flows. Angular folds in piston cores collected by Jacobi (1976) are consistent with this interpretation. Lake Tahoe piston cores from sediment mounds interpreted by Hyne et al (1973) as dewatering centers contain pebbly muds which are usually associated with debris flows. Debris flows are thought to behave as highly concentrated dispersions and exhibit the properties of plastico-viscous (Coulomb-viscous

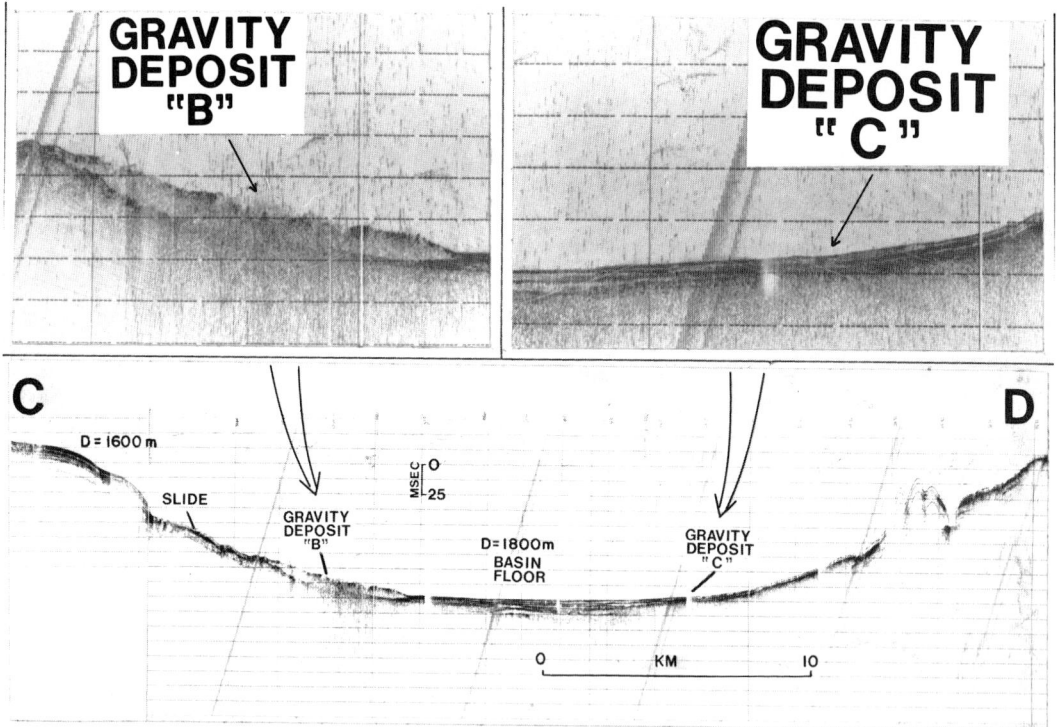

FIG. 4.—High resolution (3.5 kHz) profile C-D showing slides, the acoustically transparent lense at the base of the slope, and thin mass flow with blunt nose on the basin floor. Origin of hyperbolic reflections at southern end of profile is unknown. Travel time is one-way.

or Bingham) or quasiplastic materials all of which possess yield strengths (Johnson, 1970; Fisher, 1971; Hampton, 1972). Thus, hyperbolae associated with slope deposits have no unique rheologic interpretation but rather may reflect a variety of both elastic and plastic emplacement mechanisms.

The third mass-transport type, gravity deposit B (Figs. 3, 4), is represented by an acoustically transparent lense-shaped deposit which is up to 15 m in thickness. The lense appears to be laterally continuous with gravity deposit A. In high resolution profiles the seafloor formed by the lense appears as a prolonged reflector. The deposit lies at the base of the slope apron (1800–1840 m) where the gradient is less than 1°. It overlies gravity displaced deposits as well as ponded basin floor deposits. Deformation of the basin floor sediments (Fig. 3) indicates that there has been continued movement of the older gravity deposits at the base of the slope. The lense in turn is onlapped locally by 1 to 2 m of basin plain (hemipelagic) sediments. This together with rates of Holocene sedimentation in Santa Cruz Basin can be used to approximate the age of the mass movement. Following the precepts outlined by Gorsline et al (1968) for the California Borderland, Barnes (1970) has developed a temporal stratigraphy for Santa Cruz Basin based on vertical fluctuations in carbonate content. By this method, sedimentation rates range from 30 mg/cm^2/yr in the central basin areas to 9 mg/cm^2/yr on the basin flanks. Assuming an average bulk density of 1.2 gm/cm^3, rates of total sedimentation would be 0.8 mm/yr and 0.2 mm/yr respectively, which are in accord with estimates of 32 mg/cm^2/yr (Emery, 1960) and 0.4 mm/yr (Emery and Bray, 1962). These data suggest that the emplacement of the lense has probably occurred within the past few thousands of years (early Holocene or late Pleistocene) and may coincide with a time of lower sea level and higher slope sedimentation rates. The lense covers about 10% of the total area affected by mass movement and has a volume of 2.3×10^8 m^3 which is comparable to major slides reported by Menard (1964), Lewis (1971), Normark (1974), and Molnia et al (1977). Acoustically similar deposits occur along the Atlantic continental margins (Damuth, 1975; Embley, 1976; Jacobi, 1976) where locally cores reveal pebbly muds, angular contacts and flow folds. These deposits have been interpreted as debris flows by Embley (1976) and

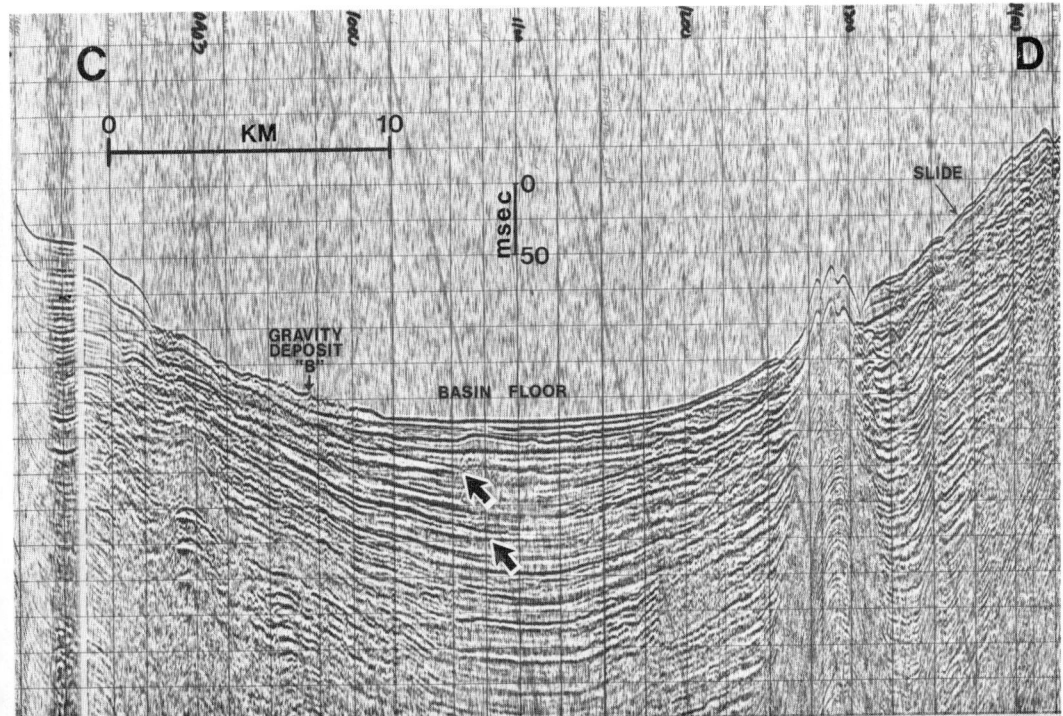

FIG. 5.—Airgun (10 in. 3) profiling transect C-D was taken simultaneous with the high resolution profile in Figure 3. Intercalations of base-of-slope gravity deposits (shown by arrows) and basin floor sediments is illustrated. Travel time is one-way.

Jacobi (1976). Using cores alone they are difficult to distinguish from the deposits associated with the diffraction hyperbolae. However, as Jacobi (1976) has suggested, the smoother surfaces of the lenses indicates that they possessed considerably less strength during emplacement. The prolonged bottom echo is attributed to seafloor irregularities too small to be resolved into discrete hyperbolae in 3.5 kHz records. The absence of subbottom reflectors within the lense may be due either to resolution limitations or preferably to high water content or deformational homogenization of the sediment mass. Damuth (1975) has correlated the prolonged echo characteristic with deposits which contain bedded, coarse, terrigenous sediment although the relationship between the two is unclear. Without piston cores showing pebbly muds we are reluctant to attribute the emplacement of the lense in Santa Cruz Basin to debris flowage although rheologically it may have behaved in a similar fashion. The Santa Cruz Basin deposit appears to have been emplaced as a sheet flow rather than as a channelized flow (Fig. 2), an important distinction in view of flow mechanics (Johnson, 1970; Lowe, 1976b) and Allen's (1971) hypothesis that sheet flows have less strength than channelized flows due to their greater efficiency in incorporating water during movement. Examination of an air gun profile (10 in^3 gun chamber) penetrating approximately 400 m of the Quaternary section (Fig. 5) reveals similarly shaped deposits intercalated with basin floor sediments. Therefore, mass emplacement, alternating through time with hemipelagic sedimentation, has been a common mode of sediment transport at the base of Santa Cruz slope for much of the Pleistocene.

In places, high-resolution records reveal deposits approximately 3 to 4 m thick which extend from the base of the slope out over the basin floor. This fourth type, gravity deposit C, typically appears to have possessed sufficient strength to rest slightly above the general level of the basin plain and terminate in blunt snouts (Figs. 3, 4) reminiscent of the debris flows described by Johnson (1970) and Cook et al (1972). This together with their apparent continuity with other gravity deposits indicates that they were emplaced in a high concentration laminar mode. The ability of a sediment mass to flow is a partial function of slope and density (or concentration) as well as thickness (Hampton, 1972), and thus the thin flows must have been markedly less viscous than the gravity deposits upslope. Survey coverage of these deposits is not sufficiently detailed to determine the shape of the deposits in plan view but their lateral extent on individual profiles suggests that they too occurred as sheet flows. The flow shown in profile A-B (Fig. 3) appears to have travelled approximately 8 km or 3/4 of the width of the basin floor. Much of the acoustic stratification beneath the basin floor (Figs. 3, 4), therefore, could be due to thin mass flows in addition to the alternation of turbidites and hemipelagic deposits.

As pointed out by Embley (1976) a brief survey of the literature which contains seismic-reflection records depicting mass movement reveals possible examples of mass flow deposits (e.g., Moore et al, 1970; Bartolini et al, 1972; Normark, 1974) that were not initially recognized as such. Submarine mass flows have also been discussed by Lehner (1969), Stanley and Silverberg (1969), Wilhelm and Ewing (1972), Stanley (1974) and Moore et al (1976) among others. These examples plus others taken from the stratigraphic record (e.g., Morris, 1971; Hendrey, 1972; Klein et al, 1972; Keith and Friedman, 1977; Stanley et al, 1978) suggest that mass flow, a common subareal process, also frequently occurs during submarine slope failure and is an important mechanism of emplacement, particularly near the base of the slope (Stanley and Unrug, 1972).

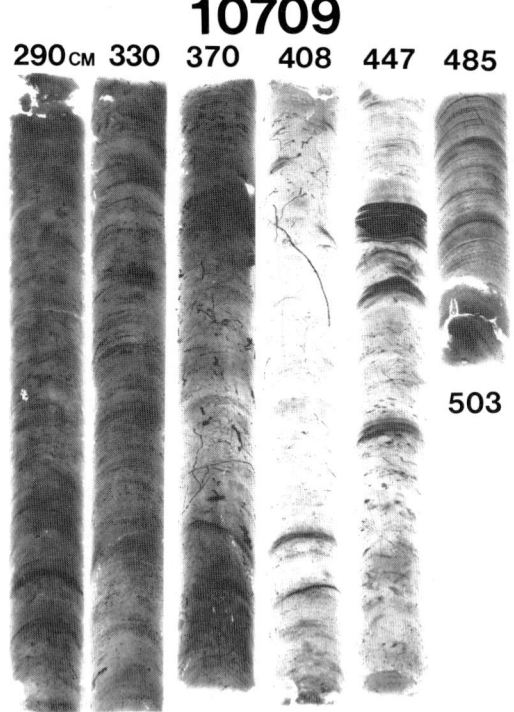

Fig. 6.—X-radiograph print of piston core AHF 10709 showing well-stratified undeformed sediments.

SEDIMENTOLOGIC EVIDENCE

Although not apparent in the seismic-reflection records, thin flows also occur in the region of general slide deposition. Figures 6 and 7 show radiographs of piston cores taken within the area of mass deposition. Core 10705 is located near the boundary between the slide deposit and gravity deposit A and shows well-stratified, internally undisturbed sediments and evidences elastic deformation in this area. Since some large masses of slope sediments move as units it is expected that random coring will not always show clear evidence of plastic deformation. However, core 10706, located in the northern portion of the slide area and well away from gravity deposits A and B shows rounded clasts, angular contacts and highly disturbed sediments (82–275 cm) sandwiched between well-stratified layers which include thin turbidites. The 1.9 meters of deformed sediments is interpreted to have once been at the surface of a slide deposit. As partially compacted sediments below moved more rigidly in the slide block, the top meter or two of sediment

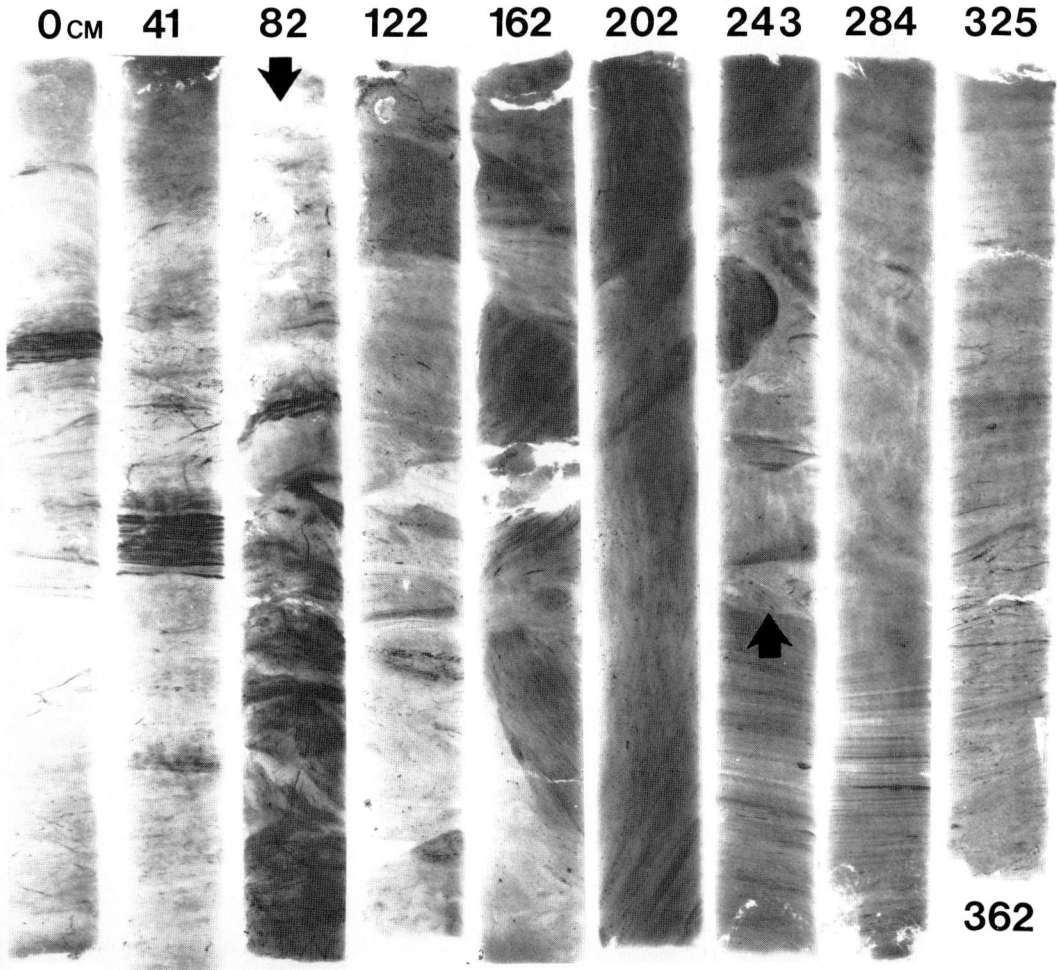

Fig. 7.—X-radiograph print of piston core AHF 10706 showing the zone of plastic deformation (bounded by arrows) between undisturbed laminated layers.

which probably possessed high water contents (contemporary basin floor sediments contain up to 70% or more in terms of wet weight) flowed across the moving block's surface when the slide activated.

Similarly, a radiograph of box core 26009 (Fig. 8) also located in the area of slide deposit, shows pebbly muds at the base of a "swirl" patterned matrix which is in abrupt contact with underlying homogeneous muds. In other box cores taken from the lower slope the sediments appear homogeneous throughout but with a distinctive swirl overprint. The clasts producing the pebbly mud texture in core 26009 appear similar to the underlying sediment suggesting that they were eroded and torn loose as the upper mass moved over the original surface. The knife-edge contact and scours between the two horizons argues for an erosive nature of emplacement.

In contrast, sediments of the basin floor typically exhibit a homogeneous character with indistinct burrow mottling although some show remnants of the swirl signature. More frequently, however, basin floor sediments are interrupted by thin turbidites (1-2 cm thick) dominated by very fine sand size carbonates swept from the ridge crest and insular shelves. Typical box core penetration in the muddy sediments of Santa Cruz Basin is of the order of 40 cm. Accepting the sedimentation rates of Barnes (1970), such cores would record 500 to 2000 years of record.

Textural analyses of surface sediments (0-5 cm) recovered by box coring shows a general fining-offshore trend (Fig. 9). The ridge crest is covered with moderately sorted fine to coarse sands with localized patches of cobbles, boulders and rock outcroppings (Fig. 10A). The upper slope (Fig. 10B) is a very poorly sorted coarse to medium silt which grades through poorly sorted fine silts on the lower slope (Fig. 10C, D) to hemipelagic very fine silts and clays on the basin floor (Fig. 10E; Edwards, 1977).

This general trend is broken along the east flank of the Santa Rosa-Cortes Ridge by a series of dislocations of relatively coarse sediment with variations in mean grain size of up to 1.4 ϕ units,

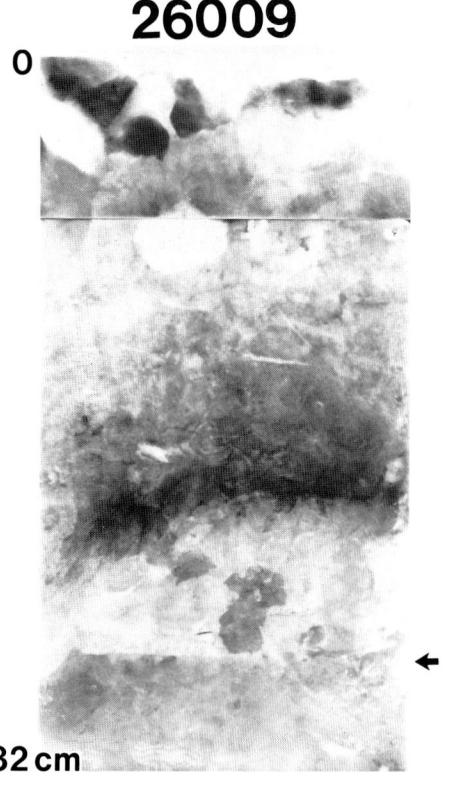

FIG. 8.—X-radiograph print of box core AHF 26009 illustrates thin mass flow deposit overlying abrupt contact (arrow) with underlying compact mud. Apparent horizon near top (8 cm) is a splice in the photo print.

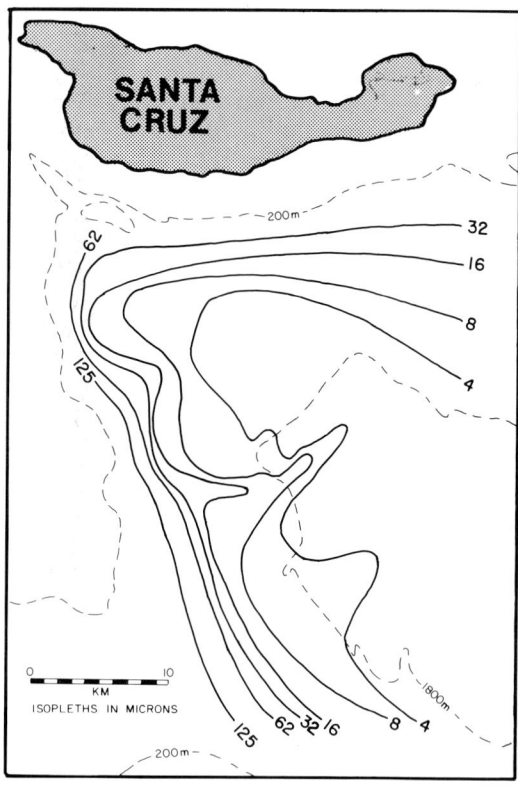

FIG. 9.—Surface sediment texture (0-5 cm) pattern showing the lobes of coarser sediment displacing the normal fining basinward trend.

well within the analytical precisions of ±0.2 φ. This, combined with displaced shallow water foraminifera, suggest recent downslope movement of upper slope sediment to the lower slope and basin floor. At present, data are too sparse to closely define the shape of the dislocations, and it remains problematical whether they result from mass movement or to the sweeping of sediments from bank and ridge tops during periods of winter storms. However, it is clear that the

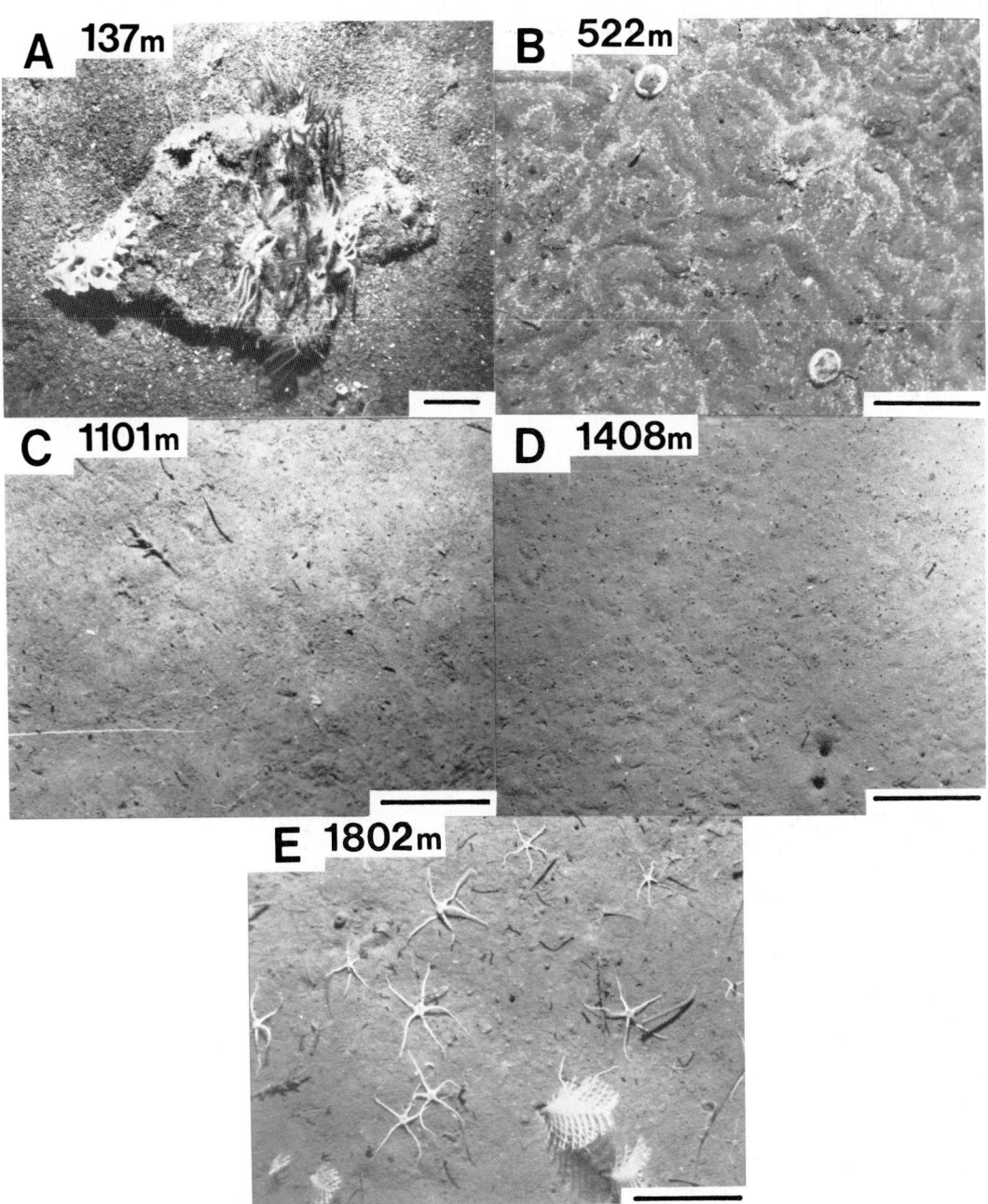

FIG. 10.—Representative bottom photographs of various environments (bar is 20 cm true scale on each photo): A) Ridge top (AFH 24300) B) Upper slope (AHF 24305) C) Lower Slope (AHF 24294) D) Lower Slope (AHF 26009) E) Basin Floor (AHF 24287)

slope is presently an active zone of sediment transport to the basin floor.

Bottom photographs are consistent with interpretation of mass movement along the slope. Because most of our photographs were taken simultaneously with coring, we were constrained to an essentially vertical orientation and relatively small areal coverage (2 m^2), with sixty-three recovered photographs showing the condition of the sediment surface at the time of core impact. The photographic evidence supports textural trends recognized by size analysis. Surface roughness is reduced with depth, grading through coarse sand on the ridge (Fig. 10A) and upper-most slope (Fig. 10B) through smoother clayey mud on the lower slope (Figs. 10C, D) and basin floor (Fig. 10E) This trend is broken by biological overprints in all environments indicating significant faunal activity.

Photographs from the lower slope show a bottom typically devoid of mobile epifauna; inferences on infaunal populations being indeterminant with photography (Owne et al, 1967). Figure 10D is typical of the lowermost slope (1000 to 1600 m) and shows a floor covered by abundant small (1 cm ±) mounds, many with apical holes. We are uncertain as to the origin of these features, but they may represent (1) burrowing and tube building by the tanaid *Apsuedes* sp. (Fauchald, pers. comm.) the dominant micro-crustacean in this environment; (2) degassing of the sediment; (3) de-watering of rapidly sedimented muds (mass flow deposits); or, (4) a combination of mechanisms.

SUMMARY AND CONCLUSIONS

Evidence for mass movement on several dimensions and time scales is present in Santa Cruz Basin. Seismic reflection records show large-scale gravity deposits of several tens of meters in thickness and many square kilometers in area which probably represent events that occurred several thousands of years before present at the time of the last glacial maximum. Box cores record small-scale evidence of contemporary mass flows within the surface few centimeters of sediment indicating a continued condition of sediment instability.

The geometry, morphology, acoustic signatures, and spatial relationships of slope deposits are believed to reflect the mechanical properties of the deposits during emplacement. Locally in Santa Cruz Basin, seismic-reflection records reveal a slope depositional sequence which suggests a range of mass emplacement modes which forms a mechanical continuum in a down-slope direction. A strikingly similar sequence has been recorded on the African Atlantic slope (Jacobi, 1976) and profiles published by Moore et al (1970), Walker and Massingill (1970), and Moore et al (1976) appear analogous. These may form the basis for a general model of gravity deposition capable of predicting the spatial relationships of slope and rise facies, particularly in areas removed from the influence of canyon-fan systems. Near the top of the slope in Santa Cruz Basin, initial slope failure occurred elastically as a slide. The toe of the slide subsequently was remobilized on the lower slope and transported plastically as a series of mass flows toward the floor of the basin. The resultant gravity deposits have been termed types *A*, *B*, and *C* in order to avoid genetic labels which imply specific rheologic behavior (e.g., debris flow). The material comprising the thin deposits on the floor (type *C*) is inferred to have been derived in part from the thicker lense-shaped unit at the base of the slope (type *B*). The lense, in turn, may have originated from the deposit represented by diffraction hyperbolae on the lower slope (type *A*). The thicknesses and morphologies of the deposits plus the gradients on which they rest suggest that the flows became increasingly less viscous as they fractionated. The sequence would seem to support the published view that jostling and remolding of material near the base of a slide is an important factor in the incorporation of water into the sediment mass which then allows it to flow with substantially less strength.

In Santa Cruz Basin, the sedimentologic evidence shows surficial downslope movement of fine sediments which may in part reflect winnowing of sediment from adjacent storm swept sills. Cores show some pebbly muds and plastically deformed zones which cannot be detected acoustically. Some of these may represent thin flows of the high water content surficial sediments.

It is also evident that sampler limitations prohibit obtaining an unambiguous picture of sedimentary structures and geometries of individual masses involved in the movements. It may be that the stratigraphic record will be the best source for evidence of the structures of the sediments in gravity deposits because of these limitations on contemporary marine data collections. Marine seismic-reflection data remain valuable because they show the large-scale three-dimensional relationships difficult to perceive in outcrop sections and define acoustic characteristics which sometimes can be used to infer physical properties as well as geologic processes.

ACKNOWLEDGMENTS

The paper has benefitted from the thoughtful critical review of Dr's. M. E. Field, E. Uchupi, and H. T. Mullins. Support for the research has been from National Science Foundation Grant OCE 78-20453.

We also acknowledge support from National Science Foundation Grant OCE76-00156 and Bureau of Land Management contracts through Science Applications, Inc. of San Diego. Ship time was funded by block grants from the National Science Foundation. Dr. P. W. Barnes kindly supplied data from his files.

REFERENCES

ALLEN, J. R. L., 1971, Mixing at turbidity current heads, and its geological implications: Jour. Sed. Petrology, v. 41, p. 97–113.

BAGNOLD, R. A., 1954, Experiments on a gravity-free dispersion of large solid spheres in a Newtonian fluid under shear: Proc. Royal Soc. London, A225, p. 49–63.

———, 1956, The flow of cohesionless grains in fluids: Phil. Trans. Royal Soc. London, A249, p. 235-297.

BARNES, P. W., 1970, Marine geology and oceanography of Santa Cruz Basin off southern California [unpubl. Ph.D. thesis]: Los Angeles, Univ. of Southern California, 175 p.

BARTOLINI, C. G., AND STANLEY, D. J., 1972, Morphology and recent sediments of the western Alboran Basin in the Mediterranean Sea: Marine Geology, v. 13, p. 159–224.

BOUMA, A. H., 1962, Sedimentology of some flysch deposits: Amsterdam, Elsevier 168 p.

CARTER, R. M., 1975, A discussion and classification of subaqueous mass-transport with particular applications to grain-flow, slurry-flow and fluxoturbidites: Earth Sci. Rev., v. 11, p. 145–177.

COATES, D. R., ed, 1977, Landslides: Geol. Soc. America Rev. in Engineering Geology, v. III.

COLEMAN, J. M., 1976, Deltas: processes of deposition and models for exploration: Champaign, Continuing Education Pub. Co., Il., 102 p.

———, AND PRYOR, D. B., 1978, Contemporary gravity tectonics—an everyday catastrophe?: Am. Assoc. Petroleum Geologists Programs with Abs., Oklahoma City, p. 56.

COOK, H. E., MCDANIEL, P. N., MOUNTJOY, E. W., AND PRAY, L. C., 1972, Allochthonous carbonate debris flows at Devonian bank ("reef") margins, Alberta, Canada: Canadian Petroleum Geologists Bull., v. 20, p. 439–497.

CROWELL, J. C., 1957, Origin of pebbly mudstones: Geol. Soc. America Bull., v. 68, p. 993–1010.

DALY, R. A., 1936, The origin of submarine canyons: Am. Jour. Sci., v. 31, p. 410–420.

DAMUTH, J. E., 1975, Echo character of the western equatorial Atlantic floor and its relationship to the dispersal and distribution of terrigenous sediments: Marine Geology, v. 18, p. 17–45.

DOTT, R. H. JR., 1963, Dynamics of subaqueous gravity depositional processes: Am. Assoc. Petroleum Geologists Bull., v. 47, p. 104–128.

ECKEL, E. B., ed., 1958, Landslides and engineering practice: Highway Research Board Spec. Rept. 29, 232 p.

EDWARDS, B. D., 1977, Animal-sediment relationships in a slope environment, southern California Continental Borderland: Geol. Soc. America Program with Abs., v. 9, p. 415.

ELLIOTT, R. E., 1965, A classification of subaqueous sedimentary structures based on rheological and kinematic parameters: Sedimentology, v. 5, p. 193–209.

EMBLEY, R. W., 1976, New evidence for occurrence of debris flow deposits in the deep-sea: Geology, v. 4, p. 371–374.

EMERY, K. O., 1960, The Sea off Southern California: New York, Wiley, 366 p.

———, AND BRAY, E. E., 1962, Radiocarbon dating of California basin sediments: Am. Assoc. Petroleum Geologists Bull., v. 46, p. 1839–1856.

———, AND TERRY, R. D., 1956, A submarine slope of southern California: Jour. Geology, v. 64, p. 271–280.

———, AND UCHUPI, E., 1972, Western North Atlantic Ocean: topography, rocks, structure, water, life and sediments: Am. Assoc. Petroleum Geologists Memoir 17, 531 p.

FELSHER, M., 1971, Physical sedimentology and bathymetry, Santa Cruz Submarine Canyon Complex, California Continental Borderland, California [unpubl. Ph.D. thesis]: Austin, Univ. of Texas, 329 p.

FISHER, R. V., 1971, Features of coarse-grained, high-concentration fluids and their deposits: Jour. Sed. Petrology, v. 41, p. 916–927.

GORSLINE, D. S., AND BARNES, P. W., 1972, Carbonate variations as climatic indicators in contemporary California flysch basins: 24th International Geol. Congr., Montreal, Sec. 6, p. 270–277.

———, AND EMERY, K. O., 1959, Turbidity current deposits in San Pedro and Santa Monica Basins off southern California: Geol. Soc. America Bull., v. 70, p. 279–290.

———, DRAKE, D. E., AND BARNES, P. W., 1968, Holocene sedimentation in Tanner Basin, California Continental Borderland: Geol. Soc. America Bull., v. 79, p. 659–674.

GREENE, H. G., CLARKE, S. H., FIELD, M. E., LINKER F. I., AND WAGNER, H. C., 1975, Preliminary report on the environmental geology of selected areas of the southern California Continental Borderland: U.S. Geol. Survey Open-file Rept. 75-596, 70 p.

HAMPTON, M. A., 1972, The role of subaqueous debris flow in the generation of turbidity currents: Jour. Sed. Petrology, v. 42, p. 775–793.

———, 1975, Competence of fine-grained debris flows: Jour. Sed. Petrology, v. 45, p. 834–844.

HANER, B. E., AND GORSLINE, D. S., 1978, Processes and morphology of continental slope between Santa Monica and Dume Submarine Canyons, southern California: Marine Geology, v. 28, p. 77–87.

HEEZEN, B. C., AND DRAKE, C. L., 1964, Grand Banks slump: Am. Assoc. Petroleum Geologists Bull., v. 48, p. 221–225.

———, AND EWING, M., 1952, Turbidity currents and submarine slumps and the 1929 Grand Banks earthquake: Am. Jour. Sci., v. 250, p. 849–873.

HENDRY, H. E., 1972, Breccias deposited by mass flow in the breccia nappe of the French pre-Alps: Sedimentology, v. 18, p. 277–292.

HOWELL, D. G., ed., 1976, Aspects on the geologic history of the California Continental Borderland: Am. Assoc. Petroleum Geologists, Pacific Sec. Misc. Paper 24, 561 p.

HSÜ, K. J., 1975, Catastrophic debris streams (sturzstroms) generated by rockfalls: Geol. Soc. America Bull., v. 86, p. 129–140.

HYNE, N. J., GOLDMAN, C. R. AND COURT, J. E., 1973, Mounds in Lake Tahoe, California-Nevada: a model for landslide topography in the subaqueous environment: Jour. Geology, v. 81, p. 176–188.

JACOBI, R. D., 1976, Sediment slides on the northwestern continental margin of Africa: Marine Geology, v. 22, p. 157–173.

JOHNSON, A. M., 1970, Physical processes in geology: San Francisco, W. H. Freeman, Inc., 571 p.

KEITH, B. D., AND FRIEDMAN, G. M., 1977, A slope-fan-basin-plain model, Taconic sequence, New York and Vermont: Jour. Sed. Petrology, v. 47, p. 1220–1241.

KELLING, G., AND STANLEY, D. J., 1976, Sedimentation in canyon, slope and base-of-slope environments, *in* Stanley, D. J., and Swift, D. J. P., eds., Marine Sediment Transport and Environmental Management: New York, Wiley, p. 379–435.

KLEIN, G. DEV., DEMELO, U. AND FAVERA, J. C. D., 1972, Subaqueous gravity processes on the front of Cretaceous deltas, Reconcavo Basin, Brazil: Geol. Soc. America Bull., v. 83, p. 1469–1492.

KUENEN, PH. H., 1937, Experiments in connection with Daly's hypothesis on the formation of submarine canyons: Leid. Geol. Meded., v. 7, p. 327–351.

———, 1951, Properties of turbidity currents of high density: Soc. Econ. Paleontologists Mineraologists Spec. Pub. No. 2, p. 14–33.

———, 1955, The difference between sliding and turbidity flow: Deep-sea Research, v. 3, p. 134–139.

———, AND MIGLIORINI, C. I., 1950, Turbidity currents as a cause of graded bedding: Jour. Geology, v. 58, p. 91–127.

LARSEN, V., AND STEEL, R. J., 1978, The sedimentary history of a debris flow dominated Devonian alluvial fan—a study of textural inversion: Sedimentology, v. 25, p. 37–59.

LEHNER, P. 1969, Salt tectonics and Pleistocene stratigraphy on continental slope of northern Gulf of Mexico: Am. Assoc. Petroleum Geologists Bull., v. 53, p. 2431–2479.

LEWIS, D. N., 1976, Subaqueous debris flows of early Pleistocene age at Motunau, North Canterbury, New Zealand: New Zealand Jour. Geol. Geophys., v. 19, p. 535–567.

LEWIS, K. B., 1971, Slumping on a continental slope inclined at 1°–4°: Sedimentology, v. 16, p. 97–110.

LOWE, D. R., 1976a, Subaqueous liquified and fluidized sediment flows and their deposits: Sedimentology, v. 23, p. 285–308.

———, 1976b, Grain flow and grain flow deposits: Jour. Sed. Petrology, v. 46, p. 188–190.

———, 1979, Sediment gravity flows: Their classification and some problems of its application to natural flows and deposits: this volume.

MENARD, H. W., 1964, Marine Geology of the Pacific: New York, McGraw-Hill, 271 p.

MIDDLETON, G. V., AND HAMPTON, M. A., 1973, Sediment gravity flows *in* Turbidites and deep water sedimentation, Soc. Econ. Paleontologists Mineraologists Pacific Sec. Short Course Notes: p. 1–38.

MOLNIA, B. F., CARLSON, P. R., AND BRUNS, T. R., 1977, Large submarine slides in Kayak Trough, Gulf of Alaska, *in* Coates, D. R., ed., Landslides: Geol. Soc. America Rev. in Engineering Geology, v. 3, p. 137–148.

MOORE, D. G., 1961, Submarine slumps: Jour. Sed. Petrology, v. 31, p. 343–357.

———, 1969, Reflection profiling studies of the California Continental Borderland: structure and Quaternary Basins: Geol. Soc. America Spec. Paper 107, 142 p.

———, Submarine slides, *in* Voight, B., ed., Rockslides and avalanches, Developments in Geotechnical Engineering, 14A, p. 563–604.

———, CURRAY, J. R., AND EMMEL, F. J., 1976, Large submarine slide (olistostrome) associated with Sunda Arc, subduction zone, northeast Indian Ocean: Marine Geology, v. 21, p. 211–226.

MOORE, T. C., VAN ANDEL, T. H., BLOW, W. H., AND HEATH, G. R., 1970, Large submarine slide off northeastern continental margin of Brazil: Am. Assoc. Petroleum Geologists Bull., v. 54, p. 125.

MORELOCK, J., 1969, Shear strength and stability of continental slope deposits, western Gulf of Mexico: Jour. Geophys. Res., v. 74, p. 465–482.

MORGENSTERN, N. R., 1967, Submarine slumping and the initiation of turbidity currents, *in* Richards, A. F., ed. Marine Geotechnique: p. 189–220.

MORRIS, R. C., 1971, Classification and interpretation of disturbed bedding types in Jackfork flysch rocks (Upper Mississippian) Ouachita Mountains, Arkansas: Jour. Sed. Petrology, v. 41, p. 410–424.

MUTTI, E., AND RICCI-LUCCHI, F., 1972, Turbidites of the northern Appennines: introduction to facies analysis: International Geology Rev. v. 20, p. 125–166.

NORMARK, W. R., 1974, Ranger submarine slide, northern Sebastian Viscaino Bay, Baja California, Mexico: Geol. Soc. America Bull., v. 85, p. 781–784.

OWEN, D. M., SANDERS, H. L. AND HESSLER, R. R., 1967, Bottom photography as a tool for estimating benthic populations, in Hersey, J. B., ed., Deep Sea Photography: Johns Hopkins University Press, p. 229–234.
PADGETT, G. R., EHRLICH, R., AND MOODY, M., 1977, Submarine debris flow deposits in an extensional setting: Upper Devonian of western Morocco: Jour. Sed. Petrology, v. 47, p. 811–818.
PIPER, D. J. W., VON HUENE, R., AND DUNCAN, J. R., 1973, Late Quaternary sedimentation in the active eastern Aleutian Trench: Geology, v. 1, p. 19–22.
REID, J. L., JR., RODEN, G. I., AND WYLIE, J. G., 1958, Studies of the California Current System: Calif. Coop. Fisheries Invest. Calif. Dept. Fish and Game, p. 27–56.
RODINE, J. D., AND JOHNSON, A. M., 1976, The ability of debris, heavily freighted with coarse clastic material to flow on gentle slopes: Sedimentology, v. 23, p. 213–234.
RONA, P. A., 1969, Middle Atlantic continental slope of the U.S.: deposition and erosion: Am. Assoc. Petroleum Geologists Bull., v. 53, p. 1453–1465.
SANDERS, J. E., 1965, Primary sedimentary structures formed by turbidity currents and related resedimentation mechanisms, in Middleton, G. V., ed., Primary sedimentary structures and their hydrodynamic interpretation: Soc. Econ. Paleontologists Mineralogists Spec. Pub. No. 12, p. 192–219.
SHARPE, C. F. T., 1938, Landslides and related phenomena: New York, Columbia Univ. Press, 127 p.
SHEARMAN, D. J., 1976, The geological evolution of southern Iran: The Geog. Jour., v. 142, p. 383–410.
SHEPARD, F. P., AND EMERY, K. O., 1941, Submarine topography off the California Coast: Geol. Soc. America Spec. Paper 31, 171 p.
STANLEY, D. J., 1974, Pebbly mud transport in the head of Wilmington Canyon: Marine Geology, v. 16, p. M1–M8.
———, AND SILVERBERG, N., 1969, Recent slumping on the continental slope off Sable Island Bank, southeast Canada: Earth and Planetary Science Letters, v. 6, p. 123–133.
———, AND KELLING, G., eds, 1978, Sedimentation in submarine canyons, fans, and trenches: Stroudsburg, Pa., Dowden, Hutchinson and Ross, Inc. 395 p.
———, AND UNRUG, R., 1972, Submarine channel deposits fluxoturbidites and other indicators of slope and base-of-slope environments in modern and ancient marine basins, in Recognition of Ancient Sedimentary Environments: Soc. Econ. Paleontologists Mineralogists Spec. Pub. 16, p. 287–340.
———, PALMER, H. D. AND DILL, R. F., 1978, Coarse sediment transport by mass flow and turbidity current processes and downslope transportation in Annot Sandstone canyon-fan valley systems, in Stanley, D. J., and Kelling, G., eds., Sedimentation in Submarine Canyons, Fans and Trenches: Stroudsburg, Dowden, Hutchingson and Ross, Inc., p. 85–126.
TERZAGHI, K., 1956, Variations of submarine slope failure: Proc. 8th Texas Soil Mechanics Found. Engr. Conf., p. 1–41.
TWENHOFEL, W. H., 1926, Treatise on Sedimentation: Baltimore, Williams and Wilkins, 926 p.
VAN DER KNAAP, W., AND EIJPE, R., 1968, Some experiments on the generics of turbidity currents: Sedimentology, v. 11, p. 115–124.
VEDDER, J. G., BEYER, L. A., JUNGER, A., MOORE, G. W., TAYLOR, J. C., AND WAGNER, H. C., 1974, Preliminary report on the geology of the Continental Borderland of southern California: U.S. Geol. Survey Misc. Field Studies Map MF-624, 34 p.
WALKER, J. R., AND MASSINGILL, J. V., 1970, Slump features on the Mississippi fan, northeastern Gulf of Mexico: Geol. Soc. America Bull., v. 81, p. 3101–3108.
———, 1967, Turbidity current sedimentary structures and their relationship to proximal and distal depositional environments: Jour. Sed. Petrology, v. 37, p. 25–43.
———, 1970, Review of the geometry and facies organization of turbidites and turbidite-bearing basins, in Lajoie, J., ed., Flysch Sedimentology in North America: p. 219–251.
———, 1973, Mopping up the turbidite mess, a history of the turbidity current concept, in Ginsburg, R. N., ed., Evolving Concepts in Sedimentology: Johns Hopkins Press, p. 1–37.
WILHELM, O., AND EWING, M., 1972, Geology and history of the Gulf of Mexico: Geol. Soc. America Bull., v. 83, p. 575–600.
ZARUBA, W., AND MENCL, V., 1969, Landslides and their control: Amsterdam, Elsevier, 205 p.

SMALL-SCALE SLUMPS AND SLIDES AND THEIR SIGNIFICANCE FOR BASIN SLOPE PROCESSES, SOUTHERN CALIFORNIA BORDERLAND

MICHAEL E. FIELD AND SAMUEL H. CLARKE, JR.
U.S. Geological Survey
345 Middlefield Road
Menlo Park, CA 94025

ABSTRACT

Large-scale sediment slides and slumps, measuring several square kilometers in area, have been described from basin slopes in southern California as well as from many other continental slopes of the world. However, little attention has been focused on smaller features that may reflect significant downslope transport of sediment by small-scale mass movements. High-resolution seismic reflection data, seafloor photographs, and side-scan sonographs obtained using a deeply-towed geophysical instrument package show that at least six such features or zones of submarine gravity transport are present in a relatively small (150 km^2) area on an intercanyon slope off San Nicolas Island in the southern California borderland. Each zone is small (hundreds of meters on a side) and somewhat different from the others in age, shape, and inferred origin. These features range from rotational slumps to chaotic slides. Photographs show rounding and extension fractures in transported clasts of sediment, and a difference in the activity of benthic organisms within and outside of the slide zones. The thickening of sedimentary units downslope, coupled with the presence of small features indicative of sediment failure, suggests that gravity-driven processes contribute significantly to the construction of basin slopes in this region.

INTRODUCTION

The importance of transport by gravity in moving large volumes of sediment on submarine slopes is well known (e.g., Heezen and Drake, 1964; Menard, 1964; Moore and others, 1970; Kelling and Stanley, 1970; Moore, 1977). Submarine slumps and slides have been identified on open slopes as low as < 1° to 4° (Lewis, 1971), as well as on steeper slopes and in canyons. Displaced sediment masses range in size from a few square meters to tens of thousands of square meters (Heezen and Drake, 1964; Moore and others, 1970; Lewis, 1971). Recent studies show that recurrent submarine mass movements of varied types are responsible for the transport of a major part of the total volume of sediment moved across the shelf and upper slope off the Mississippi delta (Coleman and Prior, 1978). In this region, which is characterized by rapid sedimentation and underconsolidation of sediments, slides have been documented on open slopes as low as 0.2°.

Submarine slumps and slides identified previously in geophysical surveys on the shelf and slope and described in the literature generally are large (several square kilometers in area) features; the absence of well-documented small features probably reflects the resolving power of conventional geophysical systems and the broad scope of most studies of the continental margin. However, recently acquired high-resolution marine geophysical data, considered in the light of recent studies of ancient deposits on land, suggest that sediment gravity transport on a much smaller scale (affecting areas up to a few hundreds of square meters) may be of volumetrically equal or greater importance in moving sediment downslope into basins of the southern California borderland, and presumably in other parts of the world ocean.

This investigation initially focused on a large sediment slide detected on 3.5 kHz seismic reflection profiles collected during the course of geologic resource and geohazard assessment studies in March 1976 aboard the U.S. Geological Survey research vessel *S. P. Lee*. The slide area is located on the northeast slope of the Santa Rosa-Cortes Ridge east of San Nicolas Island at a depth of 250 to 550 m. Subsequently, this area was surveyed in a greater detail aboard the Scripps Institution of Oceanography research vessel *Melville* using the Scripps Marine Physical Laboratory (MPL) deep-tow instrument. This vehicle is an unmanned instrument package equipped for this study with up-, down-, forward-, and side-looking sonars, a 4 kHz seismic reflection profiling system, a snapshot television camera, and bottom camera (Spiess and others, 1976). Towed at speeds of 2 to 3 knots and at heights above the bottom ranging from 7 to about 80 m, the deep-tow vehicle records detailed information on bottom topography, shallow subbottom structure, and surface sediment character. A newly developed real-time computer processing technique to enhance the resolution of 4 kHz seismic reflection profiles was successfully tested during this study. Precision navigation for the survey was provided by a shore-based transponder system; because the instrument package was towed at relatively shallow depths, its position at a given time approximated that of the ship, and ocean bottom transponders were not used. The accuracy of ship positioning is estimated to be ± 30 m.

The 36-hour survey obtained acoustic profiles from over 120 km of trackline (Fig. 1), and more than 700 bottom photographs were taken. These data provide evidence for substantial mass sediment transport as a result of failures on the submarine slopes flanking the San Nicolas Island platform.

SETTING

The southern California continental borderland is a submerged, tectonically formed basin and ridge province characterized by shallow banks and ridges interspersed with basins ranging in depth from about 800 m to 2100 m. Evidence of large-scale submarine mass transport on basin slopes throughout this region is well documented (e.g., by Moore, 1961; Greene and others, 1975; Haner and Gorsline, 1978; Field and Richmond, in press). Located in the western part of the borderland, the San Nicolas Island platform is a complexly faulted anticlinorium that forms a northwest-trending salient in the central part of the Santa Rosa-Cortes Ridge, a prominent linear ridge that extends from Santa Rosa Island approximately 210 km south-southeast to Cortes Bank (Vedder and Norris, 1963; Vedder and others, 1974). Eocene thin- to very thick-bedded sandstone, shale, and minor conglomerate are exposed in the core of this anticline, which underlies San Nicolas Island and the southwest part of the platform (Cole, 1970). Miocene mudstone and shale with subordinate sandstone, limestone, and conglomerate make up the northeast part of the platform and the surrounding slopes (Uchupi, 1961; Vedder and others, 1974). Pliocene and younger sediments are present on the lower slopes and in the basins flanking the platform. The ridge top and upper part of the adjacent slopes are veneered by Pleistocene and Holocene authigenic, biogenic, and terrigenous sand and mud a few tens of meters thick; the adjacent basins contain correlative deposits of mud and fine sand as much as 165 m thick (Vedder and others, 1974; Greene and others, 1975).

The submarine slope surrounding the San Nicolas Island platform is incised by numerous gullies and small canyons, similar to slopes described elsewhere in the borderland and along the mainland shelf (Buffington and Moore, 1963). Smaller gullies typically are 50 to 100 m deep, whereas larger gullies and small canyons are 200 to 300 m deep. Declivity of the basin slope in the area surveyed averages 6°, ranging from about 2° to 16°, although side slopes within gullies locally exceed 20°. Rates of deposition estimated for interbasin areas in this region are low, ranging from 10–20 cm/1000 yr (Emery and Bray, 1962; Moore, 1969), but resedimentation by wave and current action of material initially deposited on bank and ridge tops may result in local oversteepening of the ridge slopes. Slide masses on these slopes are at least 50 m thick and are believed to be composed chiefly of Miocene sedimentary rocks and the overlying Quaternary deposits.

RESULTS

Six zones of submarine sliding were observed in the 150 km² study area (Fig. 1). Four of these (along lines, D, E, L, and N) were identified from 4-kHz seismic reflection and side-scan sonar records, and the remaining two (on lines O and Q) from bottom photographs. The slide zones defined by acoustic techniques differ from one another in size, morphology, and inferred history. These differences suggest that, even within a small area, the mechanisms of downslope sediment transport are varied, and no single process predominates.

Deep-tow line E was run adjacent to *Lee* line 115, which previously had recorded a large slide at depths of 250 to 550 m (Fig. 2a). Preservation of stratification within the slide and apparent absence of rotation suggest that this mass may be a debris slide, involving movement along a relatively planar slide surface (Hampton and Bouma, 1977; Varnes, 1958). On deep-tow line E, a large submarine slide characterized by hummocky surface topography and internal deformation of acoustic reflectors was observed at depths exceeding 750 m (Fig. 2b). The slide, which lies at the base of a 3° slope, has positive relief of

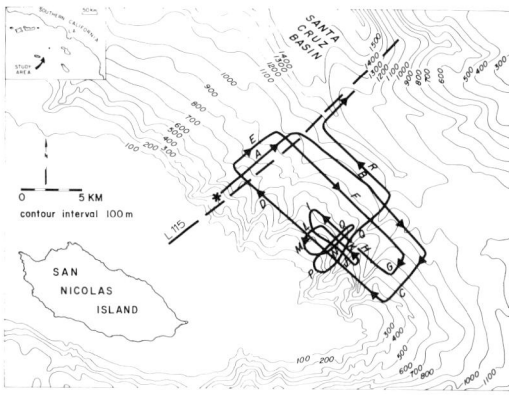

FIG. 1.—Map showing bathymetry and tracklines over slope east of San Nicolas Island. Dashed tracklines located surface ship data (R/V *Lee*); solid lines show deep-tow profile lines (R/V *Melville*).

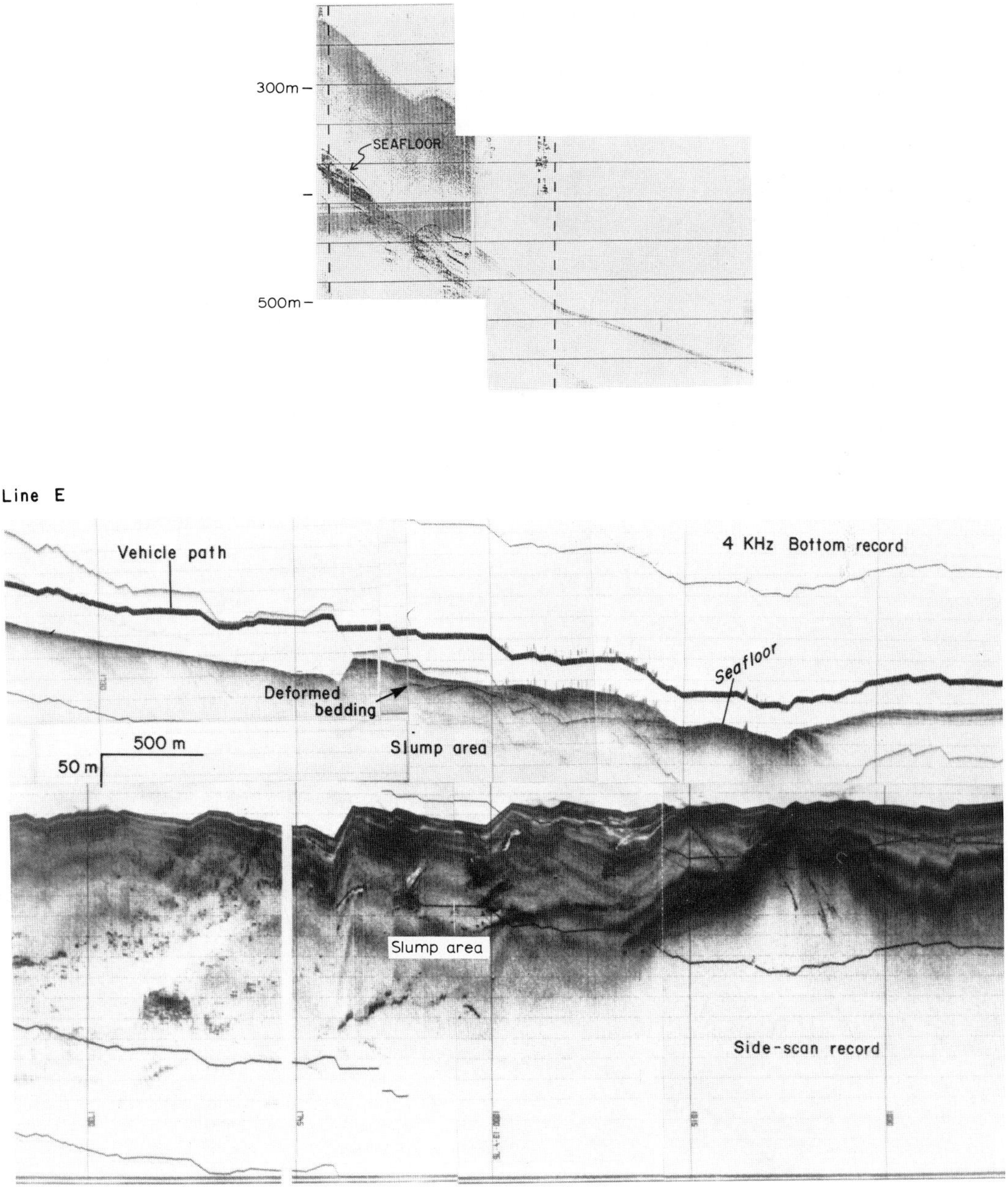

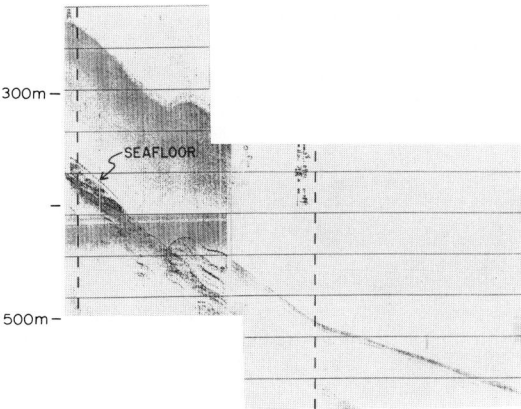

FIG. 2.—a) High-resolution seismic record (3.5 kHz) obtained from surface vessel along dashed trackline in Figure 1. b) Deep-tow high-resolution seismic record (4 kHz) and right channel of side-scan record across slide zone on line E. Height of towed instrument package above seafloor is indicated as vehicle path.

50 m and is more than 75 m thick. The zone of hummocky topography associated with this slide extends for at least 1600 m downslope; its breadth is not known, but features that may be part of the slide are identifiable on side-scan sonographs as far as 200 m from the line. The hummocky surface of this mass suggests that it was emplaced by downslope sliding or flowage;

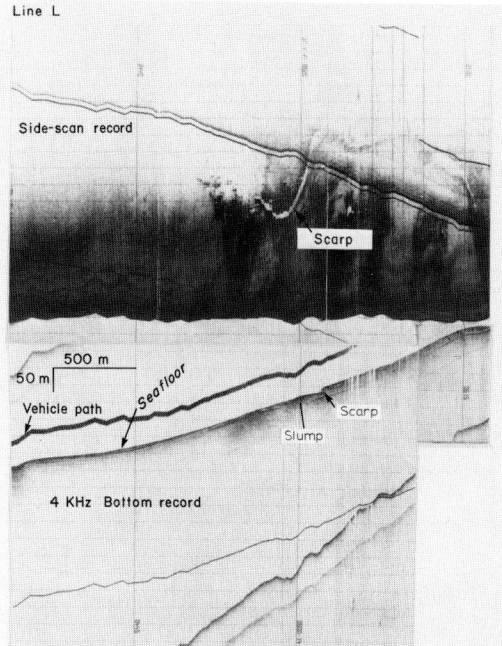

FIG. 3.—a) Deep-tow high-resolution seismic record (4 kHz) and right channel of side-scan record over rotational slump on line L. b) Interpretation of 4-kHz subbottom record over rotational slump, line L.

the absence of an identifiable scarp upslope along line E indicates that it must have originated above the 700 m isobath and (or) out of the plane of the profile line.

One small rotational slump on a slope of about 5° was identified on line L (Fig. 3a). The 4-kHz record reveals a scarp that is about 10 m high and a slip plane estimated to be about 15 to 20 m below the seafloor. Acoustic reflectors within the slump are back-rotated to a nearly horizontal position (Figs. 3a, b). Undulations in near-surface reflectors upslope from this scarp may indicate failure from downslope creep without detachment. The side-scan sonar record shows a long sinuous scarp that lies about 100 m south of the trackline and meanders from a predominantly north-south to east-west orientation. Although this scarp does not connect directly with the scarp identified on the 4-kHz profile, their proximity and morphologic similarity suggest that they are related by time and mode of origin. Because the scarps on both records show little evidence of erosion or burial, these failures are probably relatively recent.

Interpretation of submarine slides from geophysical evidence on lines D and N is more subjective than for lines E and L. Acoustic records from line D, run parallel to slope (Fig. 1), show a zone of deformed and disrupted subsurface reflectors nearly 1 km in breadth (Fig. 4a). The extent of this zone downslope was not determined; however, large (150 m × 300 m) masses of sediment at the surface appear from side-scan sonographs to extend at least 350 m downslope from the line (Fig. 4a). These sediment masses are irregular in shape and are discontinuous; they are interpreted to reflect submarine slumping. The side-scan sonar record between 1430 and 1445 reveals a horseshoe-shaped feature having a scarp (light area) facing downslope that may be a single, recent rotational slump. The 4-kHz profile indicates that the surface above this zone of disrupted reflectors is smooth, showing only a slight, 5-m-high bulge in the sea floor. The zone of internal deformation, which is located between a canyon and gulley, appears at least 50 m thick and may reflect long-term failure. The estimated minimum volume of sediment associated with this zone is 14×10^6 m^3. Side-scan sonar records along line N reveal large (approximately 100 m × 350 m) isolated bodies interpreted as displaced masses of sediment (Fig. 4b). The 4-kHz subbottom profile shows disrupted reflectors and a change in acoustic character in the vicinity of the displaced masses; however, surficial features that can be related to failure are absent.

More than 700 photographs were taken during two continuous photographic runs to study the nature of the sea floor and surface sediments in the vicinity of slides noted on geophysical records. Evidence of sliding was noted in two areas, one on line O at a depth of about 400 m and one on line Q at a depth of about 850 m (Fig. 1). The character of surface sediments and evidence for sliding are similar in the two areas. The sea floor adjacent to these areas appears

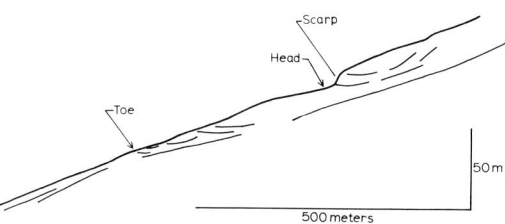

FIG. 4.—Deep-tow high-resolution seismic (4 kHz) and side-scan sonar record. a) Depicts slope-parallel line (D) and contains right channel of side-scan sonograph (slope is toward reader). b) Depicts slope-normal line (N) and shows varied side-scan sonographs.

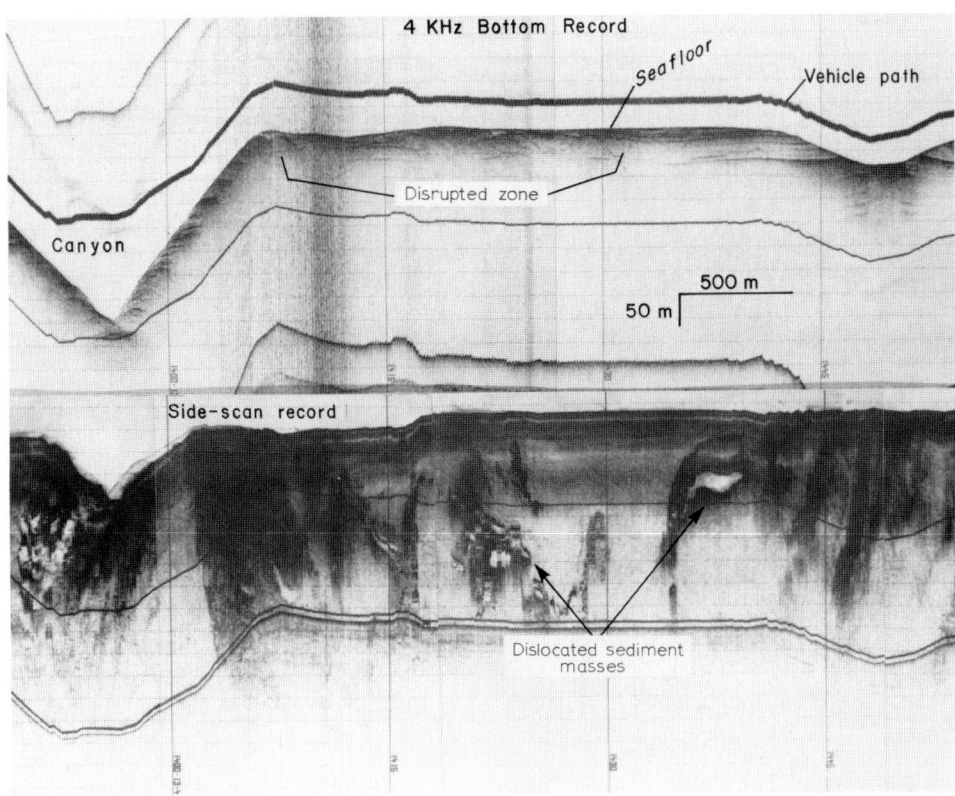
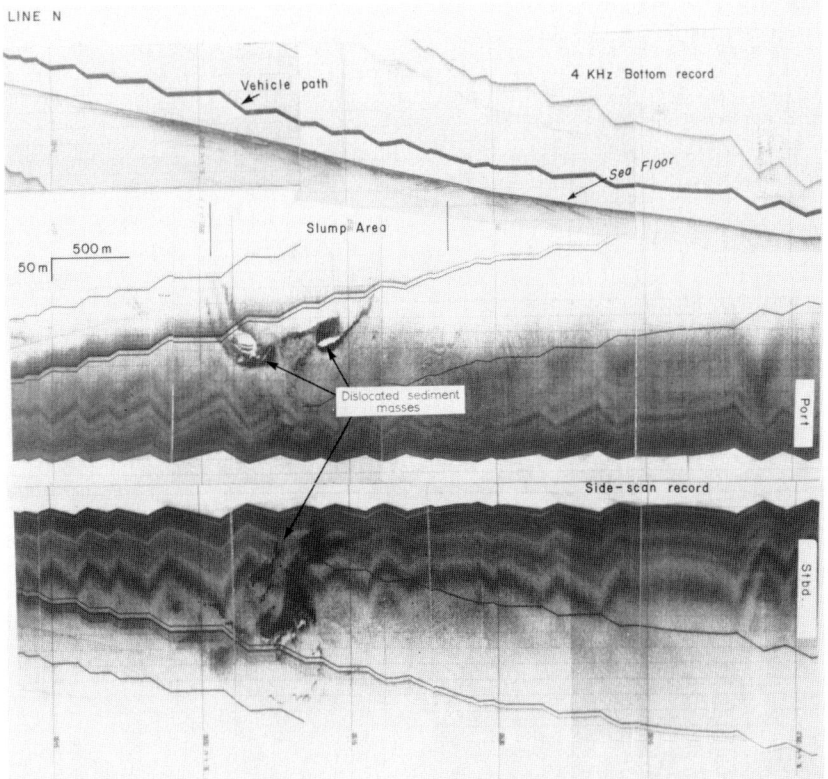

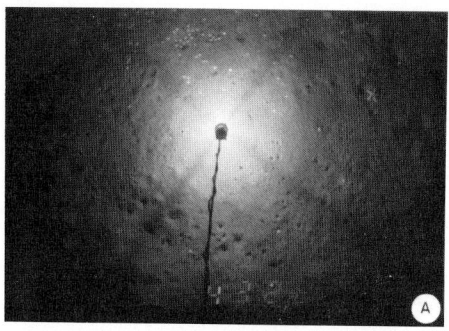
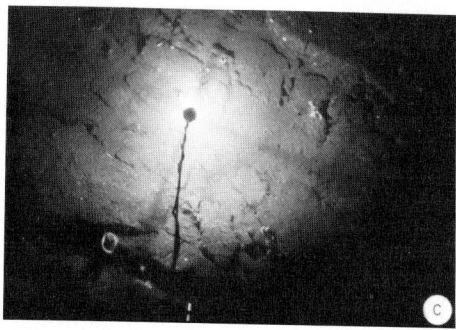

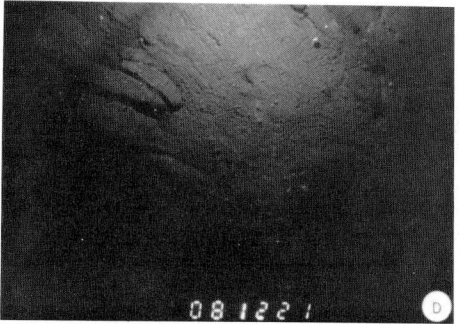

Fig. 5.—Photographs showing characteristics of slide zones and adjacent undisturbed parts of slope off San Nicolas Island. a) Seafloor is characteristically densely burrowed in area that has not been disturbed by gravity transport; b) Tracks and trails are dense, but burrows rare in vicinity of slide zones; c and d) Photographs showing seafloor character in sediment slide zone. Note sag fractures and rounded noses of clasts and density of trails made by holothurians and echinoids.

to be covered by sandy mud that has been extensively bioturbated by benthic organisms. Burrows greatly predominate over trails; holothurians, echinoids, starfish, and brittle stars are distributed sparsely on the sea floor, and echinoids locally form dense clusters (Fig. 5a). In contrast, the sea floor in the slide areas appears to be covered by muddy sand having a smooth surface covered by tracks and trails of holothurians for a distance of several hundred meters downslope, and in some areas directly upslope from the slide as well (Fig. 5b). Burrows are not evident, and the density of feeding holothurians is commonly as much as one per m^2 (Fig. 5b). Photographs from slide areas show grooves and lineations oriented downslope, extension fractures in sediment mounds, and clasts of sedimentary rock (Figs. 5c, d). Between clasts near the terminus of the slide, the bottom is sandy and covered by trails. Rounding of sediment mounds is common and may be indicative of transport direction (Fig. 5d).

A series of alongslope and downslope profiles characterizing the sea floor topography and shallow subsurface reflectors is shown in Figure 6. There is a progressive downslope change both in canyon morphology and sediment thickness (J, H, B, Fig. 6); canyons on the upper slope are V-shaped with older rocks cropping out in the canyon walls. Downslope, the canyons become U-shaped and relatively young sediment covers the valley floor. The form of the canyon appears to be influenced principally by erosion on the upper slope, whereas on the lower slope, levee construction and channel filling, coupled with channel widening, are dominant. The basin slope between canyons, as seen in profile lines A and

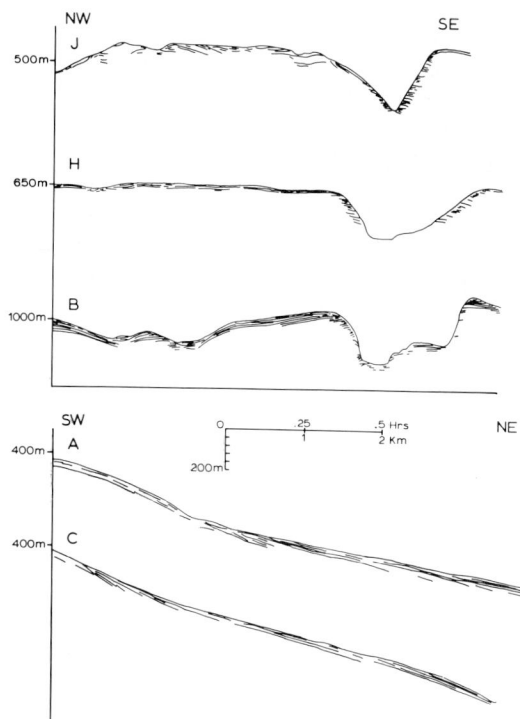

FIG. 6.—Comparison of line drawings of interpreted records from slope-parallel profiles (*J*, *H*, and *B*) taken at increasing depths and from downslope profiles (*A* and *C*).

C (Fig. 6), is smooth and underlain by basinward dipping sediments. The slight thickening of some acoustic units in the downslope direction may reflect the activity of gravity-controlled sediment transport in slope construction.

DISCUSSION

Most studies of submarine slides have focused on large features measuring tens to hundreds of square kilometers in area and as much as two or three hundred meters thick (e.g., Heezen and Drake, 1964; Lewis, 1971; Hampton and Bouma, 1977; Carlson and Molnia, 1978). Studies by Cook (this volume) of Cambrian slope deposits in Nevada and by Piper and others (1976) of Plio-Pleistocene slope sedimentary rocks in northern California indicate that small-scale mass movements, many too small to be resolved by conventional geophysical profiling techniques in the marine environment, make up a volumetrically important part of the suite of rocks representing this environment. Our results suggest that small-scale submarine slumps and slides are common on the flanks of ridges in the southern California borderland, and they may play an important part in ridge-to-basin sediment transport and shaping of the intervening slope.

Six zones of mass failure were identified in this study within a 150-km^2 area on an intercanyon slope. These failures include both rotational slumps and slides. Bottom photographs along two slope-normal transects show that the substrate on undisturbed areas of the slope is highly burrowed, in contrast to a dense covering of trails and the absence of burrows in the immediate vicinity of slides. This contrast suggests that the slides are relatively young and that some factor (e.g., release of organic material or modification of surface texture) associated with the failure may stimulate grazing activity by benthic organisms.

The basinward thickening of some sedimentary units observed on these slopes suggests that gravity transport may be important in slope construction, and mass failures on slopes may give rise to gravity-flow deposits that form a part of the fill in adjacent basins. Although most basin fill is commonly thought to be moved through large canyons, it appears probable that significant quantities of sediment also may be transported downslope between canyons.

The southern California borderland is a tectonically active area, and it seems probable that sediment failure in this region results principally from earthquake shocks, although dynamic loading from storm waves passing over the 100-m platform off San Nicolas Island could be a significant triggering agent. Nine earthquakes of magnitude 4 to 6 have occurred in the 3000-km^2 region surrounding the study area within the 40 year period from 1932–1972 (Hileman and others, 1973). Many of the numerous smaller earthquakes that occurred during that period may have been capable of causing failure in unconsolidated sediments. It is noteworthy that the frequency of earthquakes and rate of sedimentation are low in the study area relative to other parts of the borderland. We anticipate that, in areas of higher seismicity and sedimentation rates, more evidence will be obtained showing that small-scale mass movement is a major process of sediment transport on basin slopes.

ACKNOWLEDGMENTS

We wish to thank F. N. Spiess and his staff at the Scripps Institution of Oceanography Marine Physical Laboratory for their cooperative efforts on this cruise. We thank M. A. Hampton, W. R. Normark, and H. G. Greene for their critical reviews of the manuscript.

REFERENCES

BUFFINGTON, E. C., AND MOORE, D. G., 1963, Geophysical evidence on the origin of gullies submarine slopes, San Clemente, California: Jour. Geology, v. 71, p. 356–370.

CARLSON, P. R., AND MOLNIA, B. F., 1978, Submarine faults and slides on the continental shelf, northern Gulf of Alaska: Marine Geotechnology, v. 2, p. 275–290.

COLE, M. R., 1970, Paleocurrent and basin analysis on San Nicolas Island, California [unpub. Masters thesis]: Athens, Ohio Univ., 110 p.

COLEMAN, J. M., AND PRIOR, D. B., 1978, Contemporary gravity tectonics—an everyday catastrophe? (abs): Am. Assoc. Petroleum Geologists Bull., v. 62, no. 3, p. 505.

COOK, H. E., 1979, Ancient continental slope sequences and their value in understanding modern slope development: this volume.

EMERY, K. O., AND BRAY, E. E., 1962, Radiocarbon dating of California basin sediments: Am. Assoc. Petroleum Geologists Bull., v. 46, p. 1839–1856.

FIELD, M. E., AND RICHMOND, W. A., in press, Sedimentary and structural patterns on the northern Santa Rosa-Cortes Ridge, southern California: Marine Geology.

GREENE, H. G., CLARKE, S. H., JR., FIELD, M. E., LINKER, F., AND WAGNER, H. C., 1975, Preliminary report on the environmental geology of selected areas of the southern California continental borderland: U.S. Geol. Survey Open-file report 75-596, 69 p.

HAMPTON, M. A., AND BOUMA, A. H., 1977, Slope instability near the shelf break, western Gulf of Alaska: Marine Geotechnology, v. 2, p. 309–332.

HANER, B. E., AND GORSLINE, D. S., 1978, Processes and morphology of continental slope between Santa Monica and Dume canyons, southern California: Marine Geology, v. 28, p. 77–87.

HEEZEN, B. C., AND DRAKE, C. L., 1964, Grand Banks slumps: Am. Assoc. Petroleum Geologists Bull., v. 48, p. 221–225.

HILEMAN, J. A., ALLEN, C. R., AND NORDQUIST, J. M., 1973, Seismicity of southern California region: California Inst. Technology, Div. of Geology and Planetary Sciences, Contr. No. 2385, 487 p.

KELLING, G., AND STANLEY, D. J., 1970, Morphology and structure of Wilmington and Baltimore submarine canyons, eastern United States: Jour. Geology, v. 78, p. 637–660.

LEWIS, K. B., 1971, Slumping on a continental slope inclined at 1–4°: Sedimentology, v. 16, p. 97–110.

MENARD, H. W., 1964, Marine geology of the Pacific: New York, McGraw-Hill, 271 p.

MOORE, D. G., 1961, Submarine slumps: Jour. Sed. Petrology, v. 31, p. 343–357.

———, 1969, Reflection profiling studies of the California continental borderland-structure and Quaternary turbidite basins: Geol. Soc. America Spec. Paper 107, 142 p.

———, 1977, Submarine slides, in Voight, B., ed, Rock slides and Avalanches, Vol. 1: Developments in Geotechnical Engineering, p. 563–604.

MOORE, T. C., JR., VAN ANDEL, T. H., BLOW, W. H., AND HEATH, G. R., 1970, Large submarine slide off north western continental margin of Brazil: Am. Assoc. Petroleum Geologists Bull., v. 54, p. 125–128.

PIPER, D. J. W., NORMARK, W. R., AND INGLE, J. C., JR., 1976, The Rio Dell Formation: a Plio-Pleistocene basin slope deposit in northern California: Sedimentology, v. 23, p. 309–328.

SPIESS, F. N., LOWENSTEIN, C. D., BOEGEMAN, D. W., AND MUDIE, J. D., 1976, Fine scale mapping near the deep-sea floor: Proceedings Oceans '76, Mar. Tech. Soc.—IEEE Ann. Mtg., p. 8A1–8A9.

UCHUPI, E., 1961, Submarine geology of the Santa Rosa-Cortes Ridge: Jour. Sed. Petrology, v. 31, no. 4, p. 534–545.

VARNES, D. T., 1958, Landslide types and processes, in Landslides and engineering practice, Highway Research Board Spec. Rept. 29, Natl. Research Council Pub. 544, p. 20–47.

VEDDER, J. G., BEYER, L. A., JUNDER, ARNE, MOORE, G. W., ROBERTS, A. E., TAYLOR, J. C., WAGNER, H. C., 1974, Preliminary report on the geology of the continental borderland of southern California: U.S. Geol. Survey Misc. Field Studies Map (MF) 624, 34 p., 9 maps.

VEDDER, J. G., AND NORRIS, R. M., 1963, Geology of San Nicolas Island, California: U.S. Geol. Survey Professional Paper 369, 65 p.

SLOPE AND BASIN BENTHIC FORAMINIFERA OF THE CALIFORNIA BORDERLAND

ROBERT G. DOUGLAS AND HAL L. HEITMAN[1]
Department of Geological Sciences
University of Southern California
Los Angeles, California 90007

ABSTRACT

The southern California borderland provides an excellent laboratory for the study of ecologic relationships of benthic foraminifera. Inner basins are shallow, anoxic and contain clastic detrital sediments that accumulate at rapid rates, and contain broad, well developed basin plains, and small areas of slope. Outer basins are deeper and relatively more oxygenated, contain sediments of largely hemipelagic origin, that accumulate at slow rates, and contain small irregular basin plains and large steep slopes.

Benthic foraminifera living in continental margin and basin environments in the southern California borderland can be divided into five distinct assemblages: Outer Shelf, Bank, Upper Slope, Lower Slope, and Basin Floor Assemblages. The biogeography of the assemblages is related to the distribution of sediments and water mass. Differences in the species composition of the assemblages in the borderland appear related to differences in the mode and rate of sediment accumulation in the nearshore and outer basins.

Analysis of bathymetric distribution of live foraminifera shows that the majority of species are heterobathyal and few species have sufficiently consistent upper depth limits to be considered isobathyal. Comparative studies of live and dead populations reveals that the live and dead assemblages have low similarity. We propose that the differences are due in part to shifts in species distributions in response to changes in shelf, slope and basin environments in the last several hundred years.

INTRODUCTION

The continental margin off southern California consists of roughly parallel rows of basins and banks or ridges which trend NE-SW and extend westward from the mainland shelf to the oceanic slope (Patton Escarpment) that descends to the deep ocean floor. This unique region, which resembles a basin and range topography, has been termed the continental borderland (Shepard and Emery, 1941). The result is a complex array of slope and basin environments ranging from shallow to deep, anoxic to oxygenated, and from high sedimentation rate basins dominated by terrigenous detritus to low sedimentation rate basins dominated by hemipelagic or biogenic debris. This array of environments provides an excellent laboratory to study the ecological relationships of benthic foraminifera.

The benthic foraminifera of the California borderland have been the object of investigation since the pioneering study of Natland (1933, 1957) who demonstrated that the bathymetric and temperature distribution of species in San Pedro Basin could be used to interpret the paleoecology of Late Neogene deposits in the Los Angeles and Ventura Basins. Most of the previous work has concentrated on the foraminifera of the mainland shelf and adjacent slope or basin (Zalesny, 1959; McGlasson, 1959; Bandy, 1953, 1964; Bandy et al, 1964a, b, c, 1965a, b; Uchio, 1960; Harmon, 1964). Information on the benthic foraminifera in the outer borderland is limited (Crouch, 1952; Resig, 1958; Kheradpir, 1970; Douglas et al, 1976; Blake, 1976).

Unfortunately, many of the earlier investigations did not distinguish the live from the dead foraminifera and all except the recent baseline studies are based on small size samples collected by gravity corers or grab samplers. These devices disturb the sediment-water interface and produce biased estimates of the live populations (Douglas, et al, 1976; in prep.).

The present study was initiated to characterize the distribution of live benthic species in slope and basin environments; to examine the bathymetric relationships of major faunal boundaries based on the upper depth limit of assemblages and important species; and to document the distribution of dead and relict populations of foraminifera with respect to present-day distribution of live populations.

Slope and Basin Environments of the Borderland

Individual basins across the borderland, and the slope-basin environments contained within them, possess characteristics that are related to 1) structural configuration (e.g., sill depth, basin

---

[1] Present Address: Union Oil Company of California, 2323 Knoll Drive, Ventura, California

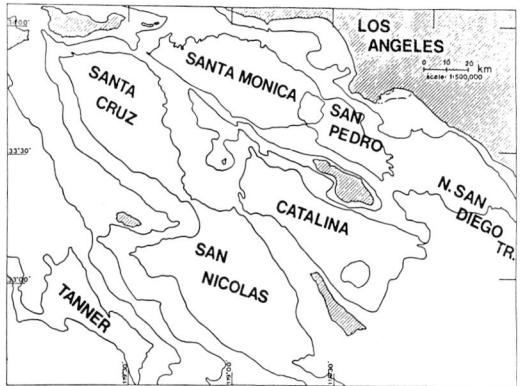

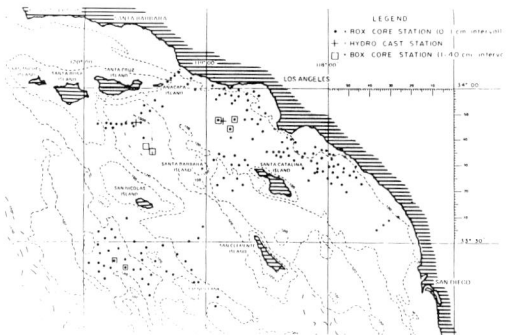

FIG. 1. A) Location of the basins in the central portion of the southern California borderland. The row of basins next to the mainland, Santa Monica, San Pedro and northern San Diego Trough, are referred to as the "nearshore" or "inner" basins; basins to the west of the islands are referred to as the "offshore" or "outer" basins. B) Location of the sample and hydrocast stations. Hydrocast data (temperature, salinity and oxygen) are shown in Figure 2.

walls, basin bottom depths); 2) sedimentary rates and processes; and 3) water quality existing at or below the sill depth.

The inner borderland basins occur east of the first line of islands and ridges, and include the northern extreme of the San Diego Trough, San Pedro and Santa Monica Basins (Fig. 1A). These basins are very broad and relatively shallow (900 meters and less) owing to higher sedimentation rates: detrital clastic material from the mainland has all but filled the inner basins, producing broad basin plains where deep basins with steep sides once existed. Because their sills are within the oxygen minimum zone, Santa Monica and San Pedro Basins are anoxic at water depths corresponding to the lower slope and basin floor. Oxygen values are at about 0.5 ml/l at 500 meter depth, and less than 0.1 ml/l on the central basin floor below the 850 meter isobath (Fig. 2). Basins of the outer borderland discussed here include Santa Cruz, Santa Catalina, San Nicolas, and Tanner Basins. These are generally much deeper basins (1300–1900 m) than those of the inner borderland and have deeper sill depths (1000–1100 m). Hemipelagic and biogenic sediments predominate in the outer basins so sedimentation rates

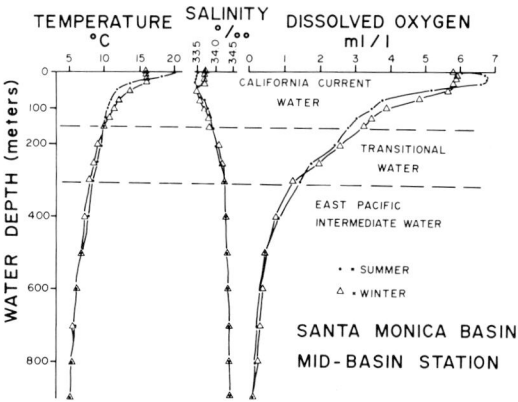

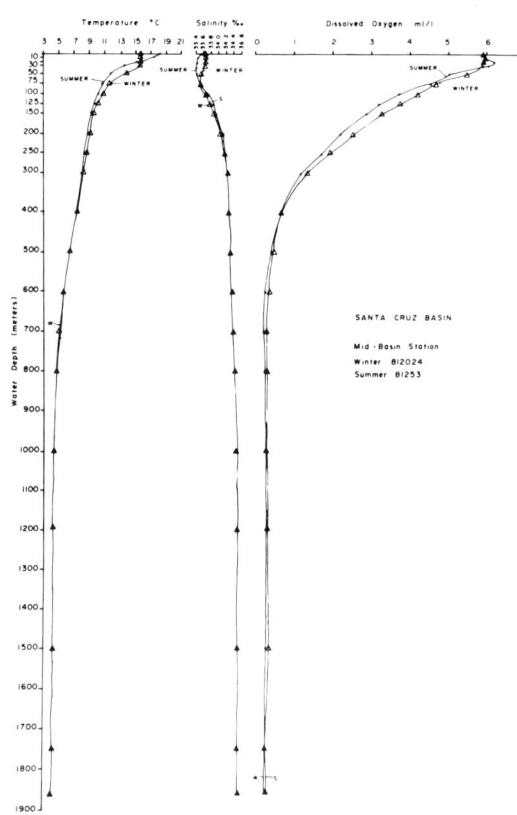

FIG. 2.—Distribution of temperature, salinity and dissolved oxygen in Santa Monica (2A) and Santa Cruz Basins (2B).

are much lower. This is, in part, responsible for their deeper, less well developed basin plains, and much larger areas of slope environment. Sediment textures of the outer basins and slopes are related to sediment source and to depositional processes. Basin bottom sediments are finest (mean phi ± 8) and are mainly hemipelagic in origin. Slope sediments are coarser owing to influx of clastic material from bank and ridge tops. Outer basin sill depths are deeper than in the inner borderland, and occur near the base of the oxygen minimum zone. Oxygen values at sill depth are, then, much higher (0.5 ml/l) and fluctuate slightly down to the basin bottoms (Fig. 2). Outer basin bottoms are, therefore, relatively more oxygenated than inner basin bottoms.

Data and Methods

Sediment samples used in foraminiferal analysis of slope and basin assemblages are subsamples of cores collected by a modified Reineck-NEL box corer, Soutar box corer, Soutar chemical grab sampler and prototype Soutar-Van Veen sampler. These devices are designed to obtain a large surface area sample without disturbing the sediment-water interface. Samples were taken from the top centimeter of the core and mixed with a solution of seawater, glutaraldehyde preservative and Rose Bengal stain. Each sample is equivalent to a square plug of sediment approximately 10 cm on a side and representing 100 cm^3 of sediment. All 0-1 cm samples were subsampled onboard ship; the remainder of the core sample was transported to the shore laboratory, extruded and subsampled at 2 cm intervals. Samples for foraminiferal analysis were wet-sieved over a 63 micron screen, dried and stored dry until counted.

A foraminifer was considered live at the time of collection if two or more chambers in the test or the area near the aperture absorbed sufficient Rose Bengal stain to turn a deep pink or rose color. Recent cytologic examination of stained specimens suggests that some stained individuals are not viable at the time of collection and that the dead protoplasm may be preserved for varying lengths of time under conditions of low oxygen and temperature. Live counts based on samples from nearshore low oxygen environments of the lower slope and basin floor are probably overestimated relative to samples from the more oxygenated and warmer environments of the upper slope and shelf.

Many of the samples used in the analysis of the pattern of live populations were originally analyzed as part of a baseline study of benthic foraminifera in the California bight (Douglas et al, 1976; Douglas, Cotton and Wall, in prep.). Details of the sample techniques, data reduction and computer analysis used is contained in these reports.

The analysis of live assemblages is based primarily upon the occurrence of numerically dominant species, that is, species which constitute more than 5% of the total live assemblage. Species having frequencies of less than 5% of the total are subject to large probable errors in sampling. Accurate documentation of their occurrence and distribution requires the use of replicate collections within the same sampling site (Douglas et al, in prep.).

LIVE ASSEMBLAGES

Living benthic foraminifera in the southern California borderland can be divided into five distinct assemblages: Outer Shelf, Bank, Upper Slope, Lower Slope and Basin Floor Assemblages. Each assemblage has been named for the physiographic habitat occupied by the dominant species.

The distribution of the assemblages is closely related to the patterns of sedimentation and water mass. Distinct differences in these regimes across the borderland are reflected in the distribution of the assemblages, and in their species composition. The Outer Shelf and Upper Slope Assemblages are restricted to terrigenous clastic substrates and limited to the mainland shelf and nearshore basins respectively, except for a few occurrences around the northern Channel Islands. The Bank Assemblage is presently restricted to the outer Borderland, west of the first line of islands and submarine ridges, whereas the Lower Slope and Basin Floor Assemblages are distributed throughout the region. There are, however, some species compositional differences between the assemblages in the nearshore and offshore basins.

Because of these biogeographic differences, the benthic faunas of the nearshore and the outer basins and slopes were analyzed separately by cluster analysis by Douglas et al, (1976), and we have followed this approach herein. The species composition and approximate bathymetric distribution of the assemblages is summarized in Table 1 and briefly discussed below.

Outer Shelf Assemblage

The shelf break in the nearshore region occurs at a water depth of about 90 meters and does not correspond to a faunal boundary: most outer shelf species also inhabit the uppermost slope, down to a depth of 120 to 140 meters.

Species characteristic of the shelf break area include *Nonionella stella*, *Bolivina acuminata*, *Bulimina denudata* and *Noninella basispinata*, which are usually dominant, and *Angulogerina angulosa*, *Epistominella sandiegoensis* and *Bolivina pacifica*, which become more important on

TABLE 1.—RECURRENT SPECIES ASSEMBLAGES PRESENT IN SLOPE AND BASIN ENVIRONMENTS IN THE SOUTHERN CALIFORNIA BORDERLAND. THE ASSEMBLAGES ARE BASED ON R-MODE CLUSTER ANALYSIS OF 192 SPECIES AT 146 STATIONS (DOUGLAS, ET AL., 1976). A SPECIES MAY OCCUR IN MORE THAN ONE ASSEMBLAGE BUT ONLY ITS MOST TYPICAL ASSOCIATION IS LISTED. ENVIRONMENTAL ATTRIBUTES ARE BASED ON PHYSICAL DATA COLLECTED AT THE SAME STATIONS AND IN NEARBY AREAS. THE MEAN RANGE OF VALUES ASSOCIATED WITH THE ASSEMBLAGE IN THE PRINCIPAL AREA OF ITS DISTRIBUTION IS GIVEN.

DR = DEPTH RANGE (METERS) OF THE ASSEMBLAGE
MUDL = MEAN UPPER DEPTH LIMIT OF THE ASSEMBLAGE (METERS)
WM = WATER MASS: CALIFORNIA CURRENT WATER, EAST PACIFIC INTERMEDIATE WATER, TRANSITIONAL WATER, I.E., WATER FORMED BY THE MIXTURE OF CALIFORNIA CURRENT AND INTERMEDIATE WATER FOUND AT DEPTHS OF 200 TO 400 METERS.
T°C = WATER TEMPERATURE
‰ = SALINITY
O_2 ml/l = DISSOLVED OXYGEN CONTENT
MΦ = MEAN GRAIN SIZE OF THE SEDIMENT, GIVEN IN PHI UNITS
%CRB = CALCIUM CARBONATE CONTENT, IN PERCENT, OF THE TOTAL SEDIMENT
%ORG = TOTAL ORGANIC CARBON CONTENT, IN PERCENT, OF THE TOTAL SEDIMENT

NEARSHORE BASINS ASSEMBLAGES	ENVIRONMENTAL ATTRIBUTES	OFFSHORE BASINS ASSEMBLAGES	ENVIRONMENTAL ATTRIBUTES
OUTER SHELF ASSEMBLAGE		**BANK ASSEMBLAGE**	
Nonionella stella Cushman and Moyer (NOL STZ)**	DR = 50–200	(*Angulogerina angulosa* (Williamson))[1]	DR = 20–400
N. basispinata (Cushman and Moyer) (NOL BSP)	MUDL = 50	*Cassidulina limbata* Cushman and Hughes	MUDL = 50
Bulimina denudata Cushman and Parker (BUM DND)	WM = Calif. Current	*C. tortuosa* Cushman and Hughes (CAS TRT)	WM = Calif. Current & Transitional
Textularia schenki Cushman and Valentine (TEX SCK)	T°C = 9–15	*C. depressa* Asano and Nakamura (CAS DPS)	T°C = 7–12
Buccella angulata Uchio (BUC AGA)	‰ = 33.5–34.0	*Cibicides fletcheri* Galloway and Wissler	‰ = 33.4–34.2
Ammotium planissimum (Cushman) (AMT PSM)	O_2ml/l = 2.5–6.5	*Islandiella subglobosa* (Brady)	O_2ml/l = 0.8–5.5
Quinqueloculina laevigata d'Orbigny (QUN LVG)	MΦ = 3.5–5	*Rotorbinella campanulata* (Galloway and Wissler) (RTL CMU)	MΦ = 2–4
	%CRB = 3–12	*Rosalina columbiensis* (Cushman) (ROS COB)	%CRB = 25–36
	%ORGC = 1.0	*Hanzawaia nitidula* Lankford (HAN NTI)	%ORGC = 1.1
BANK ASSEMBLAGE		*Ehrenbergina compressa* Cushman (EHR CPE)	
Angulogerina angulosa (Williamson) (ANG ANO)	DR = 50–200	*Astrononion stellatum* Cushman and Edwards (ASN STL)	
Cassidulina limbata (Cushman and Hughes) (CAS LMB)	MUDL = 50	*Spirillina vivipara* Earland (SPR VVP)	
Cibicides fletcheri Galloway and Wissler (CIB FLC)	WM = Calif. Current		
Islandiella subglobosa (Brady) (ISL SGB)	T°C = 9–15		
Cancris auricula (Fitchel and Moll) (CAN AUR)	‰ = 33.5–34.0		
Trochammina pacifica Cushman (TRO PAC)	O_2ml/l = 2.5–6.5		
Pullenia salisburyi Steward and Steward (PUL SLB)	MΦ = 3		
	%CRB = 5–12		
	%ORGC = 0.9		
UPPER SLOPE ASSEMBLAGE			
Bolivina pacifica Cushman and McCulloch (BOL PAC)[2]	DR = 85–450		
Globobulimina pacifica Cushman (GLB PAC)	MUDL = 125		
Suggrunda eckisi Natland (SUG ECK)[3]	WM = Transitional		
	T°C = 7–10.5		

UPPER SLOPE ASSEMBLAGE

Balivina humilis Cushman and McCulloch (BOL HML)

Not present as a recurrent assemblage; isolated occurrences of the starred (*) species found around the channel islands.

Environmental parameters:
- ‰ = 33.8–34.2
- O_2 ml/l = 0.5–3.0
- MΦ = 4–6
- %CRB = 1–5
- %ORGC = 1–3

Species:
- *Alliatina primitiva* (Cushman and McCulloch) (ALT PMV)
- *Epistominella sandiegoensis* Uchio (EPL SDG)
- *Alveolophragmium columbiense* (Cushman) (AVM COB)
- *Uvigerina juncea* Cushman and Todd (UVG JNC)
- *U. excellens* Todd (UVG ECL)
- Bulimina cf. *B. denudata* Cushman and Parker (BUM CDN)
- Cassidulina depressa Asano and Nakamura (CAS DPS)

LOWER SLOPE ASSEMBLAGE (Group I MUDL = 350, Group II MUDL = 500)

Environmental parameters:
- DR = 400–950
- MUDL = 450
- WM = E. Pac. Inter.
- T°C = 4.8–7
- ‰ = 34.2–34.4
- O_2 ml/l = 0.1–0.8
- MΦ = 5–7.5
- %CRB = 4–9
- %ORGC = 2–6

Group I Species
- *Bolivina argentea* Cushman (BOL ARG)
- *Loxostomum pseudobeyrichi* (Cushman) (LOX PDB)[4]
- *B. spissa* Cushman (BOL SPS)
- Bolivinita minuta Natland (BOT MNT)
- Cassidulina delicata Cushman (CAS DLC)
- C. subcarinata Uchio (CAS SCI)

Group II Species
- *Epistominella smithi* (Steward and Steward) (EPL SMT)
- *Uvigerina curticosta* Cushman (UVG CTA)
- U. peregrina Cushman (UVG PRG)
- Valvulineria araucana d'Orbigny (VOL ACN)

Less Common, Wide Spread Species
- Ammodiscus hoeglundi (Uchio) (AMD HGD)
- Eponides leviculus (Resig) (EPO LVI)
- Bolivina bicostata Cushman (BOL BCA)
- Karreriella parkerae Uchio (KAR PKR)

LOWER SLOPE ASSEMBLAGE (Group I MUDL = 400, Group II MUDL = 900)

Environmental parameters:
- DR = 300–1900
- MUDL = 400
- WM = E. Pac. Inter.
- T°C = 4.1–8
- ‰ = 34.3–34.4
- O_2 ml/l = 0.3–1.1
- MΦ = 5–8
- %CRB = 1–9
- %ORGC = 1–6

Group I Species
- *Bolivina argentea* Cushman
- *B. spissa* Cushman
- Bolivinita minuta Natland
- *Cassidulina delicata* Cushman
- *C. subcarinata* Uchio
- Astrononion incilis Lankford (ASN INC)
- Globobulimina spinifera Cushman (GLB SIF)
- Valvulineria glabra Cushman (VAL GAA)
- Chilostomella ovoidea Reuss (CHL OVD)[5]
- Haplophragmoides quadratus Uchio (HAP QDR)

Group II Species
- *Epistominella smithi* (Steward and Steward)
- *Uvigerina curticosta* Cushman
- U. peregrina Cushman
- Valvulineria araucana d'Orbigny
- *Eponides leviculus* (Resig)
- *Gyroidina io* Resig (GYR IO)
- Bulimina mexicana Cushman (BUM MXC)
- *Hoeglundina elegans* (d'Orbigny) (HOG ELG)
- Haplophragmoides neobradyi Uchio (HAP NBD)[5]

Less Common Species
- Ammodiscus hoeglundi (Uchio)
- Karreriella parkerae Uchio
- Trochamminina globigeriniformis (Parker and Jones)[5] (TRO GBG)
- Cancris inequalis (d'Orbigny) (CAN IQL)
- T. charlottensis Cushman (TRO CLT)
- Proteonina atlantica Cushman (PRN ALI)

TABLE 1.—CONTINUED

BASIN FLOOR ASSEMBLAGE		BASIN FLOOR ASSEMBLAGE	
*Fursenkoina apertura (Uchio) (FRS APT)	DR = 550–950 M	*Fursenkoina apertura (Uchio)	DR = 1200–1900
*F. bramlettei (Galloway and Morey) (FRS BMT)	MUDL = 750	F. bramlettei (Galloway and Morey)	MUDL = 1800
*F. seminuda (Natland) (FRS SMD)	WM = E. Pac. Inter.	F. seminuda (Natland)	WM = E. Pac. Inter.
*Cassidulinoides cornuta (Cushman) (CSN CRN)	T°C = 4.8–6.1	*Cassidulinoides cornuta (Cushman)	T°C = 4.1–6.2
Globobulimina affinis (d'Orbigny) (GBL AFN)	‰ = 34.3–34.4	*Loxostomum pseudobeyrichi (Cushman)[4]	‰ = 34.2–34.4
G. hoeglundi Uchio (GBL HGD)	O_2 ml/l = 0.08–0.9	*Buliminella tenuata (Cushman)	O_2 ml/l = 0.25–0.5
*Textularia earlandi Parker (TEX ELD)	MΦ = 5–7	*Chilostominella ovoidea Reuss[5]	MΦ = 6–8
*Buliminella tenuata (Cushman) (BUL TAA)	%CRB = 10–15	Alveolophragmium evolutum (Natland)	%CRB = 20–25
Chilostominella ovoidea Reuss (CHL OVD)	%ORGC = 6–8	Globobulimina hoeglundi Uchio	%ORGC = 5–8
Alveolophragmium evolutum (Natland) (AVM EVL)		Reophax horridus Cushman	
A. advena (Cushman) (AVM ADV)		R. micaceous Earland	
Reophax horridus Cushman (REO HRD)		Saccammina longicollis (Wiesner) (SAC LCO)	
R. micaceous Earland (REO MIC)		*Bolivina pacifica Cushman[2]	
		Haplophragmoides neobradyi (Uchio)[5]	
		Trochammina globigeriniformis (Parker and Jones)	

Explanation

*Species that are numerically dominant (greater than 5% of the total assemblage) or have consistent occurrences at greater than 1% of total.
**Species abbreviation, in parntheses, are used in Figure 3.
+Physical environmental data from several sources, including CALCOFI, Bureau of Land Management, S. Calif. Baseline study, 1975–1978, and Natural Science Foundation supported study of the California Borderland, 1976–1978, unpublished.

1. *Angulogerina angulosa* is more common in the insular shelf assemblage in the outer borderland.
2. *Bolivina pacifica* is an opportunistic species that occurs commonly in areas of sediment disturbance; in the outer borderland it is most characteristic of the basin floor.
3. *Suggrunda eckisi* may occur in considerable abundance in the basins and is a dominant species of the Basin Floor Assemblage in Santa Barbara Basin. However, to the south, it is more typical of the upper slope.
4. *Loxostomum pseudobeyrichi* becomes a Basin Floor species in the outer basin though it may occur on the slope.
5. These species cluster with the Lower Slope Assemblage but are typical taxa in the basin floor.

the upper slope (Fig. 3). (Note: See Appendix A for complete listing of species names). Several agglutinated species are associated with the Outer Shelf Assemblage, particularly *Eggerella advena, Eggerella scrippsi, Textularia schencki* and *Alveolophragmium columbiense*, that are most numerous in summer months (Douglas et al, in prep.).

Occurrences of the more abundant species of the assemblages, such as *N. stella, E. advena,* and *B. denudata,* are commonly found (sometimes as numerically dominant species) in slope and basin samples and attest to the constant transport of live shallow water taxa into deeper water (Fig. 3).

The Outer Shelf Assemblage is widely distributed on the mainland shelf, except in areas of relict sediments and outcrops, and around the Channel Islands where the substrate is predominantly detrital sediment. The lower depth limit of the assemblage coincides with the bottom of the seasonal layer and the lower boundary of pure California Current Water (defined at approximately the 10° C isotherm and 33.8 ‰ isohaline).

Bank Assemblage

The fauna occurring on the bank tops and ridges of the outer borderland and a few areas on the mainland shelf is dominated by *Cassidulina limbata, C. tortuosa, Islandella subglobosa* (and subspecies), *Angulogerina angulosa,* and smaller populations of *Cancris auricula, Buccella angulata* and *Cibicides fletcheri.* Three other species, *Ehrenbergina compressa, Cassidulina depressa,* and *Hoeglundina elegans* are common or numerically dominate in the outer banks area but are rare or absent in the mainland shelf.

Occurrences of the assemblage on the mainland shelf are restricted to relict sediments and outcrops in Santa Monica Bay and in San Pedro Bay, although dead occurrences of the assemblage are widespread over most of the mainland shelf and uppermost slope. For this reason, we believe that the assemblage is presently relict in the nearshore region and was more abundant and widespread in earlier Holocene time. Further, the assemblage was prevalent in the nearshore region during Pleistocene time, as evidenced by the occurrence of species of the Bank assemblage in the Lomita Marl, and other outcropping deposits (Blake, 1976).

In the greater part of its biogeography, the Bank Assemblage is associated with carbonate-rich substrates composed of either bioclastic debris or relict foraminiferal sand with glauconite and basalt fragments. Around the larger islands such as Santa Rosa and Santa Cruz Islands, where the shelf is covered by clastic sediment, the assemblage is replaced by an insular equivalent of the Outer Shelf Assemblage.

The Bank Assemblage extends from bank top depths, often less than 50 meters, to depths in excess of 300 meters. The lower depth limit of the assemblage is difficult to pinpoint because of the scarcity of sediment cover (which makes sampling difficult) on the slopes and ridges in the outer borderland, but it is probably less than 400 meters. The lower limit is approximately coincident with the top of the East Pacific Intermediate Water (defined by the 8° C isotherm and 34.3 ‰ isohaline).

Upper Slope Assemblage

The most distinctive association in the nearshore basin slopes is an assemblage dominated by *Epistominella sandiegoensis, Uvigerina juncea, U. excellens, Globobulimina pacifica* (and in some locations, *G. hoeglundi*), *Suggrunda eckisi* and *Alliatina primitiva.* Various agglutinated taxa are usually associated with the assemblage, the most notable being species of *Reophax* and *Alveolophragmium columbiense.* Abundant occurrences of key species such as *E. sandiegoensis* and *Bolivina pacifica* may occur on the outer shelf during summer months when dissolved oxygen at these depths decreases by more than 1.0 ml of O_2/l.

The upper depth limit of the assemblage is between 125 and 140 meters and coincides with the top of the transitional zone between California Current Water and East Pacific Intermediate Water (Fig. 4). Abundant occurrences of the assemblage is limited to depths of less than about 300 meters in northern San Diego Trough and to about 250 meters in San Pedro and Santa Monica Basins, although several species range into deeper water. Distribution of the assemblage is closely associated with the band of active sedimentation on the upper slope; in areas of slumping (as indicated by high resolution seismic reflection profiles) *Bolivina pacifica* and *Globobulimina pacifica* become the most abundant species.

The Upper Slope Assemblage is geographically restricted to the mainland slope except for rare occurrences of a few of the key species (e.g., *E. sandiegoensis, U. juncea, G. pacifica*) at shelf-break and slope depths around the northern Channel Islands. Several species of the assemblage have disjunct distributions. For example, *Suggrunda eckisi* and *Globobulimina pacifica* are important species of the upper slope. They are also present on the basin floor in Santa Monica and San Pedro Basins but are connected by few or no live occurrences on the lower slope in these basins.

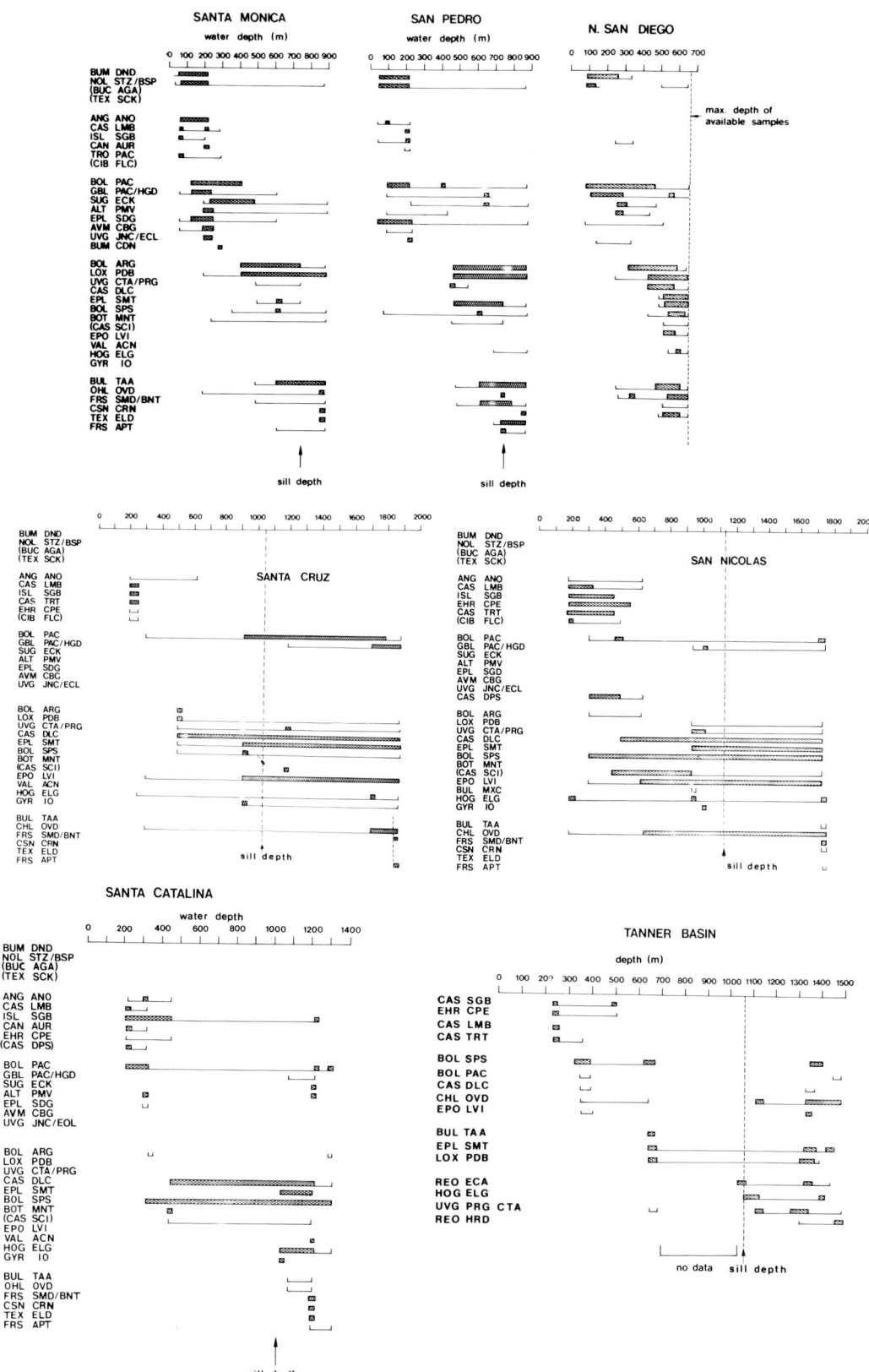

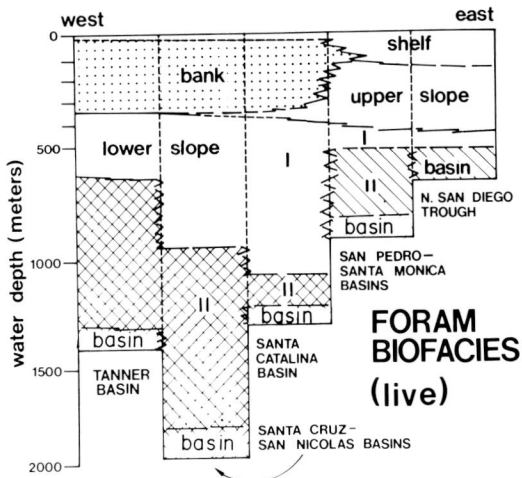

FIG. 4.—Schematic diagram showing the distribution of the recurrent assemblages in an east-west profile across the borderland. Group II species of the Lower Slope Assemblage are indicated by pattern: diagonal lines indicate Group II species in the nearshore basins; crosshatched indicates Group II species in the outer basins. See Table 1 for the species in the recurrent assemblages. Note that in the northern San Diego Trough, the distribution of the Group II, Lower Slope Assemblage and the Basin Floor Assemblage overlap. In the outer basins, species of the Lower Slope Assemblage are commonly mixed with species of the Basin Floor assemblage, probably because of the downslope transportation of sediment.

Lower Slope Assemblage

On the basin slopes at a depth of between 300 and 400 meters, a faunal change is apparent, that is marked by the first appearance of *Bolivina argentea, Loxostomum pseudobeyrichi* and *Bolivina spissa*. This change marks the upper depth limit of the Lower Slope Assemblage.

Species diversity within this habitat is high, as more than 90 species have been recorded. However, most of them have patchy distributions and small populations so that about nine (9) abundant and recurrent species best characterize the assemblage of the lower slope. Species differences between the top and bottom of the lower slope allows the assemblage to be subdivided into two groups. Species differences also exist between the nearshore and outer basins. For this reason, the assemblage is discussed separately in terms of the two regions.

Nearshore Basins

On the inner slopes of the nearshore basins the benthic foraminiferal community is characterized by large populations of *Bolivina argentea, Bolivina spissa, Loxostomum pseudobeyrichi* and smaller numbers of *Uvigerina curticosta, Buliminella tenuata, Suggrunda eckisi, Chilostomella ovoidea, Bolivina pacifica* and species of *Globobulimina*. Agglutinated foraminifera include species of *Reophax, Alveophragmium* and *Karreriella*. These species are referred to as the Lower Slope Assemblage, Group I.

Group I species are widely distributed on the mainland slope between 300 and 450 to 500 meters. (Many also occur in basin depths (Fig. 3A, B). In the outer borderland, the group is less well developed and joined by species not present in the nearshore basins. Group I species inhabit the zone lying immediately above the oxygen minimum layer where oxygen values range from 1.5 ml/l to less than 0.5 ml/l. In this zone, organic carbon content of the sediments increases to about 2-3%, approximately twice the amount of organic matter occurring in upper slope and outer shelf sediments.

At depths of 450–500 meters, a subtle change occurs in the assemblage. *Epistominella smithi* and *Valvulineria araucana* appear, *Buliminella tenuata* increases in abundance and *Suggrunda eckisi, Bolivina pacifica* and *Bolivina seminuda* disappear or become very rare. These changes define the Lower Slope Assemblage, Group II (Fig. 3A, B). In Santa Monica and San Pedro Basins, *Uvigerina curticosta* and *Cassidulina delicata* are associated with Group I but they are common elements of Group II in northern San Diego Trough. Most species in the group range into basin depths, approximately 900 meters.

The upper depth limit of the Group II species is closely associated with the top of the East Pacific Intermediate Water in the nearshore basins and coincident with the depth at which the slope tends to broaden and become less inclined. Oxygen values range from less than 0.5 ml/l to less than 0.2 ml/l where the slope merges with the basin floor.

Outer Basins

In the outer basins of the borderland, the species composition and bathymetric range of the Lower Slope Assemblage, including both Group I and

FIG. 3.—Bathymetric distribution of numerically dominant or key species in the recurrent assemblages. Thin line indicates the total depth range; heavy bar indicates the depth range where the species is 5% or more of the total live assemblage. Key to species abbreviations is in Table 1 and Appendix A. Sill depth is shown by dashed line.

Group II species, is different than in the nearshore basins (Fig. 3C, D, 4).

The most abundant Group I species include *Bolivina spissa* and three species that are rare or missing in the nearshore region: *Cassidulina delicata*, *C. subcarinata* and *Eponides leviculus*. *Bolivina argentea* and *Loxostomum pseudobeyrichi* are rare, and *Uvigerina curticosta*, generally associated with Group I in the nearshore basins, is found in deeper water. However, in Santa Cruz and San Nicolas Basins, in the depth range between 300 and 900 meters, few samples contain abundant live foraminifera, and there is evidence of extensive slumping (Mitsushio et al, 1978). Group I species of the nearshore region then, may be more common than our data suggests.

Group II species are well represented in the outer basins and include, in addition to the species in the nearshore region, *Hoeglundina elegans*, *Eponides leviculus*, *Gyroidina io*, *Bulimina mexicana*, *Uvigerina curticosta* and *U. peregrina*.

The upper depth limit of Group II species is close to sill depths in San Nicolas (900 meters), Santa Cruz (900 meters) and Santa Catalina (1000 meters) Basins, but is at 600 meters in Tanner Basin. The lower limit of the Lower Slope Assemblage is in excess of 1200 meters in Tanner and Santa Catalina Basins and 1800 meters in San Nicolas and Santa Cruz Basins. Group II species inhabit a depth interval of 200 to nearly 900 meters in the outer borderland.

Except for the close proximity to sill depth in the three central basins (Santa Catalina, San Nicolas and Santa Cruz), there is no obvious correlation of any environmental feature (sediment or water quality) and the boundary between Groups I and II of the Lower Slope Assemblage in the outer basins.

Temperature and salinity are essentially constant below sill depth and a small variation in oxygen in San Nicolas and Santa Cruz Basins occur below the sill depths and are not coincident with the faunal change. However, a shift in sediment texture from fine sandy silt to silt and clay at sill depths may signal the presence of subtle sedimentological processes that are important in controlling species distributions. The standing crop of foraminifera and most macrobenthos (Fauchald and Jones, 1976) is higher above sill depth (Figure 5).

Basin Floor Assemblage

A faunal change occurs at the point where the slope merges with the floor of the basin: several slope species, previously occurring only in low numbers, increase in abundance and are joined by new species to form the Basin Floor Assemblage.

Important members of the assemblage are *Buli-*

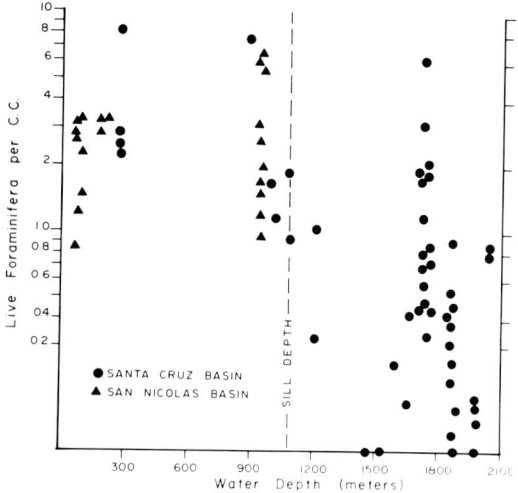

Fig. 5.—Variation in the standing crop (number of live individuals/cc) in basins of the outer borderland. Note the significant decrease in the standing crop below the sill depth.

minella tenuata, *Chilostomella ovoidea* and species of *Globobulimina*, all inherited from the slope, and the species: *Cassidulinoides cornuta*, *Fursenkoina apertura*, *F. bramletti* and *F. seminuda*. In the nearshore basins these species occur with *Textularia earlandi*, which is commonly the dominant species. In the outer basins, *T. earlandi* is absent and replaced by other agglutinated taxa, in particular *Haplophragmoides neobradyi*, *Ammodiscus pacifica* and *Trochammina globigeriniformis*.

In the central broad plains of the nearshore basin, especially in the areas of smooth topography characterized by hemipelagic sedimentation, the Basin Floor Assemblage is the dominant group of foraminifera present. These areas, below sill depth, have the lowest dissolved oxygen values in the borderland: less than 0.1 ml/l. The low oxygen values reduce the macrobenthos, particularly subsurface and burrowing types (Hartman, 1956, 1966; Fauchald and Jones, 1976), and result in either slightly mottled or laminated sediments. Foraminiferal densities are very high in the laminated sediments (exceeding 100 individuals/cc) and lower in the mottled sediments found towards the edge of the central plain. A complex interaction of topography, sedimentation, oxygen values, and macrobenthos appear to control the occurrence of the Basin Floor Assemblage.

Away from the central basin area, and in the proximity of the basin sides, the Basin Floor Assemblage becomes mixed with species from the lower slope; the proportion of slope species depends upon the amount and origin of sediments transported into the basin. The character of the basinal foraminiferal assemblage may vary considerably. In several of the outer basins there is evidence of large scale slumping of slope sediment (Mitsushio et al, 1978; unpub. data). Basin Floor species are then a small component in an assemblage composed mostly of species from slope and bank habitats. The standing crop of foraminifera in the outer basin averages less than one individual per cm^3, but varies considerably from sample to sample. Samples from the areas of slumping on the west slope of Santa Cruz Basin contains few or no living individuals.

Bathymetric Distribution

The depth range of live foraminifera in the borderland were compared in order to better understand the use of benthic foraminifera as indicators of water depth. In determining depth range, consideration was given to sample spacing and the pattern of species abundance. Greatest reliance was placed on species occurrence that exceeded 5% of the total assemblage (Fig. 3A, B, C,).

Because foraminifera vary considerably in time and space, determination of whether a species (or an assemblage) is isobathyal (same upper depth limit (UDL) in different locations) or heterobathyal (different UDL in different locations) (Bandy and Chierici, 1966), is in part dependent on the season in which samples are collected. For example, populations of *Bolivina argentea* and *Epistominella smithi*, both of which have more or less consistent UDL and are considered isobathyl in southern California waters, vary in numbers by a factor of 3 to 10 between winter and summer seasons in deeper slope stations (Douglas et al, in prep.). These deeper dwelling species can also occur in shallower water during summer months, when oxygen levels in shallow water are depressed. Less common species may have even larger seasonal fluctuation in abundance.

A further difficulty in defining depth ranges occurs with species having discontinuous distributions. High density sampling in the nearshore basins indicate that several species, e.g., *Suggrunda eckisi*, occur in well defined zones on the upper slope and in the basin, but few or no live occurrences connect the patches.

Uvigerina curticosta and *U. peregrina*, which occur in mid-slope depths (400–600 meters) in the Santa Monica Basin and the Northern San Diego Trough, are restricted to below sill depths (greater than 1000 meters) in San Nicolas and Santa Cruz and below 600 m in Tanner Basin (Fig. 3B, C). This pattern is of interest because *Uvigerina peregrina* is considered an isobathyal species of the upper slope (200 meters) in the Gulf of Mexico (Pflum and Frerichs, 1976).

Several conclusions can be drawn from the depth range data:

1. The majority of species are heterobathyal and of the 100 or so more common taxa, only eight have sufficiently consistent upper depth limits to be considered isobathyal. The more important species are indicated in Figure 3.

2. Species that are isobathyal in the inner basins, such as *Bolivina argentea*, *Loxostomum pseudobeyrichi*, *Epistominella smithi*, *Valvulineria araucana* and *Globobulimina pacifica* have different depth limits in the outer basins. The reason for the difference is unknown but it illustrates again that the factor(s) that regulate bathymetric habitats are not controlled by water depth per se. Faunal-environmental correlation suggest that species distribution are most closely related to water mass and substrate (Douglas et al, 1976).

3. The UDL of the Upper Slope (120–140 meters) and the Lower Slope (400–500 meters) Assemblages define fairly consistent depth horizons. However, the UDL of the Lower Slope, Group II species, and the Basin Floor Assemblage occur at water depths differing by nearly 900 meters between the inner and outer basins (Fig. 6).

DISTRIBUTION OF RELICT SPECIES AND THEIR PALEOECOLOGICAL IMPLICATIONS

Comparative studies of the distribution of live and dead assemblages of benthic foraminifera reveals that, except for marginal marine environments (tidal marsh and lagoon), live and dead assemblages have a low degree of resemblance (Murray, 1976; Uchio, 1960; Douglas et al, 1976). Uchio (1960), for example, found in slope and shelf environments off San Diego (Calif.) that surface samples contain dead species not presently extant in the area and that the distribution of live assemblages differed significantly from that of the total (live plus dead) assemblages.

Several reasons have been offered to explain the lack of resemblance, including postmortem transportation, introduction of fossil species, and preservation (Murray, 1976). Here we consider the possibility that the differences may be due to recent shifts in species distribution.

Our analysis is based on the foraminifera contained in the top centimeter of sediment in 108 widely distributed box cores and in down-core samples from selected cores in Santa Monica, Santa Cruz and Tanner Basin.

In each sample, the equivalent of the live assemblage was subtracted from the dead assemblage and the remaining species considered a residue assemblage. Species whose proportions between live and dead differed by more than a

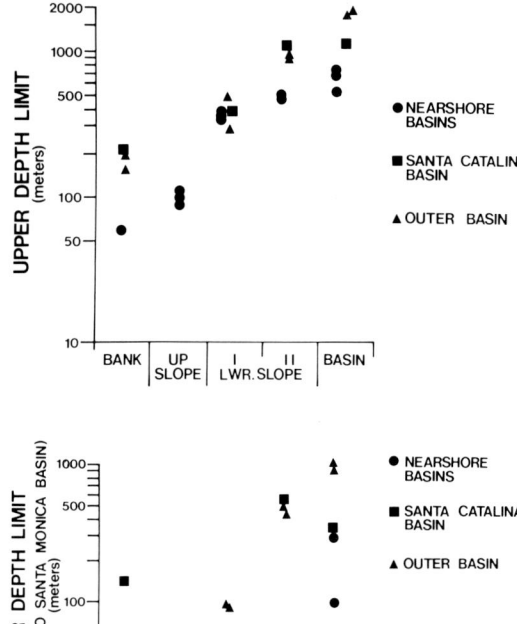

FIG. 6.—A) Upper depth limit of the major recurrent assemblages in the different basins of the borderland. B) Upper depth limit of the major recurrent assemblages relative to their distribution in Santa Monica Basin. Note the wide difference in upper depth limit for Lower Slope and Basin Floor Assemblages.

factor of 10 were included in the residue. For example, a live species that constituted 1% of the total and was represented in the dead assemblage by 10% or more of the total was included as a residue species.

The residue assemblage consists of species of three (3) different origins:

1. *Displaced species.* Species endemic to shallower water habitats, mostly shelf, which had been transported to the slope or basin.

2. *Fossil species.* Species from early Pleistocene, Pliocene and Miocene deposits that are not now living in the borderland.

3. *Relict species.* Species presently living in the borderland that are either rare (generally less than 1% of the assemblage) or not live at the time of collection although the species constitutes more than 5% of the dead assemblage in the sample. In many cases, relict species are the dominant taxa of the dead assemblage.

In a few samples, live species were found that were not represented in the dead assemblage, but they form no apparent pattern of distribution and appear to be the result of random dispersal.

The relict species form recurrent groups with distinct patterns of geographic distribution and can be easily identified ecologically in terms of present-day live assemblages. These groups are summarized in Table 2.

In outer shelf and upper slope samples of the mainland margin the most important relict species are *Cassidulina limbata, C. tortuosa, C. translucens,* and in some locations, *C. subglobosa, Cibicides fletcheri* and *Angulogerina angulosa.* Except for scattered live occurrences of these species in Santa Monica Bay and San Pedro Bay, they are found only in the dead assemblage. They are

TABLE 2.—RELICT SPECIES PRESENT IN SAMPLES FROM THE OUTER SHELF, SLOPE AND BASIN ENVIRONMENTS. RELICT SPECIES ARE DEFINED AS TAXA NUMERICALLY DOMINANT IN THE DEAD ASSEMBLAGE BUT NOT PRESENT OR VERY RARE IN THE LIVE ASSEMBLAGE FROM THE SAME SAMPLE.

Relict Species*	Outer Shelf	Upper Slope	Lower Slope	Basin Floor	Geol. Equival.
Cassidulina limbata	+	+			BANK
C. tortuosa	+				BANK
C. translucens	(+)				BANK
Cibicides fletcheri	+				BANK
Islandiella subglobosa	+				BANK
Angulogerina angulosa	+	+			BANK
Bolivinita minuta		+	+		LOWER SLOPE
Cassidulina subcarinata			+		LOWER SLOPE
C. delicata			+	+	LOWER SLOPE
C. depressa			+		LOWER SLOPE
Eponides leviculus			++		LOWER SLOPE (outer Basin)
Gyroidina io			++		LOWER SLOPE (outer Basin)
Bulimina mexicana			++		LOWER SLOPE (outer Basin)
Valvulineria araucana			++		LOWER SLOPE (outer Basin)

also present as a prominent part of the fossil assemblage found in samples below the sediment interface (Douglas et al, 1976). However, these relict species are presently living in the bank and ridge environments of the outer borderland and include four of the most characteristic species of the Bank Assemblage.

There is no doubt that the relict species in some of the surface sediment and the subsurface fossil samples are of pre-Holocene age and have either not been buried or are being "leaked" to the surface by bioturbation. However, it seems unlikely that this is the general case because: 1) the species occur at most stations and 2) high resolution seismic reflection profiles reveal that Holocene sediments are greater than 1.5 m thick over most of the Santa Monica and San Pedro shelves (Nardin and Henyey, 1978).

The relict species, once well established, and an important part of mainland shelf communities have, in late Holocene time, become restricted to west of the Channel Islands.

In the slope and basin environment of Santa Monica and San Pedro Basins, four species, *Bolivinita minuta, Cassidulina subcarinata, C. delicata* and *C. depressa,* are relict in surface samples but are living in southern San Diego Trough (Uchio, 1960) and the outer basins. However, all of these species are common in down-core samples in Santa Monica and indicate that they were formerly more abundant and widespread in the nearshore region. It appears that the species have restricted their biogeography to the western basins in recent time.

Lower Slope Assemblage, Group II species in the outer borderland includes six species that are either absent or very rare in the nearshore basins: *Gyroidina io, Eponides leviculus, Bulimina mexicana, Hoeglundina elegans, Valvulineria araucauna* and *Gyroidina altiformis.* However, these species are present in down-core samples from Santa Monica Basin, and indicate that they were inhabiting the nearshore region until recently. Based on age estimates derived from the laminated cores (Malouta, 1977) most of the taxa were living in Santa Monica until about 300 year BP.

Lastly, a comparison of the pattern of down-core benthic assemblages for cores from Santa Monica, Santa Cruz and Tanner Basins shows that there has been a recent shift in the proportions of Lower Slope to Basin Floor species in these basins over a period of the last several hundred of years (Fig. 7). In each case, the Basin Floor Assemblage has increased relative to Lower Slope species, and in Santa Monica Basin, now dominates the foraminiferal assemblage in the central basin plain. In part, the increase can be related to diagenetic destruction in the subsurface samples of Basin Floor species such as *Textularia*

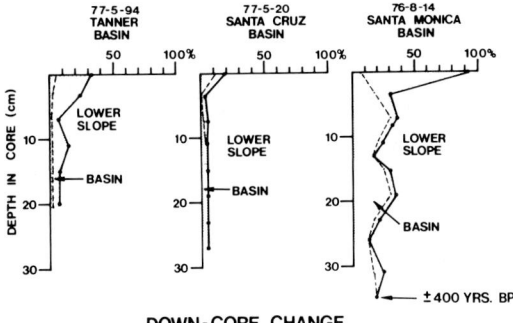

FIG. 7.—Down core variation in the proportion of Slope and Basin Assemblages in cores from Santa Monica, Santa Cruz and Tanner Basins. Dashed line indicates the down core change in species with preservation resistant shells; solid line is for total species. Several of the important species of the Basin Floor Assemblage are fragile (e.g., *Textularia earlandi, Suggrunda eckisi*) and their absence in down core samples is in part due to poor preservability.

earlandi, Suggrunda eckisi and *Cassidulinoides cornuta,* whose delicate tests are easily destroyed.

The analysis of relict species in core top and down-core samples in the borderland produces a consistent picture of faunal shifts in benthic foraminiferal assemblages from bank, slope and basin floor environments. Species that are presently residing in the outer borderland were, until some time in the last several hundred years, more widely distributed in the nearshore basins of Santa Monica, San Pedro and northern San Diego Trough. The pattern of faunal shift is the same in each case, to the west (Fig. 8). It is possible that the migration is yet incomplete and that the occurrences of relict species in isolated patches in the nearshore region are evidence that the benthic fauna are in transition.

The pattern of occurrence of the affected species, nearly all of which are now restricted to the outer borderland, suggests that the faunal shift is related to a change in the mode and rate of sedimentation in the nearshore basins. Note that for the bank and slope species involved, the range in variation of water quality with depth (i.e., temperature, salinity, oxygen) is the same throughout the region. The most important differences in the benthic environments between the inner and outer borderland are related to sedimentation. The nearshore region is dominated by high rates of terrigenous clastic sedimentation whereas the region west of the first line of ridges and islands is dominated by lower rates of hemipelagic sedimentation (Gorsline and Emery, 1959). It seems likely that several hundred years ago the nearshore basins had depositional envi-

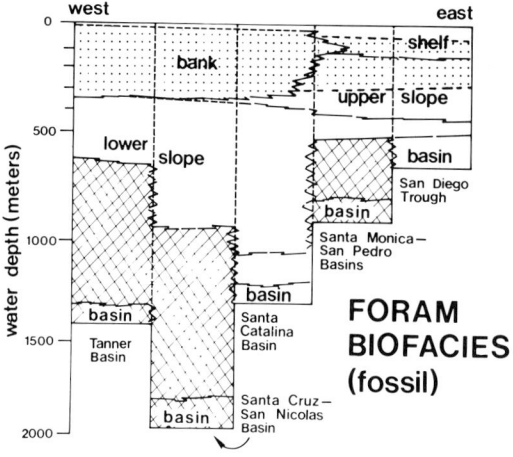

FIG. 8.—Schematic diagram showing the former distribution of assemblages in the borderland, based on the occurrence of relict and fossil species. The Bank Assemblage (dotted pattern) was until recently a well established part of mainland shelf communities. It is presently restricted to the outer borderland. Group II species of the Lower Slope Assemblage (cross-hatched) that are presently restricted to the outer basins were formerly living in Santa Monica Basin. No patterns are shown for San Pedro and northern San Diego Trough as cores from these basins have not been investigated. Compare distributions to Figure 4 which shows the distribution of live assemblages.

ronments more similar to those which currently prevail in the outer borderland: lower sedimentation rates related to an increase in the hemipelagic input component and decreases in nepheloid and turbidity flow components. Such changes are probably related to small scale climatic fluctuations.

SUMMARY AND CONCLUSIONS

Live benthic foraminifera in the southern California borderland can be grouped into five recurrent assemblages. In general, there is agreement between the assemblages and major bathymetric habitats but faunal boundaries tend to change with sediment characteristics and/or water quality. There is a bathymetric order to the occurrence of species and assemblages, and a preferred depth range in a particular habitat, but neither the order nor range is fixed.

The variation in species and assemblage depth limits illustrates that water depth per se is not an important factor in controlling distribution. In the silled basins of the borderland there is essentially no change in water quality with increasing water depth below the sill (Emery, 1960; Fig. 2). Yet, despite the increase in water depth, amounting to nearly 1000 meters in Santa Cruz and San Nicolas Basins, there is no change in the foraminiferal fauna. If water depth is an important variable in species distributions, one would expect a change in species distribution below the sill in these basins.

Differences in the mode and rate of sedimentation between the nearshore and offshore basins is reflected in variations of species composition within and between assemblages. This relationship is reflected in faunal differences occurring in the shallow water environments of the mainland shelf and the offshore banks. The Bank Assemblage is best developed in offshore areas of carbonate-rich substrates whereas the Outer Shelf and Upper Slope Assemblages occur only in areas of clastic terrigenous sediments next to the mainland. Significant variation also occurs in the Basin Floor Assemblage that is usually admixed with slope species introduced by slumping or sliding of sediments from the sides of the basins. Basin Floor species vary from less than 20% of the total fauna in areas of sediment instability to 100% of the total fauna in ponded sediments, distal areas or areas that are "shadowed" from downslope movement.

Many of the important species of the Lower Slope, Group II Assemblage occurring in the outer basins do not presently live in the nearshore basins, although downcore evidence indicates that they did until recently. Distribution of these species appears related to clastic sediment rates. Low occurrences of these outer basin species can be found in surface sediments in nearshore basins south of San Diego (Uchio, 1960, unpub. data), basins which also have low sediment rates (Gorsline and Prensky, 1975).

Comparative studies of live and dead assemblages from core top samples suggest that one of the reasons for the lack of resemblance is a recent shift in species distribution. Species have maintained their same bathymetric range but have changed their geographic range. The evidence points to a restriction of biogeographies to the west, a pattern that has persisted since the early Pleistocene (Blake, 1976).

The distribution of slope and basin benthic foraminifera in the borderland, especially the variation in species depth limits and the recent change in species geography, offers a complex picture of faunal variation in space and time. In a sense, the borderland provides a multidimensional view of faunal-environmental relationships while typical continental slope settings, which have a linear arrangement of depth related environments, provide a two dimensional view. The ecological perspective provided by borderland foraminifera suggests that the distribution of deep water benthic species is more dynamic than has been generally recognized. While this may complicate the application of empiric paleoecologic models such as Upper Depth Limits (Bandy and Chierici, 1966), in the long term, it enhances the potential of benthic foraminifera as indicators of paleoenvironment.

Acknowledgments

The authors wish to acknowledge Carolyn Walch, Gregg Blake, and Michael Abrams who contributed in various phases of the sample analysis and the officers and crew of the R/V *Velero IV* for their assistance in collecting the core samples. Many of the samples used in this study were collected under U.S. Department of the Interior, Bureau of Land Management Contracts no. 480-16 and 697-15, the study was supported by a grant from the National Science Foundation, Oceanography Section, OCE 76-00156. We wish to thank Drs. James Ingle, Johanna Resig, and George Lynts who read early versions of the manuscript and offered many helpful comments and criticisms.

REFERENCES

BANDY, O. L., 1953, Ecology and paleoecology of some California foraminifera Part I. The frequency distribution of Recent foraminiferas off California: Jour. Paleontology, v. 27, p. 161-182.

———, 1964, Foraminiferal trends associated with deep-water sands, San Pedro and Santa Monica basins, California: Jour. Paleontology, v. 38, p. 138-148.

———, AND CHIERICI, M. A., 1966, Depth-temperature evolution of selected California and Mediterranean bathyal foraminifera: Marine Geology, v. 4, p. 259-271.

———, INGLE, J. C., AND RESIG, J. M., 1964a, Facies trends, San Pedro Bay, California: Geol. Soc. America Bull. v. 75, p. 403-424.

———, ———, ———, 1964b, Foraminifera trends, Laguna Beach outfall area, California: Limnol. Oceanography, v. 9, p. 112-123.

———, ———, ———, 1964c, Foraminifera, Los Angeles County outfall area, California: Limnol. Oceanography, v. 9, p. 124-137.

———, ———, ———, 1965a, Foraminiferal trends; Hyperion outfall, California: Limnol. Oceanography, v. 10, p. 314-332.

———, ———, ———, 1965b, Modification of Foraminiferal distribution by the Orange County outfall, California: Washington D.C., Ocean Science and Ocean Engineering, Trans. Joint Conf. Marine Tech. Soc. Jour., Amer. Soc. Limnol. Oceanography, p. 55-76.

BLAKE, G. H., 1976, The distribution of benthic foraminifera in the Outer Borderland and its relationship to Pleistocene facies [unpub. Master's thesis]: University of Southern California, p. 135.

CROUCH, R. W., 1952, Significance of temperature on foraminifera from deep basins of southern California coast: Am. Assoc. Petroleum Geologists Bull., v. 36, p. 807-841.

DOUGLAS, R. G., WALCH, C., AND BLAKE, G., 1976, Benthic microfauna community studies, *in* Southern California Baseline Studies and Analysis: U.S. Department of the Interior, Bureau of Land Management, BLM Contract 08550-CT-5-52, 183 p.

———, COTTON, M. L., AND WALL, L., in prep., Benthic foraminifera in the southern California Bight as indicators of environment: U.S. Department of the Interior, Bureau of Land Management, Contract AA 550-CT7-44(696-15).

GORSLINE D. S., AND PRENSKY, S., 1975, Paleoclimatic inferences for late Pleistocene and Holocene from California Continental borderland sediments: Roayl Soc. New Zealand Bull. v. 13, p. 147-154.

———, AND EMERY, K. O., 1959, Turbidity-Current Deposits in San Pedro and Santa Monica Basins off southern California, Geol. Soc. America, Bull., p. 279-289.

HARTMAN, O., 1956, Contributions to a biological survey of Santa Monica Bay, California: University of Southern California Technical Report, submitted to Hyperion Engineers, Inc., 161 p.

———, 1966, Quantitative survey of the benthos of San Pedro Basin, southern California: Part I, final results and conclusions, Allan Hancock Pacific Expeditions, v. 19(2), p. 186-456.

HARMON, R. A., 1964, Distribution of foraminifera in the Santa Barbara Basin, California: Micropaleontology, v. 10, no. 1, p. 81-86.

KHERADPIR, A., 1970, Foraminiferal trends in the Quaternary of Tanner Basin, California: Micropaleontology, v. 16, p. 102-116.

MALOUTA, D. N., 1978, Holocene Sedimentation in Santa Monica Basin: California [unpub. Master's thesis]: Univ. Southern California, 79 p.

McGLASSON, R. H., 1959, Foraminiferal biofacies around Santa Catalina Island, California: Micropaleontology, v. 5, p. 217-240.

MITSUSHIO, H., GORSLINE, D. S., EDWARDS, B. D. AND YASUDA, Y., 1978, Bottom Sediments of San Nicolas Basin, California: Am. Assoc. Petroleum Geologists Bull., v. 62 p. 544.

MURRAY, J. W., 1976, Comparative studies of living and dead benthic foraminiferal distributions, *in* Hedley, R. H., and Adams, C. G., eds, Foraminifera, v. 2: New York, Academic Press, p. 226.

NARDIN, T. P., AND HENYEY, T. L., 1978, Pliocene-Pleistocene diastrophism of Santa Monica and San Pedro shelves, California Continental Borderland: Am. Assoc. Petroleum Geologist Bull., v. 62, p. 247-272.

NATLAND, M. L., 1933, The temperature and depth distribution of some Recent and fossil foraminifera in the

Southern California region: Scripps Institute of Oceanography Bull., Tech. Ser., v. 3, no. 10, p. 225–230.

———, 1957, Paleoecology of West Coast Tertiary Sediments: Treatise on Marine Ecology and Paleoecology, v. 2, p. 543.

PFLUM, C., AND FRERICHS, W., 1976, Gulf of Mexico deep water foraminifers: Cushman Foundation Foram Research Spec. Pub. no. 14, 125 p.

RESIG, J. M., 1958, Ecology of foraminifera of the Santa Cruz Basin, California: Micropaleontology, v. 4, p. 287–308.

SHEPARD, F. P., AND EMERY, K. O., 1941, Submarine topography off the California coast: Canyons and tectonic interpretations: Geol. Soc. of America Spec. Paper 31, 171 p.

UCHIO, T., 1960, Ecology of living benthonic foraminifera from the San Diego, California area: Cushman Foundation Foraminiferal Research Spec. Pub. no. 5, p. 72.

APPENDIX A

ALTPMV	*Alliatina primitiva* (Cushman and McCulloch)
ANGANO	*Angulogerina angulosa* (Williamson)
BOLARG	*Bolivina argentea* Cushman
BOLPAC	*B. pacifica* Cushman and McCulloch
BOLSPS	*B. spissa* Cushman
BOTMNT	*Bolivinita minuta* Natland
BUCAGA	*Buccella angulata* Uchio
BULTAA	*Buliminella tenuata* (Cushman)
BUMCDN	*Bulimina* cf. *denudata* Cushman and Parker
BUMDND	*B. denudata* Cushman and Parker
CANAUR	*Cancris auricula* (Fitchel and Moll)
CASDLC	*Cassidulina delicata* Cushman
CASLMB	*C. limbata* Cushman and Hughes
CASSCI	*C. subcarinata* Uchio
CHLOVD	*Chilostomella ovoidea* Reuss
CIBFLC	*Cibicides fletcheri* Galloway and Wissler
CSNCRN	*Cassidulinoides bradyi* (Norman)
EPLSDG	*Epistominella sandiegoensis* Uchio
EPLSMT	*E. smithi* (Steward and Stewart)
EPOLVI	*Eponides leviculus* (Resig)
FRSAPT	*Furenkoina apertura* (Uchio)
FRSSMD/BMT	*F. seminuda* (Natland)
	F. bramlettei (Galloway and Morey)
GBLPAC/HGD	*Globobulimina pacifica* Cushman
	G. hoeglundi Uchio
GYRIO	*Gyroidina io* Resig
HOGELG	*Hoeglundina elegans* (d'Orbigny)
ISLSGB	*Islandella subglobosa* (Brady)
LOXPDB	*Loxostomum pseudobeyrichi* (Cushman)
NOLSTZ/BSP	*Nonionella stella* Cushman and Moyer
	N. basispinata (Cushman and Moyer)
SUGECK	*Suggrunda eckisi* Natland
TEXELD	*Textularia earlandi* Parker
TEXSCK	*T. schenki* Cushman and Valentine
TROPAC	*Trochammina pacifica* Cushman
UVGCTA/PRG	*Uvigerina curticosta* Cushman
	U. peregrina Cushman
UVGJNC/ECL	*U. juncea* Cushman and Todd
	U. excellens Todd
VALACN	*Valvulineria araucana* d'Orbigny

QUATERNARY SEDIMENTATION ON THE TECTONICALLY ACTIVE OREGON CONTINENTAL SLOPE

L. D. KULM AND K. F. SCHEIDEGGER
School of Oceanography
Oregon State University
Corvallis, Oregon 97331

ABSTRACT

The nature and distribution of Quaternary lithofacies on the active Oregon continental margin are controlled largely by the interaction of tectonism, sea level changes, sediment supply, and dispersal processes. Convergence of the oceanic plate with the North American plate produces a rapidly evolving subduction complex on the continental slope. This complex, consisting of deep water silt and sand turbidites, has been accreted to the lower slope and has formed a dynamic structural setting of ridges with intervening basins, steep escarpments, benches, and marginal plateaus. Some of the Quaternary basins are filling with silt turbidites interbedded with hemipelagic muds or hemipelagic muds alone if topographic barriers preclude turbidity current deposition. A mud facies occurs on the escarpments and on the upper and lower slope benches and plateaus. Glauconitic and foraminiferal sands and thin sand-silt layers may be interbedded with the mud. A thick sand-silt facies may also occur on upper slope benches surrounding outer shelf submarine banks. The Astoria and Rogue submarine canyons have been filling with a predominant mud facies during the Holocene because the present high sea level stand has isolated canyon heads on the outer shelf precluding direct access to coarse-grained littoral or fluvial sediment. Off southern Oregon upper slope deposits are accumulating more slowly (10 to 14 cm/1000 years) than are lower slope deposits in the south and north (20 to 65 cm/1000 years). Sedimentation rates in the canyons are quite variable ranging from 10 to 78 cm/1000 years.

Variations in sediment texture along the continental slope are related to the nature and volume of sediments supplied by coastal rivers as well as to the margin transport and depositional processes. Holocene and late Pleistocene mud facies show significant variations in sand-silt-clay content with upper slope muds of central Oregon containing more silt than those off southern Oregon. Lower slope muds contain more clay to the south and north than in the central region. Lower slope muds always have a mean diameter finer than 6 phi whereas upper slope muds appear to be coarser although there is considerable overlap between the two areas. Transport and deposition of these fine-grained muds are controlled largely by unidirectional currents in the water column, bottom nepheloid layer transport and topography.

INTRODUCTION

The Oregon continental margin is classified as an active margin because it forms one edge of the convergence zone between the Gorda-Juan de Fuca Plate (Farallon Plate) and North American Plate (Fig. 1). A value of 2.6 cm/yr is calculated for the rate of convergence between these two plates off the Oregon region (Silver, 1969). The Oregon margin reflects the effects of convergence and is characterized by large-scale deformational features. The nature and distribution of sedimentary deposits within this complex structural framework reflect these rapid tectonic movements. In a temporal and spatial sense, deformation is most noticeable on the continental slope, especially the lower slope. While the bulk of the deposits on the continental slope are terrigenous in character, sedimentation patterns are extremely variable due to the tectonic processes, eustatic sea level changes and sediment supply.

In this report, data from studies of particular portions of the Oregon continental slope (Maloney, 1965; Carlson, 1968; Spigai, 1971; Kulm and Fowler, 1974), the results of Leg 18 of the Deep-Sea Drilling Project (Kulm, von Huene et al, 1973), and the unpublished work of M. Dinkleman are integrated to produce a more comprehensive picture of the interrelationships of tectonism and sedimentation along this active plate margin. Although the data coverage is not extensive or complementary in all areas, they do give a good indication of the predominant sediment types and characteristics found in continental slope environments. The principal objective of this paper is to describe the various sedimentary facies of the continental slope and relate these to the sedimentary and tectonic processes that produce and modify these facies.

MORPHOLOGY AND STRUCTURAL FRAMEWORK

The morphology of the Oregon continental margin (shelf and slope) is characterized by a large variety of features that are largely the result of vertical tectonic movements generated by the convergence of the Gorda-Juan de Fuca plate and the North American plate. A substantial amount

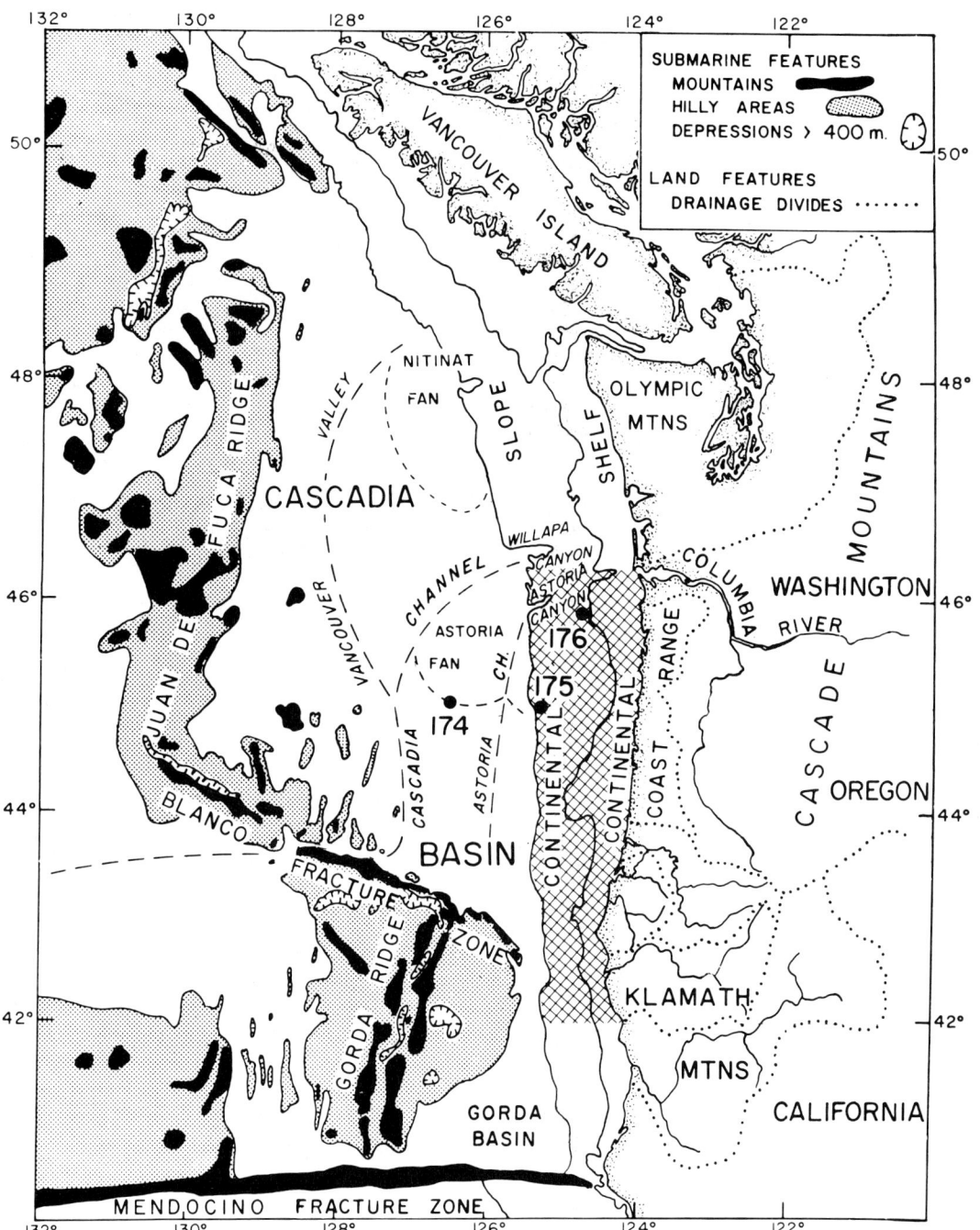

FIG. 1.—Location map of the Oregon continental shelf and slope (cross hatch) with surrounding abyssal features on oceanic plate. Note locations of Deep Sea Drilling Site 174 on the distal portions of Astoria Fan and Site 175 in a lower slope basin.

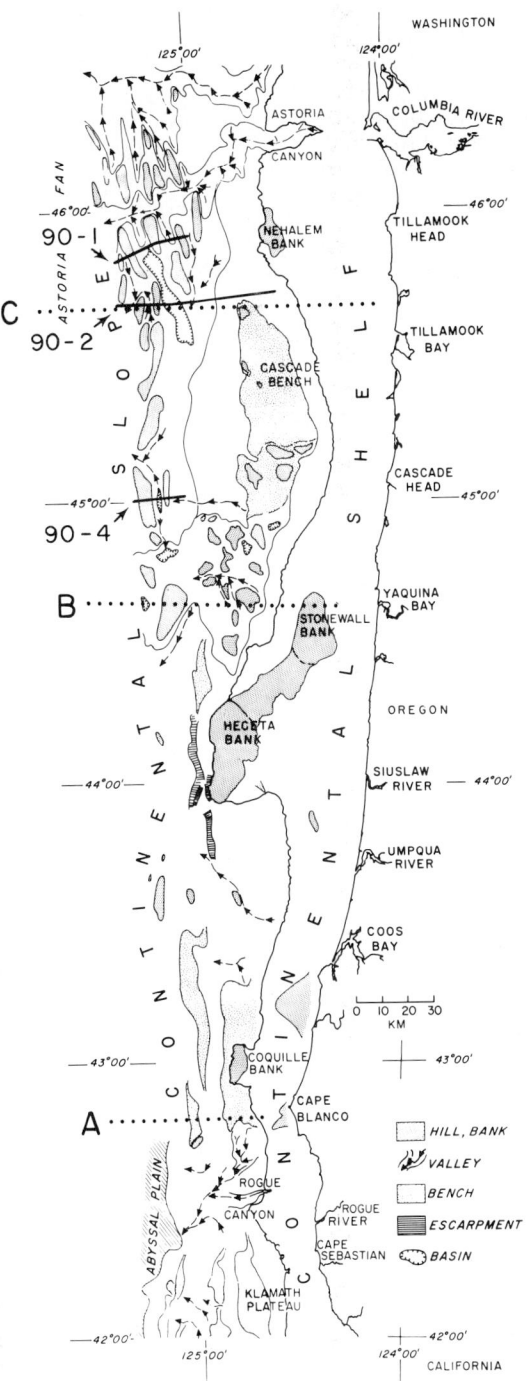

of the deep water terrigenous material from the oceanic plate is accreted to the Oregon margin producing the complicated structural framework of the subduction complex, or the accretionary prism (Kulm and Fowler, 1974; Seeley et al, 1974; Carson, 1977). The features described below for the Oregon margin are also typical of most active continental margins that surround the Pacific Ocean.

The Oregon shelf is narrow with the submarine banks (i.e., Nehalem, Heceta-Stonewall, and Coquille Banks) along the outer shelf creating a sinuous shelf break (Fig. 2). Neogene and Quaternary deep water sediments are exposed in the folded structures of these banks and reflect uplift of as much as 1000 meters (Kulm and Fowler, 1974). Pliocene and Pleistocene sediments have been deposited in a series of interconnecting synclines that occupy the central part of the shelf landward of these outer shelf structures.

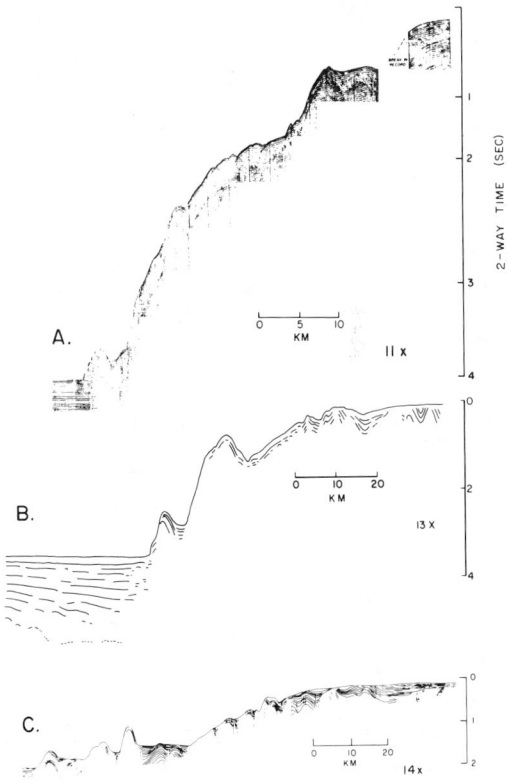

FIG. 2.—Detailed morphology of the Oregon continental shelf and slope (after Kulm and Fowler, 1974). Lines A, B, and C denote locations of seismic reflection profiles shown in Figure 3 and lines 90-1, 2, 4 in Figure 15.

FIG. 3.—Seismic reflection records of the Oregon continental slope. Southern Oregon (A), Central Oregon (B), and Northern Oregon (C). Seismic profiles were obtained by a 5,000 joule sparker (A) and 40 cubic inch air gun (B, C).

The Oregon continental slope is characterized by numerous structural benches, hills, ridges and steep escarpments (Figs. 2, 3). Three types of morphotectonic settings are recognized on the slope: (1) a region consisting of marginal plateaus, benches and intervening escarpments between the California border to the Umpqua River (Fig. 3A); (2) a region of steep escarpments and small equidimensional hills and basins from the Umpqua River to Cascade Head (Fig. 3B); and (3) a region characterized by a large structural bench (i.e., Cascade Bench) which forms the upper slope and a series N-NW trending, linear ridges and intervening basins which form the lower slope between Cascade Head and the Columbia River (Fig. 3C). Accreted deposits of former abyssal plain and continental slope environments create the structural foundation of the benches and plateaus and are exposed on the sea floor or lie just below the sea floor in many localities (Fig. 4, Kulm and Fowler, 1974). Sediments are ponded behind these tectonic ridges or hills on the continental slope, creating nearly flat benches on the upper slope (Fig. 3). The Klamath Plateau off southern Oregon is the northward continuation of the same feature off northern California (Fig. 3A, Spigai, 1971; Silver, 1971). Similarly, the prominent linear ridges off northern Oregon are continuous with analogous features on the southern Washington margin. The basins or troughs between the ridges are also being deformed as shown in Figure 3C.

Two submarine canyons, Astoria and Rogue Canyons, have their heads near edge of the continental shelf and cut across the continental

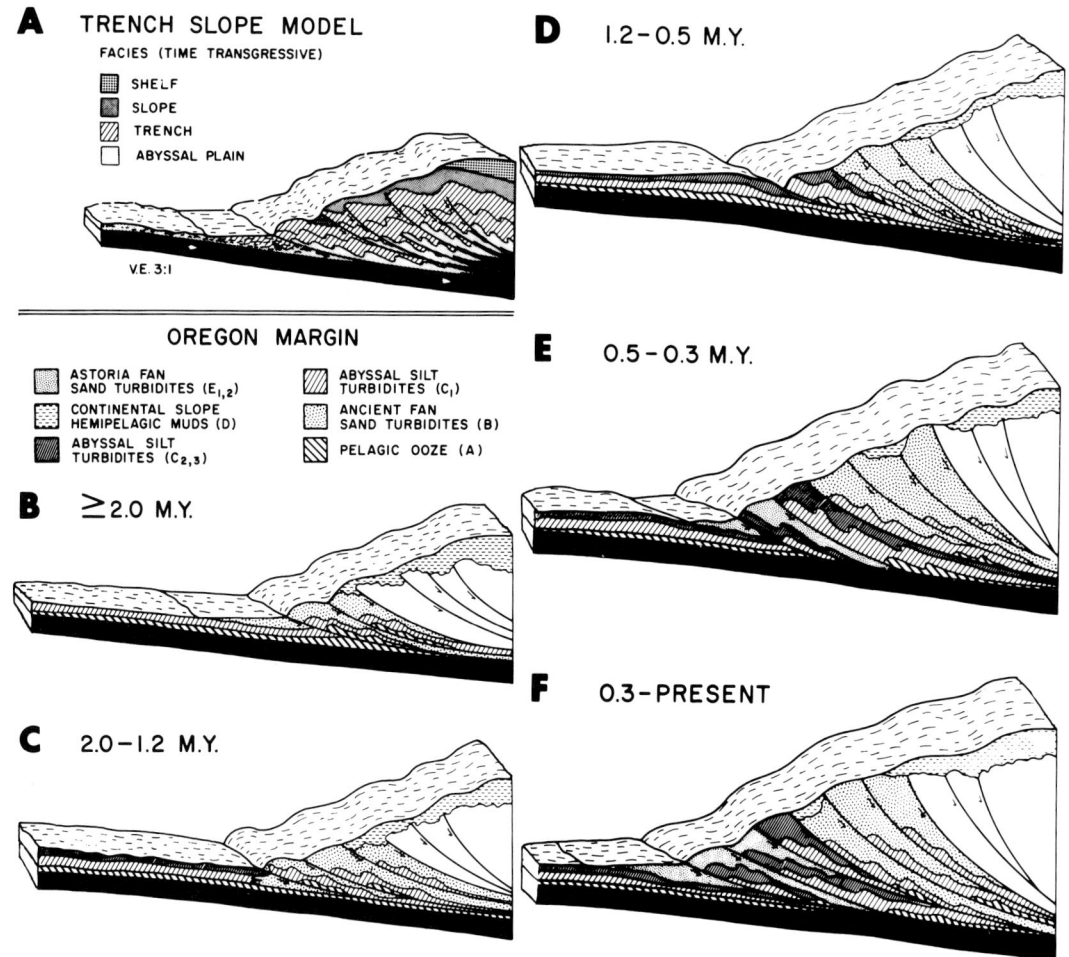

FIG. 4.—Structure and sedimentary facies of the north-central Oregon continental slope. Interpretative sections located at 44°39′N latitude (see Fig. 2 for general location). Trench slope model of Seely et al, (1974) is shown (A). Late Cenozoic chronology of postulated imbricate thrusting with sedimentary facies relationships (B-F).

slope to connect with Astoria Fan in the former case and to merge with the abyssal plain in the latter case (Fig. 2). The course of each canyon appears to be controlled by the various structural features of the continental slope. Buried channels between the present river mouths and the canyon head indicate that the two features were connected during the low sea level stands of the Pleistocene.

This diverse structural framework of the continental slope produces a variety of topographic depressions that serve as basins for sediment accumulation. Such basins formed on top of the subduction complex are called accretionary basins (Seely and Dickinson, 1977). These structural depressions generally are small, being a few tens of kilometers long and a few kilometers wide. Some depressions are even smaller and roughly equidimensional. The temporal and spatial development of these accretionary basins and their relation to sedimentation will be discussed later.

STRATIGRAPHY

Where present, Mazama ash layers are the best indicators of age in the Holocene deposits on the continental slope and in the submarine canyons. Mount Mazama, Crater Lake, in southern Oregon erupted 6,600 years ago and spread two distinct ash layers over the Northwest Pacific landmass (Fryxell, 1965). The ash was washed from the continental drainage basins and carried by rivers to the ocean where it was either deposited on the shelf through suspension in the water column or deposited as bottom-seeking turbidity current flows in certain areas of the continental slope, canyons and in channels on the adjacent abyssal plain (Nelson et al, 1968). These layers have been dated by the carbon-14 method as well as several other deposits above and below the ash layers. Total carbon was used in the determination of the C-14 dates.

Microfossils provide the standard means of age control on the stratigraphic units. Radiolarians and diatoms are abundant in the Oregon slope sediments (Kling, 1973; Schrader, 1973) and give the best resolution for dating deposits of Pleistocene age.

LITHOFACIES

Definition of Lithofacies

Six distinct lithofacies are recognized within the continental slope and within the adjacent abyssal plain. All of these lithofacies may be present in the deposits of the Oregon slope because of the transfer of abyssal deposits to the slope through the subduction processes noted previously.

Calcareous Pelagic Ooze.—This pelagic facies, which consists of calcareous muds and nannofossil ooze, contains more than 30% calcium carbonate. It occurs near the base of the sedimentary section at Site 174 and directly overlies the oceanic basalts of the Gorda-Juan de Fuca Plate (Fig. 4). A substantial amount of terrigenous clay is included with the ooze which indicates the close proximity to continental sediment sources (Hayes, 1973).

Silt Turbidites with Interbedded Muds.—The silt turbidite facies generally consists of thin silt layers typical of the "D" interval of the Bouma Sequence (Bouma, 1962) with a thick homogeneous silty clay and clayey silt "E" interval above. This overlying pelitic interval is variable in thickness and consists largely of silty clay. A hemipelagic interval above "E" represents deposition between episodic turbidity current events. This facies characterizes the abyssal plain deposits (Kulm, von Huene et al, 1973), the interchannel regions of Astoria Fan (Nelson and Kulm, 1973) and certain continental slope basins (Kulm, von Huene et al, 1973, this paper) (see Fig. 4, silt turbidites C_1, C_2 and C_3).

Sand Turbidites.—The bulk of the sand turbidites, with thin interbedded mud layers, is found in the Astoria Fan which rests upon the abyssal plain deposits (Figs. 1, 4). The fine to medium-grained sand turbidites at Site 174 (Fig. 1) commonly attain a maximum thickness of two to seven meters between relatively thin mud interbeds. This drill hole is located on the seaward edge of the suprafan where the turbidite sands should reach their maximum thickness (Nelson and Kulm, 1973).

Mud Facies.—Olive gray, olive green, dark green and gray muds characterize the continental slope deposits and are found as interbeds in abyssal plain and submarine fan deposits. The bulk of these deposits consist of clayey silt and silty clay and are deposited by hemipelagic sedimentation (i.e., from suspension in the water column). Diatoms are a common constituent of these muds. This mud facies may attain thicknesses of more than 100 meters in continental slope basins as determined by drilling. In certain depositional environments, these muds may have been resuspended by slumping and carried to lower bathymetric levels of the seafloor. Benthic Foraminifera, displaced from depositional environments which have shallower water depths than the present depositional site, are frequently found in the deposits.

Glauconite Facies.—Glauconite is a common constituent of outer continental shelf and upper continental slope deposits. It usually occurs as thin layers which are interbedded with the mud facies or the transgressive and regressive sands of the shelf. Some glauconitic layers have a substantial admixture of mud. Deposits containing greater than 30% glauconite are defined as the glauconite facies.

Sand-silt Facies.—Very fine to coarse-grained sand is the dominant facies on the inner continental shelf and portions of the outer shelf and upper slope benches (Kulm et al, 1975; Spigai, 1971). While the inner shelf sands are transgressive sands deposited during the post-glacial rise of sea level, certain patches of outer shelf sands and those of the upper slope benches are believed to be relict materials deposited during the glacial regression of the strandline. These deposits also contain various admixtures of silt and clay.

Distribution of Lithofacies on the Continental Slope

Subduction Complex.—Accreted deposits from the oceanic plate form a portion of the subduction complex beneath the Oregon continental slope as shown in Figure 4 (Kulm and Fowler, 1974). Silt and sand turbidite lithofacies have been dredged or drilled on the slope and were originally deposited in water depths typical of the adjacent abyssal plain with subsequent uplift to their present position on the slope. Because the pelagic ooze facies is only a few tens of meters thick on the oceanic plate, and because it has not been recovered in the slope, it is problematical as to

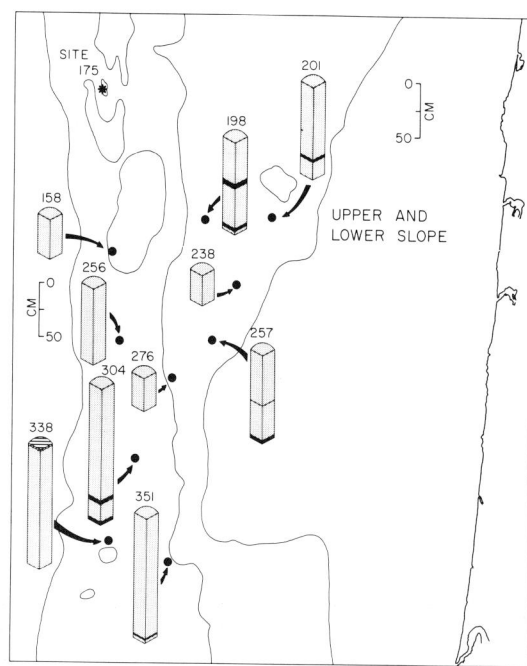

Fig. 6.—Lithofacies of the upper and lower continental slope off central Oregon. Note length of gravity cores given in centimeters. Drill Site 175 shown on lowermost slope. See Figure 5 caption for explanation and legend. See Figure 2 for location of Heceta Bank and coastal features. (Data compiled from Maloney, 1965).

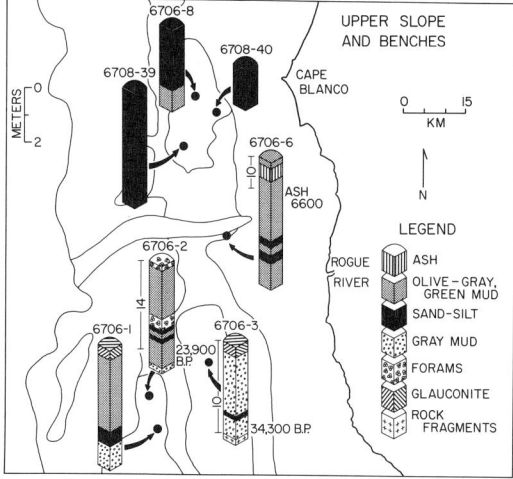

Fig. 5.—Lithofacies of the upper continental slope, including benches and Klamath Plateau off southern Oregon. See Figure 2 for location of morphologic and geographic features. Number of each piston core indicated above core with length given in meters, Carbon-14 dates given on right side of core in years Before Present, Mazama ash (6600 years B.P.) layer indicated where present, and sedimentation rates given on left side of core in centimeters per thousand years. Explanation of lithofacies given in legend (Diagram modified from Spigai 1971, Fig. 29).

whether it is incorporated into the subduction complex. If imbricate thrusting is the chief mechanism for the emplacement of these lithofacies, then they would be distributed within the structural framework shown in Figure 4.

Benches and Escarpments.—Holocene and late Pleistocene deposits of the upper slope benches off southern Oregon are characterized by a sand-silt, glauconite or olive gray to gray mud facies. The sand-silt facies is most prominent on the Cape Blanco bench which surrounds the seaward extremities of Coquille submarine bank on the outer edge of the shelf (cores 6706-8, 6708-39, 40, Figs. 2, 5). Glauconite apparently covers a portion of the Klamath Plateau (core 6706-3) as well as the bench seaward of the plateau (core 6706-1). The lack of terrigenous mud deposition also is seen in core 6702-2 on the same feature because planktonic and benthonic foraminiferal sands are accumulating at the top of the core.

Farther to the north off central Oregon olive green mud is the dominant facies over the entire slope and escarpment seaward of Heceta Bank (Figs. 2, 6). Occasional sand-silt layers are interbedded with these muds. Glauconite occurs as a layer in one core (338), but it is more commonly disseminated throughout the cores on the upper slope (see Sand Fraction Components). It should be noted that these gravity cores do not penetrate as deep into the surface sediment as the piston cores to the north and south and probably are sampling only the Holocene deposits. Other gravity cores taken in this area recovered Pliocene-Pleistocene consolidated siltstones with only a few centimeters of covering unconsolidated sediment (Maloney, 1965).

Submarine Canyons.—The bulk of the Holocene deposits of the Astoria and Rogue submarine cnayons consist chiefly of the olive gray or green and gray muds with occasional interbeds of silt and sand associated with turbidites (Figs. 7, 8). One exception, core 6706-7 at the head of the Rogue Canyon, contains coarse silt and very fine sand and is associated with the sand-silt facies that occur on the outer edge of the continental shelf (Fig. 8, Spigai, 1971). These deposits apparently were laid down at the end of the late Pleistocene regression because they have an iron oxide coating typical of relict shelf deposits (Spigai, 1971). The glauconitic facies at the top of core 6706-4 also suggests that other areas of the Rogue Canyon have not been sites of mud deposition during all of the Holocene. On the other hand, the olive gray and gray mud facies, with interbedded sand-silt layers, indicate a much more uniform and thicker mud facies was deposited

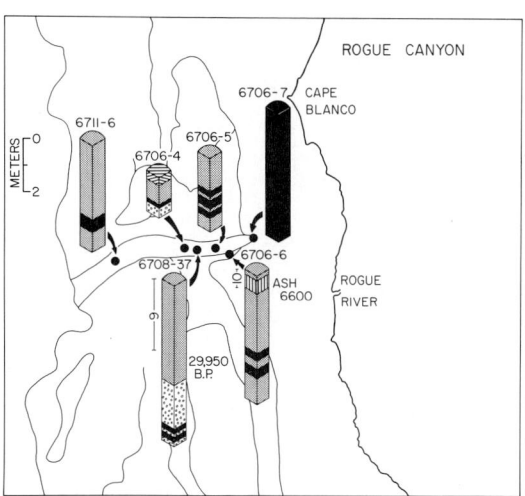

Fig. 8.—Lithofacies of the Rogue Canyon off southern Oregon. Length of piston cores given in meters. See Figure 5 caption for explanation and legend. Figure 2 shows geographic location of Rogue Canyon on continental slope. (Diagram modified from Spigai, 1971, Fig. 31).

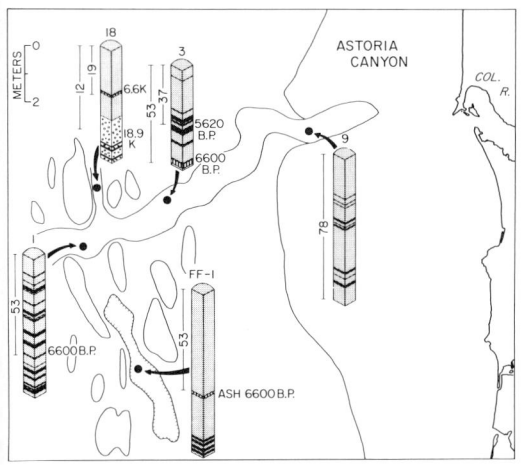

Fig. 7.—Lithofacies of Astoria Canyon off northern Oregon. Length of piston cores given in meters. See Figure 5 caption for explanation and legend. Location of Astoria Canyon and Columbia River shown in Figure 2. (Data compiled from Carlson, 1968 and Dinkleman, unpublished).

in Astoria Canyon than Rogue Canyon during the Holocene (Fig. 7).

Accretionary Basins.—During the Holocene, olive green mud was deposited in the basins developing on the lowermost slope off northern Oregon (core FF-1, Fig. 7). Sand-silt turbidites occur near the bottom of the core and no doubt extend into the late Pleistocene deposits in the same basin since they display prominent reflecting horizons (Fig. 3C).

The lower slope basin deposits at Site 175 (Fig. 1) are characterized by dark greenish gray mud interbedded with rare fine sand or coarse silt layers during the past 300,000 years (Fig. 9). This is the approximate length of time that the basin has existed at its present depth on the slope (Kulm, von Huene, et al, 1973). Deeper in the core this dominant mud facies grades into a dominant mudstone facies with an increasing frequency of interbedded silt turbidites in lower parts of the core. Both the turbidites and the mudstones were originally deposited on the adjacent abyssal plain and were uplifted, with consequent partial consolidation, to their present position during the past 600,000 years (Kulm, von Huene et al, 1973).

The olive gray mud facies on the lower slope off southern Oregon (Fig. 10) is similar to that found in the accretionary basins forming on top of the subduction complex off northern Oregon. While the sedimentary deposits are similar, the lower slope basins off southern Oregon have not

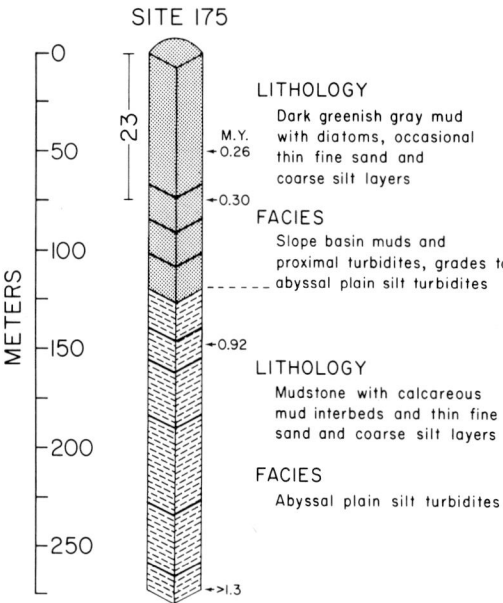

Fig. 9.—Lithofacies of drill Site 175 on lowermost continental slope off north-central Oregon. Ages of the deposits indicated on right side of core (according to Kling, 1973; Schrader, 1973) and sedimentation rate (cm/1000 yrs) given on left side of core. Dashed line at 120 m designates boundary between accreted abyssal deposits of subduction complex below and accretionary basin mud above. Former lithofacies are designated continental slope hemipelagic mud (D) and latter lithofacies, the youngest abyssal silt turbidites, (C_3) in Figure 4. See Figures 1 and 6 for location of Site 175. (Data from Kulm, von Huene et al, 1973).

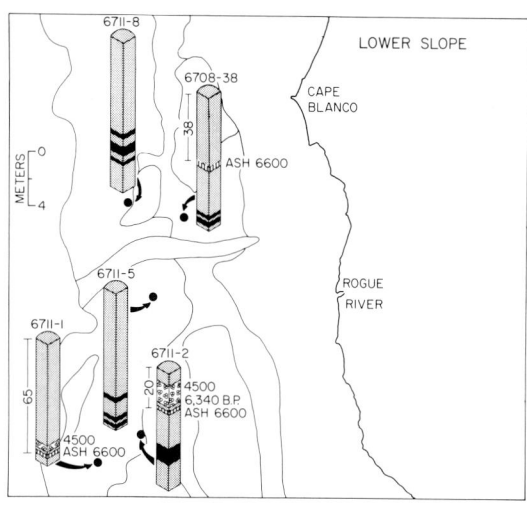

Fig. 10.—Lithofacies of the lower continental slope off southern Oregon. Length of piston cores given in meters. See Figure 5 caption for explanation and legend. Figure 2 shows geographic location and morphologic features of the region. (Diagram modified from Spigai 1971, Fig. 30).

yet developed to the point where extensive sediment ponding is occurring in a structural depression on the slope.

Composition of Continental Slope Lithofacies

Sand Fraction Components.—The composition of the sand fraction (>62 μm) was determined with a binocular microscope for the surface and subsurface sediments on the continental slope and in the submarine canyons. Because the sand fraction commonly represents such a small percentage of the total sediment (<5%) and the composition of the mud facies is rather uniform for a given core, we have averaged the data over the entire length of core in the southern Oregon and in the Astoria Canyon areas.

The bulk of the olive gray, olive green and gray muds in the lower slope deposits off southern Oregon contains less than 2% sand-sized material. This fraction is composed almost exclusively of terrigenous and biogenous particles with varying admixtures of these two components (Fig. 11).

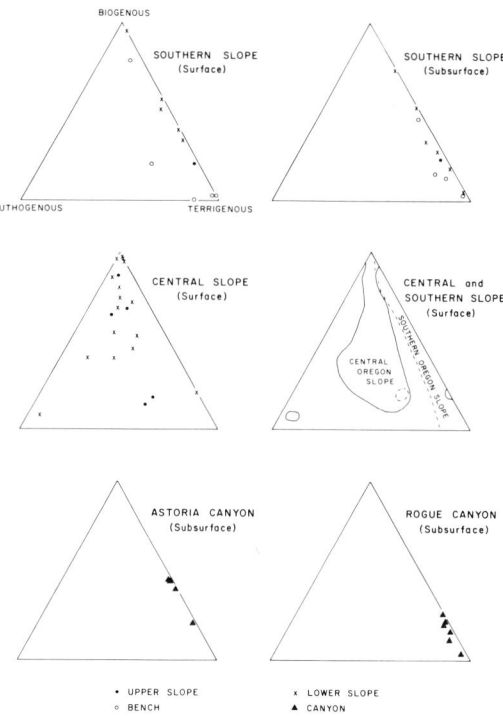

Fig. 11.—Composition of sand fraction (>62 μ) of mud facies in continental slope environments. Symbols denoting type of environment given at bottom of diagram. Surface sediment analyses only indicated (surface). Both surface and subsurface indicated as subsurface.

The authigenic glauconite and pyrite are either absent or occur in very minor amounts. Radiolarians predominate in the Holocene while planktonic Foraminifera are the major constituents in the glacial Pleistocene. Diatoms are common throughout these periods of time. Benthic Foraminifera are present, in a few percent, in all lower slope deposits. Although there is no comparable data for Site 175 and Core FF-1 off northern Oregon, smear slide analysis of the total sediment indicates that these sediments are composed largely of biogenous and terrigenous particles. Diatoms account for 5-20 percent of these dark green and olive green muds.

Upper slope muds in the south reflect a somewhat larger authigenic component which is the mineral glauconite and have a dominantly terrigenous sand fraction (Fig. 11). The quartz and feldspar grains resting upon the benches off southern Oregon are also oxidized.

The composition of the sand fraction of the central Oregon slope muds represent a radical departure from those to the north and south. The olive green muds on the escarpments contain from 1 to 14% sand-sized particles and a much larger concentration of authigenic constituents, chiefly glauconite (Fig. 11). They also have a predominantely biogenous component which consists of diatoms, radiolarians, planktonic Foraminifera and siliceous sponge spicules. The rocky outcrops of the escarpments are a favorite habitat of sponges. No real distinction can be made in the

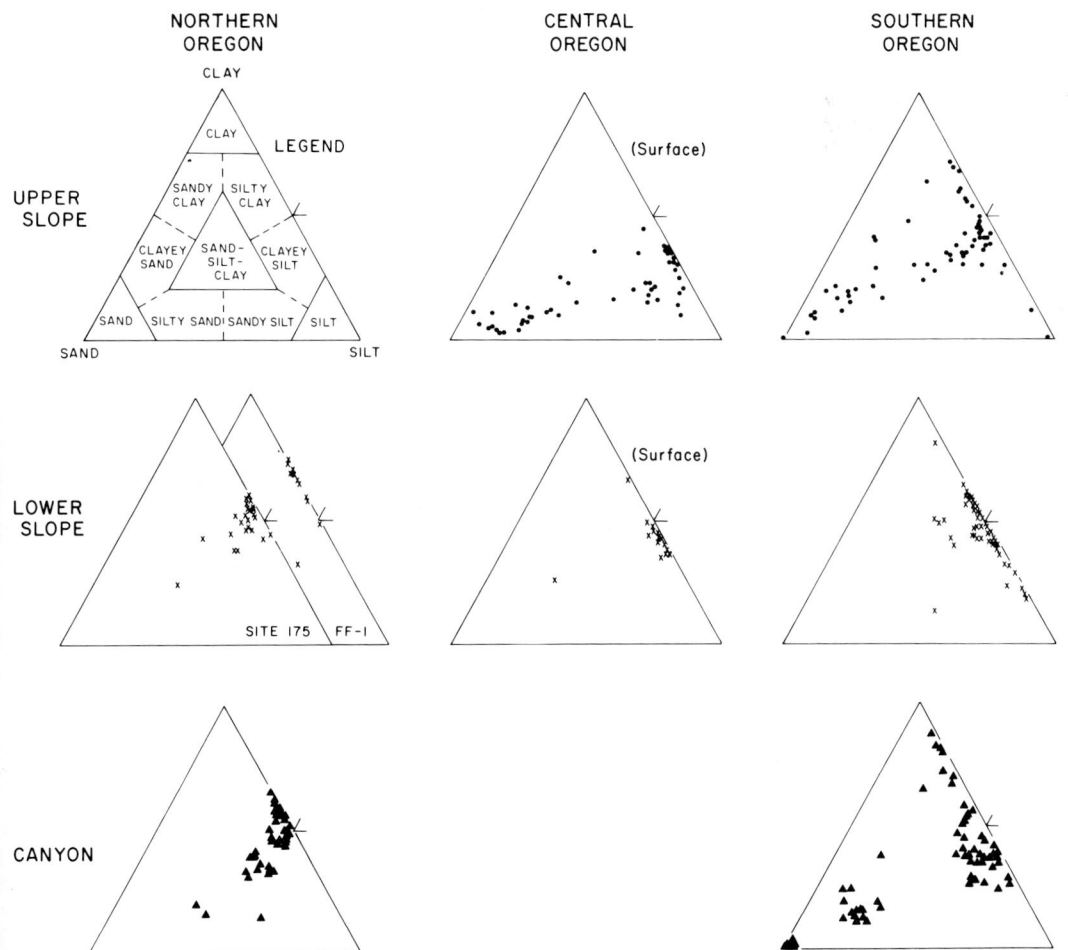

Fig. 12.—Sediment types of continental slope and submarine canyon deposits. Nomenclature according to Shepard (1954). All analyses include both surface and subsurface deposits with the exception of central Oregon where only the surface sediments were analyzed (Data compiled from Maloney, 1965). Submarine canyon off northern Oregon is Astoria Canyon and off southern Oregon, Rogue Canyon. Drill Site 175 shown in Figure 9 and piston core FF-1 in Figure 7.

composition of any of the upper and lower slope environments off Oregon.

Submarine canyon deposits are characterized by their dominant terrigenous content although Astoria Canyon appears to have a somewhat higher concentration of biogenous components than the Rogue Canyon.

Texture.—The most easily determined textural parameters of unconsolidated and consolidated sediments are the percents of sand, silt and clay. These have been widely used to group continental margin sediments into sediment types (Shepard, 1954). All of the continental slope facies described herein have been analyzed in this fashion.

As shown in Figure 12, the Oregon slope deposits have been divided into the upper slope, lower slope and submarine canyon environments for the northern, central, and southern Oregon regions. Both surface and subsurface deposits were analyzed for sand, silt and clay content, except off central Oregon where only the surface sediments were studied. No sediment data are available for the upper slope deposits off northern Oregon.

Upper slope mud, sand-silt and glauconite lithofacies show the greatest variation in texture. They consist of silty clay, clayey silts which are typical of the mud facies, and sand-silt-clay, clayey sand, silty sand and sand which is characteristic detrital sand and glauconite facies (Fig. 12). The mud facies generally contains less than 5% sand unless there are significant admixtures of authigenic sand as in the case of the central Oregon region (Fig. 11). Figure 13 shows that the upper slope deposits tend to be coarser-grained off central than southern Oregon although there is overlap of sediment types of the two regions. More clay is being deposited to the south in the vicinity of the Rogue River, which has a large drainage basin in the Klamath Mountains, than opposite the much smaller drainage basins of the central Oregon Coast Range.

The most distinctive sediment types are the clayey silts and silty clays of the lower continental slope in all three regions (Figs. 12, 13). These olive gray, olive green, dark green and gray muds generally contain less than 5% sand with a large number having less than 1% sand. Interestingly, all three regions exhibit a rather distinctive clustering of sediments between 75% clay and 75% silt with the northern region displaying more clay and the central region more silt. Lower slope sediments in the southern region have the complete spectrum of silt and clay noted in the other two regions (Fig. 13). Drill site 175 exhibits a rather close clustering of sediment with respect to silt and clay content over the past 300,000 years (Fig. 12). Core FF-1 just to the north has a slightly higher clay content; both cores are located in accretionary basins.

Submarine canyon deposits show a broad range of sediment types because they contain both mud and sand-silt turbidite facies. Astoria Canyon muds group in the silty clay category which is typical of the lower slope deposits to the north (Fig. 12). Rogue Canyon muds are texturally similar to the adjacent lower slope muds except that some sediments are classified as clay which makes them the only deposits containing dominantly clay on the entire continental slope. The sand and silt turbidites of both canyons fall into the largely silt and sand categories with a distinct clustering of Rogue Canyon deposits near the sand type.

More detailed textural information was obtained from the southern Oregon slope deposits (Spigai, 1971). The Inman (1952) statistical parameters are shown as mean diameter versus skewness and deviation (sorting) in Figure 14. Olive gray and

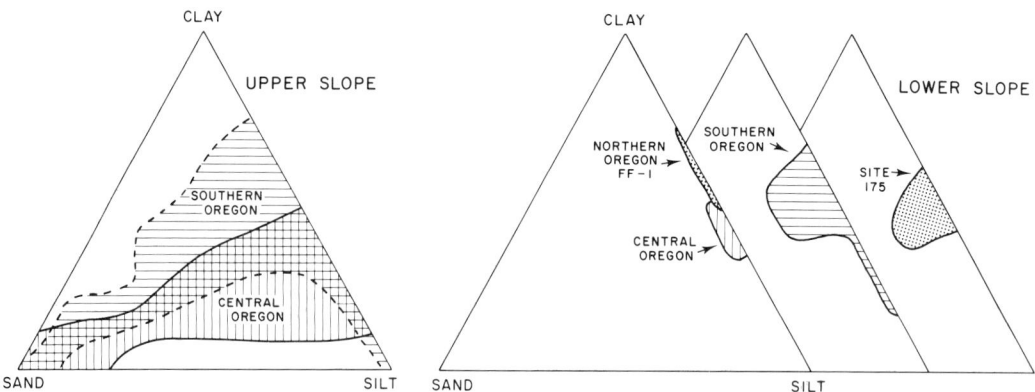

Fig. 13.—Comparison of sediment types on the upper and lower continental slope given in Figure 12. Explanation given in text.

gray muds have similar textural characteristics (see Fig. 14, solid line enclosing muds) and are inseparable even though the former represents deposition during the Holocene transgression and the latter during the glacial regression or early part of the transgression (Fig. 5). The bulk of these muds have mean diameters between 6 φ and 12 φ, which is the fine silt to clay size classes. These muds are poorly sorted and may have either a negative or positive skewness. Shells of planktonic organisms apparently produce the negative skewness (Spigai, 1971), particularly in the lower slope sediments which tend to have the largest percentage of biogenic constituents in the sand fraction (Fig. 11). The olive gray muds of the lower slope exhibit the poorest sorting (Spigai, 1971) which may reflect additions of slumped mud debris from shallower water depths.

All facies are plotted according to the type of continental slope environment (Fig. 14). Sand-silt facies and mixed sand and mud facies of the upper slope and upper slope benches are characterized by the coarsest grained deposits. Some of these bench sands exhibit good sorting values, similar to the adjacent continental shelf sands, and are believed to be relict deposits (Spigai, 1971). The sand fraction of the upper slope sediments off southern Oregon is largely terrigenous in origin, whereas the lower slope sediments can be either terrigenous or biogenous (Fig. 11).

Comparing the statistical parameters in Figure 14, one can conclude that lower slope muds always

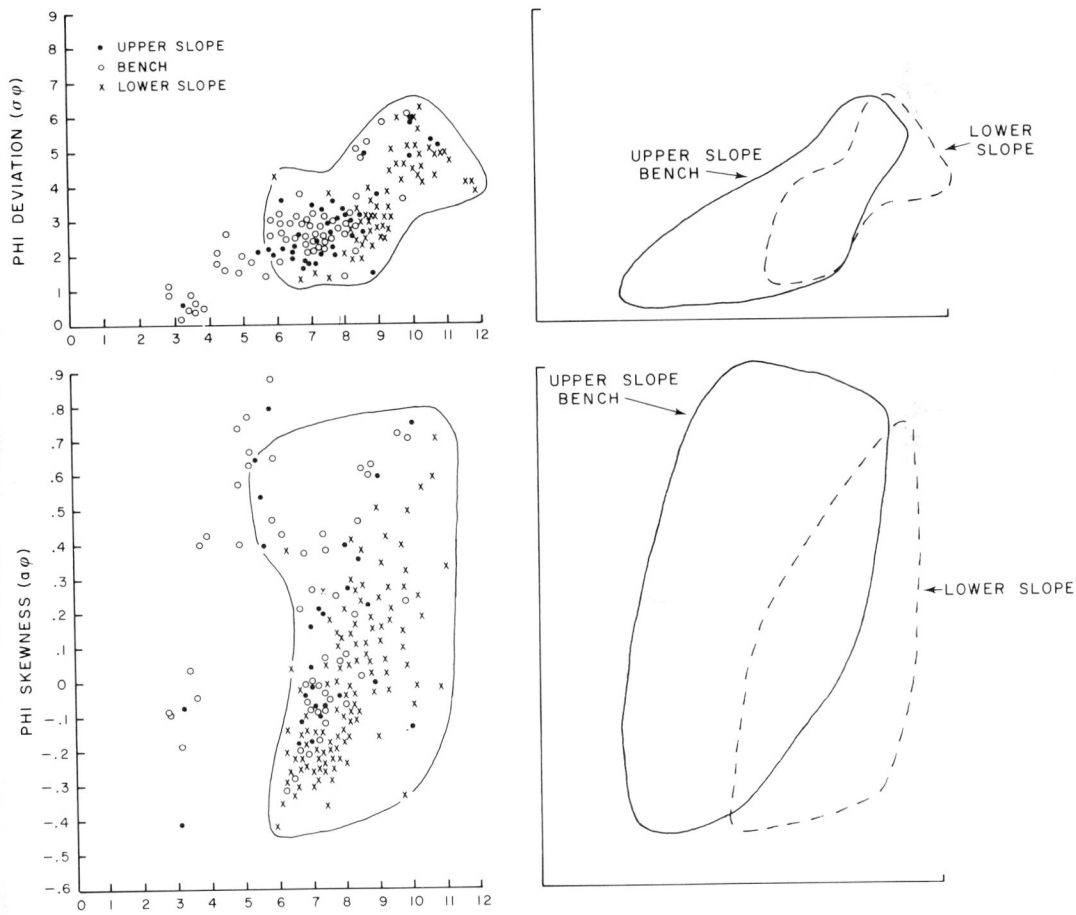

FIG. 14.—Textural analysis of continental slope deposits off southern Oregon (i.e., cores shown in Figs. 5, 10). Statistical parameters according to Inman (1952). Circled region on left pair of plots indicates olive gray and gray muds which are indistinguishable on the basis of texture. Right pair of plots show degree of separation and overlap in texture between upper slope—bench and lower slope deposits.

RATES OF SEDIMENTATION

Holocene

Holocene rates of sedimentation on the Oregon continental slope are calculated on the basis of the Mazama ash horizons in the slope deposits which are dated at 6,600 years B.P. (Table 1). Lower slope mud facies have rates ranging from 20 to 65 cm/1000 years (Figs. 7, 9, 10). The highest rates are found off southern Oregon on a lower slope escarpment (core 6711-1, 65 cm/1000 years, Fig. 10) and off northern Oregon in a large basin in the ridge and basin province (core FF-1, 53 cm/1000 years; Fig. 7). Higher up on the lower slope off southern Oregon the sedimentation rates in the mud facies decrease to 20 and 38 cm/1000 years (cores 6711-2 and 6708-38, respectively, Fig. 10).

The Astoria submarine canyon off the Columbia River exhibits equally high post-Mazama rates ranging from 37 to at least 78 cm/1000 years in the canyon axis (cores 1, 3, 9) and 19 cm/1000 years in one of the tributaries (core 18) (Fig. 7). This canyon is back filling with mud with the highest rates of Holocene deposition near the canyon head (Carlson, 1968). Much lower rates of mud deposition (10 cm/1000 years) are observed in the head of the Rogue Canyon off the Rogue River. Sand, probably relict from the last lowering of sea level, occurs at the most headward part of the Rogue Canyon.

Upper continental slope sedimentation rates probably do not exceed 10 cm/1000 years as noted in core 6706-6 recovered on the wall of the Rogue Canyon (Fig. 8). No other Holocene datums could be recognized in the remainder of the cores in this region.

Pleistocene

Latest Pleistocene.—Cores obtained from the Klamath Plateau and upper slope benches exhibit the lowest slope sedimentation rates of 10 and 14 cm/1000 years, respectively (cores 6706-3 and 6706-2, Fig. 5, Table 1) between 34 and 24,000 years ago. These gray and olive gray muds are accumulating at even slower rates in the recent past as evidenced by the glauconite facies and planktonic forams in the top of these cores. Terrigenous detrital sands on these upper slope benches are oxidized which suggests they are relict deposits that have not been covered with muds or reworked by currents since the Holocene rise of sea level (Spigai, 1971).

Late Pleistocene.—Drill hole Site 175 gives the best long term rates of sedimentation for an accretionary basin on the lower continental slope. Using diatom and radiolarian biostratigraphy, rates of 23 and 24 cm/1000 years are calculated for the past 300–400,000 years, respectively (Fig. 9). Dark greenish gray silty clays and clayey silts dominate the sedimentary section with minor fine sand and silt beds scattered throughout the upper 100 meters of the section but exhibit no size grading or other structures typical of turbidity current deposits. However, a portion of these muds have been carried to this depositional site from adjacent shallower environments since they contain benthic foraminiferal faunas indicative of shallower depths (Kulm, von Huene, et al, 1973).

FACTORS INFLUENCING CONTINENTAL SLOPE SEDIMENTATION

Tectonism

As discussed previously, the interface between the Oregon continental slope and Cascadian Basin is the subduction boundary between the Juan de Fuca and North American Plates (Fig. 1). Convergence of these two plates produces a deformational front that is migrating westward (Silver, 1972; Carson et al, 1974; Kulm and Fowler, 1974) as the sand and silt turbidite facies of abyssal plain become incorporated into the ridges (accreted to the margin) and escarpments of the lowermost continental slope (Fig. 4). These features are being uplifted, relative to the adjacent plain, at minimum rates ranging from 550 to 1000 m per million years (Kulm and Fowler, 1974, their Fig. 12). The uplift is not uniform along the slope in a temporal and spatial sense which creates a number of structural basins of various sizes. The rapidly rising deformational front continually modifies these basins and tends to produce a number of smaller sized accretionary basins, at least off Oregon, as subduction continues.

Ridges and hills on the continental slope (Figs. 2, 3) form structural barriers to turbidity current

TABLE 1.—SEDIMENTATION RATES FOR OREGON CONTINENTAL SLOPE AND SUBMARINE CANYON DEPOSITS IN CENTIMETERS PER THOUSAND YEARS

	North	Central	South
Upper Slope			
Holocene	—	—	10
L. Pleistocene	—	—	10–14
Lower Slope			
Holocene	53	—	20–65
L. Pleistocene	23*	—	—
Canyon			
Holocene	19–78	—	10
L. Pleistocene	12	—	9

*Site 175 represents 0.3 million years of sedimentation.

flows and can isolate lower slope basins for several hundred thousand years or more from this type of deposition. Mainly hemipelagic muds accumulate during these periods of isolation (e.g., Site 175, Fig. 9) with slumping of hemipelagic muds from surrounding highs adding a significant amount of debris to the basin deposits. Rates of deposition in this basin were calculated at 23 cm/1000 years (Table 1). Turbidites may be deposited on the lower slope initially while the basin is still in its early formative stage and there is a connection from shallower slope environments by bottom seeking currents. Rates of deposition are much higher for turbidites but cannot be calculated from the data that are available. On the other hand, steep escarpments with no intervening basins or structural benches (Figs. 2, 3) allow turbidity currents flowing down the continental slope to carry their sediment load directly to the abyssal plain before deposition occurs.

Sedimentation on the structural benches and plateaus off southern Oregon is more complicated than the lower slope basins off northern Oregon. Piston Core 6706-3 (Fig. 5) penetrated the entire late Pleistocene and Holocene sediment cover and recovered mid-Pliocene siltstone and mudstone from the Klamath Plateau (Spigai, 1971). A major unconformity occurs between these deposits and apparently represents a structural-depositional contact rather than erosional contact (Spigai, 1971). Structural benches may have a thin sediment cover as noted on seismic reflection records (Fig. 3A).

Mass movement of late Quaternary deposits is inferred from seismic reflection records obtained over the Oregon and Washington continental slopes (Carson, 1977; Fig. 15). Acoustically transparent sedimentary masses occur at the base of the steep escarpments along the north-south trending ridges forming the lower slope. Gravity coring shows that the upper few meters of these deposits consist of olive gray or green muds. Displacement of these muds from shallower to deeper water depths has not yet been confirmed from the sediments seen in the reflection records although displaced benthic foraminifera in the deposits drilled in the basin at Site 175 do show significant down slope movement of muds from the adjacent ridge escarpments (Kulm, von Huene et al, 1973). Large scale slumps may be expected on this tectonically active margin, but unlike most convergent continental margins, the Oregon-Washington margin exhibits only minor earthquake activity which may preclude large scale mass movement of sediment.

These rapidly evolving tectonic features undoubtedly affect deposition of suspended material in the water column, particularly when unidirectional currents flow along or across the slope.

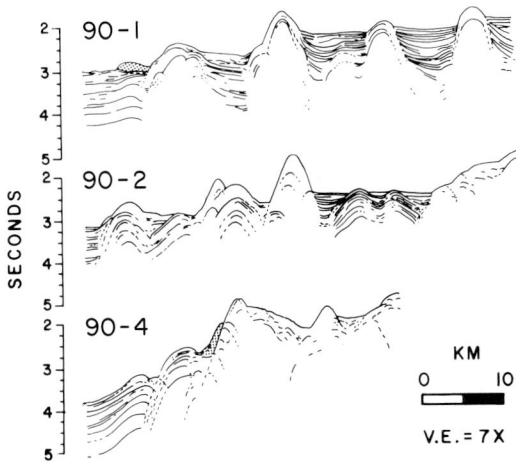

FIG. 15.—Line drawings of seismic reflection profiles across the central and northern Oregon continental slope. Stippled areas indicate slump or sediment mass movement (modified after Carson, 1977, his Fig. 2). See Figure 2 for location of profiles.

Knowledge of such currents is sparse in the northeast Pacific, but preliminary current measurements across the Oregon-Washington slope suggest that a prominent poleward flowing (south to north) undercurrent exists in this region (R. Smith and A. Huyer, 1978, pers. comm.). Furthermore, the clay mineral distribution patterns in the surface sediments of the Oregon shelf and slope indicate that clay-sized material is being transported northward from southern Oregon and northern California sources to the vicinity of the Columbia River, an observation consistent with the postulated flow direction of the undercurrent (Karlin, 1978). This contour-like current probably impinges on the bottom whose topography is constantly being modified through the tectonic events noted previously.

The distribution of tectonic ridges and hills that characterize portions of the slope will greatly influence the movement of bottom seeking turbid layers laden with silt and clay (see section on Sediment Supply and Dispersal Processes).

Sea Level Fluctuations

Sea level fluctuations on the continental margin, whether they be due to eustatic causes or to regional tectonic (uplift or subsidence) causes, have an important influence upon sedimentation on the continental slope, especially where the continental shelf is narrow as off Oregon. The present high strandline restricts the movement of modern sand-sized particles coarser than very find sand (>125 μ) to the shoreline and nearshore regions (Kulm et al, 1975). Abyssal plain lithofacies show that there have been no turbidity current events, containing sand, generated on the Oregon margin since approximately 10–12,000 years ago or since the beginning of the Holocene transgression (Duncan, 1968). However, there has been a continual supply of very fine sand (125–62 μ) and finer particles from the Columbia River to Willapa Canyon (Fig. 1), which is the source of material for turbidity current flows carried though Cascadia Deep Sea Channel at about 400 year intervals during the entire Holocene time (Royce, 1967; Griggs and Kulm, 1970). Transport of these finer grained sediments northwest from the Columbia is described in the next section.

The present high stand of sea level is also responsible for the accumulation of the mud facies on the Oregon continental shelf and portions of slope because the coastal estuaries and continental shelf act as a filter for most coarse-grained sediments that would produce a silt or sand turbidite facies on the slope (Kulm et al, 1975). Accumulation rates of slope muds range from 10 to 53 cm/1000 years for the post-6600 year period (Table 1).

During lower sea level stands like the glacial Pleistocene, the rivers may be connected, via channels cut into the shelf, to submarine canyons which head near the outer edge of the shelf (e.g., Columbia and Rogue systems). Large quantities of sand and gravel are transported through the canyons to the continental slope and abyssal regions at this time (Duncan, 1968; Griggs et al, 1970). Where the Astoria Canyon passes through the ridges of the northern Oregon slope, some of the turbidity flows are diverted southward into the evolving accretionary basins on the lower slope (e.g., see Fig. 7, core FF-1 with Mazama ash and sand turbidites near bottom of core). The rapid evolution of the accretionary basins and the accompanying changes in surrounding topography may be responsible, in certain areas, for the development of largely mud facies in these slope basins. They become isolated from turbidity currents very early in their formation and they may not come under their influence until a much later stage of development and possibly not even at this time, depending upon the rate of uplift and deformation surrounding the basin.

The upper slope sand-silt facies was deposited off southern Oregon during a low stand of the sea when the shoreline was located approximately 6 km from the shelf break (Fig. 5, cores 6706-8, 6708-39, 40). Very fine sand and coarse silt from the surf and nearshore zone must have diffused seaward, analogous to the modern setting, and then deposited on the bench. Coquille bank on the outer edge of the shelf was a nearby source of material at this time (Fig. 2). On the other hand, chiefly gray mud accumulated on the Klamath Plateau during the glacial Pleistocene record shown in core 6706-3 (Fig. 5). Cores 6706-2 and 6706-1 do not contain abundant sand either.

From these limited number of cores and the brief geologic record it is difficult to determine the relative importance of the sand facies in these slope deposits. However, in this southern region it appears that few sands coarser than 125 μm were deposited on the upper or lower slope even during the latter part of the glacial Pleistocene.

Sediment Supply and Dispersal Processes

Bottom Turbid Layer.—The numerous coastal rivers (e.g., Columbia, Umpqua, Rogue) drain regions of mountainous relief and rainfall is moderate to high, especially in the coastal regions. Maximum discharge occurs during the winter months (October to March) in the coastal rivers while peak discharge occurs during May and June in the Columbia River. Although the coastal estuaries may act as filters (sediment traps), a certain amount of sediment is transported into the open shelf environments (Kulm et al, 1975). Patchy accumulation of mud on the shelf, (concentrated in the vicinity of rivers with the highest discharge), plumes of turbid water emanating from the river mouths onto the adjacent shelf, and the texture of the shelf sediments indicate that material finer than 125 μm (very fine sand, silt and clay) is presently escaping these estuarine systems (Kulm et al, 1975).

According to the work of Pak and Zaneveld (1977) and Harlett and Kulm (1973), some of these fine-grained sediments diffuse through the turbulent surf zone and are advected seaward to form a bottom nepheloid or turbid layer. Bottom photographs show the nearshore area is a zone of extreme turbidity most of the year. If seaward advection is an effective process of generating the bottom nepheloid layer, it must be an effective means of horizontal transport of suspended particles (Pak and Zaneveld, 1977). In addition, resuspension of the transgressive mud facies on the continental shelf is documented in bottom photographs which show oscillating ripples formed by orbital motion of surface waves, and this may provide another means of enhancing bottom nepheloid layer development. As a consequence, a well-developed bottom nepheloid layer occurs over the bottom, especially during the winter when long period waves agitate the bottom to water depths of 200 meters (Komar et al, 1972), and sediment supply from rivers is at a maximum. Unidirectional bottom currents associated with surface and internal tides and wind stress on the shelf commonly have velocities of 10 cm/sec (Harlett and Kulm, 1973) with storm induced velocities up to 60 cm/sec (Smith and Hopkins, 1972). The direction of flow is variable but commonly oriented subparallel to the offshore region. Bottom photographs over the middle and outer shelf region show that clearest water is found over submarine banks and the most turbid water is found in the swales or small valleys of the outer shelf and uppermost slope. Bottom turbid layers apparently seek the low areas of the shelf and slope and transport suspended material to the topographically accessible regions of the continental slope.

Some submarine canyons are avenues of dispersal for fine-grained material on both the Oregon and Washington slope with deposition in the canyon axis as well as on the submarine fans that connect to the canyons (e.g., Stokke et al, 1977; Baker, 1976; Plank et al, 1974; Spigai, 1971; Royce, 1967; Carlson, 1968). Astoria Canyon exhibits a high rate of sedimentation (19-78 cm/1000 yrs) during the Holocene (Fig. 7, Table 1). A study of suspended matter in Astoria Canyon (Plank et al, 1974; Fig. 1) shows a bottom nepheloid layer is present within the axis of the canyon and is associated with a mud facies (Fig. 7). The upper reaches of the canyon (about 200-1350 m) are characterized by large spatial and temporal gradients in light scattering and suspended matter, and have lead Plank et al, (1974) to suggest that this was a region of bottom erosion and near bottom turbulent transport in a channelized setting. However, as seen in Figure 7, this region has the highest mud sedimentation rates (>78 cm/1000 yrs) in the canyon so deposition rather than erosion is the dominant long-term process. In the lower reaches of the canyon (1350-2000 m) gradients are not as large in optical properties and suspended matter as the upper reaches which suggest a quieter zone of deposition (Plank et al, 1974). Sedimentation rates are about 53 cm/1000 yrs. here.

Interestingly, the deposits displaying the highest clay content are observed in the muds in the axis of the Rogue Canyon off southern Oregon, whereas, the surrounding upper and lower slope muds have less clay. We suggest that the especially fine-grained canyon muds may be preferenally concentrated and deposited by nepheloid layer transport down the axis of the canyon in a manner similar to Astoria Canyon although muds sampled in that canyon are not as fine-grained (see Fig. 12).

Particle Settling.—Terrigenous suspended matter which diffuses through the surf zone may be also transported over the shelf in the surface turbid layer or in mid-water turbid layers (Harlett and Kulm, 1973; Pak et al, 1970) to be eventually deposited on the continental slope in the typical hemipelagic sedimentation process (i.e., particle by particle settling in the water column). This dispersal process may be modified by strong unidirectional currents which occur in the water column over the shelf and slope. As noted previously, the distribution pattern of clay mineral assemblages in the slope muds indicate south to north transport of clay-sized particles along the Oregon slope, an observation consistent with a poleward flowing undercurrent. Rivers in northern California and southern Oregon are the terrigenous sources of the clay-sized sediments.

Biogenous material is derived through primary productivity in the water column overlying the continental shelf. Diatom production along the Oregon coast is high, particularly during the summer months, due to upwelling and subsequent phytoplankton blooms (Small and Ramberg, 1971; Small and Curl, 1968). Areas high diatom productivity vary in intensity and are rarely found more than 20 km from the coast and yet diatoms constitute a fairly high percentage of the sand fraction and, no doubt, smaller sized particles in the continental slope mud facies. The processes responsible for the apparent seaward displacement of these siliceous components, either in the water column or by resuspension and turbid layer transport of previously deposited material, is not well understood at this time.

CONCLUSIONS

The Oregon continental slope is characterized by a variety of structural features that result from the convergence of the adjacent oceanic plate and the continental block. The subduction complex produced by this process consists of accreted ocean plate and trench deposits which are uplifted rapidly to form the ridge and basin topography off northern Oregon, the steep escarpments off central Oregon and the benches and plateaus off southern Oregon (Fig. 2). Numerous small sedimentary basins are evolving within this structural framework.

Six distinct lithofacies are recognized on the abyssal plain and adjacent continental slope off Oregon. They include: (1) calcareous pelagic ooze, (2) silt turbidites with interbedded muds, (3) sand turbidites, (4) mud facies, (5) glauconite facies, and (6) sand-silt facies. Silt and sand turbidite lithofacies from the abyssal plain and Astoria Submarine Fan are incorporated into the slope in the form of ridges with intervening accretionary basins of various sizes (Fig. 4). Early deposition in these newly formed basins may consist of a silt turbidite sequence followed by a mud facies when turbidity current deposition is precluded in the basin by the rapidly developing topographic (structural) barriers. A mud facies may develop directly upon the subduction complex as in the case of drill Site 175 on the lowermost continental slope off north-central Oregon (Fig. 9).

A mud facies is accumulating at a rather slow

rate (10 to 14 cm/1000 yrs.) on the upper slope benches and plateaus (Fig. 5), while the bench and basin features on the lower slope are experiencing more rapid mud deposition at rates of from 20 to 65 cm/1000 years (Fig. 10). A thin glauconite and sand silt facies also occur on these upper slope features with mud below, particularly during the past few thousand years. A thin mud layer is present on the steep escarpments of the upper and lower slope off central Oregon with occasional rock outcrops of older Quaternary and Neogene deposits exposed on the slope (Fig. 6). On the Klamath Plateau a thin layer of latest Quaternary muds lies unconformably upon Pliocene mudstones.

Astoria and Rogue submarine canyons off northern and southern Oregon, respectively, have been filling with mud during much of Holocene time (Figs. 7, 8). Rates of deposition range from 10 to greater than 78 cm/1000 yrs. with the highest rates apparently occurring near the heads of the canyons located at the outer edge of the continental shelf. The present high sea level stand, which separates the present heads of the canyons from their river sources, is allowing only mud deposition in these two canyons. The Astoria Canyon displays a bottom nepheloid layer which suggests that mud is being deposited, in part, from this layer of suspended material.

The Holocene and late Pleistocene mud facies show significant variations in sand-silt-clay content along the continental slope (Fig. 13). For example, upper slope muds off central Oregon contain more silt than those off southern Oregon. On the lower slope, the muds to the south and north contain more clay than those off central Oregon, but southern Oregon muds also may contain less clay. Using statistical parameters (phi mean diameter, skewness, and sorting), the upper slope muds appear to be coarser than 6 ϕ while the lower slope muds are finer-grained, although there is considerable overlap in the upper and lower slope textural parameters (Fig. 14). The sediments supplied from the various drainages appear to exert a strong control on the texture of the slope mud facies with unidirectional currents in the water column and nepheloid layer transport along the bottom further influencing the texture of these muds.

REFERENCES

BAKER, E. T., 1976, Temporal and spatial variability of the bottom nepheloid layer over a deep-sea fan: Marine Geology, v. 21, p. 67–69.

BOUMA, A. H., 1962, Sedimentology of some flysch deposits, a graphic approach to facies interpretation: Amsterdam, Elsevier, 168 p.

CARLSON, P. R., 1968, Marine geology of Astoria Submarine Canyon [unpub. Ph.D. thesis]: Corvallis, Oregon State Univ., 259 p.

CARSON, B., YUAN, T. W., MYERS, P. B., JR., AND BARNARD, W. D., 1974, Initial deep-sea sediment deformation at the base of the Washington Continental slope: a response to subduction: Geology, v. 2, p. 561–564.

———, 1977, Tectonically induced deformation of deep-sea sediments off Washington and northern Oregon: mechanical consolidation: Marine Geology, v. 24, p. 289–307.

DUNCAN, J. R., 1968, Postglacial and late Pleistocene sedimentation and stratigraphy of deep-sea environments off Oregon [Ph.D. thesis]: Corvallis, Oregon State University, 179 p.

FRYXELL, R., 1965, Mazama and Glacier Peak volcanic ash layers: relative ages: Science, v. 147, p. 1288–1290.

GRIGGS, G. B., AND KULM, L. D., 1970, Sedimentation in Cascadia Deep-Sea Channel: Geol. Soc. America Bull., v. 81, p. 1361–1384.

———, KULM, L. D., AND FOWLER, G. A., 1970, Deep-Sea gravels from Cascadia Channel: Jour. Geol., v. 78, p. 611–619.

HARLETT, J. C. AND KULM, L. D., 1973, Suspended sediment transport on the northern Oregon continental shelf: Geol. Soc. America Bull., v. 84, p. 3815–3826.

HAYES, J. B., 1973. Clay petrology of mudstones, Leg 18, Deep Sea Drilling Project, in Kulm, L. D., von Huene, R., et al, Initial reports of the Deep Sea Drilling Project, Washington D.C., U.S. Government Printing Office, v. 18, p. 903–914.

INMAN, D. L., 1952, Measures for describing the size distribution of sediments: Jour. Sed. Petrology, v. 22, p. 125–145.

KARLIN, R., 1978, Sediment sources and clay mineral distributions off the Oregon coast: evidence for a poleward slope undercurrent [M.S. thesis]: Corvallis, Oregon State Univ., 88 p.

KLING, S. A., 1973, Radiolaria from the eastern north Pacific, Deep Sea Drilling Project, Leg 18, in Kulm, L. D., von Huene, R., et al, Initial reports of the Deep Sea Drilling Project: Washington, D.C., U.S. Govt. Printing Office, v. 18, p. 617–671.

KOMAR, P. D., NEUDECK, R. H., AND KULM, L. D., 1972, Observations and significance of deep-water oscillatory ripple marks on the Oregon continental shelf, in Swift, D., Duane, D. B., and Pilkey, O. H., eds., Shelf sediment transport: process and pattern: Stroudsburg, Dowden, Hutchinson and Ross, p. 601–619.

KULM, L. D., VON HUENE, R., et al, 1973, Initial Reports of the Deep Sea Drilling Project, v. XVIII: Washington, D.C., U.S. Govt. Printing Office.

———, AND FOWLER, G. A., 1974, Oregon continental margin structure and stratigraphy: a test of the imbricate thrust model, in Burk, C. A. and Drake, C. L., eds., The Geology of Continental Margins: New York, Springer-Verlag, p. 261-283.

———, Roush, R. C., Harlett, J. C., Neudeck, R. H., Chambers, D. M., and Runge, E. J., 1975, Oregon continental shelf sedimentation: interrelationships of facies distribution and sedimentary processes: Jour. Geol., v. 83, p. 145-175.

Maloney, N. J., 1965, Geology of the continental terrace off the central coast of Oregon [unpub. Ph.D. thesis]: Corvallis, Oregon State Univ., 233 p.

Nelson, C. H., Kulm, L. D., Carlson, P. R., and Duncan, J. R., 1968, Mazama ash in the northeastern Pacific: Science, v. 161, p. 46-49.

———, and Kulm, L. D., 1973, Submarine fans and deep-sea channels: Soc. Econ. Paleontologists Mineralogists Pacific Sec., Short course lecture notes, Anaheim, Turbidites and deep water sedimentation, p. 39-78.

Pak, H., Beardsley, G. B., Jr., and Smith, R. L., 1970, An optical and hydrographic study of a temperature inversion off Oregon during upwelling: Jour. Geophys. Research, v. 75, p. 629-638.

———, and Zaneveld, J. R., 1977, Bottom nepheloid layers and bottom mixed layers observed on the Continental Shelf off Oregon: Jour. Geophys. Res., v. 82, p. 3921-3931.

Plank, W. S., Zaneveld, J. R. V., and Pak, H., 1974, Temporal variability of suspended matter in Astoria Canyon: Jour. Geophys. Res., v. 79, p. 4536-4541.

Royce, C. F., Jr., 1967, Mazama ash from the continental slope off Washington: Northwest Science, v. 41, p. 103-109.

Schrader, H. -J., 1973, Cenozoic diatoms from the northeast Pacific, Leg 18, in Kulm, L. D., von Huene, R., et al, Initial Reports of the Deep Sea Drilling Project: Washington, D.C., U.S. Govt. Printing Office, v. 18, p. 673-797.

Seely, D. R., Vail, P. R., and Walton, G. G., 1974, Trench slope model, in Burk, C. A. and Drake, C. L., eds., The Geology of Continental Margins: New York, Springer-Verlag, p. 249-260.

———, and Dickinson, W. R., 1977, Structure and stratigraphy of forearc regions, in Geology of Continental Margins, Am. Assoc. Petroleum Geologists Continuing Education Course Note Series, No. 5, p. C1-C23.

Shepard, F. P., 1954, Nomenclature based on sand-silt-clay ratios: Jour. Sed. Petrology, v. 24, p. 151-158.

Silver, E. A., 1969, Late Cenozoic underthrusting of the continental margin off northern California: Science, v. 166, p. 1265-1266.

———, 1971, Tectonics of the Mendocino Triple Junction: Geol. Soc. America Bull., v. 82, p. 2965-2978.

———, 1972, Pleistocene tectonic accretion of the continental slope off Washington: Marine Geology, v. 13, p. 239-249.

Small, L. F., and Ramberg, D. A., 1971, Chlorophyll a, carbon, and nitrogen in particles from a unique coastal environment: Costlow, Z. D., ed., Fertility of the Sea: Gordon and Breach, p. 475-492.

———, and Curl, H. C., 1968, The relative contribution of particulate chlorophyll and river tripton to the extinction of light off the coast of Oregon: Limnol. Oceanog. v. 13, p. 84-91.

Smith, J. D., and Hopkins, T. S., 1972, Sediment transport on the continental shelf off of Washington and Oregon in light of recent current measurements, in Swift, D., Duane, D. B., and Pilkey, O. H., eds. Shelf sediment transport: process and pattern: Stroudsburg, Pa., Dowden, Hutchinson and Ross, p. 143-180.

Spigai, J. J., 1971, Marine geology of the continental margin off southern Oregon [unpub. Ph.D. thesis]: Corvallis, Oregon State Univ., 214 p.

Stokke, P. R., Carson, B., and Baker, E. T., 1977, Comparison of the bottom nepheloid layer and late Holocene deposition on Nitinat Fan: implications for lutite dispersal and deposition: Geol. Soc. America Bull., v. 88, p. 1586-1592.

SEDIMENTATION ON THE ANTARCTIC CONTINENTAL SLOPE

JOHN B. ANDERSON
DENNIS D. KURTZ
Department of Geology
Rice University at Houston, Texas 77001
AND
FRED M. WEAVER
Exxon Production Research
Houston, Texas 77001

ABSTRACT

The Antarctic continental margin is presently characterized by three different glacial regimes. These include: continental margins where the ice sheet is grounded on the continental shelf, those whose inner shelves are covered by floating ice shelves, and those which bound mountainous coasts where valley glaciers debouch directly into the sea. Piston cores from all three types were examined.

The West Antarctic sectors of the Ross and Weddell Seas are characterized by broad continental shelves, relatively smooth, gentle slopes and vast floating ice shelves. Glacial sedimentation occurs well inland of the ice shelf front and the important influence of waves and wind generated currents is lacking. Consequently, sediment supply to the continental slope is primarily restricted to that of contour currents. Piston cores from the continental slopes of both the Ross and Weddell Seas penetrated similar sedimentary sequences. Surficial sediments consist of current derived silts and clays which contain from 9% to 44% ice-rafted detritus. These sediments overlie well sorted, laminated silts and clays containing only minor amounts of ice-rafted material. Laminated sediments overlie debris flow deposits which exhibit sharp upper contacts, remarkable textural homogeneity and contain displaced benthonic foraminifera.

The East Antarctic ice sheet is aground over much of the continental margin. In the northeastern Weddell Sea the weight of the ice sheet depresses the continental shelf in the form of proglacial basin with the shelf sloping toward the continent. Sediment transport to the continental slope is presently restricted, and strong contour currents are winnowing upper slope deposits. Sedimentation on the lower slope is primarily by mass flow processes.

The northwestern Ross Sea bounds the mountainous coast of Victoria Land where valley glaciers flow directly into the sea. Much of the inner shelf is free of significant ice cover during summer months. Sediments of the continental shelf have been winnowed and redistributed so that sands and gravel blanket bathymetric highs and muds fill depressions. Sediment transport to the continental slope is primarily via turbidity currents.

INTRODUCTION

During the past decade a comprehensive circum-Antarctic marine geology survey was conducted using the USNS *Eltanin,* ARA *Islas Orcadas,* and USCGC *Glacier.* The geologic samples taken during this program number in the hundreds.

The primary research objective of the Antarctic marine geology program has been to reconstruct that continent's glacial history from the marine sedimentary record. The most direct marine evidence for Antarctic glacial conditions is that of ice-rafted sediments. On the continental shelf these deposits are quite conspicuous, being virtually indistinguishable from continental tills. Unfortunately, the Antarctic continental shelf has been severely eroded by advancing ice shelves, so the record there is incomplete.

Glacial marine sedimentation in Antarctica is dominated by large ice shelves which deposit most of their sediment on the continental shelf. Relatively small amounts of debris are rafted out to sea by floating ice (Warnke, 1970). On the continental slope, ice-rafted sediments are incorporated into and interbedded with sediments deposited by currents and mass flow processes. Hence, while the stratigraphic record probably becomes more complete with increasing distance from the shelf break, the sedimentary evidence for glaciation becomes less conspicuous. Proper interpretation of the sedimentary record can only be accomplished once we are able to distinguish bottom current, mass flow and ice-rafted components of Antarctic marine sedimentary sequences. The same is true for ancient rock sequences of supposed marine glaciogenic origin, such as those of the Middle and Late Precambrian.

Of the hundreds of sediment cores collected to date around Antarctica, only a small percentage (42 cores in all) are from the continental slope. Most of these were taken in the Ross and Weddell Seas. This sample density is too small to construct sediment distribution maps. What can be gained from studying these cores is an understanding of the dominant physical processes affecting sedimentation around Antarctica. The variability of modes of deposition is reflected in the considerable diversity of sediments present, most cores having penetrated several distinct units. The purposes of this paper are: 1) to emphasize those aspects of the Antarctic continental margin which make it unique as a sedimentary environment, 2) to describe those sediments which have been cored on the continental slope, 3) to present our interpretations concerning the sedimentary processes that form these deposits, and 4) to present criteria that may be useful in distinguishing ancient glaciogenic, bottom current, and mass flow deposits.

THE ANTARCTIC CONTINENTAL MARGIN

Before beginning our discussion of Antarctic continental slope sedimentation, it is important that we emphasize those unique aspects of the continental margin which influence sedimentation.

1) Most of the Antarctic continental margin has had a passive tectonic history since at least Late Paleozoic time (Weissel et al, 1977; Craddock, in press). The Bellingshausen and Amundsen margins (Fig. 1) are the only known exceptions (Weissel, et al, 1977).

2) The continental margin of East Antarctica is characterized by a relatively narrow shelf, steep slope, and broad rise (Fig. 1). The West Antarctic margin has broad shelves, and slope gradients tend to be more gentle than those of East Antarctica (Fig. 2).

3) Erosion on the continent is limited to glacial processes. Chemical erosion is insignificant (Angino, 1966).

4) The continental margin of Antarctica is characterized by three basic types of glacial setting. In most of East Antarctica the ice sheet is grounded on the continental shelf and the ice profile steepens abruptly over the margin. The West Antarctic Ice Sheet is largely situated below sea level, and is characterized by vast floating ice shelves such as the Ross and Ronne Ice Shelves (Fig. 1). Along the mountainous coasts of Antarctica, valley glaciers debouch directly into the sea or coalesce to form small ice shelves.

5) The maximum seaward limit of the grounded ice sheet is the continental shelf break. Since most glacially transported sediment is carried at or near the ice-substrate interface (Boulton, 1975), deposition of this sediment is confined, for the most part, to the continental shelf. These sediments consist of non-sorted highly compacted ice-contact till (Anderson et al, 1977). Only those sediments carried at higher levels in the ice (englacial sediment) will be transported out to sea by floating ice. The amount of sediment transported englacially is probably quite small in relation to that which is carried subglacially (Boulton, 1975). Hence, sediment supply directly by ice shelves to the deep-sea floor may be small in relation to that transported down slope by bottom currents and mass flow. Valley glaciers may carry considerable englacial debris and, hence, their bergs may be laden with sediment (i.e., Ovenshine, 1970). Continental shelves which bound mountainous terrains may receive large amounts of sediment directly form floating ice as well as from meltwater streams. These sediments will more likely be influenced by marine agents than will those deposited subglacially.

Carey and Ahmad (1961) and Boulton (1972) have emphasized the importance of subglacial conditions, that is wet-base versus dry-base conditions, in influencing glacial erosion and transport. Hughes (1973) has argued that there are probably major differences in the glacial regimes of the East and West Antarctic ice sheets, the former having a dry-base regime and the latter a wet-base regime.

6) The Antarctic continental shelf is subject to considerable erosion by grounded ice shelves, ice streams and subglacial meltwater streams. Large submarine valleys, some over 1,000 meters deep, are widespread on the continental shelves of Antarctica. These features are believed to be formed by ice streams during major glacial advances (Hughes, 1977). Where the continental shelf is narrow, such as in the western Ross Sea, these valleys connect with submarine canyons.

7) Most of the Antarctic coastal area is covered

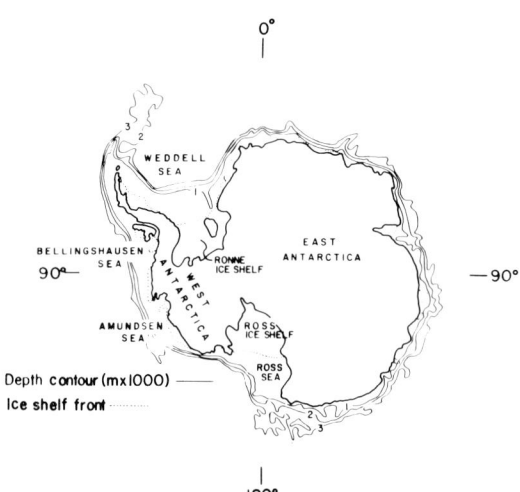

FIG. 1.—Antarctic geography and bathymetry.

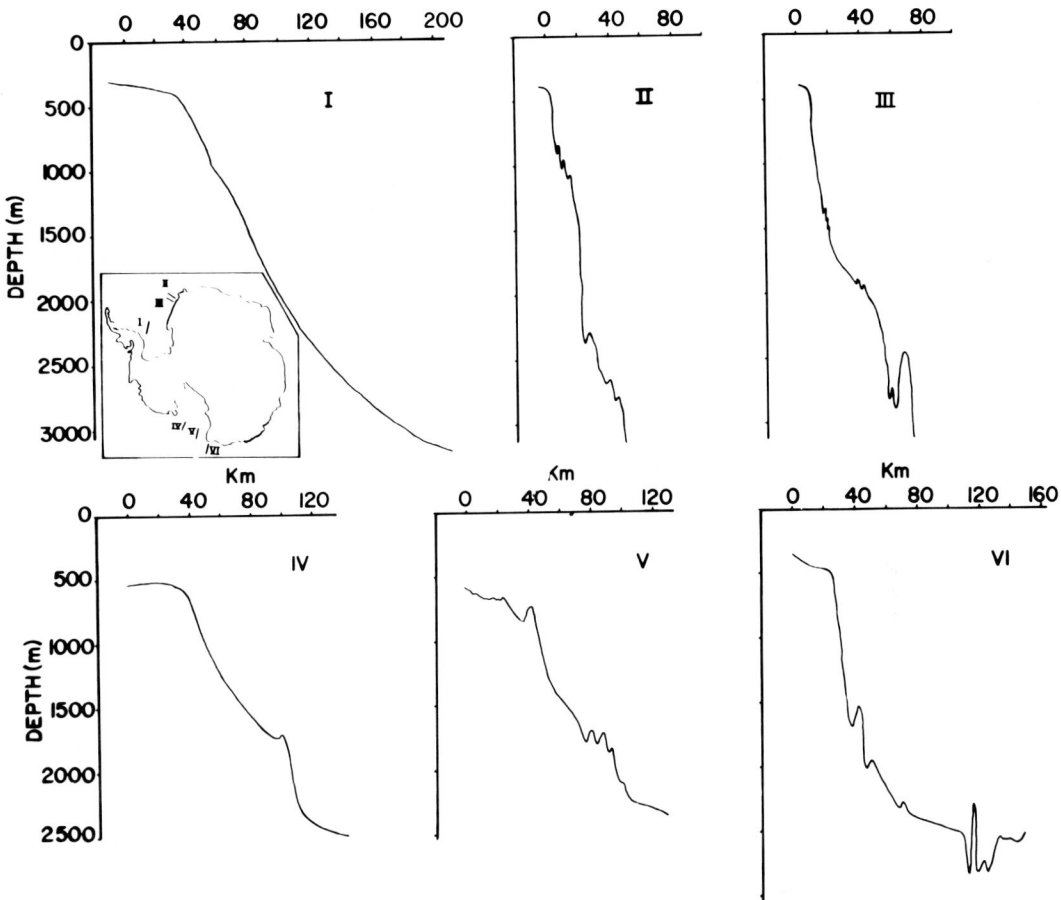

FIG. 2.—Representative bathymetric profiles of the Antarctic continental slope. Vertical exaggeration = 80:1. Inset shows profile locations. Profiles I, IV, V—West Antarctica; profiles II, III, VI—East Antarctica.

by perennial ice, either pack ice or ice shelves. Hence, the important sedimentologic influence of waves and wind generated currents is greatly retarded.

8) Because in most areas the Antarctic ice sheet extends onto the continental margin, the weight of the ice, and thus isostatic influences, are felt there. Presently, the Antarctic continental shelf break varies from 300 to 540 meters, as compared to an average of 132 meters for the other continental shelves of the world (Shepard, 1948). These differences are attributed to isostatic downwarping of the Antarctic continental shelf (Thiel, 1962). The continental slope probably serves as the hinge zone between the isostatically active continent and continental shelf and the ocean basin (Bloom, 1967; Brotchie and Silvester, 1969; Lliboutry, 1971; Walcott, 1970).

During a period of glacial retreat from the continental shelf, rates of isostatic uplift are probably on the order of 5 to 10 cm/yr^{-1}, as evidenced by rates of isostatic response to deglaciation in Canada and Greenland (Andrews, 1968; Muller and Barr, 1966; Marcel and Fairbridge, 1978; Ten Brink, 1974). Such rapid movements could influence slope instability and mass movement on the continental slope.

Due to the interaction between glacio-isostatic downwarping and the flexural rigidity of the lithosphere, a proglacial depression will form for some distance beyond the ice sheet grounding line (Walcott, 1970). For an ice sheet with dimensions and surface profile similar to that of East Antarctica, the proglacial depression will be some 180 kilometers wide and almost 300 meters deep (Broecker, 1966; Walcott, 1970) (Fig. 3a). Consequently, an ice sheet grounded on the continental shelf or within 200 kilometers of the shelf break

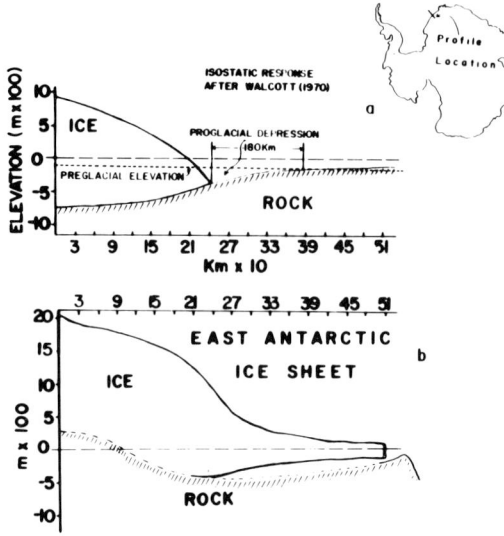

FIG. 3.—Isostatic depression of the Antarctic continental shelf. a) Depression of a flat continental shelf beneath an ice sheet with dimensions and surface profile approximately that of the East Antarctica (after Broecker, 1966 and Walcott, 1970). b) Islas Orcadas Cruise 1578 profile of the northwestern Weddell Sea continental margin showing proglacial isostatic depression. Glacier profile and subglacial topography from Bentley (1964).

will cause the shelf to slope toward the continent. This situation is probably common in East Antarctica (Fig. 3b), although bathymetric profiles of the East Antarctic margin are scarce. On the other hand, the West Antarctic ice sheet extends across the continental shelf in most areas as a relatively thin floating ice shelf, so that depression of the sea floor is not as pronounced.

9) Water mass properties and circulation on the Antarctic continental slope are greatly influenced by surface freezing and melting on the continental shelf. Freezing produces dense shelf waters which, as they flow onto the continental slope, displace and mix with Circumpolar Deep Water, the dominant water mass of the Antarctic continental slope (Gordon, 1971). This mixing results in the formation of Antarctic Bottom Water which appears to have a greater downslope component of flow than does Circumpolar Deep Water (Anderson, 1972). Mixing at the continental slope greatly influences surface productivity and hence, the supply of biogenic sediment to the continental slope (Anderson, 1975a).

METHODS

Sedimentary units were initially defined using information gained from x-radiographs and core descriptions. Parameters used to identify different units from x-radiographs include: sharpness of sedimentary contacts, textural grading, the presence or absence of stratification and sedimentary structures, and the presence or absence of pebbles. Samples were taken from those units distinguished on this basis.

Sediment samples (approximately 1.5 cms^3) were dried after a small subsample was removed for fine silt-clay analysis. Sand samples were analysed using a large (1.5 m long) settling tube and coarse silt (16 μ to 63 μ) sample using a small (25 cm long) settling tube. The 1.0 μ to 16.0 μ fraction was analysed by hydrophotometer (Jordan et al, 1971). A wet sample was used for the fine silt-clay size analysis. The settling tubes used are totally automated; size data are calculated from Gibb's equation (Gibbs et al, 1971) for the large tube and Oden's formula (Krumbein and Pettijohn, 1938) for the small tube. Data output is in the form of cumulative and frequency curves, moment measures of mean grain size ($M\phi$) and standard deviation ($\sigma\phi$), and graphic measures of skewness (SK_i) and kurtosis (K_G) (Krumbein and Pettijohn, 1938; Folk and Ward, 1957).

The benthonic foraminiferal assemblage in each sample was examined to determine if the fauna had been displaced. Displaced foraminiferal assemblages are generally quite conspicuous on the Antarctic continental slope because the shallow CCD (Kennett, 1966; Anderson, 1975b) limits the depth of distribution of most shelf forms. Depth zonations used in this investigation for the recognition of displaced faunas were taken from Kennett (1968), Fillon (1974), and Anderson (1975b).

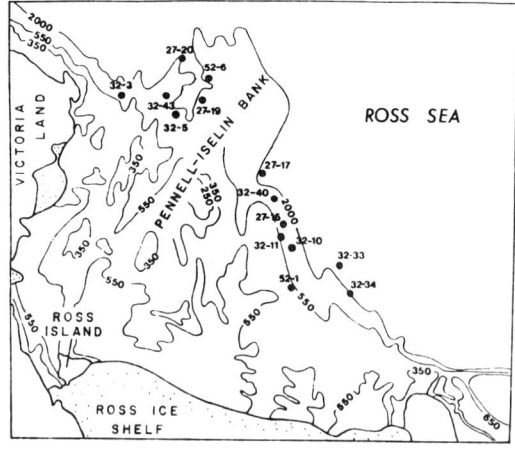

FIG. 4.—Ross Sea geography, bathymetry and core locations.

RESULTS

Ross Sea Sediments

In the Ross Sea the ice drainage divide for East and West Antarctica coincides approximately with the Pennell-Iselin Banks (Fig. 4). These banks also separate margins with very different physiographies. In the eastern Ross Sea the continental slope has a relatively straight southeast to northwest trend and is unbroken by submarine canyons (Fig. 4). The adjacent continental shelf is characterized by large northeast-to-southwest trending ridges and troughs. Troughs slope toward the continent and are separated from the slope by a sill (at a depth of approximately 550 meters).

Bottom currents on the continental shelf flow toward the continent (Jacobs et al, 1970). Sediments of the eastern shelf consist of marine tills overlain by a thin layer of diatomaceous mud (Fillon, 1972; Chriss and Frakes, 1972). The glacial regime of the eastern Ross Sea is dominated by the Ross Ice Shelf which extends over 300 kilometers onto the continental shelf (Hughes, 1973).

West of the Pennell-Iselin Banks the continental slope is steep, quite sinuous and is cut by large submarine canyons (Fig. 4). On the adjacent shelf, bathymetric highs are capped by coarse sand and gravel deposits, and in some areas by calcareous shell hash, while depressions are floored by muddy deposits (Chriss and Frakes, 1972). The

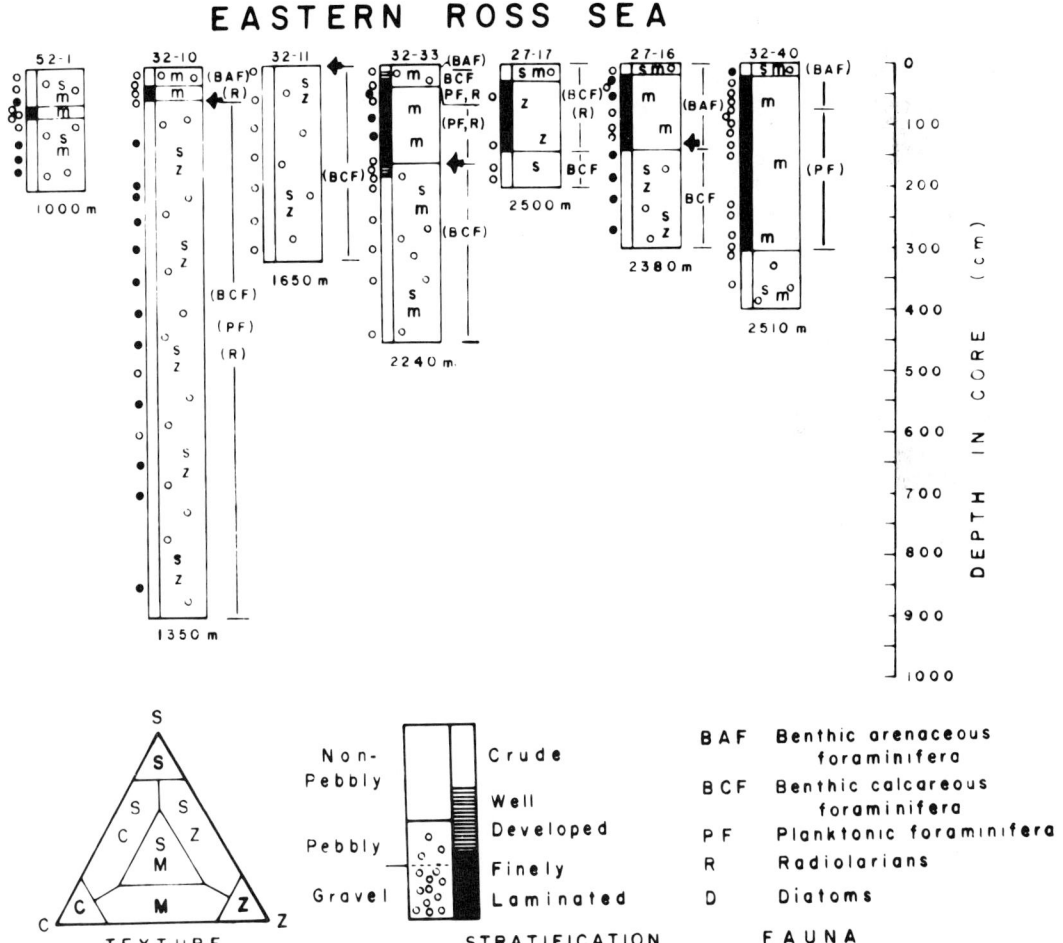

FIG. 5.—Sediment descriptions for piston cores from the Ross Sea continental slope east of Iselin Bank. Faunal designations in parentheses ( ) indicate very low abundance, otherwise fauna is common. Solid circles to the left of core logs indicate samples for which only sand-silt-clay fractions were determined, open circles represent those samples for which total size-analyses were conducted. Arrows mark Brunhes-Gauss boundary (Fillon, 1972, 1975).

glacial regime of the western Ross Sea is one in which valley glaciers from the Transantarctic Mountains flow directly into the sea. The mountainous coastline is exposed over large areas.

As might be expected from this comparison of the eastern and western Ross Sea, piston cores from either side of the Iselin Bank reflect major differences in slope sedimentation. Figure 5 shows sediment descriptions for piston cores collected on the continental slope east of Iselin Bank.

The microfossils in these cores (excluding core 52-1) were examined by Fillon (1972, 1974, 1975). He concluded that cores 32-10, 32-11, 32-33 and 27-16 all contain unconformable surfaces separating sediments of Gauss age ($t > 2.4$ m.y.) from a thin surficial layer of Brunhes sediments ($t < 0.7$ m.y.). This unconformable surface is shown by arrows in Figure 5. Note that 32-11 consists entirely of Gauss sediments. Core 27-17 penetrated a graded unit of sands and silts containing displaced benthonic foraminifera (Fillon, 1972). This core was taken along the steep flank of the Iselin Bank.

Cores 52-1, 32-10, 32-33, 27-16, 27-17, and 32-40 (Fig. 5) all contain a thin surface layer of non-stratified gravelly muds which contain arenaceous benthonic foraminifera. Cumulative curves for representative samples from this unit are shown in Figure 6. Samples 52-1 (9-11 cm) and 27-16 (9-11 cm) show little sorting, whereas in the other samples the coarse, non-sorted, ice-rafted mode is easily distinguished from silts which show varying degrees of sorting. These sediments resemble modern deposits of the adjacent continental shelf and reflect a combination of ice-rafting

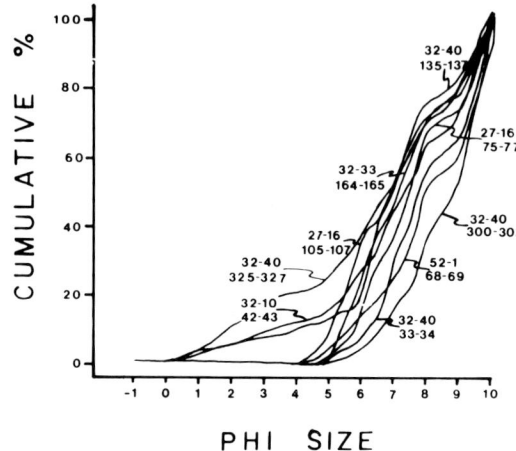

FIG. 7.—Cumulative curves for representative samples from laminated muds in eastern Ross Sea continental slope cores (Type 2, Table 1).

and bottom current deposition (Chriss and Frakes, 1972).

The surface gravelly mud units in cores 52-1, 32-10, 32-33, 27-16 and 32-40 overlie laminated clayey silts and silty clays (Fig. 5). The contact between these units is fairly sharp. Mean grain sizes range from 6.8 ϕ to 8.4 ϕ, standard deviations from 1.1 ϕ to 2.3 ϕ, skewness values from 9.4 to −0.5 and kurtosis values from 0.5 to 1.8. Figure 7 shows cumulative curves for several samples from within this laminated unit.

Laminations are readily observable in x-radiographs, but are difficult to discern in core sections. Individual laminae average 0.5 to 1.0 cm in thickness. Laminated units contain up to 13% (by weight) ice-rafted material which is generally concentrated within thin (<1 cm thick) layers. Ice-rafted grains are seldom larger than 2 mm in diameter. Fossils are rare to absent in laminated deposits.

The lower boundary of the laminated mud unit corresponds to Fillon's (1972, 1975) Gauss-Brunhes unconformity (Fig. 5). This surface marks a very sharp and dramatic change in sediment type (Fig. 5, cores 32-33, 32-10, 27-16 and 32-40). A thin (<10 cm thick) layer of sandy gravels, with sands coarser than 2.0 ϕ, marks this boundary (Fig. 8).

Supposed Gauss sediments in cores 52-1, 32-10, 32-11, 32-33, 27-16 and 32-40 consist of gravels dispersed in a very poorly sorted sand-silt-clay unit. Grain-size analyses of 25 samples from this unit generated sorting values (σϕ) which range from 2.3 ϕ to 3.2 ϕ, negative skewness values (0.0 to −0.6) and kurtosis values from 0.6 to 1.4. The most striking feature of these deposits is their

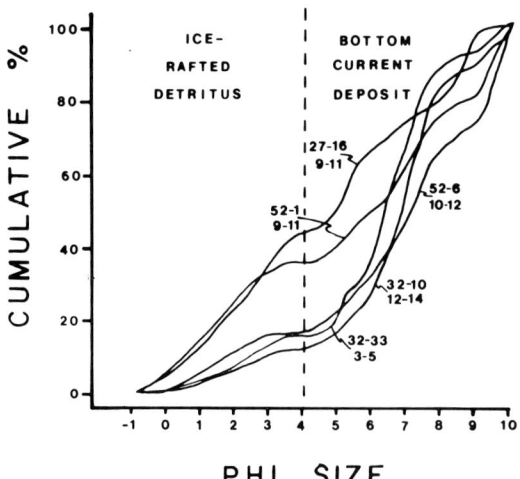

FIG. 6.—Cumulative curves for representative samples from surficial gravelly muds in eastern Ross Sea continental slope cores (Type 1, Table 1).

remarkable vertical textural homogeneity. To illustrate this homogeneity sand-silt-clay concentrations of samples taken at various levels in cores 32-10, 32-11, and 27-16 are plotted in Figure 9. Fossils are rare to absent in sediment samples taken from this unit. Those few calcareous benthonic foraminifera that do occur are indicative of shallower depths and are probably displaced.

Piston cores collected on the Ross Sea continental slope west of Iselin Bank are described in Figure 10. Cores 32-5 and 32-43 contain graded calcareous hash, and calcarenite units consisting of mollusk, bryozoan and echinoid fragments and foraminiferal tests. Core 27-19 penetrated three distinct units in which quartz sands grade upward into foraminiferal sands. Interbedded quartz sands (>85% quartz) in these cores show a very limited (2.0 ϕ to 3.0 ϕ) size range. All of these sediments contain dispersed ice-rafted clasts.

Core 52-6, collected on the western flank of Iselin Bank (Fig. 4), penetrated two meters of gravelly mud in which the upper 80 cm is crudely stratified and the lower part of the sequence non-stratified. Core 32-4 penetrated a similar sequence. These sediments are similar to the surficial gravelly muds of the eastern slope and consist of two components, ice-rafted sands and gravels and well-sorted current derived silts (Fig. 6).

Core 32-3 penetrated one of the most interesting sediment sequences yet encountered on the continental slope. Figure 11 shows representative cumulative curves for the different sediment types in this core. A surficial, thin sand layer (0 to 10 cm) consists of quartz sand ranging in size from 2.0 ϕ to 3.0 ϕ. This unit has a sharp basal contact but is not graded.

A thin gravelly mud (10 to 20 cm) separates this sand unit from a 15 cm thick gravelly sand

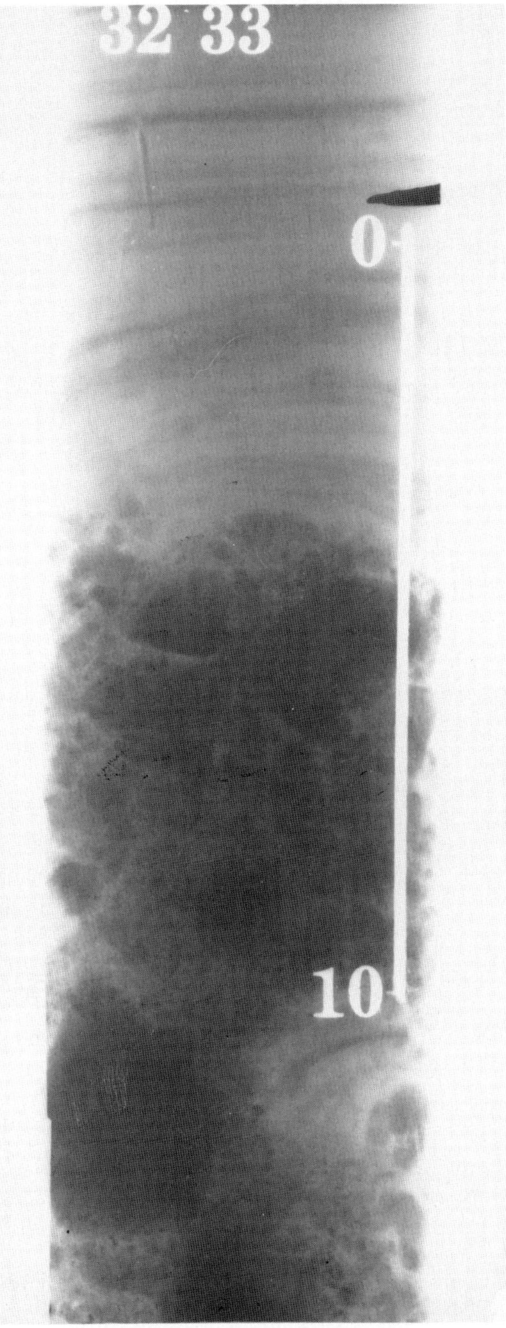

FIG. 8.—X-radiograph showing sharp contact between laminated muds and sandy, muddy gravels in core 32-33 from the eastern Ross Sea continental slope (Scale in cm).

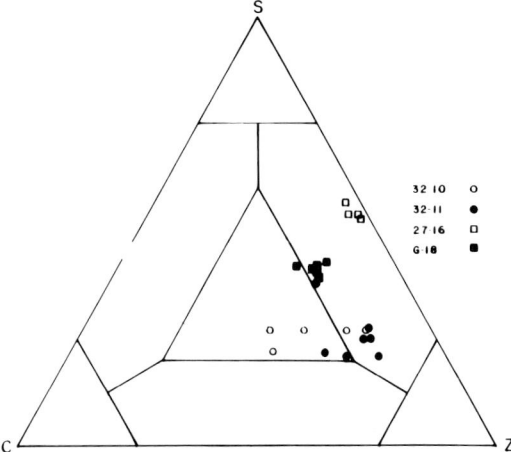

FIG. 9.—Ternary plot of sand-silt-clay percentages of samples from basal gravelly deposits of cores 32-10, 32-11, 27-16, and G18 to show textural homogeneity of these deposits.

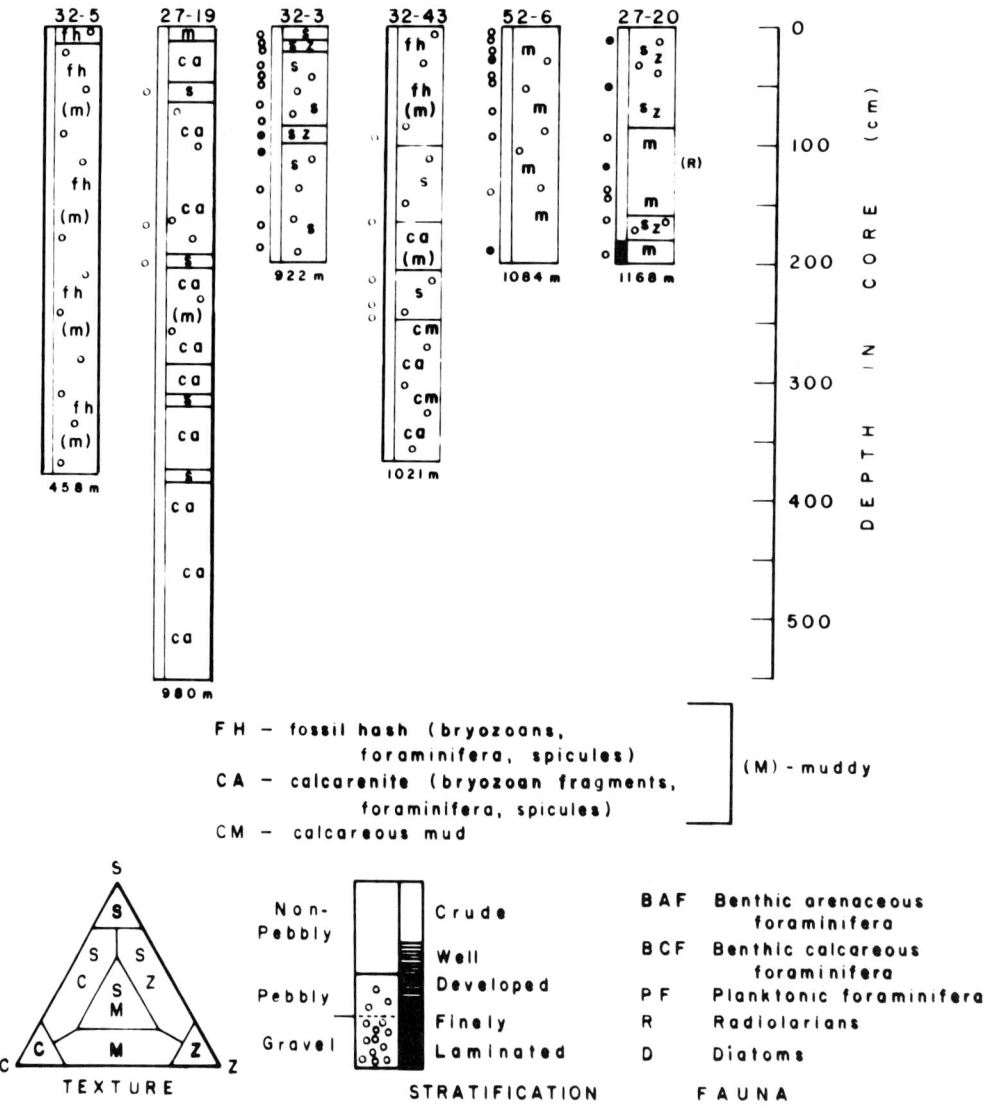

Fig. 10.—Sediment descriptions for piston cores collected on the Ross Sea continental slope west of Iselin Bank. Faunal designations in parentheses ( ) indicate very low abundance, otherwise fauna is common. Solid circles to the left of core logs indicate samples for which only sand-silt-clay fractions were determined, open circles represent those samples for which total size-analyses were conducted. See Table 2 for grain-size data.

unit. Sands of the latter unit are predominantly (>85%) quartz, and are coarser (0.0 ϕ to 3.0 ϕ) than the upper sand unit, but still possess a strong 2.0 ϕ to 3.0 ϕ mode. Silts comprise less than 13% and clays less than 3% of the smaller than 2 mm fraction. The amount of gravel decreases abruptly in the upper 10 cm of the unit. A similar gravelly sand unit occurs at 45 to 100 cm depth in this core with sands consisting of 0.0 ϕ to 4.0 ϕ quartz grains with strong modes at 0.5 ϕ and 2.5 ϕ (Fig. 11). Sands in the lower-most gravelly sand unit display considerable sorting

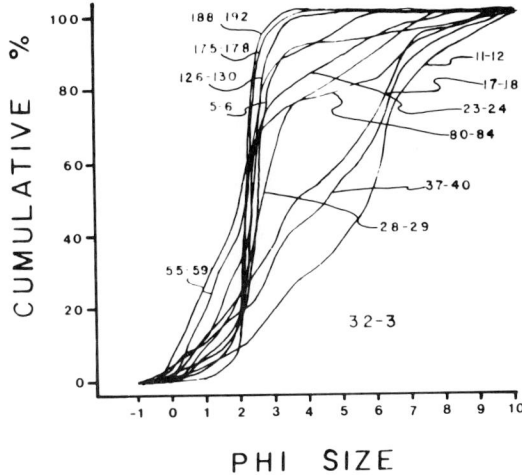

Fig. 11.—Cumulative frequency curves for sediment samples from core 32-3 from the western Ross Sea continental slope.

(Fig. 11) and are predominantly (>85%) quartz. This unit is almost a meter thick and contains virtually no silt or clay-sized particles.

Weddell Sea Cores

Like the Ross Sea, the Weddell Sea is a major ice drainage basin for both East and West Antarctica. The east and west Antarctic margins are separated by a deep, linear trough (Crary Trough) whose axis trends from approximately 75° S and 30° W southwestward beneath the ice front to approximately 80° S and 40° W (Fig. 12). The West Antarctic margin (west of Crary Trough) is characterized by a broad continental shelf and gentle slope (Fig. 12). The glacial regime of the margin is dominated by the vast Ronne Ice Shelf which extends over 250 kilometers out onto the continental shelf (Behrendt, 1962). The outer shelf and slope are covered by perennial sea ice.

The most comprehensive marine geologic survey of the east Antarctic continental margin has been in the eastern Weddell Sea. Here the depth of the continental shelf break varies from approximately 300 meters to just over 400 meters from south to north along the Weddell Sea margin. This northward depression of the shelf break appears to correspond to an increasing thickness of the marginal ice sheet in that direction.

In the northeastern Weddell Sea, the ice shelf extends to within a few tens of kilometers of the shelf break. The continental shelf appears to be depressed near the ice sheet (Fig. 3b). The shelf is floored by marine tills capped with a thin sandy mud layer (Anderson, 1972). The continental shelf in the southeastern Weddell Sea is largely free of significant ice cover. The shelf slopes away from the continent and is covered by sands and gravels (Anderson, 1972).

Figure 13 shows core descriptions for sediment cores collected on the western Weddell Sea continental slope.

All of the cores from the western slope contain surficial non-stratified to crudely stratified, clayey silts with scattered ice-rafted clasts. Pebble-size clasts are common within this unit in cores G18, G19 and G20. Ice-rafted debris in those cores taken farther west on the slope (021, 018, 019 and 026) consist of finer (<2 mm) grains. The silts in this surface unit have been sorted to varying degrees ($\sigma\phi$ from 1.5ϕ to 2.4ϕ) and hence, reflect bottom current influence (Fig. 14). Planktonic foraminifera are abundant in these surficial deposits and indicate high biological productivity due to mixing along this portion of the slope (Anderson, 1975a).

The surficial non-stratified clayey silt units in cores G18 and G19 overlie crudely stratified, gravelly, sand-silt-clay units (Fig. 13). A displaced shelf-endemic foraminiferal assemblage occurs in these sediments. Also, size-analysis data for samples from this unit in core G18 show remarkable similarity and hence indicate homogenization, probably through down slope movement.

Core G20 penetrated two distinct sandy units, one at 25 to 58 cm and another at 80 to 97 cm (Fig. 13). Both units have sharp basal contacts. The upper sandy unit in core G20 contains abun-

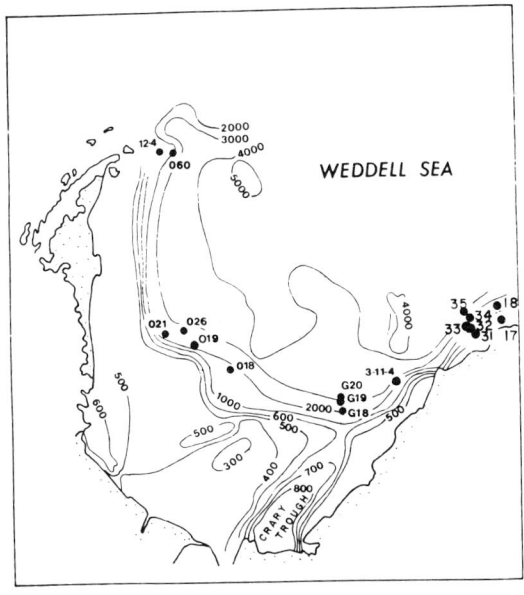

Fig. 12.—Weddell Sea geography, bathymetry and core locations.

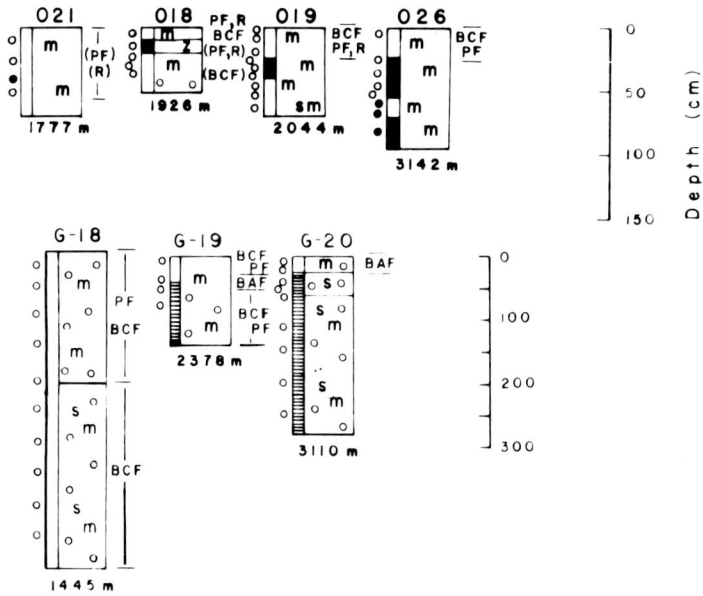

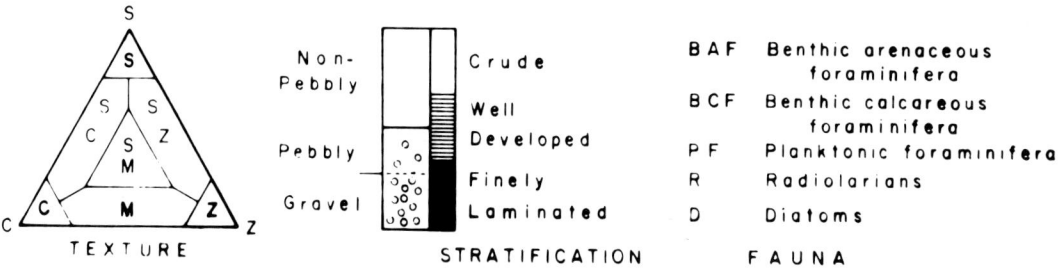

Fig. 13.—Sediment descriptions for piston cores and Phleger cores from the western Weddell Sea continental slope. Faunal designations in parentheses ( ) indicate very low abundance, otherwise fauna is common. Solid circles to the left of core logs indicate samples for which only sand-silt-clay fractions were determined, open circles represent those samples for which total size-analyses were conducted.

dant pebbles. It is not graded and sorting decreases upward in the unit.

Basal sediments (below 100 cm) in core G20 are similar to those in the lower part of core G19. They are very poorly sorted ($\sigma\phi = 2.3\phi$ to 2.8ϕ), contain displaced foraminifera, and display textural homogeneity.

The subsurface deposits in cores 018, 019 and 026, unlike those of G18, G19 and G20, contain distinctly laminated muds similar to those of the eastern Ross Sea (Fig. 13). Individual laminae average 0.5 to 1.0 cm in thickness and consist of well sorted silts. These sediments become finer and better sorted down slope. Fossils are rare to absent.

Stratified clayey silts in cores 018, 019 and 026 sharply overlie non-stratified muds containing scattered ice-rafted grains which are generally smaller than 2 mm. Fossils are rare to absent in this deeper unit; otherwise it is similar to the surficial sediments in these cores.

During *Islas Orcadas* Cruise 1578, in January and February, 1978, a number of closely spaced piston cores were acquired along two transects of the Weddell Sea continental margin north of 72° S latitude (Fig. 12). Preliminary analyses of

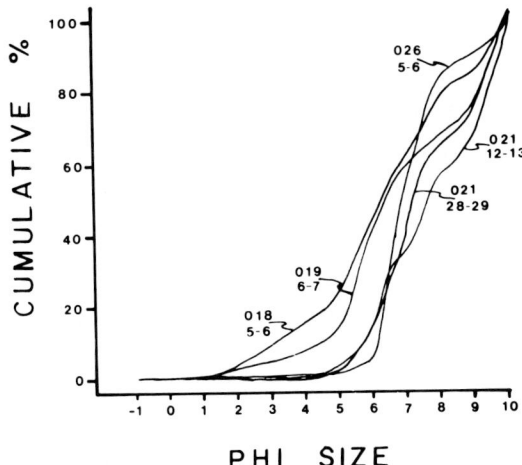

Fig. 14.—Cumulative curves for representative samples from surficial clayey silt deposits of the western Weddell Sea continental slope (Type 1, Table 1).

these cores has begun and generalized descriptions are shown in Figure 15. Core 3-11-4, collected in 1970 in the same area, is also illustrated.

Piston cores 17, 18, and 31 were taken above 1200 meters depth on the upper continental slope (Fig. 12). These cores penetrated muddy, sandy gravels. A sieve analysis of gravels in core 17 revealed no sorting or grading. Several samples of the sandy matrix material in this core were also analyzed. This sandy matrix consists of very poorly sorted ($\sigma\phi$ = 2.9ϕ to 3.7ϕ), coarse (larger than 3.0ϕ), lithic sand with average sand-silt-clay concentrations of 75%, 13% and 12%, respectively. Benthonic foraminifera in this sediment are displaced. Size-analysis data show that a clay-rich layer at 55 to 65 cm depth in core 17 separates this sandy gravel sequence into at least two units. Core 31 contains a number of interbedded gravelly deposits with marked difference in the sand, silt and clay percentages of their matrix. These units are water saturated and separated by sharp contacts.

Core 18 consists of 110 cm of sandy gravel which is overlain by, and appears to be gradational with, an overlying lithic sand unit which displays grading and improved sorting upward in the unit (Fig. 16). Core 32 penetrated two distinct graded sand units which also show considerable sorting upward in the unit. A large clay clast was penetrated near the base of this core.

Core 33 is one of the more interesting cores studied to date. The upper 150 cm of this core consists of interbedded muddy gravels, muddy sands, and pebbly muds separated by sharp contacts. Muddy gravels at 60 to 83 cm, 83 to 100 cm and 130 to 150 cm are inversely graded (Fig. 17). This upper gravelly sequence overlies a 20 cm thick graded sand-silt unit with sharp, angular, basal contact. This unit overlies a three meter thick, highly compacted pebbly clay with large clay clasts which sets atop a thin, water-saturated gravelly layer.

Core 34 penetrated three meters of crudely stratified gravelly mud overlying a 2.5 meter thick layer of interbedded laminated muds and crudely stratified gravelly muds. At 560 to 725 cm depth in this core an overcompacted clay with scattered granules overlies a thin water-saturated gravelly layer.

Core 35 penetrated 60 cm of crudely stratified, gravelly muds overlying a three meter thick, laminated mud containing widely dispersed gravels.

One other core (core 3-11-4) has been collected on the eastern Weddell Sea slope. Although taken some distance south of the *Islas Orcadas* 1578 cores (Fig. 12), this core penetrated through surficial gravelly muds into an overcompacted clay unit, which overlies a water-saturated gravelly mud unit. Preliminary age determinations for compacted clay units in cores 33, 34 and 3-11-4 indicate a Pliocene age for these deposits.

DISCUSSION

Most of the continental slope sediments of the west Antarctic sectors of the Ross and Weddell Seas can be catagorized into three distinct types (Table 1).

Surficial sediments in both areas consist of fine (Mϕ 4.6ϕ to 7.3ϕ) poorly sorted ($\sigma\phi$ 1.8ϕ to 3.3ϕ) muds which lack distinct stratification. They can be characterized by crudely bimodal grain-size distributions. Their coarse (sand- to pebble-size) fraction displays little if any sorting and includes angular, faceted, mineral and lithic grains. This material comprises the ice-rafted component of these sediments whose concentration varies from 9% to 44%. The fine mode displays varying degrees of sorting and is, in general, negatively skewed. This component reflects sedimentation by slowly moving bottom currents (Figs. 6, 14). Benthonic and planktonic foraminifera are abundant.

A second group of sediments are distinctly laminated. They are typically finer (Mϕ = 6.2ϕ to 8.4ϕ) and better sorted ($\sigma\phi$ = 1.1ϕ to 2.3ϕ) than surficial muds in these areas. Their better sorting is largely due to a smaller quantity of ice-rafted debris with relation to surficial sediments. Ice-rafted debris is concentrated within thin (<1 cm thick) widely spaced layers and is seldom larger than 1 centimeter. Individual silt laminae are well sorted and average 0.5 cm in thickness. The grain-size of individual silt laminae

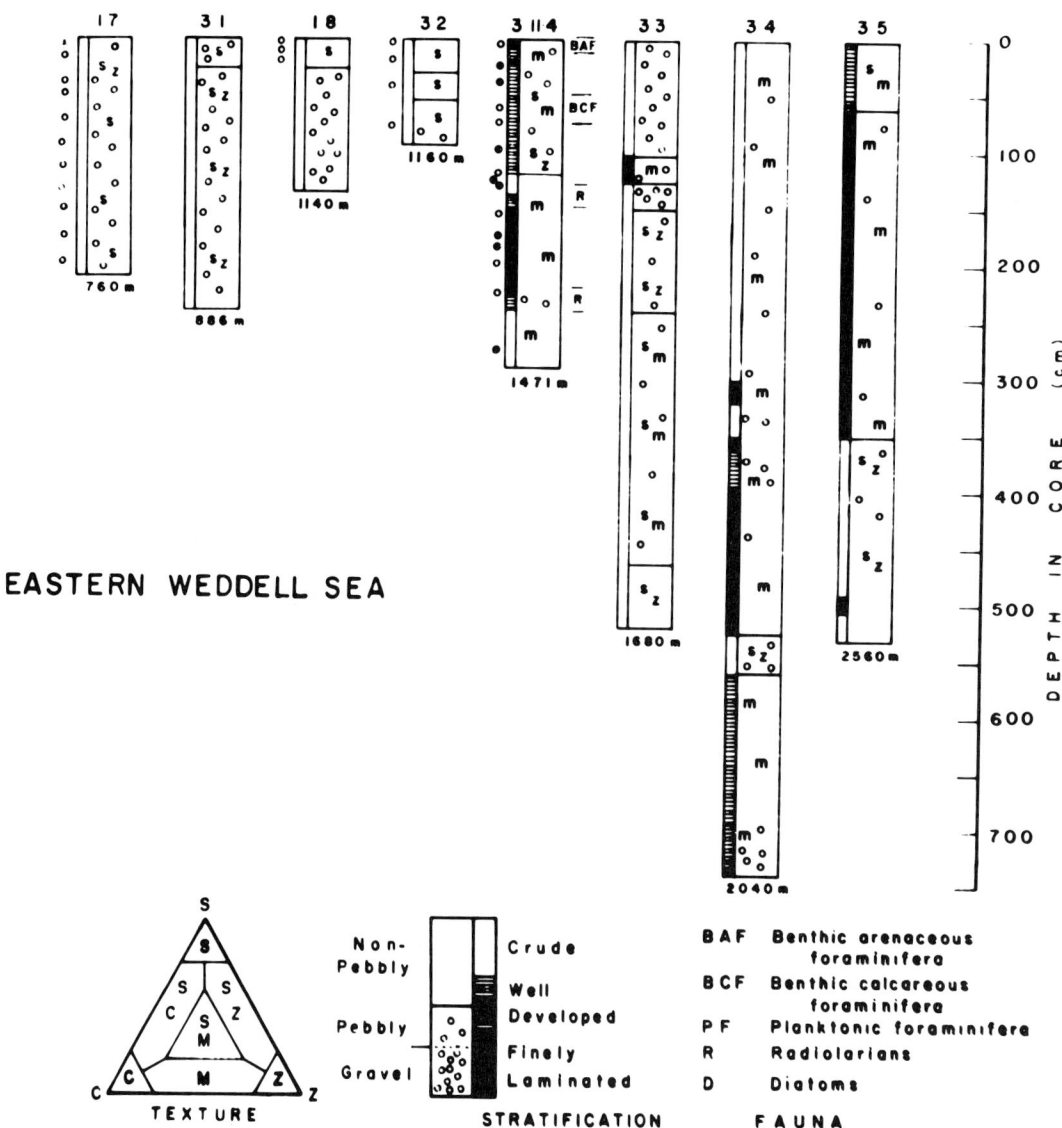

Fig. 15.—Sediment descriptions for piston cores collected on the continental slope of the eastern Weddell Sea. Faunal designations in parentheses ( ) indicate very low abundance, otherwise fauna is common. Solid circles to the left of core logs indicate samples for which only sand-silt-clay fractions were determined, open circles represent those samples for which total size-analyses were conducted.

TABLE 1.—SUMMARY OF GRAIN-SIZE DATA FOR THE THREE MOST COMMON SEDIMENT TYPES IN

Sediment Type	Stratification	Mean Grain Size		Sorting ($\sigma\phi$)	
		Average	Range	Average	Range
1	Crude to Absent	6.3	(4.6–7.3)	2.5	(1.8–3.3)
2	Laminated	7.5	(6.2–8.4)	1.5	(1.1–2.3)
3	Crude to Absent	5.7	(4.4–6.6)	2.8	(2.3–3.2)

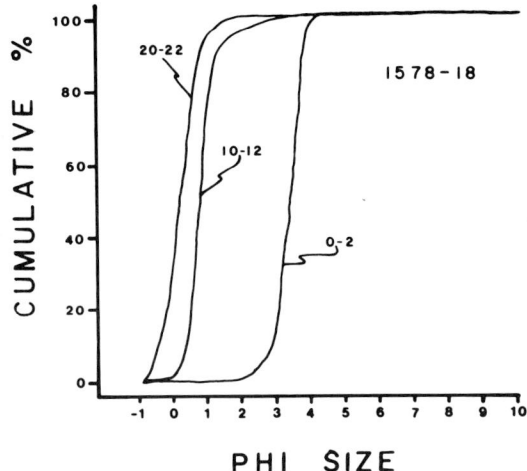

Fig. 16.—Cumulative curves for samples from a graded sand unit in core 18 of the northwestern Weddell Sea.

decreases and their sorting increases down slope. Also, the thickness of laminated silt beds increases down slope. Fossils are rare to absent. Figure 18 shows a scatter diagram of mean grain size (Mϕ) versus sorting ($\sigma\phi$) for laminated and non-laminated mud samples.

The origin of lamination in deep sea sediments has been a subject of interest for some time. A summary of this work is presented by Bridge (1978). Early models call upon velocity pulsations and variations in current direction (Kuenen, 1966) to explain lamination. Stow and Bowen (1978) argue that lamination can result from the depositional sorting of silt and clay particles by increased shear near the base of the bottom boundary layer. The resulting depositional unit would be graded as well as laminated (Stow and Bowen, 1978).

Laminated muds of the Antarctic continental slope may be separated into three categories on the basis of grain-size distributions (Fig. 19). These include samples with distinct, well-sorted, silt laminae (Fig. 19a), those consisting of graded silt laminae (Fig. 19b), and those which are characterized by more poorly sorted, normally distributed silt layers (Fig. 19c). All three types occur within any given laminated unit, with no consistent vertical relationships. Deposition of laminated mud, therefore, appears to result from a variety of depositional processes. For this reason, we have chosen to avoid the use of the term "contourite."

The source of silts and clays in both laminated and non-laminated slope sediments of the eastern Ross Sea and western Weddell Sea is questionable, particularly in the Ross Sea. Piston cores from the outer continental shelf of the eastern Ross Sea penetrated marine till units whose matrix is enriched in silt and clay relative to other Antarctic marine tills (Anderson et al, in press). Based on those sites drilled in the Ross Embayment during DSDP Leg 28, these muddy shelf sediments comprise the entire glacial sequences (Barrett, 1975). Also, the supply of terrigenous sediments from the eastern Ross Sea continental shelf must be greatly restricted as the shelf slopes away from the continental slope. Today, currents on the continental shelf flow toward the continent, while continental slope circulation is dominated by strong (up to 15cm s^{-1}) contour currents which flow from east to west along the slope (Jacobs et al, 1970).

Hence, the source of silt and clay rich continental slope deposits in the eastern Ross Sea may be to the east along the poorly explored Bellingshausen-Amundsen continental margins. Sedimentation of these silts and clays on the eastern Ross Sea continental slope may have been in response to thermohaline mixing which would disrupt circumpolar flow. This possibility is supported by the apparent distribution of surficial slope deposits in the Weddell Sea.

The eastern continental slope of the Weddell Sea is affected by strong (15 cm s^{-1}) contour currents (Klepikov, 1963; Foster and Carmack, 1976—indirect measurements) and sediments of the upper continental slope have been strongly winnowed. Current-winnowed silts and clays are probably entrained by contour currents and transported south and then west along the continental slope. Bottom photographs from the upper Weddell Sea slope indicate turbid bottom conditions (Hollister and Elder, 1969).

THE WEST ANTARCTIC SECTORS OF THE ROSS AND WEDDELL SEAS. DATA ARE IN ϕ UNITS.

Percent Sand		Percent Silt		Percent Clay		Number of Samples Analyzed
Average	Range	Average	Range	Average	Range	
17	(9–44)	46	(36–65)	27	(7–40)	20
4	(0–15)	61	(39–85)	35	(15–58)	35
32	(21–43)	48	(34–66)	20	(2–34)	29

The distribution of non-laminated silts and clays on the southwestern slope of the Weddell Sea coincides with the region of pronounced mixing

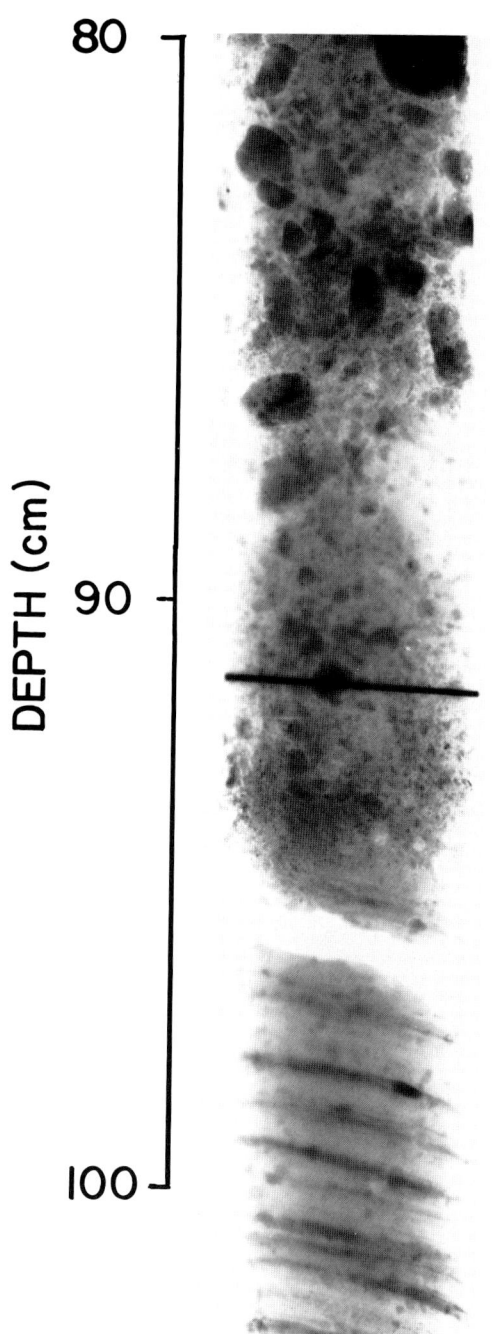

FIG. 17.—X-radiograph showing inverse grading in muddy gravel unit in core 33 from the northeastern Weddell Sea.

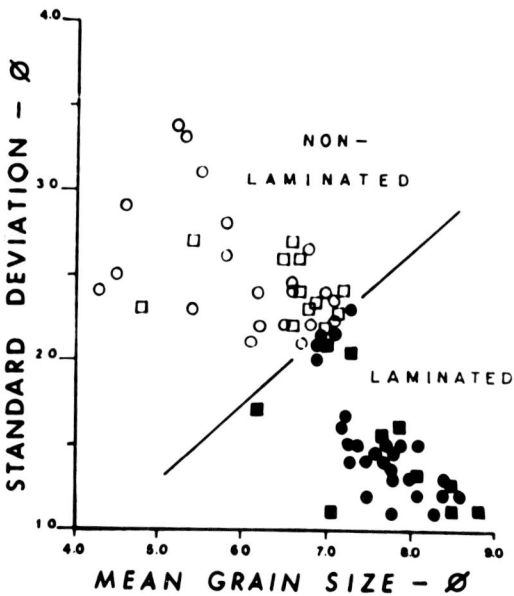

FIG. 18.—Scatter diagram of mean grain size (Mϕ) versus sorting ($\sigma\phi$) in laminated and non-laminated muds of the West Antarctic sectors of the Ross and Weddell Seas. Circles represent Ross Sea samples; squares represent Weddell Sea samples. Solid symbols represent samples from laminated units.

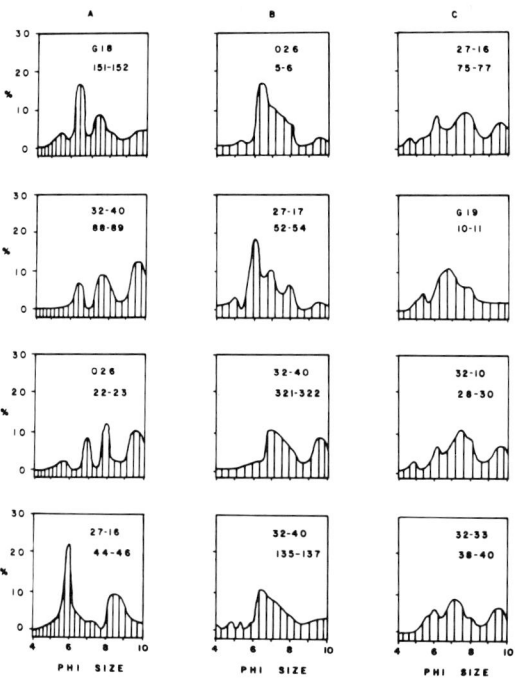

FIG. 19.—Representative frequency curves for three types of laminated sediments: (a) samples with distinct, well-sorted silt laminae, (b) samples consisting of graded silt laminae, and (c) samples of poorly sorted, normally distributed silt layers.

between shelf and slope waters. Here contour currents are disrupted by the down-slope flow of Antarctic Bottom Water (Anderson, 1972).

A third group of sediments encountered of the eastern Ross Sea and western Weddell Sea continental slopes display very different properties from Type 1 and Type 2 sediments. Type 3 sediments are either crudely stratified or non-stratified. They are virtually non-sorted ($\sigma\phi = 2.3$–3.2), and their grain-size distribution is polymodal and typically lacking a distinct fine, current-derived mode. These sediments are interpreted as debris flows because they exhibit:

1) sharp upper and lower contacts, commonly without lag deposits;

2) remarkable textural homogeneity within a given unit;

3) shallow-water (displaced) benthonic foraminifera; and

4) no indication of sorting by bottom currents.

They are identical to marine tills of the continental shelf. However, glacial deposits of the continental slope, because they were deposited by floating ice, tend to exhibit changes in texture and fossil content within a given unit, as for example in type 1 deposits. These changes result from variations in glacial sediment supply, bottom current activity and surface productivity. The most striking feature of debris flow units is their textural homogeneity. This homogeneity implies mixing within the flow, and that relatively little sediment was lost from the flow during movement. Most debris flow samples analyzed contained greater than 50% silt-sized material and greater than 10% clay-sized particles. Hence, these sediments possessed ample strength to carry the largest measured pebbles in suspension (Hampton, 1972). This appears to be the most common mode of sediment mass movement on those parts of the West Antarctic slope studied.

In the Ross Sea, the supposed Gauss-Brunhes unconformity on the continental slope (Fillon, 1972, 1974, 1975) corresponds to the contact between an apparently widespread debris flow unit and overlying laminated muds. We point this out in order to emphasize the need for caution in interpreting the glacial record from continental slope deposits. Previously, these debris flows were interpreted as representing a period of pronounced ice-rafting during Gauss time (Fillon, 1972, 1975). Obviously our interpretation of this sequence has considerably less climatic implication.

Piston cores from the western Ross Sea continental slope contain very different sediments from those collected on the slope east of Iselin Bank. Three cores (32-5, 27-19, 32-43, Fig. 4) penetrated graded sequences consisting of fossil hash, calcarenites and quartz sands, with sharp basal contacts. These sediments are identical to those which blanket the banks in the northwestern Ross Sea and were apparently transported down slope by turbidity currents.

These turbidites are interlayered with muddy calcareous sediments with dispersed gravels. The calcareous debris in these deposits consist of the same fossil fragments found in associated turbidites and hence, also reflect downslope movement. The high mud content of these sediments suggest that the movement was in the form of debris flows.

Core 32-3, collected in the northwestern corner of the Ross Sea (Fig. 4), penetrated three texturally distinct gravelly sand units ranging from 15 cm to 1 m in thickness separated by thin silty sand units. The lower gravelly sand body consists of well-sorted, quartz sands and gravel. They lack silts and clays. Sorting within the sand fraction decreases upward but a strong $2.0\ \phi$ to $3.0\ \phi$ sand mode is dominant throughout this one meter thick unit (Fig. 11). The middle gravelly sand unit samples are distinctly bimodal (Fig. 11). Those of the upper unit are poorly sorted, negatively skewed, and contain up to 20% non-sorted silt and virtually no clays. Again, 2.0ϕ to 3.0ϕ size sands dominate these units. Basal contacts for all units are sharp.

Core 32-3 is capped by a 10 cm thick, well-sorted ($M\phi = 1.4\ \phi$) quartz sand unit. Given the unusual textural properties of these gravelly sands in core 32-3, their mode of emplacement is impossible to determine using the existing criteria for recognition of mass flow deposits (Middleton and Hampton, 1973). The predominance 2.0ϕ to 3.0ϕ size quartz sands in cores 32-43, 27-19 and 32-3 (Table 2) suggests a similar source for these sediments.

The unusual textural and mineralogic maturity of these sands is uncommon for the Antarctic continental margin. They are probably derived from relict deposits on the continental slope. Local derivation of these sands and deposition by traction currents coupled with ice-rafting would explain the unusual textural properties of gravelly sand bodies in cores 32-3 and 32-43.

Piston cores from the northeastern Weddell Sea appear to have penetrated a number of transitional phases of sediment mass movement. Cores 33, 34 and 3-11-4 all contain thick (2.0 to 3.0 meters), overcompacted clay units containing numerous clay clasts within some intervals and resting on water saturated gravelly sediments (Fig. 15). Sharp angular contacts occur within the clay unit in core 33. Preliminary age determination for these clay units indicate a Pliocene age for these deposits. They apparently represent slump deposits which have, for the most part, maintained their textural integrity during down-slope movement.

Cores 17 and 18 penetrated thick (up to 2.5 meters) sandy gravel deposits (Fig. 15). Both the

TABLE 2.—Grain-size data for sediment samples from piston cores of the western Ross Sea continental slope

Depth in Core (cm)	Mϕ	$\sigma\phi$	Sk_i	K_G	%S	%Z	%C
			Core 27-19				
65–68	2.9	0.5	0.0	1.2	>90		
177–179	2.4	0.9	−0.5	1.2	>90		
193–197	3.0	0.4	0.0	1.2	>90		
			Core 32-3				
5–6	2.8	1.4	0.2	4.7	80	4	16
11–12	5.4	2.3	−0.4	1.0	30	61	9
17–18	4.1	2.4	0.1	0.7	52	44	4
23–24	2.9	1.9	0.3	2.1	82	15	3
28–29	3.2	1.9	0.3	1.7	78	21	1
37–40	4.5	2.5	−0.1	0.8	44	50	6
55–59	2.3	1.3	−0.1	1.2	90	10	0
80–84	2.9	2.5	0.4	1.2	76	19	5
126–130	2.4	0.7	−0.2	2.0	>90		
175–178	2.4	0.4	0.0	1.2	>90		
188–192	2.2	0.5	0.2	1.8	>90		
			Core 32-43				
94–98	3.1	0.6	0.2	0.9	>90		
165–169	3.1	0.4	0.0	1.2	>90		
211–216	2.6	0.6	−0.3	1.2	>90		
235–239	2.9	0.4	0.0	1.2	>90		
244–248	2.7	0.7	0.0	1.2	>90		
			Core 52-6				
10–12	6.9	2.4	−0.2	1.0	12	56	32
14–16	7.1	2.2	−0.2	0.9	10	56	34
27–29	5.5	3.1	−0.3	0.8	27	53	20
44–46	6.6	2.2	0.1	1.0	11	65	24
47–49	5.8	2.8	−0.1	0.9	25	52	23
75–77	5.8	2.6	0.1	1.2	24	55	21
93–95	6.3	2.2	0.0	1.2	16	65	19
138–140	6.5	2.6	−0.2	0.8	23	48	29
			Core 27-20				
98–99	6.5	1.6	0.5	1.0	3	62	35
140–141	6.6	2.4	−0.2	1.4	14	63	23
144–145	6.7	2.1	0.0	1.5	12	69	19
163–164	4.2	1.7	0.4	1.3	59	29	2
195–196	5.6	0.7	0.2	1.1	0	96	4

gravels and their sandy matrix are non-sorted. Strong currents have removed most of the fine (<3.0ϕ) material. The matrix in core 17 has average silt and clay concentrations of 13% and 12%, respectively.

The uppermost part of the gravel deposit in core 18 appears to be gradational upwards into an overlying graded sand unit (Fig. 16). Core 32, taken at about the same level on the continental slope to the south, contains two distinct graded sand bodies (Fig. 15). These sands represent turbidites that are perhaps transitional with associated gravel deposits.

From the previous discussion it should be evident that mass flow deposits are indeed widespread on the Antarctic continental slope. Mass flow deposits of the West Antarctic slopes consist primarily of muddy debris flows. In contrast, the East Antarctic slope is floored by a variety of different mass flow deposits which, with the exception of those slump deposits described, show greater sorting. The reasons for these differences lie in the physiographic and glacial settings of these areas.

The West Antarctic slope bounds a broad, ice covered margin. Sediment supply to the continental slope by currents is greatly restricted because the ice shelf covers the inner continental shelf.

Most of the sediment delivered to the shelf and slope consists of very poorly sorted glacial debris. The textural properties of Ross Sea and Weddell Sea debris flows indicate that this depositional process does little to change the original character of these sediments.

In the western Ross Sea, sediment supply to the continental shelf is by valley glaciers and meltwater streams. The shelf is free of ice cover during the austral summer so that waves and currents have ample opportunity to winnow these glacial sediments. Thus, sediments supplied to the outer shelf-slope of the western Ross Sea have already experienced considerable sorting. The most common mode of mass flow deposition in this area is by turbidity currents.

The sediments of the northeastern Weddell Sea continental shelf consist of non-sorted marine tills. On the upper continental slope, thick gravelly deposits, such as those in cores 17, 18, and 31, indicate strong winnowing; their $<3.0\phi$ fractions have been removed. Down-slope movement of these sediments appears to occur as debris flows which may, through liquefaction (Lowe, 1976) or some other process, generate turbidity currents. Turbidites in cores 18 and 32 may have been derived in this manner.

CONCLUSIONS

Although sediment transport to the Antarctic continental margin is predominantly by glaciers, sediments of the continental slope are highly varied and commonly bear only subtle evidence of their glacial origin. The supply of sediment to the continental slope is controlled by glacial regime, distance of the ice sheet grounding line from the shelf break, the extent of floating ice cover on the margin, and glacial stability. Three basic types of glacial marine sedimentary environments exist in Antarctica:

1) those which support a large grounded ice sheet,
2) those whose continental shelf is largely covered by floating ice shelves, and
3) those which bound mountainous coasts where valley glaciers extend into the sea.

A large stable ice sheet, such as the East Antarctic ice sheet, may ground itself on the continental shelf resulting in the development of a proglacial isostatic basin up to 180 kilometers wide and 300 meters deep. The continental shelf then slopes toward the ice shelf and sediment supply to the continental slope is greatly reduced. The northeastern Weddell Sea appears to typify this situation. Piston cores from the continental slope in this area penetrated debris flows which are overlain by, and appear to grade laterally into turbidites. Cores from the lower slope contain slump deposits consisting of Pliocene clays.

In the eastern Ross Sea and western Weddell Sea, large floating ice shelves, which represent the seaward extension of the West Antarctic Ice Sheet, cover the inner shelf. Most glacial sedimentation occurs well inland of the floating icefront near the grounding line and sediment transport to the continental slope by bottom currents is greatly restricted.

Surficial slope deposits in the West Antarctic sectors of the Ross and Weddell Seas contain from 9 to 44% ice-rafted detritus which is dispersed in non-stratified poorly sorted muds. These deposits rest atop laminated, well sorted silt units which contain very little ice-rafted debris. They were possibly derived from some easterly source and transported to these sediment starved margins by contour currents. Several modes of deposition are implied by size-analysis data.

Debris flows are widespread on the Weddell Sea and Ross Sea slopes. These deposits are texturally indistinguishable from marine tills of the continental shelf and consequently, could lead to erroneous interpretations with regard to glacial histories. They are nonstratified, non-sorted and have sharp upper and lower contacts. Their textural homogeneity, and in some cases displaced shelf faunas, distinguish them from floating ice deposits. These deposits have moved down slope over 1,000 meters without losing their textural integrity.

The western Ross Sea represents a third type of glacial marine environment, one in which mountain glaciers debouch directly into the sea. The continental shelf area is floored by well washed sands and gravels. Piston cores from the continental slope penetrated graded fossil hash, calcarenite and quartz sand units with sharp basal contacts and debris flows containing similar coarse materials with a muddy matrix.

ACKNOWLEDGMENTS

Financial support for this investigation was provided through grants from the American Chemical Society—Petroleum Research Fund (#11101-AC2) and the national Science Foundation—Division of Polar Programs (DP 77-26407). We would also like to thank Sue Davis for assistance in the preparation and analysis of samples; and Dennis Cassedy of the Antarctic Marine Geology Research Facility, Florida State University for his assistance. Stan Jacobs was kind enough to review the manuscript and made helpful comments concerning physical oceanography. Andy Gumbos provided assistance with biostratigraphic determinations.

REFERENCES

ANDERSON, J. B., 1972, Marine geology of the Weddell Sea: Florida State University Sed. Res. Lab. Rept. No. 35, 222 p.

———, 1975a, Factors controlling $CaCO_3$ dissolution in the Weddell Sea from foraminiferal distribution patterns: Marine Geology, v. 19, p. 315-332.

———, 1975b, Ecology and distribution of foraminifera in the Weddell Sea of Antarctica: Micropaleontology, v. 21, p. 69-96.

———, CLARK, H. C., AND WEAVER, F. M., 1977, Sediments and sediment processes on high latitude continental shelves: Proc. Offshore Tech. Conf., p. 91-95.

ANDREWS, J. T., 1968, Postglacial rebound in Arctic Canada: similarity and prediction of uplift curves: Canadian Jour. Earth Sciences, v. 5, 39-47.

ANGINO, E. E., 1966, Geochemistry of Antarctic pelagic sediments: Geochim. Cosmochim. Acta, v. 30, p. 939-961.

BARRETT, P. J., 1975, Textural characteristics of Cenozoic preglacial and glacial sediments at site 270, Ross Sea Antarctica, in Hayes, D. E., Frakes, L. A., et al, Initial Reports of the Deep-Sea Drilling Project, v. 28, p. 757-767.

BEHRENDT, J. C., 1962, Geophysical and glaciological studies in the Filchner Ice Shelf Area of Antarctica: Jour. Geophys. Res., v. 67, p. 221-234.

BENTLEY, C. R., 1964, Ice thickness and physical characteristics of the Antarctic Ice Sheet: New York. Antarctic Map Folio Series, Folio 2, American Geogr. Soc.

BLOOM, A., 1967, Pleistocene Shorelines: A new test of isostasy: Geol. Soc. America Bull., v. 78, p. 1477-1494.

BOULTON, G. S., 1972, Role of thermal regime of ice shelves in glacial sedimentation: Instit. British Geographers Spec. Pub. No. 4, p. 1-19.

———, 1975, Processes and patterns of subglacial sedimentation: a theoretical approach, in Wright, A. E., and Moseley, F., eds., Ice Ages: Ancient and Modern: Liverpool, Seel House Press, p. 7-42.

BRIDGE, J. S., 1978, Origin of horizontal lamination under turbulent boundary layers: Sedimentary Geology, v. 20, p. 1-16.

BROECKER, W. S., 1966, Glacial rebound the the deformation of the shorelines of proglacial lakes: Jour. Geophys. Res., v. 71, p. 4777-4783.

BROTCHIE, J. F., AND SILVESTER, R., 1969, On crustal flexure: Jour. Geophys. Research, v. 74, p. 5240-5252.

CAREY, S. W., AND AHMAD, N., 1961, Glacial marine sedimentation, in Raasch, G., ed., First International Symposium on Arctic Geology, Proceedings: Toronto, Univ. of Toronto Press, v. 2., p. 865-894.

CHRISS, T., AND FRAKES, L. A., 1972, Glacial marine sedimentation in the Ross Sea, in Adie, R., ed., Antarctic Geology and Geophysics: Oslo, Comm. Antarctic Res., p. 747-762.

FILLON, R. H., 1972, Late Cenozoic geology, paleo-oceanography and paleoclimatology of the Ross Sea, Antarctica [Ph.D. thesis]: Kingston, Rhode Island, University of Rhode Island, 183 p.

———, 1974, Late Cenozoic foraminiferal paleoecology of the Ross Sea, Antarctica: Micropaleontology, v. 20, p. 129-151.

———, 1975, Late Cenozoic Paleo-oceanography of the Ross Sea, Antarctica: Geol. Soc. America Bull., v. 86, p. 839-845.

FOLK, R. L., AND WARD, W. C., 1957, Brazos River Bar: A study in the significance of grain size parameters: Jour. Sed. Petrology, v. 27, p. 3-26.

FOSTER, T. D., AND CARMACK, E. C., 1976, Frontal mixing and Antarctic Bottom Water formation in the southern Weddell Sea: Deep-Sea Research, v. 23, p. 301-317.

GIBBS, R. J., MATTHEWS, M. D., AND LINK, D. A., 1971, The relationship between sphere size and settling velocity: Jour. Sed. Petrology, v. 41, p. 7-18.

GORDON, A. L., 1971, Oceanography of Antarctic waters, in Reid, J. L., ed., Antarctic Oceanology I, Antarctic Res. Ser., American Geophys. Union, Washington, D.C., v. 15, p. 169-203.

HAMPTON, M. A., 1972, The role of subaqueous debris flow in generating turbidity currents: Jour. Sed. Petrology, 42, p. 775-793.

HOLLISTER, C. D., AND ELDER, R. B., 1969, Contour currents in the Weddell Sea: Deep-Sea Res., v. 16, p. 99-102.

HUGHES, T., 1973, Is the West Antarctic ice sheet disintegrating?: Jour. Geophys. Res., v. 78, p. 7884-7910.

———, 1977, West Antarctic ice streams: Rev. Geophys. and Space Physics, v. 15, p. 1-46.

JACOBS, S. S., AMOS, A. F., AND BRUCHHAUSEN, P. M., 1970, Ross Sea Oceanography and Antarctic Bottom Water Formation: Deep-Sea Res., v. 17, p. 935-962.

JORDAN, C. F., JR., FRYER, G. E., AND HEMMEN, ELZE, 1971, Size analysis of silt and clay by hydrophotometer: Jour. Sed. Petrology: v. 41, p. 489-496.

KENNETT, J. P., 1966, Foraminiferal evidence of shallow calcium carbonate solution boundary, Ross Sea, Antarctica: Science, v. 153, p. 191-193.

———, 1968, The fauna of the Ross Sea, Part 6—ecology and distribution of Foraminifera: New Zealand Dept. Sci. Indust. Res., Bull. 186, p. 1-47.

KLEPIKOV, V. V., 1963, Weddell Sea hydrology: Works of the Soviet Antarctic Expedition, hydrology of Antarctic Waters, v. 17, p. 45-93.

KRUMBEIN, W. C., AND PETTIJOHN, F. J., 1938, Manual of Sedimentary Petrology: New York, N.Y., D. Appleton—Century Co., 549 p.

LLIBOUTRY, L. A., 1971, Rheological properties of the asthenosphere from Fennoscandian data: Jour. Geophys. Research, v. 76, p. 1433-1480.

MARCEL, C. H., AND FAIRBRIDGE, R. W., 1973, Isostasy and eustasy of Hudson Bay: Geology, v. 6, p. 117–122.
MIDDLETON, G. V., AND HAMPTON, M. A., 1973, Sediment gravity flows: mechanics of flow and deposition: Soc. Econ. Paleontologists Mineralogists short course, Pacific Sec., 1973, p. 1–38.
MULLER, F., AND BARR, W., 1966, Postglacial isostatic movement in northeastern Devon Island: Arctic, v. 19, p. 263–269.
OVENSHINE, A. T., 1970, Observations of iceberg rafting in Glacier Bay, Alaska and identification of ancient ice-rafted deposits: Geol. Soc. America Bull., v. 81, p. 891–894.
SHEPARD, F. P., 1948, Submarine Geology: New York, Harper and Row, 348 p.
STOW, D. A. V., AND BOWEN, A. J., 1978, Origin of lamination in deep sea, fine-grained sediments: Nature, v. 274, p. 324–328.
TEN BRINK, N., 1974, Glacio—isostasy: new data from West Greenland and geophysical implications: Geol. Soc. America Bull., v. 85, p. 219–228.
THIEL, E. C., 1962, The amount of ice on planet Earth: Antarctic Research: Washington, D.C., American Geophys. Union, pub. no. 1036, Geophys. Monograph no. 7, p. 172–175.
WALCOTT, R. I., 1970, Isostatic response to loading of the crust in Canada: Canadian Jour. Earth Sciences, v. 7, p. 716–727.
WARNKE, D. A., 1970, Glacial erosion, ice-rafting, and glacial-marine sediments: Antarctica and the Southern Ocean: Jour. Sci., v. 269, p. 276–294.
WEISSEL, J. K., HAYES, D. E., AND HERRON, E. M., 1977, Plate tectonic synthesis: The displacement between Australia, New Zealand, and Antarctica since the Late Cretaceous: Marine Geology, v. 25, p. 231–277.

PART III
ANCIENT CONTINENTAL SLOPES

… SEPM Special Publication No. 27, p. 287–305, August 1979 …

ANCIENT CONTINENTAL SLOPE SEQUENCES AND THEIR VALUE IN UNDERSTANDING MODERN SLOPE DEVELOPMENT

HARRY E. COOK
U.S. Geological Survey
Menlo Park, California 94025

ABSTRACT

A north-trending continental slope existed in central Nevada during the Late Cambrian and earliest Ordovician. The regional depositional regime changes from a deep-water slope in the west, represented by 150 m of hemipelagic limestone, to a shoal-water shelf 170 km to the east, composed of 600 m of coeval limestone. Slope and shelf environments prograded seaward with time. Seaward of this relatively thin slope sequence, 1,000 to 2,000 m of sediment formed a submarine fan.

Slope development involved both depositional and erosional processes. Up to 50 percent of the slope sequence is composed of slide, slump, debris-flow and turbidity-flow deposits. These deposits reflect extensive mobilization of semiconsolidated to unconsolidated sediment from both slope and shoal-water shelf sites to the east. The upper slope is characterized by translational and rotational slides up to about 10 m thick and 400 m wide and massive boulder-bearing debris-flow deposits that occur in channels up to 10 m deep and 300 to 400 m wide. The zones of basal shear and thin-tapering margins of the slides were transformed into conglomeratic debris and turbidity flows through mechanical shock, strain, and incorporation of water. Mass-transport deposits lower on the slope consist of slides less than a meter thick and a few tens of meters wide, and numerous conglomeratic debris and turbidity-flow sediment in channels 0.2 to 3.5 m deep and 20 to 100 m wide. The remainder of the section shows no evidence of remobilization and consists of dark, fine-grained, thin bedded, and laminated lime mudstone which forms the in situ slope deposits. This limestone represents hemipelagic sedimentation.

It is reasonable to speculate that resedimentation processes may be more significant on modern continental slopes than is presently recognized on the basis of abundant small-scale mass-transport deposits less than a few meters thick in this and similar ancient slope settings. The smooth undisturbed nature of some modern slopes as seen on continuous seismic-reflection profiles may be in part a function of the limited resolving power of the seismic equipment. Slump or flow features less than a quarter of a kilometer long and less than 10 m thick may not be recognized by present seismic-reflection techniques, especially with commonly used low frequency seismic systems. The implications of this are pertinent to both exploration for petroleum within continental slope sequences and the recognition of geologic hazards on modern slopes: (1) Seismically undetectable coarse-grained debris-flow and turbidity-flow deposits are potential conduits for migration of hydrocarbons upslope to shoal-water reservoirs or for forming stratigraphic traps on the slope, and (2) a modern slope undergoing numerous small-scale mass movements poses potential environmental hazards.

INTRODUCTION

With the need to expand our search for energy resources into deeper marine environments, it is increasingly important to utilize a variety of approaches to increase our knowledge of these settings. Geologists working on modern continental slopes have recognized that these are often sites of seafloor instability, especially during the past five to ten years (Carlson and Molnia, 1977; Embley, 1976; Field and Clarke, 1977; Hampton and Bouma, 1977; Kelling and Stanley, 1976; Lewis, 1971; Stanley and Silverberg, 1969; Stanley and Unrug, 1972). Likewise geologists working on land have recognized that ancient continental slopes, in a variety of tectonic and sedimentologic settings, were sites of extensive submarine mass movement (Cook and Enos, 1977; Cook and Chamberlain, 1978; Cook and Taylor, 1977; Thomson and Thomasson, 1969). These studies have demonstrated that some of our previous ideas about slope stabilities need to be revised. Indeed, many slopes, even ones with angles of less than one degree, seem to be areas of extensive seafloor instability. Thus, there is a need to better understand the scale, frequency and relative importance of the different types of mass movement processes. Because there are relatively few published data demonstrating that mass flows can be generated from submarine slides and slumps, there is a need to better document the genetic link and transitional characteristics between slides and associated mass-flow deposits. There is a need also to determine whether or not the smooth undisturbed nature of some slopes as seen on continuous seismic reflection profiles may be in part a function of the limited resolution of commonly used seismic systems. The purpose of this paper is to present field evidence from an ancient slope that bear on these problems.

This study summarizes the anatomy of a pro-

Copyright © 1979, The Society of Economic Paleontologists and Mineralogists

grading continental slope sequence, discusses how its evolution may help us gain insight into the nature and development of modern slopes, and describes the types of potential hydrocarbon reservoirs that could occur in similar settings. Focus is on the description and interpretation of submarine mass movements by sliding, slumping, and sediment gravity flow. Data are presented to show that some of the sediment gravity flows were generated from slides. Of particular importance is the fact that up to 50 percent of the ancient slope described in this paper is made up of slide, slump, and gravity-flow deposits, all of which occur in units less than 10 m thick. If mass-movement processes of this small scale are common to modern slopes, their products may be beyond the resolving limits of much seismic-reflection equipment. Continuous seismic surveys are one of the fundamental tools in marine geology; however, in many instances slump or flow features less than 250 m long and 10 m thick may not be recognized, especially with commonly used low-frequency seismic systems. High-resolution seismic systems are capable of resolving thin beds, however, there is difficulty in distinguishing between thin-bedded deposits of mass-movement origin and in situ pelagic and hemipelagic sediment, because both give acoustically transparent signatures. Thus, a seismic-reflection profile across a modern slope having a history of small-scale mass movements might appear as a series of smooth, undisturbed reflectors.

Studying ancient slope sequences allows us to examine the development of a slope in space and time, bed by bed. Where outcrops permit, the changes that occur in slides, slumps, and mass-flow deposits during their initiation and subsequent transportation and deposition can be examined. Caution must be used before an ancient slope model can be used to help decifer a modern slope sequence; however, the rock record can often allow a more complete awareness of what processes are possible, the scale and periodicity of these processes and the nature of their products. In essence, the additional perspective of ancient studies can sharpen our abilities to make more accurate interpretations and stratigraphic predictions in these frontier areas.

REGIONAL STRATIGRAPHIC SETTING

Lower Paleozoic rocks of the Great Basin reflect a broad continental shelf to the east and either a marginal ocean-basin and volcanic-arc system (Burchfiel and Davis, 1972; Churkin, 1974) or an open oceanic basin to the west (Dietz and Holden, 1966; Stewart, 1972; Stewart and Poole, 1974). Recent field work on well-exposed Upper Cambrian and lowermost Ordovician marine carbonate rocks in the Egan Range (Whipple Cave

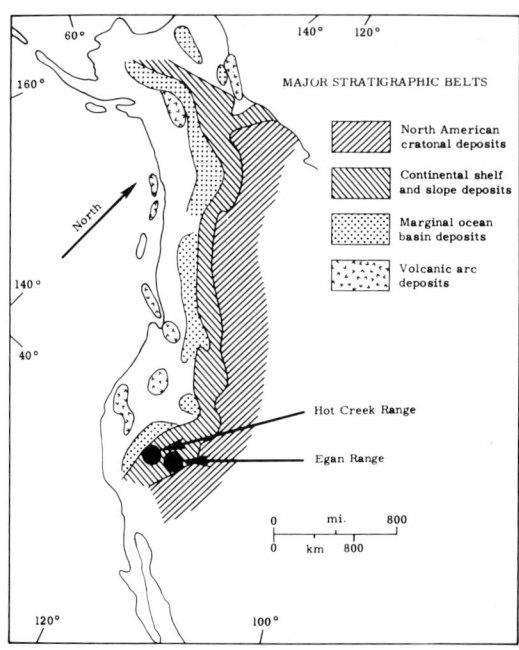

FIG. 1.—General location of sections in the Hot Creek Range and central Egan Range, Nevada, in relation to major regional stratigraphic belts (from Cook and Taylor, 1977, Fig. 1; map generalized from Churkin, 1974).

Formation and lower House Limestone) and Hot Creek Range (Hales Limestone) of Nevada (Fig. 1) show that the Egan Range section represents shoal-water deposition on a shallow carbonate shelf, whereas 170 km to the west coeval limestone in the Hot Creek Range was deposited on a continental slope (Fig. 2; Cook and Taylor, 1975, 1977). Sedimentation from the base of each section upward records a progressive shallowing of water that reflects a westward progradation of the continental shelf and slope. The Egan Range section displays deposition that ranged from shallow subtidal, low-energy open shelf environments at the base to tidal-flat settings at the top. Environments in the Hot Creek Range ranged from deep-water slope settings at the base of the section to settings higher on the slope or slope-shelf break at the top of the section (Cook and Chamberlain, 1978). Lithologic, faunal, and field characteristics that contrast the shelf and slope sediment have recently been described (Cook and Taylor, 1975, 1977; Taylor, 1976, 1977; Taylor and Cook, 1976). Farther seaward of the Hot Creek Range slope section, thick accumulations of coeval shale and sandstone with minor amounts of conglomerate and limestone, interpreted to be part of a prograding submarine fan (Rowell and others, 1979), make up the continental rise (Stewart and Poole, 1974).

Fig. 2.—View north at Upper Cambrian and lowermost Ordovician prograding slope sequence, Hot Creek Range, Nevada. Arrows bracket the interval measured. Many of the resistant ledges are slide, slump, and sediment gravity flow deposits. First large ledge above the white arrow is a translational slide 10 m thick and 400 m wide (outlined by black lines) which moved obliquely out of the photo to the left (southwest) (modified from Cook and Taylor, 1977, Fig. 22).

Variation in sediment thickness across this ancient continental margin parallel trends in sediment thickness across modern margins (Emery and others, 1970). About 540 m of limestone accumulated in shoal-water shelf environments during the Late Cambrian, whereas only about 130 m formed in deeper water slope settings. Farther west, coeval submarine fan deposits reach thicknesses on the order of 1,000 to 2,000 m (Rowell and others, 1979).

GENERAL SLOPE CHARACTERISTICS

Much of the lower slope sequence is dark gray to black limestone, mostly lime mudstone (Fig. 3). It is further characterized by its dark color, thin bedding to millimeter-thick laminae, and scarcity of apparent burrowing (Cook and Taylor, 1977, Figs. 25, 26). These beds exhibit contacts that range from planar and nearly parallel and continuous for tens of meters to more undulatory and discontinuous. Particles in the sediment are composed largely of fine-grained carbonate (and spar) and minor amounts of pyrite and terrigenous siliciclastics of clay and silt size. The associated fauna consists of abundant sponge spicules and deep-water trilobites (Cook and Taylor, 1977, Fig. 26). This dark laminated limestone is interpreted to represent hemipelagic sedimentation by suspension and other normal pelagic processes (Cook and Taylor, 1977). Upper slope deposits are lighter in color, coarser grained, and thicker bedded than lower slope sediment.

Approximately 40 to 50 percent of the slope sequence consists of two major types of gravity-displaced sediment. These are slide and slump masses and sediment gravity-flow (Middleton and Hampton, 1973, 1976) deposits. Extensive thick translational slides and rotational slides (slumps) are restricted to the upper part of the slope, whereas small slides occur on the lower parts of the slope. The shear plane(s) of translational slides is predominantly along planar or gently undulating surfaces parallel to the underlying beds, whereas the shear plane of a slump or rotational slide is concave upward (Varnes, 1958). Translational slides on this slope exhibit a wide range of internal deformation, whereas slumps show very little disruption of bedding. Sediment gravity-flow deposits occur throughout the slope; however, different types characterize different parts of the slope. Thick massive boulder-bearing conglomerate occurs on the upper part of the slope in relatively deep wide channels. This conglomerate is in contact with large slides and slumps. Finer grained thin-bedded conglomerate and sand is most abundant on the lower slope, where it occurs in shallow, narrow channels. In contrast to the thick massive conglomerate, thinner bedded conglomerate shows a variety of primary sedimentary structures including cross-bedding, normal grading, imbrication and parallel alignment of clasts.

The regional paleoslope was to the west and had a north-south trending strike (Fig. 1). Paleoslope directions in the Hot Creek Range were determined by plotting the orientation and sense of rotation of fold axes within the translational slides on a Schmidt net. Hansen (1966) discusses the general solution for determining the slip line of slides and slumps. Paleocurrent directions were determined for the sediment gravity-flow deposits by plotting the dip direction of imbricated tabular

Fig. 3.—In situ hemipelagic deposits of black lime mudstone and wackestone. Bedding planes are parallel and fairly planar; individual beds can be traced for a few tens of meters (from Cook and Taylor, 1977, Fig. 24).

TABLE 1.—GENERAL CHARACTERISTICS OF SLIDES*

Type and Stratigraphic Position	Sediment Moved	Maximum Thickness	Lateral Dimension	Nature of Lower Contact	Nature of Upper Contact	Degree of Internal Deformation	Mechanics of Movement	Generation of Mass Flows
Translational (Upper slope)	Hemipelagic lime mudstone and wackestone Figs. 4, 7	10 m Fig. 7	400 m Fig. 2	Basal shear plane subparallel to enclosing beds; local scouring up to 1.5 m when in contact with hemipelagic sediment and up to 3 m when in contact with conglomeratic channel deposits Figs. 4, 5	Hummocky, slightly convex-up	Most intense at base and lateral margins; intermediate within zone above the base and below the top; least intense at top Figs. 4, 6, 7, 10	Dominantly elastic at top; plastic in the interior; plastic and plastico-viscous at base and thin lateral margins Figs. 4, 6, 7, 10	At base and lateral margins; tabular and equant shaped clasts developed; debris flow type fabric Figs. 6, 10
Translational (Upper slope)	Same as above Figs. 11, 12	3.5 m Fig. 12	100 m	Basal shear plane subparallel to enclosing hemipelagic sediment Fig. 11	Same as above	Most intense at base and lateral margins; least intense above the base Figs. 11, 13–16	Same as above Figs. 11, 13–16	Same as above Fig. 16
Rotational (Upper slope)	Same as above Fig. 18	About 10 m Fig. 18	At least 100 m Fig. 18	Basal shear plane concave-up incised into hemipelagic sediment Fig. 18	Sheared flat by an overlying translational slide Fig. 18	Very little throughout Fig. 18	Dominantly elastic with local plastic at base Fig. 18	None apparent
Translational (Lower slope)	Same as above Fig. 24	About 1 m Fig. 24	About 30 m	Subparallel to enclosing hemipelagic sediment Fig. 24	Subparallel to enclosing hemipelagic sediment Fig. 24	Moderately intense throughout Fig. 24	Dominantly plastic with local plastico-viscous at hinge axes of overfolds Figs. 24, 25	None apparent

*Figure numbers refer to these characteristics.

TABLE 2.—GENERAL CHARACTERISTICS OF SEDIMENT GRAVITY-FLOW DEPOSITS*

Stratigraphic Position	Range in Maximum Thickness	Lateral Dimension	Shape	Lower Contact	Upper Contact	Clast Types	Maximum Clast Size	Stratification	Grading
Upper slope	6–12 m Figs. 21, 23	400 m	Channel	Erosional in central part of channel with channel walls cut 25° into enclosing hemipelagic sediment; subparallel to enclosing hemipelagic sediment where deposit thins laterally to less than a few meters; local scouring up to a few meters Figs. 19, 20	Hummocky, convex-up in central part of channel; subparallel to enclosing beds away from channel axis; locally incised by overlying slides Fig. 5	Hemipelagic lime mudstone and wackestone; dominantly tabular shapes Figs. 22, 23	3 × 15 m; most commonly 2 to 5 cm by 10 to 20 cm Figs. 21, 22	Massively bedded, clasts randomly oriented in central part of channel; clasts oriented subparallel where deposit thins laterally to less than a few meters Figs. 22, 23	None apparent
Lower slope	0.2–3.5 m Figs. 26–28, 30	100 m	Channel Figs. 26–28	Erosional, concave-up, incised into hemipelagic sediment Figs. 26, 28	Flat or slightly hummocky, convex-up Figs. 26, 27	Hemipelagic lime mudstone and wackestone dominantly tabular shaped or a mixture of hemipelagic lime mudstone and shoal-water derived limestone Figs. 28, 29	10 × 50 cm; most commonly 2 × 10 cm Figs. 28, 29	Massively bedded; imbricated; crossbedded; parallel orientation of clasts Figs. 28–31	Normal in some Figs. 29, 30

*Figure numbers refer to these characteristics.

clasts, the orientation of imbricated clasts on bedding planes and the orientation of tabular clasts on bedding planes that are aligned subparallel to bedding. This paleoslope and paleocurrent data indicate that the strike of the slope trended in a northwest to northeast direction. Because the exposurees in the Hot Creek Range trend almost due east-west, the shapes and dimensions of the slides and sediment gravity-flow deposits can be closely defined.

Slides of different sizes show diverse characteristics and thus are seaparated into three different groups based on thicknesses of less than 1 m thick, 2 to 4 m and 8 to 10 m. Sediment gravity-flow deposits also can be grouped into two types—those less than 3.5 m thick that commonly contain abundant primary sedimentary structures and those 6 to 12 m thick that are massive deposits. General characteristics of these slides and sediment gravity-flow deposits are summarized in Tables 1 and 2.

DESCRIPTION OF THE UPPER SLOPE SEQUENCE

Translational Slides (8–10 m Thick)

The thickest translational slides occur within the stratigraphically highest part of the slope sequence. One slide is described here. This slide moved a mass of black, thin-bedded hemipelagic limestone. It is about 400 m wide and 8 to 10 m thick in its central part. This slide maintains a fairly uniform thickness of 8 to 10 m but tapers out gradually along its margins. Within 100 m of each margin, it is 3 m thick and gradually pinches out with no obvious major channeling at its base. Its upper surface may be slightly convex-up as suggested by the geometry of a pebbly sediment gravity-flow deposit which caps the top of the slide. This pebbly conglomerate is 10 cm thick in the central part of the slide and thickens to 2 m at both margins of the slide, suggesting the slide had some topographic relief which was at least partially reduced by deposition of the pebbly conglomerate.

Part of this slide moved across thin-bedded hemipelagic limestone, whereas most of it slid across a previously deposited thick conglomeratic debris-flow deposit. The basal shear plane is parallel to the underlying beds, where it slid over hemipelagic limestone, with only local scouring up to 1.5 m deep (Fig. 4). Its movement across the conglomerate, however, was more complex. The slide embedded itself several meters down into the conglomerate indicating that the conglomerate was probably unconsolidated at the time the slide was emplaced (Fig. 5).

The degree of deformation within the slide, although locally chaotic, is somewhat predictable. Deformation is most intense within the basal 0.5

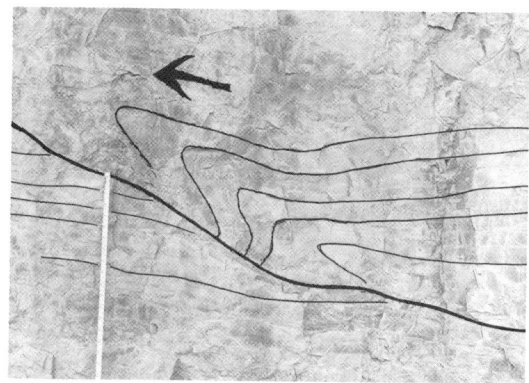

Fig. 4.—Contact between basal shear plane of 10 m thick translational slide and underlying in situ hemipelagic limestone. Top of 50 cm long tape is on the shear plane. Overfolds within the basal shear zone are confined to the lower 30 cm of the slide. Arrow points in direction slide moved. The shear plane is parallel to the in situ beds at the right then abruptly cuts up-section to the left.

Fig. 5.—Black lines mark the contacts between in situ hemipelagic beds (1), conglomeratic channel deposit (2), and translational slide (3). At this location the contact between the slide and conglomerate is very irregular where a finger-like lobe of the slide embedded down into the conglomerate several meters. Outcrop shown is about 5 to 7 m high.

to 1.0 m and at the lateral margins of the slide where it thins to less than 2 m. The least amount of deformation occurs within the upper 1 to 2 m of the slide and intermediate degrees of deformation are found between the base and top.

The intensely deformed basal zone is characterized by tight overfolds. The development of

FIG. 6.—Base of 10 m thick translational slide. Note the original bedding above the tape has broken into tabular clasts on the order of 2 × 5 cm. Lime mud fills interclast space. Tape is 22 cm long (from Cook and Chamberlain, 1978, Fig. 7).

FIG. 7.—Interior part of 10 m thick translational slide showing large open overfolds developed in the semiconsolidated hemipelagic limestone.

FIG. 8.—Interior part of 10 m thick translational slide showing the development of tabular clasts along the axis of an overfold. Clasts are dominantly tabular with either angular or subrounded terminations. Interclast space is filled with lime mud. Pencil is 8 cm long.

clasts in this zone is common and appears to be largely the result of overfolds or near-horizontal beds breaking into tabular fragments (Fig. 6). Intermediate degrees of deformation in the interior part of the slide are characterized by large open overfolds that still clearly retain original bedding (Fig. 7; Cook and Taylor, 1977, Fig. 28). Clast abundance in most cases here is related to the tightness of the folds with clasts being more abundant in areas of tight folds (Fig. 8). Parallel bedding in this interior zone is often partially broken into tabular clasts (Fig. 9). In both bases the clasts exhibit a jigsaw pattern whereby their displacement from the original bedding is only slight.

In the lateral margin of this slide where it is less than 2 m thick, original bedding in tight overfolds has totally broken into clasts. These clasts are randomly oriented and set within a pervasive dark lime mudstone matrix. The texture of this conglomerate is virtually identical to that

Fig. 9.—Interior part of 10 m thick translational slide showing near parallel beds partly broken into tabular clasts. The clasts are dominantly rounded at their ends. Some of the clasts have undergone partial rotation. Lime mud fills interclast space. Lens cover is 5 cm wide.

of a debris-flow deposit (Fig. 10). The least deformed zone appears to be at the top of the slide where bedding is parallel to the enclosing facies and clast development is minor.

Translational Slides (2–4 m Thick)

The upper part of the slope has several slides that are 2 to 4 m thick. Like the thicker slides these slides moved black, thin-bedded hemipelagic sediment. One well-exposed example is used to illustrate characteristics of these slides. This slide is 100 m wide and is 3.5 m thick in its central part. It maintains a fairly uniform thickness of 2.5 to 3.5 m throughout most of its width, then abruptly pinches out within 10 m on either margin, which appears to reflect the shape of the basal shear plane. The shear plane is approximately parallel to the underlying beds (Fig. 11), except at the margins of the slide where the shear plane is concave up and cuts a few meters diagonally upsection.

Internal deformation of the slide exhibits a fairly consistent pattern. The bulk of the slide shows very little disruption of original bedding planes, especially in the middle, thickest part of the slide (Fig. 12). Deformation is common within the basal 0.1 to 1.0 m where overfolds and local development of tabular clasts took place (Fig. 13). Formation of tabular clasts occurs along the axial planes of overfolds and by the local breakup of horizontal beds (Fig. 14). Deformation is most intense within the marginal 10 m of each side of the slide where the slide abruptly thins. Here overfolds and the beginning stages of clast development are common throughout the thickness of the slide (Fig. 15).

Fig. 10—Completely conglomeratic texture developed at the base of translational slide. Clasts are set within a pervasive lime-mud matrix. Note that the texture of the conglomerate is virtually identical to mass-flow deposits commonly inferred to be products of debris flows. Tape is 84 cm long.

Fig. 11.—Basal shear zone in 3.5 m thick translational slide. Here the shear plane is parallel to the underlying in situ hemipelagic beds and the shear zone is 30 to 40 cm thick. The shear folds are overturned in the direction of sliding (to the left). Pen is 10 cm long.

In the marginal 2 m of the slide, where the thickness is 1 m, the original bedding has changed into a conglomerate with tabular clasts. These clasts are normally graded and imbricated in an upslope direction (Fig. 16). This fabric transition at the margins of the slide, from overfolds, to overfolds associated with clasts, to a completely conglomeratic texture, takes place within a lateral distance of only 5 m (Fig. 17).

Rotational Slides (Slumps)

Only one rotational slide or slump has been identified. It occurs on the face of a vertical cliff. Its base is concave-up and is truncated nearly flat by an overlying translational slide (Fig. 18). This slump moved a mass of black, thin-bedded hemipelagic limestone that had a stratigraphic thickness of about 10 m and a lateral dimension

FIG. 12.—3.5 m thick translational slide showing only minor deformation of parallel bedding in slide. Slide has moved hemipelagic limestone.

FIG. 14.—Base of 3.5 m thick translational slide showing near horizontal bedding breaking up into tabular clasts. Lime mud fills interclast space. Tape is 40 cm long.

FIG. 13.—Base of 3.5 m thick translational slide showing basal shear folds, developed in semiconsolidated sediment, breaking up into tabular clasts. Tape is 45 cm long.

FIG. 15.—Upper part of lateral margin of 3.5 m thick translational slide with abundant overfolds and partial development of clasts. Tape is 40 cm long (from Cook and Taylor, 1977, Fig. 29).

Fig. 16.—Extreme lateral margin of 3.5 m thick translational slide showing the top of the slide where all original bedding has broken up into tabular clasts. Clasts are normally graded and imbricated in an upslope direction. Flow westerly downslope from right to left. Tape is 40 cm long (from Cook and Chamberlain, 1978, Fig. 6).

Fig. 18.—10 m thick rotational slide. Black lines show the concave-up basal shear plane and the upper contact with an overlying translational slide.

Fig. 17.—Extreme lateral 5 m of 3.5 m thick translational slide where it terminates. Figure 15 is from the left side of photo and figure 16 is from the right side of photo.

Fig. 19.—Pick lies parallel to erosional contact between 12 m thick conglomeratic channel deposit and underlying is situ hemipelagic limestone. Here channel axis is 12 m deep with channel walls having erosional slopes up to 25°.

of at least 100 m. Original bedding in the slumped pelagic limestone is clearly evident, and deformation appears to be largely confined to the basal shear zone.

Sediment Gravity Flow Deposits (6–12 m Thick)

This conglomerate occurs in broad shallow channels that have a width of 275 to 400 m. The axial part of the channels are about 5 to 10 m deep and 30 to 50 m wide. These channels shallow markedly away from this central axis. Conglomerate in these channels is up to 12 m thick but on either side of the narrow channel thins to 2 to 4 m.

Basal erosion is greatest in the thickest central parts, with up to 5 to 12 m of channeling, and channel walls having slopes of up to 25° (Fig. 19). Basal contacts, where the conglomerate is relatively thin, are usually nearly parallel with the underlying sediment. Locally, however, this contact is nearly vertical with up to 2 m of erosion of the hemipelagic sediment (Fig. 20) or the hemipelagic sediment is partly deformed and overturned in the direction the conglomerate was flowing.

The upper surfaces of the thinner, lateral parts of the deposits are typically flat or slightly hummocky, whereas the thick, narrow part of the channel-fill shows a convex-up mounded surface. In several instances the upper surface is highly

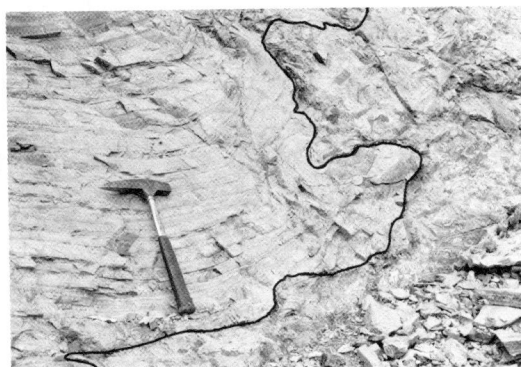

Fig. 20.—Black line marks the near vertical erosional contact between a conglomeratic channel deposit, on the right, and the in situ hemipelagic beds on the left. Note how the in situ beds are sharply upturned at the contact.

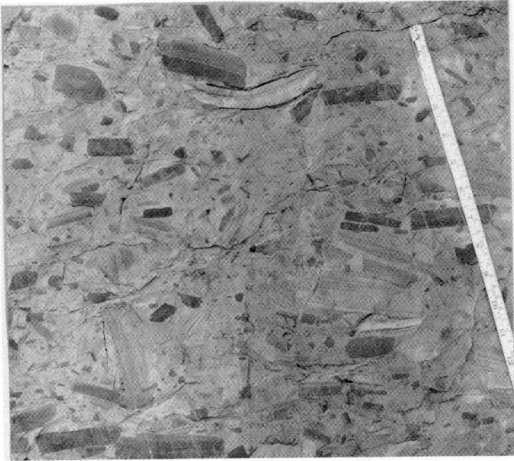

Fig. 22.—2 m thick lateral part of the 12 m thick conglomeratic channel deposit showing subparallel orientation of tabular clasts. Clasts are all hemipelagic limestone. Note that the clasts have blunt or slightly rounded ends identical to the clasts generated within the slides. Tape is 30 cm long.

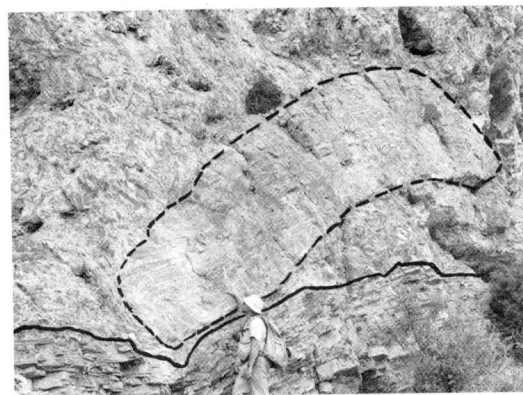

Fig. 21.—Black solid line shows the base of 12 m thick conglomeratic channel deposit and the underlying in situ hemipelagic limestone. Top of channel not shown. Dashed line encloses a 3 × 15 m clast of hemipelagic limestone surrounded by smaller randomly oriented tabular clasts.

irregular where part of an overlying slide has embedded itself down into the conglomerate (Fig. 5).

Clasts in these deposits are all dark laminated lime mudstone and wackestone that contain deep-water fossils. They are embedded in a pervasive dark, lime mudstone matrix that also contains deep-water fossils. The lithology of these clasts is identical to that of the enclosing hemipelagic limestone. Clasts are virtually all tabular and as large as 3 by 15 m (Fig. 21) and often have clasts 1 by 2 m in cross section. Boulders of this size are only found in the axial parts of the channels.

The most abundant clast size is on the order of 2 to 5 cm by 10 to 20 cm in cross section (Fig. 22) similar in size and shape to those generated within the slides. Occasionally these clasts are pillow-shaped where hemipelagic limestone clasts have been deformed into overfolds.

Clasts within the thick central parts of the channels most commonly are randomly oriented, giving the conglomerate a massive appearance (Fig. 23). Occasionally tabular clasts are subparallel within a thin interval near the upper and lower contacts of the conglomerate. Subparallel orientation of tabular clasts is abundant where the conglomerate thins to less than a few meters (Fig. 22).

DESCRIPTION OF THE LOWER SLOPE SEQUENCE

Translational Slides (0.2–1.0 m Thick

The lower part of the slope is characterized by small slides that moved semiconsolidated, black, thin-bedded hemipelagic limestone. Internal bedding is typically deformed throughout the thickness of the slide into tight recumbent overfolds (Fig. 24). These small slides have either a basal shear plane or shear zone a few centimeters thick. Their upper surface is approximately parallel to overlying, undeformed sediment. These semiconsolidated beds are commonly broken into tabular clasts with blunt or rounded terminations along the axial planes of the overfolds but apparently these small-scale slides were frozen before a complete conglomeratic texture could

FIG. 23.—Channel axis of 12 m thick conglomerate. Black line at contact of conglomerate and hemipelagic limestone. Clasts are all hemipelagic limestone, dominantly tabular shaped, randomly oriented above the base, subparallel at the base and floating in a lime-mud matrix.

develop (Fig. 25). Dark lime mud fills the interclast spaces.

Sediment Gravity Flow Deposits (0.2–3.5 m Thick)

This conglomerate forms thin channels whose maximum lateral dimensions are on the order of 100 m. They exhibit distinct scouring at their base and either have a nearly flat upper surface (Fig. 26) or a hummocky convex-up surface (Fig. 27). There are two distinct types of clasts in these deposits. One is clasts that are virtually all tabular and consist only of dark, laminated lime mudstone and wackestone identical to that of the enclosing hemipelagic beds (Fig. 28). Interclast matrix is dark lime mudstone admixed with deep-water fossils. The other type is a mixture of the above clast type and light-colored mud and grain-

FIG. 25.—Close-up of overfold axis of translational slide shown in figure 24 showing tabular clasts starting to develop. Tape is 18 cm long.

FIG. 24.—Small-scale 50 cm thick translational slide that moved hemipelagic limestone. Tape is 40 cm long.

FIG. 26.—30 cm thick conglomeratic channel deposit showing erosional base and fairly flat upper surface. Tape is 33 cm long.

Fig. 27.—1 m thick conglomeratic channel deposit with slightly erosional base and hummocky convex-up top.

Fig. 28.—1.5 m thick conglomeratic channel deposit. Clasts are all hemipelagic limestone, tabular shaped and oriented randomly to subparallel to bedding. This conglomerate has a high-mud matrix.

Fig. 29.—Upper part of a 1.5 m thick conglomeratic channel deposit. Clasts are a mixture of hemipelagic slope limestone and shoal-water derived limestone. Clasts are normally graded, imbricated in an upslope direction at the top of the channel, oriented subparallel below the top of the channel. Note the thin crossbedded calcarenite at the top of the channel. This is a typical low-matrix clast-supported conglomerate. Tape in inches (from Cook and Taylor, 1977, Fig. 38).

supported clasts of tabular and ovoid shape whose fossils and rock types are like that from the shoal-water environments to the east (Fig. 29). Matrix in these deposits is a mixture of dark lime mudstone and deep-water and shallow-water fossils. The largest clasts in these relatively thin channel deposits are 10 by 50 cm in cross section, however, the most common size is about 2 by 10 cm or less. Ovoid clasts reach a maximum size of about 10 cm.

There appears to be a relationship between the amount of mud matrix in this conglomerate and the presence or absence of stratification, clast orientation, and overall shape of the deposit. Conglomerate with a high mud matrix commonly has clasts randomly oriented or oriented subparallel to bedding (Fig. 28). These flow units either have a flat upper surface or a mounded convex-up surface (Figs. 26, 27). Conglomerate with a low mud matrix usually exhibits imbrication and normal grading of tabular clasts or subparallel alignment of clasts that may be normally graded (Figs. 29, 30). In this low matrix conglomerate a thin cross-bedded clacarenite layer is common at the top of the bed (Fig. 29).

Clasts whose long dimensions are parallel to bedding also have the long dimension of their shape parallel to the paleoflow direction. Imbricated clasts dip up the paleoslope and their long dimensions on bedding surfaces are transverse to the paleoflow direction (Fig. 31). Only rarely

FIG. 30.—50 cm thick conglomeratic channel deposit. Clasts are normally graded and contain a mixture of slope and shoal-water derived types. Tape is 22 inches long (from Cook and Taylor, 1977, Fig. 36).

FIG. 31.—Upper surface of a 50 cm thick conglomeratic channel deposit. Bedding view shows tabular clasts imbricated in an upslope direction. Bedding plane surface shows the longer dimension of tabular clasts oriented perpendicular to the paleoflow direction. Paleoflow was downslope from right to left. Tape is 17 cm long.

is the long dimension of imbricated clasts parallel to the paleoflow direction as seen on bedding surfaces. When subparallel orientation and imbrication of clasts occur in the same bed, the imbricated clasts are at the top of the bed and clasts with subparallel orientation are at the base of the bed. It is unusual to find clasts larger than 2 by 10 cm imbricated; clasts larger than this are either subparallel or randomly oriented in respect to bedding contacts.

ORIGIN OF THE MASS-TRANSPORT DEPOSITS

Provenance

All of the slides and slumps originated on the slope or possibly near the slope-shelf margin. This is indicated by the fact that the rocks and fauna that comprise these bodies are identical to that of the enclosing undeformed hemipelagic slope limestone. All of the conglomerate that is 3.5 m thick or greater appears to have also originated on or near the slope, as its clast lithology and faunal content is the same as the surrounding hemipelagic sediment. The matrix surrounding these clasts is a dark lime mudstone with deep-water fossils.

Many of the conglomeratic sediment gravity-flow deposits less than 3.5 m thick apparently originated in a shoal-water setting and incorporated a variety of material during their transport oceanward across the shelf and down the slope. This conglomerate contains a mixture of shoal-water and deep-water clasts and fauna.

Mechanics of Movement

The mechanics of movement during sliding and slumping was complex and apparently included elastic and plastic deformation, and plastico-viscous flow. There may be a relation between the thickness of a slide and the dominant manner in which movement took place. Slides whose maximum thickness is 1 m or less are usually intensely deformed into overfolds throughout their width and thickness (Fig. 24). The yield limit of the mass in these cases was exceeded and they moved plastically along a thin basal layer. In some instances, the rupture strength of these small slides was exceeded and local development of clasts occurred (Fig. 25). Slides 2 to 4 m thick and 8 to 10 m thick exhibit elastic and plastic deformation, and plastico-viscous flow behavior. Movement along the basal shear plane caused overlying plastic deformation as exhibited by overfolding. A basal zone immediately above the shear plane has a conglomeratic texture (Fig. 6). Here the rupture strength of the slide material was exceeded and movement was by plastico-viscous flow. This conglomerate has a fabric similar to debris-flow deposits. The clasts are mud supported and are randomly oriented or are subparallel to the base of the slide. The thick interior parts of these slides retain parallel bedding or large open overfolds where the slide moved plastically and only occasionally exceeded its rupture strength (Figs. 7, 8, 9). The rupture strength was exceeded at the margins of slides, and previously folded or horizontal bedding is completely replaced by a conglomerate. In one instance, tabular clasts are graded and imbricated with long axes dipping up the paleoslope (Fig. 16). In another instance, tabular clasts are randomly oriented in a mud matrix (Fig. 10). Clearly, in these marginal, thin parts of the slides, the rupture strength was exceeded, a total clastic texture developed, and the mass moved by plastico-viscous flow. Movement was probably dominantly as debris flow (Johnson, 1970) where the clasts are randomly oriented and floating in a pervasive lime mudstone matrix. A high-density turbidity flow may have been at least partly involved for those parts of the slide where the conglomerate is clast supported, and the clasts are normally graded and imbricated.

Mass-flow channel deposits up to 5 to 10 m in thickness probably moved mainly by debris flows. They consist of disorganized conglomerate in which the flow unit is massively bedded and the clasts are randomly oriented and up to 3 by 15 m in size (Fig. 21). Clasts are floating in a matrix of fine-grained, dark, lime mudstone (Figs. 22, 23). These types of flows may have moved sluggishly along and came to a halt before any significant degree of turbulent flow could be attained to sort the clasts or produce other sedimentary structures and fabrics.

Mass-flow deposits less than a few meters thick commonly exhibit a variety of primary sedimentary structures. Normal grading, imbrication and parallel orientation of clasts, crossbedding and a clast-supported fabric are common in these thin flow units (Figs. 29, 30). These conglomerates probably were emplaced by turbidity-current flow. Other thin conglomerates are massively bedded with the clasts floating in a lime mudstone matrix (Fig. 28). These deposits probably represent debris flows.

Generation of Mass Flows from Slides

There are relatively few published data demonstrating that sediment gravity flows can be generated from slides and slumps. Hampton (1972) observed the subaerial formation of debris flows by remolding of slides and slumps and entrainment of water and suggested a transition from debris flow to turbidity current flow. Field and Clarke (1977) suggest that small slides and slumps on modern slopes of the southern California borderland are a major process for generation of sediment gravity flows. The present study shows that once a slide is in motion its base and thin margins are capable of transforming into mass flows. Probably because of a variety of factors including strain, mechanical shock and incorporation of water, the rupture strength of these parts of the slide were exceeded and movement by plastico-viscous flow resulted. As pointed out by Middleton and Hampton (1973, 1976), if subaerial slides undergoing strain and mechanical shock are capable of generating debris flows, then submarine slides should be even more capable of forming debris flows owing to the incorporation of water which reduces the strength of the sediment.

Debris-flow movement is suggested for those conglomeratic parts of the slide where the fabric is totally clastic, tabular clasts are randomly oriented or subparallel to the base of the slide, and the clasts are floating in a lime mudstone matrix. In one instance, the margin of a slide transformed into a conglomerate whose tabular clasts are both imbricated and normally graded. This suggests that a high-density turbulent flow was involved for a sufficient period of time to size-sort the clasts. A continuum from debris flow to turbidity flow is not seen in any part of a slide or in any conglomeratic channel deposit. However, as different channel deposits and conglomeratic parts of different slides exhibit debris- and turbidity-flow characteristics, it suggests that a progression of flow processes was a likely event on this slope. The probability of this type of continuum is discussed by Morgenstern (1967) and Hampton (1972).

The deep-water derived conglomerate that occurs in broad channels and exhibits fabrics suggesting emplacement by debris flow, turbidity flow, or a combination of flow processes cannot be directly traced to any possible parent slide. However, because similar conglomerate is generated from slides, it is reasonable to infer that much of it also originated from slides (See also Embley, 1976.).

DEPOSITIONAL MODEL

It is proposed that during the Late Cambrian and Early Ordovician a north-trending continental slope existed in central Nevada (Fig. 32). This part of the continental slope was prograding seaward. Progradation of the shelf lying 170 km to the east is well documented for this time (Cook

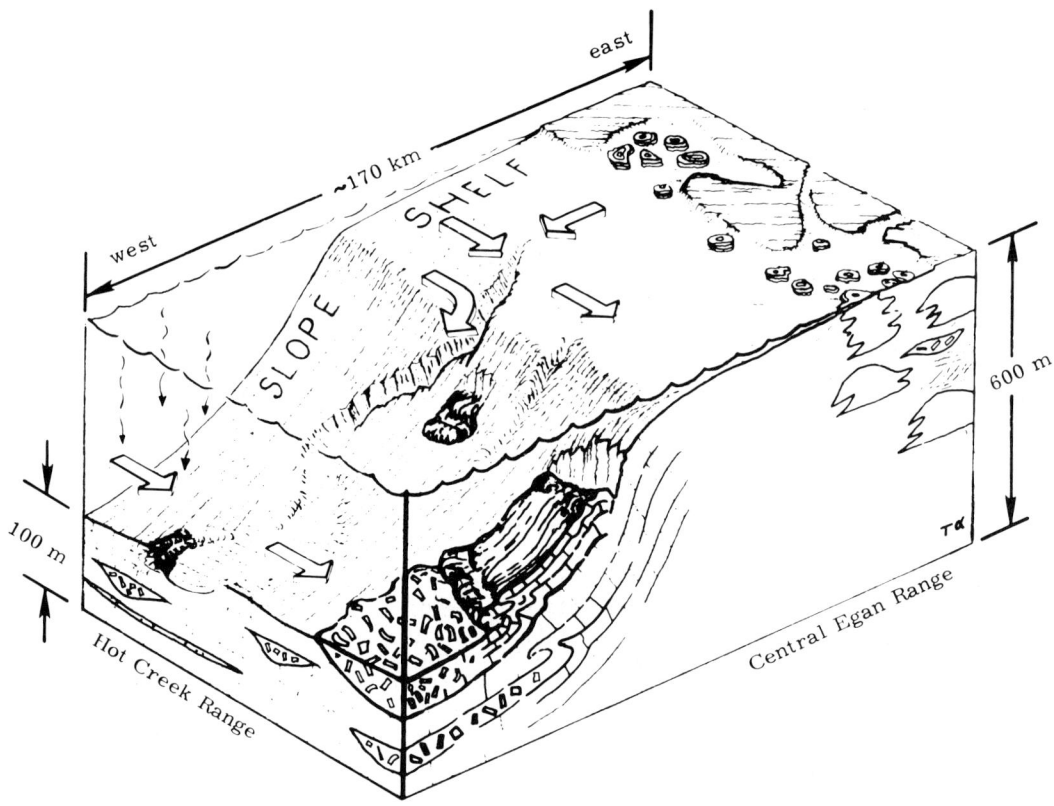

Fig. 32.—Graphic model of inferred shelf-slope transition in the Late Cambrian and earliest Ordovician of Nevada (modified from Cook and Taylor, 1977, Fig. 45). Drawn by Tau R. Alpha.

and Taylor, 1977). The character of the in situ hemipelagic limestone in this section also suggests progradation of the slope. Toward the top of the section the beds become thicker and lighter in color and textures change from virtually all lime mudstone at the base to lime wackestone at the top. Underlying formations consist of fine-grained limey shale and spicular chert of probable deep-water origin. Thus, it is likely that one is progressing up the paleoslope as the section is climbed.

The evolving model is one in which slope sedimentation involved an interplay between depositional and mass-wasting processes. The dark, laminated, thin-bedded spicule-rich mudrock in the lower part of the section largely represents hemipelagic sedimentation by suspension and normal pelagic processes, whereas the lighter-colored, thicker-bedded wackestone near the top of the section represents sediment formed higher on the slope or near the shelf-slope break. Submarine gravity movements, although not of a large scale, were a major process in the development of the slope. About half of the section is composed of mass-transport material. Sedimentary processes on the upper slope included large scale sliding, slumping, and debris flows. This is demonstrated by the presence of the large slides and slumps and the channels containing boulder-bearing, massive thick debris-flow deposits at the top of the section. Also, the deepest channels occur and the lateral dimensions of all mass-transport deposits are greatest on the upper part of the slope. Probably large portions of these slides and slumps changed into debris-flow masses that were transported farther down the slope. Some of these debris flows may have turned into turbidity flows. Thus, farther down the slope, mass-transport deposits are characterized by abundant shallow channels containing debris- and turbidity-flow deposits. The characteristics of many of these deposits suggests that turbulent motion developed and was sustained for a sufficient period of time to result in normal grading of clasts and other primary sedimentary features. Sliding on the lower part of the slope was a minor event and only involved sediment thicknesses of a meter or less. Downslope movement of these slides was probably quite limited.

VALUE OF ANCIENT SLOPE STUDIES TO
PETROLEUM EXPLORATION

General

The most obvious advantage of studying ancient slope sequences is that we are able to see what the deposits are, map their geometries, observe transitional relations in mass-transport materials, and gain a better understanding of what processes can operate in a slope environment. This exercise can be especially fruitful in a prograding slope sequence, because we can develop a depositional profile depicting processes operating on different parts of the slope. Ancient slopes cannot be directly compared to modern slopes, but by actually examining the rocks one can gain a better appreciation of the types and scales of events, and gain some knowledge of their periodicity.

Seismic Resolution Problems

How would the depositional profile of this slope appear if it existed on a modern continental slope? Seismic profiles obtained with a high energy system probably would be unable to resolve the 10 m-thick slides and debris-flow deposits on the upper part of the slope. The lower part of the slope, where thin (0.2 to 1 m) slide, debris- and turbidity-flow deposits are abundant, probably would appear as a smooth undisturbed sequence which might be interpreted to be stable in situ sediment. In reality, mass-movement events were a major depositional process. The problems of seismic resolution can at least be brought into geologic perspective if we entertain the possibility that downslope from slides and slumps, thin, relatively narrow channels filled with sediment gravity-flow deposits may be more abundant than a seismic record indicates. Thus, if acoustically disrupted reflectors, interpreted to represent slides, are recognized on some parts of the slope, it is quite possible that the lower parts of the same slope may contain mass-flow deposits generated by these slides.

Potential Petroleum Reservoirs

If modern slope environments are similar to ancient slopes such as those of the Hot Creek Range, it seems likely that mass-transport processes are more common on modern slopes than we currently recognize. The implications are pertinent to petroleum exploration within continental slope sequences. Coarse-grained mass-flow deposits are potential reservoirs on slopes. Several types of reservoirs could exist. Once a submarine slide is in motion, its base and lateral margins can be transformed into coarse-grained conglomerate. Thus, potential reservoirs may exist in direct contact with slides and slumps. These reservoirs may be sheet-like where they occur at the base of slides and more shoestring in shape at the margins of the slide. Thick-bedded conglomerate in deep, broad channels may be stratigraphically and genetically closely associated with slides. A seismically smooth, seemingly undisturbed sequence downslope from slides may have abundant small-scale channels containing thin-bedded mass-flow conglomerate and sandstone of both a deep water and shoal-water origin. Collectively, they could represent a favorable exploration target. Under the right depositional and diagenetic situations, porosity in conglomerate may be retained or enchanced. Such appears to be the case for deep-water conglomerate reservoirs in the Permian of west Texas (Cook and others, 1972), Cretaceous of Mexico (Enos, 1977) and elsewhere. In addition, mass-flow deposits on slopes could conceivably form a built-in porous plumbing system for transmitting hydrocarbons upslope to shoal-water reservoirs.

Environmental Hazards

A modern slope undergoing numerous small-scale mass-movement events poses potential geologic hazards that could influence decisions on the placement of offshore drilling platforms and subsequent installation of production facilities. Some of the sedimentary processes operating on slopes that may be responsible for a slope sequence being an attractive exploration target are at the same time those processes that can threaten the safe exploitation of a slope. Once mass-movement deposits come to rest, are they stable or do they continue to creep for long periods of time? Are they intervals of weakness that could be triggered into motion during the exploitation of a slope? These are questions that in specific slope settings may prove difficult to answer but knowing that a slope sequence may not be smooth and undisturbed can at least make us more cautious.

CONCLUSIONS

(1) During the early Paleozoic in Nevada a shoal-water carbonate shelf existed to the east and adjacent deeper water slope environments to the west. Sediment accumulating on this continental margin prograded seaward.

(2) Slope sedimentation included both depositional and erosional processes. In situ hemipelagic sediment is characterized by its dark color, thin laminated beds, and a deep-water fauna. This sediment formed by suspension and other pelagic and hemipelagic processes. Sediments formed by submarine gravity movements constitute up to 50 percent of the slope section and include slide, slump, and debris- and turbidity-flow deposits.

(3) Large-scale translational slides and slumps are confined to the upper slope and may extend

to the slope-shelf break in this ancient continental margin. In this part of the slope, thick massive boulder-bearing debris-flow deposits occur in broad channels. The lower part of the slope is characterized by thin debris- and turbidity-flow deposits which formed in shallow, narrow, channels. Minor small-scale sliding occurred on the lower part of the slope.

(4) Movement of large translational slides was by elastic and plastic deformation, and plastico-viscous flow. The basal shear zone and thin lateral margins of slides were transformed into debris and turbidity flows. These flows, once initiated, traveled farther downslope than the slide from which they formed. Some coarse-grained conglomerate found on the slope was derived from shoal-water shelf environments.

(5) On the basis of abundant small-scale slides, slumps, and sediment-gravity flows that operated in this and similar ancient slope settings, it is reasonable to speculate that similar processes and their deposits may be more significant on modern slopes than we recognize. The smooth undisturbed nature of some modern slopes as seen on continuous seismic-reflection profiles may be, in part, a function of the limited resolving power of much seismic equipment, especially low frequency, deep penetration systems.

(6) If modern slopes are undergoing more mass-movement events than we are currently aware, then the implications are at least twofold. Acoustically undetectable sediment gravity-flow deposits are potential conduits for transmitting hydrocarbons upslope to porous shallow-water reservoir or for forming stratigraphic traps on the slope. Potential reservoirs on the slope occur at the base and lateral margins of slides where the slide can be transformed into conglomeratic debris- and turbidity-flow deposits. Conglomerate in deep, wide channels may also occur in close stratigraphic proximity to slides. Thin-bedded conglomerate in sheets or shallow narrow channels may be abundant especially farther downslope. A modern slope repeatedly undergoing small-scale mass movements could pose environmental hazards to the safe exploration and exploitation of this deep-water environment.

ACKNOWLEDGMENTS

Arnold Bouma, Michael Field, James Gardner, and Monty Hampton critically reviewed the manuscript and suggested several improvements. The author very much appreciates the clerical help of Terry Coit and the drafting services of Tau Rho Alpha, Phyllis Swenson, and Lee Bailey.

REFERENCES

BURCHFIEL, B. C., AND DAVIS, G. A., 1972, Structural framework and evolution of the southern part of the Cordilleran orogen, western United States: Am. Jour. Sci., v. 272, p. 97–118.

CARLSON, P. R., AND MOLNIA, B. F., 1977, Submarine faults and slides on the continental shelf, northern Gulf of Alaska: Marine Geotechnology, v. 2, p. 275–290.

CHURKIN, MICHAEL, JR., 1974, Paleozoic marginal ocean basin-volcanic arc systems in the Cordilleran foldbelt, in Dott, R. H., Jr., and Shaver, R. H., eds., Ancient and modern geosynclinal sedimentation: Soc. Econ. Paleontologists Mineralogists Spec. Pub. No. 19, p. 174–192.

COOK, H. E., AND ENOS, PAUL, eds., 1977, Deep-water carbonate environments: Tulsa, Oklahoma, Soc. Econ. Paleontologists Mineralogists Spec. Pub. No. 25, 336 p.

———, MCDANIEL, P. N., MOUNTJOY, E. W., AND PRAY, L. C., 1972, Allochthonous carbonate debris flows at Devonian bank ('reef') margins Alberta, Canada: Canadian Petroleum Geology Bull., v. 20, p. 439–497.

———, AND TAYLOR, M. E., 1975, Early Paleozoic continental margin sedimentation, trilobite biofacies, and the thermocline western United States: Geology, v. 3, p. 559–562.

———, 1977, Comparison of continental slope and shelf environments in the Upper Cambrian and lowest Ordovician of Nevada, in Cook, H. E., and Enos, Paul, eds., Deep-water carbonate environments: Tulsa, Oklahoma, Soc. Econ. Paleontologists Mineralogists Spec. Pub. No. 25, p. 51–81.

———, AND CHAMBERLAIN, W. W., 1978, Ancient continental slope sequences and their usefulness in understanding modern slope development: Houston, Texas, Offshore Technology Conference Proc., p. 547–553.

DIETZ, R. S., AND HOLDEN, J. C., 1966, Miogeoclines (miogeosynclines) in space and time: Jour. Geology, v. 74, p. 566–583.

DOTT, R. H., JR., 1963, Dynamics of subaqueous gravity depositional processes: Am. Assoc. Petroleum Geologists Bull., v. 47, p. 104–128.

EMBLEY, R. W., 1976, New evidence for occurrence of debris flow deposits in the deep sea: Geology, v. 4, p. 371–374.

EMERY, K. O., UCHUPI, ELAZAR, PHILLIPS, J. D., BOWIN, C. O., BUNCH, E. T., AND KNOTT, S. T., 1970, Continental rise off eastern North America: Am. Assoc. Petroleum Geologists Bull., v. 54, p. 44–108.

ENOS, PAUL, 1977, Tamabra Limestone of the Poza Rica Trend, Cretaceous, Mexico, in Cook, H. E., and Enos, Paul, eds., Deep-water carbonate environments: Tulsa, Oklahoma, Soc. Econ. Paleontologists Mineralogists Spec. Pub. No. 25, p. 273–314.

FIELD, M. E., AND CLARKE, S. H., JR., 1977, Submarine slumps and sediment flows off San Nicolas Island, southern California (abs.): Am. Assoc. Petroleum Geologists Bull., v. 61, p. 783.

HAMPTON, M. A., 1972, The role of subaqueous debris flow in generating turbidity currents: Jour. Sed. Petrology, v. 42, p. 775-793.
———, AND BOUMA, A. H., 1977, Slope instability near the shelf break, western Gulf of Alaska: Marine Geotechnology, v. 2, p. 309-331.
HANSEN, EDWARD, 1966, Methods of deducing slip-line orientations from the geometry of folds: Washington D.C., Carnegie Inst. Ann. Rept. Geophys. Lab. p. 386-410.
JOHNSON, A. M., 1970, Physical processes in geology: San Francisco, California, Freeman, Cooper and Co., 577 p.
KELLING, GILBERT, AND STANLEY, D. J., 1976, Sedimentation in canyon, slope, and base-of-slope environments, in Stanley, D. J., and Swift, D. J. P., eds., Marine sediment transport and environmental management New York, Wiley p. 379-435.
LEWIS, K. B., 1971, Slumping on a continental slope inclined at $1°-4°$: Sedimentology, v. 16, p. 97-110.
MIDDLETON, G. V., AND HAMPTON, M. A., 1973, Sediment gravity flows: Mechanics of flow and deposition, in Middleton, G. V., and Bouma, A. H., eds., Turbidites and deep-water sedimentation: Soc. Econ. Paleontologists Mineralogists Pacific Sec., Los Angeles, California, p. 1-38.
———, 1976, Subaqueous sediment transport and deposition by sediment gravity flows, in Stanley, D. J., and Swift, D. J. P., eds., Marine sediment transport and environmental management: New York, Wiley, p. 197-218.
MORGENSTERN, N. R., 1967, Submarine slumping and the initiation of turbidity currents, in Richards, A. F., ed., Marine Geotechnique: Urbana, Illinois, Univ. of Illinois Press, p. 189-220.
ROWELL, A. J., REES, M. N., AND SUCZEK, C. A., 1979, Margin of the North American Continent in Nevada during Late Cambrian time: Am. Jour. Sci., v. 279, p. 1-18.
STANLEY, D. J., AND SILVERBERG, N., 1969, Recent slumping on the continental slope off Sable Island bank, southeast Canada: Earth and Planetary Sci. Letters, v. 6, p. 123-133.
———, AND UNRUG, R., 1972, Submarine channel deposits, fluxoturbidites and other indicators of slope and base-of-slope environments in modern and ancient marine basins, in Rigby, J. K., and Hamblin, W. K., eds., Recognition of ancient sedimentary environments: Soc. Econ. Paleontologists Mineralogists Spec. Pub. No. 16, p. 287-340.
STEWART, J. H., 1972, Initial deposits in the Cordilleran Geosyncline: Evidence of a Late Precambrian (850 m.y.) continental separation: Geol. Soc. America Bull., v. 83, p. 1345-1360.
———, AND POOLE, F. G., 1974, Lower Paleozoic and uppermost Precambrian Cordilleran Miogeocline, Great Basin, western United States, in Dickinson W. R., ed., Tectonics and sedimentation: Soc. Econ. Paleontologists Mineralogists Spec. Pub. No. 22, p. 28-57.
TAYLOR, M. E., 1976, Indigenous and redeposited trilobites from Late Cambrian basinal environments of central Nevada: Jour. Paleontology, v. 50, p. 668-700.
———, 1977, Late Cambrian of western North America: Trilobite biofacies, environmental significance, and biostratigraphic implications, in Kauffman, E. G., and Hazel, J. E., eds., Concepts and Methods of Biostratigraphy: Stroudsburg, Penn., Dowden, Hutchinson and Ross, Inc. p. 397-425.
———, AND COOK, H. E., 1976, Continental shelf and slope facies in the Upper Cambrian and lowest Ordovician of Nevada, in Robinson, R. A., and Rowell, A. J., eds., Cambrian paleontology and environments of western North America, a symposium: Brigham Young Univ. Geol. Studies, v. 23, pt. 2, p. 181-214.
———, AND HALLEY, R. B., 1974, Systematics, environment, and biogeography of some Late Cambrian and Early Ordovician trilobites from eastern New York State: U.S. Geol. Survey Professional Paper 834, 38 p.
THOMSON, A. F., AND THOMASSON, M. R., 1969, Shallow to deep water facies development in the Dimple Limestone (Lower Pennsylvanian), Marathon region, Texas, in Friedman, G. M., ed., Depositional environments in carbonate rocks: Soc. Econ. Paleontologists Mineralogists Spec. Pub. No. 14, p. 57-78.
VARNES, D. J., 1958, Landslide types and processes, in Landslides and engineering practice: Highway Res. Board Spec. Rept. 29, Nat. Res. Council Pub. 544, p. 20-47.

DEPOSITIONAL MECHANICS OF THICK-BEDDED SANDSTONES AT THE BASE OF A SUBMARINE SLOPE, TOURELLE FORMATION (LOWER ORDOVICIAN), QUEBEC, CANADA

RICHARD N. HISCOTT[1] AND GERARD V. MIDDLETON
Department of Geology
McMaster University
Hamilton, Ontario, Canada L8S 4M1

ABSTRACT

The Arenigian Tourelle Formation is a well exposed flysch on the south shore of the St. Lawrence Estuary in the Gaspé Peninsula of Quebec. It was derived from a tectonic landmass lying southeast of the present outcrop area, in contrast to underlying formations derived from the Canadian shield to the northwest. The Tourelle Formation consists mainly of shales and sandstones that show Bouma sequences typical of "normal" turbidites, but it includes also two types of sandstones that do not show these sequences: sandstones in layers thicker than two meters, generally found within channels at the base of thinning-upward sequences, and slurry sandstones. Thick sandstones show two end-member types: (i) massive sandstones, with poor, or inverse-to-normal grading, and variable grain orientation and imbrication; most of these lack flutes and internal lamination, and a few contain large indurated blocks of shale, dolomitic siltstone or chert; and (ii) stratified sandstones, which have large flutes and grooves, inversely graded stratification bands several centimeters thick, internal scour surfaces, distribution grading, and good grain orientation and imbrication. Slurry sandstones are beds about 40 cm thick, composed of fine to coarse sand dispersed in an abundant muddy matrix, with abundant shale chips and many larger shale fragments; they are randomly distributed throughout the formation.

The slurry sandstones and some of the massive sandstones are interpreted as submarine debris flow deposits, with debris strengths (τ_c) of the order of 10^2 to 10^4 dynes/cm^2, deposited on slopes of the order of 0.01 to 0.05. For large flows composed of such strong debris, the criterion for turbulence appears to be a critical value (about 1000) of the dimensionless number $\rho U^2/\tau_c$ (where ρ is the density and U is the speed of the flow). This number is here named the Hampton number.

Most of the thick sandstone layers were probably deposited by high density turbidity currents, which were probably about 10 to 20 meters thick, had shear velocities of the order of 20 to 30 cm/sec (corresponding to mean velocities of the order of 5 m/sec) and flowed down slopes of the order of 0.01 to 0.005. Such flows were generally supercritical. At the top of a few beds, however, there are sets of trough cross-stratification which were formed by migration of dunes. These structures indicate subcritical flow and were probably produced by the more dilute tail of some turbidity currents.

INTRODUCTION

The Lower Ordovician Tourelle Formation crops out on the Gaspé Peninsula (Fig. 1) in the allochthonous external domain of the Quebec Appalachian orogenic belt. For a summary of the stratigraphy and structure of the Quebec Appalachians, see St-Julien and Hubert (1975).

In this part of the Northern Appalachians, there is abundant evidence that the older Paleozoic clastics (sandstones, and limestones conglomerates and breccias, of Cambrian to Lower Ordovician age) were derived from the northwest and deposited on southeast-facing slopes that probably constituted the continental slope formed along the trailing, eastern margin of North America during the expanding stages of the Proto-Atlantic, or Iapetus, Ocean (Wilson, 1966; Rodgers

[1] Present address: Department of Geology, Memorial University of Newfoundland, St. John's, Newfoundland, Canada A1B 3X5

1968; Hubert et al, 1970; Lajoie et al, 1974; Davies and Walker, 1974).

The Tourelle Formation conformably overlies such rocks, but consists of at least several hundred meters of interbedded sandstones and shales that were derived from a tectonic landmass to the southeast, as indicated by a study of its paleocurrents and provenance (Hiscott, 1979). Shortly before the deposition of the Tourelle, therefore, there must have been a reversal in paleoslope. This reversal in slope, and the deposition of the Tourelle Formation, is one of the earliest recorded effects of a change from a stage of spreading and expansion, to one of consumption and contraction of the Proto-Atlantic Ocean.

The Tourelle Formation is well exposed on a broad shore platform in the intertidal zone of the south shore of the St. Lawrence estuary. Measurement of numerous closely-spaced sections has permitted detailed analysis of the sequence of facies types and reconstruction of the

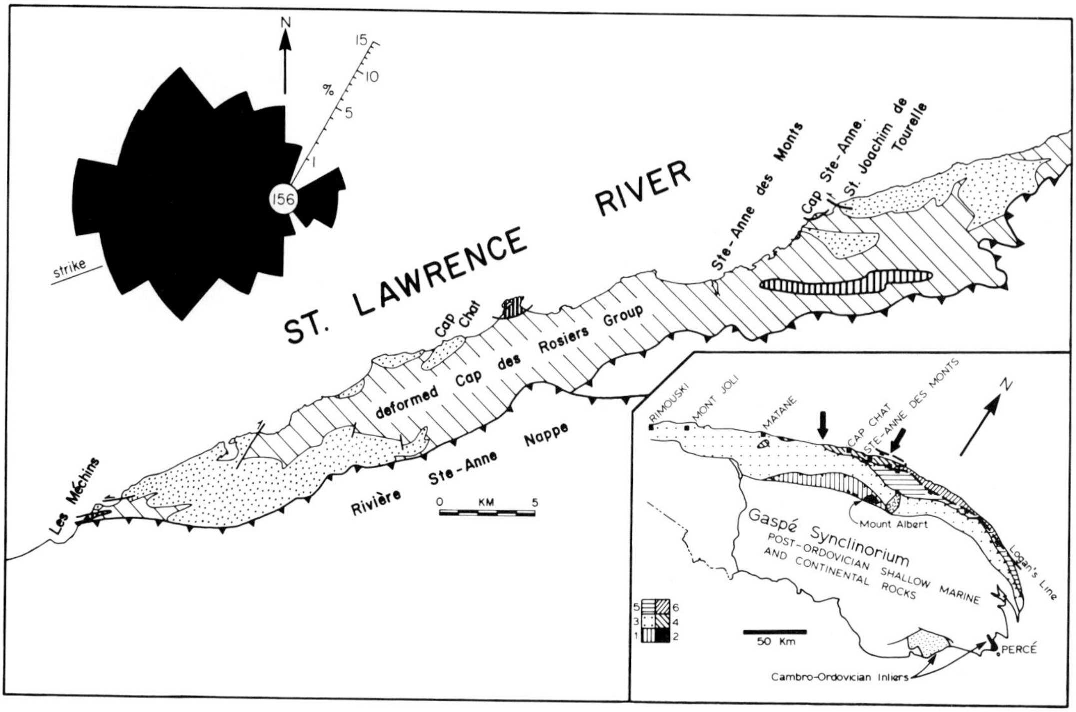

FIG. 1.—Location of the Tourelle Formation, and summary of paleocurrent data. In the main map, the Tourelle is stippled. Oblique rule indicates the Cap des Rosiers Group (or "Cap Chat Mélange") and heavy vertical rule indicates areas of Middle Ordovician klippen (After Biron, 1972, 1974). In the smaller index map, unit 4 includes the Tourelle Formation and deformed Cap des Rosiers Group, unit 6 is the Cloridorme Formation, a Middle Ordovician flysch described by Enos (1969), and unit 5 is allochthonous Middle Ordovician flysch. Stippled areas (unit 3) are allochthonous Cambrian to Lower Ordovician clastics. Unit 1 contains both arkose and mafic lavas of the ?Cambrian Shickshock Group, and unit 2 is an ophiolite slice obducted during closing of the Iapetus Ocean. Bold arrows indicate the location of the main map.

geometry of sandstone layers, at least in two dimensions.

Facies and sequence analysis, which will be the subject of a subsequent paper, indicates that the Tourelle Formation was deposited on the channelized mid-fan region of a series of submarine fans. The unit is characterized by a high proportion of thick-bedded, coarse, graded, massive sandstones (Fig. 2). Individual sandstone layers may be tens of meters thick with channelled bases, and a few contain large slabs of shale, siltstone, or bedded chert enclosed in sand. Rare cross-stratification and unusual near-horizontal stratification are the only internal depositional structures in the thick sandstones. Post-depositional liquefaction of some layers is indicated by the presence of coarse fluid-escape pillars that record elutriation of fine matrix material. The assemblage of fluid-escape structures, however, does not include dish structure.

The purpose of the present paper is to describe two of the unusual facies types in the Tourelle

FIG. 2.—Thick amalgamated sandstone beds below the Cap Chat lighthouse. Two thinning-upward sequences are visible in the cliffs. Note figure for scale.

Formation, to review some of the current theories about the mechanism of deposition of thick, coarse sand beds in submarine environments, and to apply these theories to an interpretation of the mechanics of deposition of the unusual facies in the Tourelle Formation. Reconstruction of the hydraulics of deposition further leads to some conclusions, necessarily speculative in character, about the magnitudes of the flows and slopes involved.

THE TOURELLE FORMATION

The Tourelle Formation crops out on the south shore of the St. Lawrence River, most prominantly in a region extending about 100 to 200 km east of Matane. The base of the formation is marked by the relatively abrupt appearance of massive graywacke-type sandstones, which conformably overlie an assemblage of gray, green, and red shales, with interbedded cherts, calcareous clastics (ranging in texture from calcisiltites to limestone conglomerates) and quartzose sandstones: this assemblage has commonly been assigned to the Cap des Rosiers Group (or Formation). The transition between the two units is generally marked by a tectonically undisturbed horizon, up to 60 m in thickness, of interbedded red shale, dolomitic siltstones and rare chert bands, but below this horizon the underlying beds of the Cap des Rosiers Group are commonly so disturbed that they have been described by Biron (1974) as the Cap Chat Mélange. The Tourelle Formation consists of several hundred (perhaps 1000) meters of sandstones and shales, generally organized into interstratified sandstone- and shale-rich "packets," each some tens of meters thick. Fossils are uncommon, lateral lithologic variability is high, and although the Tourelle Formation is not as intensely deformed as the underlying Cap des Rosiers Group, it has been strongly folded and cut by numerous faults. The upper boundary of the formation is everywhere erosional or tectonic. Thus, it is possible that the original thickness of the formation was considerably in excess of 1000 m.

On the basis of graptolite fauna, the Tourelle Formation has been assigned a late Arenigian age. The underlying Cap des Rosiers Group ranges in age from Arenigian to Tremadocian or older.

The Tourelle Formation has the general aspect of a flysch, and somewhat resembles the Cloridorme Formation, described by Enos (1969). It occupies a different position both tectonically (it is separated from the Cloridorme by a major thrust) and stratigraphically (the Cloridorme is Caradocian, i.e., substantially younger than the Tourelle). The Tourelle is also much coarser and thicker bedded (more "proximal" in character) than the Cloridorme Formation.

Hiscott (1977, unpub. Ph.D. thesis, McMaster University) has classified the lithologies observed in the Tourelle Formation into 9 major facies. Most of these are similar to those described from many other flysch formations and fit the conventional "turbidite" models of deposition. In the present paper we wish to analyse mainly the unusual Thick Sandstone Facies and the Slurry Sandstone Facies, because these two facies display features distinctly different from those that characterize "normal" turbidites, and thus may have been deposited from sediment gravity flows that differed from normal turbidity currents.

THICK SANDSTONES

Thick sandstones are arbitrarily defined as layers (= individual sedimentation units) greater than two meters thick, and as such compose 35% of the total thickness of the Tourelle Formation. These layers tend to occur in amalgamated sandstone packets which are separated by thick units of fine sediment (Fig. 3). These fine-grained rocks consist of shale, ripple-drifted siltstones, and thin turbidite sandstones. Also included in amalgamated sandstone packets are thinner (<2 m) graded sandstone layers. Only 10 percent of 247 measured thick sandstone layers can be fitted into the Bouma (1962) model of turbidites. Fifty percent are massive (lack internal structures) and 40 percent show structures not normally found in turbidites, such as large-scale cross-stratification (3%), near-horizontal stratification (23%; plus 2% having both of these structures) and internal scour surfaces (12%).

Layer Thickness

It would be unreasonable to examine layer thickness of only those sandstones greater than two meters thick, even though they form a convenient facies for field study. Instead, a cumula-

Fig. 3.—Airphoto of the intertidal platform at Cap Ste-Anne, showing segregation of sandstones into thick amalgamated packets (labelled 1 to 6). Note channellized base of packet 5 (arrows). Beds are upright and dip 80° to the north (top of photo).

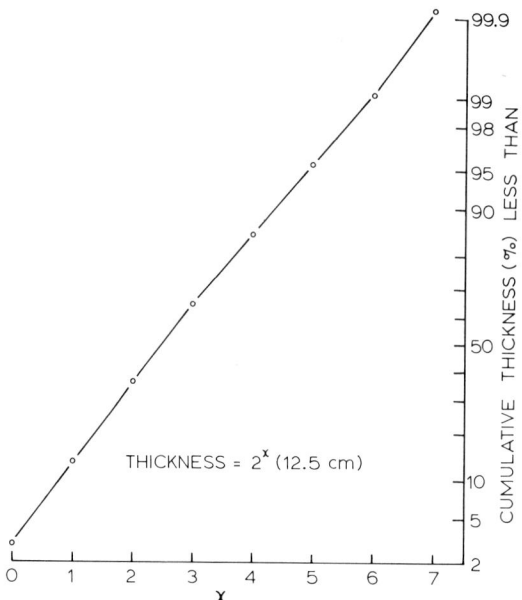

FIG. 4.—Cumulative thickness distribution for all sandstone layers, based on over 4,300 m of measured sections and 1,380 measurements of layer thickness. Thickness is given as a product of (12.5 cm) and 2 rised to the power x. Cumulative percentage is plotted on a probability scale. Note the close approximation to a log-normal distribution.

FIG. 5.—50 cm thick set of cross-stratification in coarse sand at top of sandstone bed, Cap Chat.

tive distribution curve for all sandstone layers in the Tourelle Formation is presented (Fig. 4). These layer-thickness data produce an unskewed, log-normal distribution, suggesting that all sandstone layers belong to the same layer-thickness population. The thinnest layers are classical turbidites (Walker, 1976). If thicker sandstones are characterized by a different depositional mechanism, then there must be a continuum between this mechanism and turbidity-current deposition to account for layers of intermediate thickness. Corbett (1972) and Moore (1973) have applied similar reasoning to suggest a common depositional process for both thin and thick deep-water sandstones.

Depositional Structures

Thick sandstone layers contain two depositional structures that are unusual or absent in classical turbidites. These are large-scale cross-stratification and near-horizontal stratification.

Large-scale cross-stratification occurs preferentially at the top of thick sandstone layers, and contains much better sorted sand than the rest of the layer (Fig. 5). Trough cross-stratification is the rule. Foreset slopes are characteristically asymptotic to bedding at their lower extremity. Figure 6 is a sketch of a form set of cross-stratification preserved on top of a massive, graded sandstone layer. The length to height ratio of 14 is in good agreement with length to height ratios of about 20 for dunes (Costello, 1974). Thickness of cross-strata sets ranges from 15 cm to greater than 100 cm. The 15 to 25 cm sets commonly occur as a coset of 2 to 4 sets, and some exhibit positive construction analogous to ripple-drift cross-lamination. The larger sets are solitary scour fills.

Near-horizontal stratification is the most common depositional structure and occurs in about 25 percent of all thick layers (Fig. 7). The particular type of stratification observed has not previously been described in the literature. Individual stratification bands are from several centimeters to greater than 10 cm thick, with many bands constituting a division of stratification. Each band consists of the following idealized upward progression: 1) a basal horizontal or near-horizontal erosion surface; 2) a subdivision of inverse-graded sand. Grading is from 2 ϕ or 3 ϕ to approximately 1 ϕ. This subdivision may constitute the entire band; 3) a subdivision of massive, -1 ϕ to 1 ϕ sand that commonly displays strong grain imbrication.

Basal scour surfaces in each stratification band may converge laterally due to differential erosion,

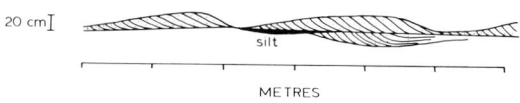

FIG. 6.—Sketch of locally preserved form sets at the top of a massive sandstone bed, section on coast near Romieu-Dalibaire Tp. Line. Note silt lens over lowest cross-stratified scour fill; this silt may have been deposited in the lee of a migrating dune. Grain size is 0-1 ϕ.

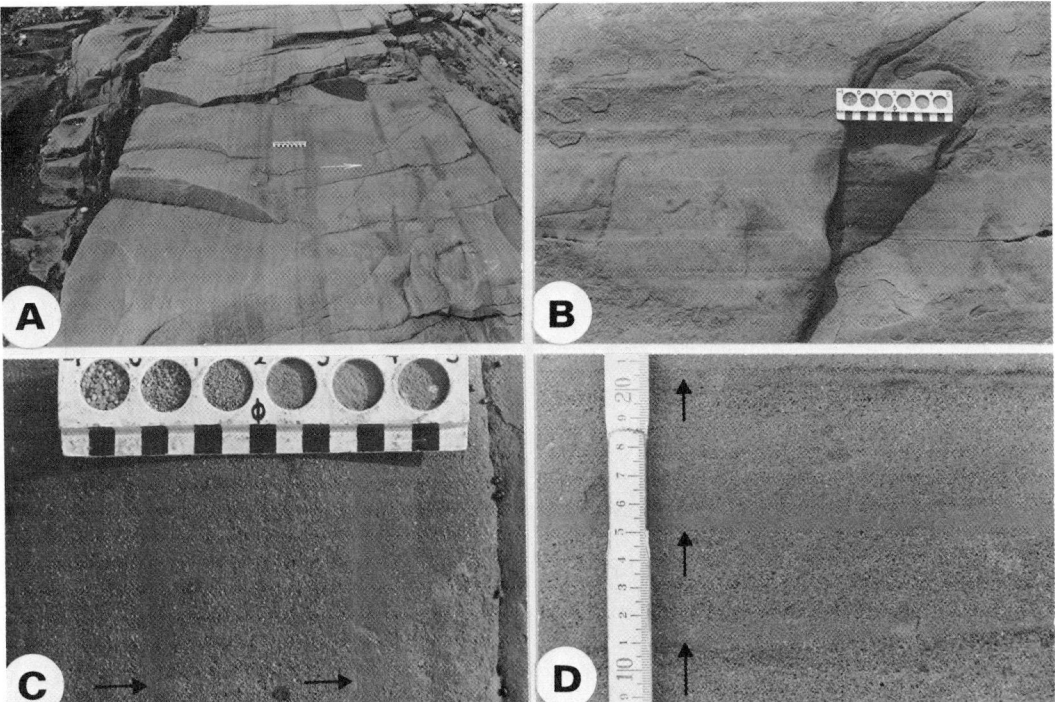

FIG. 7.—Near-horizontal stratification. A) thick graded layer formed almost entirely of unusual coarse stratification, and found at the top of the Cap Ste-Anne channel fill. Stratification bands become thinner with height above base of the layer. Scale 15 cm, top to right. B) coarse stratification, overlying massive sandstone, Romieu-Dalibaire Tp. Line Scale 15 cm. C) close-up of stratification shown in B, displaying basal erosion surfaces (arrows) and internal inverse grading. Scale in cm, top to right. D) another example showing inverse grading above basal erosion surfaces (arrows), West Baie des Capuncins. Scale in cm.

although in most cases this structure appears on the outcrop scale as plane, horizontal stratification. Careful tracing of individual stratification band boundaries in one layer, however, shows the degree to which individual bands can deviate from original horizontality (Fig. 8). The mechanism of formation of this unusual stratification will be discussed in a later section.

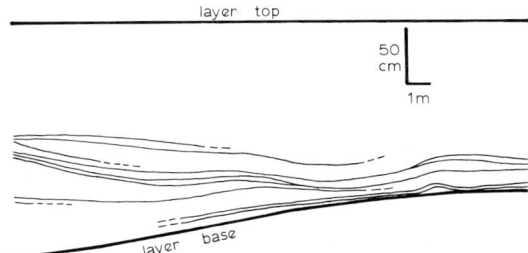

FIG. 8.—Variation in position of near-horizontal stratification surfaces, traced within a single sandstone layer over a horizontal distance of 19 m. Section at coast near Romieu-Dalibaire Tp. Line.

Fluid-Escape Structures

About 40 percent of all thick sandstones contain dewatering structures (Fig. 9). Pillars and sheet structures (Laird, 1970) are the types of escape structure present in the Tourelle Formation. Dish structures were not observed. There is no evidence (such as truncation of pillars by lamination) that these fluid-escape channels formed during deposition. Some fluid-escape pillars extend through more than one layer, requiring post-depositional pore-fluid escape in the lowest layer (Fig. 9B).

Although formation of fluid escape structures may have taken place after deposition of the sand layers, their presence does suggest that sands were emplaced by rapid, mass deposition which trapped pore fluids and produced an unstable grain packing. If this sediment is then disturbed, perhaps as a result of earth tremors or shearing and loading under the weight of later currents, liquefaction may occur. This liquefaction would result in pore-fluid expulsion, compaction, and perhaps some disturbance of depositional structures or

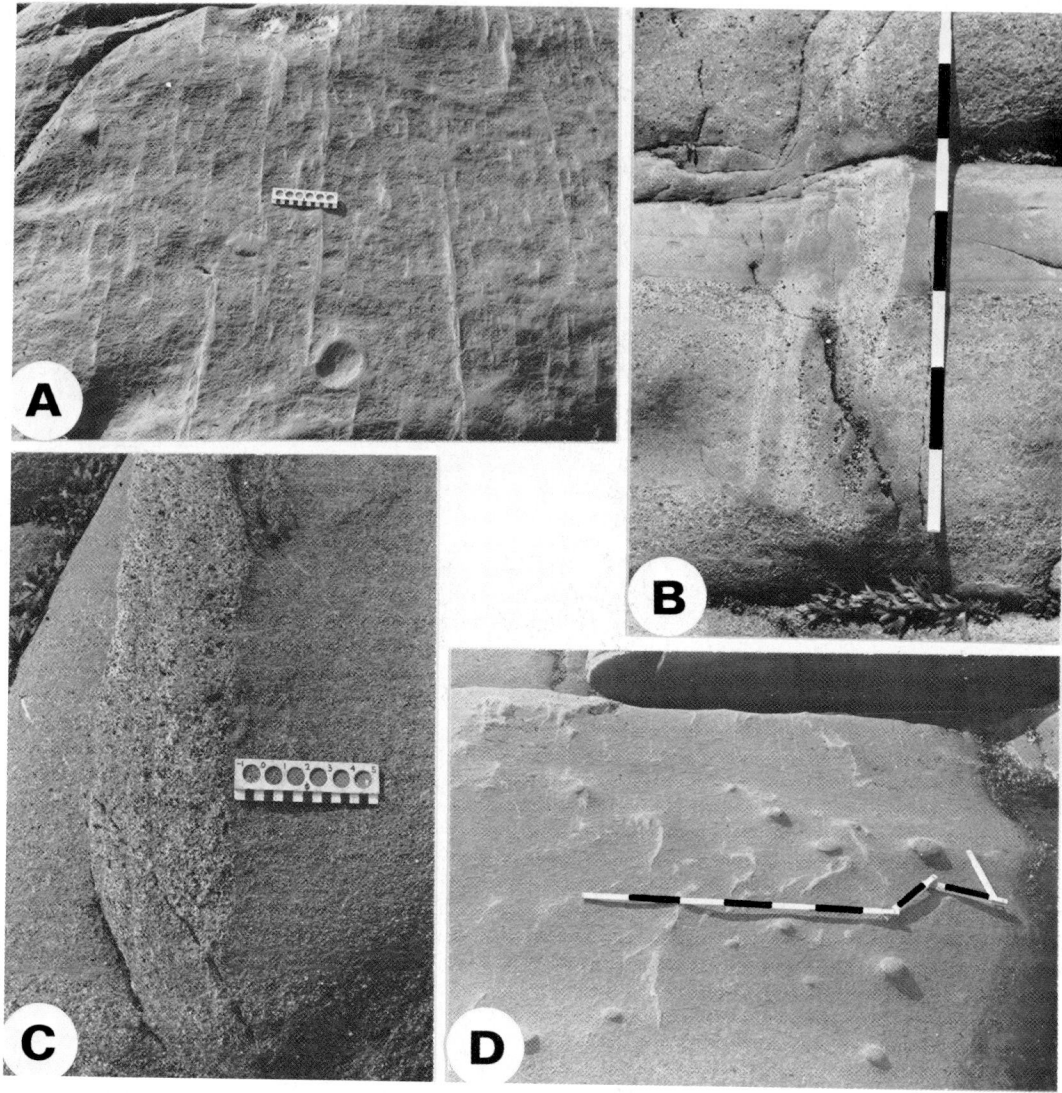

Fig. 9.—Coarse fluid-escape pillars. A) Closely spaced, long straight pillars in a thick sandstone near Ruisseau à Fournier. Scale 15 cm. B) Broad complex pillar which cuts across an amalgamation surface, west of Cap Chat. Scale bars are 10 cm. This is not a clastic dike because most grains are very near their original position of deposition—the major effect has been elutriation of fine-grained matrix. C) Thick pillar, within which coarse granular sand has been transported up from the base of the layer, west of Cap Chat. Scale in cm. D) Sinuous pillars in massive sandstone at Anse à Carlot (west). Scale bars are 10 cm.

fabric. Spectacular loading at the interface between some massive sandstones supports this hypothesis. The loading occurred beneath the weight of the current which deposited the upper layer.

Sole Marks

Tool marks (grooves) are much more abundant than scour marks (flutes). Both grooves and flutes may, however, attain large dimensions (Fig. 10A, B, D). The bases of layers are characteristically irregular, although some very thick layers appear to have completely flat bases. Basal irregularity can have a wave-like geometry reminiscent of beds molded by standing waves in open channels. These features are associated with discontinuous internal scour surfaces (Fig. 11) which are delineated by sharp grain-size variations.

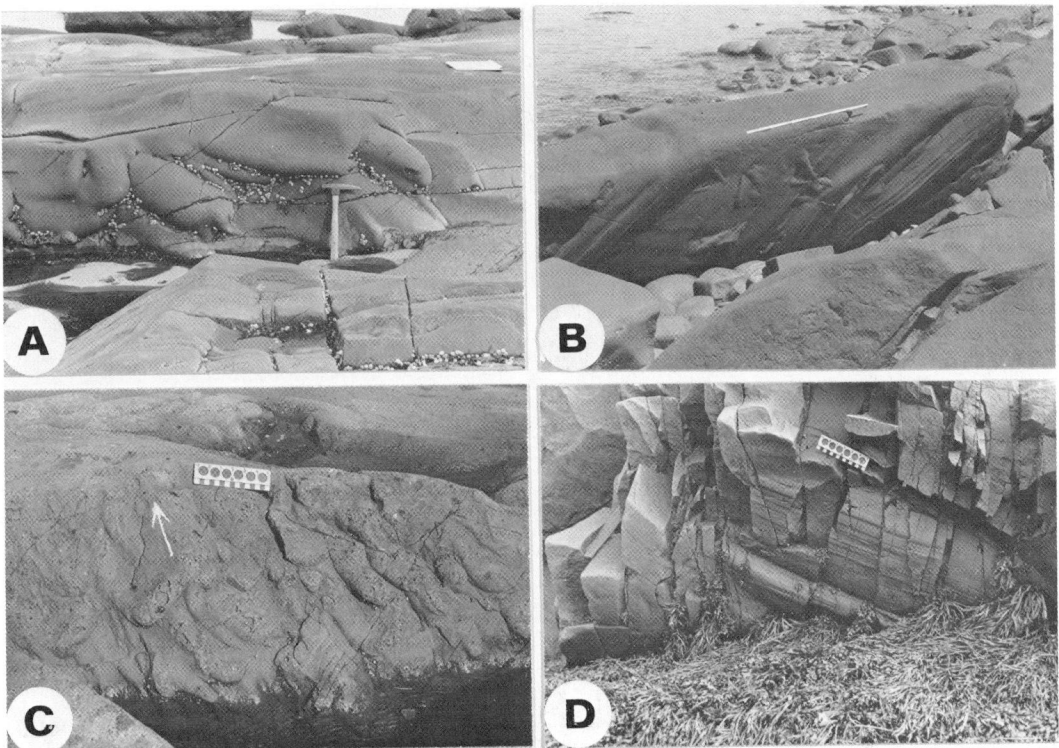

FIG. 10.—Sole marks on thick sandstone layers. A) Large bulbous flutes on the base of a layer 435 cm thick, east of Romieu-Dalibaire Tp. Line. B) Striated grooves on base of a stratified sandstone layer 2.2 m thick, Cap Ste-Anne. Scale 1 m. C) Flutes on a graded sandstone layer 10.5 m thick. The flutes are filled with coarse sandstone containing limestone pebbles. Anse à Carlot (east). D) Large striated groove on base of inverse-to-normally graded sandstone layer, 2 cm thick, Cap Chat. Scale 15 cm.

Outsize Clasts and Included Blocks

Rip-up clasts of shale are quite common in these sandstone layers, and are also present in other coarse, deep-water sandstone units (Nilsen and Clarke, 1975). Of much greater importance, however, are larger blocks of shale, siltstone, and bedded chert that range in maximum size from 1 m to greater than 15 m, and that appear to be suspended in sand (Fig. 12). These blocks were probably plucked from the walls of upslope submarine canyons. Similar processes have been observed in modern canyons (Shepard and Dill, 1966, p. 106–107; Palmer, 1976).

Grading

Normal grading is the rule, with only a few layers displaying basal inverse grading before reverting to normal grading. Modal basal grain size is $-1\ \phi$ to $0\ \phi$. In the upper part of the bed the mode lies between $1\ \phi$ and $2\ \phi$ for layers with truncated tops, and between $3\ \phi$ and $4\ \phi$ for layers with top preserved. A few layers of exceptional thickness (4 to 22 m thick) are essentially ungraded, and suggest a rather unusual depositional process.

The size distribution of three thick sandstone beds was studied in thin section (Fig. 13). One of the beds (1d) showed near-horizontal stratification, and size analysis of this bed showed consistent distribution grading, with size distributions that were either normal, or positively skewed (basal samples) as a result of truncation of the coarse tail. The other two beds were massive. One showed inverse to normal grading, and distributions that were coarse-truncated (positively skewed). The other showed very little variation in grain size over a total thickness of 14.25 m, but although the decrease was so small (0.6 phi) it was consistent from base to top of the bed and therefore it probably has some geological significance. In the two massive beds, petrographic "matrix" increased from about 7–14 percent near the base, to about 18–24 percent near the top of the bed.

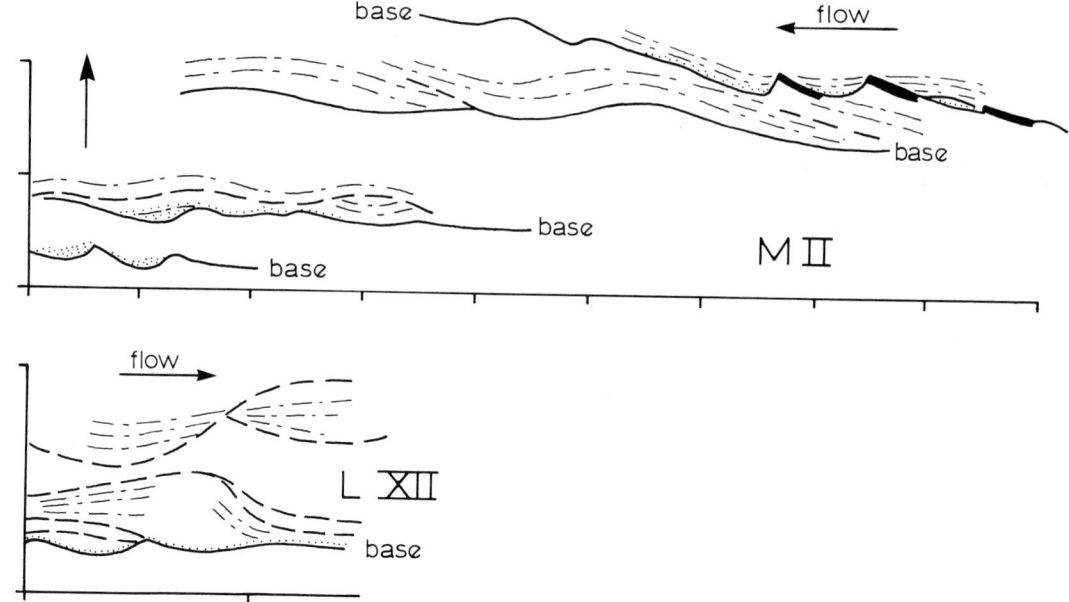

Fig. 11.—Sketch of internal scour surfaces in amalgamated sandstone beds. Layer bases indicated by full lines, light dashed lines indicate faint laminations, heavy dashed lines are internal scour surfaces, stipple indicates concentrations of coarse grains. Anse à Carlot. All scale divisions in meters.

Structure Sequences

In order to analyse and interpret sedimentary structure sequences within individual layers, transitions between internal divisions were studied using a Markov chain model. The method of analysis was similar to that of Miall (1973): details of the calculated transition, probability and difference matrices are given in the thesis by Hiscott. The following states were recognized: scour surface; 1) massive or graded division; 2) internal scour surfaces; 3) upper flow regime plane lamination; 4) ripples or climbing ripples; 5) convolution; 6) large-scale cross-stratification; 7) muddy lamination (Bouma division d); 8) near-horizontal stratification. Results are summarized graphically in Figure 14.

Examination of this figure indicates that the internal depositional structures of thick sandstones may be classified as traction structures, or as structures formed during mass deposition.

Traction Plus Fall-Out Structures.—These correspond to Bouma's divisions b (state 3), c (states 4 and 5) and d (state 7), plus large-scale cross-stratification (state 6). Convolution almost invariably occurs below a scour surface and is not formed by deformation of ripples, but can probably be attributed to liquefaction and flowage caused by shearing and loading of a sand layer by a sediment gravity flow passing over it.

Most large-scale cross-stratification occurs at the top of thick sandstone layers. Several layers containing cross-stratification are generally found in close stratigraphic proximity, suggesting (i) that cross-stratification may be the result of reworking of the tops of massive sandstones by strong currents after deposition, or (ii) that local variation in hydraulic parameters, such as slope, resulted in formation of dunes beneath several successive currents.

Depositional structures formed by traction are found at the top of thick sandstone layers. Ripples, convolution, cross-stratification, and muddy lamination all have a high probability of being succeeded by shale or a scour surface.

Structures Formed During Mass Deposition.—These structures make up the greatest proportion of thick sandstone layers and are succeeded upward by traction structures. Both massive sand (state 1), and massive sand with discontinuous internal scour surfaces (state 2), are interpreted to result from rapid deposition: (i) by fall-out without traction from a dense turbulent suspension, or (ii) by freezing of a sediment gravity flow previously supported by matrix strength, grain interaction, and possibly turbulence.

The Markov analysis clearly groups near-horizontal stratification (state 8) with massive sands, rather than with tractional structures as might have been expected. Near-horizontal stratification is characterized by inversely graded strata

Fig. 12.—Large blocks within thick sandstone layers. A) 30 × 300 cm shale raft in 5.2 thick sandstone layer, Capucins. Top to left. Scale 1 m. B) 2.3 × 25 m dolomitic siltstone block in the top of a 4 m thick sandstone layer, Pointe de la Tourelle. Top to left. Scale 1 m. C) 2.3 × 17 m bedded chert block, which originally projected from the top of a 4.3 m thick sandstone layer. The block settled into the top of the layer some time after deposition of an overlying thin sandstone. Top to right. Pointe de la Tourelle.

lying above internal scour surfaces: inverse grading is generally thought to be produced by grain interaction during deposition from highly concentrated flows, which is consistent with the results of Markov analysis. Two possible mechanisms for the production of this structure will be discussed in a subsequent section.

Fabric

Detailed studies were made of the fabric (dimensional orientation of quartz grains) in ten thick sandstone layers using techniques similar to those described by Parkash and Middleton (1970). These studies will be reported in full in a subsequent paper: the results are summarized below, because they are relevant to the discussion of depositional mechanics that follows.

Thick sandstone layers can generally be classified as (i) stratified and well graded, or (ii) massive and poorly graded, although both massive and stratified divisions may be present in the same layer.

Stratified layers have strong and consistent (both vertically and areally) preferred grain fabric with long-axes parallel to flow and imbrication upcurrent commonly at angles in excess of 20°. In some layers, a-axis orientation is also consistently rotated from sole markings. Vector magnitude has a minimum immediately above the base and again decreases at the top of layers. Within individual stratification bands, vector magnitude is strongest in fine subdivisions.

Massive layers have strongest grain orientations at the base and at the top. Basal samples may be bimodal. The middle portions of these layers are characterized by isotropic or bimodal fabrics. Grain orientation in each layer is generally highly variable, both vertically and areally. Imbrication also shows pronounced variability, and

316 RICHARD N. HISCOTT AND GERARD V. MIDDLETON

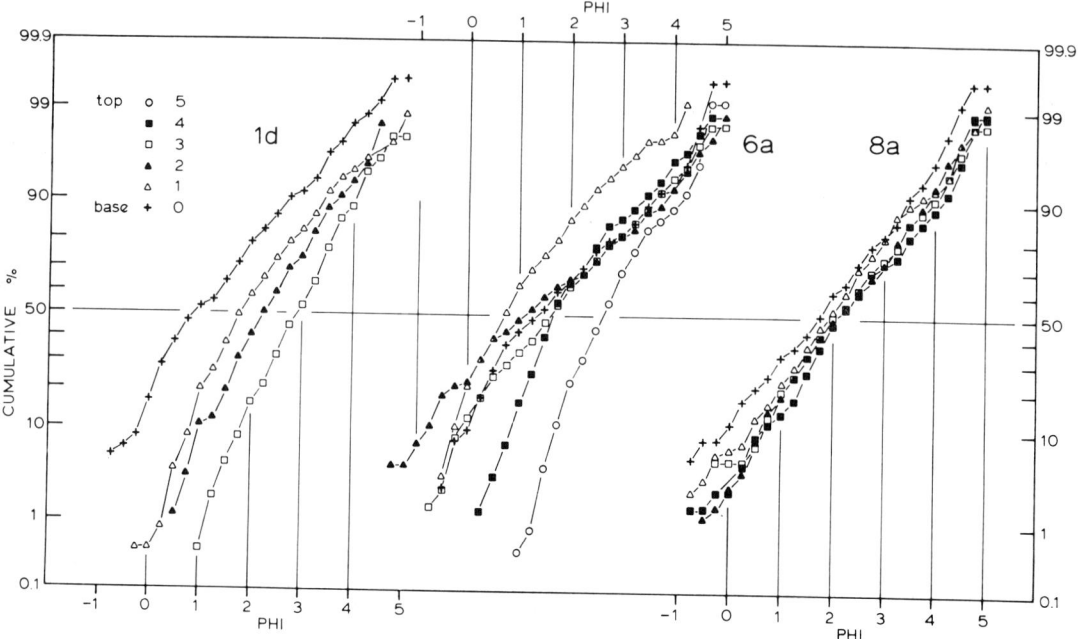

FIG. 13.—Cumulative grain size distributions for three thick sandstone layers. Distributions were determined from thin section and are uncorrected for effect of sectioning. Layer 1d was stratified (only the coarse parts of each stratum were sampled); the other two layers were massive. See text for discussion.

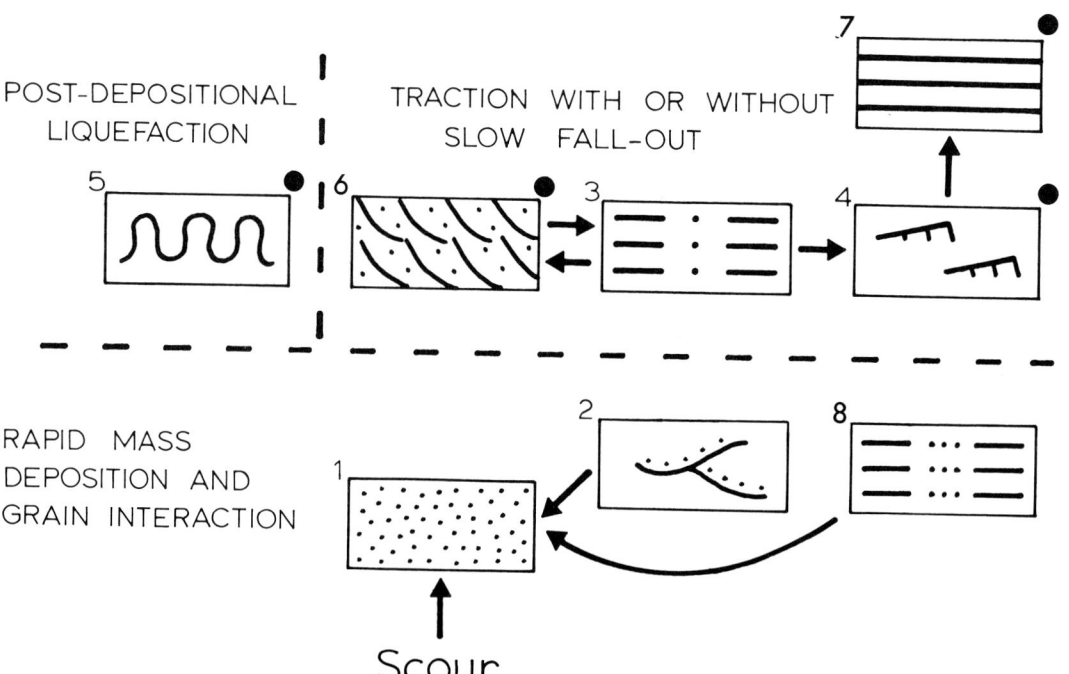

may be variously upcurrent or downcurrent, with downcurrent imbrications being associated with low vector magnitudes.

Bimodality and isotropic fabrics in both stratified and massive layers can be traced to the presence of small regions (micro-domains, a few grains in diameter) with a common long axis orientation. Orientation differs from one micro-domain to another. In bimodal samples, micro-domains are present with long axis orientation both parallel and transverse to flow. Similar clustering in flume experiments was noted by Schwarzacher (1951).

Parkash and Middleton (1970) reviewed theoretical arguments concerning production of primary grain fabric. They concluded that strong unimodal orientations with a-axis parallel to flow can be produced by several mechanisms: (i) grain interaction during deposition from grain flow (Rees, 1968) or debris flow; (ii) slow deposition from suspension (Schwarzacher, 1951); and (iii) deposition from traction in the upper flow regime (Allen, 1964). Flow-parallel, flow-normal, and isotropic fabrics can be expected (i) in debris flow deposits, particularly within the rigid plug; and (ii) in sands deposited rapidly from suspension and quickly buried (Schwarzacher, 1951; Rukavina, 1965).

Imbrication angle for upper flow regime plane bed is generally less than 15° (Allen, 1964; Martini, 1965; Taira, 1976a). Steeper imbrications are associated with rapid deposition from suspension (Schwarzacher, 1951) and high flow velocities (Johansson, 1976), and are common in massive deepwater sandstones (Parkash, 1969; Taira, 1976b). Taira (1976a) showed experimentally that imbrication angles of 20° to 35° are characteristic of grain flow deposits.

Stratified layers in the Tourelle Formation are characterized by steep upcurrent imbrication (11° to greater than 45°), whereas massive layers have extremely variable upcurrent imbrication angles from 0° to 42°. These data strongly suggest that grain interaction was responsible for the depositional fabric of both stratified layers (or divisions) and some samples from massive layers. Inverse grading in stratification bands and at the base of some massive layers also suggests grain interaction.

Bimodal and isotropic fabrics with variable imbrications in the middle of massive layers can be explained by rapid deposition from suspension, or by deposition from a viscous (debris) flow. A choice between these two mechanisms can only be attempted after considering evidence presented in later sections.

Micro-domains, or grain clusters, are interpreted to result from jostling of several elongate grains into parallelism within a dense dispersion under shear. The a-axis orientations within micro-domains might be dispersed about the sample mean to produce an overall unimodal fabric. Both flow-parallel and flow-normal micro-domains would contribute to overall bimodality. The presence of flow-normal clusters would depend on the abundance of grains with transverse orientations within the depositing flow. Transverse orientation is possible in viscous (debris) flows (Jeffreys, 1922; Taylor, 1923) and, arguing on the basis of symmetry, should be second in importance to flow-parallel orientation in all flows.

SLURRY SANDSTONES

This facies consists of beds, generally about 40 cm thick, of fine to coarse sand dispersed in an abundant muddy matrix, with abundant shale chips. The chips are subangular to rounded, generally 1 to 2 cm in diameter and have an apparently random orientation (Fig. 15A). The beds also contain larger shale and chert clasts (some rounded), and slabs of shale and calcisiltite or siltstone up to several meters long are not uncommon (Fig. 15B). The bases of the beds are flat, the tops grade up into shale, and 80 percent of the beds show no internal sedimentary structures. Many of the beds weather down and are easily mistaken for shale. Twenty percent of the beds have rip-up clasts 10 to 100 cm long, and 6 percent have slabs more than 1 m long. Swirled internal structure, distorted fragments, convolute bedding, and pseudonodules are found in some beds.

Wood and Smith (1959, p. 172–173) described "fragmented beds" that showed all transitions into "slurried beds" in the Aberystwyth Grits of Wales. At the base of these beds was a thin graded layer, with larger grains than in the rest of the bed, and above this were numerous clasts of mudstones and graywacke, set in a sandy-silty matrix with irregular coarse sandy patches, without precise boundaries. Burne (1970, p. 221–226,

---

FIG. 14.—Transition diagram, based on Markov analysis, for structural divisions within thick sandstone layers. All arrows indicate transitions significant at greater than the 99% probability level. The various states are: 1) massive or graded division; 2) internal scour surfaces; 3) upper flow regime plane lamination; 4) ripples or climbing ripples; 5) convolution; 6) large-scale cross-stratification; 7) muddy lamination (Bouma division d); and 8) near-horizontal stratification. Black dots indicate that transitions to the next scour surface are significant at greater than the 95% probability level. There are no preferred transitions from states 1, 2, or 8 to tractional structures. Near-horizontal stratification is clearly associated only with structures indicating rapid mass deposition.

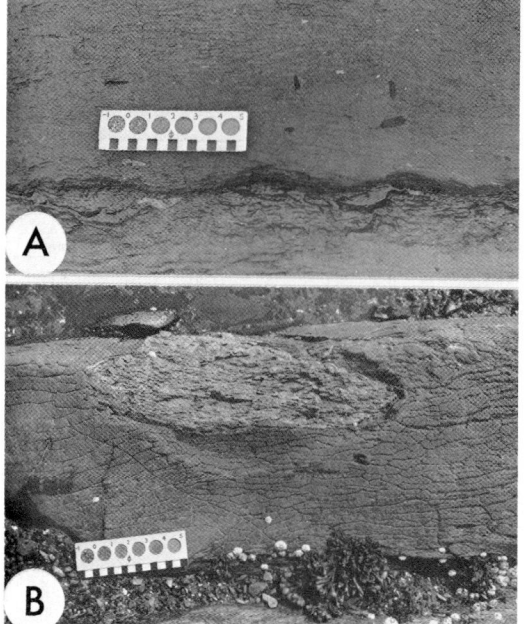

Fig. 15.—Slurry sandstone layers: A) Amalgamation surface between two slurry sandstone layers. Note abundant unoriented shale chips and dispersed coarse sand grains. Scale in cm. Petits Méchins. B) Slurry sandstone with large block of green shale. Scale in cm. Anse à Carlot (west).

see especially his Fig. 8) described "slurried beds" from the Bude Formation (Carboniferous of southwest England) which are very similar to those of the Tourelle Formation. Burne recognized three divisions within his beds: the lowest division is sandier, has few large fragments of mudstone, and grades up into the middle division, which is silty, has large mudstone clasts, fluid escape structures and pseudonodules. Burne recognized an upper thin unit of fine sand or silt, with symmetrical cross-laminations interpreted as formed by sand volcanoes. This unit appears not to be present in the Tourelle Formation—at least, no sand volcanoes were observed. Morris (1971, p. 418) used the term "slurried bedding" as "the general term . . . for all slumped masses of friable sand and varying amounts of mud, not always of the same consistency." The main difference between the beds in the Tourelle and those described by Wood and Smith and by Morris, is that the latter contain sandstone fragments, which are rare in the Tourelle Formation.

Slurry sandstones of the type seen in the Tourelle Formation appear quite common in some other Cambro-Ordovician flysch units in the northern Appalachians, and have been observed in the St-Roch Formation near St-Jean-Port-Joli and in the Humber Arm Supergroup of western Newfoundland. They are also similar in some respects to some of the Type 2 and Type 3 graywackes of the Cloridorme Formation (Enos, 1969, p. 26–27; Skipper and Middleton, 1975). The beds in the Cloridorme also show abundant shale clasts and dispersed coarse sand whose abundance decreases upwards, particularly near the top of the bed. The Cloridorme beds differ in that they are generally thicker, have flutes on their soles, are better sorted near the base, show a well defined sequence of structural divisions in the lower part of the bed, and have a more extensive development of pseudonodules in the upper part of the bed.

Sequence analysis has shown that in the Tourelle Formation the slurry beds occur in near-random fashion. They are not strongly associated with any other facies, and do not occur preferentially in thinning-upward sequences (i.e., in channels) or in thickening-upward sequences. The absence of scouring at the base of the beds, the absence of internal structures produced by traction, the poor sorting, the abundance of matrix and large fragments, and the near-random orientation of shale chips within the bed, lead us to interpret these beds as the result of submarine debris flows formed by the slumping of mud (of varying degrees of consolidation) and unconsolidated sand on a submarine slope (see also Burne, 1970, p. 223–224). Larger slumps probably produced flows that travelled farther out into the basin of deposition. Such flows, which probably entrained more water and consequently became fully turbulent, might have deposited beds of the types seen in the Cloridorme Formation.

DEPOSITIONAL MECHANICS

Introduction

The majority of the sandstone beds in the Tourelle Formation show features that are now generally accepted as indicating deposition from turbidity currents. These include a variety of sole marks produced by scour, grading, and the sequence of internal sedimentary structures first systematized by Bouma (1962). As described above, there are many beds, found in close association with the "normal" turbidites, that show distinctively different features, which suggest that their formation may have involved different mechanisms of transportation or deposition. Before considering the origin of these specific beds it is necessary to review what other mechanisms are theoretically possible.

Considering only flows, not slides or slumps, and only those flows in which mass movement is produced by gravity acting on the sediment (*sediment gravity flows*), Middleton and Hampton (1976) suggested a theoretical classification of flow mechanisms into four types. For a flow to be possible at all, the internal friction of the sediment-water dispersion must be kept low by separating the grains from each other, that is, by raising the grains up against the action of gravity. There seem to be only four possible support mechanisms: *turbulence* (producing *turbidity currents*), *upward flow of intergranular fluids* (producing *liquefied sediment flows*), *grain interactions* (producing *grain flows*), and *matrix-strength* (producing subaqueous *debris flows*). It is emphasized that these are theoretical distinctions, and that in real flows more than one mechanism may operate, either at the same time or sequentially. Only two of the mechanisms have been well documented by observations on large flows in modern environments: turbulence, and matrix strength. In neither case has the documentation come from observations of natural subaqueous sediment gravity flows. Almost all the information comes from observations made on rivers (turbulence) and subaerial debris flows (matrix strength), supplemented by observations on small-scale experiments.

Criteria for Turbulence

As fluid turbulence is by far the most important mechanism for the transport of sediment in both subaerial and subaqueous environments, it is worth reviewing what conditions produce turbulence in a flow, of whatever type. The criterion depends on the rheological properties of the material that is flowing, and two important cases can be distinguished: simple (Newtonian) fluids, and ideal (Bingham) plastics.

For simple Newtonian fluids, such as air or water, the criterion is well-established and is the Reynolds number:

$$Re = \frac{U d \rho}{\mu}$$

where U is the average speed, d is the thickness of the flow, and ρ is the density, and μ the dynamic viscosity of the fluid. For open channel flows, turbulence develops for Reynolds numbers greater than about 2000. For density currents composed of salt water, experiments suggest that the critical Reynolds number is closer to 500 (Ippen and Harleman, 1952; unpublished data of experiments by Middleton). It has been demonstrated experimentally that mixtures of sand and water behave as Newtonian fluids up to concentrations of about 30 percent by volume (Daily and Chu, 1961; Lee, 1969). The presence of sand increases the bulk viscosity of the mixture, by as much as two orders of magnitude, up to values of the order of one poise (Lee, 1969). Even if very high concentrations are assumed (corresponding to flow densities of about 1.5 grams per cm^3) application of the Reynolds number criterion indicates that a flow one meter thick would become turbulent at very low speeds, of the order of 20 cm/sec (see also Lowe, 1976). This implies that although liquefaction may be important in generating sand flows on slopes, and is probably important in the last stages of deposition of concentrated sand flows, large sand flows moving down relatively steep slopes will be fully turbulent: in other words, they will generally be turbidity currents rather than liquefied sediment flows or grain flows.

For Bingham plastics, the Reynolds number alone is not an adequate criterion of turbulence. Hampton (1972) has reviewed the experimental evidence that indicates that the criterion for turbulence is both the Reynolds number and the Bingham number:

$$B = \frac{\tau_c d}{\mu U}$$

where τ_c is the strength of the plastic. (The Hedstrom number, equal to the product of the Reynolds and Bingham numbers, is employed instead of the Bingham number by some authors: e.g., Hanks and Pratt, 1967; Govier and Aziz, 1972, p. 212-216). The experimental data are not extensive. They have been replotted on Figure 16: in this figure, after Hampton (1972), the scales have been adjusted to the correct value for sediment gravity flows using thickness of the flow as the characteristic length (rather than pipe diameter). It can be seen from this figure (see also the discussion by Thomas, 1963; Govier and Aziz, 1972, p. 216; Enos, 1977) that, at relatively large values of either Bingham or Reynolds number, a conservative criterion for turbulence is:

$$Re \geq 1000\, B$$

which is equivalent to the criterion:

$$\frac{\rho U^2}{\tau_c} \geq 1000$$

As this dimensionless number does not appear to have a name, we here propose that it be called the *Hampton number*.

The significance of these results can be seen by estimating the probable maximum values of

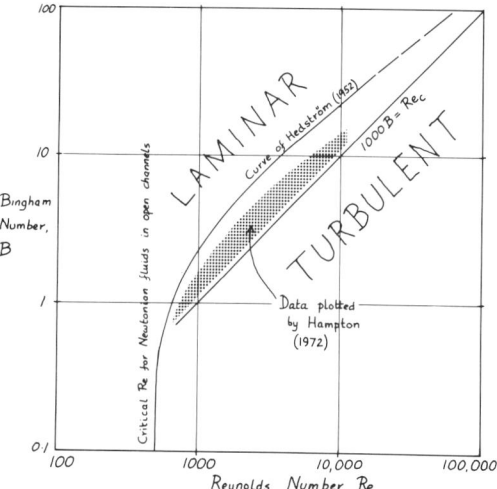

Fig. 16.—Relation of Bingham number to critical Reynolds number for turbulence in a Bingham plastic. Experimental data are for pipe flow but scales have been adjusted to the correct values for two dimensional turbidity currents, using the thickness of the flow as the length scale. From Middleton and Southard (1977, p. 8.16) based on Hampton (1972).

strength and density of natural flows. The only data come from subaerial debris flows. Johnson (1970) has shown that the strength of subaerial debris flows is of the order of 10^4 dynes/cm^2. A flow with this strength, and a density of 2 grams/cm^3, would need to reach an unrealistically high speed of some 20 m/sec before it became turbulent. This prediction is in agreement with direct observations that indicate that even large, fast subaerial debris flows (with speeds up to one meter per second) do not appear to be fully turbulent.

Submarine debris flows may be expected to differ from subaerial flows by including more water, the water being derived both from the high water content of the mud that slumped or liquefied to produce the flow, and from mixing of seawater into the flow as it moves. A small increment of water in muddy debris may produce a large reduction in strength (Johnson, 1970): therefore the possibility exists that some large subaqueous debris flows may become turbulent. Full assessment of this possibility requires data on the strength of debris-water mixtures, and these data are at present scarce.

Slurry Sandstones

These beds are believed to have been emplaced by a debris flow mechanism. Evidence for this is largely negative: the absence of any indication that the flow was turbulent, such as scouring of the mud below the bed, or development of internal sedimentary structures indicating traction (except, in some cases, at the top of the bed, where they could have been produced by the movement of an associated, turbulent, entrained layer of fluid). Positive evidence is the presence of abundant muddy matrix and the fact that the large slabs of sediment observed within these beds might more easily have been supported by matrix strength than by turbulence.

Johnson (1970) has suggested two different ways in which the strength of a debris flow may be estimated: (i) from the thickness of the flow, (ii) from the maximum size of clasts supported by the flow. The first method is based on the fact that a debris flow can move only so long as the shear stress at the bed ($\tau_o = \gamma dS$, for two-dimensional flows) is greater than the strength of the debris. In the case of a subaqueous debris flow γ is the submerged specific weight of the debris, and S is the slope. Although γ is not known precisely, it can be estimated to considerably better than an order of magnitude: it must generally be in the range of 500 to 1300 grams/cm^2 sec^2. For slopes typical of the upper parts of submarine fans (0.01) this implies that in order to deposit a bed 40 cm thick, debris flows must have had a strength of the order of 5×10^2 dynes/cm^2. Similar calculations have been published by Gonzales-Bonorino and Middleton (1976).

The second method is more difficult to use, because it depends on the submerged weight of the clast in the flow. In the case of slurry sandstones, the largest blocks were certainly not fully consolidated at the time the flow took place, and it is difficult to estimate their density accurately. As there is also no way in which the density of the matrix can be accurately estimated, a large degree of uncertainty is introduced into the calculation.

The value of 500 dynes/cm^2, which is an order of magnitude less than the strength of many subaerial debris flows, seems not unreasonable for these submarine debris flows. Using the Hampton number criterion, this implies that these submarine debris flows should not become turbulent below speeds of about 5 m/sec. It is probable that speeds were well below this limit. Reduction of the strength by another order of magnitude, by further mixing with water, would permit similar but somewhat larger and more dilute flows to become turbulent at speeds of the order of 1 m/sec.

Massive Thick Sandstones with Large Blocks

A few beds in the Tourelle Formation are massive, contain large, apparently indurated blocks of shale, siltstone or chert, show poor or inverse-to-normal grading, and variable grain orientation and imbrication. Flutes are uncommon on the base of massive thick sandstones, and are not known on those beds that contain large blocks. Is it possible that these beds also were deposited by non-turbulent sandy debris flows?

Petrographic analysis indicates that many of the thick sandstones contain abundant (20 to 30 percent) matrix. The following evidence suggests that much of this matrix was primary (original mud) and not formed by diagenetic breakdown of rock fragments and feldspar:

(1) Cross-stratified divisions and fluid-escape pillars are expected to have had a low content of primary matrix due to winnowing or elutriation of fines. Sand within both of these structures generally contains much less than 10 percent petrographic matrix, and is calcite cemented. Immediately adjacent sand may contain as much as 25 to 30 percent petrographic (chloritic) matrix, and little if any calcite cement. The matrix-rich sands, however, are not relatively depleted in feldspar or lithic grains, as would be necessary if their matrix were formed diagenetically by alteration of such grains. These sands were not cemented by calcite because available pore space was filled with a primary mud matrix.

(2) The maximum variability in the total of unstable volcanic fragments and feldspar is only 7%, regardless of matrix content. A comparison of percent unstable components (all rock fragments plus feldspars) with matrix content indicates that samples containing 30% petrographic matrix only contain about 10% fewer unstable components than do matrix-free sandstones. For this reason, diagenetic growth of matrix at the expense of unstable grains cannot be responsible for more than 10% matrix in any sample, and may account for much less than this maximum value. All other interstitial chloritic material (10 to 15 percent) must, therefore, be recrystallized primary detrital matrix.

Rodine and Johnson (1976) present a graphical procedure to calculate the volume of void space which must be exceeded and filled with cohesive matrix before grain interlocking ceases and debris flow can begin. During flow, the entire mass behaves as a dense viscous fluid. The graphical procedure requires plotting of cumulative grain size on Rosin's Law probability paper: to do this for the Tourelle sandstone samples, for which grain size was determined in thin section, the data were first adjusted using the empirical curve of Friedman (1958). The analysis indicated that only 8 to 16 volume percent cohesive matrix (clay + water) would have been necessary for these Tourelle sands to have been capable of flow. Thus all samples of massive beds (and also one of a slurry sandstone) that were analysed contained sufficient matrix to expand the sands so that they could flow.

This analysis does not prove that some massive beds were deposited by debris flow, but only that such flow was possible. The debris flow hypothesis is particularly attractive for layers containing large blocks of bedded siltstone and chert. By making use of the two methods outlined above, and by making various assumptions about the original density of the blocks and the flow, it is possible to estimate the strength of the debris. Estimates were made for three of the largest blocks observed: (1) a block of bedded chert 230 cm thick, in the top of a sandstone bed 420 cm thick, (2) a rounded block of indurated siliceous siltstone, 100 cm thick, within a sandstone bed 200 cm thick, and (3) a block of dolomitic siltstone 230 cm thick within a sandstone 400 cm thick. The estimates are shown in Figure 17.

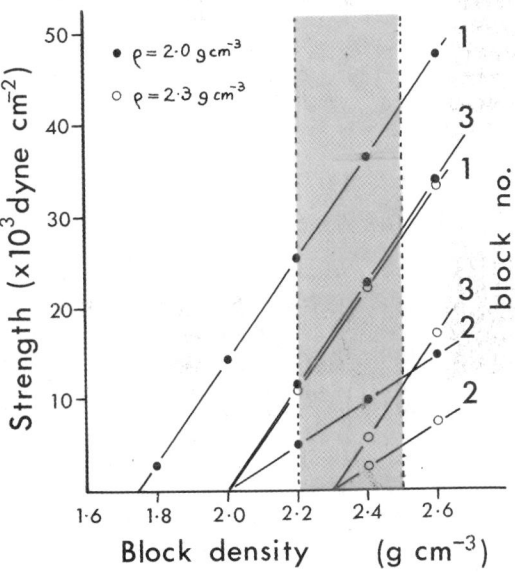

FIG. 17.—Estimates of debris strength as functions of flow and block densities, for three large blocks. Strength is given by $(h/4)\,[\rho_s - (\rho/n)]\,(g)$ where ρ_s and h are the density and thickness of the block, ρ is the overall density of the flow, and n depends on the original depth of penetration of the block into the top of the flow. If $1/4$ of the block is submerged, $n = 4$. See Johnson (1970) for details. Because all blocks have angular, brittle-fractured boundaries, they were probably at least partially lithified before transport. For this reason, it is probable that their densities were in the range of 2.2 to 2.5 g/cm^3 (shaded in diagram). If blocks were of lower density, they might have been supported mainly by buoyancy, with little or no matrix strength. Note that block 1 originally projected out of the top of the flow.

These estimates of strength may be used to estimate slope, using the equation relating strength to density, thickness and slope, described above. If we consider only those strengths within the stippled area of Figure 17, and if we assume that the depositional slope for Tourelle flows was about the same in all cases, then the slope must have been between 0.02 (for flows with a density of 2.3 g/cm^3) and 0.065 (for flows with a density of 2.0 g/cm^3). These slopes seem rather high.

The strength estimates are large (of the order of 10^3 to 10^4 dynes/cm^2) indicating that the flows would not have been turbulent. A possible alternative is to suppose that the strength was lower (of the order of 10^2 dynes/cm^2) and that some combination of strength and turbulence was responsible for the support of the clasts during transport. As speed decreased just before final deposition, turbulence would disappear but settling of the large clasts would not take place because of the rapidly increasing density, viscosity and strength of the muddy sand as it consolidated. If the clasts were partly supported by turbulence (or some other, unknown, mechanism) and the debris strengths were less than the rather high values estimated above, the estimates of slope would also be reduced to more likely values.

Stratified Thick Sandstones

Most of the thick sandstones in the Tourelle Formation show clear evidence of having been deposited by turbulent flows. This applies particularly to those showing internal stratification and scour surfaces, large flutes, well-developed distribution grading, and good grain orientation and imbrication.

Generally, the largest grains present in the sandstones are quartz or angular chert grains 3 mm in diameter. As these grains were presumably supported by turbulence, it is possible to calculate the shear velocity of the flows, making use of the criterion that suspension of grains requires shear velocities at least as large as the grain's settling velocity (Middleton, 1976). Application of this criterion can lead to only a very rough estimate, because the criterion was not developed for concentrated suspensions, and because the effective settling velocity of the largest grains depends strongly on the concentration of the smaller (and particularly clay-size) grains in the flow. Neglecting the effect of clay, the effective settling velocity, w, may be estimated roughly by using the expression given by Richardson and Zaki (1954) for the effect of concentration on the settling velocity of suspensions of coarse sands:

$$w/w_o = (1 - C)^{2.4}$$

where C is the volume concentration, and w_o is the settling velocity of the grains as $C \to 0$. w_o is about 37 cm/sec for sand 3 mm in diameter. The estimate of shear velocity given by the settling velocity may then be converted to an average velocity by using the usual expression for flow over a rough boundary, using 0.3 cm as a rough estimate of the equivalent sand roughness (k_s) of the boundary. Such an estimate can only be very approximate because it is very difficult to estimate what effect the high concentration of sediment might have on the roughness: the roughness of a plane bed composed of moving grains is not the same as that of one composed of fixed grains of the same diameter. High sediment concentration certainly tends to reduce von Karman's κ. Setting $\kappa = 0.25$ to allow for this, the equation is:

$$\bar{u}/u_* = 5.76 \log(3d/k_s)$$

where d is the depth of the flow. Strictly, d is not the total depth of flow, but the depth from the bed to the level of velocity maximum, because of the effect in a turbidity current of friction at the upper interface. The total depth of channelized flow can be roughly estimated from the thicknesses of thinning-upward sequences in the Tourelle Formation, which are observed to average about 20 meters. Flows which deposited thick beds in the base of channels were no doubt thicker than this, but how much thicker is difficult to judge. If the flows were supercritical (as the calculations suggest) there would also be considerable friction at the upper interface, so the depth from the bed to the maximum velocity level would be much less than the total depth. As a first guess, estimates are calculated for values of d from 10 to 20 meters (Table 1).

The shear velocity can be related directly to the slope by the equation

$$u_* = \sqrt{g'dS}$$

where $g' = \Delta\rho g/\rho$. The slope values (Table 1) range from 0.01 to less than 0.0005. The slopes of the mid regions of modern fans are in the range of 0.01 to 0.001. Estimates of shear velocity, density and depth that lead to estimates of slope lower than 0.001 are probably unrealistic: it is possible, for the reasons noted above, that slopes in Table 1 have been underestimated by overestimating the value of d. Estimates of densiometric Froude number indicate that for all reasonable slopes, flows would have been supercritical.

Two hypotheses may now be suggested for the origin of the near-horizontal stratification characteristic of stratified layers:

TABLE 1.—Estimates of u_*, $\bar{u}$, S, and Froude number (Fr) for flows carrying 3 mm sand in suspension and rolling 10 cm arenaceous clasts on the bed. Density (ρ) and flow depth (d) have been chosen to bracket inferred sediment gravity flow characteristics

(1) ρ (g/cm^3)	(2) C	(3) u_* (cm/sec)	(4) d (m)	(5) $\bar{u}$ (m/sec)	(6) S	(7) u_{*_c} (cm/sec)	(8) Fr
1.1	0.059	32.0	10	7.4	0.0115	29.2	2.5
			15	7.7	0.0077		2.1
			20	8.0	0.0057		1.9
1.3	0.176	23.3	10	5.4	0.0024	25.2	1.1
			15	5.6	0.0016		1.0
			20	5.8	0.0012		0.9
1.5	0.294	16.0	10	3.7	0.0008	21.7	0.7
			15	3.9	0.0006		0.6
			20	4.0	0.0004		0.5

Column (3) from $u_* = 37(1 - C)^{2.4}$; column (5) from $\bar{u} = u_* \cdot 5.76 \log (3d/k_s)$; column (6) from $S = (u_*^2/gd)(\rho/\Delta\rho)$; column (7) from application of Shields' criterion to clasts 10 cm in diameter; column (8) from Fr $= \bar{u}/\sqrt{(\Delta\rho/\rho)gd}$.

(i) Individual bands may represent stratification produced by migration and aggradation of long-wavelength antidunes beneath a supercritical flow. According to Hand (1974), the wavelength of antidunes in flows of about 10 m depth would be about 130 m. Scour on the downstream faces of these low features would produce the erosion surface at the base of a stratification band. Inverse grading would be produced by intense shearing of a basal layer of grains over the bed, which would then be buried by periodic rapid fall-out of grains from suspension into the zone of basal shearing. The thickness of bands would then depend on the intensity of deposition from suspension. Lulls in deposition, perhaps resulting from location on the migrating antidunes, would allow a basal scour surface and flowing grain layer to be reestablished.

(ii) Stratification bands may have resulted from progressive "freezing" of the base of the current and upward migration of the sediment bed, as suggested by Carter (1975, p. 159). As competence and capacity decreased, grains would settle to the base of the flow and be driven along by shear transmitted from the overlying turbidity current, producing a flowing grain layer (Sanders, 1965). Inverse grading and high imbrication angle in this layer would be produced by grain interaction. Possibly the high concentration at the base of the flow resulted in the formation of a pseudo-viscous sublayer: assuming effective viscosities about 100 times that of water, and shear velocities of the order of 25 cm/sec, the thickness of such a layer ($12\nu/u_*$, where ν is the kinematic viscosity) would be of the order of one cm. Perhaps within such a layer coarse grains would be maintained in dispersion by grain interaction (dispersive pressure) but fine grains would not. At some critical thickness, it is possible that the entire flowing carpet of grains would collapse and be deposited by "freezing." The entire process would then start again. Each stratification band would preserve the fabric which had developed during shearing above the bed. The thickness of stratification bands has been observed to diminish as mean grain size decreases from the base to the top of stratified layers. This is interpreted to indicate decreasing current velocity throughout deposition, with the thickness of the flowing grain carpet being a function of flow strength.

At any time during the deposition of coarse stratification, prolonged rapid fall-out from suspension would produce a thick massive division. As current velocity and density decreased, tractional structures would begin to form. The most prominent of these are the sets of cross-stratification, found at the tops of some thick sandstone layers. Some of these formed as scour fills, but most probably resulted from the migration of dunes. A dune origin is indicated by the presence of several (in some cases climbing) sets, and by the trough shape of the cross stratification. In one case, a form set showed a length to height ratio of about 14, close to the value of 20 typical of small dunes (Fig. 6). Formation of dunes implies a subcritical value of the densiometric Froude number, and this is consistent with a lower density and shear velocity of the flow that deposited the upper part of the bed, together with a much higher value of the friction factor than that characteristic of a plane bed or a bed covered by low-relief antidunes.

Concretion Conglomerate

At two outcrops, compact bands or broad lenses of imbricated, arenaceous clasts are found at the base of coarse to granule-grade sandstone layers (Fig. 18). These clasts have nominal diameters of about 10 cm, and are similar in lithology to

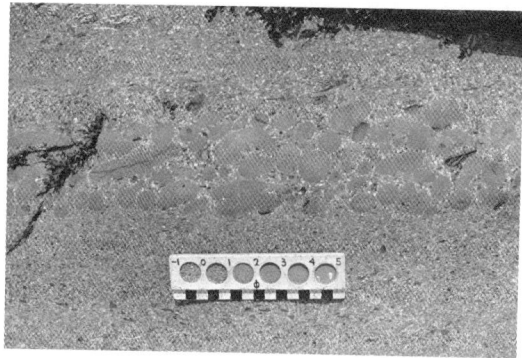

Fig. 18.—Concretion conglomerate with granular sand matrix, resting on an amalgamation surface, west of Cap Chat. Note shale nuclei in some of the transported concretions. Scale is 15 cm.

(though smaller than) the calcareous concretions common in many of the Tourelle sandstones. It seems probable that the clasts are small concretions which were eroded from sands deposited earlier on the upper parts of Tourelle fans, and moved as bed load to their present position. Somewhat similar resedimented concretions have been described by Osborne (1953) from the Cambrian Charny Sandstone near Quebec City.

The value of shear velocity necessary to initiate rolling of the clasts can be calculated from the Shields criterion, and ranges from 22 cm/sec (for a flow of density 1.5 g/cm^3) to 29 cm/sec (for a flow of density 1.1 g/cm^3). These values are in good agreement with the values of u_* estimated independently from the suspension criterion for 3 mm sand (Table 1).

CONCLUSIONS

General theoretical considerations suggest that, in subaqueous environments, thick, fast-flowing sand-water mixtures will be fully turbulent, and must therefore generally be categorized as turbidity currents, rather than as liquefied sediment flows or grain flows. Large flows of debris containing substantial amounts (more than 10%) of mud may or may not be turbulent, depending largely on the strength of the debris, which in turn depends critically on the amount of water in the flow. Strengths in subaqueous flows should generally be less than those, of the order of 10^4 dynes/cm^2, estimated for subaerial debris flows. For muddy debris with relatively high strengths, the criterion for turbulence is a critical value (about 1000) of the dimensionless number $\rho U^2/\tau_c$, which is here named the *Hampton number*.

Within the Tourelle Formation, two types of sandstone may have been deposited from non-turbulent debris flows: slurry sandstones, and a few thick massive sandstones with poor or disorganized fabric and containing large slabs of indurated chert, siltstone or shale. Debris strengths are estimated to be in the range of 10^2 to 10^4 dynes/cm^2. The larger values imply slopes of the order of 0.01 to 0.05, and may not be realistic if the large slabs originally were of lower density than estimated, or were not fully supported by debris strength.

Most thick sandstone layers, including most massive layers, were probably deposited from fully turbulent turbidity currents. The currents were probably about 20 meters or more thick, had shear velocities of about 20 to 30 cm/sec, corresponding to mean velocities of about 5 m/sec, were supercritical, and flowed down slopes of the order of 0.01 to 0.005. Densities were probably in the range of 1.1 to 1.5 g/cm^3 (perhaps relatively high in the flows that deposited massive beds and relatively low in the flows that deposited stratified beds). The cross-stratified tops to some massive beds were probably deposited from the less dense, subcritical "tails" of some flows.

Slurry sandstones appear to represent a distinct facies type, which in the Tourelle Formation was always deposited from debris flows. In some other formations there is evidence of a transition to beds with some similar characteristics which were deposited from turbulent flows.

Within the thick sandstone facies of the Tourelle Formation, however, there appear to be all transitions from massive, poorly graded beds with large blocks to stratified, well graded sandstones. The latter must certainly have been deposited from turbulent, high density turbidity currents, but the former may have been deposited from dense, non-turbulent, sandy debris flows.

ACKNOWLEDGMENTS

We thank Monty Hampton, Orrin Pilkey, and Roger Walker for comments, and National Research Council of Canada for financial assistance.

REFERENCES

ALLEN, J. R. L., 1964, Primary current lineation in the lower Old Red Sandstone (Devonian), Anglo-Welsh Basin: Sedimentology, v. 3, p. 89–108.
BIRON, S., 1974, Géologie de la région des Méchins: Quebec Dept. Nat. Resources Prelim. Rept. (and map), open file, 12 p.
———, 1972, Géologie de la région de Ste-Anne des Monts: Quebec Dept. Nat. Resources Prelim. Rept. (and map), open file.

Bouma, A. H., 1962, Sedimentology of some Flysch Deposits; a Graphic Approach to Facies Interpretation: Amsterdam, Elsevier, 168 p.
Burne, R. V., 1970, The origin and significance of sand volcanoes in the Bude Formation (Cornwall): Sedimentology, v. 15, p. 211–228.
Carter, R. M., 1975, A discussion and classification of subaqueous mass-transport with particular application to grain-flow, slurry-flow, and fluxoturbidites: Earth Sci. Rev., v. 11, p. 145–177.
Corbett, K. D., 1972, Features of thick-bedded sandstones in a proximal flysch sequence, Upper Cambrian, southwest Tasmania: Sedimentology, v. 19, p. 99–114.
Costello, W. R., 1974, Development of bed configurations in coarse sands: Cambridge, Mass., Mass. Institute Technology Expt. Sedimentology Lab. Rept. 74-2, 120 p.
Daily, J. W., and Chu, T. K., 1961, Rigid particle suspensions in turbulent shear flow—some concentration effects: Cambridge, Mass., Mass. Institute Technology Hydrodynamics Lab. Tech. Rept. 48, 51 + 16 p.
Davies, I. C., and Walker, R. G., 1974, Transport and deposition of resedimented conglomerates: the Cap Enragé Formation, Cambro-Ordovician, Gaspé, Quebec: Jour. Sed. Petrology, v. 44, p. 1200–1216.
Enos, P., 1969, Cloridorme Formation, Middle Ordovician Flysch, northern Gaspé Peninsula, Quebec: Geol. Soc. America Spec. Paper 117, 66 p.
―――, 1977, Flow regimes in debris flow: Sedimentology, v. 24, p. 133–142.
Friedman, G. M., 1958, Determination of sieve-size distribution from thin section data for sedimentary petrological studies: Jour. Geology, v. 66, p. 394–416.
Gonzales-Bonorino, G., and Middleton, G. V., 1976, A Devonian submarine fan in western Argentina: Jour. Sed. Petrology, v. 46, p. 56–69.
Govier, G. W., and Aziz, K., 1972, The flow of complex mixtures in pipes: New York, Van Nostrand Reinhold, 792 p.
Hampton, M. A., 1972, The role of subaqueous debris flow in generating turbidity currents: Jour. Sed. Petrology, v. 42, p. 775–793.
Hand, B. M., 1974, Supercritical flow in density currents: Jour. Sed. Petrology, v. 44, p. 637–648.
Hanks, R. W., and Pratt, D. R., 1967, On the flow of Bingham plastic slurries in pipes and between parallel plates: Soc. Petroleum Engineers Jour., v. 7, p. 342–346.
Hiscott, R. N., 1977, Sedimentology and Regional Implications of Deep-Water Sandstones of the Tourelle Formation, Ordovician, Quebec [Ph.D. thesis]: Hamilton, Ont., McMaster Univ., 542 p.
―――, 1979, Provenance of Ordovician deep-water sandstones, Tourelle Formation, Quebec and implications for initiation of the Taconic orogeny: Canadian Jour. Earth Sci., (in press).
Hubert, C., Lajoie, J., and Éonard, M. A., 1970, Deep sea sediments in the Lower Paleozoic Quebec Supergroup: Geol. Assoc. Canada Spec. Paper 7, p. 103–125.
Ippen, A. T., and Harleman, D. R. F., 1952, Steady-state characteristics of subsurface flow: U.S. National Bureau Standards Circ. 521, p. 79–93.
Jeffreys, G. B., 1922, The motion of ellipsoidal particles immersed in a viscous fluid: Proc. Royal Soc. London, Ser. A, v. 102, p. 161–179.
Johansson, C. E., 1976, Structural studies of frictional sediments: Geography Annales, Ser. A, No. 4, p. 201–301.
Johnson, A. M., 1970, Physical Processes in Geology: San Francisco, Freeman, Cooper and Co., 577 p.
Laird, M. G., 1970, Vertical sheet structure—a new indicator of sedimentary fabric: Jour. Sed. Petrology, v. 40, p. 428–434.
Lajoie, J., Hérous, Y., and Mathey, B., 1974, The Precambrian Shield and the Lower Paleozoic shelf: the unstable provenance of the lower Paleozoic flysch sandstones and conglomerates of the Appalachians between Beaumont and Bic, Quebec: Canadian Jour. Earth Sci., v. 11, p. 951–963.
Lee, D. I., 1969, The viscosity of concentrated suspensions: Trans. Soc. Rheology, v. 13, p. 273–288.
Lowe, D. R., 1976, Subaqueous liquefied and fluidized sediment flows and their deposits: Sedimentology, v. 23, p. 285–308.
Martini, I. P., 1965, The Sedimentology of the Medina Formation Outcropping Along the Niagara Escarpment (Ontario and New York State) [Ph.D. thesis]: Hamilton, Ont., McMaster Univ., 420 p.
Maill, A. D., 1973, Markov chain analysis applied to an ancient alluvial plain succession: Sedimentology, v. 20, p. 347–364.
Middleton, G. V., 1976, Hydraulic interpretation of sand size distributions: Jour. Geology, v. 84, p. 405–426.
―――, and Hampton, M. A., 1976, Subaqueous sediment transport and deposition by sediment gravity flows, in Stanley, D. J., and Swift, D. J. P., eds., Marine Sediment Transport and Environmental Management. New York, Wiley, p. 197–218.
―――, and Southard, J. B., 1977, Mechanics of Sediment Movement: Soc. Econ. Paleontologists Mineralogists Short Course 3, 250 p.
Moore, J. C., 1973, Cretaceous continental margin sedimentation, southwest Alaska: Geol. Soc. America Bull., v. 84, p. 595–614.
Morris, R. C., 1971, Classification and interpretation of disturbed bedding types in the Jackfork flysch rocks (Upper Mississippian), Ouachita Mountains, Arkansas: Jour. Sed. Petrology, v. 41, p. 410–424.
Nilsen, T. H., and Clarke, S. H., Jr., 1975, Sedimentation and tectonics in the early Tertiary continental borderland of central California: U.S. Geol. Survey Professional Paper 925, 64 p.
Osborne, F. F., 1953, Concretion conglomerate in the Charny Sandstone, Quebec: Royal Soc. Canada Trans., v. 47, sec. 4, p. 55–60.

PALMER, H. D., 1976, Erosion of submarine outcrops, La Jolla submarine canyon: Geol. Soc. America Bull., v. 87, p. 427–432.

PARKASH, B., 1969, Depositional Mechanism of Greywackes, Cloridorme Formation (Middle Ordovician), Gaspé, Quebec [Ph.D. thesis]: Hamilton, Ont., McMaster Univ., 238 p.

———, AND MIDDLETON, G. V., 1970, Downcurrent textural changes in Ordovician turbidite graywackes: Sedimentology, v. 14, p. 259–293.

REES, A. I., 1968, The production of preferred orientation in a concentrated dispersion of elongated and flattened grains: Jour. Geology, v. 76, p. 457–465.

RICHARDSON, J. F., AND ZAKI, W. N., 1954, Sedimentation and fluidization: London, Inst. Chem. Engineers Trans., v. 32, p. 35–53.

RUKAVINA, N. A., 1965, Particle Orientation in Turbidites; Theory and Experiment [Ph.D. thesis]: Rochester, N.Y., Rochester Univ., 57 p.

RODGERS, JOHN, 1968, The eastern edge of the North American continent during the Cambrian and Early Ordovician, in Zen, E-an et al, eds., Studies of Appalachian Geology: Northern and Maritime: New York, Wiley, p. 141–149.

RODINE, J. D., AND JOHNSON, A. M., 1976, The ability of debris, heavily freighted with coarse clastic materials, to flow on gentle slopes: Sedimentology, v. 23, p. 213–234.

SANDERS, J. E., 1965, Primary sedimentary structures formed by turbidity currents and related resedimentation mechanisms: Soc. Econ. Paleontologists Mineralogists Spec. Pub. 12, p. 192–219.

SCHWARZACHER, W., 1951, Grain orientation in sands and sandstones: Jour. Sed. Petrology, v. 12, p. 162–172.

SHEPARD, F. P., AND DILL, R. F., 1966, Submarine canyons and other sea valleys: Chicago, Rand McNally and Co., 381 p.

SKIPPER, K., AND MIDDLETON, G. V., 1975, The sedimentary structures and depositional mechanics of certain Ordovician turbidites, Cloridorme Formation, Gaspé Peninsula, Quebec: Canadian Jour. Earth Sci., v. 12, p. 1934–1952.

ST-JULIEN, P., AND HUBERT, C., 1975, Evolution of the Taconic orogen in the Quebec Appalachians: Amer. Jour. Sci., v. 275-A, p. 337–362.

TAIRA, A., 1976a, Grain Orientation and Depositional Processes—Fabric Analyses of Modern and Laboratory Flume Deposits [Ph.D. thesis]: Dallas, Univ. of Texas, p. 234–309.

———, 1976b, Settling Velocity Distributions, Magnetic Fabrics and Sedimentary Structures of the Pleiocene Pico Formation, Ventura Basin, California: Implications for the Depositional Processes of Turbidites and Associated Deposits [Ph.D. thesis]: Dallas, Univ. of Texas, p. 310–365.

TAYLOR, G. L. 1923, The motion of ellipsoidal particles in a viscous fluid: Proc. Royal Soc. London, Ser. A, v. 103, p. 58–61.

THOMAS, D. G., 1963, Non-Newtonian suspensions, Part I, Physical properties and laminar transport characteristics: Ind. Eng. Chem., v. 55, no. 11, p. 18–29.

WALKER, R. G., Facies models, 2, Turbidites and associated coarse clastic deposits: Geoscience Canada, v. 3, p. 25–36.

WILSON, J. TUZO, 1966, Did the Atlantic close and then re-open? Nature, v. 211, p. 676–681.

WOOD, A., AND SMITH, A. J., 1959, The sedimentation and sedimentary history of the Aberystwyth Grits (Upper Llandoverian): Geol. Soc. London Quart. Jour., v. 114, p. 163–195.

SAND TRANSPORT THROUGH CHANNELS ACROSS AN EOCENE SHELF AND SLOPE IN SOUTHWESTERN OREGON, U.S.A.

R. H. DOTT, JR. AND KENNETH J. BIRD
University of Wisconsin, Madison, Wisconsin 53706 and
U.S. Geological Survey, Menlo Park, California 94025

ABSTRACT

Transport of sand across shelves to deeper water still poses questions in spite of the acknowledged importance of submarine canyons as accessways, especially because canyons are little recognized in the ancient record. Eocene strata in southwestern Oregon, U.S.A. contain many small-scale channels to 100 m wide and 25 m deep that acted as conduits of much sand from a sandy littoral and deltaic zone across a narrow shelf and slope to feed deeper marine turbidity currents and other gravity flows, which built deepsea fans.

The middle Eocene Elkton Siltstone Member of the Tyee Formation (500–600 m thick) is transitional stratigraphically from thick-bedded, mid-fan sandy turbidites also of the Tyee Formation beneath to the coal-bearing deltaic Coaledo Formation above. Foraminifers suggest depths of upper bathyal at the base to inner neritic at the top of the generally fine-grained Elkton; megafauna is extremely sparse although trace fossils are common. Some channels are filled with laminated mudstone-siltstone identical with surrounding material. Many other channels, however, are filled with massive to faintly parallel-laminated and rarely graded light-colored sandstone lacking fauna; spectacular mudstone intraclast conglomerate lenses are associated. At least one small channel levee is identifiable. Sedimentary structures in the channel sands suggest gravity-flow transport and considerable post-depositional deformation. Rare thin Bouma T_a and T_{ab} graded beds in the slope mudstones attest to occasional overbanking or levee breaching by gravity flows. Symmetrical ripples and hummocky cross stratification at the top of the Elkton, together with changes in foraminifers, indicate a shoaling trend. In sharp contrast, overlying Coaledo sandstones are coarser, show large-scale cross bedding, much of it contorted, and contain abundant wood, coal, zones of shallow-marine megafossils, and trace fossils. These deposits occur in a series of coarsening-upward cycles that reflect episodic shoreline progradation.

Formerly we interpreted the channels as short delta distributary extensions onto the shelf. Recent investigations suggest channels formed deeper and farther from delta fronts, apparently as an array of sea gullies crossing the shelf and slope toward deeper water where they fed sand to deepsea fans. Modern slopes probably have many such channels that are not resolvable by conventional profiling techniques.

INTRODUCTION

In a study 10 years ago of the Coaledo Formation (Fig. 1), a late Eocene deltaic complex in southwestern Oregon, Dott observed thin-bedded turbidites in the predominantly fine silt-clay sequence of the underlying Elkton Siltstone Member (then known locally as the Sacchi Beach beds; Dott, 1966), the upper member of the Tyee Formation. Inland, the Elkton is seen to grade downward into the underlying part of the Tyee, which consists of thickly bedded sandy turbidites containing foraminifers indicating deposition at bathyal depths (> 200 m). The Elkton contains many spectacular channels filled either with mudstone or sandstone. In sea cliffs near Coos Bay, a channeled sequence is overlain abruptly by thick cliff-forming crossbedded sandstone containing much coarse plant debris and coal seams. The Coos Bay succession, as a whole, was interpreted as prograding bottomset and foreset deltaic muds and fine sands with distal distributary channels grading upward into topset distributary and swamp deposits, respectively. Paleocurrent and facies patterns indicate a northwestward progradation of a well-vegetated, humid, subtropical shoreline (Dott, 1966).

The earlier interpretation still seems basically correct and is consistent with faunal and sedimentological evidence of a regional middle and late Eocene shoaling-upward trend from bathyal to neritic and littoral depths. Subsequent research, however, allows an environmental reassessment of the Elkton and basal Coaledo strata. Bird (1967) showed in a detailed analysis of the microfauna of the pre-Coaledo strata that the Elkton around Coos Bay was deposited at mid-neritic depths, that is, about 100 m on the average, but farther eastward as deep as 300 m. This seems surprisingly deep for sandy prodeltaic sediments, indicating an apparent bathymetric conflict between faunal and sedimentological interpretations. Today significant amounts of sand are not generally found on delta fronts at depths much in excess of 10 or 20 m. For examples, even directly in front of distributary mouths of the Mississippi Delta sandy silt extends to only −40 m (Gould, 1970), and on the sandy Niger delta, a better analogue for the Coaledo, very fine sand extends only

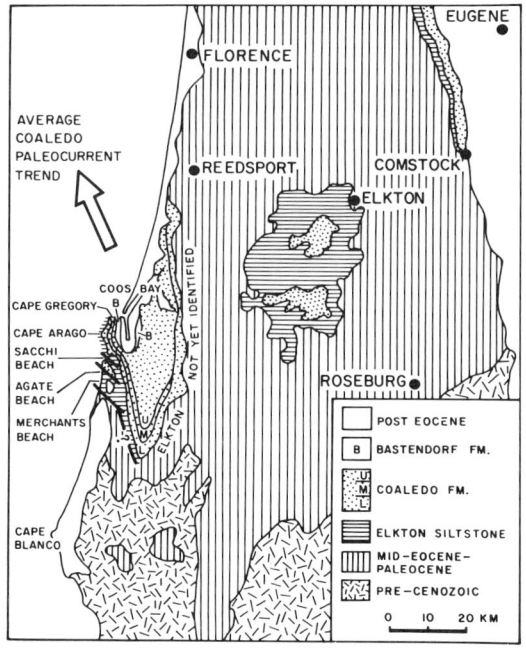

FIG. 1.—Index map of key localities for Coos Bay and inland regions showing approximate distribution of Eocene strata. The lower Tyee Formation lies at the top of the undifferentiated "Mid-Eocene-Paleocene" division.

locally to −40 m (Allen, 1965).

Recent investigations of modern subaqueous fans and fan channels, together with well-documented ancient examples, allow us to reinterpret the Coos Bay strata. Similarities are striking between channels in the Elkton Siltstone Member and proven deepsea fan-channel sands in the Miocene and Pliocene Capistrano Formation in southern California (Bergen, 1971; Piper and Normark, 1971) and Miocene deposits of the Apennine Mountains of Italy (Mutti and Ricci Lucchi, 1972). Also, there is a striking similarity between the Tyee Formation turbidites and documented deepsea fan deposits (Nelson and Nilsen, 1974). Reinterpretation of the Elkton Siltstone as a shelf-slope sequence transected by channel systems that fed deepsea fans resolves the bathymetric dilemma, providing a more plausible setting than Dott's original interpretation as a simple distributary association of graded turbidites, channels, and channel sandstones with parallel stratification.

In this paper we reexamine these rocks in the light of submarine channel and fan models. The vertical sequence is inferred to represent middle Eocene deepsea fan turbidites (lower Tyee Formation), channeled slope and shelf muds (Elkton Siltstone), and littoral and deltaic sands (Lower Coaledo). These three facies generally prograded northwestward within the southern part of a forearc basin during this time. Brief transgressive episodes are recorded in the Middle Coaledo and Bastendorff Formations, which may reflect sudden tectonic subsidence according to Rooth (1974).

There has been some confusion about the stratigraphy of the rocks at Coos Bay. Dott (1966) included the upper part of what is herein referred to as the Elkton Siltstone within the lower sandstone member of the Coaledo Formation because of thick sandstone lenses within the Elkton. We now agree with placement of the boundary at the top of the thick siltstone sequence because the Coaledo's lower sandstones are very different from the thick lenses within the underlying siltstone sequence.[1] Coaledo sandstones are coarser (even conglomeratic), cross bedded, and contain abundant coarse wood debris. Locally, the base of the Coaledo is discordant due to scouring of underlying siltstones. Baldwin (1974, 1975) infers a regional unconformity at the base of the Coaledo, and there is some unconformable overlap toward the southeast margin of Coaledo deposition. We believe, however, that the Tyee-Coaledo succession (Fig. 2) is a progradational one of regional extent, and that this discontinuity was very brief. There is no faunal or petrographic break, and the upward trend of sedimentological changes is not disrupted.

MIDDLE AND LATE EOCENE PALEOBATHYMETRY OF SOUTHWESTERN OREGON

The Eocene sequence of interest in southwestern Oregon comprises three stratigraphic units. The Tyee Formation (predominantly sandy) including its Elkton Siltstone Member (generally fine-grained) is late middle Eocene. The overlying Coaledo Formation is generally sandy, but has a medial shaly member and is entirely upper Eocene (Dott, 1966; Bird, 1967). This sequence of middle Eocene sandstone and siltstone (Tyee Formation) and Upper Eocene sandstone (Coaledo and other formations) is present from the Coos

---

[1] Baldwin has suggested (1974, 1975) two new formations in the interval discussed. These units have been proposed so recently that they have not yet been fully evaluated. For purposes of environmental interpretation, it seems satisfactory and preferable to use the older threefold nomenclature. At least some of the stratigraphic anomalies that Baldwin has noted may reflect diachrony of the litho-stratigraphic units due to lateral facies changes and local channeling at various stratigraphic levels for which there is abundant evidence. According to U.S. Geological Survey policy, the Elkton Siltstone should continue to be regarded as the upper member of the Tyee Formation rather than as a separate formation.

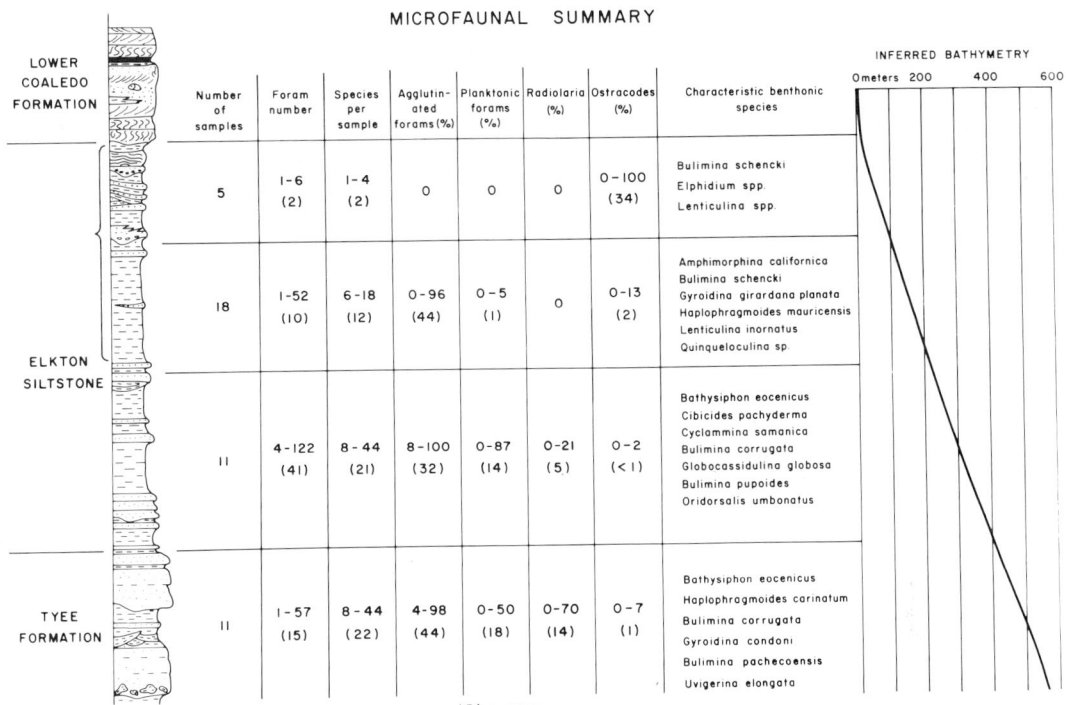

Fig. 2.—Summary of microfaunal characteristics, dominant species, and inferred bathymetry for the lower Tyee Formation and lower Elkton Siltstone Member from inland areas and the upper Elkton from the coastal Cape Argo area (shown by the bracketed interval). Data are from Bird (1967) and this study. "Number of samples" indicates those which generally contained 50 or more specimens.

Bay area northeastward for more than 130 km to the Eugene area (Fig. 1). The upper Tyee microfauna has been studied in numerous places along this trend (Bird, 1967). In all sections studied, an upward-shallowing trend is indicated by microfauna as well as lithologies. Microfaunal data (Bird, 1967) both from inland areas and coastal areas (Coos Bay) have been combined in Figure 2 to show the relationship of the upper Elkton Siltstone at Coos Bay (indicated by the bracketed interval) to the overall succession described here.

A total of 27 samples from measured sections at north Sacchi Beach, South Cove, and North Cove (Figs. 3, 4) were collected for microfossils from the channeled facies and the upper 100 m of the unchanneled facies of the Elkton in the coastal area of Coos Bay. Additional samples from this area have been studied by Rooth (1974) and to the south of this area by Bird (1967). The microfauna was analyzed quantitatively in order to compare it to similar numerical data from studies of Holocene foraminifers. The method of quantification followed uses between 200 and 300 grams of each sample disaggregated and washed on a 200 mesh screen (0.72 mm openings). The washed residue was subdivided via a microsplitter into smaller but statistically significant subsamples. Where faunal abundance permitted, about 200 specimens of foraminifers were counted. Associated ostracodes and radiolaria were also counted, while other microfaunal constituents were recorded only on the basis of presence or absence. Of the 27 samples, 17 were barren of foraminifers and ostracodes (13 of 18 samples were barren in the channeled facies, and 4 of 9 samples were barren in the unchanneled facies).

Microfauna

Microfaunal details in the coastal area are summarized according to their occurrence in the channeled or unchanneled facies of the Elkton Siltstone in Figure 2, revealing considerable difference in faunal characteristics. The channeled facies shows a lower number of foraminifers per gram of sediment (foraminiferal number), number of species per sample, and percentage of agglutinated and planktonic foraminifers. The percentage of ostracodes is higher. In contrast to the lower part of the section inland, no radiolaria were

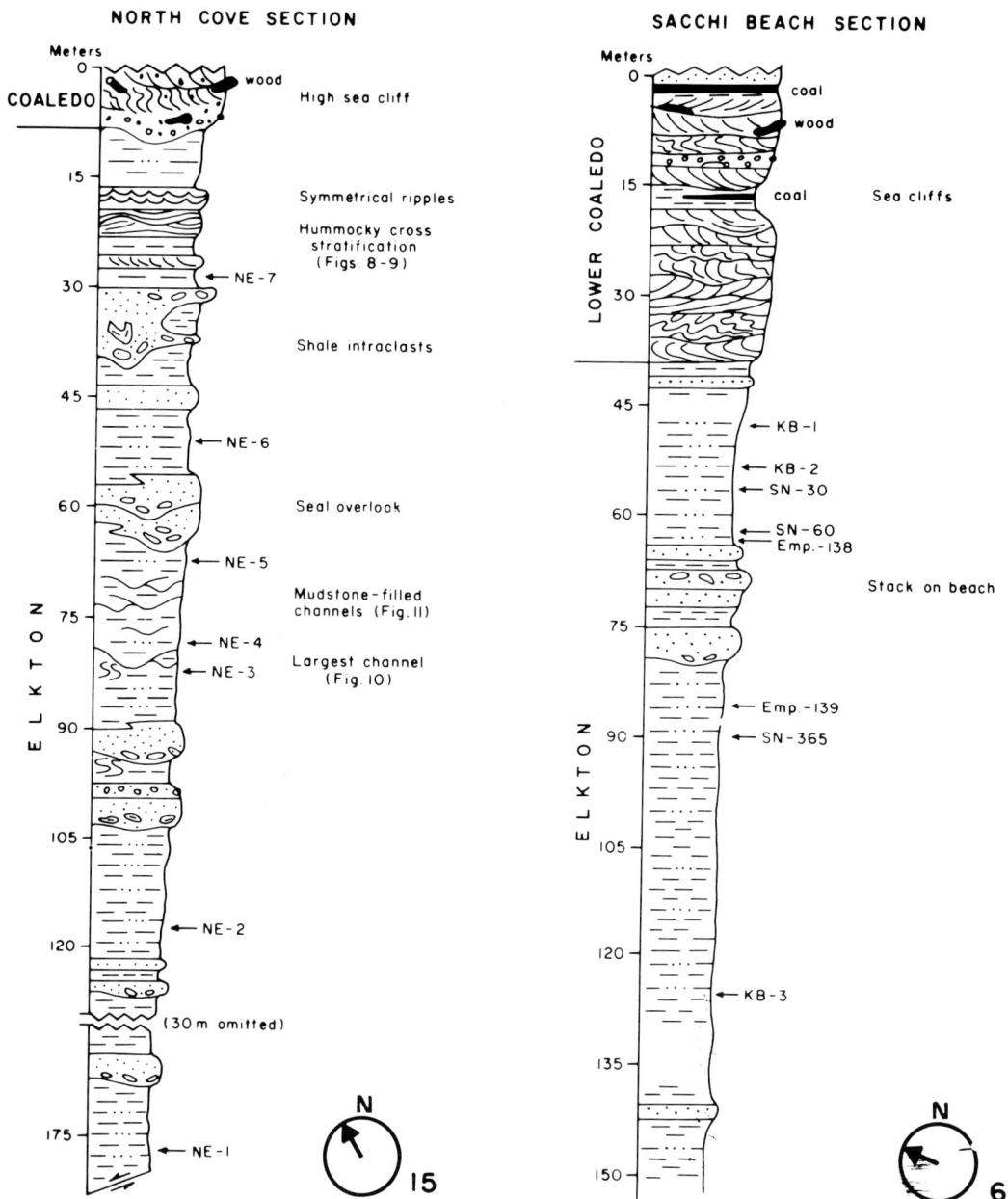

Fig. 3.—Columnar sections of the Elkton Member of the Tyee Formation and the Lower Member of the Coaledo Formation at North Cove on the north side of Cape Arago and at Sacchi Beach showing lithologies and microfossil samples. Note little channeling within the Elkton at Sacchi Beach in contrast with North Cove (compare Fig. 4). Paleocurrent summaries at bottom.

observed. *Bulimina schencki* is the only species which is dominant in both facies.

Bathymetric Interpretation

Depth estimates for the Elkton microfauna are based on comparisons with recent data on general faunal characteristics and upper depth limits of key species.

Quantitative studies of Holocene foraminifers have shown that general aspects of the fauna, such as foraminiferal number, diversity, and radiolarian number, have trends that correlate with

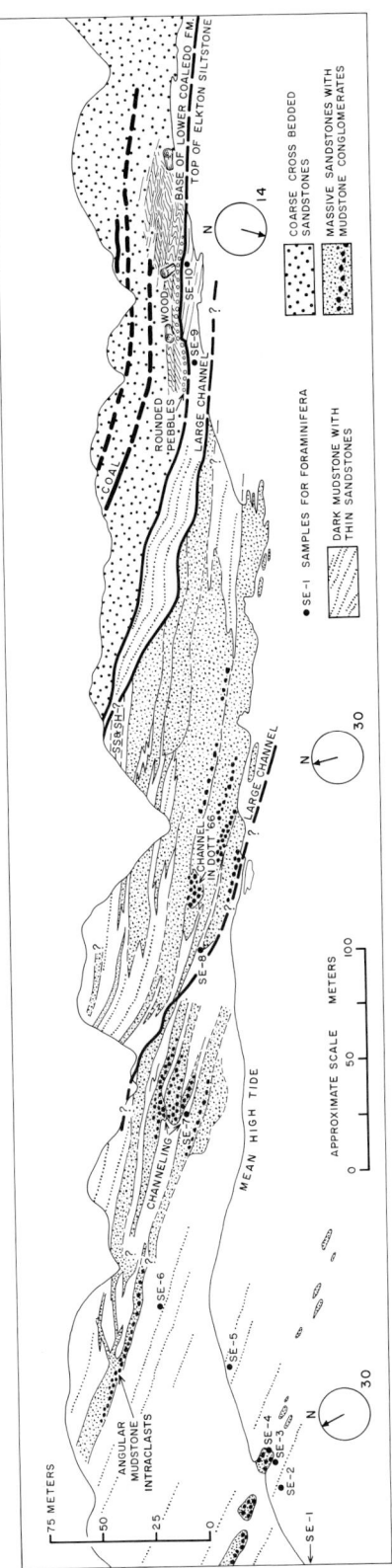

FIG. 4.—Panoramic diagram of sea cliff on east side of South Cove (south side of Cape Arago) showing location of microfossil samples in Elkton Siltstone and complexity of sand-filled channels in contrast with two large mudstone-filled ones; also note basal Coaledo Formation. Both mudstone-filled channels are of the same general dimensions as that at North Cove (Fig. 10). Hummocky and symmetrical-ripple stratification occur between SE-9 and SE-10. Paleocurrent summaries in circles. (Constructed from low-angle oblique air photos and detailed field observations).

water depth in many areas (Anderson, 1963; Bandy and Arnal, 1960; Bandy, 1963; Cockbain, 1963; Fowler and others, 1966; Frerichs, 1970; Pflum and Frerichs, 1976; Phleger, 1964; Uchio, 1960; Walton, 1955). This method of paleoecologic analysis is not dependent upon species composition and, therefore, is useful in analysis of early and pre-Cenozoic faunas where many of the species are now extinct. The following trends have been observed in many areas. The foraminiferal number and the number of species per sample reach maximum values on the outer shelf and upper and middle slope. Planktonic foraminifers and radiolarians occur in relatively low abundance on the shelf, but increase offshore and reach maximum abundance on the lower slope. Porcellaneous foraminifers are most common on the shelf and their greatest abundance occurs on the inner shelf. High percentages of agglutinated foraminifers are found below the calcite compensation depth (CCD) (about 4,000 m present day) and in very shallow marsh and lagoonal environments; although faunas dominated by agglutinated foraminifers may also occur at intermediate depths (Anderson, 1963; Cockbain, 1963). Ostracodes relative to foraminifers are reportedly rare except in lagoonal environments, where they may comprise about 10 percent of the fauna (Bandy, 1963, 1964; Fowler and others, 1966).

Frerichs (1970) has shown in a study of the Holocene foraminifers of the Andaman Sea that significant differences in faunal characteristics may occur at similar depths in different environments. A comparison of an outer shelf near a major delta (Irrawaddy) with an open shelf having no delta (Mergui Platform) indicates that delta-front faunas are generally lower in foraminiferal number, species number, and percent planktonic foraminifers. Percentages of agglutinated and porcellaneous foraminifers as well as species are fairly similar in both environments.

A similar study in the Gulf of Mexico (Pflum and Frerichs, 1976) shows that off the Mississippi Delta, the upper-depth limits of some species are depressed, whereas others are elevated relative to nondeltaic areas. Also in the area influenced by the delta, some species become unusually abundant, some are smaller than normal, and planktonic abundance and trends for benthic specimens per species show lower values.

On general faunal characteristics, the unchanneled facies of the Elkton Siltstone in the Coos Bay area appears to have been deposited at "shelf" depths (probably less than 200 m). This is inferred from the low foraminiferal number, low species number, and low percent planktonic foraminifers relative to parts of the Elkton studied farther northeast (Bird, 1967) where bathyal depths are inferred. The channeled facies may represent conditions even shallower than for the unchanneled facies, or, in light of data from the Andaman Sea (Frerichs, 1970), this facies may represent water depths similar to the depth of unchanneled facies but closer to a delta front.

Individual species are useful bathymetric indicators, assuming that their depth preferences have not changed with time. Because of downslope transport, the lower depth limit of a species is regarded as less reliable than its upper limit even in areas of nonturbidite deposition. Ingle (1967) interprets the foraminiferal fauna at any given horizon within the Pliocene Repetto Formation (former usage) at Malaga Cove, California, as being composed of 95 percent displaced neritic to middle bathyal specimens and only 5 percent indigenous lower bathyal specimens.

Some of the dominant species in the Elkton Siltstone are extinct and have no modern counterparts, and others have modern counterparts with depth ranges too broad to be useful. A most significant species, and one of the most abundant in the unchanneled facies of the Elkton, is *Gyroidina girardana planata*. This species is extinct, but has modern homeomorphs in *Gyroidina altiformis*, *G. cushmani*, *G. orbicularis*, and *G. soldanii*. The observed upper depth limits of these species are generally between 100 m and 200 m (Bandy, 1953, 1961; Frerichs, 1970; Natland, 1933; Phleger, 1960; Uchio, 1960). The shallowest reported occurrence is 59 m for *G. cushmani* in the Andaman Sea (Frerichs, 1970). Bandy and Kolpack (1963, p. 153) report a present-day trend of specimens of *Gyroidina* spp. less than 0.5 mm diameter to be found in middle and upper bathyal depths, whereas specimens larger than 0.5 mm are found in lower bathyal and abyssal depths. All of the Elkton specimens are less than 0.5 mm in diameter. The stratigraphically highest occurrence of *Gyroidina* in the Elkton in this area is near the base of the channeled facies about 50 m or more below the base of the Coaledo Formation. The absence of deep-water benthonic foraminifers, the scarcity of planktonic foraminifers, and absence of radiolarians suggest that this is an occurrence of *Gyroidina* near its upper depth limit.

Other explanations that must be considered for the occurrences of *Gyroidina* include reworking from older deep-water sediments into the Elkton and shallower water depth preference in the past. Neither is probable. A reworked origin is considered improbable because of the restricted occurrence of this genus below the channeled facies, the similarity of its preservation to other elements of the microfauna, and the absence of other deep-water species that should have been reworked along with *Gyroidina*. The association of *G. girardana planata* with other foraminifers,

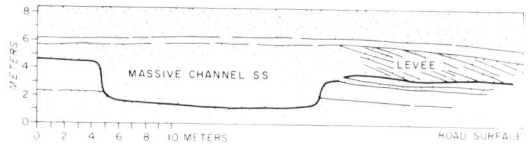

Fig. 5.—Rare channel in Tyee Formation just below Elkton Siltstone Member in roadcut on northeast side of Loon Lake, approximately 20 km southwest of Elkton in Sec. 12, T.23S., R.10W. (see Fig. 1). Like many in the Elkton Siltstone, this channel was scoured in cohesive mudstones; note inclined levee bedding formed by overbanking. Thin sandstone below appears to be an injection.

indicative of bathyal depths in the middle Eocene in the Santa Ynez Mountains (Bandy and Kolpack, 1963), and the occurrence of the homeomorph *G. soldanii*, associated with other foraminifers indicative of abyssal depths in the Miocene of the San Joaquin Valley (Bandy and Arnal, 1960), suggest that the depth preference of *Gyroidina* has not changed appreciably since the Eocene.

In summary, the unchanneled facies of the Elkton has the general faunal characteristics of shelf depths (less than about 200 m) while the abundance of specimens of *Gyroidina* indicate depths no shallower than about 100 m. Faunal differences in the channeled facies may indicate deposition in less than 100 m of water or deposition at similar depths but nearer to a delta.

MIDDLE AND UPPER EOCENE STRATA

Tyee Formation (1500–2000 m Thick)

The middle Eocene Tyee Formation found northeast of Coos Bay represents a well known turbidite succession (Snavely and others, 1964; Lovell, 1969). At the general latitude of Reedsport (Fig. 1), it is typified by thick massive amalgamated micaceous sandy turbidite units, although interstratified zones of mudstone also occur; channels are observed only rarely (Fig. 5). We interpret this phase as mid-fan deposits accumulated at upper bathyal depths. Farther north the Tyee is characterized by thinner graded sandstones and a much lower sand-shale ratio, implying outer fan deposition at slightly greater depths. To the south, however, Lovell (1969) recorded fewer turbidites and some slumps in a position he interpreted as "shelf edge or slope."

Elkton Siltstone Member (500–900 m Thick)

The Elkton Siltstone Member is a distinctive transitional fine interval between the Tyee turbidites below and the upper Eocene sandy paralic sediments typified by the Lower Member of the Coaledo Formation above (Dott, 1966). At Coos

FIG. 6.—Laminated fine sandstone, siltstone and mudstone typical of Elkton channeled phase with sporadic contorted zones and thicker sands with small mudstone intraclasts. Mica and macerated plant detritus also typify this lithology. (Cape Arago area.)

FIG. 8.—Laminated fine sandstones in upper Elkton Siltstone showing hummocky cross stratification of Harms and others (1975). Note continuity of laminae across low-angle antiformal and synformal features, which are about 0.5 m high. (Small cove 1.5 km NE of Cape Arago; also visible at North and South Coves.)

Bay, a section at least 500 m thick of dark gray mudstone and siltstone in centimeter-scale layers is well exposed in sea cliffs and on intertidal benches, especially around Cape Arago (Figs. 3, 4). Scattered beds of fine sandstone up to 10 to 15 cm thick also are present, and become more abundant upward. Fine mudstone intraclast pebbles occur sporadically within these sandstones (Fig. 6). Locally, coarser micaceous sandstones occur in beds 5–8 m thick that generally contain larger mudstone intraclasts; a few also contain pebbles of older rocks. Where exposures are good, these sandstones prove to be large lenses (Fig. 4).

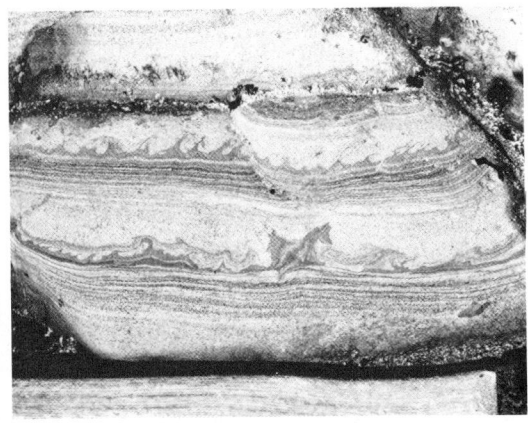

FIG. 7.—Thin graded sandstones with spectacular load structures within generally fine, laminated Elkton Siltstone near Cape Arago (hammer handle provides scale). Such rare units are interpreted as due to overbanking of sediment gravity flows from channels cutting more typical fine-grained shelf-slope deposits.

The typical fine to very-fine sandstones show both parallel and small-scale cross lamination, climbing ripples, and rare convolute lamination (Fig. 6; also Dott, 1966, Pl. 3). These probably represent Bouma T_{cd} units. Diverse load structures and thin zones of intricately contorted stratification are common. Together with rare sandstone dikes up to 0.5 m wide and 5 m long, these features attest to occasional mobility of sediments. Clearly graded beds constitute only a few percent of the Elkton strata, and these are concentrated within the more channeled phases. The grading is of Bouma T_a and T_{ab} types (Fig. 7). Typical fine-grained Elkton lithologies generally were extensively burrowed. The traces are small vertical to slightly oblique tubes averaging 1 cm long, 1 mm in diameter, and spaced about 0.5 cm apart. They penetrate both mudstone and fine sandstone indiscriminately.

In the upper 20 m of the Elkton at Coos Bay, a subtle change of sedimentary structures long puzzled us. Conspicuous rhythmically stratified fine, very micaceous, parallel-laminated sandstones occur in units typically 0.2–1 m thick interstratified with mudstone. Trace fossils are uncommon. Low-angle inclined structures superficially resembling trough cross bedding are common (Fig. 8). A paucity of truncations precludes ordinary cross bedding, and suggests instead some sort of channel fillings. This feature is identical with *hummocky cross stratification* of Harms and others (1975), which is interpreted as a lower shoreface feature produced by storm waves. Less common in the same stratigraphic interval are symmetrical ripples in fine micaceous sands (0.15–0.20 mm). Ripple wave lengths averaged 10–15 cm and amplitudes are about 1 cm. Locally,

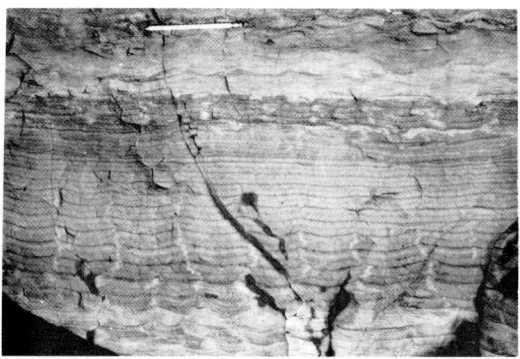

FIG. 9.—Symmetrical ripple cross lamination between zones of asymmetrical (current-modified) ripple cross sections in upper Elkton Siltstone at North Cove, all apparently wave-formed. Note the climbing or standing-wave aspect of the symmetrical ripples, which also show a striking rhythmicity.

dence suggests their formation at depths from 300 m to less than 30 m. Bird measured channels inland from Coos Bay of 4 to 30 m wide and 1 to 5 m deep. Most are filled with the same thinly stratified, generally fine sediment as they these ripples occur in a remarkable climbing or standing wave fashion (Fig. 9), which probably reflects draping over wave ripples by fallout from tidal plumes. Together with the microfossil evidence, these features indicate shallowing during late Elkton deposition at Coos Bay. Analysis of the ripples by a method suggested by Clifton (1976) indicates formation by short-period (probably 2 to 4 second) waves in water no deeper than 20 m.

Channels and Intraclast Conglomerates

Channeling characterizes the Elkton Siltstone throughout its outcrop area, which extends 80 km northeast from Coos Bay. Bird (1967) found that the channels show neither any stratigraphic nor paleobathymetric restriction; microfaunal evi-

FIG. 10.—Largest channel seen in Elkton Siltstone (approximately 100 m wide and 25 m deep). This channel is filled with fine sediments similar to those into which it was cut. (East side of North Cove approximately 0.6 km northeast of Cape Arago).

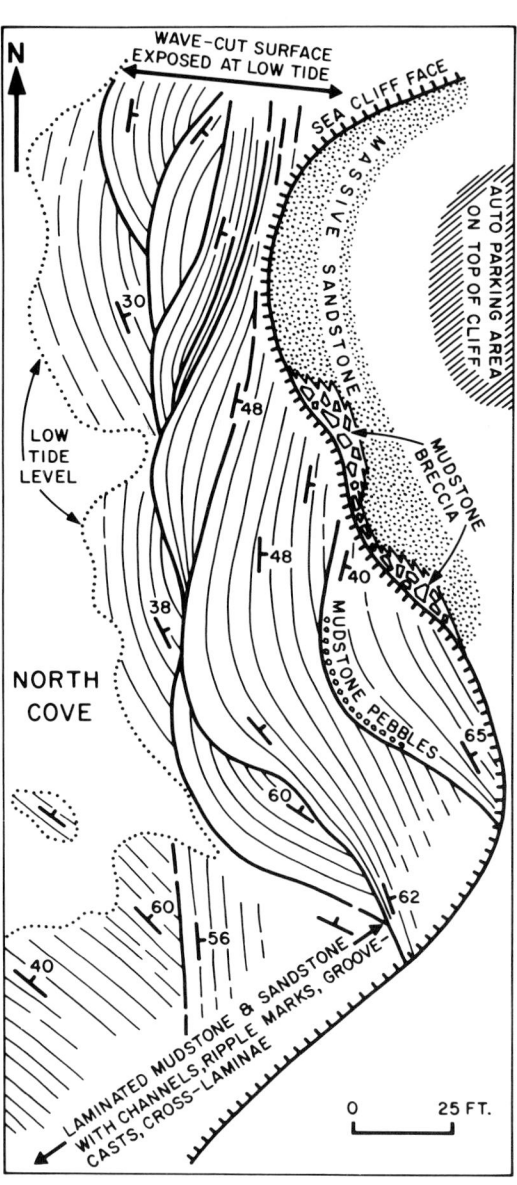

FIG. 11.—Pace and compass map of the northeast part of North Cove below Seal Overlook parking lot, showing a cross section of composite or nested mudstone-filled channels. Just north of area of Figure 10; corresponds with the middle of North Cove section (left column, Fig. 3). Note general draping by thinning with concordance of marginal versus central channel fill.

were cut into. A few, however, are filled with thick sandstone and mudstone intraclast conglomerate.

Channels are conspicuous in coastal exposures at Cape Arago near Coos Bay (Fig. 4), but they are much less abundant 5 km farther south at Sacchi Beach (Fig. 3). Around Cape Arago, the total exposed stratigraphic interval displaying channels in Elkton strata is at least 300 m thick. The largest channel is in the sea cliff just north of Cape Arago; it is 100 m wide and 25 m deep (Fig. 10). Most of the channels have width:depth ratios of about 3:1. Some channels are filled mostly with fine, thin-bedded sediments like the surrounding material, but many exposed in the sea cliffs are filled with medium to coarse, moderately sorted, light-colored sandstone and spectacular mudstone intraclast conglomerate. Channels filled with fine sediments tend to have smooth, low-angle, symmetrical bottoms, whereas ones filled with sandstone and conglomerate generally have much more irregular margins that may even show vertical and overhanging contacts (Figs. 3, 4, 10; also Dott, 1966, Pl. 2A). Complexly superposed or composite channels are the rule (Figs. 4, 11). Channel-axes where discernible in the sea cliffs, are northwest-trending, or about 30°–40° oblique to the present coastline (Fig. 1).

In the more regularly-shaped channels containing fine sediments, the strata tend to overlap and wedge out laterally from the axis regardless of scale of the channel. Even very thin laminae tend to be remarkably persistent and drape entirely across these channels (Fig. 11), suggesting deposition largely from suspension. Although channel fills are broadly lenticular in overall form, parallel lamination is the prevailing internal bedding style (perhaps 60–80%). Scattered layers show micro-cross-lamination, and, less commonly, load structures and contorted zones 1 to 50 cm thick. There seems to be no consistent restriction of these structures to either channel-fill or interchannel deposits.

Sand-filled channels have the same general dimensions as those already described, but are otherwise much more variable. In some transitional cases, sandstone beds from about 10 cm to 1 m thick are separated by laminated mudstone zones of similar scale, producing a sand:shale ratio of about 1. The more characteristic examples, however, contain much thicker sandstones with thin (if any) mudstone interbeds (Figs. 3, 4). Many of the sandstone bodies can be seen to thin and grade laterally into deposits of typical fine Elkton lithology. The sandstones appear massive from a distance, but close-up are generally seen to be very faintly parallel-laminated, especially in the finer sizes; graded bedding is only rarely evident (Fig. 12), and no dish structures

Fig. 12.—Rare graded bedding in coarse channel sandstone at Cape Arago. Note pebbles of quartz and chert as well as larger mudstone intraclasts in this Bouma T_a unit. (South Cove between samples SE-3 and SE-4).

have been observed. Subtle low-angle wedging and truncations are common as are discrete zones of contorted stratification. Evidence of burrowing is evident only as grazing traces on mudstone-sandstone interfaces.

The single most striking feature of the sand-filled channel deposits is the presence of many zones of coarse mudstone intraclasts. These intraclasts are identical with typical interstratified Elkton fine sediments from which they clearly were derived as cohesive chunks torn from newly scoured channel walls (Figs. 3, 4, 13). Contorted mudstone layers below some channel floors also attest to sudden disturbances (Dott, 1966, Pl. 1D). Most intraclasts are elongate parallel to their internal lamination; they vary from 1 cm to 1.5 m long (average 12–20 cm) and are subangular to subrounded. The mudstone conglomerate zones vary in thickness from 1 or 2 cm to as much as 4 or 5 m. Clast packing density varies from a few scattered fragments in sandstone to zones

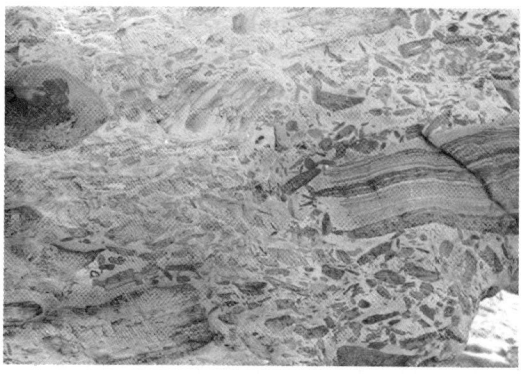

Fig. 13.—Typical coarse mudstone intraclast conglomerate within a sandstone-filled channel at Cape Arago (South Cove). Note lamination within the clasts and variation of sizes and shapes. Large clast at right is 30 cm thick.

that contain about 75% intraclasts. Except in zones of greater density, few clasts touch (Dott, 1966, Pl. 2C). Most of the thicker intraclast zones occur as lenses at scoured bases of sandstone bodies, but some occur higher within those bodies (Fig. 4). In many cases, the boundaries between individual sedimentation units within the massive sandstones are evident only where such a zone separates two sandstones. The conglomerates tend to grade both laterally and vertically into clast-free sandstone, indicating a great deal of amalgamation. No consistent fining- or coarsening-upward, thinning- or thickening-upward trends (Mutti and Ricci Lucchi, 1972), nor systematic lateral channel shifts such as Walker reports at San Clemente, California (1975) are evident here. Instead, the patterns for change within and among channels seem to have been random.

Lower Member of the Coaledo Formation (400–500 m Thick)

The Lower Member of the Coaledo Formation has a greater variety of lithologies than the Elkton, for it includes cross-bedded coarse sandstone and some pebble conglomerate together with finer sandstone, some interstratified mudstone, and coal. Mineralogically, however, these sandstones are virtually identical with those of the Tyee Formation (Dott, 1966). At Sacchi Beach, the base is concordant, appears gradational with the underlying siltstones, and only two thick sandstone lenses are seen within the Elkton (Fig. 3). In contrast, farther north in the vicinity of Cape Arago, the base shows scouring of as much as 4 or 5 m with many sandstone lenses in the Elkton below (Fig. 4). As noted above, this discontinuity is considered to represent insignificant geologic time.

The lower 50 to 60 m of the Lower Coaledo forms a very massive, yellowish-weathering cliff composed entirely of slightly conglomeratic, very coarse to coarse micaceous sandstone with medium- to large-scale cross bedding (average thickness of sets 0.3–0.5 m; Dott, 1966, Pl. 2B). Where the sets were not penecontemporaneously deformed, planar parallel truncations are characteristic. Most of the cross bedding is oversteepened or completely contorted, attesting to rapid deposition in a soft, open-packed state very susceptible to liquefaction by sudden increase of current shear over the bed (Fig. 14; also Dott, 1966, Pl. 4). Besides pebbles up to 5 cm in length, these coarse sandstones also contain abundant plant debris. Whereas macerated plant detritus together with mica is common throughout the entire Eocene sequence, large chunks of altered wood are especially characteristic of the Lower Coaledo. Smaller log fragments have been more or less flattened, but larger ones have not. Many have been thoroughly bored by organisms such as *Teredo*; the borings are filled with sandstone. Channels are not obvious within the basal sandstones, but close examination reveals many shallow scour surfaces.

The bulk of the Lower Member is only accessible north of Cape Arago (especially around Shore Acres, Sunset Bay, and Cape Gregory). This member comprises about a dozen coarsening-upward sequences, each typified by lower mudstone (more or less burrowed), followed by mudstone interstratified with tan, fine sandstone beds containing shallow scours and low-angle bedding that represents hummocky cross stratification. The tops of some of these sandstones are burrowed, and molluscan coquinas occur in some. Rare zones of homogenized sandstone occur that contain deformed laminated xenoliths liquefied after

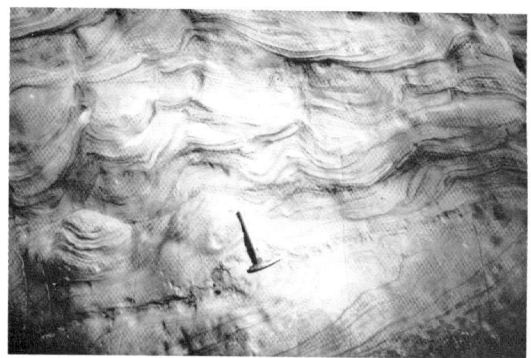

Fig. 14.—Typical deformed cross bedding in coarse sandstone of basal Coaledo Formation at Cape Arago (South Cove near SE-10 of Fig. 4). (Area of view approximately 2 m wide.)

Fig. 15.—Bioturbated very coarse sandstone at top of a coarsening-upward sequence within upper part of Lower Coaledo Formation, 1.2 km south of Cape Gregory (Norton Cove). Stratification was almost completely destroyed by large burrowing organisms; some clam shells are present.

burial. Most sequences are topped by thick beds of brown, coarse, commonly slightly conglomeratic sandstone with prominent, more or less deformed cross bedding like that described below (Fig. 14). The topmost of these sandstones is typically so intensely bioturbated that bedding may be completely destroyed (Fig. 15). The trace fossils include *Ophiomorpha;* shells are common as well. Lignitic to sub-bituminous coal seams also occur within the Lower Coaledo in some areas. They are only a few centimeters thick north of Cape Arago, but are up to a meter and are more numerous to the south near Sacchi Beach (Fig. 3). As many as seven seams are reported a few kilometers inland (Allen and Baldwin, 1944).

The basal Coaledo sandstone sequence is interpreted as representing prograding of delta distributary sands from the southeast into the Coos Bay area. This inference is based upon the lithologic characteristics described above—especially the abundance of wood—together with the lack of marine fauna. *Teredo*-bored wood, however, indicates final deposition within tidal depths. The numerous coarsening-upward sequences in the remainder of the Lower Member are interpreted as resulting from successive progradations of distributary, barrier and lagoon deposits peripheral to the delta complex. Lower Coaledo faunal evidence suggests maximum depths generally in the range 40–80 m (Rooth, 1974).

MODERN DELTA-FRONT SUBMARINE CHANNELS AND FANS

The morphology and sedimentology of modern deepsea fans are now well known (Shepard and Dill, 1966; Shepard and others, 1969; Normark, 1970, 1974; Haner, 1971; Nelson and Nilsen, 1974). While the best-known submarine fans lie beyond the mouths of deep canyons, fans also form in other settings. Seaward of deltas, for example, fans tend to be fed not by canyons but by much shallower channels or gullies that cross the lower delta fronts. According to Shepard and Dill (1966), *sea gullies* as differentiated from submarine canyons and other large valleys, are small channels generally less than 100 m deep (Buffington, 1951). They tend to be straighter and to have fewer tributaries than the larger valleys, and to terminate on the prodelta slopes at widely varying depths. Some are filled with the same fine sediments as flank them, whereas others contain coarser clastic sediments. Pilkey has suggested (pers. comm., 1977) that small channels, probably analogous to those we see in the Elkton Siltstone, are also numerous on the Atlantic upper and middle continental slope in the vicinity of Hatteras canyon, where they transect silty muds. These channels appear to "coalesce into larger entities that are part of the Hatteras canyon system." Cook and Taylor (1977, also Cook, this symposium) argue that many features associated with resedimentation (e.g., small-scale slumping) probably are missed by conventional profiling techniques because resolution of "features less than a quarter of a kilometer long and ten meters thick" is virtually impossible; in the future, new deep-tow techniques may make their detection possible. Thus, counterparts of the Elkton channels may be abundant and characteristic of slopes whether or not deltas are present. Regardless of exact type, any channel can act as a conduit for bypassing sand from the littoral zone to deepwater, leaving predominantly fine, shelf-slope sediments between.

The best-documented example of a complete modern delta-channel-fan system is that in front of the Rhone River delta in Lake Geneva, Switzerland. Descriptions by Shepard and Dill (1966) and Houbolt and Jonker (1968) compare very closely with the Oregon strata. The Rhone delta topset sediments are almost entirely medium and fine sand. At least seven channels or gullies cross the upper Rhone delta foreslope in Lake Geneva, but only one is significantly active today (Fig. 16). It heads beyond the present mouth of the river at a depth of 10 or 20 m, and is about 10 to 12 km long. The channel is about 15 m deep and 200 m wide and has natural levees that are well developed down to about −200 m. Beyond this depth, the channel becomes more and more diffuse, disappearing entirely at −280 m. From about −200 m, a fan-shaped body 10 km long and averaging 5 km wide spreads out beyond the channel and extends to the edge of the deep, central lake plain to −309 m (Fig. 16).

In 1960, Dill dived on the upper Rhone delta, and observed bottom-hugging currents. "Small ripple marks, one inch high with a wave length

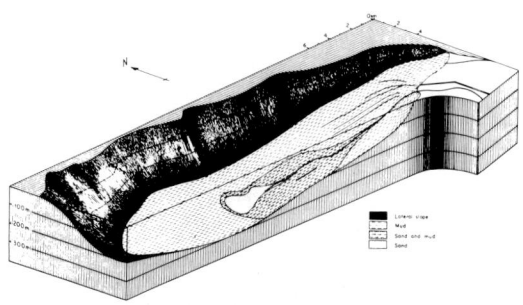

Fig. 16.—Perspective diagram of subaqueous channels and fan of Lake Geneva formed in front of Rhone River delta (right end) (after Houbolt and Jonker, 1968).

of 4 in., were moved along the bottom at an average rate of 0.1 knots" (5 cm per sec) (Shepard and Dill, 1966, p. 255). The cold, muddy river water followed the bottom down to the thermocline at about −30 m, and there "the fine-grained material spread out along the thermocline as a turbid cloud. The coarser sands settled to the bottom, forming ripples that moved downslope" (Shepard and Dill, 1966, p. 256). Within the upper part of the channel between levees, Dill observed migrating asymmetrical sand waves up to 0.5 m high. The sand was very loosely packed and flowed readily when touched.

Extensive coring by Houbolt and Jonker (1968) revealed the character of deeper sediments. Most of the delta foreslope also has some thicker and coarser medium-sand turbidites with essentially complete Bouma sequences. Through most of its length, the channel sand has parallel lamination and graded bedding; small-scale cross lamination is rare. Its levees are composed of silts and very fine, graded sand layers. The central fan consists of horizontally bedded sands with common medium- to fine-grained graded units. These sands are slightly finer and less common than in the channel sands. Small-scale cross lamination and convolute lamination are more common and parallel lamination less common. The fan margin has fine to medium sands interbedded with silt and clay; graded bedding is common. These sediments grade laterally into the normal central plain silt and mud with fewer, thinner, and finer graded sand layers, typically only 2-3 cm thick.

It is clear that cold Rhone River water regularly forms turbidity currents on the delta, and these currents carry fine and medium sands down the delta-front channel to the subaqueous fan as in deepsea fan systems. Given a larger river and delta, seemingly, one could expect correspondingly larger delta-front channel and fan systems.

INTERPRETATION OF THE COOS BAY EOCENE SEQUENCE

Coal and mollusc-bearing pebbly and cross bedded strata define sea level deposition for much of the Lower Member of the Coaledo Formation at Coos Bay, Oregon, whereas microfossils in the lower part of the Tyee Formation indicate deposition at depths greater than 300 m. The intervening mudstones of the Elkton Siltstone Member were deposited at intermediate depths, mostly between 100 and 200 m, with the upper part comprising a sequence dominated by parallel stratification and channeling and a microfauna suggesting depths less than 100 m. Rather than simply being distal delta distributary deposits as postulated before (Dott, 1966), the Elkton channel sands are now thought to have been deposited on a very narrow shelf and slope in deeper water and farther beyond the advancing Coaledo delta within a prodelta channel (or sea gully) and upper submarine fan setting (Fig. 17). The microfaunal evidence, sporadic graded bedding, and predominance of parallel stratification in these rocks all point to deposition below wave base except for the uppermost 20 m or so of the Elkton, which contains wave-formed structures. Narrowness of the shelf helps to explain the movement of sand across a shelf without the presence of a large submarine canyon.

Modern delta-front channel and fan systems provide useful general analogues, which can be applied to the Eocene vertical sequence by invoking Walther's Principle. The main Rhone delta-front channel of Lake Geneva even compares closely in dimensions (15 m deep and 200 m wide) with channels exposed at Cape Arago, Oregon. Some upper Redondo submarine fan channels are also about the same size (Haner, 1971). The least sandy, most thinly laminated phases of the Elkton are taken to be the typical Eocene shelf and slope deposits; they are compatible with the southern Lake Geneva shelf sediments and those documented on many modern ocean shelves and slopes. The lower Rhone channel and upper fan sands are very similar to the parallel-laminated and sporadically graded sandstones in the Elkton Siltstone. Both also contain small-scale cross lamination and convolute lamination. Box cores of graded sands from the La Jolla submarine fan channel (Shepard and others, 1969, Figs. 18, 19, 21) are identical with the thicker graded sandstones contained within the Elkton mudstones; some even have identical-appearing mud intraclasts. Graded sands of the Rhone fan are considered to be analogous to the Tyee's turbidites.

Apparently Eocene channels were cut to depths of at least 200 m by bottom-hugging, gravity-driven currents related indirectly to the advancing Coaledo delta. The mudstone-filled channels either were formed by submarine gravity sliding

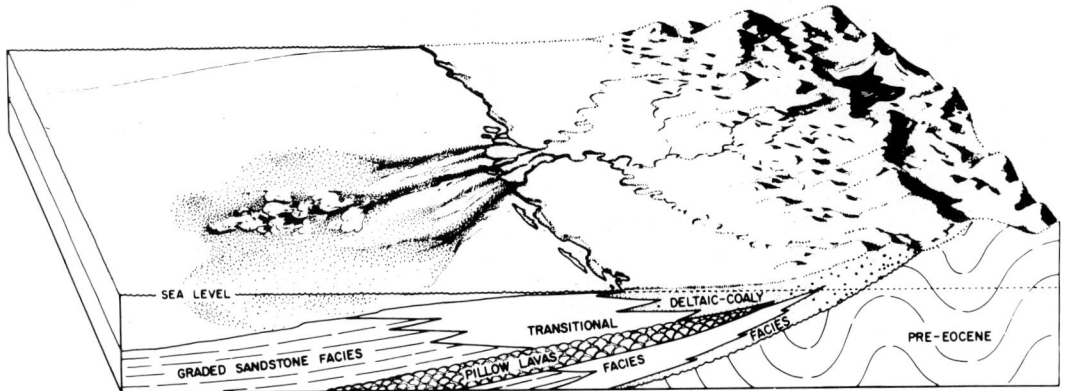

Fig. 17.—Perspective restoration of inferred relations of middle and late Eocene lithofacies in southwestern Oregon showing channels seaward of a delta and cutting across a narrow continental slope to feed deep sea fans; north to left. (Modified from Dott, 1966, Fig. 16).

or were cut by catastrophic flow events, and filled later by relatively slow sedimentation. Apparently these channels were chiefly transportive conduits where we now see them exposed; much sand may have been bypassed through them repeatedly. The sand-filled channels with their characteristic intraclast conglomerates, on the other hand, record both catastrophic scouring and rapid filling in the upper flow regime by turbidity currents and other sediment gravity flows. The cutting and filling may have been separated in time, however. Interbedded mudstone zones attest to periods of slower, fine sedimentation and occasional small gravity flows separating episodes of rapid sand deposition. The sparsity of burrowing in the channel sands and of microfauna in interstratified mudstones also are thought to reflect rapid accumulation.

The composite nature of all channel structures indicates repeated flow events down each conduit. Overbanking by large flows is envisioned as the cause of rare graded sandstones and more common Bouma T_{cd}-type beds within the normal fine Elkton lithology. Some of the margins of channel sand bodies shown in Figure 4 may represent very proximal overbank wedges. Levee morphology has only been recognized in one case (Fig. 5), however, suggesting that it is not very preservable in an identifiable condition. The Eocene channels (e.g., Fig. 4) are presumed to represent the uppermost elements of submarine fan systems. We infer that channel-filling sands were coextensive downslope with Tyee-type thick sandy turbidites deposited on mid-fan or suprafan lobes just beyond the lower ends of the channels. Thinner turbidites formed farther out on the peripheries of such fans. The contrast of thick massive channel sandstones with progressively thinner and better graded nonchannel sandstones is exactly that documented by Nelson and Nilsen (1974) as characteristic of fan systems. The Elkton channel sands with faint horizontal lamination and rare grading are typical "proximal" turbidites that belong to Facies B of Mutti and Ricci Lucchi (1972). The thick Tyee turbidites belong to their Facies C and the thinner turbidites to Facies D. To transport the coarsest sand and fine pebbles found in the channels, bed shear velocities must have reached at least 10–15 cm per sec; mean velocities would have been much greater. Fan sands that could constitute potential petroleum reservoirs are expected to occur in the subsurface offshore from Coos Bay.

The coarse sandstones with considerable contorted cross bedding, which make up the basal part of the Coaledo's lower Member and occur within coarsening-upward sequences in the rest of this member, seem to have their approximate analogue in the upper end of the Rhone prodelta channel where loosely packed sand waves occur. Accordingly, most of the basal Coaledo coarse sandstones are interpreted as distributary and distributary-mouth bar deposits perhaps as much as 2–3 km wide. Bed shear velocities here must have reached at least 20–25 cm per sec. Coal-forming swamps lay between the distributaries, and, as the distributaries, shifted, coal developed over former channel deposits. Presumably waves, tides, and longshore drift modified the delta front as well (Dott, 1966). As the Coaledo delta prograded during early late Eocene time, the various facies overlapped one another northwestward.

Relatively more cross-bedded, coal-bearing topset delta deposits occur in the southern Coos Bay region, and little, if any, coal in the Lower Member north of Cape Arago where there also seems to be more interbedded mudstone associated with at least a dozen coarsening-upward progradational sequences. While shifts of active distributary mouths may explain some of these sequences within the Lower Member, Rooth (1974) has argued convincingly from microfossil bathymetry that crustal subsidence is required to explain larger-scale changes reflected by alternations of thick, fine-grained intervals with sandy and coaly higher stratigraphic zones of the Coaledo and post-Coaledo strata. Rooth found that the fine-grained Middle Member of the Coaledo contains bathyal foraminifers but is succeeded by another shallow marine sandstone, the Upper Member. He concludes that sudden crustal subsidence of at least 300–1,000 m produced transgressions followed by filling and delta progradation. We cannot rule out lesser magnitudes of tectonic subsidence as factors affecting the progradational events recorded within the Lower Coaledo.

Thoroughly documented ancient sediments may offer as good analogues for interpretation of less known deposits as modern sediments. For modern examples, there is an enormous advantage of *knowing* what the environment is, but it is not always obvious what features of the modern case are preservable and which are most diagnostic of that particular environment. Here the outstanding ancient example adds essential insight and also provides a much better view of features too large to be displayed in box cores but too small to be resolved at all (or only as lines and shadows) on conventional subbottom profiles. It is self-evident, therefore, that the best sedimentary models are constructed both from outstanding modern *and* ancient examples. The Eocene strata at Coos Bay, Oregon, are remarkable in that their perfection of exposure in sea cliffs, their great variety of sedimentary structures, and available microfossil bathymetric control combine to offer valuable insights into pro-deltaic shelf and slope deposits. Other notable examples of similar ancient deposits have been described by Walker (1966), Carter and Lundqvist (1975), and McCabe (1978). Comparatively, the well-exposed and thoroughly-documented Miocene Doheny and San Clemente fan-channel deposits in southern California (e.g., Bergen, 1971; Piper and Normark, 1971; Walker, 1975) seem to represent a position that would be intermediate between relatively unchanneled mid-fan sandy turbidites typical of the Tyee Formation and the Elkton channels in Oregon. Whereas practically all Elkton channels were cut into slope mudstones, the Miocene examples show much more nesting of channels within fan-head sands; thus a distinct upper fan channelized facies is not clearly exposed in southwestern Oregon. A complete and idealized succession offshore is expected to show: Coaledo-type deltaic deposits; Elkton-type shelf-slope mudstones with scattered channels; San Clemente-Doheny-like upper fan channeled sandstones; Tyee-type mid-fan, nonchanneled, thick sandy turbidites; and distal, outer-fan thin Tyee-type turbidites. On the other hand, it may well be that in a moderate-sized basin with a narrow shelf and a large sand supply, the distinction of outer shelf, slope, and upper fan deposits may be nearly impossible. Indeed, the large Holocene channel-fan systems of continental rises facing major ocean basins may be misleading as predictors for ancient cases formed in smaller basins.

Accepting that the past can provide important keys to analyzing the present, we suggest that continental slopes have many more small channels than have been recognized, especially in proximity to deltas. In aggregate, these may even account for more total resedimentation of sand to the continental rise than do canyons. The Coos Bay strata also provide an expectation of frequent penecontemporaneous deformation and liquefaction of rapidly deposited shelf-slope deposits.

ACKNOWLEDGMENTS

Our investigations in southwestern Oregon have been supported over the years by grants from funds given by the Wisconsin Alumni Research Foundation to the University of Wisconsin Graduate School. We have profited by discussions with G. H. Rooth, E. M. Baldwin, Samuel Boggs, H. E. Clifton, R. E. Hunter, C. H. Nelson, and Orrin H. Pilkey and the helpful suggestions of A. H. Bouma and W. V. Sliter. Completion of the manuscript was accomplished during Dott's tenure at the U.S. Geological Survey in Menlo Park, California, under a National Science Foundation Science Faculty Fellowship.

REFERENCES

ALLEN, J. E., AND BALDWIN, E. M., 1944, Geology and coal resources of the Coos Bay Quadrangle, Oregon: Oreg. Dept. Geology Mineral Industries Bull. 27, 160 p.

ALLEN, J. R. L., 1965, Late Quaternary Niger Delta and adjacent areas: sedimentary environments and lithofacies: Am. Assoc. Petroleum Geologists Bull., v. 49, p. 547–600.

ANDERSON, G. J., 1963, Distribution patterns of recent Foraminifera of the Bering Sea: Micropaleontology, v. 9, no. 3, p. 305–317.

BALDWIN, E. M., 1974, Eocene stratigraphy of southwestern Oregon: Oreg. Dept. Geology Mineral Industries Bull. 83, 40 p.
———, 1975, Revision of the Eocene stratigraphy of southwestern Oregon, In Weaver, D. W., Hornaday, G. R., and Tipton, A., eds., Symposium on the Paleogene of the Pacific Coast: Am. Assoc. Petroleum Geologists and Soc. Econ. Paleontologists Mineralogists Pacific Sec., 629 p.
BANDY, O. L., 1953, Ecology and paleoecology of some California Foraminifera: Part I. The frequency distribution of recent Foraminifera off California: Jour. Paleontology, v. 27, no. 2, p. 161–182.
———, 1961, Distribution of Foraminifera, radiolaria and diatoms in sediments of the Gulf of California: Micropaleontology, v. 7, no. 1, p. 1–26.
———, 1963, Dominant parallic Foraminifera of southern California and the Gulf of California: Contr., Cushman Lab. Foram. Research, v. 14, pt. 4, p. 127–134.
———, 1964, Foraminiferal biofacies in sediments of Gulf of Batabano, Cuba and their geologic significance: Am. Assoc. Petroleum Geologists Bull. 48, p. 1666–1679.
———, and ARNAL, R. E., 1960, Concepts of foraminiferal paleoecology: Am. Assoc. Petroleum Geologists Bull., v. 44, no. 12, p. 1921–1932.
———, and KOLPACK, R. L., 1963, Foraminiferal sedimentological trends in the Tertiary section of Tecolote Tunnel, California: Micropaleontology, v. 9, no. 2, p. 117–170.
BERGEN, F. W., (chairman), 1971, Geologic guidebook—Newport Lagoon to San Clemente, Orange County, California: Soc. Econ. Paleontologists and Mineralogists Pacific Sec., 88 p.
BIRD, K. J., 1967, Biostratigraphy of the Tyee Formation (Eocene), southwest Oregon [Ph.D. thesis]: Univ. Wisconsin, 209 p.
BUFFINGTON, E. C., 1951, Gullied submarine slopes off southern California: Geol. Soc. America Bull., v. 62, p. 1497.
CARTER, R. M., AND LUNDQVIST, J. K., 1975, Sealers Bay submarine fan complex, Oligocene, southern New Zealand: Sedimentology, v. 22, p. 465–483.
CLIFTON, H. E., 1976, Wave-formed sedimentary structures—a conceptual model, in Davis, R. A., Jr., and Ethington, R. L., eds., Beach and nearshore sedimentation: Soc. Econ. Paleontologists Mineralogists Spec. Pub. No. 24, p. 126–148.
COCKBAIN, A. E., 1963, Distribution of Foraminifera in Juan de Fuca and Georgia Straits, British Columbia, Canada: Cushman Lab. Foram. Research Contr., v. 14, pt. 2, p. 37–57.
COOK, H. E., AND TAYLOR, M. E., 1977, Comparison of continental slope and shelf environments in the Upper Cambrian and lowest Ordovician of Nevada, in Cook, H. E., and Enos, P., eds., Deep-water carbonates: Soc. Econ. Paleontologists Mineralogists Spec. Pub. No. 24, p. 51–81.
DOTT, R. H., Jr., 1966, Eocene deltaic sedimentation at Coos Bay, Oregon: Jour. Geology, v. 74, p. 373–420.
FOWLER, G. A., HUNGER, A. A., AND MANSKE, D. C., 1966, Foraminiferal trends in marginal marine environments of Oregon: Geol. Soc. America Spec. Paper 101 (abs.), p. 72.
FRERICHS, W. E., 1970, Distribution and ecology of benthonic Foraminifera in the sediments of the Andaman Sea: Contr. Cushman Found. Foram. Research, v. 21, pt. 4, p. 123–147.
GOULD, H. R., 1970, The Mississippi delta complex, in Morgan, J. P., ed., Deltaic sedimentation: Soc. Econ. Paleontologists Mineralogists, Spec. Pub. No. 15, p. 3–30.
HANER, B. E., 1971, Morphology and sediments of Redondo submarine fan, southern California: Geol. Soc. America Bull., v. 82, p. 2413–2432.
HARMS, J. C., SOUTHARD, J. B., SPEARING, D. R., AND WALKER, R. G., 1975, Depositional environments as interpreted from primary sedimentary structures and stratification sequences: Soc. Econ. Paleontologists Mineralogists Short Course No. 2, Lecture Notes, 161 p.
HOUBOLT, J. J. H. C., AND JONKER, J. B. M., 1968, Recent sediments in the eastern part of the Lake of Geneva (Lac Leman): Geologie en Mijnbouw, v. 47, p. 131–148.
INGLE, J. C., Jr., 1967, Biofacies trends in the Late Cenozoic of Southern California, in Paleoecology, Short course lecture notes: Am. Geological Institute, Washington, D.C., pages JI-B 3–18.
LOVELL, J. P. B., 1969, Tyee Formation: undeformed turbidites and their lateral equivalents: mineralogy and paleogeography: Geol. Soc. America Bull., v. 80, p. 9–22.
MCCABE, P. J., 1978, The Kinderscoutian delta (Carboniferous) of northern England: a slope influenced by density currents, in Stanley, D. J., and Kelling, G., eds., Sedimentation in submarine canyons, fans, and trenches: Stroudsburg, Dowden, Hutchinson, and Ross, p. 116–126.
NATLAND, M. L., 1933, The temperature- and depth-distribution of some recent and fossil Foraminifera in the southern California region: Scripps Inst. Oceanography Bull., Tech. Ser., v. 3, no. 10, p. 225–230.
NELSON, C. H., AND NILSEN, T. H., 1974, Depositional trends of modern and ancient deep-sea fans, in Dott, R. H., Jr., and Shaver, R. H., eds., Modern and ancient geosynclinal sedimentation: Soc. Econ. Paleontologists Mineralogists Spec. Pub. No. 19, p. 69–91.
PFLUM, C. E., AND FRERICHS, W. E., 1976, Gulf of Mexico deep-water foraminifers: Cushman Foundation for Foram. Research Spec. Pub. No. 14, 125 p.
PHLEGER, F. L., 1964, Patterns of living benthonic foraminifera, Gulf of California, in van Andel, T., and Shor, G., eds., Marine geology of the Gulf of California: Am. Assoc. Petroleum Geologists Memoir 3, p. 377–394.
SHEPARD, F. P., AND DILL, R. F., 1966, Submarine canyons and other sea valleys: Chicago, Rand McNally, 381 p.

———, ———, AND VON RAD, R., 1969, Physiography and sedimentary processes of La Jolla submarine fan: Am. Assoc. Petroleum Geologists Bull., v. 53, p. 390–420.
SNAVELY, P. D., JR., WAGNER, H. C., AND MACLEOD, N. S., 1964, Rhythmic-bedded eugeosynclinal deposits of the Tyee Formation Oregon Coast Range, in Merriam, D. F., ed., Symposium on cyclic sedimentation: Kansas Geol. Survey Bull. 169, p. 461–480.
UCHIO, T., 1960, Ecology of living benthonic Foraminifera from the San Diego, California area: Cushman Lab. Foram. Research Spec. Pub. no. 5, 72 p.
WALTON, W. R., 1955, Ecology of living benthonic Foraminifera, Todos Santos Bay, Baja California: Jour. Paleontology, v. 29, no. 6, p. 952–1018.
WALKER, R. G., 1966, Deep channels in turbidite-bearing formations: Am. Assoc. Petroleum Geologists Bull., v. 50, p. 1899–1917.
———, 1975, Nested submarine-fan channels in the Capistrano Formation, San Clemente, California: Geol. Soc. America Bull., v. 86, p. 915–924.

CATSKILL DELTA SLOPE SEDIMENTS IN THE CENTRAL APPALACHIAN BASIN: SOURCE DEPOSITS AND RESERVOIR DEPOSITS

J. DOUGLAS GLAESER
J. D. Glaeser Associates, Consultant Geologists, 36 Riverside Drive, New York, NY 10023,
and
Geology Department and Marine Laboratory
Duke University, Durham, NC 27708

ABSTRACT

Organic-rich shales at the seaward margin of the Catskill delta have stratigraphic and sedimentary environments similar to those of Cretaceous and younger clastic wedges of North America's eastern continental margin. The Acadian-Caledonian retroarc basin wedge rests upon a shelf carbonate platform (Onondaga) and is comprised of several depositional systems tracts, assemblages of offshore shales and siltstones, nearshore sandstones and alluvial deposits of the subaerial delta plain. Overstepping of the hydrocarbon-bearing source-bed shales deposited at the seaward ends of each depositional systems tract results in enveloping successively younger slope shales with coarser-grained shoreward lithologies. Delineation of both regional and local tracts shows that spatial and temporal arrangements of delta lobes have primary influence upon the fates of the anoxic mud build-ups. In places, bioherm carbonate shoals grade laterally into these slope muds. Ideal reservoir conditions are developed where delta lobe sandstones, siltstones and limestone shoals are juxtaposed to slope deposits. These sedimentary associations have analogies in the Lower and Middle Cretaceous climatically influenced anoxic muds. Both the Catskill delta slope shales and the Cretaceous anoxic muds are hemiplagic having "linear" sources. Differing thicknesses and extents of up-from-the-basin shale tongues are controlled primarily by progradation or abandonment of Catskill lobes. Despite tectonic differences between the Catskill delta slope and the sedimentary wedges of the eastern continental margin, the organic-rich shales maintain their signatures of regional anoxic conditions across much of the Middle Devonian-Lower Mississippian craton. Catskill delta slope development has sufficient numbers of sedimentary analogues to the constructional sequences within the continental slope of eastern North America that the abundant surface and subsurface data from this ancient clastic wedge provide invaluable keys to the continent margin's sedimentary history.

INTRODUCTION

Ancient clastic wedge deltas and their modern counterparts have sufficient similarities to the constructional deposits of continental slopes that close comparison is possible between their organic-rich slope shale source deposits and nearshore coarser detrital reservoir sediments. In fact, along trailing continent margins, deltaic slope muds probably comprise a significant portion of the constructional deposits within the continental slope. A useful approach in defining the time-related components of prograded deltas and their ancient counterparts is to view both the nearshore and offshore deposits as parts of depositional systems tracts (Brown and Fisher, 1977). This approach can best be applied to ancient prograded clastic wedges because of intensive study of their stratigraphic organization in which the processes controlling sediment-body geometry must be inferred. In contrast, the majority of studies of modern delta slopes and continental slopes stresses the observable processes controlling the present depositional framework.

This study focuses upon the offshore marine sediments of the Catskill delta (Middle and Upper Devonian and Lower Mississipian) in the Central Appalachian basin which were deposited on a constructional slope in advance of prograding delta front lobes. The analogy between continental slopes and the prograding fronts of deltas was drawn by Emery (1968) who described constructional continental slopes as those having a depositional origin (Fig. 1). The size and areal distribution of the Catskill delta is such that a number of analogies can be drawn between it and the continental slope of eastern North America. To be sure, the ancient Catskill delta was built entirely upon shelf and cratonic sedimentary sequences well beyond a major juncture between oceanic and continental crust and, therefore, differs tectonically from the continental slope of eastern North America.

Organic content commonly is inversely related to grain size of terrigenous detritus so that the fine silts and clays on constructional slopes contain abundant unoxidized organic matter. As Emery (1968, p. 1381) suggested, fine sediments carried beyond the slope probably become increasingly oxidized because of slower deposition of hemipelagic muds. Shelfward, they contain lesser amounts of organics because of both grain-size increase and likelihood of oxidation. The bulk

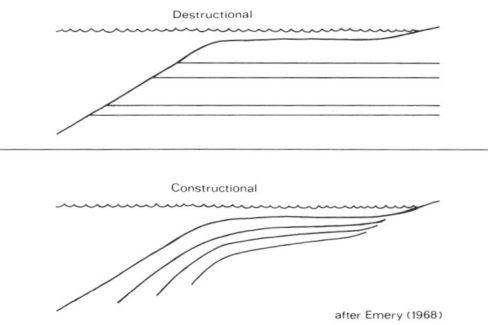

FIG. 1.—Constructional and destructional components of continental slope development (Emery, 1968). Constructional aspects are analogous to development of delta slopes.

of the Catskill delta's slope deposits are organic-rich black, gray and brown shales.

The economic significance of the Middle and Upper Devonian and Lower Mississippian offshore marine shales is evident in recent estimates of hydrocarbon reserve potential. These Appalachian basin shales contain an estimated 460 quadrillion cubic feet of natural gas (Brooks, 1972) of which about 285 quadrillion cubic feet can be extracted using present technology (Energy Research and Development Administration, 1976). These facts take on added significance because the associated shallower marine Catskill delta siltstones and sandstones continue to produce petroleum more than a century after Colonel Drake's 1859 first discovery at Titusville, Pennsylvania. More important, perhaps, is the fact that these organic-rich slope deposits act as both source and reservoir for major amounts of remaining hydrocarbons in the Central Appalachian basin. Thus, the voluminous stratigraphic information from the Catskill deposits can be an important guide in unravelling the stratigraphic relationships of source and reservoir beds in eastern North America's prograded Cretaceous and younger deposits.

GEOLOGIC SETTING OF CATSKILL DELTA

Sedimentary and Stratigraphic Framework

The essential elements of the intertonguing of offshore, nearshore and continental facies within the Catskill have been recognized since the time of Barrell (1913) who correctly interpreted the progradational nature of these deltaic deposits of the Central Appalachian basin. Recognition of gross contemporaneity among three of its intertongued facies (Portage, Chemung, and Catskill) has been a cornerstone in the history of Appalachian geology and influencal upon stratigraphic thinking about similar prograded sequences.

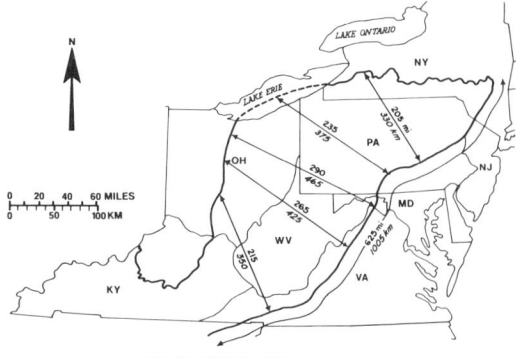

FIG. 2.—Areal extent of the Catskill delta. The delta configuration involves more than "Catskill Formation" or simply Middle and Upper Devonian rocks. Figure 3 shows most of the commonly used formation terms included here in Catskill delta.

Cracking the genetic code of the Catskill delta required identification of the types of lithologic associations present, the processes which controlled their accumulation and the physical environments (geomorphic settings) in which these processes operated. Glaeser (1973a, b, and 1974) recognized several lithologic associations within the Catskill delta* which can be readily distinguished both in outcrop and subsurface sections.

The stratigraphic and sedimentologic complexities of the Catskill delta deposits are not yet clearly understood, in part, because of its enormous size. It is roughly similar in scale to the Middle Atlantic Bight of eastern North America. A volume approximation of 288,000 cu km for the preserved Catskill sedimentary wedge was suggested in Dott and Batten (1975, p. 268–269).

Coastline length, overall distance of progradation and thickness of the sedimentary wedge (Fig. 2) give a clearer view of the Catskill delta's geometry. The preserved extent of its progradation exceeds 465 km (Fig. 2). It is 4,000 m thick along its eastern (sourceward) margin in northeastern Pennsylvania. Figure 2 data exceed those of Ayrton (1963) whose isopach-lithofacies map was confined to Upper Devonian rock units. The delta is comprised of rock units above and below

---

*The term "Catskill delta" applies here to the entire prograded sequence of marine and nonmarine rocks. Despite the fact that the term "Catskill facies" or "Catskill Formation" applies to the Middle and Upper Devonian nonmarine red-bed portion of the sequence, the writer feels no need to unscramble this dual terminology here and will use "Catskill delta" as distinct from the largely nonmarine sandstones and associated red-beds of the Catskill Formation. A review of the terminology problem appears in Glaeser (1974) and in Epstein, Sevon, and Glaeser (1974).

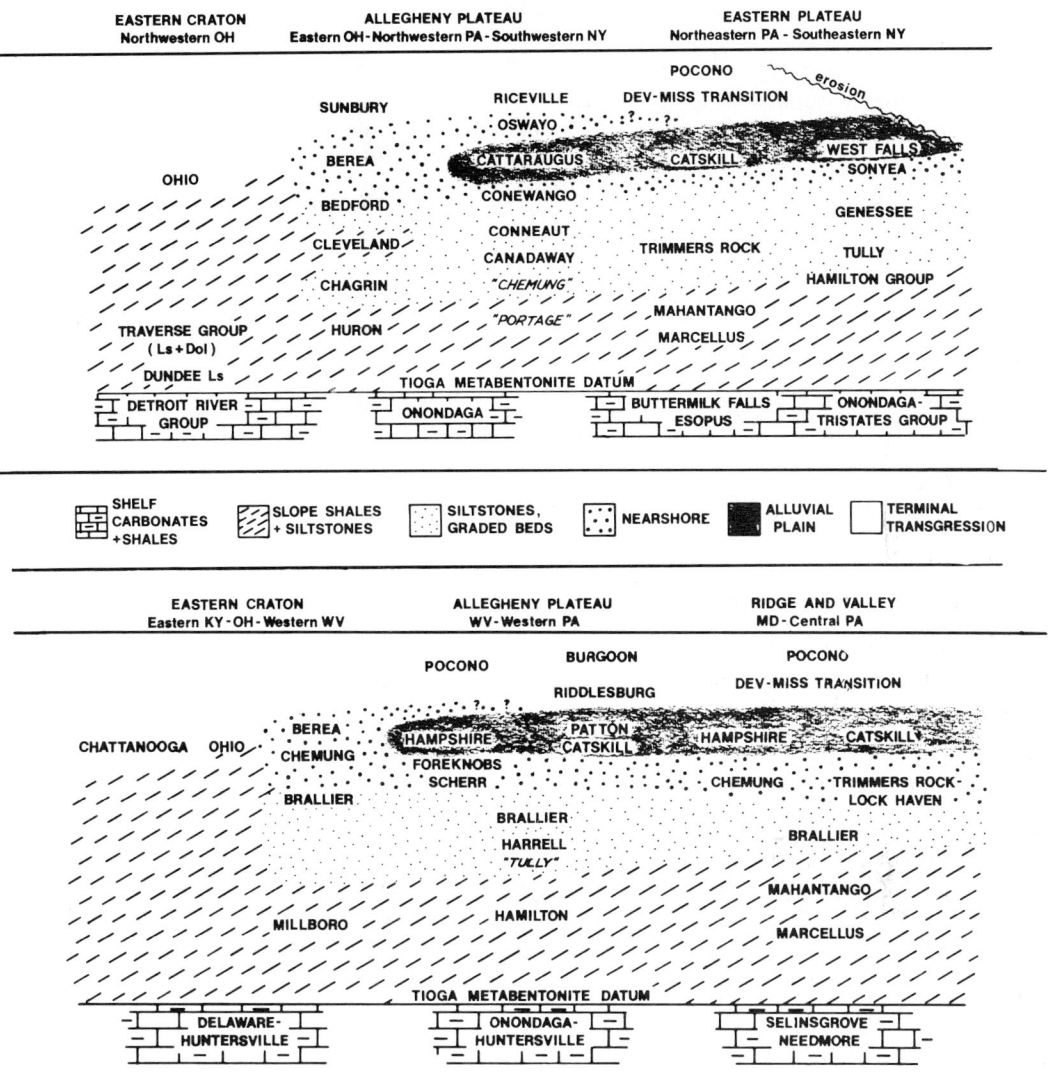

Fig. 3.—Names and general depositional settings of stratigraphic units included in the Catskill delta. Top half lists units along line from southeastern New York to northeastern Ohio. Bottom part shows units of Valley and Ridge Province to eastern Kentucky. Rocks involved in the terminal transgression (chart equivalents of Devonian-Mississippian transition) are shown in Figure 9.

the Upper Devonian time boundaries. Figure 3 is a general guide to those major rock and facies names commonly used in context to the Catskill delta wedge.

The initial size of the delta during its development is not fully depicted by the outcrop length and basinward limits of progradation (Fig. 2). The northern limit of the Catskill delta defined by the east-west line through New York state's mid-section (Fig. 2) is erosional so that all lithofacies boundaries abut that edge (Rickard, 1974; Ayrton, 1963, p. 4). Along the eastern outcrop belt, the sourceward edge of the Catskill delta, its greatest thickness is preserved. It is here that at least 30 m of offshore marine, organic-rich shales lie beneath prodelta siltstones and alluvial plain deposits. Thus, organic-rich muds had clearly accumulated in an offshore environment at the present outcrop margin inferring that nearshore and subaerial supply sites lay well to the east and southeast. Nowhere along the eastern and southeastern outcrop belt does Catskill alluvial material lie directly upon the shelf carbonates (Onondaga, Selinsgrove, etc., of Fig. 3). Thus, with the initiation of basin progradation marked by the black shales (Hamilton-Marcellus and their depositional equivalents, Fig. 3), the entire deltaic sequence of the Catskill in the Central Appalachian basin conform to Ferm's (1970, p. 246) strikingly vivid description of a delta:

> "Marine deltas form when sediments, carried by rivers into relatively large bodies of water, accumulate at the river mouth until the surface of the sediment pile reaches sea level. The emergent portion comprises the subaerial expression of the delta and is the site of most recent delta studies. Delta growth continues as sediment-laden streams pass over the emergent surfaces and deposit sand, silt and clay over the frontal delta slope. As this process of building new land at the delta margin continues, the delta is said to prograde, and the product of progradation can be thought of as the typical delta sequence."

Figure 4 graphically summarizes the essential feature of Ferm's definition.

DISTRIBUTION OF CATSKILL DELTA SLOPE DEPOSITS

Datum—Shelf Carbonate Platform

The uppermost surface of the Onondaga limestone and its correlatives (Fig. 3) in the Valley and Ridge Province is an ideal datum. This shallow water carbonate and its correlatives occur nearly everywhere throughout the Central Appalachian basin *and* in its upper beds is the widespread

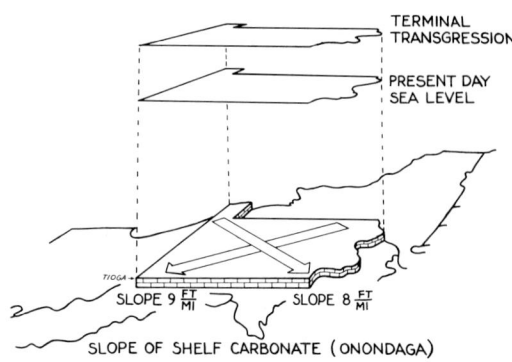

FIG. 5.—Cartoon illustrating the southeastward and southwestard components of slope on the carbonate platform upon which Catskill delta prograded. The carbonate surface slope is defined in terms of both the transgression which terminated progradation regionally and present sea level.

Tioga metabentonite described by Dennison and Textoris (1970). Thus, the limestone's upper surface can be considered virtually parallel to a time plane upon which delta growth began, signalled by the first influx of organic-rich black shales to the offshore prodelta environment (Marcellus, Mahantango, Millboro, etc. of Fig. 3).

A general picture of the carbonate platform surface upon which the Catskill delta was built can be "cartooned" (Fig. 5). If one asks, "What is the departure from parallelism between the carbonate platform and the transgressive surface terminating the Catskill delta's progradation?", the cartoon suggests that the two surfaces may be surprisingly parallel.

This inference comes from the fact that by rotating the transgressive unit* atop the Catskill delta to a horizontal position, the carbonate platform slopes at eight to nine feet per mile southward and southeastward, and eight feet per mile from

---

*A polymictic diamicitite, caused by subaqueous mudflows restricted to drowned alluvial valleys of the Catskill delta's alluvial plain in northeastern Pennsylvania, was described by Sevon (in Epstein, Sevon, and Glaeser, 1974). The sedimentary sequence which he recognized includes pebbly mudstone, laminites comprised of clay laminae associated with coarse siltstone and very fine grained sandstone laminae. Scattered sand grains and pebbles disrupt the laminae producing a "dropped in" appearance. The Riddlesburg shale (Laird, 1941) and other non-red mudstone horizons which separate the Catskill delta deposits from the distinctive overlying Pocono (Burgoon) sandstones were cited by Glaeser (1973b) as evidence of a regional transgression of the entire Catskill delta. Detailed mapping by members of the Pennsylvania Geological Survey substantiated this transition (Berg, 1979).

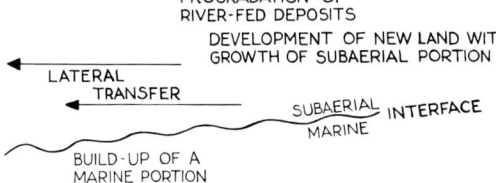

FIG. 4.—Graphic representation of components of delta building described by Ferm (1970).

the Allegheny Front to the eastern edge of the Valley and Ridge Province (Fig. 5). This "rough" estimate suggests that quantitative data on three originally horizontal surfaces (the carbonates, the transgression terminating the delta and present day sea level) can provide a way for delineating primary dips among slope shales and the enclosing shallower water reservoir rocks in this basin.

Depositional Systems Tracts of the Catskill Delta

Depositional systems tracts are defined here as the three dimensional assemblages of lithofacies genetically linked by active (modern) or inferred (ancient) processes and environments. Laterally associated lithofacies can be portrayed by surfaces (active systems) or lines (ancient systems) showing broadly temporal associations among deposits and their environments of accumulation. This definition is a combined product of the terms "depositional systems" of Fisher and McGowen (1967) and "systems tract" of Brown and Fisher (1977). Lines portraying depositional systems tracts in the Catskill delta delineate surfaces of coexistence among upper and lower delta plain, delta front lobes, graded fine prodelta sediments and the dark brown, gray and black delta slope shales. Inevitably, where depositional systems tracts can be sharply delineated, the tracts define paleoslope from the subareal portion of the delta through surf zone and ultimately into deeper water (Fig. 6).

Descriptions and observations of many hundreds of surface and subsurface sections within the Catskill delta show that most sections have a predictable vertical sequence of lithofacies. The Catskill delta's lithofacies sequence is;

Facies 7— Subaerially deposited upper deltaic
(Top) plain medium to coarse sandstones, some conglomerates and interbeds of red siltstones and shales. Fining upward cycles (meandering river depos-

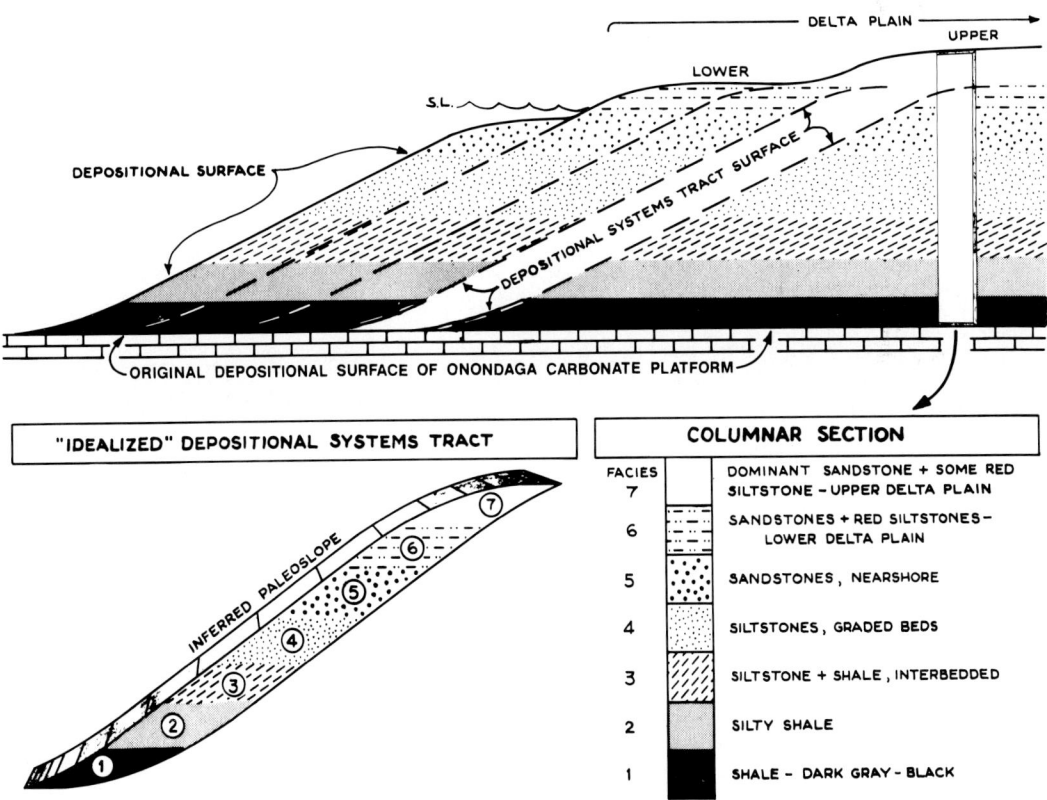

Fig. 6.—Upper portion of figure is a reinterpretation of Scruton's (1960) delta model. Present version incorporates all principal facies of the Catskill delta (see columnar section in lower right of figure), and replaces Scruton's time lines with depositional systems tract surfaces. The "idealized" tract based on columnar section is shown in lower left of figure. Facies 1, 2 and 3 are shown as slope shales and siltstones in Figure 3. Facies 6 and 7 are patterned as "alluvial plain" in Figure 3.

its) and thick sandstone sequences not interrupted by redbeds (braided river deposits).

Facies 6—Subaerially deposited lower deltaic plain sediments (commonly red siltstones and very fine sandstones).

Facies 5—Nearshore sandstones including sediments from surf zone seaward to limits of sand transport and wave or current reworking.

Facies 4—Fine to coarse siltstones commonly in graded beds. Shells occur as "conglomerate" lenses or as scattered whole and unabraded fossils.

Facies 3—Interbedded siltstones and shales (medium to dark gray). Commonly not graded and lithologies alternate vertically every few cm to tens of cm.

Facies 2—Silty shale (dominantly dark gray). Absence of well-organized interbedding and, thus, of sorting.

Facies 1—Black to dark gray and brown shales. Abundant traces of plant debris on bedding planes. Regionally persistent dark gray Purcell limestones (Dennison, 1977) several meters above base. General absence of shells throughout basal black shale unit.

The above sequence rests upon the Onondaga carbonate platform with the Tioga metabentonite near its upper surface.

The downward lithofacies sequence is thought to represent increased distance and water depth from rivers along the landward margin of the nearshore sand deposits. Major delta front lobes were outlined in Pennsylvania by Willard (1936) and extended southward by Dennison and others (1972). Sevon (1979) has recognized other input centers along the basin margin.

Depositional systems tracts are guides to correlation within this deltaic system. The conceptual model which results from using depositional systems tracts within the Catskill delta is a restatement of Walther's (1894) Law of Facies applied to this specific rock sequence. It is most easily cast as an extension of Scruton's (1960) delta model (Fig. 6). These elaborations on his model become extraordinarily powerful because of their direct links to the columnar section (Fig. 6) of the seven Catskill delta facies described above. The columnar section is highly representative of the bulk of the delta's sections. It is a simple step to then recognize the associations among the lithofacies in terms of an "idealized" depositional systems tract drawn in the lower part of Figure 6. This "idealized" tract, taken directly from the columnar section in the lower right of Figure 6, gives substantial control in making both region or local detailed correlations. Thus, the columnar section and the "idealized" depositional systems tract (Fig. 6) represent the premise underlying the correlations presented below. These correlations exemplify the association between source beds (facies 1, 2, 3) formed on the Catskill prodelta slope and the reservoir beds (facies 3, 4, 5) deposited on the upper prodelta slope and adjacent nearshore marine environments. There is actually no clear-cut distinction between source and reservoir beds in the Catskill delta from facies 1 through 4.

Principle Features Illustrated by Stratigraphic Sections Within the Catskill Delta

Eleven surface and subsurface stratigraphic columnar sections are used in this paper to illustrate four important facets of delta slope development which are quite apparent in the Catskill delta. Figure 7 is an index map of the eleven sections used along with the sources of data in the environmental interpretations and correlations. Spacing between sections is far too great for use as exploration guides, yet their spatial arrangement highlights the fundamental internal organization of this delta.

The regional correlations presented in Figure 8 and Figure 9 suggest some of the important uses of depositional systems tracts in outlining the economically important organic-rich slope source-bed shales and their associated reservoir siltstones and sandstones. Equally, or more important, are their uses in unravelling the complex geologic history buried within these classic deposits of the Catskill delta. A detailed example

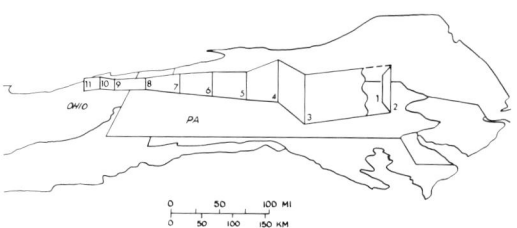

Fig. 7.—Index map of the eleven surface and subsurface stratigraphic sections shown in detail in Figures 8, 9, 10, and 11. Section name and data source: (1) #1 L. Richards (Glaeser, 1974, P. 74); (2) Lehigh River (Epstein, Sevon and Glaeser, 1974, pp. 413–439); (3) West-central part, southern anthracite field (Wood, Trexler and Kehn, 1969); (4) Lock Haven Quadrangle (Taylor, 1977); (5) #1 A. Pardee Estate, Cameron County (Fettke, 1961, p. 295); (6) Warrant 3653, #1 Pennsylvania Game Lands, Elk County (Fettke, 1961, p. 467); (7) #1 E. Collins, Forest County (Fettke, 1961, p. 263); (8) #1 E. McKnight, Mercer County (Wagner, 1958); (9) #1 Reynolds, Medina County; (10) #1 Eshtruth, Ashland County; (11) #1 Willet, Huron County (Sections 9, 10 and 11 from Lewis and Schwietering, 1971).

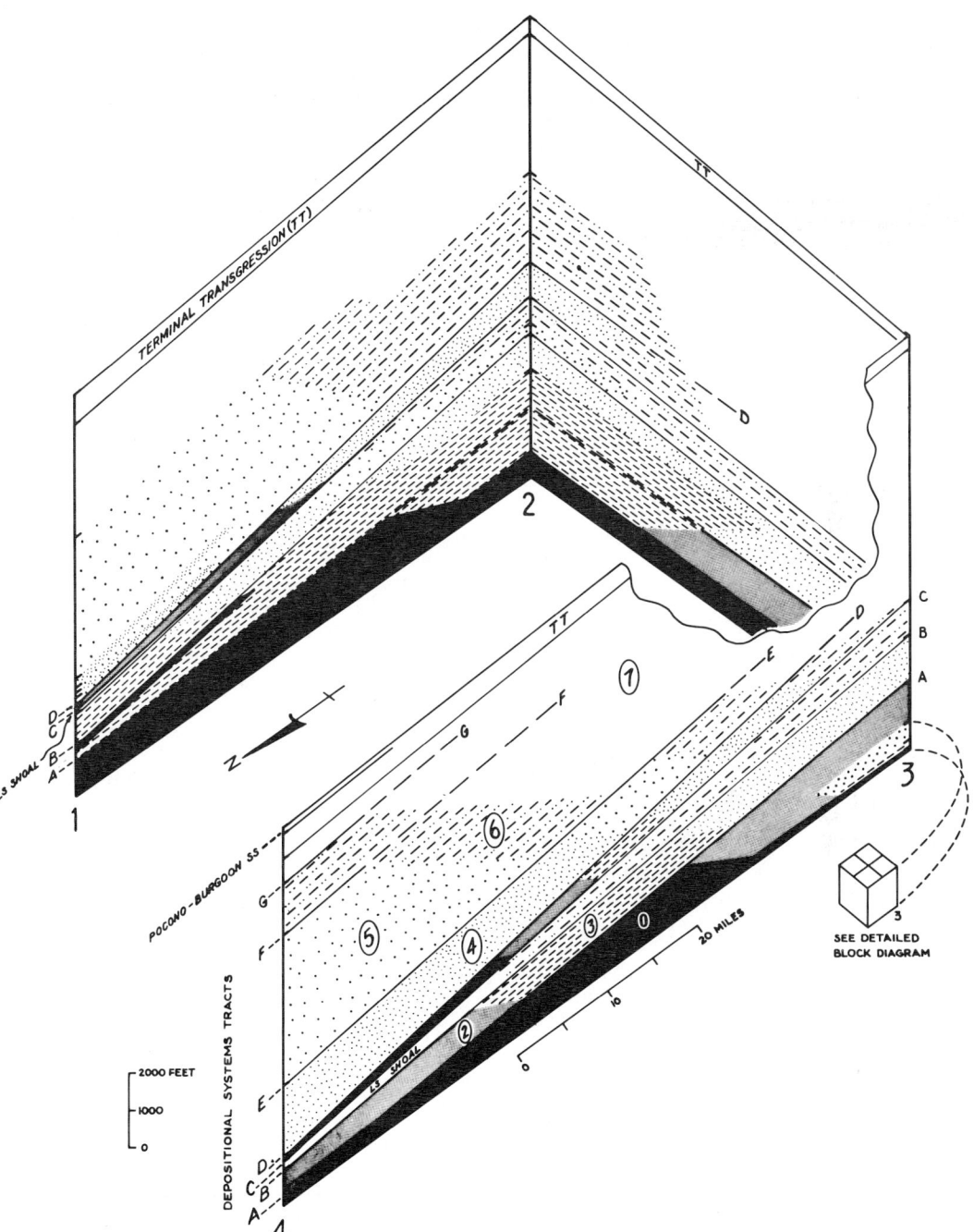

FIG. 8.—Detailed correlation diagram from southeastern margin of Valley and Ridge Province (panel connecting sections 2 and 3) to basinward sections. The panel between section 1 and 2 lies beneath Pocono Plateau. Section 3 to 4 extends to Allegheny Front. Locations and section names given in Figure 7. Detailed block diagram at base of section 3 illustrated in Figure 11. Facies types given in Figure 6.

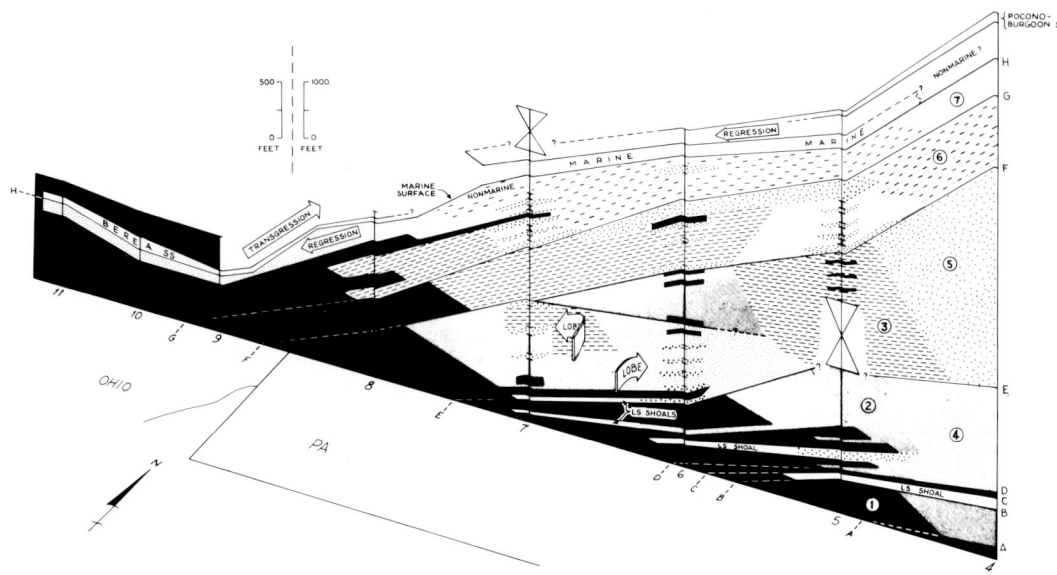

Fig. 9.—Basinward correlation diagram and designated depositional systems tracts (A through H). Diagram indicates marine transgression over most or all of the preserved delta sequence. Facies 1 through 7 illustrated in Figure 6. Outcrop and subsurface section sites and names given in Figure 7.

of potential uses of the broad associations described here is given later in the block diagram and depositional systems tract maps (Figs. 10, 11).

Four important elements of delta slope deposits are described and illustrated in the virtually complete stratigraphic sections selected (Figs. 8, 9, 10, 11).

1. Depositional systems tracts can be readily distinguished at the southeastern-eastern outcrop margin of the delta which extend westward into Ohio.

2. The slope deposits of organic-rich black shales lie at the downdip ends of the depositional systems tracts. The slope shales are practically or completely enveloped by coarser material which accumulated updip contemporaneously with the black mud and successively overstepped the mud as the delta prograded.

3. Several delta lobes can be delineated along the line of sections from near the sourceward margin to beyond the Allegheny Front basinward. Main portions of lobes, interlobe areas and "feather edges" of lobes are "exposed" in the panels illustrated.

4. Both spatial and temporal persistence of delta lobes have direct influence upon development of prodeltaic facies spectrum. Thickness of the black shales is clearly greater in interlobe areas. Where nearshore sands were deposited directly upon these thicker muds, ideal source to reservoir conditions are juxtaposed. These four main points just summarized are described and illustrated in the following sections.

Depositional Systems Tracts.—Figure 8 illustrates depositional patterns and changes along the southeastern edge of the Valley and Ridge (sec-

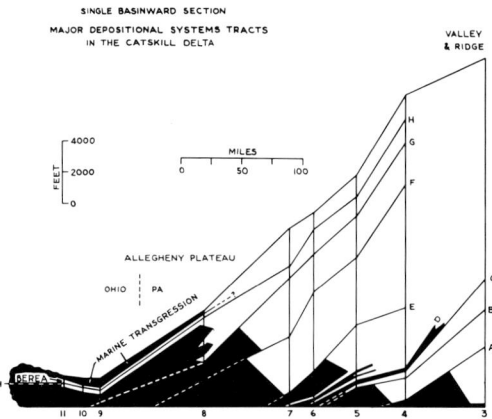

Fig. 10.—Reiteration of Figure 9 emphasizes only the shapes and extents of the delta slope black shale deposits. Depositional systems tract surfaces A through H separate slope shales into progressively younger shale bodies basinward along line of sections. Sites and names of section 4 through 11 given in Figure 7.

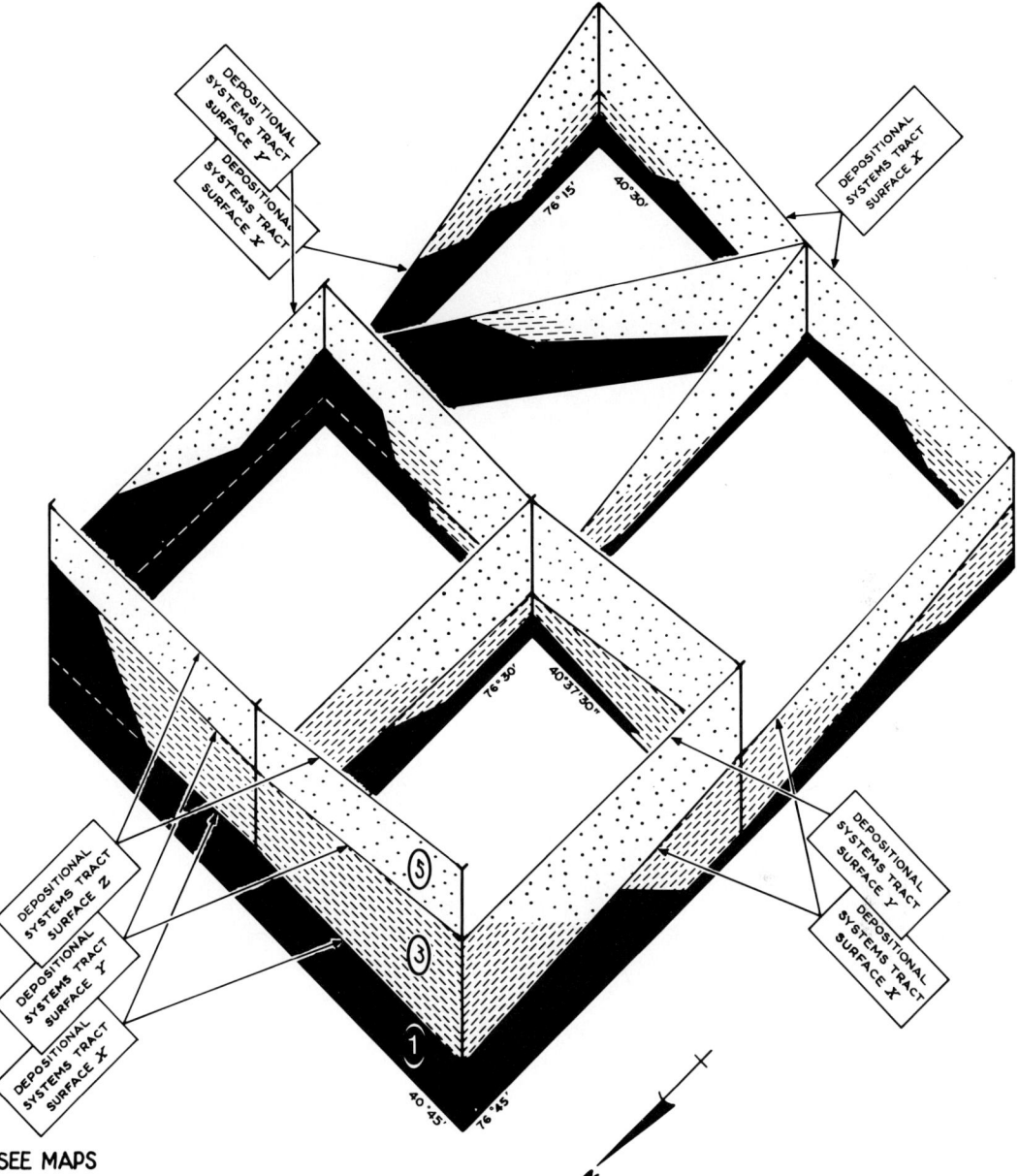

Fig. 11.—Detail of base of section 3 (Fig. 8). Block diagram from detailed measured sections, base of Marcellus through top of Montebello Formations, described by Wood and others (1969). Facies 1, 3 and 5 descriptions given in Figure 6. Note to "See Maps" refers to Figure 12.

tions 2 and 3) and into the subsurface beneath the Pocono Plateau (sections 2 to 3) and across the Valley and Ridge Province (section 3 to 4). Figure 9 (section 4 in common with Fig. 8) extends from the Allegheny Front (section 4) to northeastern Ohio (sections 5 through 11). Map distances among sections are not reconstructed palinspastically.

Eight depositional systems tracts lines (A through H) can be recognized on the correlation panels (Figs. 8, 9). The tracts shown are pertinent only to the line of sections drawn. They represent depositional surfaces for which the spectrum of deltaic facies can be demonstrated (alluvial sandstones facies 7 to black slope shales facies 1). It is important to note that, as far west as section 7 in northwestern Pennsylvania (Fig. 9), the vertical succession of facies is virtually that shown in the idealized columnar section in Figure 6. In addition, tracts A through D at section 3 (Fig. 8) mark the tops of successively higher facies in the deltaic column (silty shale, graded beds of siltstone, lower and upper deltaic plain). Section 3 shows also one example of the earlier noted fact that distal prodelta shales and siltstones are preserved at the sourceward edge of the basin.

Depositional systems tracts D through H all are within the upper deltaic plain facies 7 at the eastern margin of the Valley and Ridge Province and pass into the black shales of facies basinward (Figs. 8, 9). In fact, at least half of the entire deltaic sequence portrayed in the correlation panel connecting sections 2 and 3 is lower and upper deltaic plain sediment changing basinward to nearshore sandstones and siltstones having graded beds (facies 5 and 4) and, beyond there, to facies 3, 2, and 1 (siltstones and shales) beneath the Pocono and Allegheny Plateaus. Thus, depositional systems tracts are very useful in correlating sedimentary units which accumulated on a presumably continuous depositional surfaces linking the subaerial and submarine sedimentary environments.

Delineation of Slope Shale Bodies.—It is quite apparent from both Figures 8 and 9 that the slope shales are overstepped by coarser-grained sediment. Slope shales are partially or entirely enveloped updip by sediments formed higher on the delta platform. The organic-rich shales of facies 1 thicken and appear to spread out downdip against the pre-existing Onondaga carbonate platform surface (Figs. 3, 8, 9). Because all the depositional systems tracts (A through H in Figs. 8, 9) project into the black shale facies, it is necessary that successively basinward portions of the black shales each be considered part of the depositional surface resulting from alluvial, nearshore delta lobe and slope sedimentary processes. The shales, therefore, each are time-bounded sediment packets.

The frequent overstepping of black shales on the delta slope produces a somewhat different view of the relations between marine shales and the coarser portions of the Catskill delta. Traditional interpretations do portray the black shales as important time-related units but are thought to be results of marine incursions during transgressions during the overall progradation of the Catskill delta. The present interpretation implies that, although all the black shales appear to be part of one stratigraphic unit basinward (Figs. 9, 10), they are discrete bodies accumulated sequentially in time at the distal portion of each depositional systems tract.

The stratigraphic summary diagram (Fig. 10) incorporates the two principal elements of Figures 8 and 9; (1) the stratigraphic positions and gradients on each of the 8 depositional systems tracts, and (2) the thickness and cross-sectional shapes of the principal slope shales deposited at the seaward end of each tract. Although the projections of each tract line in Figures 8 and 9 are arbitrary extensions from higher portions of the delta (especially those defining the tops of depositional systems tracts F and G), the tacit assumption is that each slope shale is a deposit having a "linear source" (Gorsline, 1978). Thus, there is little reason to designate any of these major slope shales as either regressive or transgressive. The shales are muds winnowed from nearer shore wave and current reworking. Their depositional origin shifts the emphasis of interpreting transgression and regression to other processes, particularly those which influence delta lobe incursions into areas of organic-rich, hemipelagic mud build-ups.

Expression of Delta Lobes in Basinward Cross Sections.—Delta lobes are best viewed in cross sections which parallel regional depositional strike. However, the basinward orientation of Figures 8, 9, and 10, more suitable for delineating depositional systems tracts and emphasizing where the slope shales are in the overall stratigraphic organization of this delta, *does* show lobe incursions which are crucial to the fates of delta slope muds. Most critical to the continued accumulations of muds is lobe persistence in time and its areal extent, ultimately controlled by the gradient of the subaerial and nearshore delta platform(s). Abandonment of lobes also results in further mud and silt accumulations (discussed in the following section). Since there are no cross sections presented here which parallel the paleo-isobaths of this prograding delta, Figures 8 and 9 will be used to determine delta lobe positions and their basinward extent, to describe the appearance of lobe margins, and to assess qualitatively delta lobe influences on thickness and extent of slope shales.

Delta lobes are the nearshore sandstones (facies 5) which result from combined effects of surf and wave action on alluvial input. Thick nearshore marine sandstones lie just basinward of lower deltaic plain deposits (facies 6), most clearly illustrated between depositional systems tracts E and F (Fig. 8).

There are no appreciable accumulations of nearshore sandstones within the study area at the southeastern margin of the Valley and Ridge outcrop belt. The correlation panel connecting sections 2 and 3 (Fig. 8) shows the minor amount of facies 5 sandstones. Graded beds of prodelta facies 4 pass upward into lower delta plain motifs as first observed in the Susquehanna Valley by Walker and Harms (1971). This lateral and vertical juxtaposition of lower delta plain sediments (facies 6) with graded siltstones occurs in only one other place; that one lies between tracts G and H at sections 7 and 8 (Fig. 9) in western Pennsylvania. Gradients and/or input rates in the southeasternmost and western regions may account for the absence of appreciable nearshore sandstones. Also, laterally there may be widely spaced, time-persistent delta lobes as described by Dennison and others (1972) which simply do not lie along the basinward transect of Figures 8 and 9. In most areas, however, there are significant build-ups of major nearshore delta lobes (Figs. 8, 9; Glaeser, 1974).

The most prominent delta lobe lies within and just west of the Valley and Ridge Province where this nearshore sandstone is part of depositional systems tract F (Figs. 8, 9). A second significant lobe lies between sections 4 and 7 within tract G. This nearshore sandstone sequence overlies, and grades basinward into, facies 3 siltstone and shale interbeds.

What are interpreted here as feather edges of delta lobes can be seen at sections 6 and 7 (Fig. 9). The input directions shown by "lobe" arrows may be from either north or south of the line of the correlation diagram. Where these lobes actually are situated is intriguing and important for at least two reasons. First, one sandstone sequence lies higher stratigraphically than the other into the basin suggesting a simple progradational overstepping. Second, both sandstone units lie beneath facies 2 and 3 (both potential seals above the sandstone reservoirs). The eastward sloping line on Figure 9 shown abutting tract E (below) and tract F (above) probably defines a depositional systems tract which would be apparent if correlations were to be constructed normal to the present cross section (Fig. 9). Sufficient stratigraphic data exist that many depositional strike sections paralleling lobe paleo-isobaths can be constructed to show the lateral extents and overstepping among these economically important nearshore sandstone bodies. Their persistence in time (gauged by lobe thickness and breadth) has a direct effect upon the presence and thickness of the more distal marine lithologies (facies 4 through 1). In the following section, one detailed example of such control is described and illustrated.

Delta Lobe Influence upon Prodelta Slope Deposits.—With the broad brush approach available in the preceding discussions of Figures 8 and 9, it is useful to look at one smaller area to show by example that closely defined control can be demonstrated between delta lobe and the development of slope shale deposits. Detailed field data of formation thicknesses in contiguous quadrangles (76°15' to 76°45' and 40°30' to 40°45') mapped by Wood and others (1969) clearly display the basinward thickening of delta slope shales in response to migration of a delta lobe. The basal portion of section 3 in Figure 8 is shown in a detailed fence diagram (Fig. 11), the panels of which represent the quadrangle boundaries. The nearshore sandstones (facies 5) are parts of the Middle Devonian Montebello Formation, an early, localized group of delta lobes described in detail by Kaiser (1972). The portions of the Montebello delta lobe in Figure 11 along the southernmost margin of the fence (latitude (40°30') are thick sandstones (facies 5) stratigraphically above and grading basinward into alternating siltstones and shales (facies 3). However, in the central portion of that same fence, relatively thick black shales (facies 1) are directly overlain by nearshore sandstones. This same sand-over-shale juxtaposition also is present along the eastern panel of the fence diagram (76°15') and at both latitudes 40°37'30" and 40°45'. There are complimentary increases in amounts of facies 3 siltstones and shales to the west and northwest (Fig. 11).

Three small depositional systems tracts are recognizable within the boundaries of the fence diagram. These tracts (designated X, Y, and Z, Fig. 11) are local, definable only in context to the Montebello delta lobes. The three Montebello depositional systems tracts can be readily translated into maps (Fig. 12) to more easily interpret the patterns of delta lobe growth and explain the interesting juxtaposition of facies 5 sandstones directly upon black shales at the three sites mentioned.

The oldest of the three local tracts, surface X, can be traced around the panels in Figure 11. An interpretation of the depositional processes producing surface X (Fig. 12) is that one, and probably two, input centers lay just to the east and west of the southern boundary of this area. Black shale in the mid-portion of the southern panel (40°30', Fig. 11) is thought to be an interlobe area which was later overridden laterally by nearshore sands as the surface lobes advanced. It is clear from Figure 11, however, that input was centered more along the southwestern part of this area. Facies 3 siltstones and shales accumulated in significant amounts along the west side (76°45').

Depositional systems tract Y (Fig. 11) confirms the interpretation that sand input was dominant along the central and western portion of the area (latitude 40°37'30") in building a major delta lobe. This lobe resulted in both increased amounts of facies 3 silt and shale depositing basinward and

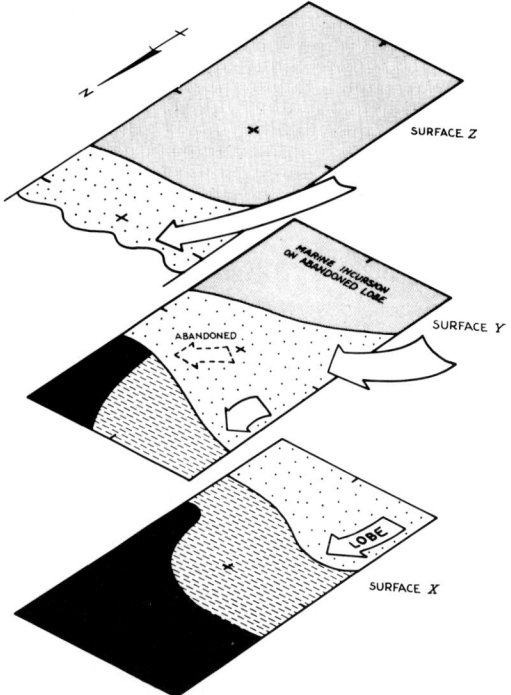

Fig. 12.—Maps interpreting three local depositional systems tracts X, Y and Z designated on Figure 11. Maps show delta lobe advancement and abandonment with ultimate incursion of marine silty shales. Facies patterns in Figure 6.

3, this marine incursion is marked by facies 2 immediately above the sandstones. These silty shales, developed upon abandoned portions of Montebello delta lobes (Figs. 11, 12), probably are not the distal prodelta sediments inferred by that facies type in Figure 6. Instead, they probably are simply poorly sorted fines expected with any local marine incursion over very low gradient lobes. When abandoned, the lobes continue subsiding and are overridden by marine muds.

The fence diagram (Fig. 11) and the interpretive maps (Fig. 12) open previously unexamined ways of deciphering the complex detail of outcrop and subsurface information which is available. The simple methods of applying the concepts in Figure 6 to determine where the depositional systems tracts are situated and portraying them as maps to interpret local paleo-geomorphic patterns and associated processes are a main point of this paper. The most interesting result of panel and map portrayal of the field data of Wood and others (1969) is the close physical contexts there between black shales and nearshore sandstones. The fences of the correlation diagram (Fig. 11) show appreciable thicknesses of potential reservoir sandstones lying immediately above black shale source beds. The intervening facies 3 in some parts of Figure 11 suggests less direct, but important, pathways between source and reservoir deposits. Because the entire Montebello delta source and reservoir beds are covered by a marine silty shale (Figs. 8, 12), the potential stratigraphic traps are apparent.

large accumulations of black shale to the east and northeast of the area enclosed by the Figure 11 fence.

The tract map of surface Y gives an interpretation for this basinward thickening of facies 3 and lateral thickening of facies 1. Apparently, as the delta lobe grew, it hugged the western margin of the area with abandonment of direct input to the east and northeast (Fig. 12). The black shale probably is hemipelagic, organic-rich muds derived from a linear source feeding winnowed fines from both the nearshore delta lobe sands and from the apparent current reworking of the silt-shale interbeds (facies 3). Depositional systems tract Z records the last phase of Montebello delta development inside these quadrangles. A relatively thick sand blanket was deposited over both the black shales and the alternating thin beds of siltstones and shales (Fig. 11).

Because the Montebello delta everywhere is overlain by the upper shale member of the Mahantango Formation (Wood and others, 1969), maps of surfaces Y and Z (Figure 12) show a marine incursion around and behind (south and southeastward) the delta lobe. In Figure 8, section

DISCUSSION

It may be possible to apply basinward correlation of depositional systems in outlining slope shales on the trailing margin of the North American continent. Regardless of the present erosion indicated by outcrops of Cretaceous and younger deposits along the continental slope, the North American eastern margin is fundamentally constructional, comprised of a series of clastic wedges built above complex Triassic-Jurassic graben structures. This Cretaceous and younger detrital blanket incorporates the sedimentary record of numerous sea level changes amidst an overall pattern of continent margin progradation.

DSDP bore holes 101 and 105 encountered about 100 m of organic-rich muds (Hollister and others, 1972) along the lower continental margin. These muds are rich in detrital plant debris and are considered a potential source for petroleum in both the Georges Bank area (Schlee and others, 1975) and the Baltimore Canyon Trough (Weed and others, 1974). Cretaceous and younger deposits are predominantly clastics with seismic, onshore and off-shore stratigraphic tests all indicating seaward thickening deltaic sequences in which potential source and reservoir beds exist. Early evaluation suggested that the 100 m dark green to black clays might extend for at least 1000 km along the continental margin of eastern North America.

Arthur and Schlanger (1979) identify Early and Middle Cretaceous as a time of increased deposition and preservation of organic-rich facies in anoxic or near-anoxic conditions. They designate two, and perhaps three, periods of sapropel and black mud accumulation. They draw attention to the contemporaneity of these oxygen-deficient muds with shallow water carbonate deposits.

Four limestone stratigraphic markers occur updip from black shales in the Catskill delta (Figs. 8, 9) suggesting a strikingly similar boundary between anoxic and carbonate environments. The limestone markers shown in Figures 8 and 9 are stratigraphic horizons including the Purcell, Centerfield and Tully limestones. They locally can contain bioherms rich in horn corals, bryozoans, crinoid columnals, along with a variety of shell material. Heckel (1969) has described the regional aspect of the Tully horizon and Dennison (1977) has shown the regional persistence of the dark gray Purcell limestone above the lowermost black shale deposits of the Marcellus and its equivalents (Fig. 3). The black shales in the Central Appalachian basin are generally devoid of fossil fauna although bedding planes are strewn with finely cominuted plant debris. Sevon (in Epstein, Sevon, and Glaeser, 1974) observed abundant organic debris and pyrite in thin sections of the Marcellus black shales.

It is probable that the slope muds of the Catskill delta owe their anoxic character to tectonic events rather than the apparent global climatic causes of the Middle Cretaceous "oceanic anoxic events." Although causes of Devonian waters devoid of oxygen can only be surmised, it is significant that the stratigraphic equivalents of the entire Catskill delta is a thin, craton-wide black shale having many anomalies which have intrigued stratigraphers for years. Rich (1951) set the stage for rigorous investigations of the Chattanooga shale and its equivalents. Conant and Swanson (1961) gave both regional and economic evaluations of this thin, broad blanket of anoxic shale. If, indeed, the anoxic black shales of the Catskill delta slope are related to climatic events, these are clearly involved with the tectonic controls the Caledonian-Acadian sedimentation of clastic wedges.

The Catskill delta can be viewed within a tectonic framework which links the prograding sedimentary wedge to crustal plate motion and interaction. The Acadian-Caledonian orogenic belt resulting in the development of the Catskill clastic wedge along the craton-margin of eastern North America was a convergent plate juncture. The orogenic setting which, at present, best defines the resultant Catskill clastic wedge is that of a retroarc basin. Such basins (Dickinson and Yarborough, 1976, p. 17) develop upon the surface of a continental block subsiding against the rear flank of an arc orogen. The ophiolitic zone lies at a greater distance from the basin than the magmatic belt of batholiths of volcano-orogenic rocks. These relationships are illustrated in Figure 13. Friedman and Johnson (1966) called the Catskill delta a "tectonic deltaic complex" because of the presence, in places, of orogenic sandstones suggesting their derivation from an active mountain front.

In summary it is quite evident that the geologic legacy preserved in ancient clastic wedges, such as the Catskill delta, has very practical applications to modern slope deposits. Ancient deltaic sedimentary sequences represent an invaluable key to present-day slope processes because the ancient sequences preserve the sum total depositional processes in a major prograding system. The facies record in the Catskill shows the important links between hydrocarbon-bearing reservoir deposits and their black shale source beds, the types of links which remain to be resolved in continental margins. As offshore exploration procedes, targeting both source and reservoir beds will in large part be accomplished by use of sedimentologic models already applied successfully to ancient constructional slopes.

This paper represents one more example that geology is largely an attempt to interpret Earth processes by the study of preserved responses in the rocks. The translation to modern slope sediments of insights gained from the studies of stratigraphic organization, sediment body geometry and facies preserved in ancient slope deposits can sharpen the quest for the really important processes and resultant responses producing constructional slope deposits. Without these insights, both scientific and economic exploration in ancient and not-so ancient slope deposits is little more than a game of chance. Costs of studies on modern slopes make it worth the risk of

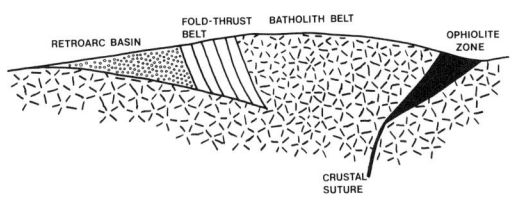

Fig. 13.—Tectonic setting of the Central Appalachians during Catskill delta progradation into retroarc basin.

inverting Hutton's principal and pursuing the point that the past is the key to the present and, therefore, a guide to future economic uses of continental slopes.

ACKNOWLEDGMENTS

This attempt to point out some uses of the ancient in understanding the modern continental slopes owes much to my participation in the activities of the Duke University Marine Laboratory since 1974, and particularly to my association and carousing with Orrin H. Pilkey, Jr. Tom Lewis, Cleveland State University, showed me the marine periphery of the Catskill delta, allowing me to look shoreward to gain a marine sedimentologist's view of a delta which I had previously visualized from upstream. Symposium co-conveners Pilkey and Larry Doyle actively encouraged completion of this manuscript and John M. Dennison critically reviewed it moments before the ax of a time deadline fell.

REFERENCES

ARTHUR, M. A., AND SCHLANGER, S. O., 1979, Middle Cretaceous "Oceanic anoxic events" as causal factors in development of reef reservoir giant oil fields: Am. Assoc. Petroleum Geologists Bull., (in press).
AYRTON, W. G., 1963, Isopach and lithofacies map of the Upper Devonian of Northeastern United States, in Shepps, V. C., ed., Symposium on Middle and Upper Devonian stratigraphy of Pennsylvania and adjacent states: Pennsylvania Geol. Survey, 4th Ser., Gen. Geol. Rept. 39, v. 1. p. 3–6.
BARRELL, J., 1913, The Upper Devonian delta of the Appalachian geosyncline, Part I, The delta and its relations to the interior sea: Am. Jour. Sci. 4th ser., v. 36, No. 1. p. 429–472; Part II, Factors controlling the present limits of the strata: v. 37, p. 87–109; Part II, The relations of the delta to Appalachia, v. 37, p. 225–253.
BERG, T. M., 1979, Upper Devonian-Lower Mississippian stratigraphic relations in north-central Pennsylvania: Geol. Soc. America Abs. with Programs, v. 11, no. 1, p. 3.
BROOKS, K., 1972, Devonian shale gas study: Devonian Shale Symposium, Columbus, Ohio.
BROWN, L. F., JR. AND FISHER, W. L., 1977, Seismic-stratigraphic interpretation of depositional systems: examples from Brazilian rift and pull-apart basins: Am. Assoc. Petroleum Geologists Memoir 26, p. 213–248.
CONANT, L. C., AND SWANSON, V. E., 1961, Chattanooga shale and related rocks of central Tennessee and nearby areas: U.S. Geol. Survey Professional Paper 357, p. 48–62.
DENNISON, J. M., 1969, Tioga bentonite and other isochronous units in the Devonian Oriskany to Tully interval of the Appalachian basin: Geol. Soc. America Abs. with Programs 1969, Part 1, p. 13.
———, AND TEXTORIS, D. A., 1970, Devonian Tioga tuff in Northeastern United States: Bulletin Volcanologique, Tome XXXIV, p. 289–294.
———, AND OTHERS, 1972, Stratigraphy, sedimentology and structure of Silurian and Devonian rocks along the Allegheny Front in Bedford County, Pennsylvania, Allegheny County, Maryland and Mineral and Grant Counties, West Virginia: Guidebooks, 37th Ann. Field Conf. Pennsylvania Geologists, Pennsylvania Geol. Survey.
DICKINSON, W. R., AND YARBOROUGH, H., 1976, Plate techtonics and hydrocarbon accumulation: Am. Assoc. Petroleum Geologists Short Course, New Orleans, May 23.
DOTT, R. H. JR., AND BATTEN, R. L., 1975, Evolution of the earth, (2nd ed.): New York, McGraw Hill.
EMERY, K. O., 1968, Characteristics of continental shelves and slopes, Am. Assoc. Petroleum Geologists Bull., v. 49, p. 1379–1384.
EPSTEIN, J. B., SEVON, W. B., AND GLAESER, J. D., 1974, Geology and mineral resources of the Lehighton and Palmerton Quadrangles, Carbon and Northampton Counties, Pennsylvania: Atlas 195 CD, Pennsylvania Geol. Survey, 4th Series, 460 p.
ENERGY RESEARCH AND DEVELOPMENT ADMINISTRATION, 1976, Outline for a project to increase the Eastern U.S. natural gas reserves from gas-bearing shales: Energy Res. and Devel. Admin., March.
FERM, J. C., 1970, Allegheny deltaic deposits: Soc. Econ. Paleontologists Mineralogists Spec. Pub., No. 15, p. 246–255.
FETTKE, C. R., 1961, Well-sample descriptions in northwestern Pennsylvania and adjacent states: Pennsylvania Geological Survey Bull. M40, 4th Series, 691 p.
FISHER, W. L. AND MCGOWEN, J. H., 1967, Depositional systems in the Wilcox Group of Texas and their relationship to occurrence of oil and gas: Trans. Gulf Coast Assoc. Geol. Socs. v. 17, p. 105–125.
FREIDMAN, G. M., AND JOHNSON, K. G., 1966, The Devonian Catskill deltaic complex of New York, type example of a "tectonic delta complex," in Shirley, M. L., ed., Deltas in their Geologic Framework: Houston Geol. Soc., p. 171–188.
GLAESER, J. D., 1973a, Delta front evolution of Catskill deposits in the Central Appalachian basin: Geol. Soc. America Abs. with Programs, v. 5, No. 2, p. 165.
———, 1973b, Pocono boundary discrepancy between eastern and western Pennsylvania: Geol. Soc. America Abs. with Programs, v. 5, No. 2, p. 166.
———, 1974, Upper Devonian stratigraphy and sedimentary environments in northeastern Pennsylvania: Pennsylvania Geol. Survey, General Geology Rept. 63, 4th Series, 89 p.
GORSLINE, D. S., 1978, Anatomy of margin basins: Jour. Sed. Petrology, v. 48, p. 1055–1068.

HECKEL, P. H., 1969, Devonian Tully limestone in Pennsylvania and comparison to type Tully limestone in New York: Pennsylvania Geological Survey, 4th Series, Inf. Circ. 60, 33p.

KAISER, W. R., 1972, Delta cycles in the Middle Devonian of Central Pennsylvania [Ph.D. thesis]: Baltimore, The Johns Hopkins Univ., 183 p.

LAIRD, W. M., 1941, The Upper Devonian and Lower Mississippian of southwestern Pennsylvania: Pennsylvania Geol. Survey, 4th Ser., Progress Rept. 126, 23 p.

LEWIS, T. L., AND SCHWIETERING, J. F., 1971, Distribution of the Cleveland black shale in Ohio: Geol. Soc. America Bull., v. 82, p. 3477–3482.

RICH, J. L., 1951, Probable fondo origin of Marcellus-Ohio-New Albany-Chattanooga bituminous shales: Am. Assoc. Petroleum Geologists Bull., v. 35, p. 2017–2040.

RICHARD, L. V., 1975, Correlation of the Silurian and Devonian rocks in New York State: N.Y. State Museum and Science Service, Map and Chart Series No. 24.

SCHLEE, J., AND OTHERS, 1975, Sediments, structural framework, petroleum potential, environmental conditions and operational considerations of the United States North Atlantic outer continental shelf: U.S. Geol. Survey Open-File Rept. 75-353, 179 p.

SCRUTON, P. C., 1960, Delta building and the deltaic sequence, in Shepard, F. P., and others, eds., Recent sediments, northwest Gulf of Mexico: Tulsa, Okla., Am. Assoc. Petroleum Geologists Bull., p. 82–102.

SEVON, W. D., 1979, Devonian sediment dispersal systems in Pennsylvania: Geol. Soc. America Abs. with Programs, v. 11. No. 1, p. 53.

TAYLOR, A. R., 1977, Geology and mineral resources of the Lock Haven 7 1/2-minute quardrangle, Clinton and Lycoming Counties, Pennsylvania: Atlas 124a, Pennsylvania Geol. Survey, 4th Series.

WAGNER, W. R., 1958, Emma McKnight No. 1 Well, Pymatuning Township, Mercer County: Pennsylvania Geol. Survey, 4th Series, Well Sample Record 40, 33 p.

WALKER, R. G., AND HARMS, J. L., 1971, The Catskill delta: a prograding muddy shoreline in Central Pennsylvania: Jour. Geology, v. 79, p. 381–399.

WALTHER, J. F., 1894, Lithogenesis der Gegenwart, Beobachtungen uber die Bildung der Gesteine an der heutigen Erdoberflache, dritten Teil einer Einleitung in die geologische Historisch Wissenschaft: Verlag von Gustav Fisher, Jena, section 8, p. 621.

WEED, G. A. AND OTHERS, 1974, Generalized pre-Pleistocene geologic map of the northern United States Atlantic continental margin: U.S. Geol. Survey Misc. Geol. Inv. Series Map I-861.

WILLARD, B., 1936, Continental Upper Devonian of northeastern Pennsylvania: Geol. Soc. America Bull., v. 47, No. 4, p. 565–608.

WOOD, G. H., JR., TREXLER, J. P., AND KEHN, T. M., 1969, Geology of the west-central part of the southern anthracite field and adjoining areas, Pennsylvania: U.S. Geol. Survey Professional Paper 602, 150 p.

SLOPE SEDIMENTS IN SMALL BASINS ASSOCIATED WITH A NEOGENE ACTIVE MARGIN, WESTERN HOKKAIDO ISLAND, JAPAN

GEORGE DEVRIES KLEIN, HAKUYU OKADA, AND KIYOHIRO MITSUI
Department of Geology, University of Illinois at Urbana-Champaign, Urbana, IL, 61801; Geosciences Institute, Shizuoka University, Shizuoka 422, Japan, and Institute of Petrology, Mineralogy and Economic Geology, Tohoku University, Sendai 980, Japan

ABSTRACT

Neogene sediments on western Hokkaido have been divided into eight formations which are organized into three depositional facies. Facies 1 consists of volcaniclastic sandstone (lithic wacke) interbedded with volcanic ash and mudstone. Ash is normally graded from coarse granule to fine sand, whereas siltstone is organized as a normally graded sequence of coarse siltstone overlain by clayey siltstone and capped by claystone. Deposition occurred by ash fall and turbidity currents on a marine slope. Facies 2 consists of mudstone and fine-grained ash containing slump blocks ranging from 20 cm to 5 m in diameter, load casts and reworked fossils. Facies 2 was derived from a volcaniclastic source and deposited on a marine slope where slumping was common. Facies 3 consists of coarsening-upward sequences from mudstone through thick, cross-stratified sandstone, produced by a prograding delta.

The lithic wackes contain more than 85 percent volcanic clasts and other lithic clasts. Clay minerals include illite, montmorillonite, chlorite and mixed layer illite-montmorillonite.

These facies were deposited in a small intraslope basin comprising part of an early Neogene active continental margin. Deposition occurred on slopes by turbidity currents and slumping. Deltaic progradation occurred during a relative lowering of sea level. It appears that slumping dominated deeper-water episodes of basin evolution, whereas sand influx occurred during lower stands of sea level associated with deltaic progradation. The primary criteria for recognition of this ancient slope sequence include paleoecology, dominance of fine-grained sediments organized into normally-graded cycles, and abundant slump features.

INTRODUCTION

Modern slope deposits have been studied extensively and much is understood about their sediment distribution and depositional processes. A paucity of examples exists, however, concerning slope sediments in ancient continental margin settings. Slope deposits were recognized as one of the depositional facies associated with the growth and development of submarine fans by Mutti and Ricchi-Lucchi (1972) in their study of Cenozoic turbidite facies in the Apennines of Italy. In their depositional model, slope deposits generally overly the coarsening-upward motif of prograding submarine fans. According to them, slope deposits consist of fine-grained sediments interbedded with thick-bedded sandstone and conglomerates of what is now known to be of debris flow origin and various types of sediment deformed by slumping. Much of this conglomerate occurs as partially-or-completely-filled submarine canyon deposits, and the preservation of such canyon fills serve as a clue to the recognition of the adjoining slope sediment. This association was recognized long ago by Whitaker (1962) in his work on the Ordovician of eastern Wales, but again he stressed the nature of the submarine canyon fills rather than the intertonguing slope sediments.

In this paper, we discuss a well-exposed ancient slope deposit on an active continental margin. The example occurs in the Neogene of western Hokkaido Island, Japan (Fig. 1). A continental slope environment is confirmed by paleoecological studies (Ujiie et al, 1977), and although their tectonic domain may differ from the better-studied slope deposits of the western Atlantic (e.g., Doyle et al, 1976, this volume), they do comprise the same depositional setting.

This paper focusses on sedimentary structures, vertical sequences, sedimentary petrology and inferred depositional processes of these Neogene sediments.

REGIONAL GEOLOGY, STRATIGRAPHY AND PALEOECOLOGY

Neogene strata, underlain by the Upper Cretaceous and locally by the Paleogene, are exposed in northwestern Hokkaido in particular, and also in smaller basins elsewhere on that island (Fig. 1). These are composed dominantly of marine sediments, and partly of brackish and non-marine sediments. Each of the basins are separated by major thrust faults with a Northwest-Southeast trend (Matsuno, 1958; Matsuno and Kino, 1960; Hata, 1961; Okada, 1978). These thrusts are considered to be of syntectonic origin during deposition.

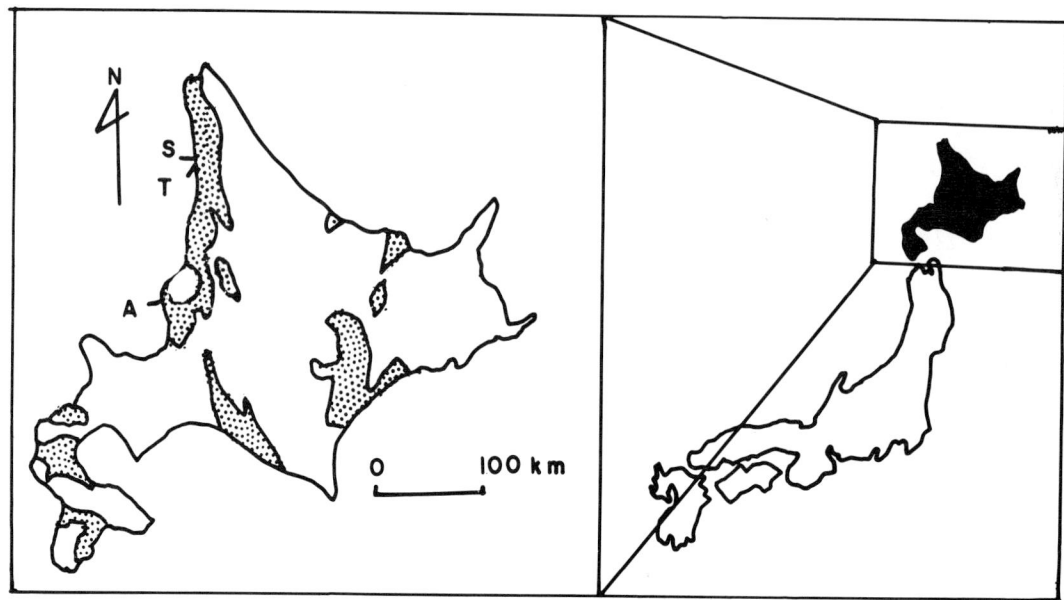

Fig. 1.—Hokkaido Island, Japan, showing major areas of outcrop of Neogene sedimentary rocks (stippled) and location of stratigraphic sections discussed in this paper (Abbreviations: A-Atsuta; S-Shosanbetsu; T-Toyosaki). Redrawn from Okada (1978).

Table 1.—Stratigraphic and Facies Summary of Slope Sediment, Atsuta, western Hokkaido Island, Japan

Age	Stratigraphic Unit	Lithology	Thickness (meters)	Facies
Late Miocene	Tobetsu Formation	Siltstone, massive, greenish-gray	more than 300	1
Late Miocene	Morai Formation	Siltstone, stratified, greenish-gray.	220–300	1
Late Miocene	Bannosawa Formation	Sandstone, green, glauconitic, massive, medium-grained.	100–350	1
Late Miocene	Atsuta Formation	Mudstone, massive, dark-gray, with calcareous conretions	260–700	1
Late Miocene	Hattari Formation	Interbedded sandstone and mudstone	not measured	

Table 2.—Stratigraphic and Facies Summary of Slope Sediment, Shosanbetsu, western Hokkaido Island, Japan

Age	Stratigraphic Unit	Lithology	Thickness (meters)	Facies
Early Pliocene	Mochikubetsu Formation	Sandstone, gray, fine-grained, pebbly at base	120	2
Early Pliocene	Embetsu Formation	Mudstone, gray to bluish gray, with blocks of sandstone, some cross-stratified.	450–1,000	2 and 3
Early Pliocene	Kinkomanai Formation	Siltstone, dark-gray, interbedded with sandstone, gray to greenish gray, fine-grained	Not measured	2 and 3

Our examples of Neogene slope deposits come from seven stratigraphic units; four are Miocene in age and three are Pliocene. These formations are exposed at two localities (Fig. 1) near the villages of Atsuta and Shosanbetsu, Hokkaido. The major litho-stratigraphic subdivisions are shown in Figures 2 and 3 and Tables 1 and 2. The stratigraphic subdivisions follow prior mapping by Tsushima et al (1956) at Atsuta and by Hata (1961) and Hata and Tsushima (1969) at Shosanbetsu. The biostratigraphy, paleoecology and magnetostratigraphy were revised recently by Ujiie et al (1977) and their stratigraphic usage is followed in this paper.

The foraminifera at both sections were studied recently by Ujiie et al (1977) and a paleo-depth determination was possible from analysis of benthonic foraminifera. At Atsuta, the presence of *Martinottiella communis* (D'Orbigny), *Goesella schencki* (Asano), and *Uvigerina akitaensis* (Asano) indicate upper bathyal depths, or an outer slope assemblage. These are also associated with beds and horizons of *Epistominella pulchella* (Husezima and Maruhasi), *E. japonica* (Asano), *Globigerina bulloides* (D'Orbigny), *G. woodi* (Jenkins), and *Globigerinita glutinata* (Egger) which are of an identical water depth assemblage. The Shosanbetsu section contains benthonic foraminifera at very few horizons but include *Cribroelphidium yabei* (Asano), *Elphidium etigoense* (Husezima and Maruhasi), *Buccella frigida* (Cushman) and *Cassidulina yabei* (Asano and Makamura) which occur in a similar slope setting (See Ujiee et al, 1977, p. 91–92). In short, the benthonic foraminifera indicate these formations to be dominantly of a slope depositional setting.

SEDIMENTARY FACIES.

Each facies recognized in our study was subdivided according to dominant lithology, grain size distribution, association of sedimentary structures, types of sedimentary sequences and biogenic components. Our approach follows that of DeRaaf et al (1965), Klein (1975), and Klein et al (1972).

Facies 1

Main Facts.—This facies occurs only at the section at Atsuta and is represented by the Tobetsu, Morai, Bannosawa and Atsuta Formations. Facies 1 consists of volcaniclastic siltstone which is light to dark gray. Interbedded in subordinate amounts are coarser-grained, sand-sized, normally graded volcanic ash beds and mudstone layers. The dominant components of each bed are pyroclastic glass, plagioclase and hornblende.

Limestone nodules are present in the upper part of Facies 1 (Fig. 4), from about 65 to 85 meters from the top of the section. These nodules are

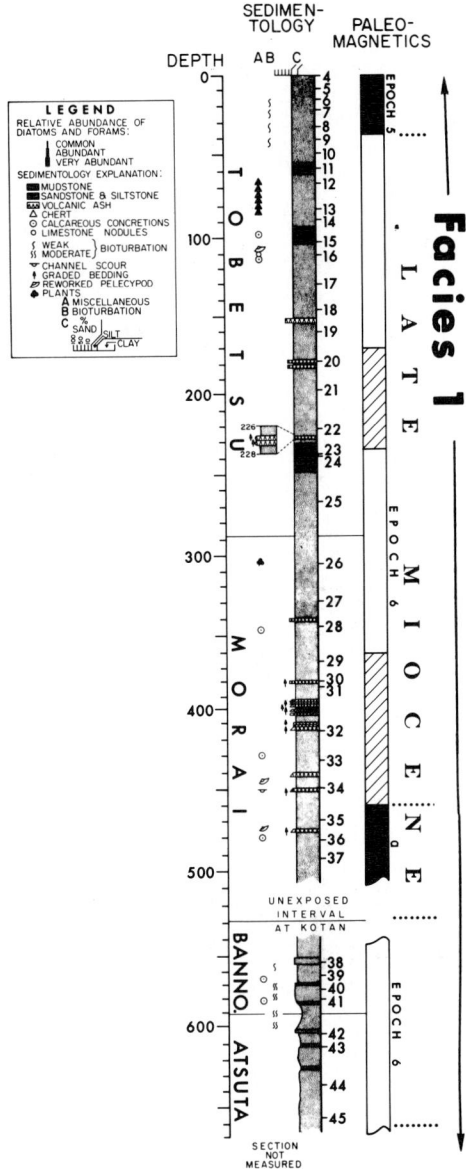

Fig. 2.—Sedimentary log and paleomagnetic log, Atsuta. Numbers refer to samples discussed in text and Tables 3, 4, and 5.

of replacement origin. Limestone nodules also occur within the middle interval of this facies (300 to 450 meters from top). These nodules are resistant to erosion and stand out in relief on sea cliffs. They range in diameter from 10 cm. to 3 m.

Sedimentary structures are rare in this facies. Many of the coarser-grained ash beds are normally graded; in some instances the graded beds are

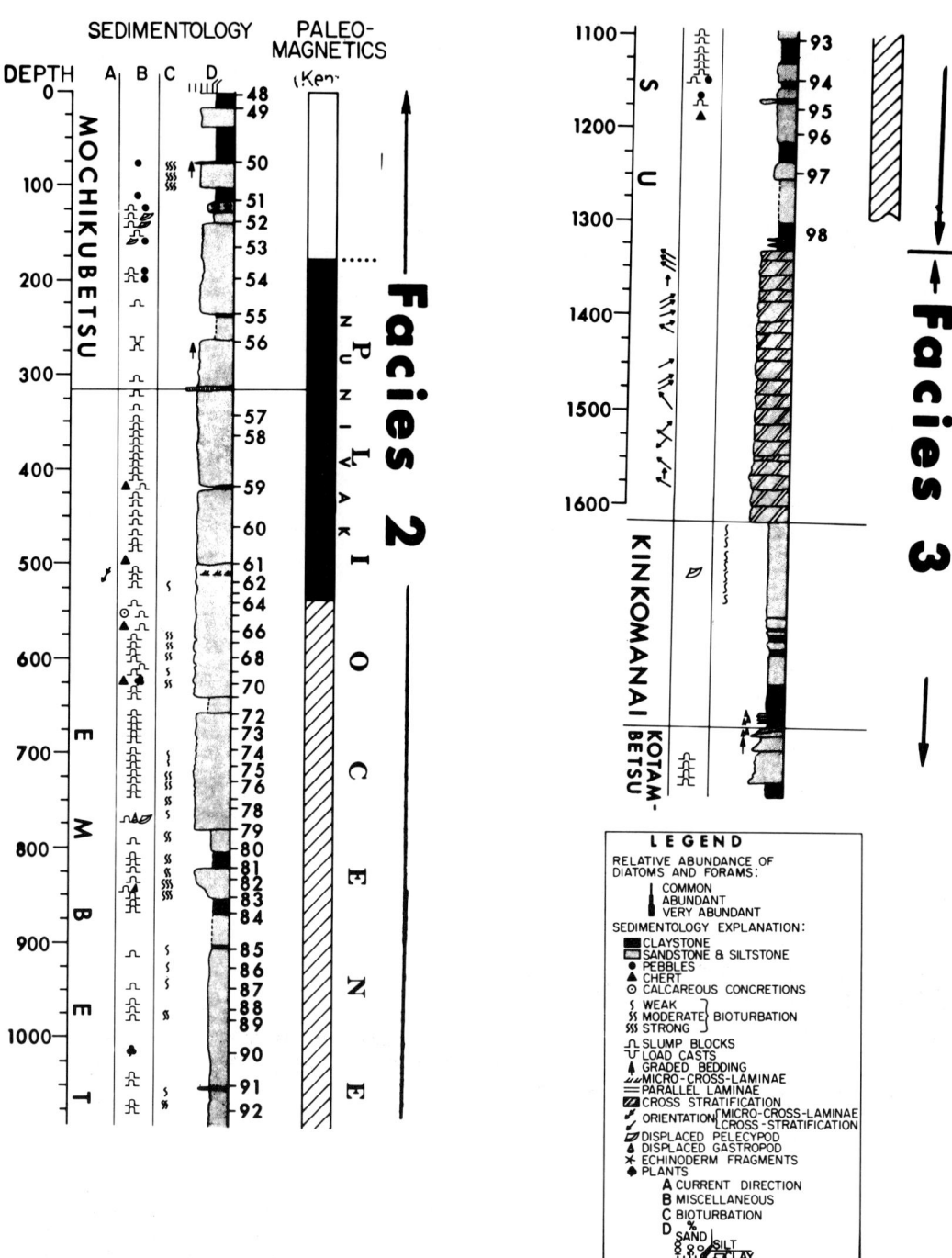

FIG. 3.—Sedimentary log and paleomagnetic log, Shosanbetsu. Numbers refer to samples discussed in text and in Tables 3, 4, and 5.

FIG. 4.—Large limestone nodule, Morai Formation, Atsuta. Scale is 2 m long.

FIG. 6.—Cyclic arrangement of coarse siltstone grading into fine clayey siltstone and mudstone, Atsuta. Scale in cm. Facies 1.

amalgamated with particle sizes changing upward from coarse granule to fine sand (Fig. 5). The siltstone beds are also normally graded and change upward into mudstone. Within this facies, a cyclic repetition of siltstone and mudstone is common. Each cycle is represented by a basal coarse siltstone which grades upward into a clayey finer-grained siltstone and is capped gradationally by a claystone. Both the lower and uppermost contact of this cycle are sharp (Fig. 6). Each such cycle is 1 m thick. This motif of sedimentation is common to the lower half to third of the section and provides a distinct bedding style similar to coarse-grained flysch deposits. In the lower third of this facies, the sand-sized content increases to 20 percent. Here, normally-graded bedding and parallel laminae are more common.

Several horizons are bioturbated with the trace fossil *Chondrites* being the most common type of burrow. Other fossils include wood fragments bored by *Teredo*, the scaphopod *Dentalium* and reworked specimens of the shallow-water pelecypod *Mactra sp*.

Interpretation.—Facies 1 was derived dominantly from a volcanic island arc source. Sedimentary processes of deposition included ash falls, turbidity currents and hemipelagic clay deposition. Both sand and silty turbidites were emplaced within a small basin on an active continental margin slope environment (Okada, 1978; Okada and Kitamura, 1978). Evidence for turbidites includes the graded ash layers, graded ashy sandstone layers and the siltstone-mudstone cycles (Fig. 6). Unusual in this facies are the silty graded cycles which, although common to several deep-water back-arc basins such as the Shikoku Basin and West Philippine Basin (Karig, Ingle et al, 1975), are rare or absent in ancient sedimentary rocks exposed on land.

Facies 2

Main Facts.—This facies is exposed at the Shosanbetsu section and characterizes the Mockikubetsu, upper Embetsu and Kotanbetsu Formations. The dominant lithology of facies 2 consists of planar-bedded dark gray mudstone containing fine-grained ash and clay minerals. Within this mudstone, however, are blocks of sandstone ranging in diameter from 20 cm to 5 m (Fig. 7). These blocks are interpreted as slump blocks (See below) because they are folded and contorted and are oriented at random, clearly indicating some type of block rotation during deposition and transport. Internally, some of these blocks, particularly the rotated ones, are parallel-laminated, micro-cross-laminated or cross-stratified (only at top of facies). The blocks show an imbricated fabric at one horizon, 900 meters below the top of the section. The direction of imbrication dips

FIG. 5.—Amalgamated, normally-graded volcanic ash beds, Atsuta. Scale in cm. Facies 1.

Fig. 7.—Deformed slump block of bedded sandstone encased in silty mudstone, Shosanbetsu. Scale in cm. Facies 2.

south to southeast which is opposite to regional paleoslope which can be inferred from cross-stratification directions shown in distributary sandstones of Facies 3 (Fig. 3).

Other sedimentary structures present include thick-bedding, demarcated by changes of the mudstone to siltier beds, and load casts at the base of clayey siltstones observed 250 m below the top of the section at Shosanbetsu.

A variety of reworked shallow-water fossils occur within this facies, particularly within the upper 300 meters. These include pelecypods (*Macoma, Mya*), gastropods (including *Neptunea vinosa*) and shallow-water foraminfera (*Cribroelphidium ezoense*). Charcoal fragments and fragments of wood bored by *Teredo* are also present.

Interpretation.—This facies was also derived from a volcaniclastic source. The environment of deposition appears to be either a continental slope in an active margin, or a delta front. Slump blocks consisting of rotated sandstones are present within the section indicating the dominance of gravity-controlled processes of deposition for this facies. The random mixing of these blocks of slump origin within thick mudstones suggests both were deposited by some sort of debris flow process. The size of the blocks is small to medium, and is similar to slump features reported by Shepard (1973) from the Magdelena Delta of Colombia and by Coleman (1976) from the Mississippi Delta. The paleoecological interpretations advanced by Ujiie et al, (1977, p. 91–92) and discussed in an earlier section, clearly indicate this slump activity and muddy debris flow activity occurred on a continental slope-type setting. This facies is a total of 1,200 meters thick, which if it were a delta front implies deltaic progradation into a very deep water environment. This excessive thickness of Facies 2 sediments is more consistent with a slope environment of deposition.

Facies 3

Main Facts.—This facies comprises the lower Embetsu and Kinkomanai formations and was observed south of the village of Toyosaki (See Fig. 1 for location). It occurs from about 1,325 to 1,750 meters below the top of the section measured at Shosanbetsu (Fig. 3). This facies consists of interbedded mudstone, mudstone and sandstone interlayers and thick-bedded sandstone. These lithologies are organized into a single coarsening-upward sequence (Fig. 8), which straddles the boundary between the Embetsu and Kinkomanai Formations (Fig. 3). The basal portion of the sequence consists of parallel-laminated claystone interbedded with siltstone. The interbedded lithologies show sharp basal and upper contacts. The middle portion of the cycle consists of interbedded thin sandstone and mudstone. This thin sandstone contains parallel laminae. The base and top of this thin sandstone is sharp.

The upper portion of the cycle consists of coarse-grained, volcaniclastic sandstone. These are thick-bedded. These beds are cross-stratified with average set thicknesses of 2.5 m (Fig. 9). The cross-strata show a dominant northwest to northeast orientation (Fig. 10). Set boundaries are sharp. Locally, the basal portion of the cross-stratified sets show channel scours cutting into the underlying beds.

Below this coarsening-upward sequence (Fig. 3) is clastic, normally-graded turbidite sandstone. This sandstone is coarse-to-medium-grained and is organized into a series of amalgamated graded beds with sharp scours. Below these turbidites is silty mudstone with slump blocks.

Fig. 8.—Coarsening-upward sequence of basal prodelta mudstone, interlayered sandstone-mudstone zone of delta front, and uppermost delta plain sandstone. Arrow shows interval represented by sequence. Facies 3. Toyosaki. Sandstone below sequence in lower right represents turbidite beds.

Fig. 9.—Thick planar and trough cross-stratification in delta plain facies of coarsening-upward sequence at Toyosaki. Facies 3. Shrine gate is 2 m high.

Interpretation.—The sequence of facies 3 is identical to similar coarsening-upward sequences described from the Mississippi Delta (Coleman, 1976) and from some ancient examples such as described by Asquith (1970), Hobday and Matthews (1975), Klein (1974), and Klein et al (1972). The basal mudstone represents the prodelta zone of such a delta, the middle interbedded sandstone-mudstone zone represents the delta front, and the thick-bedded, cross-stratified sandstone represents a bar-finger sand facies of a delta plain; this bar-finger sandstone is clearly fluvial-dominated similar to those from modern deltaic environments.

It is our interpretation that the basal turbidites below the coarsening-upward sequence may have been deposited during seasonal flooding in the river systems that supplied the sand to form the bar finger sandstone at the top of the sequence. During such seasonal flooding, the density contrast between the river effluent and marine waters would change so that the density of the effluent exceeds that of the ocean waters of the depositional basin, and thus generates a seasonal system of turbidity currents that supplies sand to the ocean floor. The density contrast interpreted here is similar to the seasonal density contrast suggested for flooding phases of the Mississippi Delta by Scruton (1956). These basal turbidites are similar to the implied deeper-water turbidites over which the modern Niger Delta is presently prograding (Burke, 1972).

The deltaic succession at Toyosaki represents a coastal margin along the edge of the depositional basin and is in a deeper-water to shallow-water transition with Facies 2 which occurs both immediately above and below the delta. This delta suggests an episode of coastal progradation from island arc sources which were possibly active and uplifted during Miocene time (Okada, 1974). The transition from deeper-water to shallower-water facies is characteristic of many so-called geosynclinal associations (Pettijohn, 1975; Pettijohn et al, 1972; Potter and Pettijohn, 1963). In our example, we observed such a transition from a

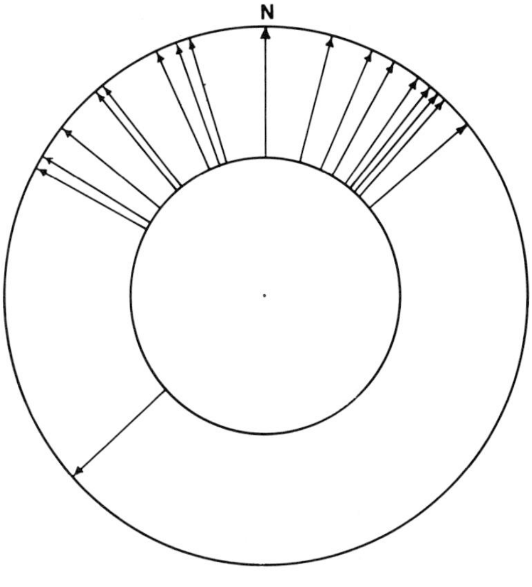

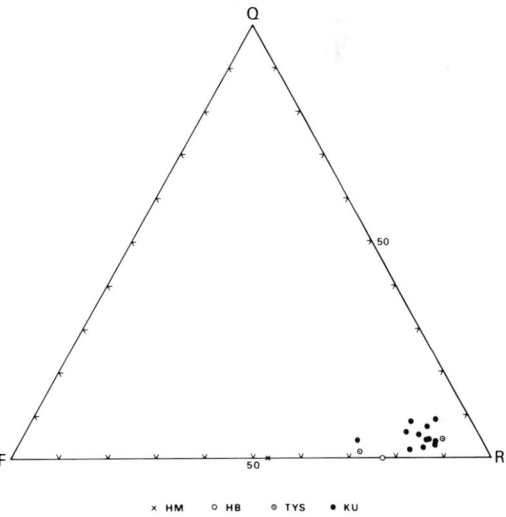

Fig. 10.—Paleocurrent orientation of cross-strata, Facies 3, delta plain sandstone, Toyosaki.

Fig. 11.—Ternary diagram showing composition of sandstone. Q-quartz, F-feldspar, R-rock fragments. HM-Morai Formation, HB-Bannosawa Formation, TYS-Embetsu Formation (Sandstone member), K-Kotanbetsu Formation.

TABLE 3.—MODAL ANALYSIS OF SANDSTONES

Specimen	Quartz	Feldspar	Rock Fragments							Opaque minerals	Serpentinite	Glauconite	Calcite cement	Matrix	Organic remains
			Chert	Slate and Sandstone	Igneous	Pumice and Glass fragments	Metamorphics	Pyroxene	Amphibole						
HM34'	0	34.99	0	0	32.42	2.40	0	4.80	0.51	+	0	0	3.95	20.93*	0
HB39	0	15.26	0.25	4.74	29.74	0.53	0	0	0	2.11	0	15.26	2.37	29.74	0
TYS1	1.19	20.44	0.85	19.59	26.58	0	2.90	1.19	0	+	0	0.34	23.34	1.02	2.56
TYS2	3.71	5.66	4.10	9.18	25.78	20.71	0.98	0.98	0	+	0	0	22.85	0	6.05
KU6501	6.74	5.29	2.37	32.24	14.03	17.49	0	0	0	+	0	0	0.16	21.68	0
KU6503	3.04	10.13	3.65	53.85	13.97	0	0.41	0	0	+	0	0	0	14.95	0
KU6504	7.19	10.68	0.97	43.89	21.75	0	0.17	0	0	+	0	0	0.39	14.96	0
KU6506	3.25	9.56	4.21	49.33	17.59	0	0	0	0	+	0	0	0.96	15.10	0
KU6510	5.81	8.30	2.29	46.89	16.39	0	1.25	0	0	+	0	0	1.66	0	17.41*	0
KU6511	3.45	23.35	0.21	6.71	56.30	0	0	0	0	+	0	0	0	2.11	7.87*	0
KU6512	3.20	8.46	3.58	55.83	15.23	0	0	0	0	+	0	0	0	0.55	13.15	0
KU6513	2.34	8.46	3.60	38.31	24.28	0	0	0	0	+	0	0	1.08	21.93	0	0
KU6514	1.33	12.40	3.14	35.54	23.97	0	0	0	0	+	0	0	0.16	23.31	0.15	0
KU6515	4.66	7.89	1.08	37.64	26.17	0	0.33	0	0	+	1.26	0	0	20.97	0	0
KU6519	4.02	9.76	5.17	29.45	28.30	0	0	0	0	+	0	0	0.16	23.14	0	0
KU6521	1.72	9.70	2.48	44.49	15.59	0	0	0	0	+	0	0	0.19	25.45	0.38	0
KU6526	3.28	7.95	1.72	40.19	19.01	0	0.78	0	0	+	1.41	0	0	25.34	0	0

N.B. HM: Morai Formation, HB: Bannosawa Formation, TYS: Sandstone member of the Embetsu Formation, KU: Kotanbetsu Formation.
+: trace, *: including glass shards

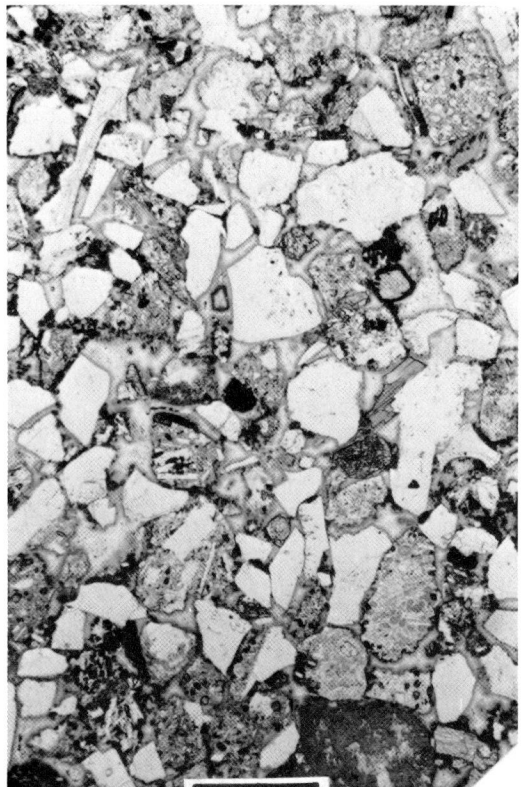

small ancient basin comprising part of an active continental margin. Similar transitions are known from the Coral Sea marginal basin where the sediment dispersal is dominated by the Fly River Delta (Galloway, 1975; Burns, Andrews et al, 1973; Andrews, Pkacham et al, 1975). Perhaps the deep-water to shallow-water transition of mobile belts represents the filling up of a series of smaller basins such as described herein, or a short-term relative lowering of sea level for reasons not fully understood.

SEDIMENTARY PETROLOGY

This section describes mineralogical and petrological characteristics of the Neogene sediments discussed above. Samples used in this study were also used for micropaleontological and paleomagnetic study (Ujiie et al, 1977); sample numbers used for all three topical studies are shown in Figures 2 and 3.

Heavy mineral analysis of sandstone and coarse siltstone were done by Okada's (1960) method. Point counting and modal analysis of sandstone followed Okada's (1971) method.

FIG. 12.—Photomicrograph of tuffaceous lithic sandstone (HM-34'), Morai Formation, Atsuta. Plain light. Scale is 0.5 mm.

Clay mineral investigations used X-ray diffraction with a JEOL diffractometer (CuKα radiation with a *Ni* filter). Identification of clay minerals was by means of ethylene glycol treatment (for montomorillonite versus haloysite) and 6 *N* hydrochloric acid treatment (for kaolinite versus chlorite). The *ZnO* interior standard method was applied to the quantitative estimation of these clay minerals (Aoyagi, 1967).

Sandstone Mineralogy

Major constituents of sandstone selected from the Morai Formation, Bannosawa Formation and the sandstone meber of the Ebetsu Formation are volcanic plagioclase, andesite clasts, older sandstone fragments, slate clasts, pumice fragments and glass shards (Fig. 11, Table 3). These sandstones are generally deficient in quartz (Figs. 12, 13, 14, 15). Rock fragments other than volcanics include chert and metamorphic rock fragments.

The sandstone composition of the Kotanbetsu Formation is characterized by abundant older sediment clasts and volcanic rock fragments. Quartz is ubiquitous but generally, a minor constituent (Figs. 16, 17). The Kotanbetsu sandstone and the lower Embetsu sandstone is classified as a matrix-poor lithic arenite (Fig. 11).

Heavy mineral investigations reveal four distinct assemblages (Table 4). These include an augite - hypersthene - hornblende-oxyhornblende - glaucophane assemblage, a zircon-garnet-augite-diopside - hypersthene - hornblende - oxyhornblende-glaucophane assemblage, an augite-hypersthene-oxyhornblende assemblage, and a zircon-tourmaline-garnet-augite-hornblende assemblage. The first assemblage characterizes the Morai Formation, the second characterizes the Mochikubetsu Formation and the mudstone member of the Embetsu Formation, and the third assemblage characterizes the sandstone member of the Embetsu Formation. The fourth assemblage characterizes the Kotanbetsu Formation. Among these assemblages, the Kotanbetsu Formation yields a sharply contrasting heavy mineral suite in that rounded zircon, tourmaline, rutile and garnet are important constituents.

FIG. 13.—Same as Figure 12. Crossed nicols. Scale is 0.5 mm.

FIG. 14.—Photomicrograph of calcareous litharenite (TYS-1), Embetsu Formation (sandstone member), Atsuta, Plain light. Scale is 0.5 mm.

TABLE 4.—HEAVY MINERALOGY OF SANDSTONE AND SANDY SILTSTONE

Mineral component		HM23	HM27'	HM32'	HM34'	HB39'	HMo48'	HMo55	HMo58'	HE67'	HE69'	HE73	TYS1	TYS2	TYS3	KU6504	KU6510	KU6526
Opaque minerals		6.03	42.31	69.07	15.15	45.45	20.60	29.70	30.95	12.20	63.60	30.00	10.89	19.25	25.37	83.08	69.23	65.77
Transparent minerals		93.97	57.69	30.93	84.85	54.55	79.40	70.30	69.05	87.80	36.40	70.00	89.11	80.75	74.63	16.92	30.77	34.23
Zircon	Euhedral colorless			6.55		1.97	1.39		0.87	0.81	4.33	2.46			0.98	25.71	21.10	7.88
	pale pink			2.80			0.46				1.44					6.43	5.96	3.45
	Rounded colorless																0.92	
	pale pink								0.43							7.14	1.83	0.99
	purple																0.46	
Tourmaline																2.14		
Rutile																1.42		
Garnet	colorless	31.11	20.00		4.39	2.46	0.46	0.44	0.43		1.44	2.46				24.29	9.18	1.97
	pale pink															10.71	2.75	1.97
	orange																	0.49
Augite				0.47			31.02	20.19	12.61	17.74	16.83	17.24	45.25	41.63	57.07	7.14	8.26	15.76
Diopside			1.46				0.46	1.31	0.43	0.81		0.49						
Hypersthene		6.67		3.74	85.37	0.49	5.56	2.52	0.87	0.40	1.44	14.78	32.58	7.69	12.20			
								6.55	10.00	6.85	15.38			2.26				8.87
Hornblende	green	62.22	39.51	7.94	7.32	88.67	37.50	48.13	49.16	49.60	35.10	44.34	20.36	41.18	24.88	2.86	7.34	
	greenish															7.88	16.46	11.82
	brown																	
Oxyhornblende	brown		20.49	77.10	1.46	0.99	11.57	9.17	11.74	16.13	15.38	16.75		3.62	1.95		21.06	33.50
			11.22		1.46	2.46	6.02	6.55	9.13	6.85	4.33	1.48	1.81	2.72	2.44			
Glaucophane			0.49	0.47			0.94	2.18	1.30	0.81	1.93							
Epidote						0.99	0.46	0.44	0.43									
Apatite									0.43									
Chlorite							0.46		0.87					0.45	0.48		1.38	
Biotite			6.83	0.93	1.48		3.24	2.52	0.43		1.92			0.45		0.71	1.92	0.49
Anatase									0.87							3.57		11.82
Spinel						0.49											1.38	0.99
Heavy mineral weight percentage		30.86	0.97	0.16	0.26	0.96	1.39	0.28	1.19	0.17	0.58	0.33	2.58	2.05	3.50	0.12	0.23	0.10

N.B. HM: Morai Formation, HB: Bannosawa Formation, HMo: Mochikubetsu Formation, HE: Embetsu Formation (mudstone member), TYS: Embetsu Formation (sandstone member), KU: Kotanbetsu Formation.

FIG. 15.—Same as Figure 14. Crossed nicols. Scale is 0.5 mm.

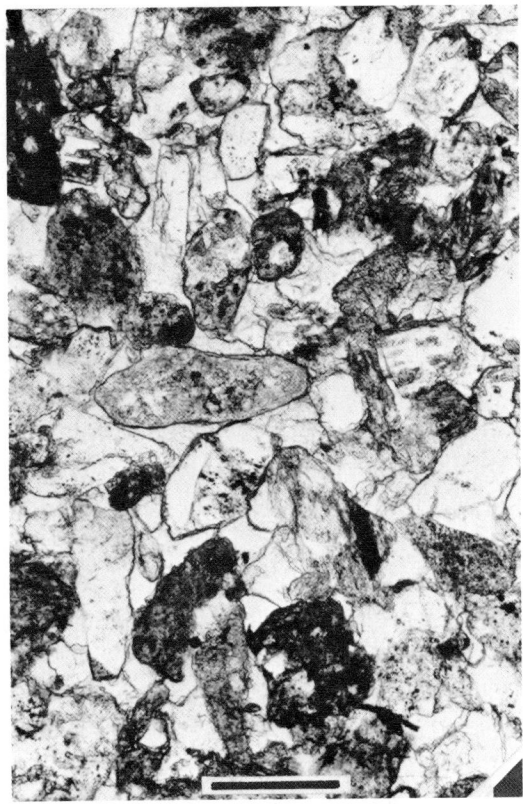

FIG. 16.—Photomicrograph of calcareous litharenite (KU-6501), Kotanbetsu Formation, Shosanbetsu. Plain light. Scale is 0.5 mm.

Clay Mineralogy.

Illite, montmorillonite, chlorite and mixed-layer illite-montmorillonite are the only clay minerals present in samples studied from the Neogene of western Hokkaido Island (Table 5). Montomorillonite is the most abundant clay mineral with chlorite of next importance. In some samples, chlorite was found to be more abundant than montmorillonite. Illite is also a widespread, but minor constituent. No stratigraphic trend was observed in the vertical distribution of clay mineral species at the two localities.

The zeolite clinoptilolite appears in the Morai Formation (Pliocene) at Atsuta, and the Embetsu Formation (Miocene) at Shosanbetsu. This clinoptilolite-bearing mudstone is assigned to the clinoptilolite zone of burial diagenesis (after Iijima and Utada, 1971). The time difference in its appearance between the two localities may be a function of differential burial depth. The mineral is an alteration product of volcanic glasses and other pyroclastics which are common in these strata.

Mineralogy of Nodules

Carbonate and cherty nodules occur in the Neogene at both localities. Most common in occurrence are greenish-gray limestone nodules (Fig. 4) which usually occur either as little elongated spheres or as thick lenses and range in their exposed longest diameter from 20 cm to three meters. The larger nodules are isolated, whereas the smaller nodules tend to occur in distinct layers (Fig. 18). The smaller nodules consist of dary gray glassy-appearing chert (Fig. 19). These smaller nodules are irregular in shape and are characteristic of the Tobetsu Formation.

Mineralogic and petrographic study of the limestone nodules of the Tobetsu Formation shows that they consist of chlorite, biotite, pyrite and diatomaceous tests (Fig. 20). One sample (HT8'-1) contains a large volume of dolomite (Table 5).

Fig. 17.—Same as Figure 16. Crossed nicols. Scale is 0.5 mm.

Fig. 19.—Chert nodules in Tobetsu Formation, Atsuta.

Fig. 18.—Bedded calcareous nodules, Tobetsu Formation, Atsuta.

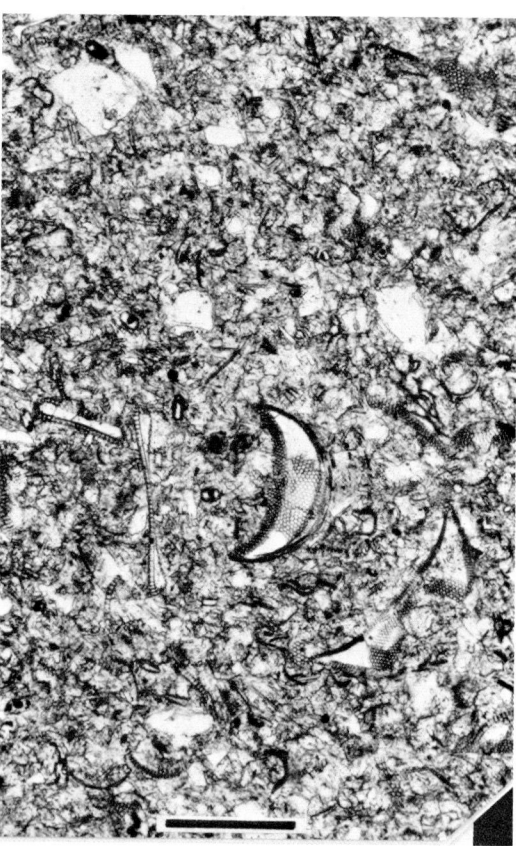

Fig. 20.—Photomicrograph of calcareous nodule (HT-12'), Tobetsu Formation, Atsuta. Plain light. Scale is 0.5 mm.

Cherty nodules contain quartz, plagioclase, diatom tests and vitric tuff (Fig. 21). One cherty nodule from the Tobetsu Formation contains a fresh core with cristobalite, and accessory quartz, feldspar, illite and chlorite, whereas the weathered rind contains a large proportion of quartz, feld-

TABLE 5.—CLAY MINERALOGY

Sample	Montmorillonite	Illite	Chlorite	Mixed Layer Illite-Montmorillonite	Clinoptilolite	Cristobalite	Quartz	Feldspar	Dolomite
HT3	•	30	+	·	·	6	40	9	
HT8'-1	•	+	·	·	·	4	9	4	36
HT8'-2	O	14	6	·	·	8	35	13	
HT11	O	14	4	·	·	6	26	6	
HT12'-A	+	4	·	6	+	30	13	12	3
HT12'-B	O	15	O	12	•	20	13	33	25
HM25'-2A	+	·	6	+	·	4	7	45	12
HM25'-2B	•	22	·	8	O	15	3	3	35
HM25'	•	30	·	3	·	4	6	2	30
HM26'	O	14	·	3	·	7	6	20	27
HM28'	·	6	·	4	O	14	+	6	35
HM29'-A	•	28	·	6	·	8	·	3	5
HA44	·	8	·	7	•	20	6	4	38
HA47	O	16	·	3	O	15	·	·	33
HMo55	O	17	·	4	O	11	·	4	27
HE62	O	14	·	5	O	14	·	6	30
HE69	·	8	·	4	·	8	3	·	27
HE72	·	9	·	6	·	9	·	·	25
HE77	O	14	·	4	·	4	·	·	30
HE83	O	14	·	3	O	18	·	·	38
HE85	⊙	17	·	4	O	10	·	·	35
HE90	•	20	·	4	·	7	3	·	25
HE96	•	20	·	5	O	14	+	·	38

N.B. •: 20–30%, O: 10–20%, ·: 3–10%, +: traceable amount. HT: Tobetsu, HM: Morai, HA: Atsuta, HMo: Mochikubetsu, HE: Embetsu Formations.

spar, montmorillonite, illite, chlorite, and a much-reduced amount of cristobalite (Samples HT12'-A and HT12'-B; Table 5).

Provenance

The marine Neogene of northwestern Hokkaido consists primarily of airborne ash and volcaniclastic sandstone of acid and andesitic composition. These volcaniclastic sandstones also contain high concentrations of pyroxene and amphibole in the various heavy mineral assemblages known to occur in them. This volcaniclastic sandstone is in agreement with Dickinson's (1974) finding that deposits in the basins associated with active continental margins contain a wide variety of pyroclastic material. The Neogene sedimentary rocks reflect active island arc magmatism at the time of deposition. The magmatism of this region is related closely to the very active Miocene igneous activity in the northwest Pacific (Matsuda et al, 1967; Oba, 1977; Klein et al, 1978). This timing of igneous activity appears to coincide with a known change in spreading direction in the North Pacific (Menard and Atwater, 1968; Jackson et al, 1972) and also with a resumption of spreading (Vine, 1966).

In addition to a volcanic arc source, a glaucophane schist source representing a buried metamorphic terrain is also indicated. The rounded garnet and zircon fragments in the Kotanbetsu Formation suggests that some recycling of older volcaniclastic sediment has occurred. The presence of clasts of older sandstone, slate and chert of the Kotanbetsu Formation suggests that the entire section in addition to having been derived from a pyroclastic source, was also derived from an arc setting exposed to the east and south of the present outcrop belt. The mudstone was probably also derived from a volcanic source.

SUMMARY AND CONCLUSIONS

The Neogene sediments of western Hokkaido Island were deposited in small basins in an active continental margin adjoining an island arc system. These sediments appear to have been derived from contemporaneous island arc volcanism, erosion of the volcanic terrain associated with an island arc and erosion of the metamorphosed root zones of this island arc system. The sources were of a multiple nature east and south of the present outcrop belt and include derivation both from the outer island arc and inner island arc zones. This

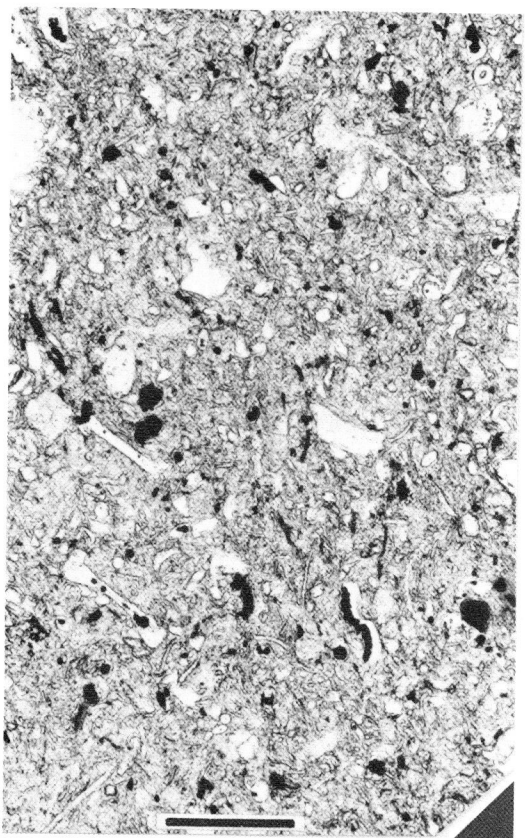

Fig. 21.—Photomicrograph of chert nodule (HT-8'-A), Tobetsu Formation, Atsuta. Plain light. Scale is 0.5 mm.

paired zonation of this island arc source contains both metamorphic and sedimentary belts, an observation consistent with prior work by Miyashiro (1961) in the Japanese Island arc system, and by Okada (1974) in Hokkaido Island.

The depositional setting of the Neogene sediments of western Hokkaido Island was clearly on a slope setting in a series of small basins. Depth of deposition was determined to be of an upper bathyal slope setting from paleoecological data. Processes of sedimentation occurring there included deposition of sediment from ash falls, deposition of sands and silts by turbidity currents, deposition of muds by hemipelagic processes and deposition of sediment by slump processes, giving rise to zones of extensive slump blocks in some formations as well as the presence of resedimented shallower water fossils. During relatively lower relative elevations of sea level, deltaic progradation occurred in this upper slope setting.

Supply of sediment to these slope settings was presumably by short length streams draining the island arc complex, which built small deltas on the shoreline. With lowering of sea level, this shoreline prograded basinward and resulted in producing a localized deltaic facies interbedded with the slope sediments. Supply of sand to the slope setting was primarily by turbidity currents. Resedimentation on the slope by slumping was common.

In recognizing these sediments to have been deposited in a slope setting, several features were found to be most critical in our interpretation. The first feature of major significance for recognition of slope settings for these sediments was paleoecological (Ujiie et al, 1977). The second feature that indicated such a setting was the presence of fine-grained clastic sediment both of hemipelagic origin, and those organized into silty graded cycles which were deposited by silty turbidity currents. The presence of rotated, isolated slump blocks of varying size set in a mudstone matrix and associated with load structures was the third criterion we used to identify a marine slope setting for these rocks. The high volcaniclastic content of the sediments indicates this depositional slope system to have been associated with an active continental margin.

Our observations and interpretations permit us to draw the following conclusions about the Neogene sedimentary rocks of northwestern Hokkaido Island:

(1) The sedimentary rocks consist mostly of mudstone, siltstone and fine-grained sandstone. The sandstone was derived from a complex island arc source. Airborne ash, the volcanic flanks and the older root zone of this island arc provided most of the volcaniclastic material.

(2) Deposition occurred by hemipelagic processes, turbidity currents and slumping. The depositional environment was an intrabasinal slope on an active continental margin.

(3) Seaward progradation of the coastline is indicated by the presence of a prograding deltaic sequence in the lower part of the section we studied. Such progradation favors an increased influx of sand into the depositional basin. The deltaic sequence occurs transitionally with deeper water sediments in a manner similar to such shoaling-upward transitions reported from mobile belt sequences.

(4) The main criteria used in the recognition of this slope sequence include paleoecology (Ujiie et al, 1977), fine-grained sediment, silty normally-graded cycles, and the occurrence of slump and load structures.

ACKNOWLEDGMENTS.

Research support for this project came from the U.S-Japan Scientific Exchange Program administered by the National Science Foundation

(Grants OIP-75-17409 and DES 75-14430) and the Japan Society of the promotion of Science and (Grant 4R028). Division of responsibility for this paper is as follows: depositional facies and processes (Klein), sandstone and siltstone petrology and heavy mineralogy (Okada) and clay mineralogy (Mitsui). The conclusions were jointly conceived. John M. Dennison and one unidentified reviewer are thanked for their constructive comments on an earlier manuscript version of this paper.

REFERENCES

ANDREWS, J. E., PACKHAM, G. H., et al, 1975, Initial Reports of the Deep Sea Drilling Project, Washington, U.S. Government Printing Office, v. 30, 753 p.

AOYAGI, K., 1967, Mineralogical study of sedimentary rocks by X-ray diffraction method and some examples of its application to petroleum geology: Jubilee Pub. Commem. Prof. Yasuo Sasa (In Japanese), p. 541–556.

ASQUITH, D. O., 1970, Depositional topography and major marine environments, Late Cretaceous, Wyoming: Am. Assoc. Petroleum Geologists Bull., v. 54, p. 1184–1224.

BURKE, KEVIN, 1972, Longshore drift, submarine canyons and submarine fans in development of Niger delta: Am. Assoc. Petroleum Geologists Bull., v. 56, p. 1975–1983.

BURNS, R. E., ANDREWS, J. E., et al, 1973, Initial Reports of the Deep Sea Drilling Project: Washington D.C., U.S. Government Printing Office, v. 21, 931 p.

COLEMAN, J. M., 1976, Deltas: Processes of deposition and models for exploration: Champaign, IL, CEPCO, 142 p.

DERAAF, J. F. M., READING, H. G., AND WALKER, R. G., 1965, Cyclic sedimentation in the Lower Westphalian of North Devon, England: Sedimentology, v. 4, p. 1–52.

DICKINSON, W. R., 1974, Sedimentation within and beside ancient and modern magmatic arcs, in Dott, R. H., Jr, and Shaver, R. H., eds., 1974, Modern and ancient geosynclinal sedimentation: Soc. Econ. Paleontologists and Mineralogists Spec. Pub. 19, p. 230–239.

DOYLE, L. E., WOOD, C. C., AND PILKEY, O. H., 1976, Sediment flux through intercanyon slope areas: U.S. Atlantic continental margin: Geol. Soc. America Abs. with programs, v. 8, p. 843.

GALLOWAY, W. E., 1975, Process framework for describing the morphologic and stratigraphic evolution of deltaic depositional systems, in Broussard, M. L., ed., Deltas, (2nd ed): Houston, Houston Geol. Soc., p. 87–99.

HATA, M., 1961, Hatsuura: explanatory text of the Geological Map of Japan: Geol. Survey of Japan (In Japanese), 60 p.

————, AND TSUSHIMA, K., 1969, Geology of the Embetsu district: Geol. Survey of Japan Quadrangle Series, 33 p.

HOBDAY, D. K., AND MATTHEWS, D., 1975, Late Paleozoic fluviatile and deltaic deposits in the northeast Karroo Basin, South Africa, in Broussard, M. L., ed., 1975, Deltas, (2nd ed.): Houston, Houston Geol. Soc., p. 457–470.

IIJIMA, A., AND UTADA, M., 1971, Present-day zeolitic diagenesis of the Neogene geosynclinal deposits in the Niigata oil field, Japan: Adv. in Chemistry Series No. 101, Am. Chem. Soc., p. 342–349.

JACKSON, E. D., SILVER, E. A., AND DALRYMPLE, G. B., 1972, Hawaiian-Emporor chain and its relation to Cenozoic Circum-Pacific tectonics: Geol. Soc. America Bull., v. 83, p. 601–618.

KARIG, D. E., INGLE, J. C. JR., et al, 1975, Initial Reports of the Deep Sea Drilling Project: Washington, D.C., U.S. Government Printing Office, v. 31, 927 p.

KLEIN, G. DEV, 1974, Estimating water depths from analysis of barrier island and deltaic sedimentary sequences: Geology, v. 2, p. 409–412.

————, 1975, Sedimentary tectonics in southwest Pacific marginal basins based on Leg 30 Deep Sea Drilling Project cores from the South Fiji, Hebrides and Coral Sea Basins: Geol. Soc. America Bull., v. 86, p. 1012–1018.

————, DEMELO, U., AND DELLA FAVERA, J. C., 1972, Subaqueous gravity processes on the front of Cretaceous deltas, Recocavo Basin, Brazil: Geol. Soc. America Bull., v. 83, p. 1469–1492.

————, KOBAYASHI, K., CHAMLEY, H., CURTIS, D. M., DICK, H. J. B., ECHOLS, D. J., FOUNTAIN, D. M., KINOSHITA, H., MARSH, N. G., MIZUNO, A., NISTERENKO, G. V., OKADA, H., SLOAN, J. R., WAPLES, D. M., AND WHITE, S. M., 1978, Off-ridge volcanism and seafloor spreading in the Shikoku Basin: Nature, v. 273, p. 746–748.

MATSUDA, T., NAKAMURA, K., AND SUGIMURA, A., 1967, Late Cenozoic orogeny in Japan: Tectonophysics, v. 4, p. 349–366.

MATSUNO, K., 1958, The depression of the sedimentary basin of the Kotanbetsu Formation: Jour. Japan. Assoc. Petroleum Technologists, v. 23, p. 130–132.

————, AND KINO, Y., 1960, "Chikubetsu-Tanko": Explanatory text of the geological map of Japan: Geol. Survey Japan, 43 p.

MENARD, H. W., AND ATWATER, TANYA, 1968, Changes in direction of sea floor spreading: Nature, v. 219, p. 463–467.

MIYASHIRO, A, 1961, Evolution of metamorphic belts: Jour. Petrology, v. 2, p. 277–311.

MORGAN, W. J., 1968, Rises, trenches, great faults and crustal blocks: Jour. Geophys. Res., v. 73, p. 1959–1982.

MUTTI, E., AND RICCI-LUCHI, F., 1972, Le torbiditi dell'Appennino settentrionale: introduzione all'analisi di facies: Mem. della Soc. Geologica Italiana, v. 11, p. 161–199.

OBA, NOBORU, 1977, Emplacement of granitic rocks in the outer zone of southwest Japan and geological significance: Jour. Geology, v. 85, p. 383–394.
OKADA, HAKUYU, 1960, Sandstones of the Cretaceous Mifune Group, Kyushu, Japan: Mem. Faculty Sci., Kyushu Univ., Ser. D., v. 10, p. 1–40.
———, 1971, Classification of sandstone: analysis and proposal: Jour. Geol., v. 79, p. 509–525.
———, 1974, Migration of ancient arc-trench systems, in Dott, R. H., Jr., and Shaver, R. H., eds., 1974, Modern and ancient geosynclinal sedimentation: Soc. Econ. Paleontologists and Mineralogists Spec. Pub. 19, p. 311–320.
———, 1978, Sedimentary patterns in apparent back-arc basins: a case study of the Neogene sequence in northwestern Hokkaido, Japan: Proc. Int. Geodynamics Conf. (in press).
OKADA, HAKUYU, AND KITAMURA, NOBU, 1978, Significance of Miocene olistostromes in apparent rear-arc belt in Hokkaido, Japan (Abs): IX. Int. Sedimentological Cong., v. 2, p. 481–482.
PETTIJOHN, F. J., 1975, Sedimentary rocks, (3rd ed): New York, Harper & Row, 628 p.
———, POTTER, P. E., AND SIEVER, R., 1972, Sand and sandstones: New York, Springer-Verlag, 618 p.
POTTER, P. E., AND PETTIJOHN, F. J., 1963, Paleocurrents and basin analysis: New York, Springer-Verlag, 269 p.
SCRUTON, P. C., 1956, Oceanography of Mississippi Delta sedimentary environments: Am. Assoc. Petroleum Geologists Bull., v. 37, p. 2119–2162.
SHEPARD, F. P., 1973, Seafloor of Magdalena Delta and Santa Maria area, Colombia: Geol. Soc. America Bull., v. 84, p. 1955–1972.
TSUSHIMA, K., KAKIMI, T., AND UEMARA, T., 1956, Atsuta: explanatory text of the Geological Map of Japan: Geol. Survey of Japan, 24 p.
UJIIE, H., SAITO, T., KENT, D. V., THOMPSON, P. R., OKADA, H., KLEIN, G. DEV., KOIZUMI, I., HARPER, H. E. JR., AND SATO, T., 1977, Biostratigraphy, paleomagnetism and sedimentology of Late Cenozoic sediments in northwestern Hokkaido, Japan: Tokyo, Nat. Sci. Museum Bull., Ser. C., v. 3, p. 51–102.
VINE, F. J., 1966, Spreading in the ocean floor: new evidence: Science, v. 154, p. 1405–1415.
WHITAKER, J. H. MCD., 1962, The geology of the area around Leintwardine, Herefordshire: Geol. Soc. London Quar. Jour., v. 118, p. 319–351.